ELECTRON PARAMAGNETIC RESONANCE

ELECTRON PARAMAGNETIC RESONANCE

Edited by S. Geschwind
Bell Telephone Laboratories
Murray Hill, New Jersey

PLENUM PRESS • NEW YORK–LONDON • 1972

Library of Congress Catalog Card Number 77-186261

ISBN 0-306-30580-1

A Division of Plenum Publishing Corporation
227 West 17th Street, New York, N.Y. 10011

United Kingdom edition published by Plenum Press, London
A Division of Plenum Publishing Company, Ltd.
Davis House (4th Floor), 8 Scrubs Lane, Harlesden, London,
NW10 6SE, England

Printed in the United States of America

Contributors

S. GESCHWIND	Bell Telephone Laboratories, Murray Hill, New Jersey
FRANK S. HAM	General Electric Research and Development Center, Schenectady, New York
E. A. HARRIS	Department of Physics, University of Sheffield, England
C. D. JEFFRIES	Department of Physics, University of California, Berkeley, California
W. B. MIMS	Bell Telephone Laboratories, Murray Hill, New Jersey
R. ORBACH	Physics Department, University of California, Los Angeles, California
J. OWEN	Clarendon Laboratories, Oxford, England
R. H. SILSBEE	Department of Physics and Laboratory of Atomic and Solid State Physics, Cornell University, Ithaca, New York
E. SIMÁNEK	Department of Physics, University of California, Riverside, California
Z. SROUBEK	Department of Physics, University of California, Los Angeles, California
H. J. STAPLETON	Physics Department, University of Illinois, Urbana, Illinois

Preface

Since its discovery by E. Zavoisky in 1944, electron paramagnetic resonance (EPR) has grown to the point where it finds application in so many different branches of the physical and biological sciences. The question may therefore be raised as to whether EPR is now a technique, or tool like a voltmeter, or a subject. The argument for it being a technique appears very compelling when one considers its use in such disparate fields as that of local moments in metals on the one hand and the study of excited triplet states of biological molecules on the other. Yet, beside the very phenomenon of electron magnetic resonance itself which is being used to explore these different subjects, they all share a body of underlying physical topics such as crystal field theory, covalency effects, sources of linewidths, hyperfine structure, relaxation, etc., which often must be understood in a fair amount of detail to reach the most meaningful interpretation of results even in very different fields.

In drawing upon such a vast field of application of EPR, matters of personal experience must be a factor which invariably enters in the selection of topics for such a volume as this. In large measure the topics treated here are related to the subject of transition-metal and rare-earth ions in solids. The recent book by Abragam and Bleaney* on EPR of transition ions is a most wonderful and comprehensive account of the subject by two people whose names are so intimately associated with so much of the pioneering and important advances in the field. The intent in this collection of articles is to supplement their treatment by a presentation of selected topics in greater detail. An effort has also been made to assemble a range of subjects of current high-interest treated by specialists in their fields that hopefully would have as wide appeal as possible.

* A. Abragam and B. Bleaney, *Electron Paramagnetic Resonance of Transition Ions*, Oxford University Press (1970).

The additional criterion of strong overlapping interest for the reader among the selected topics seemed to naturally lead to the extension of the subject matter covered beyond that of transition-metal ions and into the field of color centers as just one example. Examples of the Jahn–Teller effect, pair interactions, relaxation questions, spectral diffusion, etc., figure prominently in color center research. As another illustration of overlapping interest influencing the selection of material, we cite how the importance of spin memory in the study of excited states of transition-metal ions in solids leads naturally to its consideration in excited triplet states of organic molecules. This aspect of relatedness among the articles in this volume appears in cross referencing between them where feasible.

However, when one attempts to justify, by reasons of limitations of space or otherwise, the inclusion of some topics and omission of others such as the latest work on EPR in semiconductors and metals, an apology is called for. It is our hope, nonetheless, that the subject matter of these articles will have a high degree of common interest and prove of value to a wide audience of research workers and students.

Stanley Geschwind
February 1972

Acknowledgment: I should like to thank Miss S. Talada for her invaluable help in typing Chapter 5 and with many other editorial chores.

Contents

Chapter 3. Dynamic Polarization of Nuclei

C. D. Jeffries

Chapter 4. Electron Spin Echoes

W. B. Mims

Chapter 1

Jahn–Teller Effects in Electron Paramagnetic Resonance Spectra

Frank S. Ham

General Electric Research and Development Center
Schenectady, New York

1. INTRODUCTION

1.1. Historical Perspective and Plan of the Chapter

Over the course of the 35 years since the discovery by Jahn and Teller (1936, 1937) of their theorem demonstrating the intrinsic instability of any nonlinear complex having an electronically degenerate state (excepting odd-electron systems having simple Kramers degeneracy), electron paramagnetic resonance (EPR) studies have played a central role in revealing the experimental consequences of the Jahn–Teller theorem. Conversely, it has become increasingly clear in that time that the Jahn–Teller "effect" influences the properties of many complexes which are studied by EPR techniques. The purpose of this chapter is to review the present state of our understanding of Jahn–Teller effects in EPR spectra, particularly in the spectra of paramagnetic ions in crystals.

Immediately upon the enunciation of the Jahn–Teller theorem, Van Vleck (1937, 1939a,b) recognized its potential importance for the behavior of paramagnetic ions in solids. Not until 1950, however, was unambiguous experimental evidence of the Jahn–Teller effect found. Then, in an EPR study of Cu^{2+} in the zinc fluosilicate crystal, Bleaney and Ingram (1950) found a spectrum which was anomalous from the standpoint of ordinary crystal-field or ligand-field theory but which could be explained, as shown by Abragam and Pryce (1950), as a dynamic manifestation of the Jahn–Teller instability for the orbital doublet ground state of the Cu^{2+} ion. This nearly isotropic EPR spectrum corresponded, according to Abragam and Pryce, to an average of the anisotropic spectra expected for the various possible Jahn–Teller distortions of the Cu^{2+} complex, and it appeared as the result of rapid reorientation of the complex (at 90°K). Shortly thereafter, Bleaney and Bowers (1952) found at 20°K the anisotropic EPR spectra of the individual static Jahn–Teller-distorted configurations of Cu^{2+} in the same salt and also in copper bismuth double nitrate and copper bromate. Over the subsequent years, many other instances have been obtained of similar EPR spectra showing dynamic and static Jahn–Teller effects for complexes in an orbital doublet electronic state in cubic or trigonal symmetry. Although this investigation of Jahn–Teller phenomena by EPR techniques has focused on the Cu^{2+} ($3d^9$) ion in octahedral coordination, such spectra have been found for a number of other ions as well. For complexes in an orbital triplet state, however, far fewer examples exhibiting Jahn–Teller effects have been found than for the doublet, but a number of such cases are now known.

Much of the theoretical work on the Jahn–Teller effect has been concerned with finding the form of the distorted configurations that correspond to stable equilibrium when the symmetric configuration is unstable as a result of the Jahn–Teller theorem. After Van Vleck's (1939b) early contribution to this investigation, a more detailed treatment of this question was given in the important paper of Opik and Pryce (1957), where it was also shown [as indicated also by earlier work by Jahn (1938)] that in certain cases, spin–orbit interaction can stabilize a symmetric configuration against weak Jahn–Teller coupling. The most extensive investigation of these structural consequences of Jahn–Teller coupling is the work of Liehr (1963), where many types of molecular structures that are subject to a Jahn–Teller instability are treated in great detail. This structural aspect of the Jahn–Teller theorem, in particular the expectation that the Jahn–Teller effect manifests itself experimentally primarily through the appearance at sufficiently low temperature of stable distorted configurations having lower symmetry than that of the host crystal (the "static" Jahn–Teller effect), has until recently provided the principal guidance for experimental investigations by EPR of systems that might possibly exhibit Jahn–Teller effects. Similar considerations have of course guided those who sought evidence for the Jahn–Teller effect from molecular structure,* and those who have sought to explain the structure of certain crystals on the basis of cooperative Jahn–Teller distortions of the constituent ions (Dunitz and Orgel, 1959; Goodenough, 1963).

A rather different and more fundamental view of Jahn–Teller effects was initiated in the work of the late W. Moffitt and his students (Moffitt and Liehr, 1957; Moffitt and Thorson, 1957) and independently by Longuet-Higgins *et al.* (1958). As a consequence of the coupling between the electronic energy states and the nuclear displacements which provides the driving force of the Jahn–Teller instability, they showed that a coupled motion of the electrons and the lattice vibrational modes occurs. For sufficiently strong coupling, the complex may exhibit a static Jahn–Teller distortion as a limiting case, but with weaker coupling there is still a profound modification of the energy levels of the coupled vibrational–electronic ("vibronic") system from what one obtains with zero coupling. The resulting vibronic wave functions have their vibrational and electronic parts inextricably mixed, and in this situation the system is said to exhibit a "dynamic" Jahn–Teller effect. Even the vibronic ground state may exhibit a dynamic character of this sort, as a result of the zero-point vibrational motion of the ions. Although the early work on dynamic Jahn–Teller effects was not evidently influenced by it, a very similar situation had been analyzed more than 20 years earlier by Renner (Renner, 1934; Herzberg, 1966) in treating the vibrational states of linear triatomic molecules in π states (in the "Renner effect," the electronic degeneracy is lifted to *second* order in the amplitude of the bending coordinates of the linear molecule, while for the Jahn–Teller effect the splitting is linear in the distortion).

Applications of the new theory of dynamic Jahn–Teller effects were for a number of years limited largely to the investigation of vibrational structure in electronic transitions in molecules (Liehr, 1961; Herzberg, 1966) and to studies of molecular infrared and Raman spectra (Weinstock and Goodman, 1965; Child and Longuet-Higgins, 1962; Child, 1963). The first application of this theory to a problem under investigation by EPR methods was provided by McConnell and McLachlan (1961) in their discussion of the EPR spectrum of the negative radical ion of benzene. In connection with the continuing interest in the EPR spectra of Cu^{2+} and similar Jahn–Teller ions in solids, O'Brien (1964) in an important paper supplemented the earlier treatments of the static Jahn–Teller effect for the case of the

* Liehr (1963) has given a thorough review of this work.

orbital doublet by analyzing the dynamic effects resulting from ionic motion in the ground state and low-energy excited vibronic states of the distorted complex. As a result of the system's ability to tunnel from one distorted configuration to another, the vibronic energy levels of the complex are split in proportion to the tunneling frequency, although the vibronic ground state remains a degenerate doublet. O'Brien showed for a strong Jahn–Teller coupling how the tunneling splittings of the low-energy vibronic states depend on the barrier height between the configurations, and how the Zeeman splitting and hyperfine interaction of these states vary with the tunneling splitting. The tunneling between the equivalent distorted configurations had been noted somewhat earlier by Avvakumov (1959), and, at the same time as O'Brien's work, Bersuker (1962) undertook an extensive investigation of dynamic effects resulting from this tunneling in strongly distorted Jahn–Teller systems.

The first recognition that the theory of dynamic Jahn–Teller effects introduced by Moffitt and Liehr (1957) and Longuet-Higgins *et al.* (1958) held widespread implications in the interpretation of EPR and optical spectra of paramagnetic complexes was in a paper by Ham (1965). This discovery had grown out of the realization that dynamic Jahn–Teller effects offered an explanation for anomalous EPR spectra obtained by Ludwig and Woodbury (1962) for a number of transition metal ions in silicon having orbital triplet states in interstitial sites of tetrahedral symmetry. These spectra showed in particular a remarkably large quenching of the orbital contribution to the *g*-factor—too large to be consistent with the previous interpretation of such "orbital reduction factors" in terms of covalent bonding—without showing any indication that an individual paramagnetic ion was subject to any departure from cubic symmetry. Ham showed that a dynamic Jahn–Teller effect can cause large changes in the matrix elements of electronic operators, in particular the orbital angular momentum operator, the spin–orbit interaction, and the operators governing the splitting of the electronic degeneracy under applied strain or other perturbations, as a result of the mixing of the vibrational and electronic wave functions that occurs even in the vibronic ground state. Only in the limiting case of strong Jahn–Teller coupling, in which some of these matrix elements are quenched effectively to zero, does the system exhibit the behavior expected for a static Jahn–Teller effect. Although these results were derived explicitly in this paper only for certain examples involving an orbital triplet state, the general nature of the dynamic quenching of electronic operators by the Jahn–Teller coupling was pointed out. Isolated examples of such effects of dynamic quenching, however, had been noted in earlier papers by a number of authors (McConnell and McLachlan, 1961; Child and Longuet-Higgins, 1962; Bersuker, 1962; Washimiya, 1963; Bersuker and Vekhter, 1963; Ballhausen, 1965). In a subsequent paper, Ham (1968) carried through a similar theoretical analysis for the case of an orbital doublet state, and he used his results to interpret experimental observations by Höchli (1967) of dynamic effects in the ground state of CaF_2:Sc^{2+} and SrF_2:Sc^{2+} which could not be explained on the basis of the theories of O'Brien (1964) or Bersuker (1962).

Some of the best examples of the dynamic Jahn–Teller quenching of electronic operators have been found in the optical spectra of paramagnetic ions in solids through the work of Sturge and others (Scott and Sturge, 1966; Sturge, 1967; Stephens and Lowe-Pariseau, 1968). Other examples known from EPR spectra will be discussed in this chapter. Since most of the work relating to these effects has been done within the last few years, our detailed understanding of them is still quite incomplete. An excellent review of Jahn–Teller effects in solids has been written by Sturge (1967), including not only a full account of the recent developments concerning dynamic Jahn–Teller effects in the optical spectra of ions in solids, but also a survey of manifestations of these effects in other solid-state phenomena. Sturge's paper is recommended to the reader for these other applications and for general background as well as for a very complete bibliography; the present chapter will make very little mention of these other areas of investigation of Jahn–Teller effects and will be limited to phenomena in the EPR spectra.

The next subsection of this chapter begins with an introduction to the general nature of vibronic coupling, the Jahn–Teller theorem, and static and dynamic Jahn–Teller effects. Since Sturge has given a good account of the background for our treatment of Jahn–Teller effects in his review (covering the general problem of transition metal ions in ionic crystals, crystal-field or ligand-field theory, the use of collective coordinates and effective Hamiltonians, etc., as well as the proof of the Jahn–Teller theorem), we shall be succinct in our discussion of these fundamentals. The bulk of the chapter is concerned with the specific form which Jahn–Teller effects take in the various cases of principal interest in EPR investigations, and with a review of the available experimental data relating to Jahn–Teller effects in EPR spectra. Our treatment will deal most thoroughly with Jahn–Teller effects in the case of an orbital doublet electronic state, since it is this system on which most of the experimental work has been done and for which the theory is most complete. This case, which we consider in Section 2, illustrates very well the phenomena which may be expected in other Jahn–Teller systems. Since past treatments of these effects have often proceeded from diverse viewpoints, it is our aim in this section to establish a common theoretical framework for the interpretation of these effects in the doublet state and thereby to show how these previous treatments are related. In Section 2.1, we first give the various terms in the effective spin Hamiltonian that one would have in cubic symmetry for the doublet if the Jahn–Teller coupling were zero, and we then show in Sections 2.2 and 2.3 how this is modified for the static Jahn–Teller effect and for the dynamic effect resulting from zero-point ionic motion. The transition from such a dynamic effect to the static effect as the strength of the coupling increases is then considered in Section 2.4, along with the often crucial role of random strain. It is shown here how this transition may equally well be viewed in terms of the ability of the system to tunnel from one distorted configuration to another. A discussion of dynamic effects resulting from relaxation and thermal excitation, such as those first recognized by Abragam and Pryce (1950), follows in Section 2.5. Spin–lattice relaxation in the presence of either static or dynamic Jahn–Teller effects is then considered in Section 2.6. Section 2 is concluded with a brief account of the modifications required in the theory of the orbital doublet when the original symmetry is lower than cubic.

In Section 3, the theory for an orbital triplet state is reviewed, although this is given in much less detail than for the doublet because there is so much less experimental information available at the present time for this case. Section 4 surveys the available experimental data on Jahn–Teller effects in EPR spectra, those for the orbital doublet in Section 4.1 and those for the triplet in Section 4.2. The examples cited have been selected to illustrate, so far as is now possible, the various aspects of the theory that has been presented. Section 4 is not intended to provide a complete listing of all experimental work using EPR in which Jahn–Teller effects have been observed or are suspected. References to additional cases are given in the reviews by Sturge (1967) and Müller (1967). The chapter is concluded in Section 5 with an assessment of the current state of understanding of Jahn–Teller effects and a discussion of a number of aspects of these and related problems which we have not been able to consider in detail. These include the largely unsolved problem of the proper treatment of Jahn–Teller effects taking account of coupling to a continuous phonon spectrum, and "pseudo-Jahn–Teller effects" and related questions involving vibronic coupling between nondegenerate states, examples of which are provided by the interesting work of Inoue (1963) and Birgeneau (1967) on the properties of certain rare earth ions.

1.2. The General Problem of Vibronic States: Coupling between Electronic Energy Levels and Lattice Vibrations

Crystal-field (or ligand-field) theory based on a static crystal lattice has been so successful in providing a phenomenological framework within which to interpret the results of EPR studies that we tend to forget in such studies that the dynamic behavior of the lattice may affect the measurements we make on the electronic part of the system even at the lowest temperatures. We are accustomed, of course, to viewing relaxation phenomena in our spin system at finite temperatures as resulting from the dynamic behavior of the lattice, but we are less frequently made to realize that even the zero-point motion of the lattice at the absolute zero of temperature can affect the electronic system. It is well known, however, from the work of Born and Oppenheimer (1927), that the vibronic eigenfunctions of the coupled system usually may be taken in the approximate form*

$$\Psi(\mathbf{q}, \mathbf{Q}) = \psi(\mathbf{q}; \mathbf{Q})\chi(\mathbf{Q}) \tag{1.2.1}$$

Here, $\mathbf{q}$ denotes the coordinates of the electrons of the ion or complex, while $\mathbf{Q}$ represents symbolically all the vibrational coordinates of the lattice; $\psi(\mathbf{q}; \mathbf{Q})$ is an electronic wave function depending parametrically on $\mathbf{Q}$ and satisfying Schrödinger's equation

$$H_e(\mathbf{q}; \mathbf{Q})\psi(\mathbf{q}; \mathbf{Q}) = E_e(\mathbf{Q})\psi(\mathbf{q}; \mathbf{Q}) \tag{1.2.2}$$

where $H_e(\mathbf{q}; \mathbf{Q})$ is the *electronic* Hamiltonian

$$H_e(\mathbf{q}; \mathbf{Q}) = \sum_i (\mathbf{p}_i^2/2m) + V(\mathbf{q}; \mathbf{Q}) \tag{1.2.3}$$

(neglecting spin–orbit coupling and other refinements for now) and $V(\mathbf{q}; \mathbf{Q})$ is the total potential energy as a function of $\mathbf{q}$ and $\mathbf{Q}$. In Eq. (1.2.2), $E_e(\mathbf{Q})$ is the electronic eigenvalue for fixed $\mathbf{Q}$, and in Eq. (1.2.3), $\mathbf{p}_i$ is the momentum conjugate to $\mathbf{q}_i$. The $\psi(\mathbf{q}; \mathbf{Q})$ is normalized under integration over $\mathbf{q}$ alone:

$$\int \psi^*\psi \, d\mathbf{q} = 1 \tag{1.2.4}$$

* A good account of the Born–Oppenheimer approximation and of the nature of the vibronic problem involved in the Jahn–Teller and Renner effects is given by Longuet-Higgins (1961).

Then, as shown by Born and Oppenheimer, the vibrational wave function $\chi(\mathbf{Q})$ satisfies the vibrational equation

$$\Big[U_e(\mathbf{Q}) + \sum_\nu (P_\nu^2/2M_\nu)\Big]\chi(\mathbf{Q}) = \mathscr{E}\chi(\mathbf{Q}) \tag{1.2.5}$$

where P_ν is the momentum conjugate to Q_ν, M_ν is the associated ionic mass, $\mathscr{E}$ is the vibronic energy eigenvalue, and $U_e(Q)$ is given by

$$U_e(\mathbf{Q}) = E_e(\mathbf{Q}) + \int \psi^*(\mathbf{q};\mathbf{Q}) \sum_\nu (P_\nu^2/2M_\nu)\psi(\mathbf{q};\mathbf{Q})\, d\mathbf{q} \tag{1.2.6}$$

The expectation value of an electronic operator $\mathscr{O}$ with respect to the eigenstate $\Psi(\mathbf{q}, \mathbf{Q})$ in Eq. (1.2.1) is therefore given by

$$\langle\Psi \mid \mathscr{O} \mid \Psi\rangle = \int \chi^*(\mathbf{Q})\mathscr{O}(\mathbf{Q})\chi(\mathbf{Q})\, d\mathbf{Q} \tag{1.2.7}$$

with $\mathscr{O}(\mathbf{Q})$ the electronic matrix element

$$\mathscr{O}(\mathbf{Q}) = \int \psi^*(\mathbf{q};\mathbf{Q})\mathscr{O}\psi(\mathbf{q};\mathbf{Q})\, d\mathbf{q} \tag{1.2.8}$$

The expectation value of $\mathscr{O}$ with respect to an eigenstate of the coupled system is thus the average of the electronic expectation value $\mathscr{O}(\mathbf{Q})$ taken with the weight function $|\chi(\mathbf{Q})|^2$. But, supposing $|\chi(\mathbf{Q})|^2$ is centered at $\mathbf{Q} = \mathbf{Q}_c$, we do *not* find that $\langle\Psi \mid \mathscr{O} \mid \Psi\rangle$ is equal to $\mathscr{O}(\mathbf{Q}_c)$, the value we would get for a static lattice having $\mathbf{Q} = \mathbf{Q}_c$, unless $|\chi(\mathbf{Q})|^2$ is strongly localized with respect to the length in Q-space over which $\mathscr{O}(\mathbf{Q})$ varies significantly. Even when the Born–Oppenheimer approximation holds, therefore, the difference between $\langle\Psi \mid \mathscr{O} \mid \Psi\rangle$ and $\mathscr{O}(\mathbf{Q}_c)$ may be significant, because of the spread in $\chi(\mathbf{Q})$ which results from the dynamic nature of the vibrational system.

A necessary condition for the validity of the Born–Oppenheimer approximation, however, is that the electronic energy levels of the system (or at least those which are coupled by changes in $\mathbf{Q}$) be widely separated in energy compared to the phonon energies. Thus we may expect further complications from the vibronic coupling when this condition is not fulfilled and the Born–Oppenheimer approximation (1.2.1) is not valid. Take, for example, the case of a system having two electronic eigenstates $\psi_1(\mathbf{q};\mathbf{Q})$ and $\psi_2(\mathbf{q};\mathbf{Q})$ of Eq. (1.2.2) with energies $E_{e1}(\mathbf{Q})$ and $E_{e2}(\mathbf{Q})$ such that $|E_{e1}(\mathbf{Q}) - E_{e2}(\mathbf{Q})|$ is not very much greater than the energies $\hbar\omega$ of the various lattice phonons. We expect, in analogy to Eq. (1.2.1), that the vibronic eigenfunctions of the system may, to good accuracy, be taken in the form

$$\Psi(\mathbf{q};\mathbf{Q}) = \psi_1(\mathbf{q};\mathbf{Q})\chi_1(\mathbf{Q}) + \psi_2(\mathbf{q};\mathbf{Q})\chi_2(\mathbf{Q}) \qquad (1.2.9)$$

but that it will not be possible in general to transform Ψ to the form of Eq. (1.2.1). Various authors (Moffitt and Liehr, 1957; Hobey and McLachlan, 1960; McLachlan, 1961; Longuet-Higgins, 1961) have shown that this is indeed the case and that, moreover, $\chi_1(\mathbf{Q})$ and $\chi_2(\mathbf{Q})$ now satisfy a pair of *coupled* differential equations in place of Eq. (1.2.5). Matrix elements of operators between vibronic eigenstates of the form of Eq. (1.2.9) now obviously have a far more complicated dependence on the properties of the $\chi(\mathbf{Q})$'s than we encounter in the simple case for which the Born–Oppenheimer approximation holds.

In studying the Jahn–Teller effect, we are concerned with an extreme example of the more general situation described in the preceding paragraph, since by definition we have two or more electronic states with an energy difference which is *zero* for some $\mathbf{Q}_s$ corresponding to a symmetric configuration and which increases linearly in the appropriate $(\mathbf{Q} - \mathbf{Q}_s)$ as we depart from this configuration. While for sufficiently strong Jahn–Teller coupling we may be able to recapture the Born–Oppenheimer approximation for the low-energy states of the static Jahn–Teller effect, we must expect in the less strongly coupled dynamic case to be dealing with vibronic wave functions in which the electronic and vibrational parts are inextricably mixed as in Eq. (1.2.9). Even if it should turn out to be possible in certain special cases to use wave functions of the form of the simple Born–Oppenheimer product in Eq. (1.2.1), we must expect that different vibrational wave functions $\chi(\mathbf{Q})$ are associated with the different electronic wave functions even in vibronic states of the same energy. This latter case occurs, for example, for a triplet electronic state when coupling to only the doublet (E) vibrational mode is taken into account (Section 3.1), and in that case the dynamic quenching of electronic operators can be visualized simply as the result of the diminishing overlap between the vibrational $\chi_i(\mathbf{Q})$ associated with the different electronic states ψ_i as the strength of the Jahn–Teller coupling increases.

We may recognize also from the preceding discussion that the dynamic Jahn–Teller effect is a special case of a much wider class of problems involving electronic states which may be nondegenerate but which are not sufficiently separated in energy compared to $\hbar\omega$ of the lattice phonons for the Born–Oppenheimer approximation to be accurate. We may consequently expect significant dynamic effects on the matrix elements of electronic operators in all such cases. Although these cases would not be said from a narrow point of view to exhibit a Jahn–Teller effect (they are often described as exhibiting a "pseudo-Jahn–Teller effect"), it seems pointless to maintain a rigid distinction between two parts of what is clearly the same problem. Furthermore, as we have seen, even in cases for which the Born–Oppenheimer approximation is valid, there may still be significant dynamic corrections to the parameters characterizing the electronic system.

1.3. The Jahn–Teller Theorem

The fundamental theorem (Jahn and Teller, 1937) of the Jahn–Teller effect may be stated as follows:

If a molecule or crystalline defect has orbital electronic degeneracy when the nuclei are in a symmetric configuration, then the molecule or defect is unstable with respect to at least one asymmetric displacement of the nuclei which lifts the degeneracy. The only exception to the rule is the linear molecule.

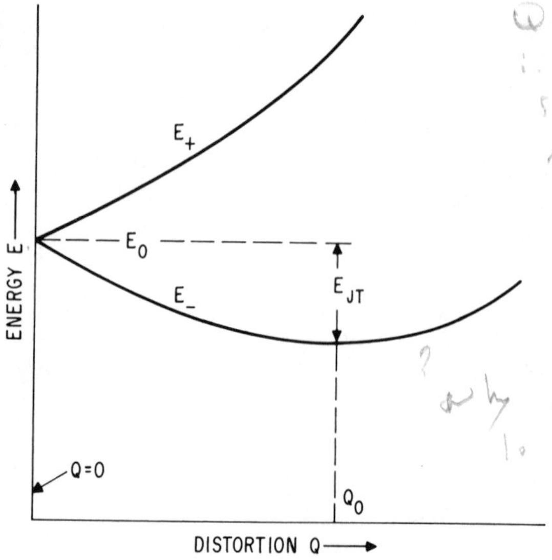

Fig. 1. Energy splitting of a doubly degenerate electronic state under a distortion that lifts the degeneracy in first order, showing the Jahn–Teller instability of the symmetric configuration.

The meaning of the theorem is illustrated in Fig. 1. Let E_0 denote the energy of a doubly degenerate electronic state in the symmetric configuration. Then, according to the theorem, there is some distortion Q that splits the electronic degeneracy and causes the lower split-off state to have an energy E_- such that $(E_0 - E_-)$ increases linearly with Q for small Q (except for the linear molecule, there is always at least one Q that produces such a linear splitting, unless of course the coupling coefficient for this mode of distortion happens accidentally to be exactly zero.) The two split electronic states are the two eigenstates of Eq. (1.2.2) that become degenerate at $Q = 0$, and their energies $E_+(Q)$ and $E_-(Q)$ are the associated eigenvalues. Since the elastic energy associated with the distortion is proportional to Q^2 (the symmetric configuration denoted by $Q = 0$ is assumed to be a configuration of stable equilibrium apart from the coupling to the degenerate electronic state), the system can lower its energy by distorting to a finite Q. The symmetric configuration is therefore unstable with respect to this distortion.

The Jahn–Teller theorem is an existence theorem concerning the availability in each case of at least one mode of distortion that can in principle cause a linear splitting, and the theorem was proved by Jahn and Teller (1937) from an examination of the properties of the point symmetry group in each possible case. The theorem says nothing about the magnitude of the coupling coefficient appropriate in any case, and of course it is the size of this coefficient that determines whether in a given situation we see a static Jahn–Teller effect, a dynamic effect, or no significant effect at all.

In the form in which we have stated the Jahn–Teller theorem above, it applies only to an electronic system having orbital degeneracy. The same theorem applies in principle to spin degeneracy as well, as Jahn (1938) has shown, with now the further exception that simple Kramers degeneracy cannot be lifted by any distortion. However, the direct coupling between the spin levels and lattice distortions is sufficiently weak so that, in practice, Jahn–Teller effects are not significant when spin degeneracy alone is present.

1.4. The Static Jahn–Teller Effect

We consider as functions of the various lattice coordinates $\mathbf{Q}$ the energy surfaces $E_e(\mathbf{Q})$ which derive from the degenerate electronic level in the symmetric configuration $\mathbf{Q} = 0$ and which are obtained as the eigenvalues of Eq. (1.2.2). As we have seen in the preceding section, these surfaces intersect with a finite slope in the symmetric configuration, which is therefore unstable. There are a number of absolute minima lying at finite $\mathbf{Q}$, which thus correspond to the configurations of stable equilibrium. In these stable, distorted configurations, the lowest electronic state has no remaining orbital degeneracy, for, by the Jahn–Teller theorem, a further distortion could otherwise lead to a still lower value of $E_e(\mathbf{Q})$. From the theoretical standpoint, investigating these energy surfaces and finding their absolute minima constitute the study of the static Jahn–Teller effect. We note that the ionic momentum operators P_ν and the ionic kinetic energy are completely omitted from Eqs. (1.2.2) and (1.2.3), which determine $E_e(\mathbf{Q})$, so that these surfaces correspond to a completely static treatment of the lattice.

The energy difference E_{JT} by which $E_e(\mathbf{Q})$ at these absolute minima (including the elastic energy) lies below the energy E_0 of the symmetric configuration, as in Fig. 1, is thus the energy of stabilization of the distorted configuration. We call this energy difference the "Jahn–Teller energy." A number of authors, notably Van Vleck (1937, 1939a,b), Opik and Pryce (1957), and Liehr (1963), have mapped these energy surfaces, found the stable configurations, and calculated E_{JT} for the different Jahn–Teller systems under a variety of assumptions. We shall consider the cases that have been of particular interest in the EPR work in the following parts of this chapter, and many others have been analyzed by Liehr (1963).

Experimentally, the static Jahn–Teller effect may be described pragmatically as any situation in which measurements indicate that an individual ion or complex has distorted spontaneously and that its properties are those corresponding to one of the stable, distorted configurations. For a Jahn–Teller effect, of course, we want this distortion to have occurred, as a result of the electronic degeneracy, from a symmetric configuration that would otherwise have been stable. We would like to exclude instances in which a distortion results from an intrinsic instability having nothing to do with the electronic state, or from the proximity in the crystal of some other defect. Needless to say, it is often difficult to distinguish these alternatives. Moreover, it is important to realize that a given system may give rise to a static Jahn–Teller effect in one experiment yet fail to do so in another. Since the system can reorient, sometimes by tunneling or else by thermal activation over a barrier, from one distorted configuration to another, a measurement made in a time short compared to the time for reorientation will indicate a static distortion, while one taking a longer time may yield an average over the configurations and thus correspond to a dynamic effect. When we do have a static Jahn–Teller effect, of course, we still have vibrational ionic motion within the well in $E_e(\mathbf{Q})$ corresponding to the distorted configuration, so that there may still be significant dynamic corrections, for example, as noted by O'Brien (1964), even from zero-point ionic motion in the ground state. The Born–Oppenheimer approximation may be applicable in this situation in describing the low-energy states in each well, even though it cannot be used with most dynamic effects.

When we include spin–orbit coupling in the electronic Hamiltonian of Eq. (1.2.3), the energy surfaces obtained from Eq. (1.2.2) are modified. Sometimes, as shown in the work of Opik and Pryce (1957) and Liehr (1963) and indicated by Jahn's treatment (1938) of the Jahn–Teller theorem for systems with spin, the spin–orbit interaction may stabilize a symmetric configuration against a Jahn–Teller instability that would otherwise be present. We illustrate such a case schematically in Fig. 2. In the absence of

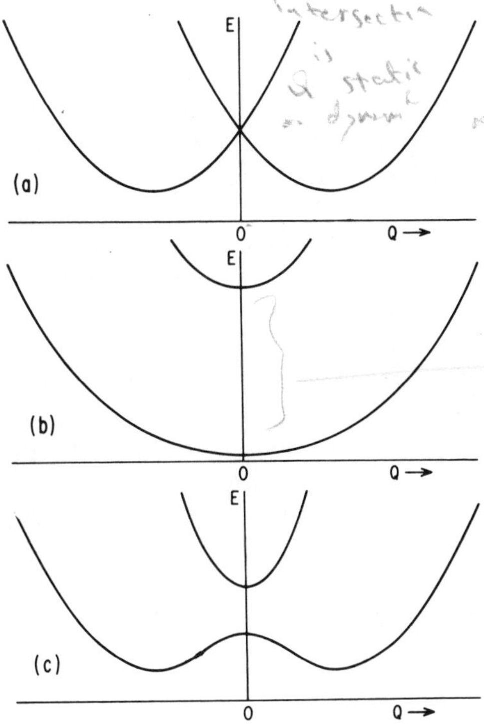

Fig. 2. Illustration of possible effects of spin–orbit interaction with respect to stabilizing a symmetric configuration against Jahn–Teller instability: (a) Jahn–Teller instability of an orbital doublet state with zero spin–orbit interaction; (b) stabilization of symmetric configuration by strong spin–orbit interaction ($S = \frac{1}{2}$); (c) instability resulting from a strong Jahn–Teller coupling despite a moderately strong spin–orbit interaction.

spin–orbit coupling, we suppose we have the energy surfaces indicated in Fig. 2(a) corresponding to a doublet orbital state in the symmetric configuration (which might be one of axial symmetry). Taking a spin $S = 1/2$, we suppose that spin–orbit coupling lifts the fourfold degeneracy in the symmetric configuration, yielding two Kramers doublets. If the spin–orbit coupling is large enough, the resulting energy surfaces would be as in Fig. 2(b), and the symmetric configuration would now be stable. On the other hand, if the spin–orbit interaction is smaller, we would have the situation shown in Fig. 2(c), where, despite the lifting of the degeneracy, the stable configurations are still distorted. From a narrow viewpoint, the spin–orbit coupling in Fig. 2(c) might still be said to have suppressed the Jahn–Teller effect, since the states are nondegenerate (apart from Kramers degeneracy) at $Q = 0$, and since, for very small Q, $E(Q)$ varies in proportion to Q^2 instead of linearly. Indeed, such a situation would traditionally be described as exhibiting a "pseudo-Jahn–Teller effect." However, in terms of the nature of the stable, distorted configurations, such a distinction between Figs. 2(a) and 2(c) is largely meaningless, and when we consider dynamic effects, we will see that we may expect rather similar such effects in all three cases in Fig. 2 unless the separation of the energy surfaces is large compared to $\hbar\omega$ of the phonons. In a similar way, low-symmetry components of the crystal field of the lattice, or those due to a nearby defect, while perhaps splitting the electronic degeneracy so as to "suppress" the Jahn–Teller effect from a narrow viewpoint, may, if weak enough, have at most a modest effect in changing the nature of the stable configurations obtained for the Jahn–Teller effect when these low-symmetry crystal-field components are ignored, and in changing the dynamic effects in the system.

1.5. Dynamic Jahn–Teller Effects

When we take account of the kinetic energy of the lattice ions, so that we are no longer treating the lattice as static, we enter upon the study of the dynamic Jahn–Teller effects. There are really two types of such effects. The first historically involves a thermally activated reorientation of a complex which in its ground state at very low temperatures exhibits a static Jahn–Teller effect. This was the type of dynamic effect first observed by Bleaney and Ingram (1950) in the EPR spectrum of Cu^{2+} in $ZnSiF_6 \cdot 6H_2O$ and interpreted by Abragam and Pryce (1950). In the second type, we are concerned with the coupling between the electronic and vibrational motions that occurs when the electronic and vibrational parts of the wave function of a single eigenstate of the coupled system are inextricably mixed, as in Eq. (1.2.9), and the Born–Oppenheimer approximation is invalid. This second type of dynamic effect, the study of which was initiated by Moffitt and Liehr (1957) and by Longuet-Higgins *et al.* (1958), can of course occur in the ground state as well as in the excited states, and it may sometimes be represented conveniently in terms of the system's ability to tunnel from one distorted configuration to another, as in the work of Bersuker (1962). In this second type, we are concerned with the properties of the individual vibronic eigenstates of the system, while in the first type the system is undergoing rapid, thermally induced transitions between different eigenstates. The first type of dynamic effect may often in fact be interpreted in terms of the conventional theory of motional averaging of resonance spectra, although tunneling in excited states may sometimes be involved.

In analyzing dynamic effects of the second type, we may no longer separate the electronic and vibrational parts of the system as we did in Eqs. (1.2.2) and (1.2.5) in connection with the Born–Oppenheimer approximation. We therefore seek to solve directly the vibronic Schrödinger equation

$$\mathscr{H}(\mathbf{p}, \mathbf{q}, \mathbf{P}, \mathbf{Q})\Psi(\mathbf{q}, \mathbf{Q}) = \mathscr{E}\Psi(\mathbf{q}, \mathbf{Q}) \tag{1.5.1}$$

where the full vibronic Hamiltonian (omitting spin–orbit interaction and other refinements for now) is obtained by adding the ionic kinetic energy operator to the electronic Hamiltonian $H_e(\mathbf{q}; \mathbf{Q})$ of Eq. (1.2.3):

$$\mathscr{H}(\mathbf{p}, \mathbf{q}, \mathbf{P}, \mathbf{Q}) = \sum_\nu (P_\nu^2/2M_\nu) + \sum_i (p_i^2/2m) + V(\mathbf{q}, \mathbf{Q}) \tag{1.5.2}$$

We shall consider various cases of Eq. (1.5.1) in what follows.

It is very important that we understand that the vibronic Hamiltonian in Eq. (1.5.2) has the full symmetry of the symmetric configuration of the complex or of the lattice site at which the Jahn–Teller ion is situated, since the total potential energy $V(\mathbf{q}, \mathbf{Q})$ has this symmetry under simultaneous transformation of the electronic coordinates $\mathbf{q}$ and the displacements $\mathbf{Q}$ of the nuclei from the symmetric configuration. As a consequence, the vibronic eigenstates of Eq. (1.5.1) may be classified according to the point symmetry group of the symmetric configuration, and states have the degeneracy of the irreducible representations of this group. We see, therefore, that *the Jahn–Teller coupling does not lift the degeneracies of the vibronic states which belong as partners to the same irreducible representation of the group of the original symmetric configuration.* Whereas the degeneracy of the electronic states which satisfy Eq. (1.2.2) is split for a fixed $\mathbf{Q}$ different from that of the symmetric configuration, there is no such splitting of the vibronic eigenstates of the coupled system. This fundamental point has often been misunderstood in discussions of the Jahn–Teller effect; for example, it has sometimes been supposed that a weak Jahn–Teller coupling can produce a small splitting in the vibronic ground state of a Jahn–Teller ion, and that this splitting might be observed by resonance techniques. We must emphasize that it is the coupled system of electrons and phonons on which experimental observations are made, and that, on grounds of symmetry, no such splitting of vibronic states belonging to the same irreducible representation of the symmetry group of the symmetric configuration can result from the Jahn–Teller coupling. If a splitting occurs, it results from some permanent lowering of the symmetry of the crystal as a result of strain or a nearby defect, but not from the dynamic Jahn–Teller effect. Even for the static Jahn–Teller effect, the degeneracy is preserved, since the several energetically equivalent distorted configurations and the associated electronic states represent vibronic states which belong as partners to one (or sometimes more) irreducible representation of the original symmetry group.

2. THEORY OF JAHN–TELLER EFFECTS IN THE EPR SPECTRUM OF AN ORBITAL DOUBLET ELECTRONIC STATE

2.1. The Hamiltonian for a Doublet State in Cubic Symmetry without Jahn–Teller Coupling

We shall first give the Hamiltonian for an orbital doublet electronic state in cubic symmetry, as given by crystal-field theory without the Jahn–Teller effect being considered. In the subsequent parts of Section 2, we shall then show how the Hamiltonian is changed by both static and dynamic Jahn–Teller effects, and how these changes affect the EPR spectrum.

An orbital doublet state in cubic (or tetrahedral) symmetry belongs to the two-dimensional irreducible representation* E of the rotation group O of the cube (or the tetrahedral group T_d). We shall label by ψ_θ and ψ_ε the two electronic states comprising the doublet and transforming, re-

* In this chapter, we use the notation (A_1, A_2, E, T_1, T_2) of Mulliken (1933) to denote the irreducible representations of O or T_d. However, in dealing with levels split by spin–orbit coupling, we will sometimes use the corresponding notation ($\Gamma_1, \ldots, \Gamma_8$) of Bethe (1929).

spectively, as $(2z^2 - x^2 - y^2)$ and $\sqrt{3}(x^2 - y^2)$, where x, y, z denotes Cartesian coordinates with respect to the cubic (fourfold) axes. We shall be particularly concerned with a ^2E state having a spin $S = 1/2$ in addition to the orbital degeneracy. A ^2E state derived from the ^2D free ion term by the cubic field splitting is, of course, the ground state for the $3d^9$ configuration (Cu^{2+}, Ni^{+}) in octahedral coordination or for the $3d^1$ configuration (Sc^{2+}, Ti^{3+}) in tetrahedral or eightfold cubic coordination. There is no spin–orbit splitting of a ^2E state in cubic or tetrahedral symmetry, since such a state belongs to the four-dimensional irreducible representation Γ_8 of the corresponding double group.

If the states of a degenerate electronic level of an ion at a site having the symmetry of the point group G belong to the representation Γ_a of G, it is well known from group theory that the only electronic operators that can have nonzero matrix elements among these states are those transforming by one of the irreducible representations of G contained in the direct product $\Gamma_a \times \Gamma_a$. Since for the representation E of the groups O or T_d we have

$$E \times E = A_1 + A_2 + E \tag{2.1.1}$$

we see (neglecting for now the spin–orbit coupling) that it is only those perturbations involving orbital operators belonging to A_2 or E that can lift the orbital degeneracy of an E doublet. We can represent any perturbation of the ion in the general form

$$\mathscr{H}' = \sum_{\Gamma,\gamma,i} V(\Gamma, i) C_\gamma(\Gamma, i) G_\gamma(\Gamma, i) \tag{2.1.2}$$

where $C_\gamma(\Gamma, i)$ is an orbital operator which, for fixed Γ and i, transforms as the γth partner belonging to the irreducible representation Γ. The $G_\gamma(\Gamma, i)$ may be functions of the components of external perturbations such as magnetic fields and strains, of other operators such as electronic and nuclear spin, and of the distortions Q_ν of the lattice; we use the index i to label the different independent choices for these functions corresponding to different physical perturbations. Choosing the $G_\gamma(\Gamma, i)$ to have the same transformation properties as the $C_\gamma(\Gamma, i)$ under cubic transformations of these various quantities, we have that the coefficients $V(\Gamma, i)$ in Eq. (2.1.2) must then be independent of γ. As a consequence of Eq. (2.1.1), for the splitting or displacement of an E doublet we need only consider terms with Γ equal to A_1, A_2, and E in Eq. (2.1.2). Moreover, because A_1, A_2, and E are each contained only once in $E \times E$, it follows from the tables of coupling coefficients for the groups O or T_d (Koster, 1958; Koster and Statz, 1959;

Koster *et al.*, 1963) that in taking matrix elements of the operators $C_\gamma(\Gamma, i)$ with respect to the doublet states ψ_θ and ψ_ε we can replace $C_\theta(E, i)$, $C_\varepsilon(E, i)$, $C_1(A_2, i)$, and $C_1(A_1, i)$ respectively by multiples of the Hermitian operators $\mathscr{U}_\theta$, $\mathscr{U}_\varepsilon$, $\mathscr{A}_2$, and $\mathscr{I}$, which are defined to have the matrix form

$$\mathscr{U}_\theta = \begin{pmatrix} -1 & 0 \\ 0 & +1 \end{pmatrix}, \qquad \mathscr{U}_\varepsilon = \begin{pmatrix} 0 & +1 \\ +1 & 0 \end{pmatrix}$$
$$\mathscr{A}_2 = \begin{pmatrix} 0 & -i \\ +i & 0 \end{pmatrix}, \qquad \mathscr{I} = \begin{pmatrix} +1 & 0 \\ 0 & +1 \end{pmatrix} \tag{2.1.3}$$

in the basis ψ_θ, ψ_ε [i.e., $\langle\psi_\theta \mid \mathscr{U}_\theta \mid \psi_\theta\rangle = -1$, etc.] Since $\mathscr{U}_\theta$ and $\mathscr{U}_\varepsilon$ belong as the θ and ε partners, respectively, to E, the proportionality factor that relates $C_\theta(E, i)$ to $\mathscr{U}_\theta$ is the same as that relating $C_\varepsilon(E, i)$ to $\mathscr{U}_\varepsilon$, so that Eq. (2.1.2) has the same form if expressed in terms of the orbital operators (2.1.3) as with the $C_\gamma(\Gamma, i)$. We note that, apart from the sign of $\mathscr{U}_\theta$, the matrices (2.1.3) are simply the Pauli matrices and the 2×2 unit matrix.

These considerations of symmetry show that, of the various lattice distortion coordinates Q_ν, it is only those that belong to the irreducible representation E of the groups O or T_d that can cause a first-order splitting of a doublet E state and thus lead to a Jahn–Teller coupling, at least so long as spin–orbit coupling is ignored. The form of this coupling follows from Eq. (2.1.2) if we set $G_\theta(E, i) = Q_\theta$, $G_\varepsilon(E, i) = Q_\varepsilon$. A linear splitting due to a Q belonging to A_2 is excluded because the Q's can all be chosen to be real, while the interaction (2.1.2) must be both Hermitian and invariant under time-reversal; these requirements cannot both be met in view of the fact that the operator $\mathscr{A}_2$ in Eq. (2.1.3) is both Hermitian and imaginary, unless the coefficient of such a term is identically zero. A distortion belonging to A_1, on the other hand, simply displaces the doublet without splitting it. In the particular case of an ion in octahedral coordination, of the even modes of distortion of the nearest neighbors given in Table I, it is thus only the E_g mode that gives rise to Jahn–Teller coupling for an E doublet state (odd modes being excluded by parity considerations when inversion symmetry is present).

In applying these symmetry arguments to obtain the form of the Zeeman, hyperfine, and strain interactions for a ^2E state, we must take into account the fact that the spin–orbit interaction, while not splitting a ^2E state, will mix it with other Γ_8 states from other levels and will thus cause it to have somewhat modified properties from those one would have for a ^2E state in the absence of spin–orbit interaction. The number of independent parameters needed to specify the splitting of an arbitrary Γ_8 state

Table I. Even Modes of Distortion[a] for an Octahedron of Six Ions,[b] Classified by the Irreducible Representation of the Symmetry Group of the Cube to Which They Belong

A_{1g}:	$Q_1 = (1/\sqrt{6})(X_1 - X_4 + Y_2 - Y_5 + Z_3 - Z_6)$
E_g:	$Q_\theta = (1/\sqrt{12})(2Z_3 - 2Z_6 - X_1 + X_4 - Y_2 + Y_5)$
	$Q_\varepsilon = \frac{1}{2}(X_1 - X_4 - Y_2 + Y_5)$
T_{2g}:	$Q_\xi = \frac{1}{2}(Z_2 - Z_5 + Y_3 - Y_6)$
	$Q_\eta = \frac{1}{2}(X_3 - X_6 + Z_1 - Z_4)$
	$Q_\zeta = \frac{1}{2}(Y_1 - Y_4 + X_2 - X_5)$

[a] After Van Vleck (1939b). For a sketch of these modes of distortion, see Fig. 2 of the chapter by Orbach and Stapleton in this volume.

[b] The ions numbered 1–6 have for their undistorted position, respectively, $(R, 0, 0)$, $(0, R, 0)$, $(0, 0, R)$, $(-R, 0, 0)$, $(0, -R, 0)$, and $(0, 0, -R)$. The displacement of the nth ion is denoted by (X_n, Y_n, Z_n).

by a given interaction is given in general by the number of times the appropriate representation is contained in the direct product $\Gamma_8 \times \Gamma_8$, for which we have

$$\Gamma_8 \times \Gamma_8 = A_1 + A_2 + E + 2T_1 + 2T_2 \qquad (2.1.4)$$

We may nevertheless take the *unperturbed* ^2E state as a prototype Γ_8 state, and, using the wave functions of this state as a basis, we may then write the general form for each type of interaction as an effective Hamiltonian having the form of Eq. (2.1.2). We shall do this using the orbital operators defined in Eq. (2.1.3) together with the electronic spin operators S_x, S_y, S_z, which suffice to represent the effect of any perturbation that lifts the spin degeneracy of a level with $S = 1/2$. The values of the necessary parameters for our ^2E level as perturbed by spin–orbit interaction are then determined from perturbation theory. The various cases are considered below.

2.1.1. Zeeman Interaction

The components of a magnetic field **H** transform as T_1. Since, according to Eq. (2.1.4), T_1 is contained twice in $\Gamma_8 \times \Gamma_8$, the linear Zeeman splitting of a Γ_8 state depends in general on two parameters which we will call g_1 and g_2. With respect to the ^2E state as basis, the Zeeman interaction for a Γ_8 state may then be written in the form of Eq. (2.1.2) as

$$\mathscr{H}_Z = g_1\beta(\mathbf{S}\cdot\mathbf{H})\mathscr{I} + \tfrac{1}{2}g_2\beta\{[3S_zH_z - (\mathbf{S}\cdot\mathbf{H})]\mathscr{U}_\theta + \sqrt{3}(S_xH_x - S_yH_y)\mathscr{U}_\varepsilon\} \tag{2.1.5}$$

where $\beta = (e\hbar/2mc)$ is the Bohr magneton.* Since the orbital angular momentum has no nonzero matrix elements within an E state, as a consequence of Eq. (2.1.1), the Zeeman interaction for a ^2E state is predominantly due to the spin-only interaction $g_s{}^0\beta(\mathbf{S}\cdot\mathbf{H})$, where $g_s{}^0 = 2.0023$. However, taking account of the spin–orbit coupling, we have for the ^2E ground state of the d^1 or d^9 configuration from crystal-field theory the values

$$g_1 = g_s{}^0 - (4\lambda/\Delta), \qquad g_2 = -(4\lambda/\Delta) \tag{2.1.6}$$

Here, $\Delta = 10\,|\,Dq\,|$ is the cubic-field splitting between ^2E and the excited $^2T_2$ term derived from the ^2D term of the free ion, and λ is the parameter such that $\lambda(\mathbf{L}\cdot\mathbf{S})$ describes the spin–orbit interaction for ^2D. The values (2.1.6) are accurate to first order in perturbation theory in the mixing by the spin–orbit interaction of the Γ_8 state from the excited term $^2T_2$ into the ^2E ground state.

Covalent bonding (Owen and Thornley, 1966) will modify the values for g_1 and g_2 given in Eq. (2.1.6), but primarily through the change it makes in λ (and Δ). However, since it is the matrix elements of the spin–orbit interaction between ^2E and $^2T_2$ that determine the corrections to both g_1 and g_2, it is the same effective ratio (λ/Δ) as modified by covalent bonding that should be taken in both parts of Eq. (2.1.6).

2.1.2. Hyperfine Interaction

Since the nuclear spin $\mathbf{I}$ also transforms as T_1, the hyperfine interaction with the central nucleus takes, in general, for a Γ_8 state the form analogous

* An alternative representation of the Zeeman interaction for a Γ_8 state may be given using a true spin-$\frac{3}{2}$ state as the Γ_8 reference state. The interaction may then be given (Koster and Statz, 1959; Ham *et al.*, 1960) in terms of the spin operators as

$$\mathscr{H}_Z = g\beta(\mathbf{S}\cdot\mathbf{H}) + u\beta[S_x{}^3H_x + S_y{}^3H_y + S_z{}^3H_z - (41/20)(\mathbf{S}\cdot\mathbf{H})]$$

where g and u are now the two independent parameters. This representation, which is mathematically entirely equivalent to Eq. (2.1.5) for a Γ_8 state, is the natural representation for a Γ_8 state derived predominantly from a true spin-$\frac{3}{2}$ state, in which case we should have $|\,g\,| \gg |\,u\,|$. For a Γ_8 state derived primarily from a ^2E level, on the other hand, Eq. (2.1.5) is the natural representation, since we have then $|\,g_1\,| \gg |\,g_2\,|$, while in the alternative representation g and u would be of comparable size.

to Eq. (2.1.5),

$$\mathscr{H}_{\mathrm{I}} = A_1(\mathbf{S}\cdot\mathbf{I})\mathscr{I} + \tfrac{1}{2}A_2\{[3S_zI_z - (\mathbf{S}\cdot\mathbf{I})]\mathscr{U}_\theta + \sqrt{3}\,(S_xI_x - S_yI_y)\mathscr{U}_\varepsilon\} \tag{2.1.7}$$

where again we use the unperturbed ^2E state for the basis. Crystal-field theory gives us by perturbation theory to first order in the spin-orbit interaction

$$\begin{aligned} A_1 &= -P[\varkappa + (4\lambda/\Delta)] \\ A_2 &= -P[6\xi + (4\lambda/\Delta) + (9\lambda/\Delta)\xi] \end{aligned} \tag{2.1.8}$$

for the ^2E ground state of a d^1 or d^9 configuration. Here, we have represented the hyperfine interaction within the ^2D term (Abragam and Pryce, 1951) by

$$\begin{aligned} \mathscr{H}_{\mathrm{I}}(^2D) = P\{&(\mathbf{L}\cdot\mathbf{I}) - \varkappa(\mathbf{S}\cdot\mathbf{I}) + \xi[6(\mathbf{S}\cdot\mathbf{I}) \\ &- \tfrac{3}{2}(\mathbf{L}\cdot\mathbf{S})(\mathbf{L}\cdot\mathbf{I}) - \tfrac{3}{2}(\mathbf{L}\cdot\mathbf{I})(\mathbf{L}\cdot\mathbf{S})]\} \end{aligned} \tag{2.1.9}$$

where $P = 2\gamma\beta\beta_N\langle r^{-3}\rangle$, $\xi = +(2/21)$, and γ is the nuclear g-factor, β_N the nuclear magneton, $\langle r^{-3}\rangle$ the one-electron average of r^{-3}, and $\varkappa$ the parameter characterizing the contact hyperfine interaction.

2.1.3. Strain

The splitting of an E state in uniform strain is described in accord with Eqs. (2.1.1) and (2.1.2) by

$$\mathscr{H}_S^{(E)} = V_2[e_\theta\mathscr{U}_\theta + e_\varepsilon\mathscr{U}_\varepsilon] \tag{2.1.10}$$

where

$$e_\theta = e_{zz} - \tfrac{1}{2}(e_{xx} + e_{yy}), \qquad e_\varepsilon = (\sqrt{3}/2)(e_{xx} - e_{yy}) \tag{2.1.11}$$

and

$$e_{ij} = \tfrac{1}{2}[(\partial u_i/\partial x_j) + (\partial u_j/\partial x_i)] \tag{2.1.12}$$

is a component of the strain tensor. V_2 is a strain-coupling coefficient, and e_θ and e_ε transform as E. Since E is contained only once in $\Gamma_8\times\Gamma_8$, Eq. (2.1.10) suffices to represent in terms of our ^2E basis the linear splitting of an arbitrary Γ_8 level by the strain components e_θ, e_ε. For the strain components e_{xy}, e_{yz}, e_{zx}, which transform as T_2, we see that, although they cannot lead to a linear splitting of an E state, they can cause such a splitting of Γ_8, since T_2 is contained twice in Γ_8 according to Eq. (2.1.4). There is, however, only one independent parameter needed in general to describe this splitting, because the second parameter is eliminated by considerations of time-reversal invariance. This interaction takes the form

$$\mathscr{H}_{\mathrm{S}}^{(T_2)} = V_3''(S_x e_{yz} + S_y e_{zx} + S_z e_{xy})\mathscr{A}_2 \tag{2.1.13}$$

in our ^2E basis. Correspondingly, there is also a strain-induced orbital Zeeman interaction which takes the form

$$\mathscr{H}_{\mathrm{SZ}}^{(T_2)} = \beta F(H_x e_{yz} + H_y e_{zx} + H_z e_{xy})\mathscr{A}_2 \tag{2.1.14}$$

For the ^2E state of the d^1 or d^9 configuration, one finds (Ham, 1968)

$$V_3'' = \lambda F = 2\lambda V_3/\Delta \tag{2.1.15}$$

to first order in the spin–orbit coupling of ^2E to $^2T_2$, where V_3 is a strain parameter giving the strength of the shear-strain coupling between ^2E and $^2T_2$.

2.2. Static Jahn–Teller Effect for a Doublet State in Cubic Symmetry

The splitting of the orbital degeneracy of a doublet E electronic state in cubic symmetry by a pair of lattice distortion coordinations Q_θ and Q_ε transforming by the representation E is described in accord with Eq. (2.1.2) by the effective Hamiltonian

$$\mathscr{H}_V = E_0\mathscr{I} + V(\mathscr{U}_\theta Q_\theta + \mathscr{U}_\varepsilon Q_\varepsilon) + \tfrac{1}{2}K(Q_\theta^2 + Q_\varepsilon^2)\mathscr{I} \tag{2.2.1}$$

which represents the effect of the electronic Hamiltonian $H_e(\mathbf{q};\mathbf{Q})$ of Eq. (1.2.3) on the degenerate electronic states ψ_θ and ψ_ε appropriate to the symmetric (cubic) configuration ($Q_\theta = Q_\varepsilon = 0$). We have included in Eq. (2.2.1) only the term describing the splitting of the degeneracy which is linear in Q_θ, Q_ε, together with the associated elastic energy $\tfrac{1}{2}K(Q_\theta^2+Q_\varepsilon^2)$, so that Eq. (2.2.1) provides the effective Hamiltonian for the static Jahn–Teller effect with only linear Jahn–Teller coupling. Thus V is the coupling coefficient for linear Jahn–Teller coupling, and E_0 is the electronic energy in the degenerate configuration.

In our treatment of Jahn–Teller effects in a doublet state, we shall consider coupling to only a single pair of lattice coordinates Q_θ, Q_ε, although in a crystal the electronic state interacts with many such modes of distortion. Our analysis is thus really more suitable for a simple molecule than it is for a complex in a crystal, and in Section 5 we shall give some discussion of the proper treatment for a crystal when the Jahn–Teller coupling is with an infinity of lattice phonons. But because this correct treatment is very complicated and has not yet been carried very far, we shall confine our analysis in all of Section 2 to the simpler case with a single pair of Q's. Although this case represents a serious oversimplification of the real situation in a crystal, it provides a tractable model which can provide at least a qualitative understanding of Jahn–Teller effects in solids, and with appropriate choices for the various parameters of the model it is quite successful in permitting a quantitative interpretation of many of the experimental results so far obtained. In our present analysis, therefore, we shall consider Q_θ, Q_ε to represent in some average sense the various lattice coordinates which belong to E and to which the Jahn–Teller coupling is most effective: Toyozawa and Inoue (1966) refer to such Q's as "interaction modes." Alternatively, we may think of Q_θ, Q_ε as representing the appropriate modes of distortion of the cluster made up of the Jahn–Teller ion and its nearest neighbors; although this cluster is principally responsible for the Jahn–Teller coupling, its modes of distortion are not normal modes of vibration of the crystal, so that we are again idealizing the situation when we represent the vibrational properties of such a mode in terms of a simple harmonic oscillator. The Q's appropriate to the nearest neighbors in such a "cluster model" are defined in Table I for an ion in octahedral coordination.

In studying the static Jahn–Teller effect for the doublet, we seek the linear combinations of ψ_θ and ψ_ε which diagonalize $\mathscr{H}_V$ in Eq. (2.2.1) for fixed Q_θ, Q_ε. The associated energy eigenvalues as functions of the Q's then provide the energy surfaces on which we seek absolute minima corresponding to stable configurations. Making the substitution

$$Q_\theta = \varrho \cos\theta, \qquad Q_\varepsilon = \varrho \sin\theta \tag{2.2.2}$$

we find for the eigenstates of $\mathscr{H}_V$ in Eq. (2.2.1)

$$\begin{aligned} \psi_-(\varrho, \theta) &= \psi_\theta \cos(\theta/2) - \psi_\varepsilon \sin(\theta/2) \\ \psi_+(\varrho, \theta) &= \psi_\theta \sin(\theta/2) + \psi_\varepsilon \cos(\theta/2) \end{aligned} \tag{2.2.3}$$

The energies corresponding to these states are

$$E_\pm(Q_\theta, Q_\varepsilon) = E_0 \pm V\varrho + \tfrac{1}{2}K\varrho^2 \tag{2.2.4}$$

These energy surfaces are shown in Fig. 3. The configurations of minimum

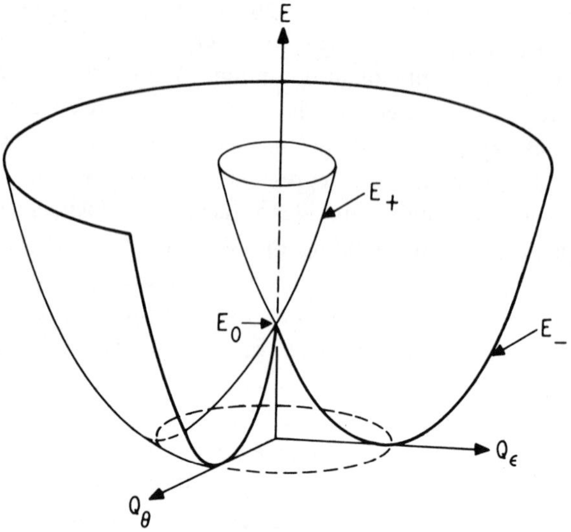

Fig. 3. Energy surfaces $E_{\pm}(Q_\theta, Q_\varepsilon)$ for the vibronic problem of the orbital doublet electronic state with linear Jahn–Teller coupling in cubic or trigonal symmetry. The surfaces have rotational symmetry about the energy axis.

energy on the lower surface are all the points on the circle $\varrho = |V|/K$, as shown originally by Van Vleck (1939b), and the corresponding energy of stabilization, or Jahn–Teller energy E_{JT}, is given by

$$E_{\text{JT}} = V^2/2K \tag{2.2.5}$$

The fact that the energies (2.2.4) are independent of θ, so that the energy surfaces in Fig. 3 have rotational symmetry about the origin, means that there are an infinite number of energetically equivalent configurations of minimum energy corresponding to all the points at the bottom of the trough in Fig. 3. Thus in this case we do not obtain a discrete set of configurations of stable equilibrium, so long as we use the Hamiltonian (2.2.1). We would expect then that, when we include the kinetic energy of the lattice ions, we would find that the system retains a dynamic behavior, corresponding to some sort of motion around the trough in Fig. 3, even for strong Jahn–Teller coupling so long as we consider only linear Jahn–Teller coupling as in Eq. (2.2.1). We shall see in Section 2.3 how such behavior affects the properties of the system.

However, Eq. (2.2.1) is not the most general form that $\mathcal{H}_V$ can take, consistent with cubic symmetry, to describe the coupling of the doublet to Q_θ, Q_ε, and when we use a more general $\mathscr{H}_V$ the rotational symmetry of the energy surfaces in Fig. 3 is removed. We may in particular add the following terms to $\mathscr{H}_V$:

$$V_q[\mathscr{U}_\theta(Q_\varepsilon^2 - Q_\theta^2) + 2\mathscr{U}_\varepsilon Q_\varepsilon Q_\theta] + V_c Q_\theta(Q_\theta^2 - 3Q_\varepsilon^2)\mathscr{I} \qquad (2.2.6)$$

The first of these (sometimes called the "quadratic Jahn–Teller coupling") changes the splitting of the electronic degeneracy by an amount quadratic in the Q's (Liehr and Ballhausen, 1958), while the second does not contribute to the splitting but does introduce a cubic anisotropy with respect to θ that is proportional to $\varrho^3 \cos 3\theta$ (Opik and Pryce, 1957). The energies of the electronic states that diagonalize the resulting $\mathscr{H}_V$ are then, in place of Eq. (2.2.4),

$$E_\pm(Q_\theta, Q_\varepsilon) = E_0 + \tfrac{1}{2}K\varrho^2 + V_c\varrho^3 \cos 3\theta$$
$$\pm[V^2\varrho^2 + V_q^2\varrho^4 - 2VV_q\varrho^3 \cos 3\theta]^{1/2} \qquad (2.2.7)$$

The bottom of the trough in the lower energy surface in Fig. 3 is now warped, with a period of $2\pi/3$ in θ. There are therefore three configurations of stable equilibrium corresponding either to $\theta = 0$, $2\pi/3$, $4\pi/3$ or to $\theta = \pi/3$, π, $5\pi/3$, depending on the signs of V_q and V_c and the relative importance of these two terms in determining the warping. We see then that, whereas linear Jahn–Teller coupling failed to yield distinct configurations of stable equilibrium, we do obtain three such energetically equivalent configurations, separated by low barriers, when the higher-order coupling terms are taken into account (Opik and Pryce, 1957; Liehr and Ballhausen, 1958). If Q_θ, Q_ε are defined as in Table I for octahedral coordination, the set of equilibrium configurations with $\theta = 0$, $2\pi/3$, $4\pi/3$ corresponds to having the octahedron of nearest neighbors elongated, respectively, along the z, x, or y axis (and symmetrically contracted in the perpendicular plane), while the set with $\theta = \pi/3$, π, $5\pi/3$ corresponds to a contraction along, respectively, the y, z, or x axis.

If the barriers between the wells are sufficiently high, the static Jahn–Teller effect for the doublet will therefore correspond to the three stable configurations at the bottoms of these wells. We must now obtain the EPR spin Hamiltonian corresponding to each of these configurations.

For the configuration characterized by a given value for θ, the electronic orbital ground state is ψ_+ of Eq. (2.2.3) if $V < 0$, or ψ_- if $V > 0$. Accordingly, the various terms in the spin Hamiltonian appropriate to this configuration are given by the diagonal matrix elements of the corresponding terms in the effective Hamiltonian of Section 2.1 with respect to ψ_- or ψ_+, as noted originally by Abragam and Pryce (1950). We obtain the following results.

2.2.1. Zeeman Interaction

We have from Eq. (2.1.5)

$$\langle\psi_\pm | \mathscr{H}_Z | \psi_\pm\rangle = g_1\beta(\mathbf{S}\cdot\mathbf{H}) \pm g_2\beta\{(\cos\theta)[S_zH_z - \tfrac{1}{2}(S_xH_x+S_yH_y)] + (\sqrt{3}/2)(\sin\theta)(S_xH_x - S_yH_y)\} \tag{2.2.8}$$

Therefore, if $\theta = 0$ is a stable configuration, we obtain from Eq. (2.2.8) the axially symmetric spin Hamiltonian for the Zeeman interaction (Abragam and Pryce, 1950; Liehr and Ballhausen, 1958)

$$\mathscr{H}_Z(\text{spin}) = g_\parallel\beta S_zH_z + g_\perp\beta(S_xH_x + S_yH_y) \tag{2.2.9}$$

with

$$\begin{aligned} g_\parallel &= g_1 + g_2 = g_s^0 - (8\lambda/\Delta) \\ g_\perp &= g_1 - \tfrac{1}{2}g_2 = g_s^0 - (2\lambda/\Delta) \end{aligned} \tag{2.2.10a}$$

if $V < 0$, where we have used Eq. (2.1.6) for g_1 and g_2 appropriate to the ^2E ground state of the d^1 or d^9 configuration, or

$$\begin{aligned} g_\parallel &= g_1 - g_2 = g_s^0 \\ g_\perp &= g_1 + \tfrac{1}{2}g_2 = g_s^0 - (6\lambda/\Delta) \end{aligned} \tag{2.2.10b}$$

if $V > 0$. The equivalent configurations with $\theta = 2\pi/3$ and $4\pi/3$ according to Eq. (2.2.8) lead to a spin Hamiltonian identical with Eqs. (2.2.9) and (2.2.10) except that the axis of symmetry is respectively the x and y axes. Alternatively, if the stable configurations are $\theta = \pi/3$, π, $5\pi/3$, we find that $\mathscr{H}_Z(\text{spin})$ has axial symmetry respectively about the y, z, and x axes, with $g_\parallel$ and $g_\perp$ given by Eq. (2.2.10b) if $V < 0$ and by (2.2.10a) if $V > 0$.

2.2.2. Hyperfine Interaction

From Eq. (2.1.7) and the expression analogous to Eq. (2.2.8), we obtain for the configuration with $\theta = 0$ the hyperfine interaction

$$\mathscr{H}_I(\text{spin}) = A_\parallel S_zI_z + A_\perp(S_xI_x + S_yI_y) \tag{2.2.11}$$

with

$$\begin{aligned} A_\parallel &= A_1 + A_2 = -P[\varkappa + 6\xi + (8 + 9\xi)(\lambda/\Delta)] \\ A_\perp &= A_1 - \tfrac{1}{2}A_2 = -P[\varkappa - 3\xi + (2 - \tfrac{9}{2}\xi)(\lambda/\Delta)] \end{aligned} \tag{2.2.12a}$$

if $V < 0$, where we have used Eq. (2.1.8), or

$$\begin{aligned} A_{\parallel} &= A_1 - A_2 = -P[\varkappa - 6\xi - 9\xi(\lambda/\Delta)] \\ A_{\perp} &= A_1 + \tfrac{1}{2}A_2 = -P[\varkappa + 3\xi + (6 + \tfrac{9}{2}\xi)(\lambda/\Delta)] \end{aligned} \tag{2.2.12b}$$

if $V > 0$. As we found for the g-tensor, the other values of θ corresponding to stable configurations lead to the same axially symmetric hyperfine interaction as for $\theta = 0$, except that the symmetry axis is along the appropriate axis x, y, or z. In all cases ($V > 0$ and $V < 0$), we find that the hyperfine parameters $A_{\parallel}$ and $A_{\perp}$ given by Eq. (2.2.12a) go with the g-tensor given by Eq. (2.2.10a), and those given by Eq. (2.2.12b) with Eq. (2.2.10b).

2.2.3. Strain

Strain does not affect the spin Hamiltonian corresponding to one of the distorted configurations, at least to the accuracy of the terms in the effective Hamiltonian considered in Section 2.1, but strain having a tetragonal component does displace the energy of the three configurations with respect to each other and thereby destroys their equivalence. From Eq. (2.1.10), we have

$$\langle \psi_{\pm} \mid \mathscr{H}_S^{(E)} \mid \psi_{\pm} \rangle = \pm V_2[e_\theta \cos\theta + e_\varepsilon \sin\theta] \tag{2.2.13}$$

so that, for example, for $\theta_1 = 0$, $\theta_2 = 2\pi/3$, $\theta_3 = 4\pi/3$, we obtain

$$\begin{aligned} \Delta E_1 &= +V_2[e_{zz} - \tfrac{1}{2}(e_{xx} + e_{yy})] \\ \Delta E_2 &= +V_2[e_{xx} - \tfrac{1}{2}(e_{yy} + e_{zz})] \\ \Delta E_3 &= +V_2[e_{yy} - \tfrac{1}{2}(e_{zz} + e_{xx})] \end{aligned} \tag{2.2.14}$$

for $V < 0$ (or the same shift but with reversed sign for $V > 0$). On the other hand, since

$$\langle \psi_+ \mid \mathscr{A}_2 \mid \psi_+ \rangle = \langle \psi_- \mid \mathscr{A}_2 \mid \psi_- \rangle = 0 \tag{2.2.15}$$

a strain having only shear components e_{yz}, e_{zx}, e_{xy} has its linear effect entirely quenched for a static Jahn–Teller effect, according to Eqs. (2.1.13) and (2.1.14).

2.2.4. Effects of Zero-Point Motion

As O'Brien (1964) has pointed out, even for a "static" Jahn–Teller distortion the vibronic wave function is not a δ-function in Q-space at a precisely defined value of θ. The wave function will instead have a finite extent within its well, and we should accordingly average over this spread in θ in obtaining the spin Hamiltonian for the distorted configuration, instead of using the value $\theta = \theta_n$ at the bottom of the well. Taking the average of $\cos(\theta - \theta_n)$ over the vibrational wave function in the nth well as a parameter u,

$$u = \langle \cos(\theta - \theta_n) \rangle_n \tag{2.2.16}$$

we find then in obtaining the g-tensor by averaging Eq. (2.2.8) over the well that Eq. (2.2.10) should be replaced by

$$g_{\parallel} = g_1 \pm u g_2, \qquad g_{\perp} = g_1 \mp \tfrac{1}{2} u g_2 \tag{2.2.17}$$

with the upper sign appropriate to Eq. (2.2.10a) and the lower sign to Eq. (2.2.10b). The hyperfine interaction is similarly altered, with Eq. (2.2.12) replaced by

$$A_{\parallel} = A_1 \pm u A_2, \qquad A_{\perp} = A_1 \mp \tfrac{1}{2} u A_2 \tag{2.2.18}$$

The finite spread of the wave function in the well thus acts to reduce somewhat the anisotropy in g and A.

A basis for an understanding of the spread of the vibronic wave function within its well is provided by the fact that, in just this case of strong Jahn–Teller coupling, we may recapture the validity of the Born–Oppenheimer approximation. Since $E_{\text{JT}} \gg \hbar\omega$ is the condition of strong Jahn–Teller coupling, in the neighborhood of a stable configuration the separation $\sim 4E_{\text{JT}}$ of the two energy surfaces in Fig. 3 is $\gg \hbar\omega$, and this is a necessary condition for the Born–Oppenheimer approximation. Accordingly, if we have sufficiently high barriers so that we can justifiably neglect tunneling between wells, we will have from Eq. (1.2.1) as the result of the Born–Oppenheimer approximation for a vibronic wave function in the nth well

$$\Psi_n(\varrho, \theta) = \psi_{\pm}(\varrho, \theta)\chi(\varrho, \theta - \theta_n) \tag{2.2.19}$$

where $\chi(\varrho, \theta - \theta_n)$ is a vibrational wave function centered at θ_n and satisfying Eq. (1.2.5). If we assume for the vibronic state of lowest energy that χ has a Gaussian dependence on θ,

$$\chi \propto \exp[-b(\theta - \theta_n)^2] \tag{2.2.20}$$

then we obtain (O'Brien, 1964) from Eq. (2.2.16)

$$u = \exp(-1/8b) \tag{2.2.21}$$

A more striking effect of the uncertainty in θ, pointed out by O'Brien (1964), concerns the hyperfine interaction with neighboring ligands. If the wave function were located precisely at $\theta = 0$, for example, we would have, from Eq. (2.2.3), $\psi_- = \psi_\theta \sim (3z^2 - r^2)$, $\psi_+ = \psi_\varepsilon \sim \sqrt{3}(x^2 - y^2)$. The state ψ_+ would in particular form σ bonds only with the four octahedrally coordinated nearest-neighbor ions lying in the x, y plane, so that, if ψ_+ is the Jahn–Teller ground state for $\theta = 0$, we would expect a superhyperfine interaction in the spectrum for this distortion involving only four of the six neighbors. However, if the spread in θ is appreciable, the ground state acquires a small admixture of ψ_θ which then introduces a small hyperfine interaction with the two neighbors previously omitted. We see thus that the dynamic nature of the Jahn–Teller problem can cause qualitative changes in the EPR spectrum even in the limiting case of a static Jahn–Teller effect.

2.3. Dynamic Jahn–Teller Effects Due to Zero-Point Ionic Motion for a Doublet State in Cubic Symmetry

The full vibronic Hamiltonian of Eq. (1.5.2) for the doublet E electronic state, to the accuracy of linear Jahn–Teller coupling, is obtained by adding the ionic kinetic energy operator corresponding to Q_θ and Q_ε to $\mathscr{H}_V$ in Eq. (2.2.1):

$$\mathscr{H} = \mathscr{H}_0 \mathscr{I} + V[Q_\theta \mathscr{U}_\theta + Q_\varepsilon \mathscr{U}_\varepsilon] \tag{2.3.1}$$

with

$$\mathscr{H}_0 = E_0 + (1/2\mu)[P_\theta^2 + P_\varepsilon^2 + \mu^2\omega^2(Q_\theta^2 + Q_\varepsilon^2)] \tag{2.3.2}$$

Here, ω is the angular frequency appropriate to the Q_θ, Q_ε mode of vibration as given by the unperturbed vibrational Hamiltonian $\mathscr{H}_0$ in the absence of Jahn–Teller coupling, and μ is the effective mass appropriate to this mode [for the cluster model, we may take $\mu = M$, where M is the mass of one of the nearest-neighbor ions (Ham, 1968, Appendix II)]. We make the assumption, as discussed in Section 2.2, that this mode is a normal mode of vibration of the lattice, although of course this represents a considerable oversimplification of the actual situation in a real crystal. In writing Eq. (2.3.2), we have replaced the force constant K in Eq. (2.2.1) by $\mu\omega^2$, so that the Jahn–Teller energy may now be expressed in the form equivalent to Eq. (2.2.5)

$$E_{\text{JT}} = V^2/2\mu\omega^2 \tag{2.3.3}$$

The exact eigenstates and energy eigenvalues of $\mathscr{H}$ in Eq. (2.3.1) cannot be obtained by analytic means. For weak Jahn–Teller coupling ($E_{JT} \ll \hbar\omega$), Moffitt and Thorson (1957) have obtained expressions for the eigenvalues by second-order perturbation theory. For stronger coupling, eigenvalues and eigenstates have been obtained numerically by Longuet-Higgins *et al.* (1958) and by Moffitt and Thorson (1958). The eigenstates Ψ_n may be expressed in terms of the electronic states ψ_+ and ψ_- of Eq. (2.2.3) as

$$\Psi_n = \chi_{n1}(\varrho, \theta)\psi_-(\varrho, \theta) + \chi_{n2}(\varrho, \theta)\psi_+(\varrho, \theta) \tag{2.3.4}$$

where we again use the definition (2.2.2) of ϱ and θ, but except in the limit of very strong coupling it is in general not possible to simplify Eq. (2.3.4) into a simple Born–Oppenheimer product. At all values of the coupling coefficient V, the ground state is a vibronic doublet belonging to E, while the first excited state is another doublet belonging to A_1 and A_2 and lying an energy Δ_1 above the ground state. [This accidental degeneracy is lifted when the warping terms as in Eq. (2.2.6) are introduced]. For weak coupling ($E_{JT} \ll \hbar\omega$), Δ_1 is given by

$$\Delta_1 = \hbar\omega[1 - (2E_{JT}/\hbar\omega)] \tag{2.3.5}$$

and for strong coupling ($E_{JT} \gg \hbar\omega$, but no warping),

$$\Delta_1 = \hbar\omega(\hbar\omega/2E_{JT}) \tag{2.3.6}$$

Thus, in the limit as $E_{JT}/\hbar\omega$ becomes very large, this excited state approaches the ground state asymptotically. Values of Δ_1 for linear Jahn–Teller coupling of intermediate strength are shown in Fig. 4.

The Hamiltonian (2.3.1) has full cubic symmetry under simultaneous transformation of both electronic and vibrational operators, so that the exact vibronic eigenstates of $\mathscr{H}$ must belong to the irreducible representations of the symmetry group of the cube and must have the corresponding degeneracies. As we have noted, the ground state is a doublet belonging to E, whatever the values of V. Thus, as discussed in Section 1.5, the twofold degeneracy of the vibronic ground state (which becomes a fourfold degeneracy when we include spin $S = 1/2$) is *not* lifted by the Jahn–Teller coupling. Neither is it lifted when the warping terms as in Eq. (2.2.6) are introduced, since no Hamiltonian of cubic symmetry can lead to a splitting of any degenerate level belonging to a single irreducible representation of the

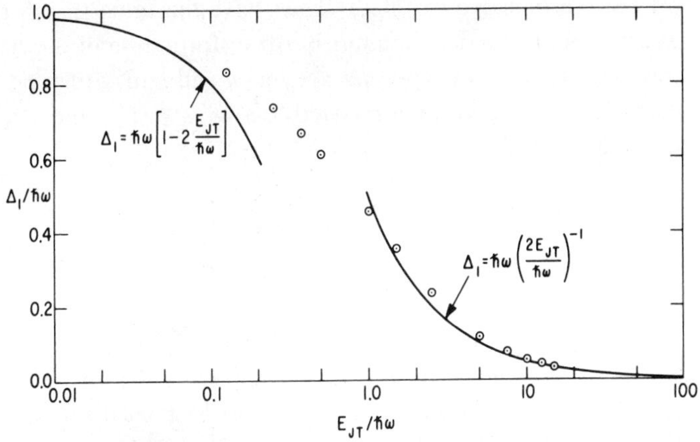

Fig. 4. Excitation energy Δ_1 of the first excited vibronic state for the case of the orbital doublet with linear Jahn–Teller coupling, as a function of the ratio of the Jahn–Teller energy E_{JT} to the mode energy $\hbar\omega$. The curves show the limiting behavior given by Eqs. (2.3.5–6), while the points are exact values from the calculations of Longuet-Higgins *et al.* (1958).

cubic group. The properties of this doublet E ground state therefore must be formally the same, so far as symmetry considerations are concerned, as those of the electronic orbital doublet state in a rigid environment of cubic symmetry when Jahn–Teller effects are ignored, at least so long as excited vibronic states are far away in energy from the ground state relative to the size of any perturbation that we apply. We expect, however, that the Jahn–Teller coupling will modify the values of the parameters that determine the size of the splitting of the ground state under a given perturbation. We shall now examine the relationship of such parameters to those of the original electronic state given in Section 2.1.

We may represent all the interactions given by Eqs. (2.1.5)–(2.1.15) in Section 2.1 in their effect on the electronic orbital doublet ψ_θ, ψ_ε by the following general operator:

$$\mathscr{V} = G_1\mathscr{I} + G_2\mathscr{A}_2 + G_\theta\mathscr{U}_\theta + G_\varepsilon\mathscr{U}_\varepsilon \tag{2.3.7}$$

Here, the G's are functions of the components of the external perturbations (magnetic field, strain, etc.) and of the other operators (e.g., electronic and nuclear spin), but they are independent of Q_θ and Q_ε; G_1 is symmetric under cubic transformation of these components, G_2 belongs to the irreducible representation A_2, and G_θ and G_ε transform as partners belonging to E. We then denote by $\Psi_{g\theta}$ and $\Psi_{g\varepsilon}$ the wave functions of the two components of the vibronic ground state which have the same transformation properties as ψ_θ and ψ_ε (under simultaneous transformation of the electronic coordinates and the Q's). Two real parameters q and p may then be defined (Ham, 1968) in terms of the matrix elements of $\mathscr{U}_\theta$, $\mathscr{U}_\varepsilon$, and $\mathscr{A}_2$ within the ground state as follows:

$$q = -\langle \Psi_{g\theta} \mid \mathscr{U}_\theta \mid \Psi_{g\theta} \rangle = \langle \Psi_{g\varepsilon} \mid \mathscr{U}_\theta \mid \Psi_{g\varepsilon} \rangle = \langle \Psi_{g\varepsilon} \mid \mathscr{U}_\varepsilon \mid \Psi_{g\theta} \rangle \quad (2.3.8a)$$

$$p = i\langle \Psi_{g\theta} \mid \mathscr{A}_2 \mid \Psi_{g\varepsilon} \rangle \quad (2.3.8b)$$

If then we let $\mathscr{U}_{g\theta}$, $\mathscr{U}_{g\varepsilon}$, and $\mathscr{A}_{g2}$ denote operators which, with respect to the basis $\Psi_{g\theta}$, $\Psi_{g\varepsilon}$, are given by exactly the same matrices as those which define $\mathscr{U}_\theta$, $\mathscr{U}_\varepsilon$, and $\mathscr{A}_2$ in Eq. (2.1.3) with respect to the basis ψ_θ, ψ_ε, we see from Eq. (2.3.8) that the operator $\mathscr{V}$ in Eq. (2.3.7) is equivalent in its effect on the ground state to the operator $\mathscr{V}_g$ given by

$$\mathscr{V}_g = G_1\mathscr{I} + pG_2\mathscr{A}_{g2} + q(G_\theta \mathscr{U}_{g\theta} + G_\varepsilon \mathscr{U}_{g\varepsilon}) \quad (2.3.9)$$

Matrix elements of the symmetric term G_1 are thus unaffected by the Jahn–Teller coupling, but those of the terms in G_2 and in G_θ and G_ε are reduced by the "reduction factors" p and q, respectively. For small coupling ($E_{JT} \ll \hbar\omega$), p and q are given by perturbation theory (Ham, 1968) as

$$p = 1 - (4E_{JT}/\hbar\omega), \qquad q = 1 - (2E_{JT}/\hbar\omega) \quad (2.3.10)$$

while for strong coupling (without warping) one obtains the asymptotic result (Williams *et al.*, 1969)

$$p \sim (\hbar\omega/4E_{JT})^2, \qquad q \sim \tfrac{1}{2}[1 + (\hbar\omega/4E_{JT})^2] \quad (2.3.11)$$

Thus, in the limit $E_{JT} \gg \hbar\omega$, we must have $p \simeq 0$, $q \simeq \frac{1}{2}$. It may be shown, moreover (Ham, 1968), that so long as only linear Jahn–Teller coupling is considered (no warping), we have the general relation

$$q = \tfrac{1}{2}(1 + p) \quad (2.3.12)$$

Figure 5 shows exact values of q and p obtained for intermediate coupling strengths from calculations by Child and Longuet-Higgins (1962). These are well approximated over the range $0.1 \leq E_{JT}/\hbar\omega \leq 3.0$ by the empirical formula

$$p = \exp[-1.974(E_{JT}/\hbar\omega)^{0.761}] \quad (2.3.13)$$

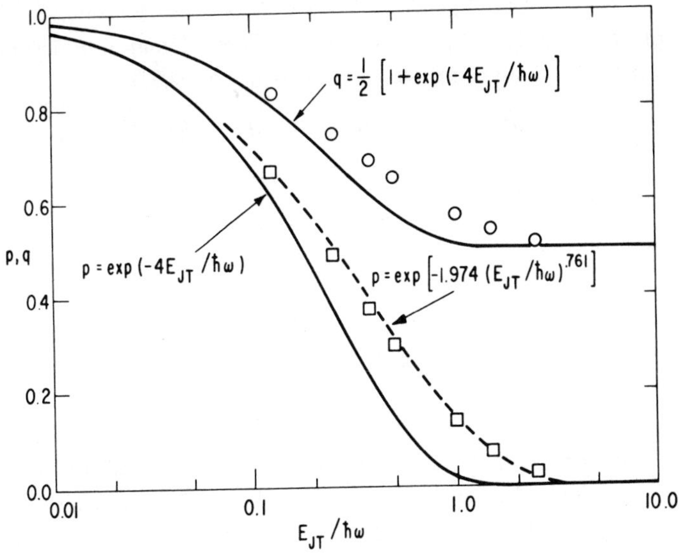

Fig. 5. Reduction factors p and q for the vibronic ground state in the case of the orbital doublet with linear Jahn–Teller coupling, as a function of the ratio of the Jahn–Teller energy E_{JT} to the mode energy $\hbar\omega$. The solid curves show the approximate expressions for q and p given by Eq. (2.3.14), which extends the perturbation results obtained for weak Jahn–Teller coupling; values for q have been related to those for p by the exact equation (2.3.12). The points are exact values from the calculations of Child and Longuet-Higgins (1962). The dashed curve is the expression given by Eq. (2.3.13), which is fitted approximately to the calculated points over the range $0.1 \leq E_{JT}/\hbar\omega \leq 3.0$.

Also shown in Fig. 5 are the curves obtained by simple extrapolation of Eq. (2.3.10) using the expression

$$p \cong \exp(-4E_{JT}/\hbar\omega) \tag{2.3.14}$$

which is a good approximation for $(E_{JT}/\hbar\omega) \lesssim 0.1$. We shall consider below how these results are modified by departures from linear Jahn–Teller coupling.

The various terms in the effective Hamiltonian for the electronic ^2E state in cubic symmetry, as given in Section 2.1 from crystal-field theory, are thus modified by a dynamic Jahn–Teller effect so that we have the following terms in the effective Hamiltonian for the vibronic ^2E ground state.

2.3.1. Zeeman Interaction

In place of Eq. (2.1.5), we now have for the Zeeman interaction in the vibronic ^2E ground state, in accordance with Eq. (2.3.9),

$$\mathscr{H}_{gZ} = g_1\beta(\mathbf{S}\cdot\mathbf{H})\mathscr{I} + \tfrac{1}{2}qg_2\beta\{[3S_zH_z - (\mathbf{S}\cdot\mathbf{H})]\mathscr{U}_{g\theta} + \sqrt{3}\,(S_xH_x - S_yH_y)\mathscr{U}_{g\varepsilon}\} \tag{2.3.15}$$

Taking ξ to be a unit vector parallel to $\mathbf{H}$ with components ζ_x, ζ_y, ζ_z, and assuming $g_1 \gg |g_2|$ so that it is a good approximation to keep only the part of the g_2 term that is diagonal in S_ζ, we find (Ham, 1968) that, *in the absence of strain*, Eq. (2.3.15) leads to two spin transitions corresponding to the anisotropic g-factors

$$g_\pm = g_1 \pm qg_2[1 - 3(\zeta_x{}^2\zeta_y{}^2 + \zeta_y{}^2\zeta_z{}^2 + \zeta_z{}^2\zeta_x{}^2)]^{1/2} \tag{2.3.16}$$

This result was originally given by Abragam and Pryce (1950) for the crystal-field case ($q = 1$) and recently by Coffman (1965, 1968) and Höchli (1967) using the theory of Bersuker (1962, 1963) for a dynamic Jahn–Teller effect resulting from tunneling ($q \simeq \frac{1}{2}$).

2.3.2. Hyperfine Interaction

In place of Eq. (2.1.7), we now have

$$\mathscr{H}_{gI} = A_1(\mathbf{S}\cdot\mathbf{I})\mathscr{I} + \tfrac{1}{2}qA_2\{[3S_zI_z - (\mathbf{S}\cdot\mathbf{I})]\mathscr{U}_{g\theta} + \sqrt{3}\,(S_xI_x - S_yI_y)\mathscr{U}_{g\varepsilon}\} \tag{2.3.17}$$

If we keep only terms diagonal in S_ζ and in the eigenvalue m of I_ζ, on combining Eqs. (2.3.15) and (2.3.17) we obtain resonance frequencies *in the absence of strain*

$$h\nu_m = (g_1\beta H + A_1m) \pm q(g_2\beta H + A_2m)[1 - 3(\zeta_x{}^2\zeta_y{}^2 + \zeta_y{}^2\zeta_z{}^2 + \zeta_z{}^2\zeta_x{}^2)]^{1/2} \tag{2.3.18}$$

2.3.3. Strain

We now have instead of Eq. (2.1.10) a splitting of the vibronic ground state under strain* given by

$$\mathscr{H}_{gS}^{(E)} = qV_2[e_\theta\mathscr{U}_{g\theta} + e_\varepsilon\mathscr{U}_{g\varepsilon}] \tag{2.3.19}$$

The strain coefficient that is directly measured experimentally with a dynamic Jahn–Teller effect is thus qV_2 instead of the electronic strain-coupling coefficient V_2.

* An alternative, but equivalent treatment of the effect of strain is given in the appendix.

For the shear strains e_{xy}, e_{yz}, e_{zx}, we replace Eqs. (2.1.13) and (2.1.14) by

$$\mathscr{H}_{gS}^{(T_2)} = pV_3''(S_x e_{yz} + S_y e_{zx} + S_z e_{xy})\mathscr{A}_{g2} \tag{2.3.20}$$

and

$$\mathscr{H}_{gSZ}^{(T_2)} = p\beta F(H_x e_{yz} + H_y e_{zx} + H_z e_{xy})\mathscr{A}_{g2} \tag{2.3.21}$$

2.3.4. Effect of Random Strain

The role of inhomogeneous strain, which may vary at random from the site of one Jahn–Teller ion in the crystal to another, and which may result from inhomogeneities, dislocations, or nearby point defects, is very often crucial in determining the EPR spectrum. Taking the local strain at the site of a given ion to be given by particular values for e_θ and e_ε, we must combine the Zeeman and hyperfine interactions in Eqs. (2.3.15) and (2.3.17) with the strain term in Eq. (2.3.19) and find the eigenstates of the resulting Hamiltonian. If the strains are small, in the sense that

$$|V_2|(e_\theta^2 + e_\varepsilon^2)^{1/2} \ll \tfrac{1}{2}|g_2\beta H + A_2 m|[1 - 3(\zeta_x^2\zeta_y^2 + \zeta_y^2\zeta_z^2 + \zeta_z^2\zeta_x^2)]^{1/2} \tag{2.3.22}$$

then the effect of random strain is simply to broaden somewhat the resonance lines as given by Eq. (2.3.18) (Ham, 1968). We note, however, that the condition (2.3.22) may be satisfied for some values of m but not for others, and that since the right-hand side of the inequality vanishes for **H** along [111] the condition will not be satisfied for any line for certain orientations of **H**. On the other hand, if the strains are sufficiently large so that the inequality (2.3.22) is reversed (in practice, this often requires only a small strain, of 10^{-4} or less), the eigenstates are determined by the strain, and the resonance frequencies are found to be given by (Ham, 1968; Chase, 1968)

$$\begin{aligned} h\nu_\pm(m) = {} & (g_1\beta H + A_1 m) \pm \tfrac{1}{2}q\,|g_2\beta H + A_2 m| \\ & \times [e_\theta(3\zeta_z^2 - 1) + e_\varepsilon\sqrt{3}(\zeta_x^2 - \zeta_y^2)](e_\theta^2 + e_\varepsilon^2)^{-1/2} \end{aligned} \tag{2.3.23}$$

If then we define angles φ and α by the relations

$$\begin{aligned} \cos\varphi &= e_\theta/(e_\theta^2 + e_\varepsilon^2)^{1/2} \\ \sin\varphi &= e_\varepsilon/(e_\theta^2 + e_\varepsilon^2)^{1/2} \\ \cos\alpha &= \tfrac{1}{2}(3\zeta_z^2 - 1)[1 - 3(\zeta_x^2\zeta_y^2 + \zeta_y^2\zeta_z^2 + \zeta_z^2\zeta_x^2)]^{-1/2} \\ \sin\alpha &= \tfrac{1}{2}\sqrt{3}(\zeta_x^2 - \zeta_y^2)[1 - 3(\zeta_x^2\zeta_y^2 + \zeta_y^2\zeta_z^2 + \zeta_z^2\zeta_x^2)]^{-1/2} \end{aligned} \tag{2.3.24}$$

we may express Eq. (2.3.23) as

$$h\nu_{\pm}(m) = (g_1\beta H + A_1 m) \pm q(g_2\beta H + A_2 m)$$
$$\times [1 - 3(\zeta_x{}^2\zeta_y{}^2 + \zeta_y{}^2\zeta_z{}^2 + \zeta_z{}^2\zeta_x{}^2)]^{1/2} \cos(\varphi - \alpha) \quad (2.3.25)$$

If the strains are random, the angle φ assumes values, at random, in the range $0 \leq \varphi \leq 2\pi$. The strain-broadened lines corresponding to $h\nu_+(m)$ and $h\nu_-(m)$ then coincide, and the extremities of the resulting line coincide with the resonance frequencies given by Eqs. (2.3.18). The absorption intensity in this line is then described by the shape function

$$g(\nu) = (1/\pi)[(\Delta\nu)^2 - (\nu - \nu_0)^2]^{-1/2} \quad (2.3.26)$$

for $|\nu - \nu_0| \leq |\Delta\nu|$ and $g(\nu) = 0$ for $|\nu - \nu_0| > |\Delta\nu|$, as shown in Fig. 6. The case of strong, but random residual strain thus leads to a spectrum exhibiting absorption peaks which coincide with those expected in the complete absence of strain. Moreover, this spectrum will be insensitive to applied stress until the resulting strain becomes comparable to the residual strain.

2.3.5. Effects of Warping

When we go beyond the approximation of linear Jahn–Teller coupling to include the warping terms (2.2.6) in our vibronic Hamiltonian (2.3.1), the principal qualitative effect on the vibronic energy levels in the dynamic

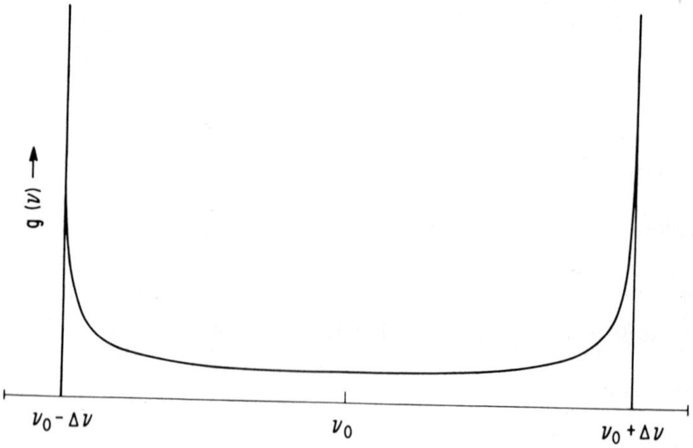

Fig. 6. Shape function $g(\nu)$ of the absorption intensity in the strain-broadened EPR lines $\hbar\nu_{\pm}(m)$, as given by Eqs. (2.3.25–26) for random strain.

regime is that the accidental degeneracy of the excited doublets belonging to A_1 and A_2 is lifted. However, the E ground state is qualitatively unaffected, and we may continue to describe its behavior in terms of Eq. (2.3.9) and the parameters q and p defined in Eq. (2.3.8). Of course, q and p will now have values which may depart somewhat from Eqs. (2.3.13) and (2.3.14), and in particular Eq. (2.3.12) will no longer hold exactly. There are no calculations available from which the dependence of q and p on the warping may be obtained in the general dynamic regime, but their values are probably given approximately by those for linear Jahn–Teller coupling so long as the warping is small compared with the Jahn–Teller energy (however, see below).

O'Brien (1964) and, more recently, several others have made calculations of the effects of warping in the limiting case of a very strong Jahn–Teller coupling ($E_{JT} \gg \hbar\omega$), when it is a good approximation to ignore the interaction between states associated with the upper and lower energy surfaces in Fig. 3. Instead of requiring the two terms involving both ψ_+ and ψ_- as in Eq. (2.3.4), the vibronic eigenfunctions with energies near the ground state now may be expressed as a single Born–Oppenheimer product,

$$\Psi_n = \chi_n(\varrho, \theta)\psi_\pm(\varrho, \theta) \tag{2.3.27}$$

with whichever of ψ_+ or ψ_- belongs to the lower energy surface. With only linear Jahn–Teller coupling, because of the rotational symmetry in Fig. 3, we have

$$\chi_n(\varrho, \theta) = (2\pi)^{-1/2} f(\varrho) \exp(in\,\theta) \tag{2.3.28}$$

Here, n must be half an odd integer in order to satisfy the requirement that $\Psi_n(\varrho, \theta)$ be single-valued as a function of θ, the functions ψ_+ and ψ_- as defined by Eq. (2.2.3) changing sign under a rotation of 2π. For states of low energy, the radial function $f_0(\varrho)$ is large only within the trough in Fig. 3, and it is, to a first approximation, independent of n for such states. The energy of the low-energy excited states relative to the ground state ($n = \pm\frac{1}{2}$) is then given by

$$\mathscr{E}_n - \mathscr{E}_{\pm 1/2} \simeq (n^2 - \tfrac{1}{4})(\hbar^2/2\mu\varrho_0{}^2) \tag{2.3.29}$$

where

$$\varrho_0 = \langle f_0 \,|\, \varrho \,|\, f_0 \rangle \simeq |\, V \,|/\mu\omega^2 \tag{2.3.30}$$

However, when the warping terms are included, the complete rotational symmetry with respect to θ is replaced by symmetry under rotation only of $\pm 2\pi/3$. Assuming the warping to be a small perturbation relative to the depth of the trough, we find on taking the diagonal matrix element of the warping terms (2.2.6) with respect to ψ_+ or ψ_- that the warping makes a contribution $(\mp V_q\varrho^2 + V_c\varrho^3)\cos 3\theta$ to the potential energy in the vibrational Schrödinger equation which $\chi_n(\varrho, \theta)$ in Eq. (2.3.27) must satisfy. The warping thus mixes states of the form (2.3.28) which differ in n by multiples of three. For low-energy states, we may then, to a good approximation, take in place of Eq. (2.3.28)

$$\chi_n(\varrho, \theta) = f_0(\varrho)\phi_n(\theta) \tag{2.3.31}$$

where $\phi_n(\theta)$ is an eigenstate of the equation

$$[-(\hbar^2/2\mu\varrho_0{}^2)(d^2/d\theta^2) + (\mp V_q\varrho_0{}^2 + V_c\varrho_0{}^3)(\cos 3\theta) - \varepsilon_n]\phi_n(\theta) = 0 \tag{2.3.32}$$

and the boundary condition

$$\phi_n(\theta + 2\pi) = -\phi_n(\theta) \tag{2.3.33}$$

for the eigenvalue ε_n. Expanding

$$\phi_n(\theta) = \sum_m a_m^{(n)} \exp(im\theta) \tag{2.3.34}$$

and defining

$$\begin{aligned} \alpha &= \hbar^2/2\mu\varrho_0{}^2 \simeq \hbar\omega(\hbar\omega/4E_{JT}) \\ \beta &= -(\mp V_q\varrho_0{}^2 + V_c\varrho_0{}^3) \end{aligned} \tag{2.3.35}$$

we obtain the recurrence relation

$$(\varepsilon_n - \alpha m^2)a_m^{(n)} + \tfrac{1}{2}\beta(a_{m-3}^{(n)} + a_{m+3}^{(n)}) = 0 \tag{2.3.36}$$

O'Brien (1964) has obtained the eigenvalues of Eq. (2.3.36) as a function of the ratio $|\beta|/\alpha$ for the states of lowest energy. Her results are shown in Fig. 7. For $\beta = 0$, the eigenvalues agree with Eq. (2.3.29). Levels with $|n| = 3/2, 9/2, 15/2, \ldots$ comprise accidentally degenerate states belonging to A_1 and A_2, and these split for nonzero β, while states belonging to E $(|n| = 1/2, 5/2, 7/2, 11/2, \ldots)$ remain doubly degenerate. The eigenstates (2.3.34) may be written for arbitrary $|\beta|/\alpha$ in the general form (O'Brien, 1964)

$$\begin{aligned} \phi_A &= a_{3/2}\cos(3\theta/2) + a_{9/2}\cos(9\theta/2) + a_{15/2}\cos(15\theta/2) + \cdots \\ \phi_{A'} &= a_{3/2}\sin(3\theta/2) + a_{9/2}\sin(9\theta/2) + a_{15/2}\sin(15\theta/2) + \cdots \\ \phi_E &= a_{1/2}\cos(\theta/2) + a_{5/2}\cos(5\theta/2) + a_{7/2}\cos(7\theta/2) + a_{11/2}\cos(11\theta/2) + \cdots \\ \phi_{E'} &= a_{1/2}\sin(\theta/2) - a_{5/2}\sin(5\theta/2) + a_{7/2}\sin(7\theta/2) - a_{11/2}\sin(11\theta/2) + \cdots \end{aligned} \tag{2.3.37}$$

If ψ_+ goes with the lower energy surface ($V < 0$), the vibronic wave functions in Eq. (2.3.27) then transform as follows: $\phi_A\psi_+ \sim A_2$, $\phi_{A'}\psi_+ \sim A_1$, $\phi_E\psi_+ \sim E_\varepsilon$, $\phi_{E'}\psi_+ \sim E_\theta$. On the other hand, if ψ_- goes with the lower surface ($V > 0$) we have $\phi_A\psi_- \sim A_1$, $\phi_{A'}\psi_- \sim A_2$, $\phi_E\psi_- \sim E_\theta$ and $(-)\phi_{E'}\psi_- \sim E_\varepsilon$. The relative energies of these states then vary as a function of $|\beta|/\alpha$

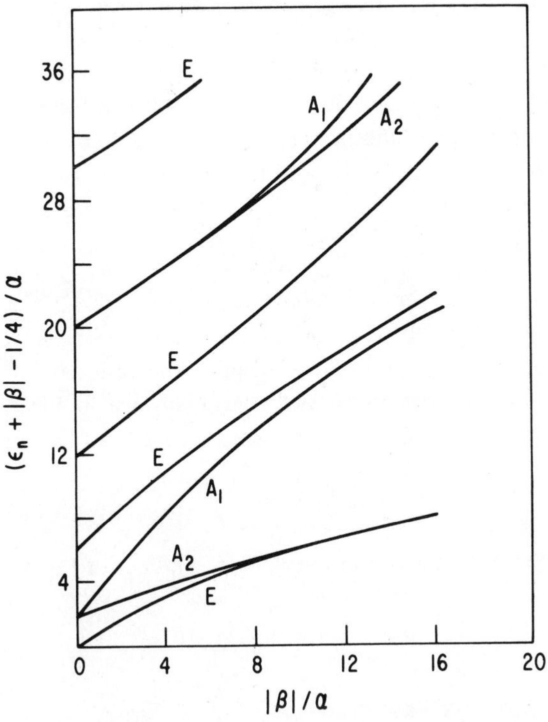

Fig. 7. Relative energies of the lowest vibronic energy levels of an orbital doublet electronic state in cubic symmetry with strong Jahn–Teller coupling ($E_{JT} \gg \hbar\omega$), including the effects of warping, as calculated by O'Brien (1964). The energies ε_n are the eigenvalues of Eq. (2.3.36) of the text, and they are plotted as functions of the ratio $|\beta|/\alpha$ of the parameters defined in Eq. (2.3.35). The height of the barrier separating adjacent wells is $2|\beta|$. The levels are labeled in the figure according to the irreducible representation of the symmetry group of the cube to which each level belongs, the labeling being appropriate to the cases (a) $V < 0$, $\beta > 0$ and (b) $V > 0$, $\beta < 0$; the labels A_1 and A_2 must be interchanged for the cases (c) $V < 0$, $\beta < 0$ and (d) $V > 0$, $\beta > 0$ [V is the Jahn–Teller coupling coefficient in Eq. (2.2.1)].

as shown in Fig. 7, where the labeling of the levels is appropriate for $\psi_+(V < 0)$ if $\beta > 0$ or for $\psi_-(V > 0)$ if $\beta < 0$. For the other two cases, $\psi_+(V < 0)$ $(\beta < 0)$ and $\psi_-(V > 0)$ $(\beta > 0)$, the energy values are the same, but the labels A_1 and A_2 in Fig. 7 must be interchanged. We see from Fig. 7 that the ground state remains an E state for all values of β/α, but that as $|\beta|/\alpha$ increases the lowest A_1 or A_2 state approaches it asymptotically. As we shall discuss in Section 2.4, this behavior represents the transition to the static Jahn–Teller limit.

We may now evaluate the parameters p and q for the ground state when warping is present, for the strong Jahn–Teller coupling case treated by O'Brien. From the definition of p in Eq. (2.3.8b) and the fact that diagonal matrix elements of $\mathscr{A}_2$ with respect to ψ_+ or ψ_- vanish, as in Eq. (2.2.15), we see that

$$p = 0 \tag{2.3.38a}$$

so long as the single Born–Oppenheimer product (2.3.27) is an accurate representation of the vibronic state. This result has been generalized by Williams *et al.* (1969) to take into account the residual interaction with the states associated with the upper energy surface in Fig. 3. They find, with warping present, that the asymptotic expression for p in Eq. (2.3.11) is changed to

$$p \sim 2(\hbar\omega/4E_{\mathrm{JT}})^2 c_3 \tag{2.3.38b}$$

where c_3 is given in terms of the solutions ϕ_E and $\phi_{E'}$ of Eq. (2.3.32) by

$$c_3 = \int_0^{2\pi} \phi_E(\theta)(\partial/\partial\theta)\phi_{E'}(\theta)\, d\theta \tag{2.3.39}$$

Using Eq. (2.3.37), we have then

$$c_3 = (\pi/2)(a_{1/2}^2 - 5a_{5/2}^2 + 7a_{7/2}^2 - 11a_{11/2}^2 + \cdots) \tag{2.3.40}$$

Numerical values for c_3 as a function of $|\beta|/\alpha$ are shown in Fig. 8 from calculations by Williams *et al.* (1969). For q, using the electronic matrix elements

$$\begin{aligned} \langle\psi_+ | \mathscr{U}_\theta | \psi_+\rangle &= -\langle\psi_- | \mathscr{U}_\theta | \psi_-\rangle = \cos\theta \\ \langle\psi_+ | \mathscr{U}_\varepsilon | \psi_+\rangle &= -\langle\psi_- | \mathscr{U}_\varepsilon | \psi_-\rangle = \sin\theta \end{aligned} \tag{2.3.41}$$

we have from Eqs. (2.3.8a) and (2.3.37) for the limiting case of the Born–Oppenheimer product (2.3.27) considered by O'Brien the result

$$q = c_2 \tag{2.3.42}$$

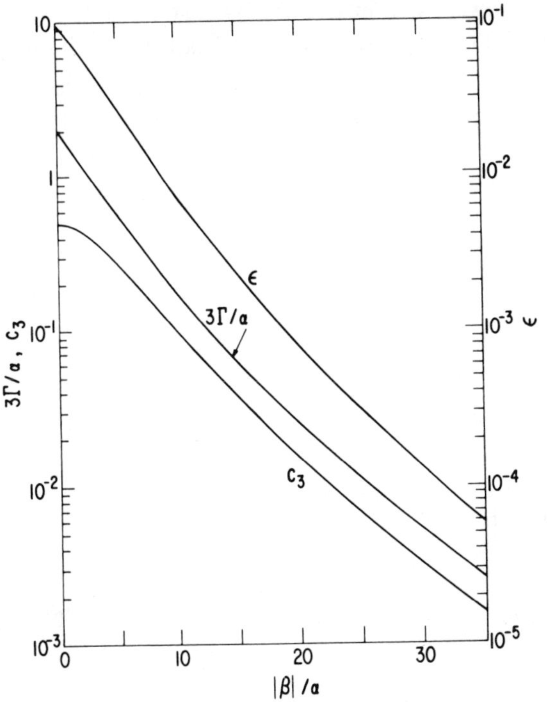

Fig. 8. Tunneling splitting 3Γ and the parameters c_3 and ε defined by Eqs. (2.3.39) and (2.4.12) for the lowest vibronic energy levels of an orbital doublet electronic state in cubic symmetry with strong Jahn–Teller coupling ($E_{\mathrm{JT}} \gg \hbar\omega$), including the effects of warping, plotted as functions of the ratio $|\beta|/\alpha$ of the parameters defined in Eq. (2.3.35). (After Williams *et al.*, 1969.)

where

$$c_2 = \int_0^{2\pi} \phi_E(\theta)\phi_{E'}(\theta) \sin\theta \, d\theta$$
$$= \tfrac{1}{2}\pi(a_{1/2}^2 + 2a_{5/7}a_{7/2} + 2a_{11/2}a_{13/2} + \cdots) \tag{2.3.43}$$

The more general asymptotic result corresponding to Eq. (2.3.38b) has been found (Williams *et al.*, 1969) to be given by

$$q \sim [1 + (\hbar\omega/4E_{\mathrm{JT}})^2]c_2 \mp (V_q\varrho_0{}^2/4E_{\mathrm{JT}})c_4 \tag{2.3.44}$$

where

$$c_4 = \int_0^{2\pi} \phi_E(\theta)\phi_{E'}(\theta)(\sin 2\theta + \sin 4\theta) \, d\theta \tag{2.3.45}$$

and the upper (lower) sign in Eq. (2.3.44) corresponds to $V < 0$ ($V > 0$). It follows from Eq. (2.3.43), the normalization condition for the a_m of ϕ_E and $\phi_{E'}$,

$$\pi(a_{1/2}^2 + a_{5/2}^2 + a_{7/2}^2 + a_{11/2}^2 + \cdots) = 1 \tag{2.3.46}$$

and the inequality

$$2\,|\,a_n\,|\,|\,a_m\,| \leq a_n{}^2 + a_m{}^2 \tag{2.3.47}$$

that

$$c_2 \leq \tfrac{1}{2} \tag{2.3.48}$$

in general, whatever the form of the warping. The value of c_2 has been calculated as a function of $|\,\beta\,|/\alpha$ by Williams *et al.* (1969) and also by Slonczewski (1969) and by Fletcher (Fletcher, 1967; Fletcher and Stevens,

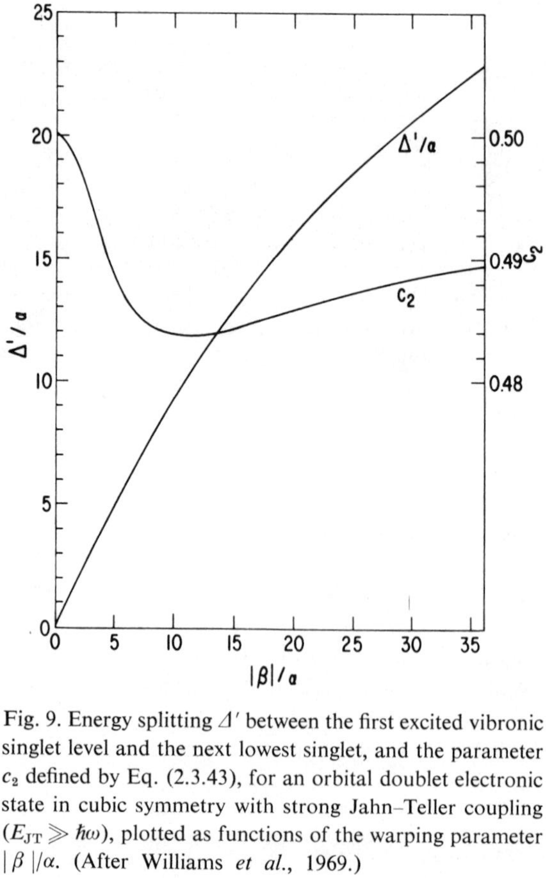

Fig. 9. Energy splitting Δ' between the first excited vibronic singlet level and the next lowest singlet, and the parameter c_2 defined by Eq. (2.3.43), for an orbital doublet electronic state in cubic symmetry with strong Jahn–Teller coupling ($E_{JT} \gg \hbar\omega$), plotted as functions of the warping parameter $|\,\beta\,|/\alpha$. (After Williams *et al.*, 1969.)

1969). The resulting values are shown in Fig. 9. Slonczewski has noted in particular that c_2 lies within the bounds

$$0.484 < c_2 \leq \tfrac{1}{2} \qquad (2.3.49)$$

for all values of β/α and therefore departs only very slightly from the value $\frac{1}{2}$ given by the upper bound (2.3.48). The same is therefore true of q, according to Eq. (2.3.42) or (2.3.44), when the Born–Oppenheimer approximation (2.3.27) is accurate, and we have $E_{JT} \gg \hbar\omega$ and $E_{JT} \gg | V_q | \varrho_0^2$, no matter what the value of the warping parameter β/α. In contrast to this result for strong coupling, we have seen, according to Eq. (2.3.12), that we must have $\frac{1}{2} \leq q \leq 1$ for a dynamic Jahn–Teller effect with only linear coupling and no warping. The value of q relative to the value $\frac{1}{2}$ therefore provides a useful criterion (Ham, 1968) as to which situation is the more appropriate in the interpretation of any given case.

2.3.6. Higher-Order Effects

Accompanying the partial quenching by the Jahn–Teller coupling of the effect of the operators in G_2, G_θ, and G_ε in Eq. (2.3.7) on the ground state, through introduction of the reduction factors p and q, there result finite matrix elements of these operators between the ground state and excited vibronic states (Ham, 1965, 1968). As a consequence, there are higher-order effects of these operators within the ground state, which involve products of G_2, G_θ, and G_ε and which may be calculated by perturbation theory. From considerations of symmetry, we may write down the form which these second-order perturbations on the ground state take in general:

$$\begin{aligned}\mathscr{V}_g^{(2)} &= a(G_\varepsilon^2 + G_\theta^2)\mathscr{I} + b[(G_\varepsilon^2 - G_\theta^2)\mathscr{U}_{g\theta} + (G_\theta G_\varepsilon + G_\varepsilon G_\theta)\mathscr{U}_{g\varepsilon}] \\ &\quad + ci(G_\varepsilon G_\theta - G_\theta G_\varepsilon)\mathscr{A}_{g2} + dG_2^2\mathscr{I} + ei[G_2 G_\varepsilon \mathscr{U}_{g\theta} - G_2 G_\theta \mathscr{U}_{g\varepsilon}] \qquad (2.3.50)\end{aligned}$$

The values which these coefficients a, b, c, ... take in several simple cases have been given elsewhere (Ham, 1968).

The perturbations on the electronic doublet which we considered in Section 2.1 have all had the form of Eq. (2.3.7) for G's which are *independent* of Q_θ and Q_ε. We must point out that other perturbations of the same form as Eq. (2.3.7), but in which the G's depend on Q_θ and Q_ε, are of course also possible. Indeed, O'Brien (1964) has given some consideration to the effect of terms of this sort on the g-factor. We will not consider these here. They may again be handled using appropriate reduction factors, although these factors of course are different from p and q, and the form of the effective interaction is not so simply related to Eq. (2.3.7) as is Eq. (2.3.9) because of the dependence of the G's on Q_θ and Q_ε. (See, however, the appendix.)

2.4. The Transition from Dynamic to Static Jahn–Teller Effects in the Limit of Strong Coupling

We have seen in connection with Fig. 7 of the preceding section that the doublet E ground state is approached asymptotically by an excited singlet state, either A_1 or A_2, when warping is introduced in addition to a strong linear Jahn–Teller coupling. We found in Section 2.2 that the warping produces three equivalent wells in the lower energy surface, and that when the barrier between the wells is sufficiently high the vibronic ground state comprises three energetically equivalent states, each of which represents the ground state of the distorted configuration corresponding to one of the wells. It is therefore clear that the approach of the singlet state to the doublet ground state as the warping is increased represents the transition from dynamic to static Jahn–Teller effect. The energy separation between the singlet and doublet, which we shall label 3Γ, thus represents the effect of the residual interaction, through the barriers, of the states of lowest energy in each well, or, in other words, of the system's ability to tunnel from one well to another. In this section, we shall consider the properties of the system in this transition region when the tunneling splitting 3Γ [Bersuker's (1962) "inversion splitting"] is fairly small but not necessarily negligible. The ratio of 3Γ to the energy parameter α of Eq. (2.3.35) is plotted in Fig. 8 as a function of $|\beta|/\alpha$, where $2|\beta|$ is the height of the barrier separating adjacent wells in Eq. (2.3.32). These plotted values from calculations by Williams *et al.* (1969) are equivalent to those of O'Brien given in Fig. 7. The energy splitting Δ' between this lowest singlet and the next lowest singlet (which of course is degenerate with the lowest singlet when $\beta = 0$, as seen in Fig. 7) is shown in Fig. 9.

When the lowest excited singlet state Ψ_{A1} or Ψ_{A2} is close to the ground state, we must clearly include it in setting up the secular equation for the effect of any perturbation on the ground state. We must first define a parameter analogous to q and p as defined by Eq. (2.3.8) but given by the matrix elements of $\mathscr{U}_\theta$ and $\mathscr{U}_\varepsilon$ between the ground-state doublet and excited singlet, namely

$$r = \langle \Psi_{A1} | \mathscr{U}_\theta | \Psi_{g\theta} \rangle = \langle \Psi_{A1} | \mathscr{U}_\varepsilon | \Psi_{g\varepsilon} \rangle \qquad (2.4.1a)$$

if the singlet is A_1, or

$$r = \langle \Psi_{A2} | \mathscr{U}_\varepsilon | \Psi_{g\theta} \rangle = -\langle \Psi_{A2} | \mathscr{U}_\theta | \Psi_{g\varepsilon} \rangle \tag{2.4.1b}$$

if the singlet is A_2. The matrix elements of any perturbation having the general form of Eq. (2.3.7) are then as follows with respect to these orbital states:

	Ψ_{A1}	$\Psi_{g\theta}$	$\Psi_{g\varepsilon}$	
Ψ_{A1}:	$3\Gamma + G_1$	rG_θ	rG_ε	(2.4.2a)
$\Psi_{g\theta}$:	rG_θ	$G_1 - qG_\theta$	qG_ε	
$\Psi_{g\varepsilon}$:	rG_ε	qG_ε	$G_1 + qG_\theta$	

or, if the singlet is A_2:

	Ψ_{A2}	$\Psi_{g\theta}$	$\Psi_{g\varepsilon}$	
Ψ_{A2}:	$3\Gamma + G_1$	rG_ε	$-rG_\theta$	(2.4.2b)
$\Psi_{g\theta}$:	rG_ε	$G_1 - qG_\theta$	qG_ε	
$\Psi_{g\varepsilon}$:	$-rG_\theta$	qG_ε	$G_1 + qG_\theta$	

In writing (2.4.2a,b), we have included the tunneling splitting 3Γ, and we have omitted terms involving G_2, since p is very close to zero in accord with Eqs. (2.3.38a,b) when we have a sufficiently strong Jahn–Teller coupling to be near the static limit.

When 3Γ is large compared with the magnitude of G_θ and G_ε, the ground state exhibits a dynamic Jahn–Teller effect as described in Section 2.3, the coupling with the singlet affecting the ground state only via the appropriate second-order terms given in Eq. (2.3.50). The excited singlet, which may be low enough in energy to be thermally populated, according to Eq. (2.4.2) has its behavior determined entirely by G_1, apart from the small second-order corrections. In particular, the spin Hamiltonian in this singlet state is isotropic and is given in accord with Eqs. (2.1.5) and (2.1.7) by

$$\mathscr{H}_A = g_1\beta(\mathbf{S} \cdot \mathbf{H}) + A_1(\mathbf{S} \cdot \mathbf{I}) \tag{2.4.3}$$

To obtain the EPR spectrum when 3Γ is of the same size as G_θ and G_ε, or smaller, we must solve the complete secular equation obtained from Eq. (2.4.2). The G's contain spin operators for electron and nuclear spin, however, so that in general this secular equation is fairly large and complicated. However, we may obtain a reasonably accurate result while simplifying the problem enormously if the anisotropic terms are small compared with the isotropic ones entering G_1, namely $|g_2| \ll g_1$, $|A_2| \ll |A_1|$. As in the derivation of Eqs. (2.3.16) and (2.3.18), we may then take ζ along **H** and keep only terms diagonal in S_ζ and I_ζ. For the Zeeman interaction alone, for example, we obtain thus from Eq. (2.4.2b) the matrix elements for the states with $S_\zeta' = +\frac{1}{2}$ or $-\frac{1}{2}$:

	$\Psi_{A2} \| S_\zeta'\rangle$	$\Psi_{g\theta} \| S_\zeta'\rangle$	$\Psi_{g\varepsilon} \| S_\zeta'\rangle$
$\Psi_{A2} \| S_\zeta'\rangle$:	$3\Gamma + g_1\beta S_\zeta' H$	$\frac{1}{2}\sqrt{3}rg_2\beta S_\zeta' H(\zeta_x^2 - \zeta_y^2)$	$-\frac{1}{2}rg_2\beta S_\zeta' H(3\zeta_z^2 - 1)$
$\Psi_{g\theta} \| S_\zeta'\rangle$:	—	$\beta S_\zeta' H[g_1 - \frac{1}{2}qg_2(3\zeta_z^2 - 1)]$	$\frac{1}{2}\sqrt{3}qg_2\beta S_\zeta' H(\zeta_x^2 - \zeta_y^2)$
$\Psi_{g\varepsilon} \| S_\zeta'\rangle$:	—	—	$\beta S_\zeta' H[g_1 + \frac{1}{2}qg_2(3\zeta_z^2 - 1)]$

(2.4.4)

O'Brien (1964) has solved the resulting secular equation to obtain the g-factors of the resonance lines in the particular case of **H** along [001] ($\zeta_z = 1$, $\zeta_x = \zeta_y = 0$) and $3\Gamma = 0$, and she has discussed how these g-factors and line intensities vary for 3Γ finite. Bersuker (1963) has given the analytical expressions for these energy levels for arbitrary values of 3Γ, also for **H** along [001], and he has discussed the variation in the g-factors as a function of the magnitude and orientation of **H**. Bersuker *et al.* (1964) have also evaluated numerically the effect of hyperfine and nuclear quadrupole interactions (for Cu^{2+}) on these EPR spectra for a number of different values of 3Γ. We will not give these interesting results of O'Brien and Bersuker here in detail, because in neither work was the effect of random strain in the crystal taken into account. As we shall discuss below, the effects of random strain almost certainly dominate the effects of anisotropy in the Zeeman and hyperfine interactions in all of the cases that have been investigated experimentally to date.

In the static Jahn–Teller limit, the states in the three wells correspond to distorted configurations of different orientation, but they are otherwise entirely equivalent. This equivalence implies a relation between the parameters q and r in this limiting case, which we will now derive. This relation will continue to hold, at least to a very good approximation, so long as the tunneling splitting is reasonably small. Using this relation, we shall be able to effect a great simplification of Eq. (2.4.2) for the case when 3Γ is finite but smaller than G_θ, G_ε. We will obtain from this analysis a condition for the experimental observation of a static Jahn–Teller effect, and we will show how a very weak static strain due to crystal imperfection can lock the system into a static Jahn–Teller effect when it would otherwise exhibit dynamic behavior.

For the limiting case of a static Jahn–Teller effect when there is negligible overlap between states in the different wells, the vibronic wave functions in each well have the form of Eq. (2.2.19). We denote by Ψ_1, Ψ_2, and Ψ_3 the ground-state wave functions in the well at $\theta_1 = 0$, $\theta_2 = 2\pi/3$, and $\theta_3 = 4\pi/3$, respectively (or $\theta_1 = \pi$, $\theta_2 = 5\pi/3$, $\theta_3 = \pi/3$). However, instead of choosing the signs of the Ψ_n exactly as given by Eq. (2.2.19), for convenience we adopt the following conventions. We take $\chi_n(\varrho, \theta) = \chi(\varrho, \theta - \theta_n)$ in all cases to be positive for $\theta \simeq \theta_n$, where $\chi(\varrho, \theta - \theta_n)$ may, for example, have approximately a Gaussian dependence on $(\theta - \theta_n)$ as given by Eq. (2.2.20). We then choose the sign of the associated electronic function, $\pm\psi_+(\varrho, \theta)$ or $\pm\psi_-(\varrho, \theta)$, such that for $\theta = \theta_n$ the dependence of Ψ_n on the electronic coordinates is given by

$$\begin{aligned} \Psi_1(\varrho, \theta_1) &= +\psi_\theta\chi_1 \sim +(3z^2 - r^2)\chi_1 \\ \Psi_2(\varrho, \theta_2) &= (-\tfrac{1}{2}\psi_\theta + \tfrac{1}{2}\sqrt{3}\,\psi_\varepsilon)\chi_2 \sim +(3x^2 - r^2)\chi_2 \\ \Psi_3(\varrho, \theta_3) &= (-\tfrac{1}{2}\psi_\theta - \tfrac{1}{2}\sqrt{3}\,\psi_\varepsilon)\chi_3 \sim +(3y^2 - r^2)\chi_3 \end{aligned} \qquad (2.4.5a)$$

when the three states belong* to $A_1 + E$, or by

$$\begin{aligned} \Psi_1(\varrho, \theta_1) &= +\psi_\varepsilon\chi_1 \sim +\sqrt{3}\,(x^2 - y^2)\chi_1 \\ \Psi_2(\varrho, \theta_2) &= +(-\tfrac{1}{2}\sqrt{3}\,\psi_\theta - \tfrac{1}{2}\psi_\varepsilon)\chi_2 \sim +\sqrt{3}\,(y^2 - z^2)\chi_2 \\ \Psi_3(\varrho, \theta_3) &= +(+\tfrac{1}{2}\sqrt{3}\,\psi_\theta - \tfrac{1}{2}\psi_\varepsilon)\chi_3 \sim +\sqrt{3}\,(z^2 - x^2)\chi_3 \end{aligned} \qquad (2.4.5b)$$

when they belong to $A_2 + E$. The normalized linear combinations of these states transforming as A_1, E_θ, E_ε for the first case are given by

$$\begin{aligned} \Psi_{A1} &= 3^{-1/2}(\Psi_1 + \Psi_2 + \Psi_3) \\ \Psi_{E\theta} &= 6^{-1/2}(2\Psi_1 - \Psi_2 - \Psi_3) \\ \Psi_{E\varepsilon} &= 2^{-1/2}(\Psi_2 - \Psi_3) \end{aligned} \qquad (2.4.6a)$$

* For the character of the reducible representation of the rotation group of the cube to which Ψ_1, Ψ_2, Ψ_3 belong, we find, in the order of the five classes E, C_3, C_2, C_4, C_2', as follows: 3, 0, 3, +1, +1 for the case (2.4.5a) and 3, 0, 3, −1, −1 for (2.4.5b). The first case thus reduces to $A_1 + E$ and the second to $A_2 + E$. We note that we get −1 rather than +1 for the character corresponding to C_4 in the second case because under, say, the rotation by $\pi/2$ about z that takes (x, y) to $(y, -x)$, the electronic function $\psi_\varepsilon \sim \sqrt{3}(x^2 - y^2)$ changes sign, while χ_1 is unchanged and Ψ_2 and Ψ_3 are interchanged. The same happens under the rotation C_2' that takes (x, y, z) to $(y, x, -z)$. In the first case, on the other hand, Ψ_1 is unchanged by these transformations.

and for the second case,

$$\Psi_{A2} = 3^{-1/2}(\Psi_1 + \Psi_2 + \Psi_3)$$
$$\Psi_{E\theta} = 2^{-1/2}(\Psi_3 - \Psi_2) \tag{2.4.6b}$$
$$\Psi_{E\varepsilon} = 6^{-1/2}(2\Psi_1 - \Psi_2 - \Psi_3)$$

Evaluating q and r from matrix elements of $\mathscr{U}_\theta$ and $\mathscr{U}_\varepsilon$ among these states, with the parameter u defined as in Eq. (2.2.16), we have from Eqs. (2.3.8a) and (2.4.1)

$$q = +\tfrac{1}{2}u, \qquad r = -(u/\sqrt{2}) \tag{2.4.7}$$

for both cases. Accordingly, we have in this limit the relation

$$r = -q\sqrt{2} \tag{2.4.8}$$

corresponding to the sign conventions in Eqs. (2.4.5) and (2.4.6).

Assuming the relation (2.4.8) to hold in the respective sets of matrix elements (2.4.2), we now transform these matrix elements (2.4.2a) by the inverse transformation of that indicated in Eq. (2.4.6a), and the matrix elements (2.4.2b) by the inverse of that in Eq. (2.4.6b). We obtain the following results:

	Ψ_1	Ψ_2	Ψ_3
Ψ_1:	$\Gamma+G_1\mp 2qG_\theta$	Γ	Γ
Ψ_2:	Γ	$\Gamma+G_1\mp q(-G_\theta+\sqrt{3}G_\varepsilon)$	Γ
Ψ_3:	Γ	Γ	$\Gamma+G_1\mp q(-G_\theta-\sqrt{3}G_\varepsilon)$

(2.4.9)

where the upper sign corresponds to Eq. (2.4.2a) $(A_1 + E)$ and the lower sign to Eq. (2.4.2b) $(A_2 + E)$.

We see therefore, when the tunneling splitting 3Γ is sufficiently small such that the relation (2.4.8) holds to a good approximation, that we can introduce linear combinations of our vibronic states Ψ_{A1} (or Ψ_{A2}), $\Psi_{g\theta}$, and $\Psi_{g\varepsilon}$, in terms of which all the anisotropic interactions involving G_θ and G_ε are diagonal. By comparing with Section 2.2, we find, moreover, that these diagonal terms in Eq. (2.4.9) are identical with the terms in the spin Hamiltonian we obtained for the three differently oriented configurations of the static Jahn–Teller effect. For example, if the G's represent the Zeeman and hyperfine interactions, and we substitute $q = \tfrac{1}{2}u$ from Eq. (2.4.7), we obtain from the diagonal term for Ψ_1 in Eq. (2.4.9) the axial spin Hamiltonian (2.2.9) and (2.2.11) with $g_\parallel$, $g_\perp$, $A_\parallel$, and $A_\perp$ given by

Eqs. (2.2.17) and (2.2.18). Similarly for Ψ_2 and Ψ_3 we get the corresponding results for the distortion with the symmetry axis along x and y, respectively.

If the off-diagonal matrix elements Γ in Eq. (2.4.9) of the tunneling interaction between the states Ψ_1, Ψ_2, and Ψ_3 are small compared with the differences in the diagonal matrix elements due to G_θ and G_ε, we see that the experimental properties of the system will be those of a static Jahn–Teller effect. This will be the case, then, when both G_θ and G_ε are nonzero, and both are large compared to Γ. We thus have a criterion for the experimental observation of a static Jahn–Teller effect.

Clearly, the above criterion for a static Jahn–Teller effect could be met if the anisotropic part of the Zeeman interaction were sufficiently large, so that it is possible in principle to have a dynamic effect for one value of H (or for appropriate orientations) and a static effect when the field strength is increased sufficiently. However, to take a typical example, we have $g_2 \simeq +0.2$ for Cu^{2+} in many crystals (Section 4.1), so that for $H \simeq 3000$ Oe these energy differences due to the anisotropic part of the Zeeman interaction are less than $\sim 0.05\ \mathrm{cm}^{-1}$. The tunneling matrix elements Γ thus have to be quite small for all dynamic effects to be suppressed by the Zeeman interaction at such field strengths. But, of greater importance, for the Zeeman interaction to be itself responsible for suppressing dynamic effects, we should require that splittings due to random strain also be less than those due to the anisotropic part of the Zeeman interaction.

We have already seen that inhomogeneous strain in the crystal can drastically affect the EPR spectrum with a dynamic Jahn–Teller effect, and we will show now that such strain can be expected usually to dominate other anisotropic affects in leading to a static Jahn–Teller effect. Since we have from Eq. (2.1.10)

$$G_\theta = V_2 e_\theta, \qquad G_\varepsilon = V_2 e_\varepsilon \tag{2.4.10}$$

for the strain interaction, we see that the contribution of the strain to the diagonal terms in Eq. (2.4.9) agrees precisely with the energy shifts for the distorted states given by Eq. (2.2.14), except that in writing this earlier equation we did not include the factor $2q = u$. For systems coupled so strongly to the lattice that they exhibit a static Jahn–Teller effect, however, we must probably always have a strain coefficient $V_2 \gtrsim 10^4\ \mathrm{cm}^{-1}$ (see Section 4.1) so that strains of order 10^{-4} lead to differences in the diagonal terms of Eq. (2.4.9) of $\sim 1\ \mathrm{cm}^{-1}$ or more. Unless crystals of sufficient perfection are available to reduce residual strains substantially below this level, it is probable that dynamic effects will be suppressed by the strain unless the tunneling splitting is greater than at least 1 cm^{-1}. Furthermore, it is usually the strain rather than, say, the Zeeman interaction that dominates the anisotropic part of the interaction.* When this is so, the eigenstates of the system may be taken to be determined by the strain, with the tunneling matrix elements taken into account if they are large enough. The spin Hamiltonian representing the effect of the Zeeman and hyperfine interactions on each of these resulting strain-split states may then be found simply from the diagonal matrix elements of these weaker interactions with respect to these states, with perturbation corrections from matrix elements to the other states included if necessary.

In Bersuker's (1962, 1963) analysis of the effects of tunneling between the distorted Jahn–Teller configurations arising from an electronic E doublet, the order of his considerations is the reverse of that we have followed in this section. Thus Bersuker starts from the states characterizing the static Jahn–Teller effect and introduces an interaction between these states corresponding to tunneling between the wells (which Bersuker calls "inversion"). In other words, Bersuker starts with the diagonal matrix elements in Eq. (2.4.9), which describe the properties of the states characterizing the static Jahn–Teller effect, and he then introduces the matrix element Γ of the tunneling interaction, exactly as in Eq. (2.4.9). Apart from detailed assumptions concerning the various matrix elements, Bersuker's theory of the "inversion splitting" of the doublet is thus formally equivalent to the theory we have outlined, at least for a sufficiently strong Jahn–Teller coupling so that Eq. (2.4.9) gives the matrix elements accurately.

In detail, however, Bersuker has made several assumptions which in practice cause his results to differ from those of the theory we have described above and which we believe is essentially correct. For one, Bersuker takes the tunneling interaction Γ to be negative, so that the singlet is the ground state of his theory, as the result of the inversion splitting, instead of the doublet as found in the theory of Moffitt and Thorson (1958), Longuet-Higgins *et al.* (1958), and O'Brien (1964). While this discrepancy is evidently the result of an error in sign in the calculation of the matrix elements of the full vibronic Hamiltonian of the crystal between the different distorted states, a more fundamental difficulty is that Bersuker did not take into account the variation with θ of the electronic part of the Born–Oppenheimer wave function (2.2.19) within each well. Thus this electronic part of the

* The dominant role of random strains in effecting the transition from a dynamic to a static Jahn–Teller effect in the EPR spectrum appears first to have been noted by Gelerinter and Silsbee (1966) in their study of the N_2^- defect in the sodium azide crystal.

wave function $\Psi_n(\varrho, \theta)$ for the well at $\theta \simeq \theta_n$ was taken to be the fixed linear combination of ψ_θ and ψ_ε appropriate for $\theta = \theta_n$ and given by Eq. (2.4.5), rather than the θ-dependent combination given by $\psi_\pm(\varrho, \theta)$ in Eq. (2.2.3). As shown by Höchli (1967) from Bersuker's (1963) results for the g-factor, this assumption leads to the value

$$q = \tfrac{1}{2}(1 + \tfrac{3}{2}\gamma) \tag{2.4.11}$$

for the reduction factor q when the vibrational overlap γ (assumed $\ll 1$) between different wells is taken into account. However, since γ is positive (Bersuker, 1963) the result (2.4.11) conflicts with the general requirement (2.3.48), proved from O'Brien's theory, that $q \simeq c_2$ must be *less* than 1/2 in the ground state under the conditions of strong Jahn–Teller coupling for which Bersuker's theory is intended to be appropriate. For strong Jahn–Teller coupling, the reduction of q due to the delocalization of the wave function within each well always dominates the increase in q resulting from overlap between different wells (Ham, 1968).*

In transforming the matrices (2.4.2a,b) to the form of Eq. (2.4.9), we assumed that q and r satisfied the relation (2.4.8). If this relation is not exact, we find that q is replaced in the diagonal matrix elements of Eq. (2.4.9) by $(1/3)(q - \sqrt{2}r)$, while small off-diagonal matrix elements given by $\pm\varepsilon q(G_\theta \pm \sqrt{3}G_\varepsilon)$ and $\pm 2\varepsilon q G_\theta$ appear, where ε is defined from the relation

$$r/q = -\sqrt{2}(1 - 3\varepsilon) \tag{2.4.12}$$

Williams *et al.* (1969) have evaluated ε numerically as a function of $|\beta|/\alpha$ from the solutions of Eq. (2.3.32), making the approximation that the small terms given by $(\hbar\omega/4E_{\mathrm{JT}})^2 c_2$ and $(V_q\varrho_0^2/4E_{\mathrm{JT}})c_4$ in q [Eq. (2.3.44)] are

* If the vibrational wave function χ_n in each well is assumed to have a Gaussian dependence on $(\theta - \theta_n)$ as in Eq. (2.2.20), the correct expression for q taking account of the θ-dependence of the electronic wave functions is given by $q = c_2 = \tfrac{1}{2}[(1 + 2\gamma)/(1 + \gamma)]e^{-1/8b}$, where $\gamma = \exp(-2\pi^2 b/9)$ is the vibrational overlap between wells. As expected, this expression satisfies the requirement $c_2 \leq \tfrac{1}{2}$. Englman and Halperin (1970) have shown that this expression [with b given by $(9|\beta|/8\alpha)^{1/2}$] gives an excellent representation of the variation of c_2 as calculated numerically (see Fig. 9) for $|\beta|/\alpha \gtrsim 10$, but that it becomes increasingly inaccurate as the barrier height diminishes and is altogether inadequate for $|\beta|/\alpha \lesssim 1$. Similar conclusions were reached for the corresponding approximate expressions for 3Γ and r. In particular, this expression for c_2 may *not* be taken to give an accurate representation in any situation in which it predicts that c_2 differs from $\tfrac{1}{2}$ by more than a few per cent. The suggestion of Zdansky (1969) that a value $q = c_2 \simeq 0.2$ could be accounted for in this way is therefore wrong.

negligible, along with the corresponding terms in the expression for r. Their results are plotted in Fig. 8; values for r equivalent to these have also been calculated by Fletcher (Fletcher, 1967; Fletcher and Stevens, 1969). Since ε is found from Fig. 8 to be <0.1 for all values of β/α and to become smaller rapidly as $|\beta|/\alpha$ increases, it is clear that off-diagonal matrix elements proportional to $\varepsilon q G_\theta$ and $\varepsilon q G_\varepsilon$ in Eq. (2.4.9) will be small compared with the differences in the diagonal matrix elements, at least for $|\beta|/\alpha \gtrsim 5$. Since in these diagonal matrix elements $\frac{1}{3}(q-\sqrt{2}r)$ equals $q(1-2\varepsilon)$ in accord with Eq. (2.4.12), we find therefore for the static Jahn–Teller effect that the parameter replacing u in Eqs. (2.2.16-18) is now

$$u' = 2q(1-2\varepsilon) \tag{2.4.13}$$

In view of the small value of ε and the narrow limits on the variation of q as given by the relation $q \simeq c_2$ and the inequality (2.3.49), we see from Eq. (2.4.13) that u' will not be smaller than $\sim$0.95 if $|\beta|/\alpha \gtrsim 10$ [assuming $E_{\mathrm{JT}} \gg \hbar\omega$ and $E_{\mathrm{JT}} \gg |V_q|\varrho_0^2$ so that the corrections to the approximate relation $q \simeq c_2$ from Eq. (2.3.44) are small]. This result therefore imposes a limit on the extent to which the finite spread in θ of the vibrational wave function within the well can be expected to reduce the anisotropy of the Zeeman or hyperfine interactions in the case of a static Jahn–Teller effect. In terms of wave functions localized in the individual wells, we may show from the work of Williams *et al.* (1969) that, to the same accuracy as in Eq. (2.3.44), we have

$$\begin{aligned} u' = {} & [1 + (\hbar\omega/4E_{\mathrm{JT}})^2]\langle\cos(\theta-\theta_n)\rangle_n \\ & \mp (V_q\varrho_0^2/4E_{\mathrm{JT}})[\langle\cos 2(\theta-\theta_n)\rangle_n - \langle\cos 4(\theta-\theta_n)\rangle_n] \end{aligned} \tag{2.4.14}$$

the averages indicated by $\langle\cdots\rangle_n$ being taken with respect to the vibrational wave function in the well at $\theta = \theta_n$. The upper (lower) sign in Eq. (2.4.14) should be taken when V and β have the same (opposite) sign.

2.5. Dynamic Effects Due to Relaxation and Thermal Excitation

Up to this point in our analysis of dynamic Jahn–Teller effects, we have been concerned primarily with those effects that are determined by the properties of the individual vibronic eigenstates which comprise the ground state of the system. We want now to consider effects resulting from thermal excitation and from rapid relaxation between these states.

We have seen that the system has various excited vibronic states belonging to A_1, A_2, and E in addition to its doublet (E) ground state. If we assume that all accidental degeneracies are lifted by the warping terms in the coupling, the lowest excited state is a singlet (A_1 or A_2) with an excitation energy which is given approximately by Fig. 4 for weak or moderately strong Jahn–Teller coupling and which may become quite small in accord with Fig. 7 if the coupling is very strong. However, so long as this separation 3Γ between this singlet and the ground state is large compared with the magnitude of the anisotropic perturbation described by G_θ and G_ε in Eq. (2.4.2), the spin Hamiltonian of the singlet is determined by the isotropic part of the perturbation and is given by Eq. (2.4.3). The same spin Hamiltonian with the same values for its parameters is appropriate to other excited singlet states as well, while for excited E states we have a spin Hamiltonian of the same form (2.3.9) as that of the ground state, but with different values for p and q. Thus we see that as the temperature rises and such excited states become thermally populated (but neglecting effects of rapid relaxation), we may expect additional EPR spectra to appear besides that of the ground state. In particular, we may expect to find an isotropic spectrum described by Eq. (2.4.3) and resulting from one or more excited singlet states. If 3Γ if fairly small (a few cm^{-1}, say), as it may be in appropriate cases, such a spectrum should be expected to appear above a few degrees Kelvin and to coexist with the spectrum of the ground state over an appreciable temperature range.

However, coupling with lattice phonons leads to thermally induced transitions between different vibronic states,* and if these transitions occur sufficiently rapidly we may also encounter relaxation effects in the EPR spectrum as the temperature is raised. In particular, the phonons directly cause transitions between vibronic states of the same spin, and such relaxation processes cause a motional averaging of the EPR spectrum. As we shall see, these effects of relaxation tend to average out the anisotropic contributions to the spin Hamiltonian. This occurs in the case of a system

* Since it is the coupling with the lattice phonons which in reality is responsible for the Jahn–Teller effect in the first place, it is not really correct to treat the phonons as independent modes of excitation, interaction with which can induce transitions between the states of the Jahn–Teller system. A fully satisfactory treatment of the interaction of a Jahn–Teller ion with a continuum of lattice phonons has not yet been given. However, if the Jahn–Teller coupling is primarily with the nearest neighbors, there is no particular difficulty in treating long-wavelength phonons as independent excitations which interact weakly with an isolated Jahn–Teller ion via the strain field which they produce.

that exhibits a static Jahn–Teller effect at low temperatures, as well as for one in which dynamic effects due to zero-point motion are present in the ground state. Although the linewidth of the resulting spectrum may show some anisotropy, the positions of the line centers (in the absence of oriented strain) depend only on the isotropic part of the spin Hamiltonian as given by Eq. (2.4.3).

Thus we see that there are two alternative mechanisms that lead to the appearance of an isotropic EPR spectrum as the temperature is raised: Either excited singlet states become thermally populated, or we have motional averaging (or both). Moreover, the same spin Hamiltonian, Eq. (2.4.3), determines the positions of the lines in the isotropic spectrum in both cases. Although there has been considerable discussion of these two alternatives (Abragam and Pryce, 1950; Bijl and Rose-Innes, 1953; Bleaney, Bowers, and Trenam, 1955; Orton *et al.*, 1961; Bersuker, 1962; Englman and Horn, 1963; O'Brien, 1964; Hudson, 1966) since the original observation of a dynamic Jahn–Teller effect of this type by Bleaney and Ingram (1950), there is only one experimental case involving the orbital doublet (Al_2O_3:Ni^{3+}) in which the nature of the transition is fairly satisfactorily understood, and the theory of this transition is by no means complete. In this section, we shall describe qualitatively the nature of the motional averaging that occurs in the different cases we have discussed previously, and we shall obtain expressions for the relevant relaxation times.

When the system exhibits a static Jahn–Teller effect at low temperature, the possibility is clearly present of a thermally activated reorientation from one of the three distorted configurations to another. This reorientation may occur because the complex absorbs one or more phonons and makes a real transition to one of the excited states in Fig. 7, whence it can decay back to the ground state in a different distorted configuration from that in which it started. Alternatively, there are one- and two-phonon processes within the ground state itself which, aided by the matrix elements of the tunneling interaction, can effect such a reorientation. If $P(a \to b)$ denotes the probability per unit time of a transition from the distorted configuration a to that at b, and if $P(a \to b)$ is approximately the same for all possible pairs of the three configurations, then the relaxation time τ for the recovery of thermal equilibrium between the populations of the differently oriented configurations is given by

$$\tau^{-1} = 3P(a \to b) \tag{2.5.1}$$

If this τ is long compared with $(2\pi \Delta\nu)^{-1}$, where $\Delta\nu$ is the frequency difference between corresponding EPR resonance lines for the different distorted configurations, the spectrum observed is the superposition of those of the individual configurations. However, if we have

$$\tau \ll (2\pi\, \Delta\nu)^{-1} \tag{2.5.2}$$

then by the theory of motional narrowing (Gutowsky *et al.*, 1953; Abragam, 1961) the resonance frequency is the average of the corresponding frequencies of the individual configurations. For the axially symmetric spectra described by Eqs. (2.2.9) and (2.2.11), we have, for **H** along [100],

$$\begin{aligned} h\,|\,\Delta\nu\,| &= |\,(g_\parallel - g_\perp)\beta H + (A_\parallel - A_\perp)m\,| \\ &= \tfrac{3}{2}u\,|\,g_2\beta H + A_2 m\,| \end{aligned} \tag{2.5.3}$$

where we have used Eqs. (2.2.17) and (2.2.18). The motionally averaged frequency is then given by

$$h\bar{\nu}_m = (\tfrac{1}{3}g_\parallel + \tfrac{2}{3}g_\perp)\beta H + (\tfrac{1}{3}A_\parallel + \tfrac{2}{3}A_\perp)m \tag{2.5.4}$$

or

$$h\bar{\nu}_m = g_1\beta H + A_1 m \tag{2.5.5}$$

in agreement with the isotropic spectrum given by Eq. (2.4.3). We note that $\Delta\nu$ in Eq. (2.5.3) depends on m, so that the transition to the motionally averaged spectrum may occur at somewhat different values of τ (and therefore different temperatures) for different hyperfine lines. On the other hand, for **H** along [111] the three distorted configurations are equivalent, and we have $\Delta\nu = 0$. Motional narrowing will thus occur at a lower temperature when **H** is near the [111] orientation than for **H** along [100]. Expressions for the theoretical line shapes resulting from rapid reorientation of the Jahn–Teller distortion have been obtained by Hudson (1966).

If the reorientation occurs via a real transition to an excited state lying below the top of the barrier between the wells, τ will be given to a good approximation by

$$\tau^{-1} = \nu_0 e^{-E/kT} \tag{2.5.6}$$

where E is the excitation energy of the excited state. This is of course the form taken by τ for an Orbach type of relaxation process (Orbach, 1961). The exponential factor reflects the relative probability of finding the system in the excited state, or equivalently, the relative probability of phonons exciting the upward transition. In the present situation, the frequency factor ν_0 is determined by the lifetime of the excited state and by the tunneling rate through the barrier in that state. Alternatively, Eq. (2.5.6) has the form of the classical rate at which the system is thermally activated over the barrier, if E is the barrier height. The validity of Eq. (2.5.6) for reorientation over or through a barrier via states lying either below or just above the barrier, together with the identification of ν_0 in various cases, has been discussed by Sussman (1967).

At temperatures below those at which τ is given by Eq. (2.5.6), the relaxation time for reorientation is dominated by one-phonon or multiphonon processes involving only the ground state in each well. We will now evaluate τ for these processes in the case in which the strain energy difference between the three wells dominates both the tunneling and other sources of anisotropy. The eigenstates of Eq. (2.4.9) are now given by perturbation theory, to first order in the tunneling matrix element Γ, as

$$\begin{aligned}\Psi_1' &= \Psi_1 - (\Gamma/\delta_{21})\Psi_2 - (\Gamma/\delta_{31})\Psi_3 \\ \Psi_2' &= \Psi_2 + (\Gamma/\delta_{21})\Psi_1 - (\Gamma/\delta_{32})\Psi_3 \\ \Psi_3' &= \Psi_3 + (\Gamma/\delta_{31})\Psi_1 + (\Gamma/\delta_{32})\Psi_2\end{aligned} \tag{2.5.7}$$

where δ_{21}, δ_{31}, and δ_{32} denote the energy differences $(E_2 - E_1)$, $(E_3 - E_1)$, $(E_3 - E_2)$ obtained from the diagonal matrix elements in Eq. (2.4.9) when the tunneling is neglected and G_θ and G_ε are given in terms of the local static strain by Eq. (2.4.10). Each of the wave functions (2.5.7) represents a state localized predominantly in one well, and we want now to obtain the rate at which phonons induce transitions between these states. Considering only acoustic phonons, and treating the phonons in the long-wavelength approximation, we take for the electron–phonon interaction Hamiltonian the strain Hamiltonian (2.1.10) with the strains e_θ and e_ε replaced by the corresponding operators of the phonon field (Stevens, 1967). This electron–phonon interaction has no off-diagonal matrix elements with respect to the states Ψ_1, Ψ_2, and Ψ_3, according to Eq. (2.4.9), so that its matrix elements between the different states Ψ_1', Ψ_2', and Ψ_3' of Eqs. (2.5.7) are nonzero only because of the admixture of states from different wells by the tunneling matrix element Γ. Assuming for simplicity an isotropic phonon spectrum, we obtain by a straightforward calculation that, for $kT > |\delta_{21}|, |\delta_{31}|, |\delta_{32}|$, the probability per unit time of a direct transition between two of the states (2.5.7) accompanied by the absorption or emission of a single phonon is given by

$$P(a \to b) = [9\Gamma^2(qV_2)^2kT/5\pi\varrho\hbar^4 s_T^5][1 + \tfrac{2}{3}(s_T/s_L)^5] \tag{2.5.8}$$

where ϱ is the crystal density and s_T and s_L are the sound velocities in the transverse and longitudinal branches, respectively. This result for $P(a \rightarrow b)$ is independent of the strain energy difference δ_{ab} (for $kT > | \delta_{ab} |$), because the factor δ_{ab}^{-2} from the coefficient Γ/δ_{ab} in the electron–phonon interaction matrix element is cancelled by a factor δ_{ab}^2 from the phonon density of states (since $\hbar\omega$ is equal to $| \delta_{ab} |$ for the energy-conserving phonon). The relaxation time τ is then obtained from Eq. (2.5.1). A result equivalent to Eq. (2.5.8) for the reorientation of the O_2^- center in an alkali halide was first given by Pirc *et al.* (1966) and has been obtained also by Sussman (1967).

The transition rate between the states (2.5.7) resulting from two-phonon Raman processes may be calculated under similar assumptions. The result is an expression varying as T^3 which was pointed out by Pirc *et al.* (1966) and is given by

$$P(a \rightarrow b) = [27\Gamma^2(qV_2)^4(kT)^3/50\pi\hbar^7\varrho^2 s_T^{10}][1 + \tfrac{2}{3}(s_T/s_L)^5]^2 \qquad (2.5.9)$$

For Eq. (2.5.9) to give a transition rate faster than that due to direct processes and given by Eq. (2.5.8), we must have T greater than a temperature T_α given by

$$T_\alpha \simeq (1/k)[10\hbar^3\varrho s_T{}^5/3(qV_2)^2]^{1/2} \qquad (2.5.10)$$

However, when this condition is satisfied there are other higher-order processes which may also contribute significantly (Gosar and Pirc, 1967; Silsbee, 1967), so that the reorientation rate may not in fact exhibit the T^3 variation for $T > T_\alpha$ that we would expect from Eq. (2.5.9) for the Raman processes alone.

If the condition (2.5.2) for motional averaging is satisfied, but the strain splittings δ_{21}, δ_{31}, and δ_{32} are not very small compared to kT, the average of the frequencies for the different distorted configurations must be taken using the respective Boltzmann factor for each state. If the strain is oriented, as in the presence of a strong, externally applied stress, the resulting motionally averaged spectrum will *not* be isotropic. On the other hand, if the strain is random, the positions of the line centers will continue to be given by Eq. (2.5.5) and will be isotropic, but there will be an anisotropic, temperature-dependent contribution to the linewidths from the asymmetric averaging. Since the linewidths also depend on both temperature and orientation through the product $\tau\,\Delta\nu$ in the theory of motional narrowing, a detailed linewidth analysis in any experimental case will undoubtedly be very complicated.

When the system exhibits a dynamic Jahn–Teller effect in its ground state at low temperature as described in Section 2.3, the effect of motional narrowing is to average the two resonance frequencies (for given m) given by Eq. (2.3.18) or Eq. (2.3.23). These two resonances correspond to spin transitions for two orthogonal vibronic states (Ham, 1968), Ψ_α and Ψ_β, between which phonons can induce transitions at a rate corresponding to a relaxation time

$$\tau^{-1} = P(\alpha \to \beta) + P(\beta \to \alpha) \tag{2.5.11}$$

For motional narrowing, τ must then satisfy the condition (2.5.2) with respect to the frequency difference between these resonances. In the case of a predominant strain-splitting corresponding to a strain characterized by the angle φ in Eq. (2.3.24) (assuming of course that 3Γ is much greater than the strain splitting, so that we can ignore mixing with the excited singlet), we have

$$\begin{aligned} \Psi_\alpha &= \Psi_{g\theta} \sin(\varphi/2) + \Psi_{g\varepsilon} \cos(\varphi/2) \\ \Psi_\beta &= \Psi_{g\theta} \cos(\varphi/2) - \Psi_{g\varepsilon} \sin(\varphi/2) \end{aligned} \tag{2.5.12}$$

with an energy splitting

$$\delta = E_\alpha - E_\beta = +2qV_2(e_\theta^2 + e_\varepsilon^2)^{1/2} \tag{2.5.13}$$

The frequency difference of the resonances is given in accord with Eqs. (2.3.23–25) by

$$h\,\Delta\nu = 2q(g_2\beta H + A_2 m)[1 - 3(\zeta_x^2\zeta_y^2 + \zeta_y^2\zeta_z^2 + \zeta_z^2\zeta_x^2)]^{1/2} \cos(\varphi - \alpha) \tag{2.5.14}$$

As in the case of the static Jahn–Teller effect, $\Delta\nu$ again depends both upon m and upon the orientation of **H**, with now an additional dependence on the strain through the angle φ. For $kT \gg |\delta|$, the motionally averaged frequency is again the isotropic value given by Eq. (2.5.5). However, if kT is not large compared to $|\delta|$, the averaging must again include the appropriate Boltzmann factors, and the motionally averaged spectrum will be anisotropic if the strain is oriented, or will show an anisotropic linewidth if the strain is random.

The calculation of the relaxation time (2.5.11) is now straightforward. In addition to an Orbach relaxation through the excited singlet, giving rise to τ of the form of Eq. (2.5.6), we have direct and Raman processes within the doublet ground state. Making assumptions for the phonon spectrum and electron–phonon interactions identical with those used in the derivation of Eqs. (2.5.8–9), we obtain for the direct process

$$\tau^{-1} = [3 \mid \delta \mid^3 (qV_2)^2 / 20\pi\hbar^4 \varrho s_T{}^5][1 + \tfrac{2}{3}(s_T/s_L)^5] \coth(\mid \delta \mid /2kT) \quad (2.5.15)$$

and for the Raman process (Ham, 1968)

$$\tau^{-1} = [6\pi(qV_2)^4(kT)^5/125\hbar^7\varrho^2 s_T^{10}][1 + \tfrac{2}{3}(s_T/s_L)^5]^2 \quad (2.5.16)$$

A rather different view of the transition from the low-temperature anisotropic spectrum to the higher-temperature isotropic spectrum, for a system which exhibits a static Jahn–Teller effect, has been proposed by Bersuker (1962, 1963). In his model, the matrix element Γ in Eq. (2.4.9) is taken to represent an average of the tunneling interaction in the various thermally populated vibrational states of each well, and because tunneling should be more rapid in the higher-energy states this average Γ is assumed to increase in magnitude rapidly with rising temperature. Thus whereas Γ is small enough at low temperatures so that the resonance spectrum is that of the static Jahn–Teller effect, at high temperature $\mid \Gamma \mid$ has increased sufficiently so that $\mid \Gamma \mid \gg \mid G_\theta \mid, \mid G_\varepsilon \mid$ and the EPR spectrum changes to that characteristic of the dynamic effect. In particular, the isotropic spectrum of the singlet (which is the ground state in Bersuker's work) should then appear, while that of the doublet is perhaps not observed because of broadening or motional narrowing. The trouble with this model is that one expects the low-energy vibrational states in the wells corresponding to the distorted configurations of the static Jahn–Teller effect to correspond roughly to the energy levels toward the right-hand side of Fig. 7, for an effective pair of parameters β and α. One would not, however, expect these parameters or the resulting tunneling splitting of the ground state to be markedly temperature-dependent, at least not for temperatures well below the Debye temperature of the crystal. Thus one would expect perhaps to see the spectra of these individual states as they become populated. On the other hand, to see a spectrum characteristic of a value for Γ averaged over these states, as Bersuker assumes, one would evidently already have to have a sufficiently rapid rate of phonon-induced transitions between these levels so that the spectra of the individual states are motionally averaged. The onset of motional narrowing, however, seems to be more appropriately described by the theory we have outlined previously than it is by one using a tunneling interaction based on an averaged Γ. Accordingly, we conclude that Bersuker's model for the transition to the high-temperature spectrum is not an appropriate one.

2.6. Spin–Lattice Relaxation

Our concern with relaxation processes in Section 2.5 was with those processes that involve transitions between different vibronic states of the same spin. The corresponding theory for spin–lattice relaxation in the presence of Jahn–Teller effects has not been developed very far, but a number of beginnings have been made in response to the stimulus of recent experimental results. In particular, it has been found in several Jahn–Teller systems that spin–lattice relaxation rates are substantially faster than in similar systems in which there is no orbital degeneracy. The principal question to which theoretical considerations have so far been addressed is whether mechanisms exist within the framework of the Jahn–Teller problem which can account for this rate enhancement.*

The first reported observation of such a spin–lattice relaxation rate enhancement in a Jahn–Teller system was that of Townsend and Weissman (1960) for the negative ion of coronene. It was proposed by McConnell (1961) that this might result from a first-order spin–orbit interaction, associated with the electronic motion around the ring, in the orbitally degenerate ground state. The Jahn–Teller coupling does not lift the degeneracy of this vibronic ground state, and since this coupling is only moderately strong for this ion the resulting dynamic Jahn–Teller effect only partially quenches this spin–orbit interaction. The enhanced relaxation rate in this case is a consequence of the spin–orbit interaction rather than of the Jahn–Teller coupling, which in fact tends to quench the spin–orbit interaction. However, for the first-order spin–orbit interaction to be present, the system must have orbital degeneracy, and such degeneracy is of course the criterion for a Jahn–Teller system, so that it is in this sense that the enhanced relaxation rate may in this case be associated with the Jahn–Teller effect.

More recently, an enhanced spin–lattice relaxation rate has been found by Breen (1966), Breen *et al.* (1969), Lee and Walsh (1968), and Lee (1969) for ions that exhibit a static Jahn–Teller effect. At liquid helium temperatures, the spin–lattice relaxation rate of Cu^{2+} in $La_2Mg_3(NO_3)_{12} \cdot 24H_2O$ was found to be four to five orders of magnitude faster than that of Cu^{2+} in potassium zinc Tutton salt (Gill, 1965), in which a strong tetragonal crystal field suppresses the Jahn–Teller effect. A relaxation rate even faster

* Some results for spin–lattice relaxation based on Bersuker's inversion model have been given by Bersuker and Vekhter (1965).

than that in the double nitrate was found for Cu^{2+} in zinc bromate (Breen, 1966). In the Tutton salt, the relaxation occurs between the component states of a single Kramers doublet, in a permanently distorted crystal field, and Stoneham (1965) has shown that Van Vleck's theory of the spin–lattice relaxation for such a doublet (Van Vleck 1940) successfully describes the data. With a static Jahn–Teller effect, however, the ground state of the Cu^{2+} is a Kramers doublet for *each* of the three energetically equivalent distorted configurations. As Williams *et al.* (1969) and Lee and Walsh (1968) have noted, as the result of tunneling matrix elements or other small residual interactions between the vibronic wave functions of the distorted configurations, in the presence of a magnetic field transitions may occur more rapidly between states of (more or less) opposite spin belonging to different configurations than between the Kramers conjugate states of a single configuration. Such spin-reversing transitions between different distorted configurations thus provide a plausible mechanism for an enhanced spin–lattice relaxation rate in the presence of a static Jahn–Teller effect. We shall now indicate how a quantitative estimate of this rate enhancement may be made.

Williams *et al.* (1969) have noted that, because of the anisotropic g-tensor given by Eq. (2.2.9) for each of the states (2.5.7) corresponding to the three distorted configurations, the spin eigenstates in a magnetic field correspond to a direction of spin quantization which is slightly different for the different distorted configurations (unless the magnetic field is along a [100] crystal axis, which is a principal axis of the g-tensor for all three configurations). As a result, the "up" spin state of one of the states (2.5.7) corresponding to one configuration is not in general orthogonal to the "down" spin state of another configuration, and the same electron–phonon interaction that leads to reorientation without spin reversal in accord with Eqs. (2.5.8–9) therefore also results in reorientation transitions which are accompanied by a spin reversal. Taking ζ along $\mathbf{H}$, we find by a straightforward calculation of the nonorthogonality of these spin eigenstates that, to lowest order in $\Delta g = (g_{\parallel} - g_{\perp})$, the relative probability* $p_r(a \to b)$ of a spin reversal in a one-phonon direct process transition between two of the

* We assume that the strain splittings δ_{21}, δ_{31}, and δ_{32} are larger than $g\beta H$, so that for the transition with spin reversal as well as for the transition without spin reversal, the energy of the phonon involved in the transition is determined principally by the strain splitting rather than by the Zeeman splitting. If $g\beta H$ were larger than δ_{ab}, the probability per unit time of a spin-reversing transition would be relatively greater because of the higher density of phonon states that would enter the transition probability.

states of Eq. (2.5.7) is given by

$$p_r(1 \to 2) = p_r(2 \to 1) = \tfrac{1}{4}(\Delta g/g_1)^2(4\zeta_x^2\zeta_z^2 + \zeta_y^2\zeta_x^2 + \zeta_y^2\zeta_z^2) \tag{2.6.1}$$

with similar results for the other pairs of states. Multiplying $p_r(a \to b)$ by $P(a \to b)$, the probability per unit time of a reorientation without spin reversal as given by Eq. (2.5.8), summing over the final states b, and averaging over the initial states a, under the condition that $kT > |\delta_{21}|$, $|\delta_{31}|$, $|\delta_{32}|$, so that the three distorted configurations are equally probable, we obtain for the direct process spin–lattice relaxation time T_1 (for $kT > |\delta_{ab}| > g\beta H$) the result

$$T_1^{-1} = \tfrac{2}{3}(\Delta g/g_1)^2(\zeta_x^2\zeta_y^2 + \zeta_y^2\zeta_z^2 + \zeta_z^2\zeta_x^2)\tau^{-1} \tag{2.6.2}$$

Here, τ is the relaxation time for reorientation without spin reversal as given by Eqs. (2.5.1), and τ is of course short compared to T_1 under the assumed conditions. When $(\Delta g/g_1)$ is as large as it is for Cu^{2+} in the double nitrate ($\Delta g/g_1 \simeq 0.17$), however, T_1 as given by this equation, say for **H** in the [111] orientation, is only about two orders of magnitude longer than τ. This mechanism can thus cause a large enhancement in the spin–lattice relaxation rate over that given by Van Vleck's theory for Cu^{2+} in the Tutton salt. However, Eq. (2.6.2) predicts a considerable anisotropy in T_1 which is not in accord with available experimental data, and this equation indicates in particular that T_1 should be very long when **H** is in a [100] direction. The anisotropy in T_1 observed by Breen *et al.* (1969) in the double nitrate was not so extreme. T_1 was longer for **H** along a principal axis of the g-tensor than for an intermediate orientation, but the difference was less than a factor of two.

The analysis leading to Eq. (2.6.2) has ignored the additional source of nonorthogonality in the spin eigenstates of different distorted configurations which results from the anisotropy in the hyperfine interaction (2.2.11) in these configurations. If $\Delta A = A_\parallel - A_\perp$ is small compared to $A_1 = \tfrac{1}{3}(A_\parallel + 2A_\perp)$, the spin–lattice relaxation time T_1 corresponding to the nuclear spin eigenstate m of I_ζ is given in place of Eq. (2.6.2) (but to a similar approximation) by

$$T_1^{-1} = \tfrac{2}{3}[(\Delta g\beta H + \Delta A m)/h\nu]^2(\zeta_x^2\zeta_y^2 + \zeta_y^2\zeta_z^2 + \zeta_z^2\zeta_x^2)\tau^{-1} \tag{2.6.3}$$

where ν denotes the EPR resonance frequency. Apart from the dependence of T_1 on m indicated by Eq. (2.6.3), this expression has essentially the same behavior as Eq. (2.6.2). However, there is an additional spin relaxation mechanism via the hyperfine interaction which involves the simultaneous flip of the electronic spin and the nuclear spin during reorientation and which remains effective when **H** is in a [100] direction. This mechanism, which again originates in the nonorthogonality of the spin eigenstates (electronic and nuclear) for the different distortions, yields the following spin transition probabilities for **H** in the [001] direction (under similar assumptions to those made previously):

$$P(+\tfrac{1}{2}\to-\tfrac{1}{2};\, m\to m+1)=(\tau^{-1}/36)(\Delta A/h\nu)^2[I(I+1)-m(m+1)] \tag{2.6.4a}$$

$$P(+\tfrac{1}{2}\to-\tfrac{1}{2};\, m\to m-1)=(\tau^{-1}/12)(\Delta A/h\nu)^2[I(I+1)-m(m-1)] \tag{2.6.4b}$$

Yet another possible mechanism for the spin–lattice relaxation with a static Jahn–Teller effect was suggested by Lee and Walsh (1968). They noted that there is a direct coupling via the phonons between states of opposite spin in different distorted configurations, as a consequence of spin–orbit coupling to excited states. If this coupling is not too strongly quenched by the Jahn–Teller effect, Lee and Walsh proposed that this mechanism might account for the spin–lattice relaxation. To estimate the value of T_1 that results, we consider the interaction given by Eq. (2.1.13), which represents the combined effect of spin–orbit interaction and the shear strains e_{xy}, e_{yz}, e_{zx}. Since the operator $\mathscr{A}_2$ (which transforms as A_2) has finite matrix elements within an E state but none between an E state and a singlet A_1 or A_2, the only effect of this interaction on the ground state $\Psi_{g\theta}$, $\Psi_{g\varepsilon}$ and first excited state Ψ_{A1} or Ψ_{A2} of the Jahn–Teller system is to introduce off-diagonal matrix elements $\pm ipG_2$ between $\Psi_{g\theta}$ and $\Psi_{g\varepsilon}$ in Eq. (2.4.2), where G_2 is given by

$$G_2 = V_3''(S_x e_{yz} + S_y e_{zx} + S_z e_{xy}) \tag{2.6.5}$$

in accord with Eq. (2.3.20). Performing the inverse of the transformation (2.4.6) to the states Ψ_1, Ψ_2, and Ψ_3, which represent the individual distorted Jahn–Teller configurations in accord with Eq. (2.4.9), we find that in terms of these states this interaction takes the form

	Ψ_1	Ψ_2	Ψ_3
Ψ_1:	0	$-ipG_2/\sqrt{3}$	$+ipG_2/\sqrt{3}$
Ψ_2:	$+ipG_2/\sqrt{3}$	0	$-ipG_2/\sqrt{3}$
Ψ_3:	$-ipG_2/\sqrt{3}$	$+ipG_2/\sqrt{3}$	0

(2.6.6)

We see therefore that this interaction has nonzero matrix elements only between states of different distorted configurations, and that these matrix elements contain as a factor the parameter p, which is very small when the Jahn–Teller coupling is strong. We have given the asymptotic expression for p in the presence of warping in Eq. (2.3.38b), and as we have seen in Section 2.3, a value for p different from zero is a measure of the inadequacy of the Born–Oppenheimer approximation (2.3.27) in describing the ground state. In particular, p is not simply an overlap integral between vibrational wave functions of different distorted configurations, as Eq. (2.6.6) might at first sight suggest, and it may be $\ll 1$ even in the absence of warping. To obtain T_1, we now replace the strains e_{xy}, e_{yz}, and e_{zx} in G_2 in Eq. (2.6.5) by the corresponding phonon operators and use Eq. (2.6.6) together with the spin operators contained in G_2 to evaluate the matrix element for the probability of a reorientation of the defect accompanied by a spin reversal and the absorption or emission of a single phonon (direct process). Using approximations and assumptions similar to those used in deriving Eq. (2.5.8), and using the value for V_3'' given by Eq. (2.1.15), we obtain by this mechanism the isotropic result for T_1

$$T_1^{-1} = (2\delta^2 p^2 kT/15\pi\hbar^4 \varrho s_T{}^5)(\lambda V_3/\Delta)^2[1 + \tfrac{2}{3}(s_T/s_L)^5] \tag{2.6.7}$$

where δ^2 denotes an average value for the squared strain energy differences δ_{21}^2, δ_{31}^2, and δ_{32}^2 and we have assumed $kT > \delta > g\beta H$.

Orton *et al.* (1961) have related the spin–lattice relaxation rate, as obtained from linewidths in the isotropic motionally averaged spectrum, to a theory given by McConnell (1956) for the spin–lattice relaxation in a molecule which undergoes a rapid tumbling motion in a liquid. Assuming an anisotropic g-factor and hyperfine interaction, McConnell used the part of the spin Hamiltonian which is explicitly time-dependent as a result of the tumbling motion to calculate the spin transition probabilities. In the notation of Eq. (2.6.3) and under the assumption $\Delta g\beta H \gg \Delta A$, his result for T_1 is

$$T_1^{-1} \approx (8\pi^2/15)(\Delta g\beta H + \Delta Am)^2(1/h^2)[\tau_c/(1 + 4\pi^2\nu^2\tau_c{}^2)] \tag{2.6.8}$$

where τ_c is the correlation time for the tumbling motion. In the limit $4\pi\nu\tau_c \gg 1$, this expression is identical with what we obtain by taking the average of Eq. (2.6.3) with respect to the orientation of $\mathbf{H}$, if we identify τ_c with the reorientation relaxation time τ. McConnell's theory represents a different approach to the same physical mechanism for the spin–lattice relaxation as that considered by Williams *et al.* (1969). The fact that Eq. (2.6.8) is isotropic, whereas Eq. (2.6.3) contains the anisotropic factor $(\zeta_x^2\zeta_y^2+\zeta_y^2\zeta_z^2+\zeta_z^2\zeta_x^2)$, reflects the ability of the tumbling molecule in the liquid to assume all possible orientations, while the reorientation in the Jahn–Teller system is between three fixed configurations. Indeed, when **H** is along a [100] direction, which coincides with a principal axis for each of these three Jahn–Teller distortions, it is clear from consideration of McConnell's theory applied to the Jahn–Teller system that the time-dependent part of the spin Hamiltonian has no matrix elements which can reverse the spin direction, except in conjunction with a nuclear spin flip as in the mechanism leading to Eq. (2.6.4) [this mechanism was not included in the derivation of Eq. (2.6.8)]. Thus, to extend McConnell's theory to the Jahn–Teller system, Eq. (2.6.8) should evidently be modified by insertion of a factor $5(\zeta_x^2\zeta_y^2 + \zeta_y^2\zeta_z^2 + \zeta_z^2\zeta_x^2)$, and the relaxation mechanism involving a simultaneous nuclear spin flip should be considered.

The same mechanisms that lead to an enhanced spin–lattice relaxation rate for a static Jahn–Teller effect are effective also for a dynamic Jahn–Teller effect. Assuming a predominant strain-splitting so that the ^2E vibronic ground state is split into the states Ψ_α and Ψ_β given by Eq. (2.5.12), we recognize that these represent two Kramers doublets separated by an energy difference $|\delta|$. Because the anisotropy in the g-tensor and hyperfine interaction is different for each doublet, there is again a small relative probability $p_r(\alpha \to \beta) = p_r(\beta \to \alpha)$ of a spin reversal in a phonon-induced transition between the doublets. Under assumptions corresponding to those used in deriving Eqs. (2.6.1) and (2.6.3), we obtain in this case as a result of this anisotropy

$$p_r(\alpha \to \beta) = [3q^2(g_2\beta H + A_2 m)^2/(h\nu)^2][\zeta_x^2\zeta_y^2 \sin^2\varphi + \zeta_y^2\zeta_z^2 \sin^2(\varphi - \tfrac{2}{3}\pi) + \zeta_z^2\zeta_x^2 \sin^2(\varphi + \tfrac{2}{3}\pi)] \tag{2.6.9}$$

If we assume a random strain distribution and replace $\sin^2\varphi$ by its average value 1/2, and if we have $kT > |\delta| > \hbar\nu$, then the average value for T_1^{-1} is given by

$$T_1^{-1} = \tfrac{3}{2}q^2[(g_2\beta H + A_2 m)/h\nu]^2(\zeta_x^2\zeta_y^2 + \zeta_y^2\zeta_z^2 + \zeta_z^2\zeta_x^2)\tau^{-1} \tag{2.6.10}$$

where τ is the relaxation time defined by Eq. (2.5.11) and given for direct and Raman processes by Eqs. (2.5.15) and (2.5.16), respectively. We must recognize, however, that we actually have a distribution of spin–lattice relaxation times for a given orientation of **H** if the strain is random, while for a definite strain we should use the appropriate value of φ in Eq. (2.6.9) in deriving T_1. Moreover, if kT is less than $|\delta|$, τ^{-1} approaches a limiting value as $kT \to 0$ in accord with Eq. (2.5.15), whereas T_1^{-1} decreases more rapidly with decreasing temperature than indicated by Eq. (2.6.10). This continued decrease in T_1^{-1} results because the lower Kramers doublet (we still assume $|\delta| > g\beta H$) becomes relatively more populated than the upper one as T decreases, and transitions to the upper doublet occur less frequently. Thus in the limit $kT \ll |\delta|$, the value of T_1 must approach that for the lower Kramers doublet taken by itself, and the enhancement of the spin–lattice relaxation rate due to transitions between the doublets is frozen out. If the magnitude of $|\delta|$ is also random, this freezeout effect provides an additional origin for a distribution in spin–lattice relaxation times when $kT < |\delta|$. A similar freezeout effect on the spin–lattice relaxation time of course also occurs for a static Jahn–Teller effect when kT is less than the strain splittings δ_{ab}.

Finally, the alternative mechanism for spin–lattice relaxation arising from the residual spin–orbit interaction, Eq. (2.1.13), leads to an expression for T_1^{-1} in the dynamic case, for $kT > |\delta| > g\beta H$, given by 3/2 times Eq. (2.6.7), where of course δ must now denote the energy difference (2.5.13). Since the parameter p is larger the weaker the Jahn–Teller coupling, this mechanism should be relatively more effective in the dynamic case than it is for a static Jahn–Teller effect.

2.7. Jahn–Teller Effects for Orbital Doublet States in Configurations with Less Than Cubic Symmetry

Many of the experimental observations of Jahn–Teller effects in the ^2E state of Cu^{2+} and related ions have been in crystals in which the site symmetry is trigonal rather than cubic. Since a doublet belonging to the irreducible representation E of the cubic group (or T_d) is not split by a trigonal crystal field, the theory we have outlined in the preceding sections is changed very little, however, if the point group symmetry is lowered to D_3, C_{3v}, or C_3. The principal change is that, in accord with Eqs. (2.1.13) and (2.1.14) (taking $e_{yz} = e_{zx} = e_{xy}$ corresponding to the trigonal axis in the [111] direction, for example), we must introduce small spin–orbit and orbital Zeeman interactions of the form

$$\mathscr{H}_{\text{SO}} = \lambda' S_t \mathscr{A}_2 \tag{2.7.1}$$

$$\mathscr{H}_{\text{OZ}} = g'\beta H_t \mathscr{A}_2 \tag{2.7.2}$$

where S_t and H_t are the components of $\mathbf{S}$ and $\mathbf{H}$ along the trigonal axis.

The first of these splits a $^2E(\Gamma_8)$ level into two Kramers doublets. But since terms of the sort (2.7.1) and (2.7.2) are quenched strongly [see Eqs. (2.3.20–21)] by Jahn–Teller coupling with the reduction factor p, which is nearly zero for strong coupling, these additions to the preceding theory will be of little importance unless the Jahn–Teller coupling is fairly weak.

The general Jahn–Teller problem for an arbitrary orbital doublet in trigonal symmetry (not only just for one arising from an E doublet in cubic symmetry) is formally identical to the theory we have outlined above. However, the resulting EPR spectrum may be quite different in detail from what we have obtained above, because of the different physical situation. In particular, the spin–orbit interaction (2.7.1) and the orbital Zeeman interaction (2.7.2) may be quite large (as they would be, for example, if the doublet results from the crystal-field splitting of a triplet T_1 or T_2 state in cubic symmetry). For all of the point groups having a threefold symmetry axis (C_3, S_6, D_3, C_{3v}, D_{3d}, C_6, C_{3h}, C_{6h}, D_6, C_{6v}, D_{3h}, D_{6h}), and for any electronic orbital doublet the degeneracy of which follows from group theory (including the time-reversal invariance of the Hamiltonian), the Jahn-Teller-active modes (the nontotally-symmetric modes of distortion which can produce a first-order splitting of the doublet) are pairs of degenerate modes. The linear Jahn–Teller coupling to one such pair of modes can in all cases be put in the form of Eq. (2.2.1), in the appropriate representation of the electronic states. Therefore the vibronic Hamiltonian for the dynamic Jahn–Teller effect with linear coupling to these modes is identical to Eq. (2.3.1), and the eigenstates and energies are the same as those in the cubic case. Moreover, the effect of any perturbation on the electronic doublet can be given in the general form of Eq. (2.3.7). The resulting reduction factors p and q that modify the respective terms of Eq. (2.3.7) in a dynamic Jahn–Teller effect are thus the same as in the cubic case, and other aspects of the theory may also be treated in a manner formally equivalent to that used in cubic symmetry. It should be noted, of course, that in the lower symmetry there may be more than one pair of Jahn–Teller-active modes of vibration of the cluster comprising a Jahn–Teller ion and its nearest neighbors. If these pairs are all equally important, the vibronic problem in the lower symmetry will therefore in actuality be much more complicated than, say, for an octahedrally coordinated ion in cubic symmetry interacting principally with its nearest neighbors. Also, if the terms (2.7.1) and (2.7.2) are large for the electronic doublet, second-order terms like those in Eq. (2.3.50) may be quite important in the ground state, giving, for example, an appreciable contribution to the g-tensor.

The theory of dynamic Jahn–Teller effects in the EPR and optical spectra of a doublet state in C_{3v} symmetry has been considered by Silsbee (1965) and Krupka and Silsbee (1966) and applied to the *R* center in KCl. In addition to calculating the reduction factor for the spin–orbit and orbital Zeeman interactions (2.7.1) and (2.7.2) from the theory of Longuet-Higgins *et al.* (1958) and evaluating the second-order *g*-shifts noted above, they have also given the form which stress splittings of the doublet must take in this symmetry. Krupka and Silsbee have also emphasized the importance of taking account of residual strain in comparing theory with experimental results.

The case of an orbital doublet in axial symmetry but with a fourfold axis (point groups C_4, S_4, C_{4h}, D_4, C_{4v}, D_{2d}, D_{4h}) gives rise to a very different Jahn–Teller problem. The Jahn–Teller-active modes are now nondegenerate, so that the coupling coefficients to each such mode and the mode frequencies are all independent parameters. The energy surfaces corresponding to linear coupling to two such modes do not then exhibit the rotational symmetry of Fig. 3 (except accidentally). This vibronic problem has been considered by Child (1960), Hougen (1964), and Ballhausen (1965) (the latter includes spin–orbit coupling). In particular, if coupling with only a single mode is important, the Jahn–Teller problem (without spin–orbit coupling) can be solved exactly [as in the case of a triplet T_1 or T_2 coupled only to *E* modes (see Section 3.1), the electronic states can be chosen such that the Jahn–Teller coupling is diagonal with respect to the states]. The only application to the theory for the EPR spectrum that has yet been made for this case is a numerical calculation by Mizuhashi (Kamimura and Mizuhashi, 1968; Mizuhashi, 1969) in proposing an interpretation of the anisotropy of the *g*-tensor in the low-spin ferrihemoglobin compounds. In this work, interaction with one mode was considered, and the spin–orbit interaction and rhombic field splitting were of comparable size to $\hbar\omega$ and the Jahn–Teller energy. The lowest vibronic eigenstate was accordingly obtained from a numerical diagonalization of the infinite-order secular equation of the vibronic problem including these other interactions. The possibility that similar vibronic corrections to the *g*-tensor are important for the O_2^- center in the alkali halides has been suggested by Shuey and Zeller (1967).

3. THEORY OF JAHN–TELLER EFFECTS IN THE EPR SPECTRUM OF AN ORBITAL TRIPLET ELECTRONIC STATE*

3.1. Jahn–Teller Effects for a Triplet State Coupled to an E Vibrational Mode

For an orbital triplet electronic state, T_1 or T_2, in cubic or tetrahedral symmetry,† the Jahn–Teller coupling to an E vibrational mode Q_θ, Q_ε is given by the vibronic Hamiltonian (Moffitt and Thorson, 1957)

$$\mathscr{H} = \mathscr{H}_0 \mathscr{I} + V_E[Q_\theta \mathscr{E}_\theta + Q_\varepsilon \mathscr{E}_\varepsilon] \tag{3.1.1}$$

where $\mathscr{H}_0$ is the same as in Eq. (2.3.2). Now, $\mathscr{I}$ is the 3×3 unit matrix, and $\mathscr{E}_\theta$ and $\mathscr{E}_\varepsilon$ are two standard electronic orbital operators belonging to E and taking the matrix form with respect to the basis ψ_ξ, ψ_η, ψ_ζ,

$$\mathscr{E}_\theta = \begin{pmatrix} +\frac{1}{2} & 0 & 0 \\ 0 & +\frac{1}{2} & 0 \\ 0 & 0 & -1 \end{pmatrix}, \qquad \mathscr{E}_\varepsilon = \begin{pmatrix} -\frac{1}{2}\sqrt{3} & 0 & 0 \\ 0 & +\frac{1}{2}\sqrt{3} & 0 \\ 0 & 0 & 0 \end{pmatrix} \tag{3.1.2}$$

Since the matrices $\mathscr{E}_\theta$ and $\mathscr{E}_\varepsilon$ are diagonal, the Hamiltonian (3.1.1) does not mix the states ψ_ξ, ψ_η, ψ_ζ for any Q_θ, Q_ε. The resulting energy surfaces describing the static Jahn–Teller effect are thus three disjoint paraboloids (Van Vleck, 1939b; Opik and Pryce, 1957; Liehr, 1963) as shown in Fig. 10 corresponding to ψ_ξ, ψ_η, ψ_ζ, with vertices lying respectively at the points

$$Q_{\theta i} = -V_E e_{i\theta}/\mu\omega^2, \qquad Q_{\varepsilon i} = -V_E e_{i\varepsilon}/\mu\omega^2 \tag{3.1.3}$$

where $e_{i\theta}$ and $e_{i\varepsilon}$ are the diagonal matrix elements, respectively, of $\mathscr{E}_\theta$ and $\mathscr{E}_\varepsilon$ corresponding to $i = \xi, \eta, \zeta$. These vertices thus correspond to the three energetically equivalent stable configurations of the static Jahn–Teller effect, and the Jahn–Teller energy is

$$E_{\rm JT} = V_E^2/2\mu\omega^2 \tag{3.1.4}$$

* The theory we have given in some detail in Section 2 for an orbital doublet state also illustrates the nature of the phenomena which are associated with the Jahn–Teller effect in other systems. Since relatively few EPR investigations to date have been concerned with Jahn–Teller effects in orbital triplet systems, we will accordingly treat these cases only briefly to summarize their most prominent characteristics.

† For the triplet state, we use as a basis the real functions $\psi_1 = \psi_\xi$, $\psi_2 = \psi_\eta$, and $\psi_3 = \psi_\zeta$, which are taken to transform respectively for T_2 (O or T_d) as yz, zx, and xy, for T_1 (O) as x, y, and z, or for T_1 (T_d) as $x(y^2 - z^2)$, $y(z^2 - x^2)$, and $z(x^2 - y^2)$.

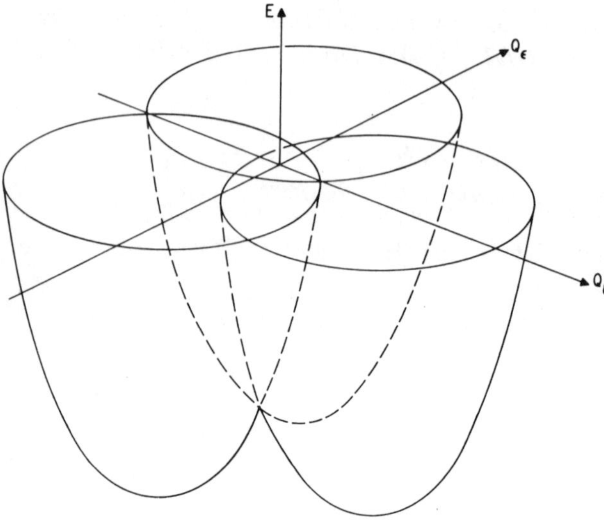

Fig. 10. Energy surfaces for the vibronic problem of an orbital triplet electronic state with Jahn–Teller coupling only with a pair of E vibrational modes, Q_θ, Q_ε. The surfaces are three disjoint paraboloids.

The stable configurations have tetragonal symmetry and as in Section 2 represent an elongation or contraction along one of the cubic axes.

The vibronic eigenstates of Eq. (3.1.1) are given exactly by (Moffitt and Thorson, 1957; Ham, 1965)

$$\Psi_{in} = \psi_i F_{n\theta}[Q_\theta + (V_E e_{i\theta}/\mu\omega^2)]F_{n\varepsilon}[Q_\varepsilon + (V_E e_{i\varepsilon}/\mu\omega^2)] \qquad (3.1.5)$$

corresponding to the energy eigenvalue

$$E_{in} = E_0 - E_{\mathrm{JT}} + (n_\theta + n_\varepsilon + 1)\hbar\omega \qquad (3.1.6)$$

where $n_\theta, n_\varepsilon = 0, 1, 2, 3, \ldots,$ and $F_n(y)$ is the one-dimensional simple harmonic oscillator wave function for the energy $(n + \frac{1}{2})\hbar\omega$. The energy spectrum (3.1.6) is thus displaced by $-E_{\mathrm{JT}}$ but otherwise is unchanged by the Jahn–Teller coupling, while the vibrational parts of the wave functions (3.1.5) are appropriate to a displaced two-dimensional harmonic oscillator corresponding to one of the paraboloids in Fig. 8. The ground state is a degenerate triplet belonging to the same representation T_1 or T_2 as the electronic triplet. In this case, the exact vibronic wave functions (3.1.5) have the form of simple Born–Oppenheimer products (1.2.1), but the vibrational factor is different for the three electronic states even for states of the same energy.

Any interaction having matrix elements within the electronic triplet may be expressed in analogy with Eq. (2.3.7) as

$$\mathscr{V} = G_1\mathscr{I} + G_\theta\mathscr{E}_\theta + G_\varepsilon\mathscr{E}_\varepsilon + G_\xi\mathscr{T}_\xi + G_\zeta\mathscr{T}_\zeta + G_\eta\mathscr{T}_\eta + G_x\mathscr{L}_x + G_y\mathscr{L}_y + G_z\mathscr{L}_z \tag{3.1.7}$$

where the standard Hermitian electronic orbital operators $\mathscr{T}_\xi$, $\mathscr{T}_\eta$, and $\mathscr{T}_\zeta$ belong to T_2, transform like yz, zx, and xy, and have the form

$$\mathscr{T}_\xi = \begin{pmatrix} 0 & 0 & 0 \\ 0 & 0 & -1 \\ 0 & -1 & 0 \end{pmatrix}, \quad \mathscr{T}_\eta = \begin{pmatrix} 0 & 0 & -1 \\ 0 & 0 & 0 \\ -1 & 0 & 0 \end{pmatrix}, \quad \mathscr{T}_\zeta = \begin{pmatrix} 0 & -1 & 0 \\ -1 & 0 & 0 \\ 0 & 0 & 0 \end{pmatrix} \tag{3.1.8}$$

while $\mathscr{L}_x$, $\mathscr{L}_y$, and $\mathscr{L}_z$ belong to T_1, transform like components of an angular momentum, and have the form

$$\mathscr{L}_x = \begin{pmatrix} 0 & 0 & 0 \\ 0 & 0 & -i \\ 0 & +i & 0 \end{pmatrix}, \quad \mathscr{L}_y = \begin{pmatrix} 0 & 0 & +i \\ 0 & 0 & 0 \\ -i & 0 & 0 \end{pmatrix}, \quad \mathscr{L}_z = \begin{pmatrix} 0 & -i & 0 \\ +i & 0 & 0 \\ 0 & 0 & 0 \end{pmatrix} \tag{3.1.9}$$

We then define reduction factors $K(E)$, $K(T_1)$, and $K(T_2)$ for the ground-state triplet Ψ_{gi} from representative matrix elements:

$$K(E) = -\langle\Psi_{g\zeta} \mid \mathscr{E}_\theta \mid \Psi_{g\zeta}\rangle \tag{3.1.10a}$$

$$K(T_1) = i\langle\Psi_{g\xi} \mid \mathscr{L}_z \mid \Psi_{g\eta}\rangle \tag{3.1.10b}$$

$$K(T_2) = -\langle\Psi_{g\xi} \mid \mathscr{T}_\zeta \mid \Psi_{g\eta}\rangle \tag{3.1.10c}$$

Within the ground-state triplet, $\mathscr{V}$ in Eq. (3.1.7) is thus equivalent to the operator

$$\begin{aligned}\mathscr{V}_g = G_1\mathscr{I} &+ K(E)(G_\theta\mathscr{E}_{g\theta} + G_\varepsilon\mathscr{E}_{g\varepsilon}) + K(T_2)(G_\xi\mathscr{T}_{g\xi} + G_\eta\mathscr{T}_{g\eta} + G_\zeta\mathscr{T}_{g\zeta}) \\ &+ K(T_1)(G_x\mathscr{L}_{gx} + G_y\mathscr{L}_{gy} + G_z\mathscr{L}_{gz})\end{aligned} \tag{3.1.11}$$

where we have assumed the G's independent of the Q's and have defined $\mathscr{E}_{g\theta}$, etc., with respect to the ground state in analogy with $\mathscr{U}_{g\theta}$, $\mathscr{U}_{g\varepsilon}$, and $\mathscr{A}_{g2}$ in Eq. (2.3.9). It follows then from Eq. (3.1.10) and the vibrational overlap integral of the wave functions (3.1.5) that we obtain the exact value (Ham, 1965)

$$K(T_1) = K(T_2) = \exp[-\tfrac{3}{2}(E_{\rm JT}/\hbar\omega)] \tag{3.1.12}$$

while $K(E) = 1$. Operators transforming as E (and of course A_1) are thus unaffected by the Jahn–Teller coupling when the coupling is solely with modes belonging to E, while Eq. (3.1.12) determines the extent of the dynamic quenching of electronic operators transforming as T_1 or T_2. In the limiting case $E_{JT} \gg \hbar\omega$, the direct effect of these latter operators is thus entirely quenched, as we expect for the static Jahn–Teller effect.

The way in which a dynamic Jahn–Teller effect affects the spin–orbit splitting and Zeeman interaction in a triplet state may now be shown. We assume that in addition to the orbital degeneracy the system has a spin **S**. The Zeeman interaction for the electronic state then has the form

$$\mathscr{H}_H = g_{\mathscr{L}}\beta(\mathscr{L} \cdot \mathbf{H}) + g_s(\mathbf{S} \cdot \mathbf{H}) \tag{3.1.13}$$

where $\mathscr{L}$ denotes the vector $(\mathscr{L}_x, \mathscr{L}_y, \mathscr{L}_z)$ and $g_{\mathscr{L}}$ and g_s are appropriate orbital and spin g-factors which include effects of covalency. Taking account of the Jahn–Teller coupling, we then have in the vibronic ground state, in accord with Eq. (3.1.11),

$$\mathscr{H}_{gH} = g'\beta(\mathscr{L}_g \cdot \mathbf{H}) + g_s\beta(\mathbf{S} \cdot \mathbf{H}) \tag{3.1.14}$$

where

$$g' = K(T_1)g_{\mathscr{L}} \tag{3.1.15}$$

The orbital part of the Zeeman interaction is thus partially quenched (Ham, 1965) by the dynamic Jahn–Teller effect through the appearance of the reduction factor $K(T_1)$. In addition, the spin–orbit interaction,

$$\mathscr{H}_{\text{so}} = \lambda(\mathscr{L} \cdot \mathbf{S}) \tag{3.1.16}$$

is equivalent within the vibronic ground state to the operator

$$\mathscr{H}_{g,\text{so}} = \lambda'(\mathscr{L}_g \cdot \mathbf{S}) \tag{3.1.17}$$

with

$$\lambda' = K(T_1)\lambda \tag{3.1.18}$$

so that the spin–orbit interaction, too, is partially quenched.* If we can

* Equations (3.1.17) and (3.1.18) are valid only if the Jahn–Teller energy E_{JT} is larger than the spin–orbit splitting of the electronic term in the absence of Jahn–Teller coupling. If these quantities are comparable, we must treat the Jahn–Teller coupling and the spin–orbit interaction on the same footing, and we must then diagonalize numerically the infinite-order secular equation for the vibronic states including the spin–orbit interaction. On the other hand, if the spin–orbit interaction dominates,

ignore second-order corrections (see below), the spin–orbit levels arising from the vibronic ground state then have energies

$$E_{0J} = E_0 - E_{\mathrm{JT}} + \hbar\omega + \tfrac{1}{2}\lambda'[J(J+1) - \mathscr{L}_g(\mathscr{L}_g+1) - S(S+1)] \quad (3.1.19)$$

where we have used Eq. (3.1.6) for $n_\theta = n_\varepsilon = 0$. In obtaining Eq. (3.1.19), we have defined an effective total angular momentum operator

$$\mathbf{J} = \mathscr{L}_g + \mathbf{S} \quad (3.1.20)$$

which is diagonal with respect to $\mathscr{H}_{g,\mathrm{so}}$ as given by Eq. (3.1.17). We have also $\mathscr{L}_g = 1$ from the definition (3.1.9). The Zeeman interaction (3.1.14) within each J level of Eq. (3.1.19) then takes the form

$$\mathscr{H}_H(J) = g_J\beta(\mathbf{J}\cdot\mathbf{H}) \quad (3.1.21)$$

with g_J given as in the derivation for the Landé g-factor as

$$g_J = \tfrac{1}{2}(g_s+g') + \tfrac{1}{2}\{[S(S+1) - \mathscr{L}_g(\mathscr{L}_g+1)]/J(J+1)\}(g_s - g') \quad (3.1.22)$$

These results in Eqs. (3.1.19) and (3.1.22) are of course valid only if the overall spin–orbit splitting of the vibronic ground state as given by Eq. (3.1.19) is small compared with $\hbar\omega$, the separation of the vibronic levels.

Other orbital interactions, such as the orbital contribution to the hyperfine interaction and the effect of strain, are also partially quenched by a dynamic Jahn–Teller effect in accord with the appropriate reduction factor in Eq. (3.1.10). For strain, the degeneracy of the electronic states $\psi_\xi, \psi_\eta, \psi_\zeta$ is lifted in accordance with the perturbation

$$\mathscr{H}_s = V_2[e_\theta\mathscr{E}_\theta + e_\varepsilon\mathscr{E}_\varepsilon] + V_3[e_{yz}\mathscr{T}_\xi + e_{zx}\mathscr{T}_\eta + e_{xy}\mathscr{T}_\zeta] \quad (3.1.23)$$

The effect of strain in lifting the orbital degeneracy of the vibronic ground state is accordingly given by

$$\mathscr{H}_{gs} = K(E)V_2[e_\theta\mathscr{E}_{g\theta} + e_\varepsilon\mathscr{E}_{g\varepsilon}] + K(T_2)V_3[e_{yz}\mathscr{T}_{g\xi} + e_{zx}\mathscr{T}_{g\eta} + e_{xy}\mathscr{T}_{g\zeta}] \quad (3.1.24)$$

---

we should treat each spin–orbit level separately in analyzing the effect of the Jahn–Teller coupling, and we may include the effect of the other levels through use of perturbation theory. In the latter case, if the lowest spin–orbit level is a singlet or Kramers doublet, or if it has only a weak coupling to the Jahn–Teller distortion as in the case of the $J = 1$ level of Fe^{2+} in MgO discussed by Van Vleck (1958, 1960) and Liehr (1960), we have an example of the spin–orbit interaction stabilizing the ground state of the system against a Jahn–Teller effect.

The splitting of the triplet by low-symmetry crystal fields may similarly be diminished by a Jahn–Teller effect, if the Jahn–Teller coupling appropriate to cubic symmetry is sufficiently strong to take preeminence over such a component of the crystal field. In particular, Jahn–Teller coupling to an E mode will tend to diminish the splitting of the triplet by a crystal field of trigonal symmetry, but it will not affect that due to a tetragonal component.

A dynamic Jahn–Teller effect also leads to second-order corrections in the ground state which arise from the matrix elements of $\mathscr{V}$ in Eq. (3.1.7) to excited vibronic states. For the Zeeman and spin–orbit interactions (3.1.13) and (3.1.16), when the Jahn–Teller coupling is only with the E modes, these corrections may be put in the form (Ham, 1965)

$$\mathscr{H}_{\rm SO}^{(2)} = -(\lambda^2/\hbar\omega)[f_a(\mathscr{L}_g \cdot \mathbf{S})^2 + (f_b - f_a)(\mathscr{L}_{gx}^2 S_x{}^2 + \mathscr{L}_{gy}^2 S_y{}^2 + \mathscr{L}_{gz}^2 S_z{}^2)] \tag{3.1.25}$$

$$\begin{aligned}\mathscr{H}_{\rm H}^{(2)} = -(\lambda/\hbar\omega)g_{\mathscr{L}}\beta\{&f_a[(\mathscr{L}_g \cdot \mathbf{S})(\mathscr{L}_g \cdot \mathbf{H}) + (\mathscr{L}_g \cdot \mathbf{H})(\mathscr{L}_g \cdot \mathbf{S})] \\ &+2(f_b - f_a)[\mathscr{L}_{gx}^2 S_x H_x + \mathscr{L}_{gy}^2 S_y H_y + \mathscr{L}_{gz}^2 S_z H_z]\}\end{aligned} \tag{3.1.26}$$

where

$$\begin{aligned} f_a &= e^{-3E_{\rm JT}/\hbar\omega} G(\tfrac{3}{2}E_{\rm JT}/\hbar\omega) \\ f_b &= e^{-3E_{\rm JT}/\hbar\omega} G(3E_{\rm JT}/\hbar\omega) \end{aligned} \tag{3.1.27}$$

$G(x)$ is defined as

$$G(x) = \int_0^x (1/u)(e^u - 1)\,du \tag{3.1.28}$$

and has the asymptotic expansion for large x

$$G(x) \sim \frac{e^x}{x}\left(1 + \frac{1}{x} + \frac{2^2}{2!x^2} + \frac{2^2 \cdot 3^2}{3!x^3} + \cdots\right) \tag{3.1.29}$$

In the limit $E_{\rm JT} \gg \hbar\omega$, the terms (3.1.25) and (3.1.26) may be shown (Ham, 1965) to agree with the second-order corrections calculated from perturbation matrix elements to the upper energy surface arising from the triplet in the distorted configuration of the static Jahn–Teller effect.

3.2. Jahn–Teller Effects for a Triplet State Coupled to a T_2 Vibrational Mode

A triplet state coupled to a set of vibrational modes Q_ξ, Q_η, Q_ζ belonging to T_2 has the vibronic Hamiltonian (Moffitt and Thorson, 1957)

$$\mathscr{H} = \mathscr{H}_0 \mathscr{I} + V_T[Q_\xi \mathscr{E}_\xi + Q_\eta \mathscr{E}_\eta + Q_\zeta \mathscr{E}_\zeta] \tag{3.2.1}$$

where $\mathscr{H}_0$ is now given by

$$\mathscr{H}_0 = E_0 + (1/2\mu)[P_\xi^2 + P_\eta^2 + P_\zeta^2 + \mu^2\omega^2(Q_\xi^2 + Q_\eta^2 + Q_\zeta^2)] \quad (3.2.2)$$

The energy surfaces in the space of the three coordinates Q_ξ, Q_η, Q_ζ which describe the static Jahn–Teller effect are now not easily visualized, but various cross sections have been illustrated by Liehr (1963). The stable configurations (Van Vleck, 1939b; Opik and Pryce, 1957; Liehr, 1963) are four in number and represent distortions of trigonal symmetry given by the coordinates

$$(Q_\xi^0, Q_\eta^0, Q_\zeta^0)_j = (2V_T/3\mu\omega^2)(m_1, m_2, m_3) \quad (3.2.3)$$

with the associated electronic states

$$\psi_j = (1/\sqrt{3})(m_1\psi_\xi + m_2\psi_\eta + m_3\psi_\zeta) \quad (3.2.4)$$

where for $j = 1, 2, 3, 4$, the set of integers (m_1, m_2, m_3) is given by $(+1, +1, +1)$, $(-1, -1, +1)$, $(+1, -1, -1)$, $(-1, +1, -1)$, respectively. The Jahn–Teller energy is

$$E_{\text{JT}} = \tfrac{2}{3}V_T^2/\mu\omega^2 \quad (3.2.5)$$

The vibronic eigenvalues of Eq. (3.2.1) for weak Jahn–Teller coupling have been obtained by Moffitt and Thorson (1957) to the accuracy of second-order perturbation theory. For stronger coupling, the eigenvalues have been evaluated numerically by Caner and Englman (1966). For all values of the coupling coefficient V_T, the vibronic ground state is a triplet belonging to the same irreducible representation T_1 or T_2 as the original electronic triplet. However, as the ratio $E_{\text{JT}}/\hbar\omega$ becomes large, an excited singlet (A_1 if the triplet is T_2, or A_2 if it is T_1) approaches the ground state asymptotically, the fourfold degeneracy in the limit corresponding to the four equivalent configurations of the static Jahn–Teller effect.

The reduction factors $K(E)$, $K(T_1)$, and $K(T_2)$ of Eq. (3.1.10) for the ground-state triplet in this case have been evaluated numerically by Caner and Englman (1966). Their results are plotted in Fig. 11, where they are compared with expressions appropriate to weak coupling given by

$$K(E) \simeq K(T_1) \simeq \exp[-(9/4)(E_{\text{JT}}/\hbar\omega)] \quad (3.2.6a)$$

$$K(T_2) \simeq \tfrac{1}{3}\{2 + \exp[-(9/4)(E_{\text{JT}}/\hbar\omega)]\} \quad (3.2.6b)$$

and obtained (Ham, 1965) from extrapolation of the results of perturbation theory. We see from Eq. (3.2.6) and Fig. 11 that, when the coupling is with T_2 modes, all types of electronic orbital operators except symmetric opera-

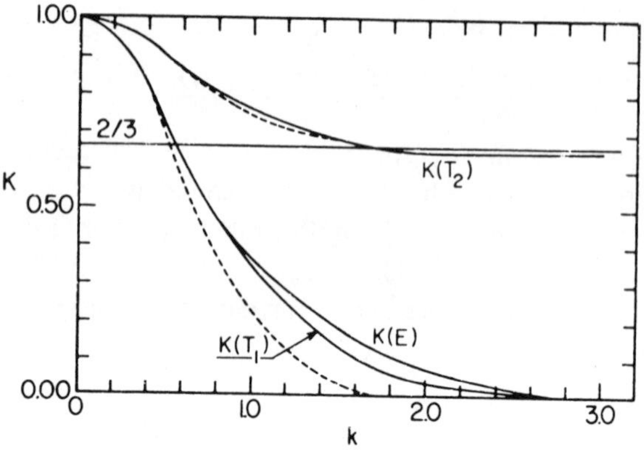

Fig. 11. Reduction factors $K(E)$, $K(T_1)$, and $K(T_2)$, as defined in Eqs. (3.1.10a–c) of the text, for the vibronic ground state in the case of the orbital triplet electronic state with Jahn–Teller coupling to a T_2 vibrational mode (from Caner and Englman, 1966). The solid lines show the results of Caner and Englman's numerical calculations, while the dashed lines show the approximate expressions given by Eqs. (3.2.6a,b), which extend the perturbation results obtained for weak Jahn–Teller coupling. The parameter k, the abscissa in the figure, is defined in terms of the ratio of the Jahn–Teller energy E_{JT} [Eq. (3.2.5)] to the mode energy $\hbar\omega$ by the relation $k^2 = \frac{3}{2}(E_{JT}/\hbar\omega)$.

tors belonging to A_1 suffer at least a partial dynamic quenching. Operators belonging to T_1 or E are entirely quenched in the limit of strong coupling, as we expect for the static Jahn–Teller effect, while those belonging to T_2 remain effective corresponding to the asymptotic value $K(T_2) = 2/3$. The fact that $K(T_2)$ is *less* than 2/3 as this asymptotic value is approached (Caner and Englman, 1966) (Fig. 11) evidently arises from the same cause that leads q to be less than 1/2 for the doublet case in strong coupling: the dependence of the electronic part of the Born–Oppenheimer product on the variation of Q within the well, and the finite spread of the vibrational wave function within the well.

When the Jahn–Teller coupling for the triplet is with both E and T_2 modes, the stable configurations of the static Jahn–Teller effect were shown by Opik and Pryce (1957) and Liehr (1963) to be either the purely tetragonal distortions given by Eq. (3.1.3), with the Jahn–Teller energy given by Eq. (3.1.4), or the purely trigonal distortions given by Eq. (3.2.3) with E_{JT} given by (3.2.5). Which of these cases is obtained depends upon the relative strength of the coupling to the two types of modes, as determined by the relative size of the Jahn–Teller energy as given by Eq. (3.1.4) or Eq. (3.2.5). In the dynamic regime, the effects of the two types of coupling should be roughly additive in diminishing the reduction factors below a value of unity when the coupling to both types of modes is weak. If either coupling is strong (or both are), the reduction factors should be those appropriate to the dominant type. The case of a dynamic Jahn–Teller effect when the coupling to both modes is almost equally strong has been considered very recently by O'Brien (1969) and applied by Hughes (1970) to propose an interpretation of the optical absorption spectrum of the F^+ center in CaO.*

4. EXAMPLES OF JAHN–TELLER EFFECTS IN EXPERIMENTAL EPR SPECTRA

4.1. Jahn–Teller Effects in EPR for Orbital Doublet States

In Table II are given examples of experimental values of g-factors and central hyperfine parameters that have been obtained from anisotropic low-temperature EPR spectra indicative of a static Jahn–Teller effect, for

* Bersuker (1962) and Bersuker and Vekhter (1963) have claimed that there is an "inversion splitting" of the orbital triplet ground state arising from a T_1 or T_2 electronic triplet in cubic symmetry, when the Jahn–Teller coupling is predominantly with the E modes but there is a weaker coupling with the T_2 modes. A singlet and a doublet state were said to result, and as a result of spin–orbit coupling a $^2T_2$ state was said therefore to split into three Kramers doublets (Bersuker and Vekhter, 1963). Such a splitting of states belonging to an irreducible representation of the cubic group is forbidden, however, on grounds of symmetry so long as the vibronic Hamiltonian has cubic symmetry. An orbital triplet belonging to T_1 or T_2 thus preserves its threefold degeneracy, while $^2T_2$ $(= \Gamma_7 + \Gamma_8)$ splits into at most a doublet and a quartet under spin–orbit interaction. Detailed consideration of the nature of the coupling to the T_2 modes, in the vibronic ground state resulting from predominant coupling to the E modes, is given by Ham (1967, Section VI), where it is shown in what sense the different tetragonally distorted configurations interact without any splitting of the threefold orbital degeneracy of the ground state. Bersuker and Vekhter (1966) have asserted, however, that symmetry considerations are in fact satisfied, because when the predominant Jahn–Teller coupling is with the E modes but there is weaker linear coupling to the T_2 modes, one obtains stable configurations having lower symmetry than tetragonal. The ground state of the system, comprising one state from each well, thus becomes 12-fold degenerate, and inversion splittings are then permitted by symmetry. However, this assertion that the stable configurations under these conditions have lower than tetragonal symmetry conflicts with the general conclusion of Opik and Pryce (1957) and Liehr (1963) and is incorrect. Wysling and Müller (1968) have recently reexamined this question explicitly and have reached the same conclusion.

Table II. Examples of Experimental g-Factor and Central Hyperfine Parameters in the Low-Temperature Anisotropic EPR Spectrum of Transition Metal Ions in Orbital Doublet States (^2E) in Cubic Symmetry[a] Which Exhibit a Static Jahn–Teller Effect

Ion	Host crystal	T,[b] °K	$g_\parallel$	$g_\perp$	$A_\parallel$, 10^{-4} cm^{-1}	$A_\perp$, 10^{-4} cm^{-1}	Ref.[c]
$Cu^{2+}(3d^9)$	$ZnSiF_6 \cdot 6H_2O$	20	2.46 ± 0.01	2.10 ± 0.01	110 ± 3	<30	A, B
$Cu^{2+}(3d^9)$	$La_2Mg_3(NO_3)_{12} \cdot 24D_2O$	20	2.470 ± 0.002	2.097 ± 0.002	−113 ± 0.5	+16 ± 4	B
$Cu^{2+}(3d^9)$	$Bi_2Mg_3(NO_3)_{12} \cdot 24H_2O$	20	2.454 ± 0.003	2.096 ± 0.003	110 ± 1	17 ± 2	B
$Cu^{2+}(3d^9)$	AgCl	77	2.30	2.07	113	45	C, D
$Pt^{3+}(5d^7)$	Al_2O_3	4.2	2.011 ± 0.006	2.328 ± 0.004	—	—	E
$Ag^{2+}(5d^9)$	KCl	77	2.193 ± 0.001	2.035 ± 0.001	40.9 ± 0.5	30.3 ± 0.5	F, G
$Ag^{2+}(5d^9)$	LiCl	4.2	2.178 ± 0.001	2.039 ± 0.001	38 ± 2	30 ± 2	G
$Ag^{2+}(5d^9)$	NaCl	4.2	2.196 ± 0.001	2.041 ± 0.001	39 ± 2	30 ± 2	G
$La^{2+}(5d^1)$	CaF_2	4	2.00 ± 0.01	1.904 ± 0.002	37 ± 9	62.6 ± 1	H
$Ni^{3+}(3d^7)$	CaO	20	2.0672 ± 0.0006	2.3828 ± 0.0006	—	—	I, J
$Ni^{+}(3d^9)$	LiF	80	2.065 ± 0.004	2.353 ± 0.004	—	—	K
$Ni^{+}(3d^9)$	NaF	60	2.067 ± 0.004	2.464 ± 0.004	—	—	K
$Y^{2+}(4d^1)$	CaF_2	4.2	2.00	1.958	—	—	L

[a] For several of the crystals, the site symmetry is lowered from cubic by a small trigonal distortion.
[b] Here, T is the temperature at which the EPR spectrum was measured.
[c] References are designated as follows: (A) Bleaney and Bowers (1952); (B) Bleaney, Bowers, and Trenam (1955); (C) Tucker (1958); (D) Burnham (1966); (E) Geschwind and Remeika (1962); (F) Delbecq *et al.* (1963); (G) Sierro (1967); (H) Hayes and Twidell (1963); (I) Low and Suss (1963); (J) Höchli *et al.* (1965); (K) Hayes and Wilkens (1964); (L) O'Connor and Chen (1964).

a number of transition metal ions having an orbital doublet (^2E) ground state in cubic symmetry (or cubic symmetry plus a small trigonal distortion). In all of these cases, this low-temperature spectrum is the superposition of three axial spectra in which the symmetry axis is one of three mutually perpendicular axes which coincide with the [100] axes for the cubic crystals or, for the crystals of trigonal symmetry, lie so as to have the trigonal axis as the body diagonal of the cube. The three superimposed spectra have equal intensity, corresponding to a random selection among these three orientations on the part of each individual paramagnetic ion. These findings agree with the predictions of Eqs. (2.2.9) and (2.2.11) concerning the symmetry expected for the g-tensor and hyperfine interaction in the configurations of stable equilibrium with a static Jahn–Teller effect.*

In each case listed in Table II (except for LiF:Ni^+), there is observed a transition with increasing temperature to an isotropic spectrum† for which g and A are given to a close approximation in terms of $g_\parallel$, $g_\perp$, $A_\parallel$, and $A_\perp$ of the low-temperature spectrum by

$$g = \tfrac{1}{3}(g_\parallel + 2g_\perp) \tag{4.1.1}$$

$$A = \tfrac{1}{3}(A_\parallel + 2A_\perp) \tag{4.1.2}$$

These data for the isotropic spectra are listed in Table III.‡ The values calculated from Eqs. (4.1.1) and (4.1.2) and the data of Table II are also given in Table III under the columns labeled g_{calc} and A_{calc}, and with only a few exceptions these agree with the experimental values to within the indicated experimental error. This transition at the higher temperature to an isotropic spectrum with g and A given by Eqs. (4.1.1) and (4.1.2) is characteristic of the sort of dynamic Jahn–Teller effect first observed by Bleaney and Ingram (1950) on Cu^{2+} in zinc fluosilicate and interpreted by Abragam and Pryce (1950). As we have seen in Section 2.5, such a transition is consistent with a motional averaging of the low-temperature spec-

* A component of apparently rhombic symmetry in the hyperfine interaction was reported (Bleaney, Bowers, and Trenam, 1955) for Cu^{2+} in $La_2Mg_3(NO_3)_{12} \cdot 24D_2O$. However, recent experiments (Breen *et al.*, 1969) have failed to confirm this finding.

† For the crystals of trigonal symmetry, there was found in some cases a small difference ($\lesssim 0.01$) between g measured parallel and perpendicular to the trigonal axis in the high-temperature spectrum, and also a similar small departure from isotropy in A. We have ignored these differences in listing the data in Table III.

‡ An isotropic spectrum attributed by Low and Rosenthal (1968) to a dynamic Jahn–Teller effect resulting from rapid relaxation in the ^2E state of Ti^{3+} in CaF_2 has since been shown by Zaitov and Stepanov (1968) to be due to $Ti^{2+}(^3A_2)$, which has $S = 1$, but no orbital degeneracy.

Table III. Experimental and Calculated[a] Values of g-Factors and Central Hyperfine Parameters in the High-Temperature Isotropic EPR Spectrum of Transition Metal Ions in Orbital Doublet States (^2E) in Cubic Symmetry[b]

Ion	Host crystal	T,[c] °K	T_T,[c] °K	g_{exp}	g_{calc}	A_{exp}, 10^{-4} cm^{-1}	A_{calc}, 10^{-4} cm^{-1}	Ref.[d]
Cu^{2+}	$ZnSiF_6 \cdot 6H_2O$	90	12–50	2.23 ± 0.01	2.22	25 ± 5	—	M, B
Cu^{2+}	$La_2Mg_3(NO_3)_{12} \cdot 24D_2O$	90	33–45	2.218 ± 0.003	2.221	28 ± 1	−27	N, B
Cu^{2+}	$Bi_2Mg_3(NO_3)_{12} \cdot 24H_2O$	90	>20, <90	2.218 ± 0.003	2.215	26 ± 1	25	B
Cu^{2+}	AgCl	400	90–300	2.15	2.15	—	—	C, D
Pt^{3+}	Al_2O_3	79	>4.2, <79	2.220 ± 0.001	2.222	—	—	E
Ag^{2+}	KCl	300	~160	2.09 ± 0.01	2.088	—	—	F, G
Ag^{2+}	LiCl	130	~40	2.085 ± 0.002	2.085	—	—	G
Ag^{2+}	NaCl	200	~95	2.092 ± 0.002	2.093	—	—	G
La^{2+}	CaF_2	20	10–12	1.937 ± 0.002	1.936	52 ± 1	54	H
Ni^{3+}	CaO	77	62–68	2.2814 ± 0.0006	2.278	—	—	I, J
Ni^{+}	NaF	150	130–230	2.332 ± 0.004	2.332	—	—	K
Y^{2+}	CaF_2	20	4–10	1.971	1.972	—	—	L
Ni^{3+}	Al_2O_3	>50	~12	2.146	—	—	—	E, O
Pd^{3+}	Al_2O_3	77	<10	2.163	—	—	—	P
Ni^{3+}	$SrTiO_3$	203	<100	2.180	—	—	—	Q
Ni^{3+}	MgO	77	<4	2.1693	—	—	—	R, J
Cu^{2+}	CaO	77	<2	2.2201 ± 0.0006	—	21.6 ± 0.3	—	I

[a] The values g_{calc} and A_{calc} are calculated from the expressions $g_{calc} = \frac{1}{3}(g_\parallel + 2g_\perp)$ and $A_{calc} = \frac{1}{3}(A_\parallel + 2A_\perp)$ using the parameters of the low-temperature spectrum of the static Jahn–Teller effect given in Table II. When no static Jahn–Teller effect has been observed, we omit g_{calc} and A_{calc}. Since the signs of $A_\parallel$ and $A_\perp$ are often not known, in these cases we have chosen the relative sign of $A_\parallel$ and $A_\perp$ that gives the best agreement between A_{calc} and the high-temperature experimental value for A when such data are available.

[b] For several of the crystals, the site symmetry is lowered from cubic by a small trigonal distortion.

[c] The temperature T is that at which the experimental spectrum was measured for which the values g_{exp} and A_{exp} are given in the table. The temperature T_T is an approximate value for the temperature at which the transition from the low-temperature to high-temperature spectrum occurs. This transition region is often quite broad; the two spectra sometimes are found to coexist within this region, while in other cases one is reported to have vanished before the other appears. For details, the original references should be consulted.

trum by rapid, phonon-induced reorientation of the distorted configuration as described by Eqs. (2.5.4) and (2.5.5). Alternatively, the appearance of the isotropic spectrum is consistent with the presence of excited singlet states which become populated with increasing temperature. Also included in Table III are parameters for the isotropic spectra of a number of ions for which published data on the low-temperature anisotropic spectrum are at the time of this writing either not available or incomplete.

Among the ions listed in Tables II and III, only in one case does sufficient information exist at the present time concerning the relaxation time τ for reorientation of the distorted configuration [Eq. (2.5.1)], for the hypothesis to be confirmed that the transition to the isotropic spectrum is explained by the theory of motional averaging. For $Al_2O_3:Ni^{3+}$, Sturge *et al.* (1967) have determined τ from the temperature dependence of the attenuation of 100- and 260-MHz acoustic waves in corundum crystals containing a few ppm of Ni^{3+}. Their results are shown in Fig. 12. Merritt and Sturge (1966) then obtained independent values for τ by fitting the theory of motional narrowing to the EPR data in the vicinity of the transition at 12°K. Since these results (the crosses in Fig. 12) agree to within the experimental uncertainties with the values of τ obtained from the acoustic attenuation data, the identification of the isotropic spectrum as resulting from the motional averaging of the low-temperature anisotropic spectrum is confirmed. Sturge *et al.* (1967) showed, moreover, that the low-temperature portion of Fig. 12 ($2°K < T < 7°K$) is consistent with τ^{-1} varying linearly with T, in agreement with the form of Eq. (2.5.8) (one-phonon direct process), while the higher-temperature data fit a formula of the form (2.5.6) (thermal activation or Orbach process) with an activation energy $E = 90\ cm^{-1}$. Another ion for which data on the reorientation time τ are available is Cu^{2+} in $La_2Mg_3(NO_3)_{12} \cdot 24D_2O$. In this case, the decay of the two-pulse electron spin echo of the Cu^{2+} has been measured (Breen *et al.*, 1969) and found to be exponential, and this decay time should approximate the time for reorientation. The reciprocal of this decay time was found to exhibit a linear temperature dependence, $\tau^{-1} \sim A + BT$, in the temperature range below 12°K with $A \lesssim 2.5\times 10^5\ sec^{-1}$ and $B \sim 5\times 10^4\ sec^{-1}\ deg^{-1}$. However, the transition to the isotropic spectrum occurs for this ion in the range 33–45°K (Table III), and the spin echo data were not extended above 12°K. Thus for this ion as well as for the others in Tables II and III (apart from $Al_2O_3:Ni^{3+}$), we have as yet no definitive evidence from independent measurement of the orientation time as to the role played by motional averaging in this transition.

From the change in the elastic constants of Al_2O_3 due to the presence

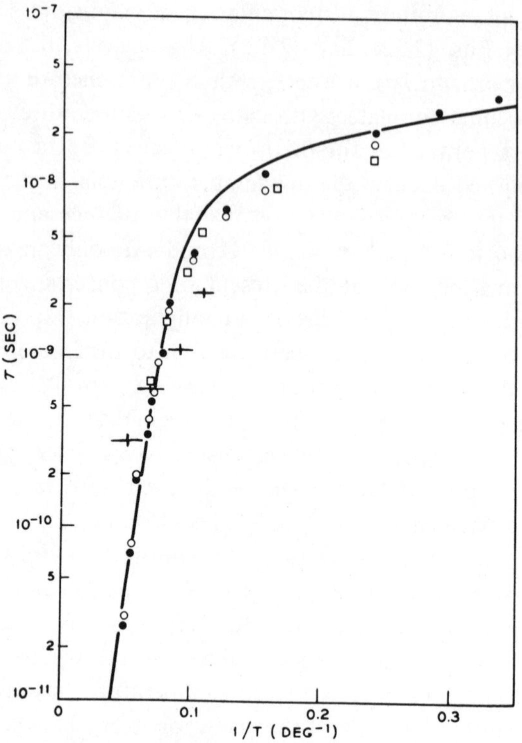

Fig. 12. Relaxation time τ for reorientation of the distorted configuration for Ni^{3+} ions in Al_2O_3 (from Sturge *et al.*, 1967). The points denote the experimental values obtained from acoustic attenuation data by Sturge *et al.* (1967), while the crosses indicate approximate values deduced by Merritt and Sturge (1966) from EPR linewidths using the theory of motional narrowing.

of the Ni^{3+} ions, as measured at 2 MHz, Sturge *et al.* (1967) were also able to determine for Al_2O_3:Ni^{3+} the coupling coefficient for the effect of strain in producing a relative shift in energy of the three distorted configurations in accord with Eq. (2.2.14). In terms of the strain coefficient V_2 that we have used throughout Section 2 [Eq. (2.1.10) or Eq. (2.2.14)], their result corresponds to a value $V_2 = 2.4 \pm 1.0 \times 10^4$ cm^{-1}. From this result, we may obtain an estimate of the Jahn–Teller energy in this system. Using Eq. (2.3.3) and the approximate relation [see the appendix, Eqs. (A.7) and (A.11)]

$$V = \sqrt{3}\, V_2/2R \tag{4.1.3}$$

between the strain coefficient V_2 and the coupling coefficient V for linear Jahn–Teller coupling, where R is the nearest-neighbor distance, and taking $\hbar\omega \simeq 200\ \text{cm}^{-1}$ for the energy of the effective E mode, we obtain from the value $V_2 = 2.4 \times 10^4\ \text{cm}^{-1}$ an estimate $E_{\text{JT}} \simeq 3000\ \text{cm}^{-1}$. This estimate is no better than as to order of magnitude, probably, because of the simplifying assumptions that had to be made in obtaining Eq. (4.1.3), as well as the uncertainty in the value of V_2. However it is gratifying that this estimate of E_{JT} agrees roughly with those made by Opik and Pryce (1957) and Liehr and Ballhausen (1958). Also, from consideration of the parameters α and β in Eq. (2.3.35), and of Figs. 7 and 8, it appears that we must have $E_{\text{JT}}/\hbar\omega > 5$ if $\hbar\omega$ is $200\ \text{cm}^{-1}$ or larger and if α is to be small enough so that a barrier height ($= 2\,|\,\beta\,|$) of 100–200 cm^{-1} will suffice to reduce the tunneling splitting 3Γ to less than 1 cm^{-1}, say. Under these conditions, for $V_2 \gtrsim 10^4$ cm^{-1}, a residual strain of $\gtrsim 10^{-4}$ will satisfy the criterion for the experimental observation of a static Jahn–Teller effect at low temperatures as discussed in Section 2.4. Thus, if $E_{\text{JT}}/\hbar\omega > 5$ and if $\hbar\omega > 200\ \text{cm}^{-1}$, to have a static Jahn–Teller effect we must have $E_{\text{JT}} > 1000\ \text{cm}^{-1}$, again in agreement with the earlier estimate for $Al_2O_3{:}Ni^{3+}$. Strain data are not available from which corresponding estimates of E_{JT} for the other ions in Tables II and III could be made.*

A final important number that can be derived from the data of Sturge *et al.* (1967) for $Al_2O_3{:}Ni^{3+}$ is the tunneling splitting 3Γ, which may be obtained from values of $(\tau T)^{-1}$, in accord with Eqs. (2.5.1) and (2.5.8), once V_2 is known (take $q \simeq \frac{1}{2}$), in the low-temperature limit in which τ^{-1} is proportional to T, $\tau^{-1} = BT$. As we have noted, Sturge *et al.* (1967) showed that their relaxation time in Fig. 12 had this form in the range $2°\text{K} < T < 7°\text{K}$, with $B = 1.1 \times 10^7\ \text{sec}^{-1}\ \text{deg}^{-1}$, and they obtained an estimate of Γ on this basis. Using our Eq. (2.5.8) instead of the expression obtained by Pirc *et al.* (1966) for a two-well tunneling problem, which Sturge *et al.* (1967) used and which differs from Eq. (2.5.8) in its numerical coefficient, we obtain $3\Gamma \simeq 1\ \text{cm}^{-1}$ for $Al_2O_3{:}Ni^{3+}$. From the spin-echo

* Borcherts *et al.* (1970) have recently reported the value $V_2 = -1.31 \times 10^4\ \text{cm}^{-1}$ from stress experiments on $NaCl{:}Cu^{2+}$, which exhibits a static Jahn–Teller effect below $\sim 90°\text{K}$. The g-factors $g_\parallel = 2.373$ and $g_\perp = 2.070$ were found to be very close to the values $g_\parallel = 2.339 \pm 0.002$ and $g_\perp = 2.070 \pm 0.002$ obtained for the similar Jahn–Teller complex $CdCl_2{:}Cu^{2+}$ by Thornley *et al.* (1961). This result for V_2 is in rather good agreement with a value obtained by Lohr (1967) from a molecular orbital calculation for the complex $CuCl_6^{4-}$. However, using this value in Eq. (4.1.3), we obtain an estimate of the Jahn–Teller energy which is considerably smaller than the value $E_{\text{JT}} = 2380\ \text{cm}^{-1}$ calculated by Lohr. This discrepancy serves as a warning that the assumptions underlying Eq. (4.1.3) may not be appropriate in all cases.

data of Breen *et al.* (1969) for Cu^{2+} in $La_2Mg_3(NO_3)_{12} \cdot 24D_2O$, in particular their value $B' \simeq 5\times 10^4$ sec^{-1} deg^{-1} for the temperature coefficient in the reciprocal decay time, we see from comparison with the value of B for Al_2O_3:Ni^{3+} and Eq. (2.5.8) that we would have a value $3\Gamma \simeq 0.1$ cm^{-1} for Cu^{2+} in the double nitrate if V_2 and the sound velocities were the same as for Ni^{3+} in corundum. However, we have no determination of V_2 in this case. For the other ions in Tables II and III, there are at present no data available from which an estimate of 3Γ can be obtained.

From the values of $g_{\parallel}$ and $g_{\perp}$ listed in Table II, we can infer whether the stable, distorted configurations are elongated or compressed along the symmetry axis. In accord with the conventions used in Section 2.2, for an ion in octahedral coordination, the configuration characterized by $\theta = 0$ ($Q_\theta > 0$, $Q_\varepsilon = 0$) represents a distortion in which the octahedron of nearest neighbors is elongated along the z axis (which passes through two of the six neighbors). We expect, then, that the Jahn–Teller coupling coefficient V in Eq. (2.2.1) will be negative for a d^9 configuration, which has one e hole in the d shell, and positive for the strong-field d^7 configuration ($et_2{}^6$). From Eq. (2.2.10), we should then have $g_{\parallel} > g_{\perp}$ (λ is negative for d^9 and d^7) for d^9 ions if the stable configuration is elongated along the symmetry axis, while for d^7 we should have $g_{\parallel} < g_{\perp}$ in the same case. We see from Table II that for all the d^7 and d^9 ions listed (with the exception of Ni^+ in LiF and NaF) the stable configuration is evidently the elongated one. While the qualitative agreement between these experimental data and Eq. (2.2.10) is evident, there are discrepancies if we try to make a quantitative fit to Eq. (2.2.10) or to Eq. (2.2.17). These discrepancies occur because, for (λ/Δ) as large as it is for the d^7 and d^9 ions of Table II, there are higher-order corrections to $g_{\parallel}$ and $g_{\perp}$ which cannot be neglected. These have been calculated by Bleaney, Bowers, and Pryce (1955) to order $(\lambda/\Delta)^2$ for d^9, also taking account of the corrections due to the splitting of the ${}^2T_2$ excited state in the distorted configuration, and these are given also by Hayes and Wilkins (1964). The exceptional cases of Ni^+ in LiF and NaF, for which $\theta = \pi$ is evidently the stable configuration, are of particular interest in this regard, because for a compression along the symmetry axis these higher-order corrections should lead to a value for $g_{\parallel}$ less than 2, in contrast to what is observed (Table II). Hayes and Wilkens (1964) have cited these data as evidence of the finite spread in θ within each well which was pointed out by O'Brien (1964), and which then leads to a value less than unity for u in Eq. (2.2.16) or u' in Eq. (2.4.13). A similar discrepancy has been observed by Pilbrow and Spaeth (1967) in the g-factors of Cu^{2+} at tetragonal sites in NH_4Cl, and these authors have also attributed this to the effects of vibrational admixtures of the electronic states. Hayes and Wilkens found separate evidence for such a spread in θ from a small but finite hyperfine interaction with the two neighboring fluorines lying on the z axis in other Ni^+ spectra in both LiF and NaF, for which an associated defect has evidently stabilized the configuration $\theta = 0$. As noted by O'Brien (1964) and discussed at the end of Section 2.2, without a spread in θ the electronic wave function would be $\psi_\varepsilon \sim (x^2 - y^2)$ and should show a hyperfine structure only for the four fluorines in the x, y plane. However, Hayes and Wilkens showed that a value $u' \simeq 0.8$ is required to account for the values of $g_\parallel$ and $g_\perp$ in the $\theta = \pi$ spectrum (Table II), while we estimated at the end of Section 2.4 that u' will not be smaller than $\sim$0.95. Some other explanation for the values of $g_\parallel$ and $g_\perp$ must therefore be found if this spectrum really results from a static Jahn–Teller effect and not from a center with some associated defect. One possibility lies in the observation that the g-factors of the actual eigenstates Ψ_1', Ψ_2', and Ψ_3' in Eq. (2.5.7) depart from $g_\parallel{}^0$ and $g_\perp{}^0$ of the "pure" states Ψ_1, Ψ_2, and Ψ_3 by corrections of order $(\Gamma^2/\delta^2)(g_\parallel{}^0 - g_\perp{}^0)$. If the strains that produce the splittings δ_{21}, δ_{31}, and δ_{32} are random, the g-factors of the strain-broadened lines should then be given, to order Γ^2/δ^2, by

$$\begin{aligned} g_\parallel &= g_\parallel{}^0 - 2\Gamma^2\langle\delta^{-2}\rangle(g_\parallel{}^0 - g_\perp{}^0) \\ g_\perp &= g_\perp{}^0 + \Gamma^2\langle\delta^{-2}\rangle(g_\parallel{}^0 - g_\perp{}^0) \end{aligned} \tag{4.1.4}$$

where $\langle\delta^{-2}\rangle$ denotes an average value. These corrections thus reduce the anisotropy in g and are therefore in the direction required to account for the LiF:Ni^+ and NaF:Ni^+ data. This hypothesis could, of course, be tested by measuring $g_\parallel$ and $g_\perp$ as functions of an applied strain; it would not apply to the data of Pilbrow and Spaeth on NH_4Cl:Cu^{2+} because of the large splitting due to the tetragonal crystal field already present at the Cu^{2+} site.

Estimates of the Jahn–Teller energy for a number of the ions in Tables II and III have been made by Höchli and Müller (1964a,b) and Höchli *et al.* (1965) on the hypothesis that the broadening of the isotropic EPR spectrum at high temperatures results from an Orbach relaxation process involving a state on the upper energy surface of Fig. 3 which is identified with the centrifugally stabilized state studied by Slonczewski (1963) and Slonczewski and Moruzzi (1967). The evidence in support of this proposal has recently been summarized by Müller (1967). However, the resulting values for E_{JT} are considerably less than 1000 cm^{-1} in a number of cases, and in view of our preceding discussion such a value would appear too small to be consistent with the observation of a static Jahn–Teller effect at low temperature. Moreover, in the theory of spin–lattice relaxation introduced in Section 2.5, it is not evident that the Slonczewski states or the upper branch of the energy surface should necessarily play an important role. The significance of the results of Höchli *et al.* (1964a,b; 1965) therefore remains to be clarified by further investigation of the various relaxation processes.

The spin–lattice relaxation time for Cu^{2+} in $La_2Mg_3(NO_3)_{12} \cdot 24D_2O$ measured by Breen *et al.* (1969) was found to be described by the formula $T_1^{-1} = CT$ for $1.3°K < T < 9°K$, with $C \approx 100\ sec^{-1}\ deg^{-1}$ for a lightly doped sample with 2×10^{17} copper ions per cm^3. Comparing this result with the values for the spin-echo decay time found in the same temperature range and described by $\tau^{-1} = A + BT$, one obtains the ratio $C/B \approx 2 \times 10^{-3}$. This ratio is roughly consistent, at least as to order of magnitude, with the value predicted by Eq. (2.6.2) for $(\Delta g/g) \simeq 0.17$ (Table II). However, as noted in Section 2.6, the observed anisotropy of T_1 is not altogether consistent with what Eq. (2.6.2) predicts. Williams *et al.* (1969) have suggested that an additional mechanism for spin–lattice relaxation may be present as a result of the trigonal crystal field at the copper site in the double nitrate. Replacing e_{yz}, e_{zx}, and e_{xy} in Eq. (2.6.5) by a parameter e_t representing the effect of this static crystal field, we would then obtain from Eq. (2.6.6) a coupling between states in different wells analogous to the tunneling matrix element Γ, but connecting states of opposite spin. This leads exactly as in the derivation of Eq. (2.5.8) to a rate for reorientation accompanied by spin reversal (except when **H** is along the trigonal axis). This contribution to T_1^{-1} remains effective when **H** is along [001], and it may be appreciable if e_t is sufficiently large and if the parameter p is not too small.

A dynamic Jahn–Teller effect resulting from zero-point motion in the ground state has now definitely been shown to occur for Cu^{2+} in MgO by Coffman (1965, 1966, 1968) and for $Sc^{2+}(3d^1)$ in CaF_2 and SrF_2 by Höchli and Estle (1967) and Höchli (1967). Instead of a low-temperature EPR spectrum given by the superposition of three axial spectra appropriate to a static Jahn–Teller effect, the spectrum at liquid helium temperature exhibits a cubic anisotropy with a g-factor described by Eq. (2.3.16). The parameters g_1 and qg_2 obtained by fitting the data to Eq. (2.3.16) are given in Table IV. For Sc^{2+} in CaF_2 and SrF_2, well-resolved resonances corresponding to both g_+ and g_- in Eq. (2.3.16) are seen, while for MgO:Cu^{2+} only the line showing the larger departure from $g = 2$ is observed because the second line is obscured by resonances due to other impurities. The hyperfine splitting due to $Sc^{45}(I = \frac{7}{2})$ for the two Sc^{2+} spectra is quite well described by Eq. (2.3.18), and the parameters A_1 and qA_2 thus obtained

Table IV. Experimental *g*-Factors and Central Hyperfine Parameters for Transition Metal Ions in an Orbital Doublet State (^2E) in Cubic Symmetry Which Exhibit a Dynamic Jahn–Teller Effect in the Low-Temperature Anisotropic EPR Spectrum As a Result of Zero-Point Ionic Motion[a]

Ion	Host crystal	T, °K	g-factor	Hyperfine interaction, 10^{-4} cm^{-1}	Ref.[b]
$Cu^{2+}(3d^9)$	MgO	77	$g = 2.192$ (isotropic)	$A = 19.0$	A, B
		1.2	$g_1 = 2.195$	$A_1+qA_2 = 63$	B
			$qg_2 = +0.108$ ($q \simeq 0.5$)		
$Sc^{2+}(3d^1)$	CaF_2	77	$g = 1.967$ (isotropic)	$A = 65.0$	C
		10	$g = 1.969$ (isotropic)	$A = 65.5$	C
		1.5	$g_1 = 1.973$	$A_1 = 65.0$	C
			$qg_2 = -0.022$ ($q = 0.75$)	$qA_2 = 24.5$	
$Sc^2(3d^1)$	SrF_2	10	$g = 1.963$ (isotropic)	$A = 67.0$	C
		1.5	$g_1 = 1.963$	$A_1 = 67.1$	C
			$qg_2 = -0.028$ ($q = 0.71$)	$qA_2 = 24.1$	

[a] Parameters of both this low-temperature spectrum and the high-temperature isotropic spectrum are listed.

[b] The references are designated as follows: (A) Orton *et al.* (1961); (B) Coffman (1965, 1966, 1968); (C) Höchli and Estle (1967).

are also given in Table IV. For MgO:Cu^{2+}, the hyperfine splitting should probably depart from Eq. (2.3.18) except when A is along [100], because of the importance of correction terms which were omitted in obtaining this equation [e.g., the assumption $|A_1| \gg |A_2|$, which is made in the derivation of Eq. (2.3.18), but which is not required if **H** is along one of the cube axes, is evidently not fulfilled for MgO:Cu^{2+}]. The data (Coffman, 1968) indicate such departures from the simple theory, and we have therefore listed for MgO:Cu^{2+} in Table IV only the value for $A_1 + qA_2$ which is obtained by fitting the hyperfine splitting of the one line observed, for **H** along [100], to Eq. (2.3.18).

However, it is likely (Ham, 1968) that what Coffman and Höchli observed was the strain-broadened spectrum described by Eq. (2.3.25) in the presence of random residual strain, rather than the no-strain spectrum. The individual hyperfine components then have the line shape given by Eq. (2.3.26) and Fig. 6 (apart from other contributions of relaxation and inhomogeneity to the line broadening), the two peaks at the extremities of each line coinciding with the resonance frequencies of the no-strain spectrum given by Eq. (2.3.18).* This proposal would offer an explanation for the remarkably sharp resolution in Höchli's Sc^{2+} spectra, despite the random strains which must be present in the CaF_2 crystals. Since an applied stress would have to produce strains comparable with the residual strains before the spectrum would be markedly changed, this proposal would also apparently account for Höchli's observation that his spectrum was unaffected when moderate stress ($\sim$250 kg/cm^2) was applied to the crystal. In MgO containing Fe^{2+}, Stoneham (1966) has shown from an analysis of EPR linewidths that typical residual strains of $\sim 2\times 10^{-4}$ are present at the sites of the Fe^{2+}. Assuming that similar strains (resulting from other defects, dislocations, etc.) are present at a typical Cu^{2+} site, and that the strain coefficient V_2 in Eq. (2.3.19) is $\sim 10^4$ cm^{-1}, we thus find that the strain splitting $2q\,|\,V_2\,|(e_\theta^2 + e_\varepsilon^2)^{1/2}$ of the orbital doublet is of order 1 cm^{-1}. The strain splitting for MgO:Cu^{2+} must therefore be expected to dominate the anisotropic part of the Zeeman and hyperfine interactions under the experimental resonance conditions. The conditions for the observation of the strain-broadened spectrum of Eq. (2.3.25), rather than the no-strain spectrum, should therefore be fulfilled for MgO:Cu^{2+}.†

* Observation of the anisotropic EPR spectrum characteristic of the dynamic Jahn–Teller effect has been reported very recently by Boatner *et al.* (1969) for $La^{2+}(5d^1)$ in $SrCl_2$ at 1.6°K, with g-factors $g_1 = 1.878$, $|\,qg_2\,| = 0.064$, indicating a value $q \simeq \frac{1}{2}$. The highly asymmetric lines were found to fit the shape function predicted by Eq. (2.3.26), confirming that it is the strain-broadened spectrum that is observed.

† When the magnetic field is in the (110) or (1$\bar{1}$0) planes, the extremities of the strain-broadened line given by Eq. (2.3.25) are due to those ions located at sites at which the local strain described by e_θ and e_ε has uniaxial symmetry with respect to the [001] direction. For these ions, we therefore have $e_\varepsilon \simeq 0$ and $\varphi \simeq 0$ or π in Eq. (2.3.24), and the doublet states $\Psi_{g\theta}$ and $\Psi_{g\varepsilon}$ are separately eigenstates of the local strain in accord with Eq. (2.5.12). Chase (1969, 1970) has pointed out that under these conditions of strain, one but only one of these states has the correct symmetry to be coupled by the strain to the excited singlet state expected from the tunneling model or from Fig. 7, as may be seen from Eqs. (2.4.2a,b). If the tunneling splitting 3Γ is small enough, a pronounced broadening of only one side of the resonance spectrum will therefore occur for these field orientations, and Chase has observed such an asymmetric broadening in the EPR spectrum of the ^2E level of CaF_2:Eu^{2+}. A similar asymmetric broadening is seen in as yet unpublished EPR data of Geschwind and Brya for MgO:Cu^{2+}. One can determine the symmetry (A_1 or A_2) of the excited singlet depending on which side of the EPR line is broadened, and these data indicate that the singlet is A_2 for MgO:Cu^{2+}. It is therefore the configuration elongated along the symmetry axis which tends to be stabilized for MgO:Cu^{2+}, in agreement with what has been found for the other cases involving Cu^{2+} listed in Table II. Chase was able to determine the tunneling splitting 3Γ to be 18 ± 4 cm^{-1} for CaF_2:Eu^{2+} from the extent of this broadening, using optical data to determine the magnitude of the random strain splitting of the ^2E level.

Also given in Table IV are the values which the data require for the reduction factor q if g_1 and g_2 are given by Eq. (2.1.6). For MgO:Cu^{2+}, $q = \frac{1}{2}$ to within the experimental uncertainties, in agreement both with our result in Eq. (2.3.11) for strong linear Jahn–Teller coupling* and with the tunneling model, as proposed by Coffman (1965, 1966, 1968), for which we have $q \simeq c_2 \simeq \frac{1}{2}$ from Eq. (2.3.49). On the other hand, the values $q = 0.75$ and 0.71 for Sc^{2+} in CaF_2 and SrF_2 represent a large departure from this limiting value $q = \frac{1}{2}$ and, moreover, are inconsistent with the requirement $q \simeq c_2 \leq \frac{1}{2}$ [Eq. (2.3.48)] for the tunneling model with strong Jahn–Teller coupling. Applying Eqs. (2.3.12) and (2.3.13) to these values for q on the assumption that this case is better described as one of only moderately strong Jahn–Teller coupling, we obtain an estimate (Ham, 1968) for the ratio $E_{JT}/\hbar\omega$, namely 0.25 and 0.34 for CaF_2:Sc^{2+} and SrF_2:Sc^{2+}, respectively. These results indicate that the Jahn–Teller coupling in these two examples may actually be quite weak (if we assume $\hbar\omega \simeq 300$ cm^{-1}, these values correspond to a Jahn–Teller energy E_{JT} of only about 60–80 cm^{-1}). For so weak a Jahn–Teller coupling, the model based on linear coupling which was described in the first part of Section 2.3 should be quite appropriate (apart from the need to generalize the theory to take proper account of the phonon continuum), so that this interpretation is self-consistent. This analysis thus supports Höchli's (1967) interpretation of his results as arising from a dynamic Jahn–Teller effect in the vibronic ground state of the Sc^{2+}, but it indicates that a model with rather weak Jahn–Teller coupling is more appropriate in this case than a strong-coupling model such as those of O'Brien (1964) and Bersuker (1962, 1963).†

At higher temperatures, these anisotropic spectra are replaced by isotropic spectra with a g-factor (Table IV) given approximately by the average (given by g_1) of g_+ and g_- in the low-temperature spectrum. The same result is found for the hyperfine interaction in the Sc^{2+} spectra, for which the data suffice to yield an unambiguous value for A_1 in the low-temperature spectrum. The appearance of the isotropic spectrum above about 6°K has been attributed by Höchli (1967) to the excited singlet state expected from the tunneling model, as in Eq. (2.4.3), and Höchli has estimated from the temperature dependence of this line that the tunneling splitting 3Γ

* Note, however, from Fig. 5 that one could have $q = 0.512$ for a value of $E_{JT}/\hbar\omega$ as small as 2.5.

† Equation (2.4.11) for q in Bersuker's tunneling model was used by Höchli to interpret his results, but for the reasons we have seen at the end of Section 2.4 this formula is not applicable.

is 10 cm^{-1} for $CaF_2:Sc^{2+}$ and 8 cm^{-1} for $SrF_2:Sc^{2+}$. Alternatively, the iso tropic line may arise from rapid relaxation within the ^2E ground state, as described in Section 2.5, and estimates of the relaxation time for the Raman process given by Eq. (2.5.16) and for the direct process given by Eq. (2.5.15), with the strain splitting of the level taken into account, are not inconsistent with this proposal (Ham, 1968). In view of the complicated roles which strain and relaxation play in determining linewidths and in causing motional averaging of the spectrum, final clarification of the nature of the transition to the isotropic spectrum will require further experimental investigation, ideally with applied stress used to override the effects of the random residual strain. Coffman (1968) found no evidence for an isotropic resonance at 1.2°K in $MgO:Cu^{2+}$; he concluded that the tunneling splitting is large enough so that the excited singlet state is not appreciably populated at this temperature. The isotropic spectrum for $MgO:Cu^{2+}$ (Table IV) has been observed at temperatures above 4°K by Orton *et al.* (1961).

Chase (1968) has recently reported observations on the EPR spectrum of the excited ^2E level of Cr^{3+} and V^{2+} in MgO, the detection of the resonance being by optical methods. This level strictly should not be said to exhibit a Jahn–Teller effect, because it couples so weakly to lattice distortions that vibronic corrections to the predictions of crystal-field theory are small enough to be unimportant. Nevertheless, the results of Section 2 for the EPR spectrum with a dynamic Jahn–Teller effect may be applied to this level if only we set $q \simeq 1$ and $p \simeq 1$ in our equations. The spectrum observed by Chase for $MgO:Cr^{3+}$ conforms to this theory very well; that of $MgO:V^{2+}$ is less well resolved. At liquid helium temperature, the $MgO:Cr^{3+}$ spectrum exhibits two peaks having an anisotropy described by g_+ and g_- in Eq. (2.3.16) with $q = 1$ and $g_2 = g_1 - 2 = -0.096$. Moreover, the spectrum is found to be broadened by random strain in just the way predicted by Eq. (2.3.25) [aside from additional sources of broadening which keep the peaks from being as sharply defined as in Eq. (2.3.26) and Fig. 6]. The strain-coupling coefficient V_2 of Eq. (2.1.10) or (2.3.19) (if $q \simeq 1$, these equations are equivalent) is known from the measurements of Schawlow *et al.* (1961) to have the value $V_2 = 660$ cm^{-1} for $MgO:Cr^{3+}$. The mean splitting of the orbital degeneracy of ^2E that results from the random strain in the crystal is then several tenths of a cm^{-1}. Although this splitting is smaller than the Zeeman splitting at 24 GHz, it is nevertheless sufficiently large compared to the anisotropic part of the Zeeman splitting, $2 \mid g_2 \beta H \mid \simeq 0.08$ cm^{-1}, for the random strains to provide the dominant splitting of the orbital degeneracy and thereby to lead to the strain-broadened spectrum of Eq. (2.3.25). The appearance of this spectrum for the ^2E level of

$MgO:Cr^{3+}$, despite the small value of V_2, thus provides further support for the proposal that it is this strain-broadened spectrum which must be observed, as in Coffman's work, for the much more strongly coupled Cu^{2+} ion in MgO.

The optical measurements of the fluorescence spectrum of $MgO:Cr^{3+}$ by Schawlow *et al.* (1961) and those of Sturge (1963) for $MgO:V^{2+}$ have also determined the value of the strain-coupling coefficient V_3'' which describes the effect of the shears e_{yz}, e_{zx}, and e_{xy} on the ^2E level as in Eq. (2.1.13) or Eq. (2.3.20) [we have $p \simeq 1$ for the ^2E $(3d^3)$ level of Cr^{3+} and V^{2+} because of the weak Jahn–Teller coupling]. The measured strain coefficients are $V_2 = 660\ \text{cm}^{-1}$ and $V_3'' = 1460\ \text{cm}^{-1}$ for $MgO:Cr^{3+}$ and $V_2 = 1080\ \text{cm}^{-1}$ and $V_3'' = 700\ \text{cm}^{-1}$ for $MgO:V^{2+}$. Chase (1968) has considered the various relaxation processes that occur for these ions as a result of this coupling to the phonons, and he has estimated that, under the conditions of his experiments, the fastest relaxation (direct process) occurs between states of opposite spin belonging to the different Kramers doublets Ψ_α and Ψ_β of Eq. (2.5.12). This result is a consequence of having V_2 and V_3'' of more or less the same size and of having a Zeeman splitting larger than the strain splitting so that the larger density of phonon states enters the spin-reversal transition. This situation contrasts with the one we have described in connection with the discussion of spin–lattice relaxation in Section 2.6, where the reverse situation was assumed, namely $|\delta| > h\nu \simeq g\beta H$. With the parameters of Cr^{3+} and V^{2+} in MgO, it is possible to show that the coupling with the shear strains via V_3'' produces a spin-reversal transition rate which is faster than that which results from the anisotropy in the g-tensor and which led to Eqs. (2.6.9–10).

Other complexes in an orbital doublet state for which dynamic Jahn–Teller effects have been identified in EPR spectra include the R center in KCl studied by Krupka and Silsbee (1966) and the negative radical ion of benzene. In the investigation of the latter, McConnell and McLachlan (1961) provided the first example of the partial quenching by a dynamic Jahn–Teller effect of an electronic orbital operator, in this case the orbital angular momentum for electronic motion of the π electron around the benzene ring. Dynamic Jahn–Teller effects have been investigated in EPR spectra in a number of other symmetric π-electron ring radicals such as C_5H_5, $C_6H_6^+$, C_7H_7, $C_8H_8^{\pm}$ and the monopositive and mononegative ions of coronene and triphenylene. A list of references to this work is given by Carter and Vincow (1967).

A number of the ions in Table III for which a static Jahn–Teller effect has not yet been found would appear to hold excellent prospects of exhibiting dynamic effects in their ground state. A low-temperature anisotropic spectrum for $CaO:Cu^{2+}$ has indeed been studied by Coffman *et al.* (1968) and interpreted in terms of a tunneling splitting $|3\Gamma| = 0.006\ cm^{-1}$. However, their analysis has not taken into account the effect of random strains, which probably cause splittings of $\sim 1\ cm^{-1}$ as in our estimate for $MgO:Cu^{2+}$.

Jahn–Teller effects in the closely related case of the ^5E state of the chromous ion Cr^{2+} ($3d^4$) have been investigated by Fletcher (1967) and Fletcher and Stevens (1969), using an analysis similar to that given for the ^2E state of Cu^{2+} by O'Brien (1964) as reviewed in Sections 2.2 and 2.3. As in O'Brien's (1964) work, strong Jahn–Teller coupling ($E_{JT} \gg \hbar\omega$) and the validity of the Born–Oppenheimer approximation (2.3.27) (in effect) were assumed. The effects of the now-familiar warping in the bottom of the trough in Fig. 3, or equivalently of the tunneling between wells, on the vibronic levels of lowest energy and on their magnetic properties were calculated, taking into account the second-order spin–orbit splitting of the ^5E state. These theoretical results were used to propose an interpretation of acoustic paramagnetic resonance measurements on $MgO:Cr^{2+}$ made by Marshall and Rampton (1968). As we would now expect, these spectra and their interpretation are complicated by the effects of random strains in the crystal.

4.2. Jahn–Teller Effects in EPR for Orbital Triplet States

Compared with the case of the orbital doublet, relatively few examples of static Jahn–Teller effects in orbital triplet states have so far been found in EPR studies. One such example is that of Cr^{2+} in CdS: A highly anisotropic EPR spectrum observed at liquid helium temperature which broadens beyond detection above 20°K has been attributed by Estle *et al.* (1963) and by Morigaki (1963, 1964a) to a static Jahn–Teller distortion of a substitutional Cr^{2+} ion, the axis of the distortion being approximately along one of the lines bisecting the angle to a pair of nearest sulfur neighbors (these would be [100] directions for perfect tetrahedral symmetry). Similar spectra have been found for Cr^{2+} in ZnSe, CaF_2, and CdF_2 (deWit, Reinberg *et al.*, 1965; Estle and Holton, 1966). Another case in which a static distortion attributed to the Jahn–Teller effect has been observed in EPR, and in which the symmetry of the distorted configuration is trigonal, is that of V^{2+} in CaF_2 first reported by Höchli (1966) and subsequently by Zaripov *et al.* (1967). Similar spectra have been observed by Zaripov

Table V. Examples of Isotropic Experimental g-Factors of Transition Metal Ions in an Orbital Triplet State in Cubic or Tetrahedral Symmetry

	Ion	g	Ref.[a]
$3d^6{:}{}^5T_2$	$MgO{:}Fe^{2+}$	3.428	A
	$CaO{:}Fe^{2+}$	3.30	B
	$Si{:}Mn^{+}$	3.01	C[b]
	$Si{:}Cr^{\circ}$	2.97	C[b]
$3d^7{:}{}^4T_1$	$MgO{:}Co^{2+}$	4.278	D
	$Si{:}Mn^{\circ}$	3.362	C[b]
	$Si{:}Fe^{+}$	3.524	C[b]

[a] References are designated as follows: (A) Low and Weger (1960); (B) Shuskus (1964); (C) Ludwig and Woodbury (1962); (D) Low (1958).
[b] These g-factors are for interstitial ions in silicon.

et al. (1967) for $CaF_2{:}Cr^{3+}$, $SrF_2{:}V^{2+}$, and $CaF_2{:}Ni^{2+}$ and by Schneider, Dischler, and Räuber (1967) for $ZnS{:}V^{2+}$. For $ZnSe{:}Cr^{3+}$, an EPR spectrum attributed to a Γ_8 level resulting from tunneling between three tetragonal distortions has been reported (at 77°K) by Rai *et al.* (1967).* Aside from this preliminary report and some effects of relaxation observed in several of the above cases, evidence for dynamic Jahn–Teller effects in these spectra has not yet been found.

The most striking examples of dynamic Jahn–Teller effects in EPR spectra of ions in orbital triplet states are the interstitial transition metal impurities Mn^+, Cr°, Mn°, and Fe^+ in silicon studied by Ludwig and Woodbury (1962). The measured g-factor for the ground state of these ions is given in Table V. For Mn^+ and Cr° the configuration is $3d^6$, the lowest term ${}^5T_2$, and the lowest spin–orbit level a triplet corresponding to $J = 1$ in Eq. (3.1.20). Substituting $S = 2$, $\mathscr{L}_g = 1$, and $J = 1$ into Eq. (3.1.22), we obtain for the g-factor of this level (apart from all second-order corrections, which we shall ignore here)

$$g_1 = \tfrac{3}{2}g_s - \tfrac{1}{2}g' \tag{4.2.1}$$

In the absence of Jahn–Teller effects, we would have from Eq. (3.1.15) and the parameters of the Zeeman interaction (3.1.13) $g_s = 2$ and $g' = g_{\mathscr{L}} = -1$, so that we would then expect $g_1 \simeq 3.5$ from Eq. (4.2.1)

* The identification of this spectrum has recently been questioned by Rai *et al.* (1969).

[for the moment, we also ignore corrections due to covalency; the result $g_{\mathscr{L}} = -1$ then follows because the true orbital angular momentum **L** is equivalent within the $^5T_2$ term of the $3d^6$ configuration to $-\mathscr{L}$, as defined in Eq. (3.1.9)]. However, if g' is reduced to nearly zero because a dynamic Jahn–Teller effect makes the reduction factor $K(T_1)$ in Eq. (3.1.15) small, we would expect $g_1 \simeq 3.0$ from Eq. (4.2.1). From Table V, we see that both Mn^+ and Cr° have isotropic g-factors close to the latter value [there is also evidence (Ludwig and Woodbury, 1962) that the spin–orbit splitting is drastically reduced from free-ion values, in accord with Eq. (3.1.18) if $K(T_1)$ is small, yet remaining larger than the Zeeman splitting as required if Eq. (3.1.22) is to hold]. So large a reduction in the orbital contribution to the g-factor could scarcely be the result of covalent bonding, while it is entirely consistent with the partial quenching due to a dynamic Jahn–Teller effect. Similarly, the ions Mn° and Fe^+ have the configuration $3d^7$, the lowest term $^4T_1$, and a Kramers doublet ($J = \frac{1}{2}$) ground state. Using $S = \frac{3}{2}$, $\mathscr{L}_g = 1$, and $J = \frac{1}{2}$ in Eq. (3.1.22), we obtain in this case

$$g_{1/2} = \tfrac{5}{3}g_s - \tfrac{2}{3}g' \tag{4.2.2}$$

For the $^4T_1$ state derived from the ^4F term of the free ion, we would have $g' = g_{\mathscr{L}} = -\frac{3}{2}$ and would expect $g_{1/2} \simeq 4.3$ from Eq. (4.2.2), while, if $g' \simeq 0$, we obtain $g_{1/2} \simeq 3.3$. As seen from Table V, Mn° has an isotropic g-factor of 3.362, while Fe^+ has $g = 3.524$, corresponding by Eq. (4.2.2) to a value of $g' \simeq -0.3$. Again these results are consistent with strong quenching of the orbital part of the g-factor by a dynamic Jahn–Teller effect (Ham, 1965).

By way of contrast, we also give in Table V the g-factors for two ions with the same ground state as those discussed above but which have been thought until recently to have properties consistent with crystal-field theory modified for a modest degree of covalency: Fe^{2+} ($3d^6$) and Co^{2+} ($3d^7$) in MgO (Low, 1958; Low and Weger, 1960). In both cases, the observed g agrees (apart from small second-order corrections) with Eq. (4.2.1) or (4.2.2) if $g' \simeq g_{\mathscr{L}}$ is given its free-ion value with a small reduction for covalency. In neither of these cases, therefore, does Jahn–Teller coupling cause a large change in the g-factor of the ground state. The absence of any evident Jahn–Teller effect in the ground state of MgO:Fe^{2+}, despite the fact that this state is a spin–orbit triplet (Γ_{5g}), was explained by Van Vleck (1958, 1960) and Liehr (1960) on the assumption that the Jahn–Teller interaction in the $^5T_{2g}$ term was weaker than the spin–orbit interaction and that matrix elements of the Jahn–Teller interaction to the higher spin–orbit levels could consequently be ignored. The matrix elements of the Jahn–Teller terms in Eq. (3.1.1) or Eq. (3.2.1) within this lowest spin–orbit triplet are found from crystal-field theory to be reduced by a factor 1/10 compared with matrix elements between the orbital states of $^5T_{2g}$. The Jahn–Teller energy evaluated for the Γ_{5g} ground-state triplet considered by itself is therefore, according to Eq. (3.1.4) or Eq. (3.2.5), only 1/100 of the value one would obtain for the orbital triplet in the absence of spin–orbit splitting. As Van Vleck and Liehr pointed out, under such circumstances, one should certainly not expect to see a Jahn–Teller distortion, and crystal-field theory (including covalency) should suffice to describe the EPR spectrum quite accurately (McMahon and Silsbee, 1964; O'Brien, 1965). For Co^{2+} in octahedral coordination, the spin–orbit ground state is a Kramers doublet, the degeneracy of which cannot be split by any distortion, so that again for this ion, if the Jahn–Teller interaction is weak compared with the spin–orbit splitting, one expects no pronounced Jahn–Teller effect in the ground state. Both Fe^{2+} and Co^{2+} in octahedral coordination thus provide examples of situations in which spin–orbit interaction tends to stabilize the ground state of the system against a Jahn–Teller instability, provided the Jahn–Teller interaction is not too strong.

It is obviously essential for the preceding argument to be appropriate that the Jahn–Teller interaction be weak compared with the spin–orbit interaction. Otherwise, if the Jahn–Teller interaction dominates, we have the situation described in Section 3, which evidently pertains to the interstitial ions in silicon listed in Table V and in which the spin–orbit interaction, the orbital contribution to the g-factor, etc., suffer a dynamic quenching. Recently, however, it has been found by Wong (1968) from far-infrared absorption studies on $MgO{:}Fe^{2+}$ that the spin–orbit splitting between the ground-state triplet and the first excited state is only 105 cm^{-1}. Crystal-field theory, on the other hand, predicts this splitting to be $\sim$200 cm^{-1}, and the reduction to 105 cm^{-1} is much too large to be the result of the modest degree of covalency expected for this ion. Ham, Schwarz, and O'Brien (1969) have now examined in some detail the question of a possible dynamic Jahn–Teller effect for $MgO{:}Fe^{2+}$, treating the intermediate situation in which the spin–orbit and Jahn–Teller interactions are of comparable size. They have shown that it is indeed reasonable to attribute this reduced spin–orbit splitting predominantly to a dynamic Jahn–Teller effect, and, if this is so, that the Jahn–Teller effect must then account for a major part of the shift in the g-factor and hyperfine field that had previously been attributed to covalency. However, despite the large change in the spin–orbit splitting, these Jahn–Teller corrections in other parameters characterizing the EPR and Mössbauer spectra of $MgO:Fe^{2+}$ are relatively much smaller and pose no inconsistency with the experimental values. It was found necessary to take account of coupling to both E_g and T_{2g} vibrational modes and also to include the coupling via the Jahn–Teller interaction with the higher spin–orbit levels of the $^5T_{2g}$ term, which had been omitted in the considerations of Van Vleck (1958, 1960), Liehr (1960), McMahon and Silsbee (1964), and O'Brien (1965). The strength of the Jahn–Teller coupling (a value for E_{JT} of the order 100 cm^{-1} was found for coupling with each type of mode) was found to be too weak to permit use of the theory outlined in Section 3 (which assumes a Jahn–Teller interaction stronger than the spin–orbit interaction and predicts equally large reductions in the spin–orbit splitting and in the orbital contribution to the g-factor and hyperfine field). The coupling was weak enough, however, to make a perturbation treatment of the Jahn–Teller coupling reasonably accurate, and the quantitative calculations of the energy level shifts and the reduction factors in the Γ_{5g} ground state were done in this way. Although the detailed calculations were made using discrete vibrational modes of each type, it was also shown, by taking advantage of the simplicity of the perturbation treatment, how in this case the calculation of the Jahn–Teller effects could be formulated when the coupling is taken to be with the continuum of lattice phonons. A closely related perturbation calculation of the g-shift for $MgO:Fe^{2+}$ as well as for several other ions, which treated the coupling with the phonon continuum while also recognizing the importance of coupling to the higher spin–orbit levels, was given by Koloskova and Kochelaev (1967). Koloskova and Kochelaev's results were found by Ham, Schwarz, and O'Brien to contain serious algebraic errors, but the conclusions of both papers were in agreement that a substantial fraction of the observed g-shift for $MgO:Fe^{2+}$ is in fact the result of the Jahn–Teller coupling. Koloskova and Kochelaev estimate that the vibronic g-shift is appreciable also for $MgO:Co^{2+}$.

The importance of the role of a dynamic Jahn–Teller effect in the properties of the octahedrally coordinated Fe^{2+} ion is given further support by the observation (Shuskus, 1964) of a smaller g-factor ($g = 3.30$) for $CaO:Fe^{2+}$ (Table V) than was found for $MgO:Fe^{2+}$ ($g = 3.428$). If the difference in g-factors of $MgO:Fe^{2+}$ and $CaO:Fe^{2+}$ (which correspond to orbital reduction factors of $\sim$0.8 and $\sim$0.6, respectively) were due to covalent bonding, one would require substantially more bonding in the latter. However, CaO has a larger lattice constant than MgO, so that one would not expect stronger bonding in CaO. The suggestion of a dynamic Jahn–Teller effect in the g-factor of $CaO:Fe^{2+}$ (Ham, 1965) has received support from Mössbauer studies by Chappert *et al.* (1967) in which an orbital reduction factor ~0.5 was obtained from the orbital contribution to the hyperfine field while the isomer shift indicated *less* covalency for Fe^{2+} in CaO than in MgO.

The Zeeman splitting of the optical absorption line believed to be due to the Cu^{2+} ion substituting for Zn^{2+} at a site of tetrahedral symmetry in cubic ZnS has been reported by de Wit (1969). The g-factor in the ground state is found to be isotropic, with the value $|g| = 0.71 \pm 0.02$. Since, from Eq. (3.1.22), we would have

$$g_{1/2} \simeq -\tfrac{1}{3}g_s + \tfrac{4}{3}g' \tag{4.2.3}$$

for the Γ_7 ($J = 1/2$) level of the ${}^2T_2$ term of Cu^{2+}, the observed g-factor apparently indicates a nearly complete quenching of the orbital contribution (we recall that $g' = g_{\mathscr{L}} = -1$ in the absence of Jahn–Teller effects or covalency), and, as de Wit suggests, this is probably the result of a dynamic Jahn–Teller effect. The g-factor of the isoelectronic ion Ni^{+} in ZnS (also isotropic) had been determined earlier (de Wit *et al.*, 1964) to be $|g| = 1.40 \pm 0.01$ from EPR. From Eq. (4.2.3), this value corresponds to taking $g' \simeq -0.55$, indicating weaker Jahn–Teller coupling for Ni^{+} than for Cu^{2+} in ZnS. Anisotropic g-factors of a large number of other paramagnetic copper centers in ZnS have been reported by Holton *et al.* (1969). Although these evidently result from copper ions in association with other defects, none of these spectra seem to be amenable to interpretation in terms of simple crystal-field theory with modest covalency corrections, and it seems quite possible that vibronic effects may be present. A dynamic Jahn–Teller effect is very probably involved in the case of $ZnO:Cu^{2+}$, for which Dietz *et al.* (1963) obtained, from EPR and optical data, anomalously small orbital reduction factors which they could not successfully account for on the basis of covalent bonding. Somewhat similar orbital reduction factors have been obtained from EPR data for the related ions $CdS:Cu^{2+}$ (Morigaki, 1964b), $CdS:Ag^{2+}$ (Morigaki and Hoshina, 1967), and also for $BeO:Cu^{2+}$ (de Wit and Reinberg, 1967), and an explanation in terms of vibronic effects will have to be investigated in these cases as well.

A dynamic Jahn–Teller effect has been proposed by Macfarlane *et al.* (1968) as offering an explanation for an anomalous g-factor observed in the ground state of $Al_2O_3:Ti^{3+}$ as well as for states of anomalously low energy observed in the far-infrared spectrum. This case is an important one, because it is the first in which a quantitative interpretation of EPR data was attempted in terms of a detailed model for a dynamic Jahn–Teller effect in a triplet state. The ground-state g-factor for $Al_2O_3:Ti^{3+}$ reported by Kornienko and Prokhorov (1960) has trigonal symmetry with $g_{\parallel} = 1.067 \pm 0.001$ and $g_{\perp} \leq 0.1$ and indicates a reduction factor ≈ 0.5 for the orbital contribution to g. This is too small a value to be plausible in terms of covalent bonding between Ti^{3+} and the neighboring oxygen ions. Moreover, far-infrared studies by Nelson *et al.* (1967) place excited states of the Ti^{3+} ion 37.8 and 107.5 cm^{-1} above the ground state. The former of these is consistent with the excitation energy inferred by Kask *et al.* (1963) from the spin–lattice relaxation time in the ground state. However, these energies are too small to be consistent with a model based on crystal-field theory alone (including effects of covalency) for any reasonable set of parameters. Macfarlane *et al.* (1968) have proposed that all of these data can be accounted for in terms of a dynamic Jahn–Teller effect if the $^2T_2$ term of $Ti^{3+}(3d^1)$ in cubic symmetry is coupled primarily with the E_g vibrational mode of the octahedron of neighboring oxygen ions. As we have seen in Section 3.1, there would then occur a dynamic quenching not only of the orbital angular momentum and spin–orbit interaction in the $^2T_2$ state, but also of the splitting of this state by a trigonal component of the crystal field (if this trigonal crystal field is not too strong). Macfarlane *et al.* have treated this dynamic quenching by perturbation theory, using the formulas for this case given by Ham (1965) and including not only the direct effect of dynamic quenching as given by the reduction factors in Eq. (3.1.12), but also the second-order corrections [similar to those in Eqs. (3.1.25–26)] arising from the spin–orbit interaction and the trigonal crystal field. They are able to fit the experimental data reasonably well for the following choice of values of their parameters: Jahn–Teller energy $E_{JT} = 200$ cm^{-1}, trigonal crystal-field splitting of the orbital triplet $v = 700$ cm^{-1}, effective E_g mode energy $\hbar\omega = 200$ cm^{-1}, spin–orbit parameter $\zeta = 120$ cm^{-1}, and orbital reduction factor due to covalency $k = 0.8$. A similar calculation was also shown to fit data on $Al_2O_3:V^{4+}$. Far-infrared spectroscopy has found for this ion the value $g_{\parallel} = 1.43 \pm 0.04$ in the ground state (Joyce and Richards, 1969), while the excited-state energies are 28.1 and 52.8 cm^{-1} (Wong *et al.*, 1968).

Use of perturbation theory to obtain the vibronic energy levels of $Al_2O_3:Ti^{3+}$ is valid, however, only if the spin–orbit and trigonal field splittings (before Jahn–Teller effects are considered) are small compared with the Jahn–Teller energy (or at least small relative to the splitting $3E_{JT}$ of the electronic states in the distorted configuration). The calculations of Macfarlane *et al.* for $Al_2O_3:Ti^{3+}$ (or $Al_2O_3:V^{4+}$) do not satisfy this requirement because their value for the trigonal crystal-field splitting $v = 700$

cm^{-1} is actually *larger* even than $3E_{JT} = 600\ cm^{-1}$. In a situation like this (which very likely may be found to be more typical in the study of dynamic Jahn–Teller effects than cases in which perturbation theory is clearly valid), one must obtain the vibronic states by calculating directly the eigenvalues and eigenvectors of the infinite-order secular equation of the vibronic problem. Preliminary results of calculations of this sort for $Al_2O_3:Ti^{3+}$ (Ham and Schwarz, 1969), however, indicate that the values of the parameters used by Macfarlane *et al.* in fact need only modest adjustment for their model to account reasonably well for the available experimental data.

Bersuker and Vekhter (1963) have proposed that a dynamic Jahn–Teller effect (or in their terms, inversion splitting) can explain the strongly temperature-dependent *g*-factors observed for Ti^{3+} in the titanium–cesium alums (Bijl, 1950; Bleaney, 1950; Bleaney, Bogle *et al.*, 1955). Probably this is the case, although, for the reasons noted in Section 3, Bersuker and Vekhter's theory of inversion splitting is not applicable in detail in explaining these data.

5. DISCUSSION

Our theoretical understanding of dynamic Jahn–Teller effects and of their relation to the static Jahn–Teller effect has advanced significantly in the last several years. In particular, recognition of the partial quenching of electronic operators by a dynamic Jahn–Teller effect has opened up many new possibilities for the experimental investigation of Jahn–Teller effects. Nevertheless, despite much detailed data on the static Jahn–Teller effect and a burgeoning interest in the dynamic effects, as documented in this chapter and in the review by Sturge (1967), in most of the cases studied experimentally values have not yet been determined for the parameters that are of crucial importance for an understanding of the role of a Jahn–Teller effect in these systems. For example, for most of the many ions in an orbital doublet state which give rise to a static Jahn–Teller effect and which have been studied by EPR over the last two decades, we do not have values for the Jahn–Teller stabilization energy E_{JT}, the Jahn–Teller coupling coefficient V, the strain coefficient V_2, the tunneling splitting 3Γ, the barrier height $2|\beta|$ between distorted configurations, the frequency ω of the effective mode (or modes) of vibration responsible for the Jahn–Teller coupling, or the relaxation time τ for reorientation. Clearly in all of these cases there is a great need for measurements such as those of

Sturge *et al.* (1967) on $Al_2O_3:Ni^{3+}$, those of Williams *et al.* (1969) on Cu^{2+} in lanthanum magnesium double nitrate, or other experiments that may be devised to determine these parameters. Similarly, experiments and analysis for ions in orbital triplet states have not yet been carried far enough to provide values for the corresponding parameters in these cases, except in the perturbation analysis of EPR and far-infrared data for Ti^{3+} and V^{4+} in Al_2O_3 by Macfarlane *et al.* (1968), in the interpretation of dynamic Jahn–Teller effects in the excited ${}^3T_2$ state of $Al_2O_3:V^{3+}$, as inferred from optical data by Scott and Sturge (1966) and Stephens and Lowe-Pariseau (1968), and in the analysis by Ham, Schwarz, and O'Brien (1969) of the far-infrared, EPR, and Mössbauer data for $MgO:Fe^{2+}$.

A major theoretical problem still awaiting solution is the proper treatment of Jahn–Teller effects when the coupling is with a continuous spectrum of lattice phonons, rather than with a discrete set of vibrational modes. Since for a complex in a crystal one must in principle have coupling with such a phonon continuum even if there are also local modes of vibration, one can fairly object that the analysis outlined in this chapter is strictly appropriate only to simple molecules and that for a crystal it represents an extreme idealization which may overlook important effects. Although a definitive answer to this objection cannot be given until better solutions are available to the continuum problem, we would like to comment here on some solutions of this problem that have been suggested, and to indicate those features of the results for the simpler problem which we believe may usefully be carried over to the more complicated case.

When the coupling is with a continuum of phonon states, we may nevertheless still classify the vibronic eigenstates of the coupled system according to the irreducible representations of the group corresponding to the original undistorted symmetry of the site of the Jahn–Teller ion, since the full Hamiltonian describing the coupling has this symmetry. As an example, for a Jahn–Teller ion in an orbital doublet electronic state at a site of cubic symmetry, the vibronic eigenstates take the form given in Eq. (1.2.9), where $\chi_1(\mathbf{Q})$ and $\chi_2(\mathbf{Q})$ are now functions of *all* of the lattice coordinates $\mathbf{Q}$. Even though the energy spectrum of these states may be continuous, we can classify these states according to the irreducible representations of the symmetry group of the cube to which they belong. It is clear, therefore, that any effective Hamiltonian describing the response of this system to an external perturbation, such as an applied magnetic field, must have cubic symmetry. In particular, if the vibronic eigenstate of lowest energy in this example is a doublet belonging to E at least up to moderately strong coupling, as we surmise from analogy with the simpler problem of Section 2.3, then we may expect in this situation that at very low temperatures (~0°K) this effective Hamiltonian of the system will have precisely the form of Eq. (2.3.9). The reduction factors p and q will now reflect the effect of coupling with the entire phonon spectrum rather than with a single pair of vibrational modes, but otherwise the behavior of the system, say in an EPR experiment, should be essentially the same as that described previously (Section 2.3) for a dynamic Jahn–Teller effect. As the temperature is increased, however, excited states of the system may become populated which give rise to effective Hamiltonians still of cubic symmetry but different from Eq. (2.3.9), so that additional lines may appear in the spectrum as the temperature is raised. Such effective Hamiltonians may differ from Eq. (2.3.9) in form, such as that corresponding to a singlet state as in Eq. (2.4.3), or merely in the values of p and q. Alternatively, if transitions occur sufficiently rapidly between the thermally populated states, we may anticipate that p and q may now exhibit a temperature dependence in analogy to the Debye–Waller factor, while a transition to an isotropic spectrum as a result of relaxation would again be expected as in Section 2.5.

For very strong coupling between our Jahn–Teller ion and the lattice phonons, on the other hand, it is quite reasonable to suppose that the system has states in which the vibrational parts of the wave function have appreciable amplitude only for lattice ions close to the Jahn–Teller ion. We surmise, in the case of the orbital doublet, that these localized states may be represented approximately in terms of the coupling of the degenerate electronic states with a single pair of localized modes of vibration belonging to E, primarily involving the nearest-neighbor lattice ions, and having some effective frequency ω. These modes would thus be linear combinations of the normal modes of vibration of the unperturbed crystal. In terms of such modes, we surmise that the approach to a static Jahn–Teller effect occurs as described in Sections 2.2–2.4, and that the residual coupling with other lattice modes is relatively weak and occurs (as in the appendix) via the strain field which they produce. Some consideration has been given by Slonczewski (1963) to the formation of such a localized vibrational mode as a result of the Jahn–Teller coupling to the phonons, but no further calculations are available from which we might determine, for example, what effective frequency we should associate with this mode in various cases. Bates *et al.* (1968) have made some further progress toward justifying the use of a "cluster model," comprising the Jahn–Teller ion and its nearest neighbors, as a good approximation to the real problem of a Jahn–Teller ion coupled to the entire lattice when the Jahn–Teller coupling is sufficiently strong. Bates *et al.* (1968) showed in particular how results equivalent to those obtained by O'Brien (1964) could be derived by transforming the vibronic Hamiltonian, in lieu of working with the wave functions of the vibronic eigenstates of the cluster model. Such operator methods should provide useful insights when the Jahn–Teller coupling is with the continuous spectrum of lattice phonons.

Similar considerations pertain to the problem of a Jahn–Teller ion in an orbital triplet state in cubic symmetry, coupled to the phonon continuum. The effective Hamiltonian of the system must again have cubic symmetry, and for moderate coupling so that we have a dynamic Jahn–Teller effect we expect that this Hamiltonian will have the form of Eq. (3.1.11) at very low temperatures (apart from higher-order effects). The reduction factors $K(E)$, $K(T_1)$, and $K(T_2)$ will then have appropriate values to take account of the coupling with the various phonons. The exact analysis for this case can be carried through very simply in one special situation, namely when the coupling is only with the normal modes of vibration belonging to E (Ham, 1965). The vibronic Hamiltonian is now given in place of Eq. (3.1.1) by

$$\mathscr{H} = E_0\mathscr{T} + \sum_q \hbar\omega_q(a_q^+a_q + \tfrac{1}{2})\mathscr{T} + \sum_q [(V_{\theta q}a_q + V_{\theta q}^* a_q^+)\mathscr{E}_\theta + (V_{\varepsilon q}a_q + V_{\varepsilon q}^* a_q^+)\mathscr{E}_\varepsilon] \tag{5.1.1}$$

Here, q denotes the propagation vector and mode of polarization of a phonon of the unperturbed crystal, ω_q its frequency, a_q and a_q^+ the corresponding annihilation and creation operators, and $V_{\theta q}$ and $V_{\varepsilon q}$ the Jahn–Teller coupling coefficients. Since we have chosen the phonons in the usual way to represent modes with a definite propagation vector rather than some combination of these modes belonging to an irreducible representation of the cubic group [in the latter case, only modes belonging to E would enter the coupling terms in Eq. (5.1.1)], we must have suitable relations among the $V_{\theta q}$ and $V_{\varepsilon q}$ so that the Hamiltonian has overall cubic symmetry. The energy spectrum of the exact eigenstates of Eq. (5.1.1) is, however, identical with that of the system when the Jahn–Teller coupling vanishes, $V_{\theta q} = V_{\varepsilon q} = 0$, except for an energy shift δE (the Jahn–Teller energy) which is common to all states and is given by

$$\delta E = -\tfrac{1}{2}\sum_q (\hbar\omega_q)^{-1}(|V_{\theta q}|^2 + |V_{\varepsilon q}|^2) \tag{5.1.2}$$

Since the matrices $\mathscr{E}_\theta$ and $\mathscr{E}_\varepsilon$ [Eq. (3.1.2)] are diagonal with respect to the electronic states $\psi_\xi, \psi_\eta, \psi_\zeta$, the exact eigenstates of Eq. (5.1.1) are given as in Eq. (3.1.5) by a product of one of $\psi_\xi, \psi_\eta, \psi_\zeta$ with a phonon state. In particular, the lowest energy state of the system is triply degenerate and belongs to the same irreducible representation of the cubic group, T_1 or T_2, as does the original electronic triplet, whatever the strength of the Jahn–Teller coupling. The effect of this coupling is thus not to split the degeneracy of this vibronic ground state [such a splitting is prevented by symmetry so long as the Hamiltonian (5.1.1) has cubic symmetry], but to associate different phonon states with each electronic state ψ_ξ, ψ_η, ψ_ζ in forming the three vibronic eigenstates that comprise this ground state. Because of this difference in the phonon part of these states, off-diagonal matrix elements of electronic operators within the ground state are reduced just as in Section 3.1. We find accordingly that the reduction factors $K(T_1)$ and $K(T_2)$ for this ground state are given by

$$K(T_1) = K(T_2) = \exp\left[-\tfrac{3}{4}\sum_q (\hbar\omega_q)^{-2}(|V_{\theta q}|^2 + |V_{\varepsilon q}|^2)\right] \qquad (5.1.3)$$

while $K(E)$ remains equal to unity.

Different conclusions have been reached by several authors concerning the influence of a dynamic Jahn–Teller effect on the EPR spectrum when the coupling is with the phonon continuum. Stevens and Persico (1966), considering the effect of such coupling on a triplet state (which they describe in terms of an effective spin $S' = 1$) in cubic symmetry, concluded that the EPR spectrum would split into two lines corresponding to a relative displacement of the ($S_z' = -1$ to $S_z' = 0$) and ($S_z' = 0$ to $S_z' = +1$) transition frequencies. Although Van Eekelen and Stevens (1967) have subsequently shown this conclusion to be in error, the nonzero splitting having resulted from an algebraic error in the calculations, it is of interest to consider in general terms why such a splitting is in conflict with symmetry requirements. A similar splitting has also recently been obtained by Böttger (1968) from a high-order diagram method applied to the same problem, but, as this result is again in conflict with symmetry, this calculation once again must have been done incorrectly.

A splitting between the ($S_z' = -1$ to $S_z' = 0$) and ($S_z' = 0$ to $S_z' = 1$) transitions when the magnetic field **H** is along the z cubic axis would indicate the presence of a term $D(S_z')^2$ in the effective Hamiltonian of the system.* The cubic symmetry of this Hamiltonian, however, then would

* Such a splitting could also arise from terms of the form $a(\mathbf{S}' \cdot \mathbf{H})^2 + b[(S_x')^2 H_x^2 + (S_y')^2 H_y^2 + (S_z')^2 H_z^2]$ which *are* allowed in cubic symmetry. However, such quadratic Zeeman energies are usually very small, and neither Stevens and Persico (1966) or Böttger (1968) were concerned with a splitting of this sort. A term of this type was obtained, however, by McMahon and Silsbee (1964).

require that the corresponding terms $D[(S_x')^2 + (S_y')^2]$ also be present, and, since $[(S_x')^2 + (S_y')^2 + (S_z')^2]$ is a constant within the manifold of states with $S' = 1$, a splitting thus cannot occur unless the symmetry of the Hamiltonian is less than cubic. In other words, since the triplet belongs to a three-dimensional irreducible representation (Γ_4 or Γ_5) of the cubic group, there can be no zero-field splitting of the degeneracy such as that given by a term $D(S_z')^2$, unless the symmetry of the Hamiltonian in the absence of a magnetic field is lower than cubic. It should be noted, of course, that the case considered by Stevens and Persico (1966) and Böttger (1968) (a triplet state such as the ground state of MgO:Ni^{2+} with a true spin 1, or the lowest spin–orbit level Γ_5 of MgO:Fe^{2+}) is not the same as the one that has been of principal concern to us in Section 3, namely an *orbital* triplet state with additional spin degeneracy. In the latter case, we have the possibility in sufficiently strong coupling of a static Jahn–Teller effect in which, in one of the distorted configurations, the orbital degeneracy is removed, but we still have an EPR spectrum corresponding to transitions among the spin levels. In the former case, on the other hand, the g-factor for the $-1 \to 0$ and $0 \to +1$ transitions is reduced by the Jahn–Teller coupling in proportion to the reduction factor $K(T_1)$ as in Eq. (5.1.3), and in the limit of strong coupling (could it be achieved without taking into account other levels of the system) g would be zero and there would be no observable resonance transitions. In the dynamic regime, however, the two cases do have similar features, since the effect of the Jahn–Teller coupling in each case can be described by the appropriate reduction factors in the effective Hamiltonian.

An earlier treatment by McMahon and Silsbee (1964) of the effect of spin–phonon interaction on the EPR spectrum of a $S' = 1$ system in cubic symmetry, taking into account the phonon continuum, did lead to results in agreement with the conclusions we have stated above. Their calculation, carried out to second order in perturbation theory, led to a reduction in the g-factor at 0°K which is consistent with Eq. (5.1.3), and the only splitting of the resonance line which they obtained is that resulting from a small quadratic Zeeman energy (as noted in an earlier footnote). Evaluating their results for a finite temperature, they found on taking a thermal average over the phonon states that the reduction in the g-factor is enhanced by a term proportional to T^2, in analogy with the expansion of the Debye–Waller factor. However, when their results were applied to the ground state of MgO:Fe^{2+} (McMahon and Silsbee, 1964; O'Brien, 1965), they found that the calculated reductions in the g-factor were too small to be observed experimentally. The generalization of this calculation for MgO:Fe^{2+} to the more realistic situation in which coupling by the

Jahn–Teller interaction to the higher spin–orbit levels is included, while still treating the phonon continuum, is given in the recent work of Ham, Schwarz, and O'Brien (1969) noted previously; however, this calculation did not include the temperature-dependent effects. A calculation somewhat similar to that of McMahon and Silsbee, but for the shift in resonance frequency of a spin-$\frac{1}{2}$ level as a result of the spin–phonon interaction, has been given by Aminov and Kochelaev (1962).

We have noted in this chapter the close relation between the problem of a dynamic Jahn–Teller effect and that of the coupling of *nondegenerate* electronic states by the vibrational motion of the lattice, including zero-point motion. It was proposed by Inoue (1963) that an effect of the latter type causes appreciable shifts in the g-factor of the Kramers doublet ground state of certain rare earth ions. Inoue showed by a perturbation calculation, using a Debye model for the phonon spectrum, that the change in g resulting from zero-point motion of the lattice was in fact in good agreement with a previously unexplained small (1.5%) but significant difference (Lewis and Sabiskey, 1963) between the observed g-factor for $CaF_2:Ho^{2+}$ and the value calculated for the ground state on the basis of a rigid lattice. More recently, an even more striking effect of this sort has been pointed out by Birgeneau (1967) for Ce^{3+} in the ethyl sulfates. In this case, a crystal-field calculation using a rigid lattice predicts a value for $g_\parallel$ which is $\sim$10% larger than the experimental value. Birgeneau has shown by a calculation with reasonable assumptions about the phonon spectrum and orbit–lattice interaction that the effect of the zero-point motion in mixing the ground-state Kramers doublet with the first excited doublet accounts very well for the magnitude of the reduction in $g_\parallel$ and also for the systematics of its variation among a number of the ethyl sulfates. The separation of these doublets is only a few cm^{-1}, so that we are in fact concerned here with a "pseudo-Jahn–Teller effect" involving two nearly degenerate electronic levels. Apart from this small separation, the vibronic problem of the coupling of these states to the lattice vibrations is identical with that of a dynamic Jahn–Teller effect for a degenerate level, and indeed Birgeneau has found that the g-shift is not sensitive to the precise value of this separation. This is what one expects from a perturbation calculation of the effect of the coupling on the ground-state g-factor, because the virtual intermediate states in the calculation have one or more phonons present, and the level separation is much smaller than typical phonon energies and therefore can be neglected to a first approximation in the energy denominators. The reduction of $g_\parallel$ due to this mixing of nondegenerate states by the zero-point motion is thus entirely analogous to the partial quenching of the Zeeman interaction by a dynamic

Jahn–Teller effect as we have described it in Section 3. Birgeneau (1967) has suggested that such vibronic corrections to static-crystal-field parameters may be quite common for rare earth ions in crystals, and that they may often mask the small effects of covalent bonding on these parameters.

Another case in which the dynamic vibronic coupling of nondegenerate electronic states affects the value of an important parameter of the EPR spectrum has been considered by Simanek and Orbach (1966).* They proposed that the explicit temperature dependence of the hyperfine-coupling coefficient for the S-state ion Mn^{2+} in MgO, observed by Walsh *et al.* (1965), resulted from dynamic mixing of excited states from the $3d^4\,ns$ configurations into the ^6S ground state of the $3d^5$ configuration. Since these excited states are 60,000 cm^{-1} or more above the ground state, and the phonon energies are negligible by comparison, the effect of the vibronic coupling on the hyperfine interaction can be treated in this case within the framework of the Born–Oppenheimer approximation, as discussed in Section 1.2 in connection with Eqs. (1.2.7–8). When the average is taken over the thermally excited phonon states, the result is a small decrease in the hyperfine coefficient which agrees well with the observations as to temperature dependence but only roughly as to magnitude. The corresponding reduction in the $T = 0°K$ hyperfine coupling from the value appropriate to the rigid lattice, as a result of zero-point motion, was calculated by Orbach and Simanek (1967) to be 1.0% for $MgO:Mn^{2+}$. This problem has subsequently attracted the attention of a number of authors.†

A class of Jahn–Teller systems (or in some cases perhaps "pseudo-Jahn–Teller systems") which has been studied extensively by EPR but which falls outside the categories we have considered in detail in this chapter includes a number of the defects observed in irradiated silicon (Watkins, 1965) and centers with somewhat similar behavior found in other crystals of the diamond and wurtzite structures. Among these are the various defects in silicon which have been identified by Watkins and his collaborators from EPR spectra as Jahn–Teller-distorted configurations involving a lattice vacancy, in particular the phosphorus–vacancy pair, or Si-E center (Watkins and Corbett, 1964), the arsenic– and antimony–vacancy pairs (Elkin and Watkins, 1968), the isolated vacancy in its singly positive and singly negative charge states (Watkins, 1963, 1965), and the divacancy in its +1 and −1 charge states (Watkins and Corbett, 1965). Other such defects include nickel in germanium (Ludwig and Woodbury, 1959), the

* See also p. 561.

† References to more recent work are given by Menne (1969).

oxygen–vacancy pair in silicon, or Si–*A* center (Watkins and Corbett, 1961), the nitrogen donor in diamond (Smith *et al.*, 1959), the halogen *A* centers in cubic and hexagonal ZnS (Schneider *et al.*, 1963), and the fluorine donor in beryllium oxide (Schirmer *et al.*, 1965; Reinberg *et al.*, 1967). The EPR data reveal that each of these centers suffers a spontaneous static distortion at sufficiently low temperature, that there are several energetically equivalent such configurations differing only in orientation, and that as the temperature is raised each individual defect reorients increasingly rapidly between these configurations as the result of a thermally induced process. However, the stabilization energy of the distorted configurations is apparently large in these systems and of the same size as the energy differences between electronic energy levels of the defect. The electronic states in the distorted configuration are therefore admixtures of states from different energy levels of the defect in the symmetric configuration, and the distortion cannot be viewed as simply the result of Jahn–Teller coupling to the lattice acting within a single degenerate electronic level. The Jahn–Teller effect in these systems is therefore much more complicated than the case of a simple orbital doublet or triplet state which we have considered in this chapter, and indeed it is not known in detail for most of these systems what form of vibronic Hamiltonian will lead to the distorted configurations that are actually found, or what states of the defect need to be considered in the coupling. However, Watkins has shown that the behavior of the various vacancy centers in silicon can be interpreted with remarkable success in terms of the formation of electron pair bonds between the atoms. Paramagnetic resonance then is attributed to an unpaired electron in a remaining orbital, and the distorted Jahn–Teller configurations correspond to the several energetically equivalent choices for the atoms which are joined by the bonds. Similar models also appear to describe the halogen *A*-centers in ZnS (Schneider *et al.*, 1963; Schneider, 1967), the nitrogen donor in diamond (Smith *et al.*, 1959), and the fluorine center in BeO (Schirmer *et al.*, 1965; Reinberg and Estle, 1967). These have been reviewed by Sturge (1967, Section 19). (See also Section 8.3 in the chapter by R. H. Silsbee in this volume for another example on the *R* center.)

The EPR spectra of all these defects exhibit dynamic effects from motional averaging as the temperature is raised, as a result of the rapid reorientation of the defect between its equivalent distorted configurations. Indeed, the experimental and theoretical investigation of motional effects carried out by Watkins and his collaborators in their papers on the various vacancy–impurity defects in silicon is the most complete analysis of such effects that has been made for any group of Jahn–Teller systems, and Watkins's work is a model for what should be done in other investigations of dynamic Jahn–Teller effects that result from motional averaging. It was found, for example, for the phosphorus–vacancy center (Watkins and Corbett, 1964) that the EPR spectrum changes drastically in the temperature range 60–150°K. At the lower temperature, the spectrum is the superposition of the spectra appropriate to the three energetically equivalent distorted configurations corresponding to each of the four possible directions for the phosphorus–vacancy axis. As the temperature rises, these lines broaden and eventually disappear, to be replaced at 150°K by motionally narrowed lines at the position of the average of the corresponding lines of the low-temperature spectra (the phosphorus–vacancy axis remains fixed in direction in this range of temperature). Watkins has used the theory of motional narrowing to obtain from the EPR linewidths the reorientation time both in the lower-temperature region, where the lines are broadening, and in the higher-temperature region, where the motionally averaged lines are narrowing. He has also determined the reorientation time at 20.4°K by using the relative intensities of the EPR lines to follow the recovery of thermal equilibrium between the populations of the different distorted configurations after the removal of an applied stress. The reorientation time τ was found to be very long at 20.4°K, namely $\tau = 1.3\times 10^3$ sec. Comparing this value with the data at higher temperature obtained from the EPR linewidths, Watkins found that his values for τ span 13 decades and that they fit a formula of the form of Eq. (2.5.6) with an activation energy $E \simeq 0.06$ eV (480 cm^{-1}), which thus provides a measure of the height of the barrier which must be surmounted by the defect in reorienting. Similar values for the activation energy for reorientation have been obtained from similar studies of the other vacancy centers in silicon, the values ranging from ~0.01–0.02 eV for the +1 charge state of the isolated vacancy to ~0.07 eV for the antimony–vacancy pair and the +1 charge state of the divacancy (a larger value of ~0.38 eV was obtained for the oxygen–vacancy pair, evidently reflecting the fact that reorienting the defect requires a substantial displacement of the oxygen atom instead of simply a switching of an electron bond accompanied by a small readjustment of the atomic positions). A similar analysis of motional averaging has been made by Dischler *et al.* (1964) for the EPR spectra of the chlorine and bromine *A* centers in cubic ZnS in the temperature range 67–121°K. The activation energy for reorientation was found to be 0.057 eV for the chlorine center and 0.058 eV for the bromine. For nickel in germanium, Ludwig and Woodbury (1959) estimated activation energies of ~0.02 eV for reorientaion by both rotation and inversion, from linewidth data in the range 20–50°K. Shulman *et al.* (1966) found for the nitrogen donor in diamond that motional averaging effects occur only above ~570°K, and they obtained the rather large activation energy of ~0.7 eV; a similar result was obtained by Loubser and van Ryneveld (1967). For the fluorine center in BeO, in contrast, Reinberg and Estle (1967) found from their analysis of motional effects on the EPR spectrum that the reorientation time seems to be described by the expression $\tau^{-1} = A + BT$, with $A = 2\times10^7\ \text{sec}^{-1}$ and $B = 3\times10^6\ \text{sec}^{-1}\ \text{deg}^{-1}$, from below 8°K to above room temperature.

In view of the uncertainties that still becloud the nature of the thermally induced dynamic Jahn–Teller effects for most of the transition metal ions in orbital doublet states listed in Tables II and III, it is of interest that, for all of the defects in silicon which have so far been studied carefully by Watkins, the change in the spectrum that occurs with rising temperature seems to be completely accounted for by the motional averaging resulting from the reorientation of the defect. It is also interesting that the reorientation times for these vacancy defects in silicon are so very long below ~20°K and that they follow an expression of the form of Eq. (2.5.6) over such an extended range of variation, corresponding to thermal excitation over a barrier or an Orbach process through an excited state. Only for the isolated vacancy, for the +1 charge state of which τ is still less than 0.1 sec at 2°K in contrast to a value $>10^{15}$ sec obtained by extrapolating the data in the range 14–20°K using Eq. (2.5.6) (Watkins, 1963), has evidence been found for any departure from this behavior such as would be expected if tunneling is important. Thus dynamic Jahn–Teller effects arising from zero-point motion have not yet been detected for any of the vacancy defects in silicon except perhaps the isolated vacancy. In particular, tunneling matrix elements between the different distorted configurations must be orders of magnitude smaller than the values so far obtained in cases of a static Jahn–Teller distortion of orbital doublet states of transition metal ions ($Al_2O_3:Ni^{3+}$, say) as discussed in Section 4.1. The linear temperature variation of τ^{-1} obtained for the fluorine center in BeO is, however, of the form expected for the one-phonon direct-process reorientation via tunneling, but it is then surprising that this mechanism appears to remain the dominant one for reorientation all the way up to room temperature.

From the relative populations of the different Jahn–Teller distortions under an applied stress, as determined in thermal equilibrium from the EPR line intensities, Watkins and his collaborators have been able to determine the coefficients for the effect of stress in producing a relative shift in the energy of the different distorted configurations for the various defects in silicon. By relating these coefficients to the strength of the Jahn–Teller coupling [in a manner analogous to that leading to Eq. (4.1.3)], Elkin and Watkins (1968) have obtained rough estimates of the Jahn–Teller stabilization energy for the phosphorus–, antimony–, and arsenic–vacancy centers in silicon. These values range from 0.6 to 1.4 eV (5,000 to 11,000 cm^{-1}). These results therefore are in support of the view that Jahn–Teller energies for these defects are very large indeed, and they illustrate the usefulness which such stress measurements will have in the study of other Jahn–Teller systems.

The nature of Jahn–Teller instabilities in cases of accidental degeneracy or near-degeneracy is the subject of a recent, very interesting, and valuable investigation by Stoneham and Lannoo (1969). The symmetry of the stable, distorted configuration for many of these cases is found to be lower than that of the purely tetragonal or trigonal distortions that result (in cubic systems) from the Jahn–Teller instability of a single orbital doublet or triplet electronic level. For example, in cubic symmetry, a doublet E and triplet T_1 or T_2 or two triplets T_1 and T_2 may interact by terms linear in the appropriate distortion coordinates (pseudo-Jahn–Teller coupling); if this coupling is sufficiently strong, the stabilization energy of the resulting stable configurations may be comparable with or larger than the original energy difference between the two electronic levels, and under these conditions, the stable distortions may be of a "mixed" type with symmetry D_{2h} (or C_{2v} or D_2 if the original symmetry is T_d). The electronic wave functions in these six distorted configurations are then linear combinations of those of the originally nondegenerate levels. Considerations such as these are clearly of major importance for the interpretation of Jahn–Teller effects in defects such as those studied by Watkins.

APPENDIX. Alternative Treatment of the Effect of Strain on a Jahn–Teller System

We have described the effect of uniform strain on a Jahn–Teller system by introducing into the vibronic Hamiltonian a term, as in Eq. (2.1.10) for the orbital doublet in cubic symmetry,

$$\mathscr{H}_S^{(a)} = V_2[e_\theta \mathscr{U}_\theta + e_\varepsilon \mathscr{U}_\varepsilon] \tag{A.1}$$

which represents a direct splitting of the *electronic* degeneracy as a result of the strain. Alternatively, we could view the effect of uniform strain in the lattice to be that of shifting the equilibrium values of the distortion modes Q_θ and Q_ε that are responsible for the Jahn–Teller coupling. Whereas n zero strain the equilibrium configuration of perfect cubic symmetry (before we consider the instability of this configuration that results from he Jahn–Teller coupling to the impurity ion) corresponds to $Q_\theta = Q_\varepsilon = 0$, a strain represented by e_θ and e_ε displaces this equilibrium position to

$$Q_\theta = Q_\theta{}^0 = De_\theta, \qquad Q_\varepsilon = Q_\varepsilon{}^0 = De_\varepsilon \tag{A.2}$$

where D is a coefficient to be discussed below. According to this view, the splitting of the electronic degeneracy in the configuration (A.2) is a consequence of the Jahn–Teller coupling and is obtained by substituting the values (A.2) into the interaction term $V[Q_\theta \mathscr{U}_\theta + Q_\varepsilon \mathscr{U}_\varepsilon]$ in Eq. (2.3.1). Corresponding to the displaced equilibrium configuration, the elastic potential energy in Eq. (2.3.2) is replaced by

$$V_{\text{el}} = \tfrac{1}{2}\mu\omega^2[(Q_\theta - Q_\theta{}^0)^2 + (Q_\varepsilon - Q_\varepsilon{}^0)^2]\mathscr{I} \tag{A.3}$$

We find on expanding V_{el} that, apart from the constant term $\frac{1}{2}\mu\omega^2[(Q_\theta{}^0)^2 + (Q_\varepsilon{}^0)^2]\mathscr{I}$, the effect of strain on the vibronic Hamiltonian (2.3.1) is now to introduce a term

$$\mathscr{H}_S^{(b)} = -\mu\omega^2(Q_\theta Q_\theta{}^0 + Q_\varepsilon Q_\varepsilon{}^0)\mathscr{I} \tag{A.4}$$

Although it appears that Eq. (A.4) represents the effect of strain on our vibronic system in quite a different way from Eq. (A.1), we will now show that in fact these representations are equivalent.

Taking the elastic potential energy in the presence of strain to be given by Eq. (A.3), we make the change of variable

$$Q_\theta' = Q_\theta - Q_\theta{}^0, \qquad Q_\varepsilon' = Q_\theta - Q_\varepsilon{}^0 \tag{A.5}$$

and the corresponding transformation of the canonical momenta, $P_\theta' = P_\theta$ and $P_\varepsilon' = P_\varepsilon$, in the vibronic Hamiltonian $\mathscr{H}$ of Eqs. (2.3.1–2). We obtain in this way

$$\begin{aligned} \mathscr{H} = E_0\mathscr{I} &+ (1/2\mu)\{(P_\theta')^2 + (P_\varepsilon')^2 + \mu^2\omega^2[(Q_\theta')^2 + (Q_\varepsilon')^2]\}\mathscr{I} \\ &+ V[Q_\theta' \mathscr{U}_\theta + Q_\varepsilon' \mathscr{U}_\varepsilon] + V[Q_\theta{}^0 \mathscr{U}_\theta + Q_\varepsilon{}^0 \mathscr{U}_\varepsilon] \end{aligned} \tag{A.6}$$

which, apart from the last term, is the same Hamiltonian in the new variables that we had in Eqs. (2.3.1–2) in the absence of strain. The last term, however, has just the form of the strain term, Eq. (A.1), used previously. We see therefore that introducing strain via the displacement in the equilibrium position of Q_θ, Q_ε leads, on the one hand, by simple expansion, to the perturbing term $\mathscr{H}_S^{(b)}$ of Eq. (A.4), and on the other hand, by the change of variable, to a perturbation given by the last term in Eq. (A.6). The latter is identical with $\mathscr{H}_S^{(a)}$ in Eq. (A.1) if we take

$$V_2 = DV \tag{A.7}$$

To first order in the strain, therefore, the effect of $\mathscr{H}_S^{(a)}$ and $\mathscr{H}_S^{(b)}$ on the energy levels of the unperturbed vibronic Hamiltonian (2.3.1) must be identical, if the strain coefficient V_2 is related to the Jahn–Teller coupling coefficient V by Eq. (A.7).

The equivalence of $\mathscr{H}_S^{(a)}$ and $\mathscr{H}_S^{(b)}$ to first order in perturbing the vibronic energy levels of the Hamiltonian (2.3.1) implies in particular that there is a relation between the matrix elements of $\mathscr{U}_\theta$, $\mathscr{U}_\varepsilon$ and those of Q_θ, Q_ε within the vibronic ground-state doublet $\Psi_{g\theta}$, $\Psi_{g\varepsilon}$ in the case of a dynamic Jahn–Teller effect as presented in Section 2.3. Since these matrix elements of $\mathscr{U}_\theta$, $\mathscr{U}_\varepsilon$ have been used previously to define the parameter q from Eq. (2.3.8a), this relation must take the form

$$\begin{aligned} -q(V/\mu\omega^2) &= -\langle \Psi_{g\theta} \mid Q_\theta \mid \Psi_{g\theta} \rangle \\ &= \langle \Psi_{g\varepsilon} \mid Q_\theta \mid \Psi_{g\varepsilon} \rangle = \langle \Psi_{g\theta} \mid Q_\varepsilon \mid \Psi_{g\varepsilon} \rangle \end{aligned} \tag{A.8}$$

Although we have not succeeded in finding a direct proof of this relation from the analytic properties of the functions $\Psi_{g\theta}$ and $\Psi_{g\varepsilon}$, we have verified that it is obeyed in the numerical examples calculated by Child and Longuet-Higgins (1962) as well as in the limits of weak ($E_{\text{JT}}/\hbar\omega \ll 1$) and strong ($E_{\text{JT}}/\hbar\omega \gg 1$) Jahn–Teller coupling. Thus Eq. (A.8) provides a convenient formula from which values of the matrix elements of Q_θ and Q_ε within the ground state may be obtained in terms of the known values of q [Eq. (2.3.13) and Fig. 5], at least so long as only linear Jahn–Teller coupling is considered.

The equivalence of $\mathscr{H}_S^{(a)}$ and $\mathscr{H}_S^{(b)}$, as given by Eqs. (A.1) and (A.4) using Eq. (A.7), may also be demonstrated in general for states of low energy in the limit of strong Jahn–Teller coupling ($E_{\text{JT}}/\hbar\omega \gg 1$) even if warping terms are present in the Hamiltonian. In this limit, the vibronic eigenfunctions of states with energies near the ground state are given as in Eq. (2.3.27) by simple Born–Oppenheimer products involving the electronic wave function ψ_+ ($V < 0$) or ψ_- ($V > 0$). From the diagonal matrix elements (2.3.39) of $\mathscr{U}_\theta$ and $\mathscr{U}_\varepsilon$ with respect to ψ_+ and ψ_- we see therefore that $\mathscr{H}_S^{(a)}$ is equivalent for such states to

$$\mathscr{H}_S^{(a)\prime} = -D \mid V \mid (e_\theta \cos\theta + e_\varepsilon \sin\theta) \tag{A.9}$$

where we have used Eq. (A.7). From the definition $Q_\theta = \varrho \cos\theta$, $Q_\varepsilon = \varrho \sin\theta$ we have, on the other hand,

$$\mathscr{H}_S^{(b)} = -\mu\omega^2 D\varrho(e_\theta \cos\theta + e_\varepsilon \sin\theta) \qquad \text{(A.10)}$$

where we have used Eq. (A.2). The equivalence of these operators (A.9) and (A.10) then follows when we note that for the low energy states in the trough in Fig. 3, ϱ has the value $|V|/\mu\omega^2$ given by Eq. (2.3.30).

The coefficient D in Eq. (A.2) is easily evaluated if Q_θ, Q_ε represent the distortion modes of six octahedrally coordinated nearest-neighbor ions as in Table I, and if the local elastic constants near the Jahn–Teller ion are taken to be the same as in the host crystal. We then obtain (Ham, 1968, Appendix II)

$$D = (2/\sqrt{3})R \qquad \text{(A.11)}$$

for the case of octahedral coordination (and similar, but numerically different results for tetrahedral or eightfold cubic coordination), where R is the nearest-neighbor distance. On the other hand, if the Q's are defined in a more complicated way to take account of the displacements of more distant ions, or if the local elastic constants are different from those of the host crystal, the value of D may be quite different from that given by Eq. (A.11).

ACKNOWLEDGMENTS

I am grateful to the following for sending me preprints of their papers in advance of publication: N. A. Blum, R. H. Borcherts, D. P. Breen, J. Chappert, L. L. Chase, R. E. Coffman, M. de Wit, R. Englman, T. L. Estle, J. R. Fletcher, R. B. Frankel, U. T. Höchli, A. E. Hughes, R. R. Joyce, H. Kamimura, D. C. Krupka, K. P. Lee, W. Low, M. Lowe, R. M. Macfarlane, S. Mizuhashi, K. Morigaki, K. A. Müller, M. C. M. O'Brien, P. L. Richards, J. Schneider, J. C. Slonczewski, P. J. Stephens, K. W. H. Stevens, A. M. Stoneham, M. D. Sturge, D. Walsh, G. D. Watkins, F. I. B. Williams, and J. Y. Wong. I am indebted to the above and in addition to the following for discussions and correspondence from which I have benefited greatly: A. Abragam, I. B. Bersuker, R. J. Birgeneau, B. Bleaney, D. C. Burnham, S. Geschwind, W. C. Holton, L. L. Lohr, G. W. Ludwig, M. H. L. Pryce, W. M. Schwarz, R. H. Silsbee, G. A. Slack, W. Thorson, W. P. Wolf, and H. R. Zeller.

REFERENCES

Abragam, A., 1961, *The Principles of Nuclear Magnetism* (Clarendon Press, Oxford).

Abragam, A., and Pryce, M. H. L., 1950, *Proc. Phys. Soc.* (*London*) **A63**, 409.

Abragam, A., and Pryce, M. H. L., 1951, *Proc. Roy. Soc.* **A205**, 135.

Aminov, L. K., and Kochelaev, B. I., 1962, *Fiz. Tverd. Tela* **4**, 1604 [English Transl.: *Soviet Phys.—Solid State* **4**, 1175 (1962)].

Avvakumov, V. I., 1959, *Zh. Eksperim. i. Teor. Fiz.* **37**, 1017 [English Transl.: *Soviet Phys.—JETP* **16**, 933 (1960)].

Ballhausen, C. J., 1965, *Theoret. Chim. Acta* (*Berlin*) **3**, 368.

Bates, C. A., Dixon, J. M., Fletcher, J. R., and Stevens, K. W. H., 1968, *J. Phys. C* (*Proc. Phys. Soc.*) [2], **1**, 859.

Bersuker, I. B., 1962, *Zh. Eksperim. i. Teor. Fiz.* **43**, 1315 [English Transl.: *Soviet Phys.—JETP* **16**, 933 (1963)].

Bersuker, I. B., 1963, *Zh. Eksperim. i. Teor. Fiz.* **44**, 1239 [English Transl.: *Soviet Phys.—JETP* **17**, 836 (1963)].

Bersuker, I. B., Budnikov, S. S., Vekhter, B. G., and Chinik, B. I., 1964, *Fiz. Tverd. Tela.* **6**, 2583 [English Transl.: *Soviet Phys.—Solid State* **6**, 2059 (1965)].

Bersuker, I. B., and Vekhter, B. G., 1963, *Fiz. Tverd. Tela.* **5**, 2432 [English Transl.: *Soviet Phys.—Solid State* **5**, 1772 (1964)].

Bersuker, I. B., and Vekhter, B. G., 1965, *Fiz. Tverd. Tela.* **7**, 1231 [English Transl.: *Soviet Phys.—Solid State* **7**, 986 (1965)].

Bersuker, I. B., and Vekhter, B. G., 1966, *Phys. Stat. Sol.* **16**, 63.

Bethe, H. A., 1929, *Ann. Physik* **3**, 133.

Bijl, D., 1950, *Proc. Phys. Soc.* (*London*) **A63**, 405.

Bijl, D., and Rose-Innes, A. C., 1953, *Proc. Phys. Soc.* (*London*) **A66**, 954.

Birgeneau, R. J., 1967, *Phys. Rev. Letters* **19**, 160.

Bleaney, B., 1950, *Proc. Phys. Soc.* (*London*) **A63**, 407.

Bleaney, B., Bogle, G. S., Cooke, A. H., Duffus, R. J., O'Brien M. C. M., and Stevens, K. W. H., 1955, *Proc. Phys. Soc.* (*London*) **A68**, 57.

Bleaney, B., and Bowers, K. D., 1952, *Proc. Phys. Soc.* (*London*) **A65**, 667.

Bleaney, B., Bowers, K. D., and Pryce, M. H. L., 1955, *Proc. Roy. Soc.* (*London*) **A228**, 166.

Bleaney, B., Bowers, K. D., and Trenam, R. S., 1955, *Proc. Roy. Soc.* (*London*) **A228**, 157.

Bleaney, B., and Ingram, D. J. E., 1950, *Proc. Phys. Soc.* (*London*) **A63**, 408.

Boatner, L. A., Dischler, B., Herrington, J. R., and Estle, T. L., 1969, *Bull. Am. Phys. Soc.* **14**, 355.

Borcherts, R. H., Kanzaki, H., and Abe, H., 1970, *Phys. Rev.* **B2**, 23.

Born, M., and Oppenheimer, J. R., 1927, *Ann. Physik* **84**, 457.

Böttger, H., 1968, *Phys. Stat. Sol.* **26**, 681.

Breen, D. P., 1966, Ph. D. Thesis, Oxford University (unpublished).

Breen, D. P., Krupka, D. C., and Williams, F. I. B., 1969, *Phys. Rev.* **179**, 241.

Burnham, D. C., 1966, *Bull. Am. Phys. Soc.* **11**, 186.

Caner, M., and Englman, R., 1966, *J. Chem. Phys.* **44**, 4054.

Carter, M. K., and Vincow, G., 1967, *J. Chem. Phys.* **47**, 292.

Chappert, J., Frankel, R. B., and Blum, N. A., 1967, *Phys. Letters* **25A**, 149.

Chase, L. L., 1968, *Phys. Rev.* **168**, 341.

Chase, L. L., 1969, *Phys. Rev. Letters* **23**, 275.

Child, M. S., 1960, *Mol. Phys.* **3**, 601.
Child, M. S., 1963, *Phil. Trans. Roy. Soc.* (*London*) **A255**, 31.
Child, M. S., and Longuet-Higgins, H. C., 1962, *Phil. Trans. Roy. Soc.* (*London*) **A254**, 259.
Coffman, R. E., 1965, *Phys. Letters* **19**, 475.
Coffman, R. E., 1966, *Phys. Letters* **21**, 381.
Coffman, R. E., 1968, *J. Chem. Phys.* **48**, 609.
Coffman, R. E., Lyle, D. L., and Mattison, D. R., 1968, *J. Phys. Chem.* **72**, 1392.
Delbecq, C. J., Hayes, W., O'Brien, M. C. M., and Yuster, P. H., 1963, *Proc. Roy. Soc.* (*London*) **A271**, 243.
De Wit, M., 1969, *Phys. Rev.* **177**, 441.
De Wit, M., Estle, T. L., Holton, W. C., and Schneider, J., 1964, *Bull. Am. Phys. Soc.* **9**, 249.
De Wit, M., and Reinberg, A. R., 1967, *Phys. Rev.* **163**, 261.
De Wit, M., Reinberg, A. R., Holton, W. C., and Estle, T. L., 1965, *Bull. Am. Phys. Soc.* **10**, 329.
Dietz, R. E., Kamimura, H., Sturge, M. D., and Yariv, A., 1963, *Phys. Rev.* **132**, 1559.
Dischler, B., Räuber, A., and Schneider, J., 1964, *Phys. Stat. Sol.* **6**, 507.
Dunitz, J. D., and Orgel, L. E., 1957, *J. Phys. Chem. Solids* **3**, 20.
Elkin, E. L., and Watkins, G. D., 1968, *Phys. Rev.* **174**, 881.
Englman, R., and Halperin, D., 1970, *Phys. Rev.* **B2**, 75.
Englman, R., and Horn, D., 1963, in *Paramagnetic Resonance*, ed. by W. Low (Academic Press, New York), Vol. I, p. 329.
Estle, T. L., and Holton, W. C., 1966, *Phys. Rev.* **150**, 159.
Estle, T. L., Walters, G. K., and de Wit, M., 1963, in *Paramagnetic Resonance*, ed. by W. Low (Academic Press, New York), Vol. I, p. 144.
Fletcher, J. R., 1967, Ph. D. Thesis, University of Nottingham (unpublished).
Fletcher, J. R., and Stevens, K. W. H., 1969, *J. Phys. C* (*Proc. Phys. Soc.*) [2], **2**, 444.
Gelerinter, E., and Silsbee, R. H., 1966, *J. Chem. Phys.* **45**, 1703.
Geschwind, S., and Remeika, J. P., 1962, *J. Appl. Phys.* **33**, 370.
Gill, J. C., 1965, *Proc. Phys. Soc.* (*London*) **85**, 119.
Goodenough, J. B., 1963, *Magnetism and the Chemical Bond* (Interscience, New York).
Gosar, P., and Pirc, R., 1967, in *Magnetic Resonance and Relaxation*, ed. by R. Blinc (North-Holland Publishing Co., Amsterdam).
Gutowsky, H. S., McCall, D. W., and Slichter, C. P., 1953, *J. Chem. Phys.* **21**, 279.
Ham, F. S., 1965, *Phys. Rev.* **A138**, 1727.
Ham, F. S., 1967, in *Optical Properties of Ions in Crystals*, ed. by H. M. Crosswhite and H. W. Moos (John Wiley and Sons, New York), p. 357.
Ham, F. S., 1968, *Phys. Rev.* **166**, 307.
Ham, F. S., Ludwig, G. W., Watkins, G. D., and Woodbury, H. H., 1960, *Phys. Rev. Letters* **5**, 468.
Ham, F. S., and Schwarz, W. M., 1969 (unpublished).
Ham, F. S., Schwarz, W. M., and O'Brien, M. C. M., 1969, *Phys. Rev.* **185**, 548.
Hayes, W., and Twidell, J. R., 1963, *Proc. Phys. Soc.* (*London*) **82**, 330.
Hayes, W., and Wilkens, J., 1964, *Proc. Roy. Soc.* (*London*) **A281**, 340.
Herzberg, G., 1966, *Electronic Spectra and Electronic Structure of Polyatomic Molecules* (Van Nostrand, Princeton).
Hobey, W. D., and McLachlan, A. D., 1960, *J. Chem. Phys.* **33**, 1695.

Höchli, U. T., 1966, *Bull. Am. Phys. Soc.* **11**, 203.
Höchli, U. T., 1967, *Phys. Rev.* **162**, 262.
Höchli, U. T., and Estle, T. L., 1967, *Phys. Rev. Letters* **18**, 128.
Höchli, U., and Müller, K. A., 1964a, *Phys. Rev. Letters* **12**, 730.
Höchli, U., and Müller, K. A., 1964b, *Phys. Rev. Letters* **13**, 565.
Höchli, U., Müller, K. A., and Wysling, P., 1965, *Phys. Letters* **15**, 5.
Holton, W. C., de Wit, M., Watts, R. K., Estle, T. L., and Schneider, J., 1969, *J. Phys. Chem. Solids.* **30**, 963.
Hougen, J. T., 1964, *J. Mol. Spectry.* **13**, 149.
Hudson, A., 1966, *Mol. Phys.* **10**, 575.
Hughes, A. E., 1970, *J. Phys. C.* (*Proc. Phys. Soc.*) [2], **3**, 627.
Inoue, M., 1963, *Phys. Rev. Letters* **11**, 196.
Jahn, H. A., 1938, *Proc. Roy. Soc.* (*London*) **A164**, 117.
Jahn, H. A., and Teller, E., 1936, *Phys. Rev.* **49**, 874.
Jahn, H. A., and Teller, E., 1937, *Proc. Roy. Soc.* (*London*) **A161**, 220.
Joyce, R. R., and Richards, P. L., 1969, *Phys. Rev.* **179**, 375.
Kamimura, H., and Mizuhashi, S., 1968, *J. Appl. Phys.* **39**, 684.
Kask, N. E., Kornienko, L. S., Mandelshtam, T. S., and Prokhorov, A. M., 1963, *Fiz. Tverd. Tela* **5**, 2306 [English Transl.: *Soviet Phys.—Solid State* **5**, 1677 (1964)].
Koloskova, N. G., and Kochelaev, B. I., 1967, *Fiz. Tverd. Tela* **9**, 2948 [English Transl.: *Soviet Phys.—Solid State* **9**, 2317 (1968)].
Kornienko, L. S., and Prokhorov, A. M., 1960, *Zh. Eksperim. i. Teor. Fiz.* **38**, 1651 [English Transl.: *Soviet Phys.—JETP* **11**, 1189 (1960)].
Koster, G. F., 1958, *Phys. Rev.* **109**, 227.
Koster, G. F., Dimmock, J. O., Wheeler, R. G., and Statz, H., 1963, *Properties of the Thirty-Two Point Groups* (M.I.T. Press, Cambridge).
Koster, G. F., and Statz, H., 1959, *Phys. Rev.* **113**, 445.
Krupka, D. C., and Silsbee, R. H., 1966, *Phys. Rev.* **152**, 816.
LaCroix, R., Höchli, U., and Müller, K. A., 1964, *Helv. Phys. Acta.* **37**, 627.
Lee, K. P., 1969, Ph. D. Thesis, McGill University (unpublished).
Lee, K. P., and Walsh, D., 1968, *Phys. Letters* **27A**, 17.
Lewis, H. R., and Sabiskey, E. S., 1963, *Phys. Rev.* **130**, 1370.
Liehr, A. D., 1960, *Bell System Tech. J.* **39**, 1617.
Liehr, A. D., 1961, *Z. Naturforsch.* **16a**, 641.
Liehr, A. D., 1963, *J. Phys. Chem.* **67**, 389.
Liehr, A. D., and Ballhausen, C. J., 1958, *Ann. Phys.* (*N.Y.*) **3**, 304.
Lohr, L. L., Jr., 1967, *Inorg. Chem.* **6**, 1890.
Longuet-Higgins, H. C., 1961, in *Advances in Spectroscopy*, ed. by H. W. Thompson (Interscience, New York), Vol. II, p. 429.
Longuet-Higgins, H. C., Opik, U., Pryce, M. H. L., and Sack, R. A., 1958, *Proc. Roy. Soc.* (*London*) **A244**, 1.
Loubser, J. H. N., and van Rynevald, W. P., 1967, *Brit. J. Appl. Phys.* **18**, 1029.
Low, W., 1958, *Phys. Rev.* **109**, 256.
Low, W., and Rosenthal, A., 1968, *Phys. Letters* **26A**, 143.
Low, W., and Suss, J. T., 1963, *Phys. Letters* **7**, 310.
Low, W., and Weger, M., 1960, *Phys. Rev.* **118**, 1119, 1130; **120**, 2277.
Ludwig, G. W., and Woodbury, H. H., 1959, *Phys. Rev.* **113**, 1014.

Ludwig, G. W., and Woodbury, H. H., 1962, in *Solid State Physics*, edited by F. Seitz, and D. Turnbull (Academic Press, New York), Vol. 13, p. 223.
Macfarlane, R. M., Wong, J. Y., and Sturge, M. D., 1968, *Phys. Rev.* **166**, 250.
Marshall, F. G., and Rampton, V. W., 1968, *J. Phys. C* (*Proc. Phys. Soc.*) [2], **1**, 594.
McConnell, H. M., 1956, *J. Chem. Phys.* **25**, 709.
McConnell, H. M., 1961, *J. Chem. Phys.* **34**, 13.
McConnell, H. M., and McLachlan, A. D., 1961, *J. Chem. Phys.* **34**, 1.
McLachlan, A. D., 1961, *Mol. Phys.* **4**, 417.
McMahon, D. H., and Silsbee, R. H., 1964, *Phys. Rev.* **135**, A91.
Menne, T. J., 1969, *Phys. Rev.* **180**, 350.
Merritt, F. R., and Sturge, M. D., 1966, *Bull. Am. Phys. Soc.* **11**, 202.
Mizuhashi, S., 1969, *J. Phys. Soc.* (*Japan*) **26**, 468.
Moffitt, W., and Liehr, A. D., 1957, *Phys. Rev.* **106**, 1195.
Moffitt, W., and Thorson, W., 1957, *Phys. Rev.* **108**, 1251.
Moffitt, W., and Thorson, W., 1958, in *Calcul des Fonctions d'Onde Moleculaire*, ed. by R. Daudel (Rec. Mem. Centre Natl. Rech. Sci., Paris), p. 141.
Morigaki, K., 1963, *J. Phys. Soc.* (*Japan*) **18**, 733.
Morigaki, K., 1964a, *J. Phys. Soc.* (*Japan*) **19**, 187.
Morigaki, K., 1964b, *J. Phys. Soc.* (*Japan*) **19**, 1240.
Morigaki, K., and Hoshina, T., 1967, *J. Phys. Soc.* (*Japan*) **23**, 820.
Müller, K. A., 1967, in *Magnetic Resonance and Relaxation*, ed. by R. Blinc (North-Holland Publishing Co., Amsterdam), p. 192.
Mulliken, R. E., 1933, *Phys. Rev.* **43**, 279.
Nelson E. D., Wong, J. Y., and Schawlow, A. L., 1967, *Phys. Rev.* **156**, 298.
O'Brien, M. C. M., 1964, *Proc. Roy. Soc.* (*London*) **A281**, 323.
O'Brien, M. C. M., 1965, *Proc. Phys. Soc.* (*London*) **86**, 847.
O'Brien, M. C. M., 1970, *Phys. Rev.* **187**, 407.
O'Connor, J. R., and Chen, J. H., 1964, *Appl. Phys. Letters* **5**, 100.
Opik, U., and Pryce, M. H. L., 1957, *Proc. Roy. Soc.* (*London*) **A238**, 425.
Orbach, R., 1961, *Proc. Roy. Soc.* (*London*) **A264**, 458.
Orbach, R., and Simanek, E., 1967, *Phys. Rev.* **158**, 310.
Orton, J. W., Auzins, P., Griffiths, J. H. E., and Wertz, J. E., 1961, *Proc. Phys. Soc.* (*London*) **78**, 554.
Owen, J. and Thornley, J. H. M., 1966, *Rep. Progr. Phys.* **29**, 675 (London: Physical Society).
Pilbrow, J. R., and Spaeth, J. M., 1967, *Phys. Stat. Sol.* **20**, 225, 237.
Pirc, R., Zeks, B., and Gosar, P., 1966, *J. Phys. Chem. Solids* **27**, 1219.
Rai, R., Savard, J. Y., and Tousignant, B., 1967, *Phys. Letters* **25A**, 443.
Rai, R., Savard, J. Y., and Tousignant, B., 1969, *Can. J. Phys.* **47**, 1147.
Reinberg, A. R., and Estle, T. L., 1967, *Phys. Rev.* **160**, 263.
Renner, R., 1934, *Z. Phys.* **92**, 172.
Rubins, R. S., and Low, W., 1963, in *Paramagnetic Resonance*, ed. by W. Low (Academic Press, New York), Vol. I, p. 59.
Schawlow, A. L., Piksis, A. H., and Sugano, S., 1961, *Phys. Rev.* **168**, 341.
Schirmer, O. F., Müller, K. A., and Schneider, J., 1965, *Phys. Kondens. Materie* **3**, 323.
Schneider, J., 1967, in *II–VI Semiconducting Compounds*, ed. by D. G. Thomas (W. A. Benjamin, New York), p. 40.
Schneider, J., Dischler, B., and Rauber, A., 1967, *Sol. State Comm.* **5**, 603.

Schneider, J., Holton, W. C., Estle, T. L., and Räuber, A., 1963, *Phys. Letters* **5**, 312.
Scott, W. C., and Sturge, M. D., 1966, *Phys. Rev.* **146**, 262.
Shuey, R. T., and Zeller, H. R., 1967, *Helv. Phys. Acta* **40**, 873.
Shul'man, L. A., Zaritskii, I. M., and Podzyarei, G. A., 1966, *Fiz. Tverd. Tela.* **8**, 2307 [English Transl.: *Soviet Phys.—Solid State* **8**, 1842 (1967)].
Shuskus, A. J., 1964, *J. Chem. Phys.* **40**, 1602.
Sierro, J., 1967, *J. Phys. Chem. Solids* **28**, 417.
Silsbee, R. H., 1965, *Phys. Rev.* **138**, A180.
Silsbee, R. H., 1967, *J. Phys. Chem. Solids* **28**, 2525.
Simanek, E., and Orbach, R., 1966, *Phys. Rev.* **145**, 191.
Slonczewski, J. C., 1963, *Phys. Rev.* **131**, 1596.
Slonczewski, J. C., 1969, *Solid State Comm.* **7**, 519.
Slonczewski, J. C., and Moruzzi, V. L., 1967, *Physics* **3**, 237.
Smith, W. V., Sorokin, P. P., Gelles, I. L., and Lasher, G. J., 1959, *Phys. Rev.* **115**, 1546.
Stephens, P. J., and Lowe-Pariseau, M., 1968, *Phys. Rev.* **171**, 322.
Stevens, K. W. H., 1967, *Rep. Progr. Phys.* **30**, 189 (London: Physical Society).
Stevens, K. W. H., and Persico, F., 1966, *Nuovo Cimento B*, **41**, 37.
Stoneham, A. M., 1965, *Proc. Phys. Soc. (London)* **85**, 107.
Stoneham, A. M., 1966, *Proc. Phys. Soc.* **89**, 909.
Stoneham, A. M., and Lannoo, M., 1969, *J. Phys. Chem. Solids* **30**, 1769.
Sturge, M. D., 1963, *Phys. Rev.* **131**, 1456.
Sturge, M. D., 1967, in *Solid State Physics*, ed. by F. Seitz, D. Turnbull, and H. Ehrenreich (Academic Press, New York), Vol. 20, p. 91.
Sturge, M. D., Krause, J. T., Gyorgy, E. M., LeCraw, R. C., and Merritt, F. R., 1967, *Phys. Rev.* **155**, 218.
Sussman, J. A., 1967, *J. Phys. Chem. Solids* **28**, 1643.
Thornley, J. H. M., Mangum, B. W., Griffiths, J. H. E., and Owen, J., 1961, *Proc. Phys. Soc. (London)* **78**, 1263.
Townsend, M. G. and Weissman, S. I., 1960, J. Chem. Phys. **32**, 309.
Toyozawa, Y., and Inoue, M., 1966, *J. Phys. Soc. (Japan)* **21**, 1663.
Tucker, R. F., 1958, *Phys. Rev.* **112**, 725.
Van Eekelen, H. A. M., and Stevens, K. W. H., 1967, *Proc. Phys. Soc. (London)* **90**, 199.
Van Vleck, J. H., 1937, *Phys. Rev.* **52**, 246.
Van Vleck, J. H., 1939a, *J. Chem. Phys.* **7**, 61.
Van Vleck, J. H., 1939b, *J. Chem. Phys.* **7**, 72.
Van Vleck, J. H., 1940, *Phys. Rev.* **57**, 426.
Van Vleck, J. H., 1958, *Disc. Faraday Soc.* **26**, 96.
Van Vleck, J. H., 1960, *Physica* **26**, 544.
Walsh, W. M., Jeener, J., and Bloembergen, N., 1965, *Phys. Rev.* **139**, A1338.
Washimiya, S., 1963, Paper presented at the autumn meeting of the Physical Society of Japan [unpublished; referred to by S. Washimiya and K. Gondaira, *J. Phys. Soc. (Japan)* **23**, 1 (1967)].
Watkins, G. D., 1963, *J. Phys. Soc. (Japan)* **18**, Suppl. II, 22.
Watkins, G. D., 1965, in *Effets des Rayonnements sur les Semiconducteurs* (Dunod, Paris), p. 97.
Watkins, G. D., and Corbett, J. W., 1961, *Phys. Rev.* **121**, 1001.
Watkins, G. D., and Corbett, J. W., 1964, *Phys. Rev.* **134**, A1359.

Weinstock, B., and Goodman, G. L., 1965, in *Advances in Chemical Physics*, ed. by I. Prigogine (Interscience, New York), Vol. 9, p. 169.

Williams, F. I. B., Krupka, D. C., and Breen, D. P., 1969, *Phys. Rev.* **179**, 255.

Wong, J. Y., 1968, *Phys. Rev.* **168**, 337.

Wong, J. Y., Berggren, M. J., and Schawlow, A. L., 1968, *J. Chem. Phys.* **49**, 835.

Wysling, P., and Müller, K. A., 1968, *Phys. Rev.* **173**, 327.

Zaitov, M. M., and Stepanov, V. G., 1968, *Fiz. Tverd. Tela* **10**, 3178 [English Transl.: *Soviet Phys.—Solid State* **10**, 2520 (1969)].

Zaripov, M. M., Kropotov, V. S., Livanova, L. D., and Stepanov, V. G., 1967, *Fiz. Tverd. Tela* **9**, 209, 2983, 2984 [English Transl.: *Soviet Phys.—Solid State* **9**, 992, 2344, 2346 (1967–8)].

Zdansky, K., 1969, *Phys. Rev.* **177**, 490.

Chapter 2

Electron Spin–Lattice Relaxation

R. Orbach*

Physics Department
University of California
Los Angeles, California

and

H. J. Stapleton

Physics Department
University of Illinois
Urbana, Illinois

1. INTRODUCTION

The problem of thermal contact between a paramagnetic spin system and a heat reservoir is central to the study of magnetic resonance. Without such contact, energy from the radiation field would soon heat the spin system to an infinite temperature, after which no further absorption would take place. The introduction of a coupling to a reservoir consisting of lattice phonons was worked out in some detail by Waller[1] well before any experiments which could detect energy loss from the paramagnetic spin system had been performed, let alone magnetic resonance. It remained for Gorter† and his school to demonstrate the presence of (Zeeman) energy transfer between the spin system and the lattice using nonresonant techniques, and then for pulsed microwave methods over the past ten years to examine the detailed dynamics of relaxation processes.

* Supported in part by U.S. Office of Naval Research [NONR 233(88)] and the National Science Foundation.

† Gorter[2] has given an excellent review of the theoretical basis of nonresonant absorption methods and of experiments using this technique.

Waller[1] considered energy transfer between the spin system and the lattice (i.e., spin–lattice relaxation) to occur when lattice vibrations (phonons) caused the static dipolar interaction to fluctuate. Explicitly, the distance between the (dipolar) interacting paramagnetic ions is expanded about the equilibrium distance in powers of the strain tensor. The first term in the expansion, linear in the strain, gives rise to single-phonon emission or absorption and a relaxation process called the "direct" process. The spin system changes its energy by a unit of Zeeman energy, $\hbar\omega_0$, the lattice energy changing by the opposite amount to ensure conservation of energy. Because the phonons that can participate in the energy transfer process must have energies equal to the Zeeman energy, they are often referred to as "speaking term" phonons.[3] The direct process leads to a linear dependence of the relaxation rate on temperature T when $T \gg \hbar\omega_0/k$ [more correctly, as $\coth(\hbar\omega_0/2kT)$], and for dipole–dipole modulation, a quadratic dependence on the Zeeman splitting. Because the single-spin Zeeman energy, for reasonable magnetic fields, is never more than a few degrees Kelvin in magnitude, only the low-energy tail of the phonon spectrum induces this process. For a simple solid, the phonon density of states in this energy region is proportional to the square of the phonon energy. Hence, only a small part of the phonon spectrum, with an even smaller density of states compared to the phonons predominant in number (with energies of the order of kT) can be effective. For this reason, at higher temperatures, Waller[1] introduced a Raman-like (two-phonon) process arising from expansion of the interspin distance to second order in the lattice strain, leading to inelastic two-phonon emission and absorption processes. A phonon is assumed to scatter off of a paramagnetic ion, the difference between initial and final phonon energies equaling the change in spin Zeeman energy. This process allows the use of phonons with thermal energies, their abundance (compared to the speaking-term phonons) more than making up for the necessity of going to second order in the strain expansion. The Raman process, again for dipole–dipole modulation, leads to a relaxation rate at temperatures below the Debye temperature proportional to the seventh power of temperature, the dependence on temperature reduces to quadratic.

The theory of Waller was directly applicable to the nonresonant experiments of Gorter,[2] but it was found that the magnitudes of relaxation rates observed experimentally were considerably greater than those predicted by him (approximately seven orders of magnitude!). Heitler and Teller[4] and Fierz[5] suggested an alternate mechanism, one which in fact appears to be dominant in almost every case of spin–lattice relaxation in solids known to the authors.* They pointed out that in an ionic solid, the paramagnetic ion is subjected to intense electric fields caused by the neighboring charged ions (or ligands). These fields, more specifically their multipole expansion, result in a Stark splitting of the orbital levels of the paramagnetic electrons. This crystalline field splitting is much larger than the dipolar interaction, and can itself be modulated by vibrations of the spin–ligand distance. However, because the perturbation is electrical in origin, it can act only indirectly on the true spin component of the paramagnetic moment, through the action of the spin–orbit coupling. This indirect character of the electrostatic field reduces its magnitude to such an extent that in the case of *S*-state ions (e.g., Mn^{++}, Gd^{3+}), it is not inconceivable that magnetic interactions (e.g., dipolar or hyperfine field modulation) might be important.[7] Modulation of electric fields also produces another difficulty. According to Kramers,[8] matrix elements of an electrostatic field vanish between time-reversed states of half-integral spin (even in the presence of orbital moment). Hence, for ions with an odd number of electrons in sites of such a symmetry that only a doublet lies lowest, it appeared that even this mechanism would be ineffective for spin–lattice relaxation. Kronig[9] noted that this difficulty could be removed by the action of an external magnetic field. The field mixes the ground doublet with excited (crystal-field split) states, and finite matrix elements of the electrostatic potential could obtain between levels of different time-reversed doublets. This necessity of admixtures increases the dependence of the relaxation rate on the magnetic field by two powers, leading to a direct process and a Raman process proportional to the fourth and second power, respectively, of the Zeeman splitting, with the same dependence on temperature as found before by Waller.

Again, the relaxation rates predicted by Kronig were too slow, especially in the liquid nitrogen range. Van Vleck,[10] in two related papers, noted that the presence of yet another relaxation term would obviate the need for magnetic field admixtures in the Raman regime. This process relies on the dynamics of phonon emission and absorption. Kramers's argument is appropriate to electrostatic fields (relying on time-reversal symmetry), but the finite frequency of the thermal phonons leads to a time-dependent term which "breaks" the selection rule. Thus, in place of Zeeman admixtures, the time-dependent phonon field itself admixes excited states to the ground doublet, and extinction of the matrix element is avoided to order

* Of course there are notable exceptions which obtain for spins having very little orbital moment present in the ground state. Thus phonon modulation of the transferred hyperfine field is now thought responsible for the spin–lattice relaxation of *F* centers. (See Deigen and Zevin.[6])

the phonon frequency divided by the excited-state crystalline field separation from the ground doublet. This (Van Vleck) cancellation leads to two additional powers of temperature, so that Van Vleck[10] obtains a Raman relaxation rate proportional to the ninth power of the temperature and independent of Zeeman splitting for relaxation transitions between time-reversed states of half-integral spin. This result is to be compared with that of Kronig,[9] who invoked the Zeeman interaction to break the cancellation. The relaxation rate he obtained was proportional to the second power of the Zeeman interaction and the seventh power of the temperature. Van Vleck's[10] and Kronig's[9] results for Kramers transition rates were proportional to non-Kramers transition rates times the diminution factors $(kT/\Delta)^2$ and $(\nu\omega_0/\Delta)^2$, respectively, where Δ is the energy to the admixed level. The two-phonon Raman process is effective generally when $kT > \hbar\omega_0$, so that it is not surprising that no unequivocal experimental evidence for Kronig's expression has been obtained. Of course, should the direct process be inhibited (see below), and should $\hbar\omega_0$ exceed kT, it could be the case that Kronig's result would become significant. The Van Vleck theory was generalized to ions with large amounts of orbital moment present in their ground state by Mattuck and Strandberg.[11] Though their formulation purports to be more general than that of Van Vleck,[10] in fact, explicit use of their formulas demonstrates that Van Vleck did find all the terms which are nonzero for the cases he considered.

There were still many discrepancies, however, between the experimental magnitudes and temperature dependences of the relaxation rates and the predictions of Van Vleck's theory. Today, many of the experimental numbers have changed (!) in such a way as to yield agreement with Van Vleck's original estimates, especially when care is taken to work with dilute spin systems in order to avoid effects of spin–spin ("cross") relaxation. There were still examples prior to 1960, especially in rare earth salts, which could not be attributed to spin–spin interactions. These were explained in part by Finn *et al.*,[12] again by examining the two-phonon process. In this case, it was noted that a low-lying excited level (common to rare earth salts) could cause a resonance in the Raman integral which would dominate the usual nonresonant two-phonon relaxation process. This resonance requires the presence of phonons whose energy is equal to the ground-state–excited-state crystalline field splitting, and therefore has an "activation"-like temperature dependence. The relaxation process is field-independent, and its rapid variation with temperature was found to explain a number of experiments in the rare earth and iron group series.[13] A further calculation[14] demonstrated that, within a multiplet of closely spaced levels (multiplicity greater than two), yet another temperature dependence for the Raman relaxation rate resulted. This process is the "inverse" of the Van Vleck cancellation, leading to a Raman-process relaxation rate proportional to the fifth power of temperature.

The multiplicity of relaxation processes, many of which are specific to a particular spin species or arrangement of crystalline field levels, requires a formulation which is sufficiently general to cope with a wide variety of circumstances. Because the basic interaction is now recognized to be a phonon modulation of ion–ligand potentials, it seems appropriate to construct an approach from a viewpoint similar to that of crystalline field theory. That is, it is useful* to utilize the point group symmetry of the paramagnetic ion in order to reduce the number of unknowns in the dynamic interaction of spins with phonons, exactly as is done for the rigid lattice in the usual crystalline field theory.[15] Such an approach was (again!) first put forward by Van Vleck[10] and then expanded somewhat be a number of authors.[†,17a–21] Basically, the technique is as follows. It is assumed‡ that only the nearest ligand neighbors contribute to the crystalline potential acting on the paramagnetic electrons. The normal modes of vibration (perforce even under inversion because of parity arguments) of the resulting "cluster" are expanded in terms of the phonons of the host crystal, and the relevant matrix elements and angular averages taken. The utility of this approach, beyond the question of minimizing the number of dynamic potential parameters, is that, upon summing over the phonon states, one is allowed to take advantage of the orthogonality between different normal modes to reduce the number of terms. This will be demonstrated in Sections 2 and 3 and represents the principal reason for normal mode expansion. If, however, the assumption of near neighbors is inappropriate, as it undoubtedably will be in many situations,[23] one must include more and more ligands in the cluster, complicating the normal mode analysis, and making

* Though by no means is it necessary. To appreciate the complexity of not doing so, see the paper by Micher[15] on the two-phonon process, and compare with the group-theoretic approach of Bridges.[51]

† The paper by Blume and Orbach[17b] is marred by a sign error in the quantity γ which measures the amount of ^4G character mixed into the $^4T_4$ excited level. Thus, its final results have the wrong sign. When corrected, they agree in sign with the experimental values for the orbit lattice coefficients as measured for Mn:MgO, though somewhat smaller in magnitude.

‡ This assumption is certainly in error for the $l = 2$ components of the orbit–lattice interaction. Such terms fall off only as $1/R^3$, and thus receive contributions from much more distant neighbors. See Section 2 of this chapter for a more complete discussion, as well as Ref. 23.

the analogy with static crystal-field potentials almost useless. This is probably the case for rare earth ions, but the dominance of covalency in the static crystalline field for iron group ions (which is important only for near-neighbor ligands) may well rescue the near-neighbor approach for such systems.

When the explicit dynamic-potential argument described above is thought to fail, it is still possible to use the point group symmetry of the ion to usefully describe the interaction of the paramagnetic ion with the host lattice phonons. A dynamic spin Hamiltonian can be introduced[19–21] which has the same analogy to the static crystalline field potential. Thus, in terms of an appropriate fictitious spin for the ground multiplet, an interaction Hamiltonian can be written down which has transformation properties appropriate to the point group symmetry of the paramagnetic ion and the associated ground level degeneracy. This method has all the advantages, and disadvantages, associated with the static spin Hamiltonian. In particular, one introduces parameters which describe the dynamic Hamiltonian appropriate to (only) the ground levels. The angular dependence of the various relaxation processes[20–22,24,25] can be found easily, and various relaxation processes can be described with a minimum number of parameters. It should be emphasized that these parameters are peculiar to the ground state in question, and are much fewer in number than the parameters describing the dynamic crystalline field potential which gives rise to them. Of course, an identical situation obtains for a comparison of the number of static crystalline field and spin-Hamiltonian parameters. The advantages of the spin-Hamiltonian description of the spin–phonon interaction are sufficiently attractive to have been used in a variety of contexts. Thus, static pressure and phonon absorption experiments have been performed to obtain the first-order spin-phonon Hamiltonian coupling constants.[23,26–29] One can use these spin–phonon constants to calculate the relaxation time T_1 and a comparison between these calculated T_1's and experiment will be made in Section 5. It is clear that in the future these spin–phonon Hamiltonian coupling constants will be tabulated in the same manner as the static spin-Hamiltonian parameters.[30] It is, of course, possible to compare these dynamic quantities with the predictions of theory[17a,17b,23,28,31–34] in the same manner as, for example, Abragam and Pryce[35] dealt with the static case. Future work should provide impetus for more calculations of the dynamic parameters based on covalency, in the same manner that the static crystalline field splitting has been treated (though this has been done already for some iron group ions.[31,33,34]) Following this brief survey, we now turn our attention to a more detailed treatment of these topics.

2. ORBIT–LATTICE COUPLING HAMILTONIAN

2.1. Construction of the Orbit–Lattice Coupling Hamiltonian

The steps to be followed in the calculation of the spin–lattice relaxation time were outlined in Section 1. We begin the calculation by constructing the orbit–lattice coupling Hamiltonian. It is simplest to begin with an octahedron of ligands surrounding the paramagnetic ion in question. This configuration has been studied extensively by Van Vleck,[10] and is pictured in Fig. 1. The static Hamiltonian for the overall system consists of three parts,

$$\mathscr{H} = \mathscr{H}_L + \mathscr{H}_{\mathrm{FI}} + \mathscr{H}_{\mathrm{CF}} \tag{2.1}$$

where $\mathscr{H}_L$ is the lattice Hamiltonian with eigenvalues

$$E_L = \sum_{\mathbf{k},s} \hbar\omega(\mathbf{k}, s)(n_{\mathbf{k},s} + \tfrac{1}{2}) \tag{2.2}$$

Here, $\omega(\mathbf{k}, s)$ is the angular frequency for a phonon of wave vector $\mathbf{k}$ and polarization index s. In (2.2), $n_{\mathbf{k},s}$ is the usual Bose factor, $\{\exp[\hbar\omega(\mathbf{k}, s)/kT] - 1\}^{-1}$, and k is the Boltzmann constant. The second term, $\mathscr{H}_{\mathrm{FI}}$, is the

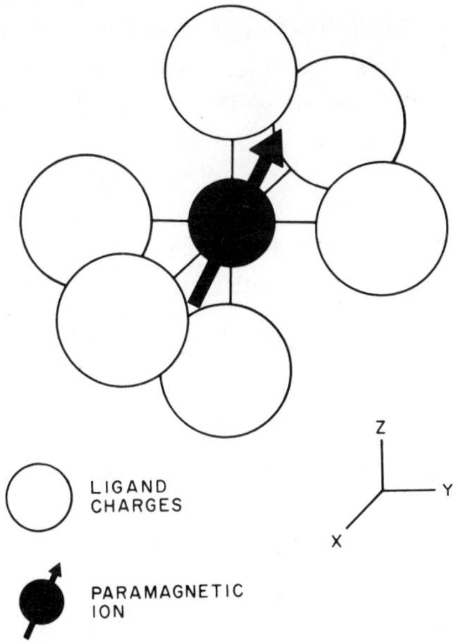

Fig. 1. Paramagnetic ion in sixfold octahedral ligand coordination.

free paramagnetic ion Hamiltonian, which, for our purposes, has eigenstates given in terms of Russell–Saunders coupling, with total spin $\mathbf{S} = \sum_i \mathbf{s}_i$ and orbit $\mathbf{L} = \sum_i l_i$. The lowercase quantities are one-electron operators, and the sum is over all electrons i. The last term in (2.1) is the static crystalline field potential. In using an electrostatic model, one is forced to neglect any overlap between the paramagnetic ion and the ligand wave functions. $\mathscr{H}_{\text{CF}}$ is evaluated by expanding the Coulomb potential between an electron at r and an infinitesimal volume d^3R of the surrounding environment at $\mathbf{R}$, with charge density $\varrho(\mathbf{R})$,

$$V(\mathbf{r}) = -\int [e\varrho(\mathbf{R})/|\,\mathbf{r} - \mathbf{R}\,|]\, d^3\mathbf{R} \tag{2.3}$$

One uses[36]

$$1/|\,\mathbf{r} - \mathbf{R}\,| = \sum_{l=0}^{\infty} (r_<^l/r_>^{l+1}) P_l(\cos\omega)$$

where ω is the angle between $\mathbf{r}$ and $\mathbf{R}$. The Legendre function is then itself expanded by use of the spherical harmonic addition theorem,

$$P_l(\cos\omega) = \sum_{m=-l} C_l^m(\mathbf{r}) C_l^{m*}(\mathbf{R})$$

where the $C_l^m(\mathbf{r}) = [4\pi/(2l+1)]^{1/2} Y_l^m(\mathbf{r})$ are the Racah normalized spherical harmonics. The arguments of the C_l^m in the above expansion refer to the angular coordinates of the electron at $\mathbf{r}$ and the environment at $\mathbf{R}$, respectively. Assuming that the charge density is finite only outside of the region where the electronic wave function is finite, (2.3) simplifies and we get

$$V(\mathbf{r}) = -\sum_l r^l C_l^m(\mathbf{r}) \left\{ e \int [\varrho(\mathbf{R}) C_l^{m*}(\mathbf{R})/R^{l+1}]\, d^3\mathbf{R} \right\} \tag{2.4}$$

Sandwiching (2.4) between the orbital part of the wave function of the paramagnetic electron, the Coulomb interaction (2.3) can be written in terms of a potential acting on the paramagnetic electrons produced by the external environment. Thus, (2.4) reduces to

$$\mathscr{H}_{\text{CF}} = -\sum_{i,l} \langle r_i^l \rangle C_l^m(\mathbf{r}_i) \left\{ e \int [\varrho(\mathbf{R}) C_l^{m*}(\mathbf{R})/R^{l+1}]\, d^3\mathbf{R} \right\} \tag{2.5}$$

It is immediately seen that a similar result could have been obtained without resort to calculation by looking for those spherical harmonics that transform as the identity representation of the point symmetry group appropriate to the paramagnetic ion and its surroundings. A glance at (2.5), however, together with the triangle rule, indicates that it is unnecessary to consider spherical harmonics with $l > 2l_i$, where l_i is the orbital moment of the

ith paramagnetic electron. Thus, for d electrons in the center of an octahedron of charges, the cubic field can be written down immediately as[16]

$$\mathscr{H}_{\mathrm{CF}}^{(C)} = A_4^{\,0}\langle r^4\rangle[C_4^{\,0} + (5/14)^{1/2}(C_4^{\,4} + C_4^{-4})] \tag{2.6}$$

with an additional term for f electrons,[16]

$$A_6^{\,0}\langle r^6\rangle[C_6^{\,0} - (7/2)^{1/2}(C_6^{\,4} + C_6^{-4})] \tag{2.7}$$

The coefficients for an octahedron of charges can be found from (2.5) and equal

$$A_4^{\,0} = 14\pi e^2/9R^5 \tag{2.6$'$}$$

and

$$A_6^{\,0} = 3\pi e^2/13R^7 \tag{2.7$'$}$$

To obtain the orbit–lattice coupling Hamiltonian, $\mathscr{H}_{\mathrm{OL}}$, the cluster in Fig. 1 is allowed to vibrate. There are 21 (7 ions $\times$ 3 degrees of freedom each) degrees of freedom, of which 6 must be subtracted out as being pure translation or rotation.* The remaining 15 can be grouped as follows:

$$\begin{array}{ll} \Gamma_{1g}\ (1) & \\ & 2\Gamma_{4u}\ (3) \\ \Gamma_{3g}\ (2) & \\ & \Gamma_{5u}\ (3) \\ \Gamma_{5g}\ (3) & \end{array} \tag{2.8}$$

where the number in parantheses refers to the degeneracy (number of subvectors) for that representation. We usually wish matrix elements of the potential produced by the vibrations (2.8) between states of the same electronic configuration. Hence, we can discard the odd [right-hand side of (2.8)] vibrations and concentrate only on the even vibrations. They are described pictorially in Fig. 2, where an algebraic representation of the subvectors in (2.8) is given.

We next imagine the vibrations grouped in (2.8) to be classified into normal modes $Q(\Gamma_{ig}, m)$, where i refers to the irreducible representation

* Though in the case of a static trigonal distortion, even uniform cluster rotation can lead to relaxation. See Ref. 10. More recently, this question has been reopened for noncubic sites by R. L. Melcher, *Phys. Rev. Letters,* **28**, 165 (1972), and shown to be quantitatively significant. An extension to cubic sites, with a detailed application to direct and "skew" transitions for hyperfine split states, has been made by A. Abragam, J. F. Jacquinot, M. Chapellier, and M. Goldman, *J. Phys. C (London)*, to be published, 1972.

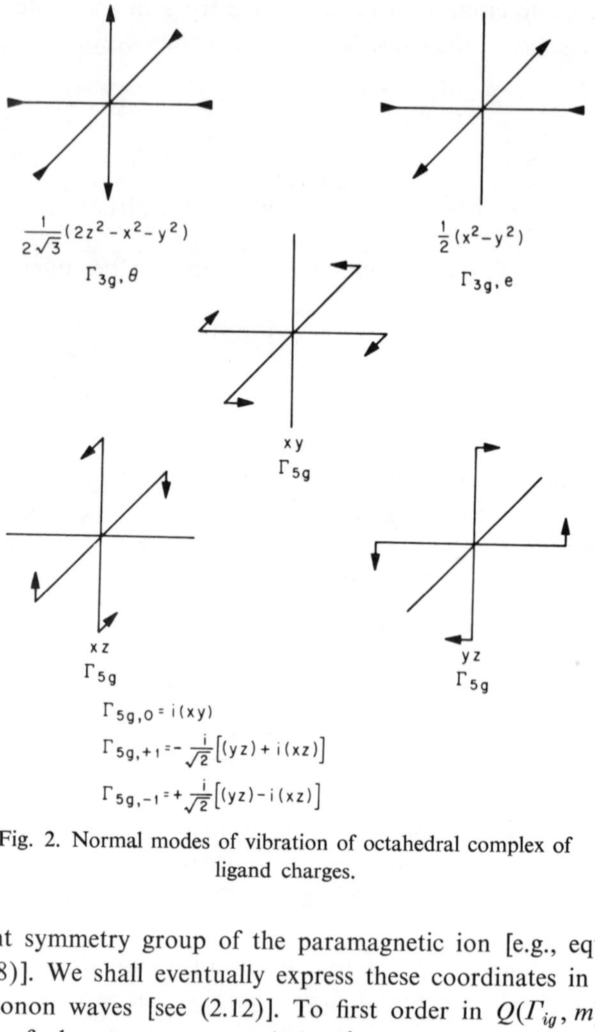

Fig. 2. Normal modes of vibration of octahedral complex of ligand charges.

of the point symmetry group of the paramagnetic ion [e.g., equals 1, 3, or 5 in (2.8)]. We shall eventually express these coordinates in terms of running phonon waves [see (2.12)]. To first order in $Q(\Gamma_{ig}, m)$, for an octahedron of charges, we can write[17a,17b]

$$\mathscr{H}_{\mathrm{OL}} = \sum_{\substack{m=0,e;\\ l=2,4,\ldots}} V(\Gamma_{3g}l)C(\Gamma_{3g}l, m)Q(\Gamma_{3g}, m)$$

$$\sum_{\substack{m=0,\pm 1;\\ l=2,4,\ldots}} V(\Gamma_{5g}l)C(\Gamma_{5g}l, m)Q(\Gamma_{5g}, -m)(-1)^m \qquad (2.9)$$

The $C(\Gamma_{ig}l, m)$ are those linear combinations of the Racah normalized spherical harmonics, summed over all electrons, that transform as the mth subvector of the ith irreducible representation and are given in Eq. (2.10).

We label the subvectors of the Γ_{3g} representation by θ, ε and of the Γ_{5g} representation by $\pm 1, 0$. The $Q(\Gamma_{5g}; \pm 1, 0)$ are the appropriate linear combinations of the Γ_{5g} normal modes shown in Fig. 2 which transform as the $C(\Gamma_{5g}, 2, m)$. The $V(\Gamma_{ig}l)$ have units of energy, and are independent of m (by the Wigner–Eckart theorem).[36] The form of (2.9) is dictated by the requirement that $\mathscr{H}_{\mathrm{OL}}$ transforms as the identity representation of the point symmetry group, i.e., that $\mathscr{H}_{\mathrm{OL}}$ be a scalar. Equation (2.9) is the most general scalar that can be built from the tensor operators C and Q.[36]

Equation (2.9) is clearly the first term in a Taylor series expansion in powers of the normal vibrations $Q(\Gamma_{ig}, m)$. A similar expression to (2.9) (with, however, two Γ_{5g} summations) obtains for eightfold coordination.[18] We have dropped the vibration transforming as Γ_{1g}. It is just the identity vibration and merely shifts all levels by the same amount. Therefore transitions *between* levels are not induced, such transitions being necessary for spin–lattice relaxation. This simplification is not always appropriate for higher-order processes. For example as discussed in Section 4, following a treatment by Walker[37] and Yafet,[38] *two*-phonon processes can obtain which are proportional to the Γ_{1g} matrix elements for one of the two levels in question, times the matrix element of the spin-flip orbit–lattice interaction *between* the same two levels.

We have already reduced the potential to only angular terms in (2.9), so that the $V(\Gamma_{ig}l)$ will be proportional to $\langle r^l \rangle$ exactly as in (2.5), (2.6′), and (2.7′). If the paramagnetic ion is taken to lie at the origin, the various quantities in (2.9) can be written as follows[17a,17b]:

$$\begin{aligned} C(\Gamma_{3g}2, e) &= (C_2^{\,2} + C_2^{-2})/\sqrt{2}\,; \\ C(\Gamma_{3g}2, \theta) &= C_2^{\,0} \\ C(\Gamma_{5g}2, 1) &= C_2^{-1}; \\ C(\Gamma_{5g}2, 0) &= (C_2^{\,2} - C_2^{-2})/\sqrt{2}\,; \\ C(\Gamma_{5g}2, -1) &= -C_2^{\,1} \end{aligned} \tag{2.10}$$

and

$$\begin{aligned} C(\Gamma_{3g}4, e) &= (C_4^{\,2} + C_4^{-2})/\sqrt{2}\,; \\ C(\Gamma_{3g}4, \theta) &= -(5/2\sqrt{3})C_4^{\,0} + (\sqrt{7}/2\sqrt{6})(C_4^{\,4} + C_4^{-4}) \\ C(\Gamma_{5g}4, 1) &= (\sqrt{7}\,C_4^{\,3} - C_4^{-1})/2\sqrt{2} \\ C(\Gamma_{5g}4, 0) &= (C_4^{\,2} - C_4^{-2})/\sqrt{2} \\ C(\Gamma_{5g}4, -1) &= (C_4^{\,1} - \sqrt{7}\,C_4^{-3})/2\sqrt{2} \end{aligned} \tag{2.11}$$

The expressions for $l = 6$, and higher, can be found in standard texts.[36]

For iron group ions, it is only necessary to consider $l = 2,4$; for rare earth ions, $l = 6$ must be included as well. The normal modes of vibration of the surrounding ligand octahedron, the $Q(\Gamma_{ig}, m)$, at the site of the jth paramagnetic ion can be expanded in terms of phonons in the following way[39]:

$$Q_j(\Gamma_{ig}, m) = \sum_{\mathbf{k},s} (i/a)[\hbar/2M\omega(\mathbf{k}, s)]^{1/2}$$
$$\times [b_{\mathbf{k},s}(\exp i\mathbf{k} \cdot \mathbf{R}_j) - b^{*}_{\mathbf{k},s} \exp -i\mathbf{k} \cdot \mathbf{R}_j]R_{\mathbf{k},s}(\Gamma_{ig}, m) \quad (2.12)$$

where

$$R_{\mathbf{k},s}(\Gamma_{3g}, \theta) = \tfrac{1}{2}[2e^z(\mathbf{k}, s)(\sin k^z a) - e^x(\mathbf{k}, s)(\sin k_x a) - e^4(\mathbf{k}, s) \sin k_y a]$$
$$R_{\mathbf{k},s}(\Gamma_{3g}, e) = \tfrac{1}{2}\sqrt{3}\,[e^x(\mathbf{k}, s)(\sin k_x a) - e^y(\mathbf{k}, s) \sin k_y a]$$
$$R_{\mathbf{k},s}(\Gamma_{5g}, +1) = -i\sqrt{\tfrac{3}{2}}\,[e^y(\mathbf{k}, s)(\sinh k_z a) + e^z(\mathbf{k}, s)(\sin k_y a) + ie^z(\mathbf{k}, s)(\sin k_x a) + ie^x(\mathbf{k}, s) \sin k_z a]$$
$$R_{\mathbf{k},s}(\Gamma_{5g}, 0) = i\sqrt{3}\,[e^x(\mathbf{k}, s)(\sin k_y a) + e^y(\mathbf{k}, s) \sin k_x a]$$
$$R_{\mathbf{k},s}(\Gamma_{5g}, -1) = i\sqrt{\tfrac{3}{2}}\,[e^y(\mathbf{k}, s)(\sin k_z a) + e^z(\mathbf{k}, s)(\sin k_y a) - ie^z(\mathbf{k}, s)(\sin k_x a) - ie^x(\mathbf{k}, s) \sin k_z a] \quad (2.13)$$

Here, $e^\alpha(\mathbf{k}, s)$ is the αth component of the unit polarization vector for the phonon of wave vector k and polarization index s, $b_{\mathbf{k},s}$ the destruction operator for the same phonon, M the mass of the entire crystal, and a the lattice constant. The operator quantities $Q(\Gamma_{ig}, m)$ are dimensionless and obey commutation relations easily deriveable from the definition of Q, (2.12), and the boson commutation relations obeyed by $b_{\mathbf{k},s}$:

$$[b^{*}_{\mathbf{k},s}, b_{\mathbf{k}',s'}] = \delta_{\mathbf{k},\mathbf{k}'}, \delta_{s,s'}$$

In general, the phonon spectrum of a real crystal is so complicated that the detailed evaluation of (2.13) is almost impossible (though Donoho[27] has worked it through for ruby in the long-wavelength limit). It is almost always assumed that the crystal can be approximated by an isotropic solid with a linear phonon dispersion curve up to the Debye cutoff k_D. Such an assumption is probably not unreasonable in the direct-process regime, but is very questionable in the two-phonon regime, where the participating phonons have energies $\gtrsim kT$. Nevertheless, since we cannot do anything else without great mathematical difficulty, we shall adopt a long-wavelength approximation and expand (2.13) accordingly. Again, this will only be valid for the direct process (at all temperatures) and the Raman process at sufficiently low temperature (depending on the details of the phonon spectrum). Performing this expansion, (2.13) reduces to

$$\begin{aligned}
R_{\mathbf{k},s}(\Gamma_{3g}, \theta) &= \tfrac{1}{2}a(2e^z k_z - e^x k_x - e^y k_y) \\
R_{\mathbf{k},s}(\Gamma_{3g}, e) &= \tfrac{1}{2}a\sqrt{3}\,(e^x k_x - e^y k_y) \\
R_{\mathbf{k},s}(\Gamma_{5g}, 1) &= -ia\sqrt{\tfrac{3}{2}}\,(e^y k_z + e^z k_y + ie^z k_x + ie^x k_z) \qquad (2.14) \\
R_{\mathbf{k},s}(\Gamma_{5h}, 0) &= ia\sqrt{3}\,(e^x k_y + e^y k_x) \\
R_{\mathbf{k},s}(\Gamma_{5g}, -1) &= ia\sqrt{\tfrac{3}{2}}\,(e^y k_z + e^z k_y - ie^z k_x - ie^x k_z)
\end{aligned}$$

A comparison of (2.9), using (2.10)–(2.14), with the corresponding terms in Van Vleck's paper[10] demonstrates the identity of the formulations. We have cast $\mathscr{H}_{\mathrm{OL}}$ into the form (2.9), rather than in a coordinate representation, because we wish to write the electron potential in terms of tabulated quantities (e.g., the matrix elements of the various C_{lm} appearing in (2.10) and (2.13) may be obtained using Racah's procedures.[36] A form similar to (2.9), but of course containing a great many more terms for symmetries lower than cubic, can be written down for any point group symmetry of interest.

The coefficients $V(\Gamma_{ig}l)$ can be evaluated immediately using a point charge model. Using (2.9) and comparing with Van Vleck,[10] we find,[17a,17b] for a single active electron of charge e on the cation,

$$\begin{aligned}
V(\Gamma_{3g}2) &= 6ee_{\mathrm{eff}}\langle r^2\rangle/a^3 \\
V(\Gamma_{5g}2) &= 4ee_{\mathrm{eff}}\langle r^2\rangle/a^3 \qquad (2.15) \\
V(\Gamma_{3g}4) &= -(5/3)\sqrt{15}ee_{\mathrm{eff}}\langle r^4\rangle/a^5 \\
V(\Gamma_{5g}4) &= -(2/3)\sqrt{15}ee_{\mathrm{eff}}\langle r^4\rangle/a^5
\end{aligned}$$

Here, e_{eff} is the effective charge on the neighboring ligands (note that $ee_{\mathrm{eff}} < 0$ for a cation and ligand of opposite charge). It must be emphasized that (2.9) is valid even if the point charge model is not, in exactly the same way as the static (parametrized) crystalline field (2.6) is valid. Both rely only on the symmetry of the surrounding cluster. However, as stated in the introduction, trouble obtains in the orbit–lattice interaction if more distant neighbors contribute [as they must for the $l = 2$ terms, since $V(\Gamma_{ig}2)$ falls off only as the cube of the distance]. The approach of counting the normal modes of vibrations and grouping them according to the irreducible representations of the point group means that a great many more terms will enter. The parametrization then becomes cumbersome, and the utility of this method is lessened. In addition, for such a spatially extended interaction, the long-wavelength approximation becomes even more suspect, since the expansion which led to (2.14) requires the phonon wavelength to be greater than the distance to the furthest contributing ion. We believe this difficulty to be a real one, making the expressions (2.15) for the $l = 2$ terms very dubious. This can be detected experimentally by a direct-process rate much different from what one would calculate from a point charge model, though the Raman rate might agree [since the shorter wavelength of the phonons in the latter process will tend to average out the long-range contributions to $V(\Gamma_{ig}2)$]. This feature may be responsible for the seemingly "good" agreement between estimated and observed two-phonon relaxation rates[13] and the relatively "poor" agreement for direct-process rates. Certainly the necessity for computing long-range contributions to $V(\Gamma_{ig}2)$ in the static limit has already been recognized.[23,31,40] One can thus appreciate the essential simplicity of a near-neighbor-only potential.

Once the form of $\mathscr{H}_{\mathrm{OL}}$ has been derived, it is a straightforward matter to evaluate the matrix elements of the electronic potentials appearing in, for example, (2.9). This can be done in a manner identical to that for the static crystalline field splitting (2.5). Note that (2.9) is applicable to any electronic system, either iron or rare earth group. It represents the one-phonon interaction between the orbital moment of the paramagnetic electron and the crystalline potential of the lattice. If one regards the quantities $V(\Gamma_{ig}l)$ as adjustable parameters, then (2.9) describes the total one-phonon "potential," even when such a quantity cannot strictly be defined (i.e., in the presence of covalency).

2.2. Dynamic Spin Hamiltonian

Sometimes, however, one is interested in a situation where the number of low-lying levels is small. It is then convenient to introduce the concept of a spin Hamiltonian. This was first carried out by Van Vleck (!),[10] rederived in detail by Pryce,[41] and developed by Abragam and Pryce[35] and by Bleaney and Stevens.[16] The essentials of the method are well known, so that we shall not discuss them here. We introduce an effective or "fictitious" spin S whose degeneracy, $2S + 1$, equals the number of lowest-lying levels. So defined, the effective spin will, of course, in general differ from the real spin, although many cases (especially in the $3d^n$ ions) are encountered where the real spin and fictitious spin are the same.* One must be careful, however, to keep under consideration a sufficient number of levels such that the effective spin S is equal to an odd (even) half-integer, depending on whether the number of paramagnetic electrons is odd (even). Otherwise, serious time-reversal symmetry difficulties arise in the construction of the spin Hamiltonian.† Once the dimensionality of the S space has been chosen, one can construct the appropriate one-phonon spin Hamiltonian using the point group symmetry of the surroundings of the paramagnetic ion to reduce the number of unknown parameters. So far, relevant experimental techniques have only been applied in the long-wavelength limit, appropriate to static uniform stress measurements or to a one-phonon (direct) spin–lattice relaxation process. The simplest case, again, is a cubic site where $S = 1$. The result can be written as[10,44]

$$\mathscr{H}_{\mathrm{SP}} = \mathbf{S} \cdot \mathsf{D} \cdot \mathbf{S} \tag{2.16}$$

where D is the second-rank tensor defined by

$$\mathsf{D} = \mathsf{G} \cdot \boldsymbol{\epsilon} \equiv G_{\alpha\beta\gamma\delta}\varepsilon_{\gamma\delta} \tag{2.17}$$

where we have adopted the convention that one sums over repeated indices. The strain tensor $\varepsilon_{\gamma\delta}$ is defined by

$$\varepsilon_{\gamma\delta} = \tfrac{1}{2}[(\partial\mu_\gamma/\partial x_\delta) + (\partial\mu_\delta/\partial x_\gamma)] \tag{2.18}$$

where $\mu(\mathbf{r})$ is the lattice displacement at the point $\mathbf{r}$. The fourth-rank tensor G can be simplified in the case of cubic symmetry, its transformation properties being similar to the usual elastic constant tensor. The antisymmetric part of D gives rise to terms linear in the spin operator. The system must be time-reversal-invariant, so that these terms vanish in the absence of magnetic field admixtures. Hence, $D_{\alpha\beta} = D_{\beta\alpha}$, and only the symmetric part of D can remain. In addition, symmetry arguments applicable to a cubic point group reduce the number of independent elements of G to only three. The requirement that Tr $\mathscr{H}_{\mathrm{SP}} = 0$, where one neglects the uniform shift of the S manifold, reduces this number to only two. We adopt

* The same symbol S will generally be used for real spin and fictitious spins, although it will be clear from the context which is being referred to.

† For example, Baker and Bleaney[42] use an $S = \frac{1}{2}$ spin Hamiltonian to describe the lowest-lying doublet of a $4f^n$ configuration (n even). Terms are exhibited which are not manifestly time-reversal invariant. This is discussed by Muller.[43]

for simplicity the Voigt notation, $xx = 1$, $yy = 2$, $zz = 3$, $yz = 4$, $xz = 5$, and $xy = 6$, so that the only nonvanishing components of **G** are

$$G_{11} = G_{22} = G_{33} = -2G_{21} = -2G_{31} = -2G_{32} = -2G_{12} = -2G_{13} = -2G_{23}$$
$$G_{44} = G_{55} = G_{66}$$

all other components vanishing. With this simplification, it is instructive to write out (2.17) in detail, first in irreducible tensor form, then (equivalently) in Cartesian coordinates, dropping terms which shift the entire S manifold uniformly:

$$\begin{aligned}\mathscr{H}_{\mathrm{SP}} = {} & G_{11}\{\tfrac{1}{2}[3S_z^2 - S(S+1)]\varepsilon(\Gamma_{3g}, \theta) + \tfrac{1}{2}\sqrt{3}\cdot\tfrac{1}{2}[S_+^2 + S_-^2]\varepsilon(\Gamma_{3g}, e)\} \\ & + G_{44}\{-\sqrt{\tfrac{2}{3}}\cdot\tfrac{1}{2}(S_zS_+ + S_+S_z)\varepsilon(\Gamma_{5g}, 1) \\ & + \tfrac{1}{3}\sqrt{3}\cdot\tfrac{1}{2}[S_+^2 - S_-^2]\varepsilon(\Gamma_{5g}, 0) \\ & - \sqrt{\tfrac{2}{3}}\cdot\tfrac{1}{2}(S_zS_- + S_-S_z)\varepsilon(\Gamma_{5g}, -1)\}\end{aligned} \tag{2.19}$$

where we have used the standard operator equivalents[45] for the spin operators, and[17a,17b] [(2.13) in the long-wave limit]

$$\begin{aligned}\varepsilon(\Gamma_{3g}, \theta) &= \tfrac{1}{2}(2\varepsilon_{zz} - \varepsilon_{xx} - \varepsilon_{yy}) \\ \varepsilon(\Gamma_{3g}, e) &= \tfrac{1}{2}\sqrt{3}\,(\varepsilon_{xx} - \varepsilon_{yy}) \\ \varepsilon(\Gamma_{5g}, 1) &= -i\sqrt{\tfrac{3}{2}}\,(\varepsilon_{yz} + i\varepsilon_{xz}) \\ \varepsilon(\Gamma_{5g}, 0) &= i\sqrt{3}\,\varepsilon_{xy} \\ \varepsilon(\Gamma_{5g}, -1) &= i\sqrt{\tfrac{3}{2}}\,(\varepsilon_{yz} - i\varepsilon_{xz})\end{aligned} \tag{2.19a}$$

In Cartesian coordinates, (2.19) is equal to

$$\begin{aligned}\mathscr{H}_{\mathrm{SP}} = {} & \tfrac{3}{2}G_{11}\{S_x^2\varepsilon_1 + S_y^2\varepsilon_2 + S_z^2\varepsilon_3\} + G_{44}\{(S_xS_z + S_zS_x)\varepsilon_5 \\ & + (S_yS_z + S_zS_y)\varepsilon_4 + (S_xS_y + S_yS_x)\varepsilon_6\}\end{aligned} \tag{2.20}$$

If (2.17) is written in terms of the stress tensor σ, instead of the strain ε (since stress is the measured quantity in a static pressure experiment), the equation analogous to (2.16) is[26]

$$\mathscr{H}_{\mathrm{SP}} = \mathbf{S}\cdot\mathbf{C}\cdot\sigma\cdot\mathbf{S} \tag{2.21}$$

where the **C** tensor has the same properties as the **G** tensor. Clearly, **G** and **C** are related by the elastic constant tensor.

In applying Eq. (2.19), it is generally assumed that the influence upon the paramagnetic ion of the strain associated with the phonons is the same as that of a corresponding static bulk strain. Note that this assumption in no way implies that the local strain at the impurity site is the same as the bulk strain, as this is not true even in the static case. All that one assumes is that the effect of long-wavelength phonons on the paramagnetic ion is the same as a static strain and that the dynamical properties of these long-wavelength phonons are not affected by the impurities. Obviously this assumption is no longer valid for the very-short-wavelength phonons, whose dynamical properties will be severely modified by the impurity. Such a situation could have profound effects on the two-phonon relaxation time* and the reader is referred to Refs. 45 and 47 for a complete discussion. In general, the frequencies of interest involving the one-phonon terms (2.17) are so low (seldom more than 1 or 2°K) that such perturbations are not usually important [though, see Ref. (46)], affecting as they do phonons with much higher frequencies in the excitation spectrum.

The example we have investigated here, that appropriate to $S = 1$, in fact has greater importance than might be apparent at first sight. An example is to be found in the iron group series when the lowest-lying level in an orbital singlet. In such a case, the effective spin S equals the true total spin S. The basic potential interaction, (2.9), affects only the orbital component of the magnetic moment, of which, by assumption, there is none to lowest order. The powers of S which appear in $\mathscr{H}_{SP}$ are then directly related to the powers of the spin–orbit coupling which admix excited crystal-line field levels (as well as other excited-term values[17a,17b,33,34]) into the ground state.[47] The admixture is linear, the matrix elements quadratic, so that in fact (2.16) applies to a wide variety of iron group ions. An example is Mn^{++} in MgO,† where the coefficients G_{11} and G_{44} were evaluated explicitly in terms of the $V(\Gamma_{ig} l)$. A point charge model was used to evaluate these parameters, as exhibited in (2.15), and both spin–orbit coupling (taken twice) and atomic spin–spin coupling (taken once) were included. Additional examples have been considered by Al'tshuler *et al.*[32] Zdansky[31] has corrected their treatment by including additional excited-term values (which contribute significantly) and covalent and spin–spin effects. His

* The predicted effects of a resonance in the lattice spectrum on the Raman rate have been observed by Feldman *et al.*[47]

† See Schawlow *et al.*[17a] and Blume and Orbach.[17b] Leushin,[48a] Kondo,[33] and Sharma *et al.*,[34] following the methods of Pryce[48b] and Watanabe,[48c] have also examined the effect of atomic spin–spin interactions in this context. Each time this interaction acts, two powers of S are obtained.

results, for V^{++}, Cr^{3+}, and Ni^{++}, together with those of Sharma *et al.* for Mn^{++} in MgO, and the experimental values of Feher[26] and Calvo *et al.*[29] are listed in Section 5.

Paramagnetic ions with $S > 1$ can, of course, exhibit terms in $\mathscr{H}_S$ higher in order than quadratic in S. Such terms, though often small (they will *not* be small when there is substantial orbital moment in the ground manifold), may be significant in experiments of high accuracy. They have been derived to order S^4 by Koloskova[25] and Calvo *et al.*[49] They are of the form

$$\mathscr{H}_{SP} = \tfrac{1}{4}(S_\alpha S_\beta + S_\beta S_\alpha)(S_\gamma S_\delta + S_\delta S_\gamma)G_{\alpha\beta\gamma\delta\xi\eta}\varepsilon_{\xi\eta} \tag{2.22}$$

where explicit expressions for the G's in a cubic environment are tabulated in both references and measured for Eu^{2+} and Gd^{3+} in CaF_2 and Gd^{3+} in CaO by Calvo *et al.*[49]

The form of the overall spin-phonon Hamiltonian, when (2.22) is added to the second-order interaction Hamiltonian, (2.20), is similar to the expansion of the one-phonon lattice potential (2.9) in that the coefficients of a given component of the strain tensor are summed over powers of S equivalent to the sum over l in the case of $\mathscr{H}_{OL}$.

There is a further complication which we have not yet considered, though this is not a real restriction on the use of (2.9), but only should one choose to use the spin-Hamiltonian form (2.16). This difficulty will be treated explicitly in Section 3, and is caused by the vanishing of the spin-phonon Hamiltonian between time-reversed states of half-integral spin [this is true for both (2.16) and (2.22)]. This is not a problem for spin systems with $S > \frac{1}{2}$, since there are sufficient other transitions within the manifold to ensure a return to equilibrium of the entire manifold.[10,50] However, for $S = \frac{1}{2}$, a direct one-phonon process is known to occur. It is necessary, if one uses the spin-Hamiltonian formalism, to construct a suitable $\mathscr{H}_{SP}$. This Hamiltonian arises from the product of the one-phonon orbit–lattice interaction and the Zeeman interaction. For a cubic environment, it is straightforward to write the interaction as[25]

$$\begin{aligned}\mathscr{H}_{SP} = {} & \mathbf{H}\cdot\mathbf{S}\cdot\tfrac{1}{3}(G'_{11}+2G'_{12})(\varepsilon_1+\varepsilon_2+\varepsilon_3)\\ &+(\mathbf{H}\cdot\mathbf{S}-3H_zS_z)\cdot\tfrac{1}{3}(G'_{11}-G'_{12})[\tfrac{1}{2}(\varepsilon_1+\varepsilon_2)-\varepsilon_3]\\ &+(H_xS_x-H_yS_y)(G'_{11}-G'_{12})\cdot\tfrac{1}{2}(\varepsilon_1-\varepsilon_2)\\ &+2G'_{44}[(H_yS_z+H_zS_y)\varepsilon_4+(H_xS_z+H_zS_x)\varepsilon_5\\ &+(H_xS_y+H_yS_x)\varepsilon_6]\end{aligned} \tag{2.23}$$

It must be emphasized that the expressions (2.16), (2.22), and (2.23) hide the real complexity of the problem. Evaluation of the various G's which are present in these expressions requires a detailed knowledge of the one-phonon lattice potential and the energy level structure of the paramagnetic ion. Nevertheless, these spin-phonon Hamiltonians are useful in the sense that they provide a framework with which to describe the angular and frequency dependences of the relaxation time.

Indeed, the various spin-Hamiltonian parameters G can be regarded as empirical parameters which completely describe the relaxation process. The forms of the expressions are model-independent, and serve as a minimum parameter fit, with complete utilization of symmetry reductions, for the relaxation Hamiltonian within the effective spin manifold. In practice, the various G's have been evaluated in precisely this manner, from uniaxial static stress measurements, or through microwave phonon attenuation. How well these empirically determined parameters G are described by a point charge model, for example, is an entirely separate theoretical question.

So far, we have concentrated solely on the one-phonon coupling Hamiltonian. There are cases, to be discussed in Section 4, when two-phonon terms in the Taylor expansion of the crystalline potential can be important. The one-phonon terms, when taken to second order, can appear as being "two-phonon-lie." However, there are dynamic features in the latter case which distinguish such terms from a first-order two-phonon interaction. It is clear when the one-phonon terms are *not* important. For example, they must vanish when taken between time-reversed states of half-integral spin. Thus, for Ti^{3+} alum, $S = 1/2$, and Van Vleck[10] demonstrated that only the one-phonon interaction in second order contributes to the two-phonon relaxation process. For Cr^{3+} alum, however, $S = 3/2$, and both second-order one-phonon and first-order two-phonon terms contributed to the Raman relaxation process (via transitions between the levels $\pm 3/2 \leftrightarrow \pm 1/2, \mp 1/2$). The two-phonon lattice potential can be calculated in a manner analogous to (2.9).[51] The analogous spin-Hamiltonian form has been derived, in the long-wavelength limit, by Koloskova.[25] She finds

$$\begin{aligned}
\mathscr{H}_{\text{SP}} = {} & \tfrac{3}{2}S_z^2\{G'_{111}[\varepsilon_3^2 - \tfrac{1}{2}(\varepsilon_1^2 + \varepsilon_2^2)] \\
& + 2G'_{112}[\varepsilon_3(\varepsilon_1 + \varepsilon_2) - 2\varepsilon_1\varepsilon_2] + 4G'_{166}[\varepsilon_4^2 + \varepsilon_5^2 - 2\varepsilon_6^2]\} \\
& + 3(S_x^2 - S_y^2[\tfrac{1}{4}G'_{111}(\varepsilon_1^2 - \varepsilon_2^2) \\
& + G'_{112}(\varepsilon_1 - \varepsilon_2)\varepsilon_3 + 2G'_{166}(\varepsilon_5^2 - \varepsilon_4^2)] \\
& + 4(S_yS_z + S_zS_y)[G'_{661}(\varepsilon_2 + \varepsilon_3)\varepsilon_4 + G'_{663}\varepsilon_1\varepsilon_4 + 2G'_{456}\varepsilon_5\varepsilon_6] \\
& + 4(S_xS_z + S_zS_x)[G'_{661}(\varepsilon_1 + \varepsilon_3)\varepsilon_5 + G'_{663}\varepsilon_2\varepsilon_5 + 2G'_{456}\varepsilon_9\varepsilon_6] \\
& + 4(S_xS_y + S_yS_x)[G'_{661}(\varepsilon_1 + \varepsilon_2)\varepsilon_6 + G'_{663}\varepsilon_3\varepsilon_6 + 2G'_{456}\varepsilon_4\varepsilon_5] \qquad (2.24)
\end{aligned}$$

In the next two sections, use wiil be made of the formalism developed in this section to compute the relaxation times in the direct process and two-phonon regimes. Time-reversal arguments applied to $\mathscr{H}_{\mathrm{OL}}$ in second order, to be detailed in Section 4, will introduce yet another "effective" two-phonon $\mathscr{H}_{\mathrm{SP}}$, involving the phonon momentum explicitly.[20] This term will give rise to the Van Vleck ninth-power temperature dependence of the Raman process for $S = 1/2$, as discussed in the introduction. We will not write this term down explicitly here, as its form will become evident in Section 4 and as it is really just a result of the dynamics of the two-phonon relaxation process.

3. THE ONE-PHONON (DIRECT) PROCESS

3.1. Evaluation of T_1 Using the Dynamic Spin Hamiltonian

In Section 2, the one-phonon orbit–lattice (or spin–phonon) Hamiltonian has been derived, both in a microscopic and a phenomenological manner. In this section, we wish to apply it to the case of spin–lattice relaxation. Its application to, for example, phonon attenuation has already been given in a number of places.[32,52] In order to give a sketch of the method, we shall first consider the simplest possible case, that appropriate in two isolated levels with $S = 1$, $M_S = \pm 1$. This is not uncommon; for example, it applies to Pr^{3+} and Ho^{3+} in ethyl sulfate.[43] The basic process is the following (Fig. 3). A spin is in the excited level $|\, S = 1, M_S = +1\rangle \equiv |+\rangle$ and the transition to $|\, S = 1, M_S = -1\rangle \equiv |-\rangle$ level occurs with the emission of a single phonon of energy $\hbar\omega(\mathbf{k}, s) = 2g\beta H_0$ (since $S = 1$), where H_0 is parallel to the z axis. We use the "golden rule" to write, for a given phonon $\mathbf{k}, s$,

$$W_{+\to-}^{\mathbf{k},s} = (2\pi/\hbar^2)\,|\,\langle\varphi'\,|\,\langle -\,|\,\mathscr{H}_{\mathrm{SP}}\,|\,+\rangle\,|\,\varphi\rangle\,|^2\,\delta[\omega(\mathbf{k}, s) - \omega_0] \qquad (3.1)$$

where $|\,\varphi\rangle$ and $|\,\varphi'\rangle$ are the state vectors of the phonon system before and after the transition, and $\omega_0 = 2g\beta H_0/\hbar$. We use (2.20) for $\mathscr{H}_{\mathrm{SP}}$, so

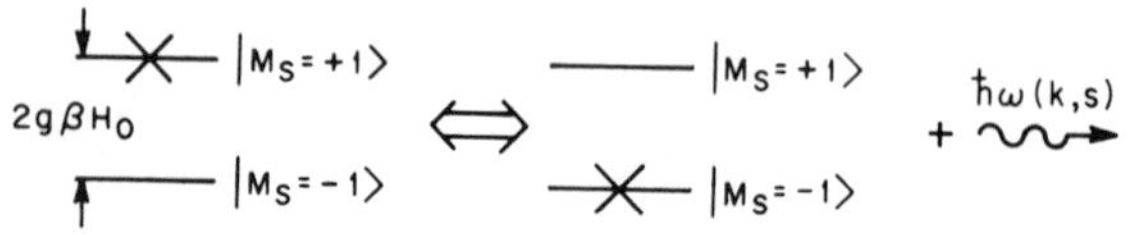

Fig. 3. Schematic illustration of direct relaxation process $\hbar\omega(\mathbf{k}, s) = 2g\beta H_0$.

hat the only relevant terms are

$$\mathscr{H}_{\text{SP}} = \tfrac{3}{2}G_{11}(S_x{}^2\varepsilon_1 + S_y{}^2\varepsilon_2) + G_{44}[(S_xS_y + S_yS_x)\varepsilon_6] \tag{3.2}$$

Then

$$\langle - \mid \mathscr{H}_{\text{SP}} \mid + \rangle = \tfrac{3}{4}G_{11}(\varepsilon_1 - \varepsilon_2) + iG_{44}\varepsilon_6 \tag{3.3}$$

The expression (3.1) must be averaged over initial states and summed over final states with care to be taken that a *net* transition probability per unit time is obtained. Because the strain appears only linearly in (3.3), $\mid \varphi' \rangle$ differs from $\mid \varphi \rangle$, according to Fig. 3, by a net increase in the occupation number $n_{\mathbf{k},s}$ by unity. Here,

$$n_{\mathbf{k},s} = 1/(e^{\hbar\omega(\mathbf{k},s/kT} - 1)$$

represents the equilibrium Bose factor with the temperature T of the "speaking-term" phonons equal to the bath temperature. This is not always the case, and when departures occur, one is said to be in the phonon bottleneck regime (see the chapter in this volume by S. Geschwind). Summing over all permissible final phonon states, and using the expansion for the strains ε analogous (2.12) in the long-wave length limit [allowable for phonons on speaking terms with the spins because $\omega_0 = \omega(\mathbf{k}, s) = v_s \mid \mathbf{k} \mid$, so that $\mid \mathbf{k} \mid / k_{\text{D}} = \omega_0 / v_s k_{\text{D}} = \omega_0/\omega_{\text{D}} \ll 1$], we obtain

$$\begin{aligned} W_{+\rightarrow-} = (\pi/\hbar M) \sum_{\mathbf{k},s} [1/\omega(\mathbf{k}, s)] \mid \tfrac{3}{4}G_{11}(e^x k_x - e^y k_y) \\ + i \cdot \tfrac{1}{2}G_{44}(e^x k_y + e^y k_x) \mid^2 (n_{\mathbf{k},s} + 1)\, \delta[\omega(\mathbf{k}, s) - \omega_0] \end{aligned} \tag{3.4}$$

The next step can be taken if we assume an isotropic phonon spectrum where $\omega(\mathbf{k}, s) = v_s \mid \mathbf{k} \mid$ where $s = l$ or (2) t, corresponding to a longitudinal and (two) degenerate transverse phonon branches. The quantities in parentheses are averaged over angles using[10]*

$$\begin{aligned} \langle (e^\alpha)^2 k_\alpha{}^2 \rangle &= k^2 \times \begin{cases} 1/5 & \text{for } s = l \\ 1/15 & \text{for } s = t \end{cases} \\ \langle (e^\alpha)^2 k_\beta{}^2 \rangle &= k^2 \times \begin{cases} 1/15 & \text{for } s = l \\ 2/15 & \text{for } s = t \end{cases} \\ \langle e^\alpha k_\alpha e^\beta k_\beta \rangle &= k^2 \times \begin{cases} 1/15 & \text{for } s = l \\ -1/30 & \text{for } s = t \end{cases} \end{aligned} \tag{3.5}$$

* Menne[53] has used the lattice symmetry to carry out these averages, without necessity of a long-wavelength approximation, though requiring isotropy of the vibrational spectrum. In Ref. 27, Donoho carried out, in the large-wavelength limit, an exact angular average for the Al_2O_3 structure.

where $\alpha, \beta = x, t, z;\ \alpha \neq \beta$. All other angular averages vanish. Lettin[g]

$$\sum_{\mathbf{k},s} = [V/(2\pi)^3] \sum_s (1/v_s^3) \int \omega^2\, d\omega\, d\Omega \tag{3.6}$$

where V is the volume of the crystal, and using (3.5), it is straightforwar[d] to evaluate (3.4). We find, introducing $\varrho = M/V$, the crystal mass density,

$$\begin{aligned} W_{+\to-} &= (3\omega_0^3/40\pi\hbar\varrho)[G_{11}^2 + (4/9)G_{44}^2] \\ &\quad \times [(1/v_l)^5 + \tfrac{3}{2}(1/v_t)^5][n(\omega_0) + 1] \end{aligned} \tag{3.7}$$

The transition probability per unit time for the inverse process, appropriate to the absorption of a phonon and the raising of a spin from $|-\rangle$ to $|+\rangle$, is given by (3.7) with $[n(\omega_0) + 1]$ replaced by $n(\omega_0)$. The change in population with time of the two spin levels is given by

$$\begin{aligned} dN_+/dt &= -W_{+\to-}N_+ + W_{-\to+}N_- \\ dN_-/dt &= -W_{-\to+}N_- + W_{+\to-}N_+ \end{aligned} \tag{3.8}$$

with $N_+ + N_- = N$, the total number of spins in the system. It is easy to show that the difference in spin population, $\Delta N = N_+ - N_-$, changes with time as[54,55]

$$d(\Delta N)/dt = -(\Delta N - \overline{\Delta N})(W_{+\to-} + W_{-\to+}) \tag{3.9}$$

where $\overline{\Delta N} = \bar{N}_+ - \bar{N}_-$, the thermal equilibrium difference in spin population. Equation (3.9) demonstrates that it is possible to define a spin–lattice relaxation time for a two level system *exactly*, namely

$$1/T_1 = W_{+\to-} + W_{-\to+} \tag{3.10}$$

We use (3.7) to arrive at a direct-process relaxation time given by

$$\begin{aligned} 1/T_{1D} &= (3\omega_0^3/40\pi\hbar\varrho)[G_{11}^2 + (4/9)G_{44}^2] \\ &\quad \times [(1/v_l)^5 + (3/2)(1/v_t)^5] \coth(\hbar\omega_0/2kT) \end{aligned} \tag{3.11}$$

This completes the calculation of the direct-process relaxation time for this very simple case. One notices some interesting features of the temperature and field (ω_0) dependences of $1/T_{1D}$. At high temperatures, $kT \gg \hbar\omega_0$, $1/T_{1D} \propto \omega_0^2 T$, whereas at low temperatures, $kT \ll \hbar\omega_0$, $1/T_{1D} \propto \omega_0^3$. Hence, we have the result that $1/T_{1D}$ becomes temperature-independent at low temperatures, reflecting the fact that relaxation occurs via spontaneous honon emission only. Its precise form, $\sim\coth(\hbar\omega_0/2kT)$ should not be ırprising, for this is a result of statistical theory for a system relaxing via ngle-quantum exchange with a randomly fluctuating heat bath. In addi-on, the inverse fifth power of the sound velocity appearing in (3.11) ıeans that the harder the crystal (larger v_l and v_t), the smaller the relaxation ıte. A factor of 2.5 change in the sound velocities changes $1/T_{1D}$ by two rders of magnitude. This factor will even be more important in the Raman-rocess regime where the relaxation rate is proportional to the inverse :nth power of the sound velocity. Indeed, other things being equal, this is he reason why Raman processes dominate the relaxation rate at lower ɜmperatures in soft crystals as compared to hard materials.

There is feature of (3.11) which should lead to a certain nervousness n the reader. This is the rather cavalier use of the phonon temperature T which appears in the "speaking-term" phonon occupation number $n(\omega_0)$. Normally one assumes that $T = T_{\mathrm{Bath}}$ with which the crystal is in contact. Van Vleck[3] first noted that if the spin system is being heated by an external ource, (3.9) will lead to the absorbed energy being fed into a narrow band of speaking-term phonons. If the contact with the heat bath of the latter s weak (as it is in most situations[3]) heating of the speaking-term phonons esults, and it will not in general be true that $\Delta N(t)$ will follow a simple xponential law implied by (3.9) for fixed T. This situation is referred to ıs a phonon "bottleneck" and striking examples of its effects have been observed in recent years[13,56,57] and will be referred to in Section 5. The phonon bottleneck is also discussed by S. Geschwind in another chapter n this volume. Clearly, one way to avoid this effect is to work at low con-centrations so as to minimize the spin heat capacity.

The next step in our examination of the direct-process relaxation time is the case of $S = 1/2$, using (2.23). The calculation proceeds analogously to that which led to (3.7) and hence (3.11). For H_0 parallel to z, identi-fying $|+\rangle = |S = 1/2, M_s = 1/2\rangle$ and $|-\rangle = |S = 1/2, M_s = -1/2\rangle$, we find

$$1/T_{1\mathrm{D}} = (\omega_0^3 H_0^2/15\pi\hbar\varrho)(G'_{44})[(1/v_l)^5 + \tfrac{3}{2}(1/v_t)^5]\coth(\hbar\omega_0/2kT) \quad (3.12)$$

This result displays an important feature of $S = 1/2$ direct-process relaxa-tion, namely at high temperatures, $1/T_{1\mathrm{D}} \propto H_0^4 T$. It is remarkable that this field dependence was unambiguously observed only as late as 1963[58] and 1964,[59] though predicted by Kronig[9] and Van Vleck[10] in 1939 and 1940!

It is possible to generate the angular dependence of the direct-process rates (3.11) and (3.12) in a straightforward manner by simple rotation of the spin operators S_α to a system of coordinates where S_z is parallel to H_0 Thus, in (3.11), the first term in parentheses is replaced by*

$$[G_{11}^2 + (4/9)G_{44}^2] \Rightarrow [G_{11}^2 + (4/9)G_{44}^2] - [G_{11}^2 - (4/9)G_{44}^2](l^2m^2 + m^2n^2 + n^2l^2) \quad (3.13)$$

where l, m, and n are the direction cosines H_0 makes with the original crystallographic x, y, and z axes. Thus, for $G_{11} \neq (4/9)G_{44}$, angular anisotropy will be present in the $S = 1$ direct-process relaxation rate.

A similar method can be applied in the case of $S = 1/2$, where (G'_{44}) in (3.12) is replaced by[25]

$$(G'_{44})^2 \Leftrightarrow (G'_{44})^2 - [2(G'_{44})^2 - \tfrac{1}{4}(G'_{11} - G'_{12})^2](l^2m^2 + m^2n^2 + n^2l^2) \quad (3.14)$$

More general angular relationships, appropriate to lower (than cubic crystallographic symmetries, are exhibited in the work of Ray *et al.*[20] and will not he repeated here.

The above demonstrates the utility of the spin-phonon Hamiltonian in generating the one-phonon relaxation rate. It is instructive to backtrack a bit and consider what happens if we use the orbit–lattice interaction (2.9 directly. Clearly, some higher-order terms must be included if, for example transitions are to obtain between time-reversed states of half-integral spin This can be seen in the following manner. Imagine that we wish to calculate the transition probability per unit time, $W_{+\rightarrow-}$, analogous to (3.12). However, we now use the orbit–lattice interaction (2.9) instead of the spin-phonon Hamiltonian (2.23) for $S = 1/2$ to effect the transition from $|+\rangle$ to its time-reversed partner $|-\rangle$. In terms of Russell–Saunders coupling where for simplicity we restrict our attention to states of given L, S, the time-reversed pair of levels can be written as

$$|+\rangle = \sum_{M_S, M_L} a_{M_S,M_L} |M_S\rangle |M_L\rangle$$
$$|-\rangle = \sum_{M_S, M_L} a^*_{M_S,M_L} (-1)^{M_S+M_L} |-M_S\rangle |-M_L\rangle \quad (3.15)$$

The form of (3.15) is quite general,[60] applying to both iron-group and rare earth ions, but neglecting magnetic admixtures [which, arising from external fields, hyperfine interactions,[61] and dipole–dipole interactions, cause admixtures which destroy the time-reversed relation between $|+\rangle$ and

* This result differs slightly from those of Refs. 10, 19, 20, and 24 because the form of H_{SP} in Eq. (2.20), is numerically different from their formulations.

$|-\rangle$ in (3.15)]. We construct the matrix elements of $\mathscr{H}_{\text{OL}}$, (2.9), between the states (3.15).* Since $\mathscr{H}_{\text{OL}}$ is independent of S, we have

$$\langle - \mid \mathscr{H}_{\text{OL}} \mid + \rangle = \sum_{M_S, M_L, M_L'} a_{M_S, M_L} a_{-M_S, M_L'} (-1)^{M_S + M_L} \langle -M_L' \mid \mathscr{H}_{\text{OL}} \mid M_L \rangle \tag{3.16}$$

Interchanging the dummy variables M_L and $M_{L'}$ and reversing the sign of the other dummy variable M_S, (3.16) becomes

$$\langle - \mid \mathscr{H}_{\text{OL}} \mid + \rangle = \sum_{M_S, M_L, M_L'} a_{M_S, M_L} a_{-M_S, M_L'} (-1)^{-M_S + M_L'} \langle -M_L \mid \mathscr{H}_{\text{OL}} \mid M_L' \rangle \tag{3.17}$$

The orbit–lattice Hamiltonian is time-even, so that

$$\langle -M_L \mid \mathscr{H}_{\text{OL}} \mid M_L' \rangle = (-1)^{M_L - M_L'} \langle -M_L' \mid \mathscr{H}_{\text{OL}} \mid M_L \rangle \tag{3.18}$$

Inserting (3.18) in (3.17), we find

$$\langle - \mid \mathscr{H}_{\text{OL}} \mid + \rangle = \sum_{M_S, M_L, M_L'} a_{M_S, M_L} a_{-M_S, M_L'} (-1)^{-M_S - M_L} \langle -M_L' \mid \mathscr{H}_{\text{OL}} \mid M_L \rangle \tag{3.19}$$

Comparing (3.19) with (3.16), we see that, for M_S equal to an odd half-integer (an odd number of electrons), the summands must vanish term by term. Hence, between time-reversed states of half-integral spin,

$$\langle 0 \mid \mathscr{H}_{\text{OL}} \mid + \rangle = 0 \tag{3.20}$$

The procedure which lifts this difficulty is clearly one of introducing a magnetic perturbation which will break the time-reversed relationship between $|+\rangle$ and $|-\rangle$. This can be achieved by considering excited time-reversed levels $|+'\rangle$ and $|-'\rangle$ at an energy Δ compared to the ground doublet. The Zeeman interaction

$$\mathscr{H}_{\text{Z}} = \beta(\mathbf{L} + 2\mathbf{S}) \cdot \mathbf{H}_0 \tag{3.21}$$

is assumed to have matrix elements between, say, $|+\rangle$ and $|+'\rangle$, and $|-\rangle$ and $|-'\rangle$, where

$$\begin{aligned} \langle +' \mid \mathscr{H}_{\text{Z}} \mid + \rangle &= -\langle -' \mid \mathscr{H}_{\text{Z}} \mid - \rangle \\ &\equiv AH_0 \end{aligned} \tag{3.22}$$

* A very clear, concise treatment of time-reversal symmetry is given by Abragam and Bleaney.[62] We shall adopt their language (hence, time-even) in our treatment of the matrix elements of the orbit–lattice operator.

The matrix element of $\mathscr{H}_{\mathrm{OL}}$ can now be taken between the admixed ground doublet states, namely

$$
\begin{aligned}
&\{\langle -\,| + AH_0/\Delta\langle -'\,|\}\mathscr{H}_{\mathrm{OL}}\{|\,+\rangle - AH_0/\Delta\,|\,+'\rangle\} \\
&\quad = -(AH_0/\Delta)\{\langle -\,|\,\mathscr{H}_{\mathrm{OL}}\,|\,+'\rangle - \langle -'\,|\,\mathscr{H}_{\mathrm{OL}}\,|\,+\rangle\} \\
&\quad = -(2AH_0/\Delta)\langle -\,|\,\mathscr{H}_{\mathrm{OL}}\,|\,+'\rangle
\end{aligned}
\tag{3.23}
$$

where the last step can be proven in a manner identical to the steps (3.15)–(3.19). The result (3.23) is the orbit–lattice equivalent of the matrix element of the spin-Hamiltonian $\mathscr{H}_{\mathrm{SP}}$ given by (2.23), within an $S = 1/2$ manifold. The matrix elements of $\mathscr{H}_{\mathrm{OL}}$ are finite only because of Zeeman admixtures, and are therefore proportional to the external field H_0 exactly as is the spin-phonon Hamiltonian $\mathscr{H}_{\mathrm{SP}}$ given by (2.23). Al'tshuler *et al.*[32] give a number of examples of the evaluation of the coefficients $G_{\alpha\beta'}$ in terms of the more fundamental orbital operators contained in $\mathscr{H}_{\mathrm{OL}}$ [i.e., the $V(\Gamma_{ig}, l)$].

3.2. Direct Processes in Multilevel Systems

For $S > 1/2$, where transitions of $\mathscr{H}_{\mathrm{OL}}$ are allowed by time-reversal symmetry, it is also possible to express $\mathscr{H}_{\mathrm{SP}}$, now given by the quadrupolar form, (2.16), in terms of matrix elements of $\mathscr{H}_{\mathrm{OL}}$.[17,24,32] This case is also interesting because of the multiplicity of relaxation paths. Consider a system where $S > 1/2$ where there are a number of relaxation transitions $W_{M_S \to M_{S\pm1} M_{S\pm2}} \cdots$. For simplicity, we shall consider iron group salts where (2.16) applies so that at most, $\Delta M_S = \pm 2$ transitions can occur. The situation is then complicated because of a multitude of relaxation paths. The first treatment of relaxation of such a multilevel system was given by Van Vleck[10] for $S = 3/2$ (chrome alum). It was rederived by Hebel and Slichter[63] and has been treated since by a number of authors.[27,50,64]

A specific case may help to make their treatments more apparent. Consider Mn^{2+} in a cubic environment; $S = 5/2$, where, ignoring the cubic field splitting, there are four different finite relaxation transition rates:

$$
\begin{aligned}
&W_{5/2\to 3/2} = W_{-3/2\to -5/2}, \qquad W_{5/2\to 1/2} = W_{-1/2\to -5/2} \\
&W_{3/2\to 1/2} = W_{-1/2\to -3/2}, \qquad W_{3/2\to -1/2} = W_{1/2\to -3/2} \\
&W_{1/2\to -1/2} = W_{-1/2\to 1/2} \cong 0
\end{aligned}
\tag{3.24}
$$

The last rate is much smaller than the other rates for sensible values of magnetic field. Of course, from (2.16), [more explicitly, (2.20)], all four nonzero terms in (3.24) are related, depending as they do on only two parameters, G_{11} and G_{44}. The rates (3.24), as well as the inverse processes, can be calculated as in (3.4). The next problem is to decide on the initial set of conditions, since there are six levels whose populations are to be calculated as a function of time. Since the sum of the populations is fixed, $\sum_{M_S} N_{M_S} = N$, the number of unknowns is reduced to five. We define the difference from equilibrium of the difference of populations for adjacent levels $(N_{M_S+1} - \bar{N}_{M_S+1}) - (N_{M_S} - \bar{N}_{M_S}) = n_{M_S+(1/2)}$. It is straightforward to write

$$\begin{aligned}
\dot{n}_2 &= -n_2\{2W_{5/2\to3/2} + W_{5/2\to1/2}\} + n_1\{W_{3/2\to1/2} + W_{3/2\to-1/2} - W_{5/2\to1/2}\} \\
&\quad + n_0\{W_{3/2\to-1/2}\} \\
\dot{n}_1 &= n_2\{W_{5/2\to3/2} - W_{5/2\to1/2}\} - n_1\{2W_{3/2\to1/2} + W_{5/2\to1/2}\} \\
\dot{n}_0 &= n_2\{2W_{5/2\to1/2}\} + n_1\{2W_{5/2\to1/2} + W_{3/2\to1/2}\} - n_0\{2W_{3/2\to1/2}\}
\end{aligned} \tag{3.25}$$

with symmetric equations for n_{-1} and n_{-2}. There are in general, subject to initial conditions, five relaxation times arising from the five linear coupled equations for the quantities $n_{M_S+(1/2)}(t)$. For our purposes, we shall suppose the spin system to have been prepared "symmetrically": $n_{-2} = n_2$, $n_{-1} = n_1$, so that only three of the five equations (3.25) remain independent and the number of relaxation times is reduced to three. This initial condition is in fact satisfied in two rather important cases: (1) the entire spin system is in thermal equilibrium in zero magnetic field and an external field is suddenly applied, and (2) the spin system is in thermal equilibrium in a finite magnetic field and is then saturated by a strong microwave field (assuming the various lines overlap with each other). Since the equations (3.25) are symmetric in $n_{M_S+(1/2)}$, $n_{-M_S-(1/2)}$, if this condition is initially fulfilled, it will so remain at all subsequent times.

There is yet another condition which can further simplify these equations: All the levels can be described by the same spin temperature, and $n_0 = n_1 = n_2$. This would be the case for the nonresonant experiments of Gorter,[2] for example, in a system with equal level spacing and rapid spin–spin relaxation. Then, a single relaxation time obtains equal to

$$1/T_1 = (4/35)\{W_{5/2\to3/2} + 4W_{5/2\to1/2} + 4W_{3/2\to-1/2} + W_{3/2\to1/2}\} \tag{3.26}$$

Under these very restrictive conditions, (3.26) can also be derived from the general multilevel relaxation formula[10,63]

$$1/T_1 = \tfrac{1}{2}\Big\{\sum_{n,m} (E_n - E_m)^2 W_{n\to}\Big\}\Big/\sum_n E_n^2 \tag{3.27}$$

where E_n is the energy of the nth level. Equation (3.26) also follows immediately from (3.27). The formulation (3.27) is quite useful because it can be recast in the form of a trace of a product of two commutators. This enables one to include[10,63,65] additional perturbation into the relaxation time; for example, spin–spin and hyperfine interactions. In general, the one-phonon expressions for $W_{n\to m}$ are themselves proportional to the frequency, so that (3.27) is a complicated result.[10,65] Thus, Huber[66] calculates modifications to (3.26) arising from the use of (3.27) in the presence of the dipolar and hyperfine fields for $S = 1/2$ ions. He needs two "effective" fluctuating fields to describe the effect of these perturbations, one entering from the energy splitting of the doublet levels, the other from the admixture effect [see (3.12)] which breaks the time-reversed character of the doublet. Van Vleck,[10] in his early treatment, had made use of a Gaussian-distributed fluctuating local field. In general, this method is not correct, though there are, special cases where it can be valid.[65] These occur for two-phonon relaxation only, when the perpendicular g value vanishes. Actually, the direct process, relying as it does on the density of states of speaking-term phonons, is probably swamped in most cases by the two-phonon process when the external field is so low as to be of the order of the internal fields. Hence, these corrections to the direct process are probably not important in general, though they can contribute in exceptional cases (e.g., the rare earth group, where the hyperfine field can be quite large).

Returning to (3.25), we now examine the more general case of unequal population differences $n_{M_S+(1/2)}$. There will now be three relaxation times, and any given population difference, n_2, n_1, or n_0, will not in general decay in a simple exponential fashion, but rather as a linear combination of three exponentials. We can compute these times by setting $n_{m_S+(1/2)} \propto e^{-\lambda t}$. Then, (3.25) generates the secular equation

$$\begin{vmatrix} \lambda-(2W_{5/2\to3/2}+W_{5/2\to1/2}) & (W_{3/2\to1/2}+W_{3/2\to1/2}-W_{5/2\to1/2}) & W_{3/2\to-1/2} \\ (W_{5/2\to3/2}-W_{5/2\to1/2} & \lambda-(2W_{3/2\to1/2}+W_{5/2\to1/2}) & 0 \\ 2W_{5/2\to1/2} & (2W_{5/2\to1/2}+2W_{3/2\to1/2}-2W_{3/2\to-1/2}) & \lambda-2W_{3/2\to-1/2} \end{vmatrix} = 0 \tag{3.28}$$

The three eigenvalues represent the three values for the relaxation time, and the respective eigenvectors that linear combination of population differences which decay according to that (single) exponential with a relaxation time equal to the inverse of the eigenvalue. It is a simple matter to use a linear combination of the three eigenvectors which satisfies the $t = 0$

conditions to find how a given population will decay in time. Thus,

$$n_m(t) \sum_{p=1}^{3} a_{mp} e^{-\lambda_p t} \tag{3.29}$$

where the λ_p are the three eigenvalues of (3.28) and the coefficients a_{mp} depend on the initial conditions and the eigenvectors of (3.28). Andrew and Tunstall[50] have analyzed these equations in considerable detail and the reader is referred to their work. For some values of the relaxation coefficients $W_{n\to m}$, which in turn are simply related to G_{11} and G_{44} from (2.20), the quantity $n_0(t)$ actually can reverse sign as a function of time in a system where initially all levels were in thermal equilibrium (at zero magnetic field). The importance of spin–spin interactions, which attempt to keep a uniform spin temperature and thus avoid the vagaries of the complicated recovery (3.29), can now be appreciated. Manenkov and Prokhorov[67] looked at multilevel systems where the spin–lattice times are comparable to spin–spin times, so that the recovery is intermediate between (3.26), where all levels are assumed to be described by the same spin temperature, and (3.29), where they are all assumed to relax independently with a separate spin temperature associated with each population difference $n_{M_S+(1/2)}$.

This completes our study of the one-phonon or direct process. We have omitted mention of the resonance process, which some authors[64,68] describe as a limiting case of the direct process, for reasons which will become apparent in the next (two-phonon) section.

4. THE TWO-PHONON PROCESS

4.1. $\mathscr{H}_{\mathrm{OL}}$ in Second-Order Time-Dependent Perturbation Theory

In the previous section, we examined the effect of $\mathscr{H}_{\mathrm{OL}}$ to first order in the quantities $Q(\Gamma_{ig}, m)$, which, in turn, represent the normal modes of vibration of the local complex. In this section, we consider terms quadratic in these operators, arising both from the first-order expression for $\mathscr{H}_{\mathrm{OL}}$, (2.9), taken to second order in time-dependent perturbation theory, and the second-order expression for $\mathscr{H}_{\mathrm{SP}}$, (2.24), taken to first order. The reason for using $\mathscr{H}_{\mathrm{OL}}$ in one case and $\mathscr{H}_{\mathrm{SP}}$ in the other is clear from the discussion of Section 3. For $S = 1/2$, matrix elements of $\mathscr{H}_{\mathrm{SP}}$ must vanish between the time-reversed doublet states, (3.20), so that it is necessary to examine directly the matrix elements of $\mathscr{H}_{OL}$ between ground and excited levels. If $S > 1/2$, $\mathscr{H}_{SP}$ can account for two-phonon (Raman) relaxation within the ground multiplet. In addition, $\mathscr{H}_{OL}$ can cause two-phonon relaxation in second order involving transitions where energy is "almost" conserved at each step of the virtual process (thus, the "resonance" relaxation process, analogous to resonance fluorescence). Yet a third process can occur for multilevel spin systems, where the "ground" and "excited" levels are less than kT apart in energy. This process yields a Raman relaxation rate proportional to T^5 as mentioned in the introduction.

In order to sort out this plethora of terms, it is advantageous to consider first the general case of a multilevel spin system, picture in Fig. 4, and treat the one-phonon interaction $\mathscr{H}_{OL}$ in second-order time-dependent perturbation theory. For the moment, we do not specify the number of paramagnetic electrons on the central ion. We shall assume for simplicity that the ground and excited levels are time-reversed doublets. This is always true for ions with an odd number of electrons. For ions with an even number of electrons, the same situation, or the simpler one of an excited singlet, could obtain. The latter simplification would only reduce our calculated relaxation rates by a factor of two.[68] and the formal expression for the spin–lattice relaxation time would be the same as for an excited doublet.

The relaxation process "begins" with the paramagnetic ion in the upper ground level, $|+\rangle$, and the phonon states with wave vectors and polariza-

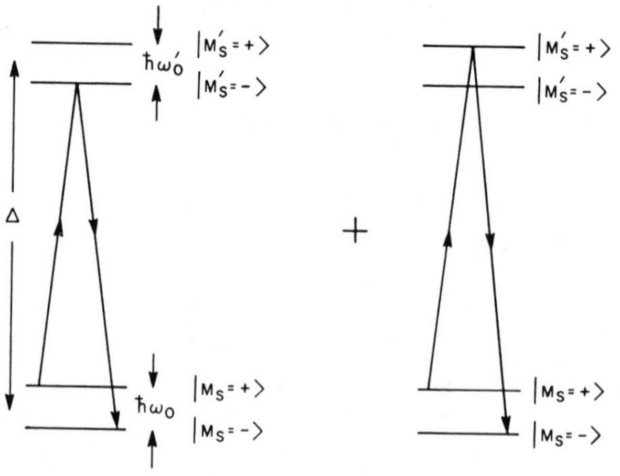

Fig. 4. Illustration of the two-phonon process. According to whether the intermediate transition to the level Δ above the ground state is virtual or real, the process is labeled Raman or resonance.

tion indices $\mathbf{k}, s$ and $\mathbf{k}', s'$ possessing equilibrium occupation numbers $n_{\mathbf{k},s}$ and $n_{\mathbf{k}',s'}$, respectively. An intermediate transition occurs to an excited ionic level $|j\rangle \equiv |\pm'\rangle$ with either a reduction in $n_{\mathbf{k},s}$, or an increase in $n_{\mathbf{k}',s'}$, by unity. According to whether this intermediate transition is virtual or real, the process is labeled Raman or resonance. In the next (and final) stage of the transition, the ion reverts to the lower ground level, $|-\rangle$, and $n_{\mathbf{k}',s'}$ is increased, or $n_{\mathbf{k},s}$ decreased, respectively, by unity. There are clearly many different paths which will lead to this final state and interference between paths is, as we shall see, a crucial factor in two-phonon transitions for ions with an odd number of electrons. For the purposes of our calculation, the orbit–lattice Hamiltonian will be taken to have the form (2.9) with the phonon operators $Q(\Gamma_{ig}, m)$ defined by (2.12). The second-order form for the orbit–lattice interaction is

$$\mathscr{H}_{\mathrm{OL}}^{(2)} = \sum_{j=\pm';\phi'} \frac{\mathscr{H}_{\mathrm{OL}}\,|\phi'\rangle\,|j\rangle\langle j|\,\langle\phi'|\,\mathscr{H}_{\mathrm{OL}}}{E_{\phi,+} - E_{\phi',j}} \tag{4.1}$$

where $|j\rangle$ represents the excited ionic level. It will be a matter of some importance later as to whether the energy difference $E_j - E_\pm = \Delta$ between this excited level (or levels) and the ground doublet is greater or less than the two characteristic energies $k\theta_{\mathrm{D}}$ and kT. The phonon intermediate states are labeled by $|\phi'\rangle$, where $|\phi'\rangle$ can be either $|\ldots, n_{\mathbf{k},s} - 1, \ldots, n_{\mathbf{k}',s'}, \ldots\rangle$ or $|\ldots, n_{\mathbf{k},s}, \ldots, n_{\mathbf{k}',s'} + 1, \ldots\rangle$, depending on the combination of phonon operators appearing in (4.1). The energies $E_{\phi,\pm}$ equal $E_\phi + E_\pm$, the sum of spin and phonon energies appropriate to the energy of the initial states. In almost every two-phonon case of interest (save the T^5 process mentioned above), we may ignore the energy difference between $|+\rangle$ and $|-\rangle$, or $|+'\rangle$ and $|-'\rangle$, compared to the energy difference $E_{\pm'} - E_\pm = \Delta$.

Using (2.9), we can rewrite (4.1) as

$$\begin{aligned}\mathscr{H}_{\mathrm{OL}}^{(2)} = -\sum_{\substack{i,l,m;\\ i',l',m';j,\phi'}} & V(\Gamma_{ig}, l)V(\Gamma_{i'g}, l')(-1)^m(-1)^{m'} \\ & \times Q(\Gamma_{i'g}, -m')\,|\phi'\rangle\langle\phi'|\,Q(\Gamma_{ig}, -m) \\ & \times C(\Gamma_{i'g}l', m')\,|j\rangle\langle j|\,C(\Gamma_{ig}l, m)(\Delta + E_{\phi'} - E_\phi)^{-1}\end{aligned} \tag{4.2}$$

When we calculate the transition probability per unit time for $\mathscr{H}_{\mathrm{OL}}^{(2)}$ to cause relaxation transitions, a delta function with argument $\hbar\omega_0 + \hbar\omega(\mathbf{k}, s) - \hbar\omega(\mathbf{k}', s')$ will be present (conservation of energy). This simply means that the difference between the initial and final phonon state energies must equal the change in ground-state Zeeman energy $\hbar\omega_0$. Usually, $\hbar\omega_0 \ll kT$, so that, to make effective use of the phonon density of states, as discussed in the introduction, it is clear that we should consider only processes where a phonon is absorbed (emitted) and another is emitted (absorbed). Double absorption would involve phonons of only half the speaking-term phonon energy, and the higher order of perturbation theory makes such processes negligible compared to the direct process. The most important contributions to $\mathscr{H}_{\mathrm{OL}}^{(2)}$ will arise from phonons whose energies ($\sim kT$) are $\gg\hbar\omega_0$ so that later we will drop the Zeeman energy in the argument of the delta function.

The easiest case to consider is when $\Delta \gg \hbar\omega_{\mathrm{D}}$, the maximum phonon energy, so that the energy denominator in (4.2) never causes trouble. This is the case treated by authors prior to 1961, and results in the usual T^7 and T^9 Raman processes. The former is appropriate to ions with an even number of electrons, the latter to ions with an odd number. The reason for this difference of behavior is of some interest, and we shall examine it in detail below.

If we carry out the sum over excited electronic states $|j\rangle$ or $|+'\rangle$ and $|-'\rangle$, we can write (4.2) out in detail for phonon emission and absorption appropriate to Raman-type processes as described above [see Eq. (2.12)]. We have

$$\begin{aligned}\mathscr{H}_{\mathrm{OL}}^{(2)} = &\sum_{\substack{\mathbf{k},s;\\ \mathbf{k}',s'}} \sum_{\substack{i,l,m,\\ i',l',m'}} V(\Gamma_{ig}l)V(\Gamma_{i'g}l') \\ &\times(-1)^m(-1)^{m'} \frac{\hbar R_{\mathbf{k},s}(\Gamma_{ig}, m)R_{\mathbf{k}',s'}(\Gamma_{i'g}, m')}{2Ma^2[\omega(\mathbf{k}, s)]^{1/2}[\omega(\mathbf{k}', s')]^{1/2}} \\ &\times C(\Gamma_{i'g}l', m')\,|+'\rangle\langle+'|\,C(\Gamma_{ig}l, m) \\ &\times\left[\frac{b^*_{\mathbf{k}',s'}b_{\mathbf{k},s}}{\Delta-\hbar\omega(\mathbf{k}, s)} + \frac{b_{\mathbf{k}',s'}b^*_{\mathbf{k},s}}{\Delta+\hbar\omega(\mathbf{k}, s)}\right] \\ &+\sum_{\substack{\mathbf{k}'',s''\\ \mathbf{k}''',s'''}} \sum_{\substack{i'',l'',m'',\\ i''',l''',m'''}} V(\Gamma_{i''g}l'')V(\Gamma_{i'''g}l''') \\ &\times(-1)^{m''}(-1)^{m'''} \frac{\hbar R_{\mathbf{k}'',s''}(\Gamma_{i''g}, m'')R_{\mathbf{k}''',s'''}(\Gamma_{i'''g}, m''')}{2Ma^2[\omega(\mathbf{k}'', s'')]^{1/2}[\omega(\mathbf{k}''', s''')]^{1/2}} \\ &\times C(\Gamma_{i'''g}l''', m''')\,|-'\rangle\langle-'|\,C(\Gamma_{i''g}l'', m'') \\ &\times\left[\frac{b^*_{\mathbf{k}''',s'''}b_{\mathbf{k}'',s''}}{\Delta-\hbar\omega(\mathbf{k}'', s'')} + \frac{b_{\mathbf{k}''',s'''}b^*_{\mathbf{k}'',s''}}{\Delta+\hbar\omega(\mathbf{k}'', s'')}\right]\end{aligned} \tag{4.2a}$$

We have kept the contributions to $\mathscr{H}_{\mathrm{OL}}^{(2)}$ arising from the two intermediate spin states separate for good reason. If we sandwich $\mathscr{H}_{\mathrm{OL}}^{(2)}$ between initial ($|+\rangle$) and final ($|-\rangle$) ground spin states between which relaxation transitions are to occur, the first term in (4.2a) contains the matrix element

$$\langle - \mid C(\Gamma_{i'g}l', m') \mid +' \rangle \langle +' \mid C(\Gamma_{ig}l, m) \mid + \rangle$$

while the second contains the matrix element

$$\langle - \mid C(\Gamma_{i'''g}l''', m''') \mid -' \rangle \langle -' \mid C(\Gamma_{i''g}l'', m'') \mid + \rangle$$

Following Ref. 62 [their Eq. (15.23b)] or, equivalently, (3.15) of this chapter, the latter can be written as

$$(-1)^{2M_S} \langle - \mid C(\Gamma_{i''g}l'', m'') \mid +' \rangle \langle +' \mid C(\Gamma_{i'''g}l''', m''') \mid + \rangle$$

Comparing with the former, we can identify the double with the single primed, and the triple with the unprimed variables. We thereby are allowed to collapse (4.2a) to read

$$\begin{aligned} \langle - \mid \mathscr{H}_{\mathrm{OL}}^{(2)} \mid + \rangle = \sum_{\substack{\mathbf{k},s \\ \mathbf{k}',s'}} \sum_{\substack{i,l,m \\ i',l',m'}} & V(\Gamma_{ig}l) V(\Gamma_{i'g}l') \\ & \times (-1)^m (-1)^{m'} \frac{\hbar R_{\mathbf{k},s}(\Gamma_{ig}, m) R_{\mathbf{k}',s'}(\Gamma_{i'g}, m')}{2Ma^2[\omega(\mathbf{k}, s)]^{1/2}[\omega(\mathbf{k}', s')]^{1/2}} \\ & \times \langle - \mid C(\Gamma_{i'g}l', m') \mid +' \rangle \langle +' \mid C(\Gamma_{ig}l, m) \mid + \rangle \\ & \times \left\{ b^*_{\mathbf{k}',s'} b_{\mathbf{k},s} \left[\frac{1}{\Delta - \hbar\omega(\mathbf{k}, s)} + (-1)^{2M_s} \frac{1}{\Delta + \hbar\omega(\mathbf{k}', s')} \right] \right. \\ & \left. + b_{\mathbf{k}',s'} b^*_{\mathbf{k},s} \left[\frac{1}{\Delta + \hbar\omega(\mathbf{k}, s)} + (-1)^{2M_s} \frac{1}{\Delta - \hbar\omega(\mathbf{k}', s')} \right] \right\} \end{aligned} \tag{4.2b}$$

It is interesting to note that (4.2b) would vanish for relaxation transitions between time-reversed ground states of half-integral spin, but for the presence of the phonon frequency factors in the denominators. This is the dynamic equivalent to the static result (3.20) that electric (time-even) operators cannot split (have matrix elements between, or differing values within) the components of a time-reversed doublet of half-integral spin. Indeed, in the static limit [all $\omega(\mathbf{k}, s)$, $\omega(\mathbf{k}', s')$ zero], (4.2b) vanishes identically for $2M_S$ odd. The cancellation has been termed the "Van Vleck" cancellation. It is also interesting to stress the off-diagonal character of the matrix elements of $\mathscr{H}_{\mathrm{OL}}^{(2)}$. If the phonon states $\mathbf{k}, s$ and $\mathbf{k}', s'$ were identical, (4.2b) would vanish for $2M_S$ odd, even in the dynamic regime.

The Raman process rate, corresponding to the absorption of the phonon $\mathbf{k}$, s and the emission of the phonon $\mathbf{k}'$, s', follows from (4.2b) by sandwiching $\mathscr{H}_{\mathrm{OL}}^{(2)}$ between the phonon states as well as the spin states. The former are chosen to be

$$|\,(\text{initial})\rangle = |\ \ldots, n_{\mathbf{k},s}, \ldots, n_{\mathbf{k}',s'}, \ldots\rangle$$
$$|\,(\text{final})\rangle = |\ \ldots, n_{\mathbf{k},s} - 1, \ldots, n_{\mathbf{k}',s'} + 1, \ldots\rangle$$

We find

$$\begin{aligned} W_{+\to-}^{\mathbf{k},s\to\mathbf{k}',s'} &= (2\pi/\hbar^2)\Bigg|\sum_{\substack{i,l,m,\\ i',l',m'}} V(\Gamma_{ig}l)V(\Gamma_{i'g}l')(-1)^m(-1)^{m'} \\ &\times \frac{\hbar[R_{\mathbf{k},s}(\Gamma_{ig}, m)R_{\mathbf{k}',s'}(\Gamma_{i'g}, m') + (-1)^{2M_S}R_{\mathbf{k}',s'}(\Gamma_{ig}, m)R_{\mathbf{k},s}(\Gamma_{i'g}, m')]}{2Ma^2[\omega(\mathbf{k}, s)]^{1/2}[\omega(\mathbf{k}', s')]^{1/2}} \\ &\times \langle -\,|\, C(\Gamma_{i'g}l', m')\,|\,+'\rangle\langle +'\,|\, C(\Gamma_{ig}l, m)\,|\,+\rangle \\ &\times \left\{\frac{n_{\mathbf{k},s}^{1/2}(n_{\mathbf{k}',s'}+1)^{1/2}}{\Delta - \hbar\omega(\mathbf{k}, s)} + (-1)^{2M_s}\frac{n_{\mathbf{k},s}^{1/2}(n_{\mathbf{k}',s'}+1)^{1/2}}{\Delta + \hbar\omega(\mathbf{k}', s')}\right\}\Bigg|^2 \\ &\times \delta[\omega(\mathbf{k}, s) - \omega(\mathbf{k}', s')] \end{aligned} \tag{4.3}$$

4.2. Raman Processes for Non-Kramers Levels

We evaluate (4.3) for the simpler case of transitions between non-Kramers levels.* This means here that $|\pm\rangle$ is a doublet (Fig. 4) formed from eigenstates of an ion with an even number of electrons. In this case, $2M_S$ is even, and since Δ is assumed $\gg kT$, the phonon frequencies in the denominator of (4.3) can be neglected compared to Δ. The delta function results in $n_{\mathbf{k},s} = n_{\mathbf{k}',s'}$ and we find, summing over all phonons $\mathbf{k}$, $\mathbf{k}'$, s, s', a

* Van Vleck[70] objects to the use of the terms introduced by one of us (R.O.) "non-Kramers" and "Kramers" ions, pointing out that Cr^{3+}, for example, has an odd number of electrons, three to be precise, but yet matrix elements of H_{OL} in the absence of magnetic mixing are allowed between states of the ground, $S = 3/2$, manifold (e.g., between $M_S = +3/2$ and $+1/2$, $+3/2$ and $-1/2$). This remark is of course completely correct, and for this reason, we have chosen to refer to non-Kramers and Kramers *transitions* in this chapter. By this, we mean transitions between time reversed states of integral and half-integral spin, respectively. Hopefully, this will clarify the unfortunate usage of these terms which one of us introduced into the literature previously.[54]

total transition probability per unit time equal to

$$W_{+\to-} = (2\pi/\hbar^2) \sum_{\substack{\mathbf{k},s \\ \mathbf{k}',s'}} \Bigg| \sum_{\substack{i,l,m \\ i',l',m'}} V(\Gamma_{ig}l)V(\Gamma_{i'g}l')(-1)^m(-1)^{m'}$$

$$\times \frac{\hbar[R_{\mathbf{k},s}(\Gamma_{ig}, m)R_{\mathbf{k}',s'}(\Gamma_{i'g}, m') + R_{\mathbf{k}',s'}(\Gamma_{ig}, m)R_{\mathbf{k},s}(\Gamma_{i'g}, m')]}{Ma^2\omega^{1/2}(\mathbf{k}, s)\omega^{1/2}(\mathbf{k}', s')\,\Delta}$$

$$\times\langle - \mid C(\Gamma_{i'g}l', -m') \mid +'\rangle\langle +' \mid C(\Gamma_{ig}l, m) \mid +\rangle \Bigg|^2$$

$$\times n_{\mathbf{k},s}(n_{\mathbf{k},s} + 1)\, \delta[\omega(\mathbf{k}, s) - \omega(\mathbf{k}', s')] \tag{4.4}$$

In order to evaluate (4.4), it is necessary to evaluate the angular sums over $\mathbf{k}$, $\mathbf{k}'$, s, and s' for the crystal of interest. In general, this is very difficult, but some simplifications can be achieved utilizing the orthogonality properties of the subvectors Γ_{ig}, m. The absolute value square in (4.4) will involve products of four R functions, but their angular average will vanish unless they can be paired in conjugate quantities. This is as far as one can go on symmetry properties alone. It is seen from (4.4) that whether one can bring outside of the square the sums over i and m depends on the specific symmetry under consideration. In general, one cannot, and the particular decision to sum first and then square, or sum the squares of the individual terms in (4.4), as suggested by Scott and Jeffries,[13] is indeterminate. To proceed further, we note that the R's in the long-wavelength limit, a requirement which is severely limiting ($T \ll \theta_D$), according to (2.14) are proportional to $ka = \omega a/v$. This means that, transforming from a sum to an integral [see (3.6)]

$$W_{+\to-} = (C/\varrho^2\,\Delta^2 v^{10}) \int d\omega\, n(\omega)[n(\omega) + 1]\omega^6 \tag{4.5}$$

where ϱ is the crystal density, v some appropriately averaged velocity of sound, and C a constant with units of (energy)^4. In the Debye approximation, (4.5) reduces to

$$W_{+\to-} = (C/\varrho^2\Delta^2 v^{10}) I^6 (kT/\hbar)^7 \tag{4.6}$$

where

$$I^n = \int_0^{\theta_D/T} dx[x^n e^x/(e^x - 1)^2] \tag{4.7}$$

Thus, at sufficiently low temperatures (the integrals I^n converge rather slowly), I^6 is independent of the upper limit only for[71] $T < 0.02\theta_D$. Numerical evaluation of the I^n integrals has been carried out by Rogers and Powell and their results are available in tabular form.[72] Even using the correct numerical values for I^n can lead to error; structure in the phonon density of states can show up in the measured relaxation rate.[73–75] For sufficiently low temperature, the upper limit in I^n may be replaced by infinity so that $I^n \to \int_0^\infty x^n e^{-x}\, dx = n!$. Therefore,

$$W^{R}_{+\to-} = [(6!)C/\varrho^2 \Delta^2 v^{10}](kT/\hbar)^7 \tag{4.8}$$

This is the usual Raman process result for non-Kramers transitions. The numerical coefficients appearing in (4.8) are large, and care must be taken to carry out the angular averages contained in the full expression (4.4). Equation (3.10) demonstrates that the inverse of the Raman spin–lattice relaxation time, $1/T_1^{R}$, is twice (4.8) when $kT \gg \hbar\omega_0$. At high temperatures ($T \gtrsim \theta_D$) and $x \ll 1$ in Eq. (4.7), $I_n \to \int_n^{\theta_D/T} x^4\, dx$ and the temperature dependence of the relaxation rate goes as T^2. The tenth power of sound velocity appearing in $W^{R}_{+\to-}$ causes a diminution in Raman relaxation rates in hard crystals, as pointed out earlier. For example, this is the reason why the resonance relaxation process (to be described below) dominates the Raman process up to 100°K for Co^{2+} in Al_2O_3.[75]

4.3. Raman Processes for Kramers Levels

For transitions in Kramers ions $2M_S$ is odd, and instead of adding, the last term in the curly brackets in (4.3) almost cancels with the first. The lack of complete cancellation is due to the dynamic character of the second-order matrix element contributing to the relaxation process, and was an important contribution of Van Vleck's[10] treatment. The resultant relaxation rate is reduced from that of non-Kramers transitions by $[\hbar\omega(\mathbf{k}, s)/\Delta]^2$ in the integrand, or by $(kT/\Delta)^2$ when the integration is carried out, giving a T^9 dependence to the relaxation rate. As noted in the introduction and in Sections 2 and 3, use of the Zeeman interaction to mix the ground $|\pm\rangle$ levels with the excited $|\pm'\rangle$ levels would also "break" the cancellation of the two terms in (4.3). The amount to which cancellation is avoided is of order $(\hbar\omega_0''/\Delta)^2$ in the calculated rate,[9] where $\hbar\omega_0''$ represents the matrix element of the Zeeman interaction between the ground and excited levels, yielding an H^2T^7 dependence of the relaxation rate. For every case studied so far, the two-phonon contributions have only been important when $kT \gg \hbar\omega_0''$, so that the more important contribution arises from the dynamical character of the two-phonon process, and *not* from static Zeeman admixtures. Finally, then, the relaxation rate for half-integral M_S becomes

$$W^{\mathrm{R}}_{+\to-} = (4\pi/\hbar^2) \sum_{\substack{\mathbf{k},s \\ \mathbf{k}',s'}} \Bigg| \sum_{\substack{i,l,m \\ i',l',m'}} V(\Gamma_{ig}l)V(\Gamma_{i'g}l')(-1)^m(-1)^{m'}$$
$$\times \frac{\hbar[R_{\mathbf{k},s}(\Gamma_{ig}, m)R_{\mathbf{k}',s'}(\Gamma_{i'g}, m') - R_{\mathbf{k}',s'}(\Gamma_{ig}, m)R_{\mathbf{k},s}(\Gamma_{i'g}, m')]}{2Ma^2\omega^{1/2}(\mathbf{k}, s)\omega^{1/2}(\mathbf{k}', s')}$$
$$\times\langle - \mid C(\Gamma_{i'g}l', m') \mid +'\rangle\langle +' \mid C(\Gamma_{ig}l, m) \mid +\rangle$$
$$\times n^{1/2}_{\mathbf{k},s}(n_{\mathbf{k}',s'} + 1)^{1/2}[\hbar\omega(\mathbf{k}', s') + \hbar\omega(\mathbf{k}, s)]$$
$$\times \{[\Delta - \hbar\omega(\mathbf{k}, s)][\Delta + \hbar\omega(\mathbf{k}', s')]\}^{-1} \Bigg|^2 \delta[\omega(\mathbf{k}, s) - \omega(\mathbf{k}', s')] \tag{4.9}$$

The same difficulty with respect to angular averaging which plagued us for non-Kramers transitions does so again here. To obtain a qualitative picure, we again expand R in the long-wavelength limit so that

$$W^{\mathrm{R}}_{+\to-} = (C\hbar^2/\varrho^2\, \Delta^4 v^{10})I^8(kT/\hbar)^9 \tag{4.10}$$

At sufficiently low temperatures [this requirement is even more severe here than for non-Kramers transitions—see discussion following (4.7)], we have

$$W^{\mathrm{R}}_{+\to-} = (8!C\hbar^2/\varrho^2\, \Delta^4 v^{10})(kT/\hbar)^9 \tag{4.11}$$

It is often stated that this rate is substantially smaller than the non-Kramers rate (4.8), but one must be quantitatively careful because of the large numerical factors involved when estimating the integrals. As with (4.8), the inverse of the Raman relaxation time is simply twice (4.10). At high temperatures, $T \gtrsim \theta_{\mathrm{D}}$, (4.10) is proportional to T^2.

4.4. Two-Phonon Interaction in First-Order Perturbation Theory

A word should be said concerning the use of the two-phonon interaction once, instead of the one-phonon interaction twice. For non-Kramers transitions, there is no essential change from (4.8), except of course that the square of the two-phonon coupling coefficient replaces C/Δ^2. As a rule of thumb, when the various terms in the spin-phonon Hamiltonian are expanded as a power series in the strain (valid at long wavelengths),

$$\mathscr{H}_{\mathrm{SP}} = GO(S)\varepsilon + G'O'(S)\varepsilon^2 + \cdots \tag{4.12}$$

where $O(S)$ is an operator acting on the spin coordinates, and the magnitudes of the coefficients G are (very) roughly equal. Since these coefficients are *also* roughly equal to the static crystalline field coefficients,[54] and these are of order Δ, the contribution of the two-phonon interaction taken once, proportional to $(G')^2$, is comparable to the one-phonon interaction taken twice, proportional to $(G')^4/\Delta^2$, i.e., the two are roughly comparable.

This makes quantitative estimates of non-Kramers relaxation rates difficult, requiring the knowledge of not only the one-phonon orbit–lattice coupling coefficients (2.16), measurable from static strain experiments, but also the two-phonon coupling coefficients, (2.24), for which at the present time no method of measurement exists. A further "in principle" difficulty exists for the determination of two-phonon coupling constants. Raman processes involve short-wave phonons [of energy $\sim kT$ or wave vector $(T/T_D)k_D$], whereas static stress measurements can only produce $\mathbf{k} = 0$ strains. Any frequency dependence (from, for example, dispersion, or the implicit frequency dependence of the Raman coupling constant in ionic crystals coming from the interplay between phonon wavelength and distant-neighbor potential contributions) of the coupling constant can thus produce serious discrepancies between static-stress and Raman-relaxation-rate parameters. For Kramers transitions, the situation is much simpler. The two-phonon spin Hamiltonian (2.24) vanishes between time-reversed half-integral states and only the dynamical character of the one-phonon spin Hamiltonian via intermediate levels allows transitions to occur. Of course, Zeeman interactions could "break" the time-reversed character of the ground levels, but this would again give rise to a rate proportional to $(G')^2(\hbar\omega_0''/\Delta)^2$, the factor to be compared with the "one-phonon twice" result, (4.9), $(G)^4(kT)^2/(\Delta)^4$. Since $G \sim G' \sim \Delta$, the "one-phonon twice" rate (4.9) dominates the "two-phonon once" rates for Kramers ions by $(kT/\hbar\omega_0'')^2$, which is $\gg 1$ when the two-phonon rate is important.

4.5. Resonant Two-Phonon Relaxation

So far, we have considered only the simplest of two-phonon relaxation processes, those for which the energy of the excited level(s) $\Delta \gg k\theta_D$, the maximum phonon energy. It is of interest to ask what would happen should $\Delta < k\theta_D$. A glance at the general expression for the two-phonon rate, (4.3), demonstrates the difficulty. The first term in the curly brackets diverges when $\hbar\omega(\mathbf{k}, s) = \Delta$. This divergence is a manifestation of a resonance in the two-phonon relaxation rate, the resonance occurring when one (and hence both) of the phonons involved in the normally virtual two-step process can satisfy energy conservation by itself [rather than as in the

Raman process, where one requires only that the difference of phonon energies, $\hbar\omega(\mathbf{k}, s) - \hbar\omega(\mathbf{k}', s')$, equal the Zeeman energy $\hbar\omega_0$]. This feature is quite common in optical spectroscopy, and is known as resonance fluorescence. The conservation of energy at *each* step in the virtual process makes one immediately wonder if a virtual process is conceptually necessary. Indeed, the resonance relaxation process which shall result from this condition is treated by many authors as just another form of the direct process in a multilevel spin system.[12,64,68] Though we shall not follow such a procedure here, Heitler[76] demonstrates that the two approaches yield nearly identical results [apart from some cross terms—see discussion after (4.23)] when the spectrum of incoming phonons (there, photons) is considerably broader than the lifetime width of the transition at energy Δ. This will always be the case in crystals [barring some catastrophic structure in the phonon density of states at $\hbar\omega(\mathbf{k}, s) = \Delta$]. We prefer to use the virtual transition (resonance fluorescence) approach because of its generality, and because in the limit of extreme dilution (where spin–spin effects are small compared to the linewidth of the excited $|\pm'\rangle$ levels due to phonon emission), it is the more appropriate method.

The method goes as follows.[76] The denominator of the first term in the curly brackets in (4.3) is kept from vanishing at resonance by recognizing that the excited spin levels, $|\pm'\rangle$, have finite lifetime against decay to the ground $|\pm\rangle$ levels because of phonon emission. The associated widths $\Gamma_{\pm'}$ are of course frequency-dependent. However, we shall assume the splitting to be much greater than $\Gamma_{\pm'}$ (and certainly greater than the energy width of the ground doublet), so that we need only consider transitions at energies Δ in our evaluation of $\Gamma_{\pm'}$. The appropriate result for $\Gamma_{+'}$ is simply [compare (3.4)]

$$\Gamma_{+'}/\hbar = (\pi/\hbar M a^2) \sum_{\substack{\mathbf{k},s\\ i,m\\ j=\pm}} \Big| \sum_l V(\Gamma_{ig} l)[R_{\mathbf{k},s}(\Gamma_{ig}, m)/\omega^{1/2}(\mathbf{k}, s)] \times \langle j |\, C(\Gamma_{ig} l, m) \,| +'\rangle (n_{\mathbf{k},s} + 1)^{1/2} \Big|^2 \delta[\omega(\mathbf{k}, s) - (\Delta/\hbar)] \tag{4.13}$$

where j denotes these levels to which transitions can take place. We are assuming here that only the four levels of Fig. 4 are relevant. Making use of the delta function and Eq. (3.6), we can simplify (4.13) to

$$\Gamma_{+'}/\hbar = \Delta[n(\Delta) + 1]/(2\pi\varrho a^2 \hbar^2) \sum_{\substack{i,m,s\\ j=\pm}} (1/v_s)^3 \times \Big[\Big| \sum_l V(\Gamma_{ig} l) R_{\mathbf{k},s}(\Gamma_{ig}, m) \langle j |\, C(\Gamma_i\, l, m) \,| +'\rangle \Big|^2 \Big]_{\mathrm{AV}} \tag{4.14}$$

The (angular) average is taken over the solid angle of **k**, and the magnitude of **k** in $R_{\mathbf{k},s}(\Gamma_{ig}, m)$ is fixed, equal to $\Delta/\hbar v_s$. The Bose factor in (4.14) is simply

$$n(\Delta) = (e^{\Delta/kT} - 1)^{-1} \tag{4.15}$$

The prescription we follow is to introduce this lifetime width into $W_{+\to-}$ by replacing $\Delta - \hbar\omega(\mathbf{k}, s)$ by $\Delta - \hbar\omega(\mathbf{k}, s) - (i/2)\Gamma_{+'}$ in (4.3). We are, of course, ignoring the real part of the dynamic corrections to the energy denominator, though one can imagine these to have been incorporated into the (much larger) splitting Δ to begin with. We shall next assume that the variation of the remaining frequency-dependent terms (including the nonresonant terms) in the summand of (4.3) are slow compared to the now Lorentzian-like variation of the first term in curly brackets. This is *not* always the case, and can result in an interference between the resonant and nonresonant terms.* This interference will be very important when kT is of the order of Γ and Δ/n, where $n = 7$ and 9 for non-Kramers and Kramers transitions, respectively. The need for such a correction does not seem to have been recognized in the literature as yet, but it can be responsible for significant departures from the normal Raman and resonance relaxation process rates. We shall neglect such effects here, as the integral in (4.3) must otherwise be done numerically. We next pull all the slowly varying frequency-dependent terms outside the summation over the magnitude of **k**, replacing $\omega(\mathbf{k}, s)$ everywhere by $\Delta/\hbar$ and making the usual spherical average. We neglect the second (nonresonant) term in the curly brackets in (4.3) for the moment, as indeed we have the nonresonant contribution of the first term. We find

$$\begin{aligned} W^{\circ}_{+\to-} = (2\pi/\hbar^2) \sum_{\substack{\Omega_{\mathbf{k}},\Omega_{\mathbf{k}'} \\ s,s'}} \Bigg| \sum_{\substack{i,l,m \\ i',l',m'}} & V(\Gamma_{ig}l)V(\Gamma_{i'g}l')(-1)^m(-1)^{m'} \\ & \times \hbar^2 n^{1/2}(\Delta)[n(\Delta)+1]^{1/2}[R_{\mathbf{k},s}(\Gamma_{ig}, m)R_{\mathbf{k}',s'}(\Gamma_{i'g}, m') \\ & +(-1)^{2M_s}R_{\mathbf{k}',s'}(\Gamma_{ig}, m)R_{\mathbf{k},s}(\Gamma_{i'g}, m')] \\ & \times (2M\,\Delta a^2)^{-1}\langle - \mid C(\Gamma_{i'g}l', m') \mid +'\rangle\langle +' \mid C(\Gamma_{ig}l, m) \mid + \rangle \Bigg|^2 \\ & \times \sum_{|\mathbf{k}|,|\mathbf{k}'|} \{[\Delta - \hbar\omega(\mathbf{k}, s)]^2 + \tfrac{1}{4}\Gamma^2_{+'}\}^{-1}\,\delta[\omega(\mathbf{k}, s) - \omega(\mathbf{k}', s')] \end{aligned} \tag{4.16}$$

We have written separately the sums over the solid angles $\Omega_{\mathbf{k}}$ and $\Omega_{\mathbf{k}'}$ from the sums over the magnitudes of the wave vectors $|\mathbf{k}|$ and $|\mathbf{k}'|$

* This is a well-known phenomenon in high-energy physics, and has been used in a separation of overlapping scattering and excitonic states in crystals by Phillips.[77]

(soon to be converted to integrals over frequencies), and have set $n_{\mathbf{k},s} = n_{\mathbf{k}',s'}$ by virtue of the delta function at the end of (4.16). The magnitudes of the wave vectors $\mathbf{k}$ and $\mathbf{k}'$ appearing in the expressions $R_{\mathbf{k},s}$ and $R_{\mathbf{k}',s'}$ are equal to $\Delta/\hbar v_s$ and $\Delta/\hbar v_{s'}$, respectively. We have neglected the effects of finite Zeeman energies in (4.16), though these will be exhibited below in a simplified result. In most cases, such modifications make no significant difference (since the process itself is usually only important at temperatures $kT \gg \hbar\omega_0$). The double sum over the magnitudes of $\mathbf{k}$ and $\mathbf{k}'$ is readily performed if we neglect the zero- and Debye-frequency cutoffs. This is valid if the resonance is sufficiently sharp and not localized near either zero or the Debye frequency. In this case,

$$\{[\Delta - \hbar\omega(\mathbf{k}, s)]^2 + \tfrac{1}{4}\Gamma_{+'}^2\}^{-1} \approx 2\pi\delta[\Delta - \hbar\omega(\mathbf{k}, s)]/\Gamma_{+'}$$

and using Eq. (3.6) to replace the sums by integrals, Eq. (4.16) becomes

$$\begin{aligned} W^{\circ}_{+\to-} = {} & (\Delta^2/4\pi^2\varrho^2\hbar^3 a^4)\{n(\Delta)[n(\Delta)+1]/\Gamma_{+'}\} \\ & \times \sum_{s,s'} (1/v_s^{\,3} v_{s'}^3) \Bigg[\Bigg| \sum_{\substack{i,l,m \\ i',l',m'}} V(\Gamma_{ig}l) V(\Gamma_{i'g}l')(-1)^m(-1)^{m'} \\ & \times [R_{\mathbf{k},s}(\Gamma_{ig}, m) R_{\mathbf{k}',s'}(\Gamma_{i'g}, m') \\ & + (-1)^{2M_s} R_{\mathbf{k}',s'}(\Gamma_{ig}, m) R_{\mathbf{k},s}(\Gamma_{i'g}, m')] \\ & \times \langle - | C(\Gamma_{i'g}l', m' | +' \rangle \langle +' | C(\Gamma_{ig}l, m) | + \rangle \Bigg|^2 \Bigg]_{\mathrm{AV}} \end{aligned} \tag{4.17}$$

where the (angular) average is taken over the solid angles of $\mathbf{k}$ and $\mathbf{k}'$. We now use (4.14) in (4.17) to write

$$\begin{aligned} W^{\circ}_{+\to-} = {} & (\Delta/2\pi\hbar^2\varrho a^2) n(\Delta) \\ & \times \Bigg\{ \sum_{s,s'} (1/v_s^{\,3} v_{s'}^{3}) \Bigg[\Bigg| \sum_{\substack{i,l,m \\ i',l',m'}} V(\Gamma_{ig}l) V(\Gamma_{i'g}l')(-1)^m(-1)^{m'} \\ & \times [R_{\mathbf{k},s}(\Gamma_{ig}, m) R_{\mathbf{k}',s'}(\Gamma_{i'g}, m') \\ & + (-1)^{2M_s} R_{\mathbf{k}',s'}(\Gamma_{ig}, m) R_{\mathbf{k},s}(\Gamma_{i'g}, m')] \\ & \times \langle - | C(\Gamma_{i'g}l', m' | +' \rangle \langle +' | C(\Gamma_{ig}l, m) | + \rangle \Bigg|^2 \Bigg]_{\mathrm{AV}} \Bigg\} \\ & \times \Bigg\{ \sum_{\substack{i,m,s \\ j=\pm}} (1/v_s^{\,3}) \Bigg[\Bigg| \sum_{l} V(\Gamma_{ig}l) R_{\mathbf{k},s}(\Gamma_{ig}, m) \\ & \times \langle j | C(\Gamma_{ig}l, m) | +' \rangle \Bigg|^2 \Bigg]_{\mathrm{AV}} \Bigg\}^{-1} \end{aligned} \tag{4.18}$$

This expression looks formidable, but it can be reduced very rapidly if one is willing to make some rather unrealistic approximations. Before doing so note that the temperature dependence is contained entirely in the Bose factor $n(\Delta)$, so that when $kT \ll \Delta$,

$$W^{\circ}_{+\to-} \propto e^{-\Delta/kT} \quad (4.19)$$

This is a very useful result. Apart from describing the temperature dependence of a specific relaxation mechanism, it can be "turned around" to yield the energy position of an excited paramagnetic level which might otherwise be unobservable. Indeed, this kind of "phonon spectroscopy" has been used on a number of occasions in just such a context.* Interestingly, there are cases when optical data are available which seem to show systematic discrepancies between the values of Δ so obtained and those obtained from relaxation time measurements using (4.19). In general, the latter always seem to lie below the former in energy†: This systematic variation has been explained recently by Young and Stapleton[80] as being due to a spatial distribution of values of the crystalline field splitting (hence, of Δ). The exponential character of the dependence of $W_{+\to-}$ on Δ then favors the contribution from these ions with smaller Δ's in the sample. Young and Stapleton hypothesize a Gaussian distribution of these quantities, and the relaxation-time-fitted widths of the distributions nicely match the observed optical widths. This is further discussed in Section 5.

The relaxation rate (4.18) appears to depend on Δ in only two ways: the direct proportionality of the first term, and the exponential contribution of the second (Bose) term. In fact, of course, the projection coefficients $R_{\mathbf{k},s}(\Gamma_{ig}, m)$ are also proportional to Δ in the long-wavelength limit [see the defining equation (2.13) and (2.14), and remember that $|\mathbf{k}| = \Delta/\hbar v_s$ in (4.18)]. The four powers in the numerator cancel against two powers in the denominator, so that in this limit

$$W^{\circ}_{+\to-}(\text{long wave}) \propto \Delta^3(e^{\Delta/kT} - 1)^{-1} \quad (4.20)$$

The long-wave approximation means here that $\Delta \ll k\theta_D$, an extremely limiting inequality for soft crystals (but adequate for hard crystal hosts—e.g., Al_2O_3). In the short-wavelength limit (when $\Delta \sim k\theta_D$), $R_{\mathbf{k},s}(\Gamma_{ig}, m)$

* For example, Castner[78] used this technique to obtain the valley–orbit splitting of P, As, Sb, and Bi in Si.

† Compare the 34 ± 0.5°K value for in cerium magnesium nitrate obtained by T_1 measurements in Ref. 12 and the value of 36.25 ± 0.4°K observed optically by Thornby.[79]

is only a weakly varying function of $|\mathbf{k}|$ (essentially independent of Δ) so that

$$W_{+\to-}^{\mathrm{O(short\ wave)}} \propto \Delta(e^{\Delta/kT} - 1)^{-1} \tag{4.21}$$

Clearly, great care must be taken when using (4.20). For example, when studying the systematics of the resonance relaxation process in isomorphic hosts, one must take into account the changes in $R_{\mathbf{k},s}(\Gamma_{ig}, m)$ as Δ and θ_{D} vary.

The result (4.17) illustrates our remarks on the applicability of the resonance fluorescence approach. It is clear that the width $\Gamma_{+'}$ present in the denominator of (4.17) is *at least* given by the phonon emission linewidth (4.14). It may, in the spirit of the approach, also be due to spin–spin interactions, or some other "detuning" or lifetime effect. The resulting relaxation rate (4.17) could then be reduced from the value we have found using (4.14). Since we have argued that the final result (4.18) ought to be identical with that obtained from the two-step direct process, the latter not affected directly by spin–spin interaction widths, an inconsistency is raised. We claim that this is only apparent for the following reason. When the (dipolar) broadening exceeds $\Gamma_{+'}$, as given by (4.14), the coherence of the excited level in the resonance process is destroyed. Our technique of treating the spin–lattice relaxation as essentially a one-step process is then no longer valid. It is now really a two-step process, and the relaxation time must be computed utilizing the two-step method.[12,64,68,81] The destruction of coherence should increase the calculated relaxation time for a one-step process, and this is apparently the case in (4.17). This increase is spurious. If we restrict $\Gamma_{+'}$ to be only that part of the width of $|+'\rangle$ given by (4.14), then (4.17) and hence (4.18) will be valid quite generally, regardless of the relative magnitude of the dipolar interaction. Just this state of affairs occurs in resonance fluorescence in gases, as was pointed out to the authors by Drs. G. W. Series and R. Hindmarsh (private communication). There, the additional broadening may be due to collision effects, and the methods of handling it are just those of which we have made use above.

It is possible to approximately connect (4.18) with the two-step relaxation rate by glancing at the expression for the transition rates which lead to $\Gamma_{+'}$. We designate the *spontaneous* part of the transition rate (4.14) as

$$\Gamma_{+'\to+}^{\mathrm{spont}} = \hbar B_1, \qquad \Gamma_{+'\to-}^{\mathrm{spont}} = \hbar B_2 \tag{4.22}$$

where, in (4.14), $j = +$ and $-$, respectively. Neglecting correlations between the phonons $\mathbf{k}$ and $\mathbf{k}'$ when performing the angular averages indicated in

(4.18), and using (4.14) and (4.22), we find

$$W^{\circ}_{+\to-} = [2B_1B_2/(B_1+B_2)]n(\Delta)$$
$$+(\Delta/2\pi\hbar^2\varrho a^2)n(\Delta)\Big\{\sum_{s,s'}(1/v_s^3v_{s'}^3)\sum_{\substack{i,l,m;\\ i',l',m';\\ \bar{l},\bar{l}'}}(-1)^{2M_s}$$
$$\times(-1)^m(-1)^{m'}V(\Gamma_{ig}l)V^*(\Gamma_{ig}\bar{l}')V(\Gamma_{i'g}l')V^*(\Gamma_{i'g}\bar{l})$$
$$\times[|R_{\mathbf{k},s}(\Gamma_{ig},m)R_{\mathbf{k}',s'}(\Gamma_{i'g},m')|^2$$
$$+|R_{\mathbf{k},s}(\Gamma_{i'g},m')R_{\mathbf{k}',s'}(\Gamma_{ig},m)|^2]_{\text{AV}}$$
$$\times[\langle-|C(\Gamma_{i'g}l',m')|+'\rangle\langle+'|C(\Gamma_{ig}l,m)|+\rangle]$$
$$\times[\langle-|C(\Gamma_{ig}\bar{l}',m)|+'\rangle\langle+'|C(\Gamma_{i'g}\bar{l},m')|+\rangle]^*\Big\}$$
$$\times\Big\{\sum_{\substack{i,m,s\\ j=\pm}}(1/v_s^3)\Big[\Big|\sum_l V(\Gamma_{ig}l)R_{\mathbf{k},s}(\Gamma_{ig},m)$$
$$\times\langle j|C(\Gamma_{ig}l,m)|+'\rangle\Big|^2\Big]_{\text{AV}}\Big\}^{-1} \tag{4.23}$$

The complexity of (4.23) is caused by cross terms which depend on the particular symmetry of the point group. For example, for Kramers transitions, they vanish for site symmetries with crystalline potentials of even modulae. They do *not* vanish for trigonal symmetry. For non-Kramers transitions, they do not vanish in general. Their existance complicates any simple relation between the one-step resonance fluorescence transition rate and the two-step direct-process rate. Lyo[81b] has shown that (4.23) reduces (remarkably!) to the form first derived by Culvahouse and Richards[68] for the resonance relaxation rate:

$$\frac{1}{T_1^0}=\frac{4B_1B_2}{B_1+B_2}\left[1-\frac{|C|^2}{B_1B_2}\,\frac{(B_1+B_2)^2}{(B_1+B_2)^2+(\omega_0')^2}\right]n(\Delta) \tag{4.24a}$$

where $\hbar\omega_0'$ is the Zeeman splitting of the excited doublet, C is always $\leq[B_1B_2]^{1/2}$, and is a rate corresponding to the product of the transition amplitudes for $|-\rangle\to|-'\rangle$ and $|-\rangle\to|+'\rangle$. The lack of energy conservation is apparently taken care of by the spontaneous widths at $|\pm'\rangle$, accounting for the Lorentzian form of the second term in (4.24a). The magnitude of C varies according to the local site symmetry and the direction of the external field, vanishing, for example, in a site of threefold symmetry with the magnetic field perpendicular to the symmetry axis. Culvahouse and Richards[68] also find a frequency (g) shift with a magnitude, under the most favorable conditions, of 1/7 the ground doublet lifetime broadened linewidth. This shift, given careful attention, should be observable, but has yet to be reported.

When the spontaneous width $\Gamma_{+'}$ is small compared to the Zeeman splittings of the excited doublet, $\hbar\omega_0'$, the second term in the square bracket of (4.24a) can be neglected, and the resonance relaxation rate takes the familiar form

$$1/T_1^0 = 4B_1B_2n(\Delta)/(B_1 + B_2) \tag{4.24b}$$

Should we relax the condition[81] that $\hbar\omega_0$ and $\hbar\omega_0'$ be small compared to kT, the Bose factor in (4.24b) should be replaced by

$$n(\Delta) \rightarrow [\cosh(\hbar\omega_0/2kT)\cosh(\hbar\omega_0'/2kT)]n(\Delta) \tag{4.25}$$

The replacement is not appropriate for use in (4.24a) because the second term in the square brackets arises from a spread of excited state energies, and one cannot utilize the energy conservation conditions which lead to (4.25).

As mentioned in Section 3, a field dependence of the two-phonon process resulting from dipole–dipole and hyperfine interactions also obtains. This dependence is usually trivial in EPR measurements where the Zeeman splitting, $\hbar\omega_0$, is much greater than the root-mean-square dipolar or hyperfine widths. However, in nonresonant methods,[2] this is not the case and these interactions are important. Orbach[65] has shown that

$$\left(\frac{1}{T_1^0}\right)_H = \left(\frac{1}{T_1^0}\right)_{H\text{ large}} \frac{H^2 + \mu H_{\text{hyp}}^2 + \frac{1}{2}\mu' H_{\text{dip}}^2}{H^2 + H_{\text{hyp}}^2 + \frac{1}{2}H_{\text{dip}}^2} \tag{4.26}$$

where H_{hyp}^2 and H_{dip}^2 are the mean-squared hyperfine and dipolar fields, respectively. The quantities μ and μ' are complicated functions in general [see Eq. (38) of Ref. 65, where it is shown that μ' may even be temperature-dependent!] but, for a vanishing perpendicular g value in the ground doublet, reduce to 1 and 2, respectively. This latter result yields an expression identical to that derived by Van Vleck[10] and demonstrates the limitation of his purely "effective field" model.

Returning to our discussion of (4.18), it should be clear that the overall relaxation rate is a sum of (4.18) together with either (4.8) or (4.11), according to whether the transition is non-Kramers or Kramers in character. This addition of relaxation rates is only valid, however, as stated earlier, when the area under the resonant part of the integrand in (4.3) is well-separated from the area under the nonresonant part. An overlap can occur when $\Delta \sim nkT$; and if $\Gamma_{+'}$ is also comparable to kT, an appreciable interference results, modifying the above remarks. A perturbation treatment of the integral was first given by Stoneham,[81c] while Lyo[81b] carried out the full numerical integration. The effect of finite excited state widths is to reduce the apparent energy separation of excited states from the ground state.

4.6. T^5 Relaxation

Finally, there is one "further" temperature dependence for the relaxation rate which can be extracted from (4.3). This result, listed in the introduction as the T^5 process, can occur when $\hbar\omega(\mathbf{k}, s) \gtrsim \Delta$, a situation common to multilevel systems (where $S_{\text{eff}} > 1/2$). It occurs for Kramers transitions if we use (4.3), but Walker,[37] and independently, Yafet,[38] have demonstrated that the use of Γ_{1g} vibrational modes leads to a similar result for non-Kramers transitions. Their scheme is illustrated in Fig. 5. The origin of the effect for Kramers transitions is straightforward. From (4.3), the term in curly brackets ($2M_S$ odd) becomes

$$-2n_{\mathbf{k},s}^{1/2}(n_{\mathbf{k}',s'} + 1)^{1/2}\hbar\omega(\mathbf{k}, s)/\{\Delta^2 - [\hbar\omega(\mathbf{k}, s)]^2\} \tag{4.27}$$

where we have made explicit use of the delta function on the energies $\hbar\omega(\mathbf{k}, s)$, $\hbar\omega(\mathbf{k}', s')$. For $\hbar\omega(\mathbf{k}, s) \gg \Delta$, the curly bracketed term reduces to

$$2n_{\mathbf{k},s}^{1/2}(n_{\mathbf{k}',s'} + 1)^{1/2}/\hbar\omega(\mathbf{k}, s) \tag{4.28}$$

This result, when compared to the opposite limit, which led to an *increase* in the power of $\omega(\mathbf{k}, s)$ appearing in the summand by two, and thence to

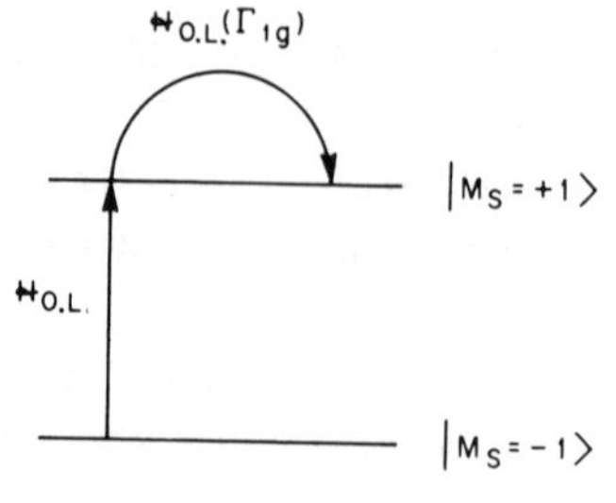

Fig. 5. T^5 relaxation in two-phonon process. A Γ_{1g} vibration is introduced in one of the transitions. See text.

(4.11), now leads to a *reduction* in the power of $\omega(\mathbf{k}, s)$ in the summand by two. We have, therefore,

$$W^{R}_{+\to-} = [C\hbar^2/(\varrho^2 \Delta^4 v^{10})] I^4 (kT/\hbar)^5 \tag{4.29}$$

The argument due to Walker[37] and Yafet,[38] appropriate to non-Kramers transitions, is almost identical, except that they introduce Γ_{1g} vibrations which do not change the state of the spin system in one of the two transition matrix elements, resulting in a "bare" $\hbar\omega(\mathbf{k}, s)$ in the energy denominator.

For an ion with unquenched orbital moment, the criterion for the importance of (4.29), as compared to the non-Kramers transition rate (4.8), is roughly that

$$\Delta < kT \tag{4.30}$$

where Δ can be a crystalline field, spin–orbit, or Zeeman splitting. This inequality obtains because the integrand is greatest when $\hbar\omega(\mathbf{k}, s)$ is of the order of kT (actually, when of the order of nkT where $n = 7$). For quenched orbital moment, the condition for the T^5 term to dominate the normal Raman term is (again roughly)

$$\lambda(\lambda/\Delta) > kT \tag{4.31}$$

where λ is the spin–orbit coupling parameter, and Δ an appropriate crystal field or Zeeman splitting. The condition (4.31) is satisfied for rare-earth *S*-state ions over a wide temperature range [$\lambda \sim 1500$ cm^{-1}, $\Delta \sim 30{,}000$ cm^{-1}, so that (4.29) dominates for $T \lesssim 110$°K]. For iron-group *S*-state ions (e.g., Mn^{2+}), $\lambda \sim 300$ cm^{-1} and $\Delta \sim 23{,}000$ cm^{-1}, leading to a condition that $T < 4$°K for the T^5 dependence to dominate over the T^7 dependence. Since at such low temperatures the predominant relaxation process is direct, the T^5 term may in fact not be observable for this system.

It should be mentioned that, again, one must be careful about interference between the T^5 and the resonance relaxation process. Whereas, in the case of T^7 or T^9 situations (non-Kramers or Kramers transitions, respectively), the interference region was approached from below (i.e., $kT \ll \Delta$), for temperatures where the T^5 dependence is important, the interference is approached from above (i.e., $kT \gg \Delta$). Hence, there may be significant departures from a T^5 law when it first becomes effective, but as the temperature continues to rise, the interference will become significantly less. Only quantitative evaluation of (4.3) under the conditions of the experiment in question can determine the importance of this effect.

This completes the discussion of theoretical models for the various relaxation processes.* In the next section, application will be made of the results of Sections 2–4 to specific materials, and experimental values obtained for the various parameters exhibited in section 2 will be given.

5. EXPERIMENTAL RESULTS

5.1. Introduction

Several experimental techniques are available for determining spin–lattice relaxation times and numerous references can be cited for each method.†

Nonresonant[82a–82c] methods have been used for about 30 years and involve measurements of magnetic absorption (χ'') and/or dispersion (χ') as a function of frequency in the audio or RF regions. Sample heating or resistive changes in a sample-loaded coil can be used to measure χ'', while χ' can be found from the change in the inductance of a sample-loaded coil as measured by frequency changes in a tuned RF circuit or by an inductance bridge. From the frequency dependence of χ' or χ'', the relaxation time T_1 can be obtained. These nonresonant methods lack any means of isolating a particular pair of magnetic states for study if more than one is involved and they require a number of measurements in order to yield one relaxation time at a given temperature and field. They are, however, simpler than resonance methods if one wishes to determine T_1 at several magnetic fields, since the same apparatus can be used.

With the development of paramagnetic resonance after World War II, three new methods of measuring relaxation times become available. These are CW saturation[82d] of an ESR line, measurements of ESR linewidths,[82e] and measurement of transient signals after strong microwave pulses.[82f–82h] The technique of CW saturation[82d] or linewidths[82e] does not involve a direct time measurement but depends instead upon a model of an ESR line. The simplest situation is a homogeneously broadened line with a Lorentzian shape. If used alone, the CW saturation method requires an

* The entire discussion has been for isolated paramagnetic centers. Relaxation of exchange coupled clusters is discussed in Section 4.6 in Chapter 6 by Owen and Harris. Spin-lattice relaxation in the presence of a Jahn–Teller effect requires special consideration and is given in Section 2.6 of Chapter 1 by Ham.

† Available techniques include those involving mutual inductance,[82a] heterodyne beat frequencies,[82b] calorimetry,[82c] CW saturation,[82d] line broadening,[82e] pulse saturation,[82f] electron spin echoes,[82g] spin population inversion,[82h] transient Faraday rotation,[82i] and fluorescence.[82j] The papers just cited list further references which the interested reader can consult.

accurate knowledge of the microwave magnetic field over the sample volume at partial saturation of the ESR line for the value of T_1 to be extracted. Alternatively, measurements of only the relative incident microwave power may be substituted in the CW saturation method if the system is calibrated at some temperature by another technique such as pulse saturation. This assumes that the Q of the microwave cavity resonance remains constant with temperature. Linewidth measurements yield T_1 values only when the ESR line is appreciably broadened by relaxation processes and this generally limits T_1 values to less than 10 nsec.

The laboratory production of fast microwave pulses became simpler in the early 1960's with the advent of solid-state microwave switches with 10-nsec switching times. A single switch typically produces by reflection a 25 dB on–off transmission ratio and the switches can be easily ganged by interposing phase shifters. By means of these switches, a single microwave source can be used to both saturate a given pair of energy levels at high power and to monitor the return of the populations to thermal equilibrium at very low power. Unless special precautions are observed, the bandwidth of the monitoring receiver system will limit the measurable values of T_1 to 10 μsec or longer. More sophisticated systems, to which we shall refer later, allow one to measure T_1 and T_2 by spin-echo techniques or to follow the recovery of an ESR signal after inversion by adiabatic fast passage.

By combining pulse saturation, CW saturation, and linewidth techniques, one can measure T_1 values in a fairly continuous manner over nine decades of time. However, in any T_1 measurement, the transient behavior of the ESR signal may not reflect the basic Van Vleck–Kronig mechanism, but may instead be determined by the effects of cross-relaxation within a multilevel spin system, cross-relaxation to a rapidly relaxing impurity or to exchange-coupled pairs, spectral diffusion within an inhomogeneously broadened line, slight admixtures of dispersion into the absorption signal, sample heating, or a phonon bottleneck. By working with magnetically dilute samples, sharp resonance lines are obtained and the probability of cross-relaxation is reduced. Care must be taken to keep rapidly relaxing impurities such as Fe^{2+} out of the host lattice. The signal recovery must be checked to see if it is truly exponential and independent of the power or duration of the microwave pulse. Of course, if the system is phonon-limited, multileveled, or inhomogeneously broadened, a single exponential recovery may not be expected, but useful information may still be obtained from the signal recovery if it becomes exponential near thermal equilibrium.

Optical methods have also been employed to measure spin–lattice relaxation. For example, transient measurements of Faraday rotation[82i] after pulsing the magnetic field yield T_1 values and also provide a usef method of measuring the magnetic field dependence of T_1. Another ma netooptical technique which is applicable to T_1 measurements in metastab excited states involves the microwave pulse saturation of two Zeema levels of the optically pumped metastable state and the optical detectio of intensity changes in one of the Zeeman components of the fluorescenc spectrum.[82j]

Selected experimental data on static strain effects and relaxation rat will be presented in the following sections. Summaries of these data ar presented in Tables I and II, respectively.

Table I. Summary of Strain Data on the ESR Spectra of Iron Group Io in MgO and CaO[a]

		Point charge	Covalent	Spin–spin anisotropy	MgO theor. total	MgO exp.	CaO exp.
V^{2+}	G_{11}	0.51	−0.07	0.12	0.56[b]	0.42[c]	1.3[d]
	G_{44}	2.16	0.97	0.00	3.13[b]	3.0[c]	—
Cr^{3+}	G_{11}	0.40	0.13	0.08	0.61[b]	0.6[e]	—
	G_{44}	2.64	2.32	0.00	4.96[b]	4.2[e]	—
Ni^{2+}	G_{11}	15	28	—	43[b]	57[e]	110[d]
	G_{44}	49	9	—	58[b]	36[e]	42[d]
Fe^{2+}	G_{11}	—	—	—	—	800[e]	—
	G_{44}	—	—	—	—	540[e]	—
Co^{2+}	G'_{11}	−642	—	—	−642[f]	−69[f]	—
	G'_{12}	321	—	—	321[f]	32[f]	—
	G'_{44}	88	—	—	88[f]	10[f]	—
Mn^{2+}	G_{11}	0.442	−0.029[i]	—	0.413[g]	1.5[h]	0.48[d]
	G_{44}	−0.216	0.042[i]	—	−0.174[j]	−0.32[h]	−0.10[d]
Fe^{3+}	G_{11}	—	—	—	—	5.5[h]	1.5[d]
	G_{44}	—	—	—	—	−0.83[h]	−0.45[d]

[a] Results are given in terms of the G_{ij} or G'_{ij} coupling tensors defined by (2.17) an (2.23), respectively. Units are cm^{-1}/unit strain for G_{ij} and (unit strain)$^{-1}$ for G'_{ij}.
[b] Reference 31.
[c] Elsa Feher, *Bull. Am. Phys. Soc.* **10**, 699 (1965).
[d] Reference 29.
[e] Reference 26, Watkins and Feher.
[f] Reference 28.
[g] Reference 34.
[h] Reference 26, Feher.
[i] Overlap only.
[j] Equation (31) of Ref. 34 is correct, but Eq. (30), which leads to it, has the incorrect sign.

Table II. Summary of the Selected Experimental Relaxation Data Discussed in the Text, Arranged According to Electronic Configuration and Ion[a]

Ion	Host	Mechanism	Comments	Text	Ref.
$(3d^1{:}Ti^{3+})$	Al_2O_3	1, 2	Direct rate proportional to $\sin^2\theta$	205	139
$(3d^3{:}V^{2+})$	Al_2O_3	2	Verified Δ^3 dependence by comparison with Cr^{3+} and Mn^{4+}	205	142
		—	Measured changes in fine structure under stress	183, 184	85, 86
$(3d^3{:}Cr^{3+})$	Al_2O_3	2	Verified Δ^3 dependence by comparison with V^{2+} and Mn^{4+}	205	142
		—	Measured changes in fine structure under stress	183, 184	85, 86
		2	Measured for metastable $\bar{E}(^2E)$ states which relax through $2\bar{A}(^2\bar{E})$	184	81
		2	Calculation from static strain data	184	87
		2	Evidence for a phonon-limited resonant process	184	89, 90
		1	Measured θ dependence at 4.2°K, 9.36 Hz	180, 181	83
		1	Measured θ dependence at 35 GHz	180	84
		1	Calculation of θ dependence from stress data	180, 181	27
		—	Static stress measurements at 10 GHz, 300°K, 0.05% Cr	179	27
	Emerald	2	Measured for the two doublets	185	91
	$ZnWO_4$	2	Measured effects due to Li^+ charge compensation	185	92
$(3d^3{:}Mn^{4+})$	Al_2O_3	2	Verified Δ^3 dependence by comparison with V^{2+} and Cr^{3+}	205	142
$(3d^5{:}Fe^{3+})$	Al_2O_3	—	Measured effects of strain on fine structure	188	100
	$K_3Co(CN)_6$	1, 3	H^4 dependence of direct process	195	58, 59
		1, 3	Angular dependence of direct process due essentially to $g(\theta)$	205	140

Table II (*continued*)

Ion	Host	Mechanism	Comments	Text	Ref.
$(3d^7:Co^{2+})$	MgO	1	Compared direct rate to calculations from strain data on Zeeman and hyperfine interactions	180	28
	$ZnSiF_6 \cdot 6H_2O$	1	Hyperfine and angular effects measured	205	19
	$La_2Zn_3(NO_3)_{12} \cdot 24H_2O$	1, 3	Hyperfine effects on T_1^D measured	205	19, 141
$(3d^9:Cu^{2+})$	Organic complex	1	Compared full temporal recovery of signal from a bottlenecked system to theory	202	126
$(4f':Ce^{3+})$	$Ce_2Mg_3(NO_3)_{12} \cdot 24H_2O$	2	AC susceptibility data yielded first observation of resonant process; $\Delta = 34°K$	190	12
		2	Field dependence of resonant process noted even at high fields	206	144
	$La_2Mg_3(NO_3)_{12} \cdot 24H_2O$	2	Pulse method used, $\Delta = 32 \pm 2°K$	190	102
		2	Spin echo method used, $\Delta = 34°K$	190	103
		1, 2	Measurements to 0.25°K indicate phonon-limited direct process	192, 193	107
		1	Phonon avalanche observed	200, 201	56
		1	Phonon bottleneck studied	202	118
		1, 2, 3	Frequency dependence of resonance process implies phonon bottleneck	206	143
		2	Discrepancy between pulse and linewidth data implies bottleneck	206	145
	CaF_2	1, 3	Tetragonal sites studied	194	98
	$LaCl_3$	1, 2	Angular dependence of direct process studied; resonant process with $\Delta = 46°K$	197, 199	121, 122
	$LaCl_3:LaBr_3$	2	Related effective Δ to Br^- concentration	199	80
	LaF_3	1, 2, 3	Six nonequivalent sites; cross-relaxation at low temperature	204	126

	$Y(C_2H_5SO_4)_3 \cdot 9H_2O$	2, 3	Direct process still too weak to observe at 1.5°K, $\Delta = 25 \pm 1$°K	202	127
	$La(C_2H_5SO_4)_3 \cdot 9H_2O$	2	$\Delta = 5.67 \pm 1$°K, T_1 measured from 1.4 to 1.8°K	190	13
	$YCl_3 \cdot 6H_2O$	1, 2, 3	Monoclinic host	204	137
	$CaWO_4$	1, 3	Spin echo technique, measured $T_1^D(\theta)$, found $T_1^R \propto T^{10.9}$	203	129
		1, 3	T_1 measured between 1.5 and 50°K, measured T_1^D at two frequencies	204	130
	$PbMoO_4$	1, 3	T_1 measured between 1.5 and 35°K, measured T_1^D at two frequencies	204	130
($4f^2$:Pr^{3+})	$La_2Mg_3(NO_3)_{12} \cdot 24H_2O$	1, 2, 3	Bottlenecked direct process	190, 192	13
		1, 2, 3	Isotropic resonant and Raman mechanisms	202	118
	$La(C_2H_5SO_4)_3 \cdot 9H_2O$	1, 2	Bottlenecked direct process	202	127
	$Y(C_2H_5SO_4)_3 \cdot 9H_2O$	1, 2	Bottlenecked direct process	202	127
	$LaCl_3$	1, 2	Bottleneck direct process	197, 198	121
($4f^3$:Nd^{3+})	CaF_2	—	Uniaxial stress measurements, UPR absorption calculation	186	95
		—	Negative UPR results	186	96
		—	T_1^D calculations from strain data	186, 187	95
		1	T_1^D measured, compared to estimates	188	98
		1, 2	Tetragonal sites studied	194	98
		1, 2	Tetragonal and rhombic sites studied	195	115
	$La_2Mg_3(NO_3)_{12} \cdot 24H_2O$	1, 2	Pulse-saturation measurements between 1.4 and 4.5°K	190, 192	13
		1, 2	Pulse saturation data down to 0.25°K	192	107
		1, 2	Tested H^4 dependence of direct process	195	61
		2	Angular dependence due to anisotropy in the excited-state g values	202	118
		1	Studied hyperfine effects with ^{143}Nd	202	118
		2	Discrepancy between pulse and linewidth data implies bottleneck	206	145
	$La(C_2H_5SO_4)_3 \cdot 9H_2O$	1, 3	No resonant process; T^9 Raman rate	190, 191	13
		1	T_1^D influenced by cross-relaxation to ground and first excited states of Ce^{3+} impurity	202	118

Table II *(continued)*

Ion	Host	Mechanism	Comments	Text	Ref.
	$Y(C_2H_5SO_4)_3 \cdot 9H_2O$	1, 3	Isotropic Raman rate, anisotropic direct process	202	118, 127
	LaF_3	1	Field dependence has H^4, H^2, and H^0 terms	195	62
		1, 2, 3	Six nonequivalent sites, concentration-dependent cross-relaxation at low temperatures	204	136
	$LaCl_3$	1, 3	Measured $T_1^D(\theta)$	197, 199	121, 122
		3	Observed anisotropy in Raman process at 10°K	199	122
	$YCl_3 \cdot 6H_2O$	1, 2, 3	Monoclinic lattice, anisotropy factor for T_1^D is about 10^3	204	137
	$CaWO_4$	1, 3	Used spin-echo techniques, found Raman rate $\propto T^{10.4}$ and slightly anisotropic	203	129
		1, 2, 3	Measured between 1.5 and 30°K at two frequencies	204	131
		3	No resonant process identified	204	133
	Yttrium gallium garnet	1, 2	$\Delta = 112$°K	196	101
	Yttrium aluminum garnet	1, 2	$\Delta = 108$°K	196	101
	$CdMoO_4$	1, 2, 3	T_1 measured between 1.5 and 30°K	204	131
	$PbMoO_4$	1, 2, 3	Measured between 1.5 and 30°K; two frequencies	204	131
	$CaMoO_4$	1, 2, 3	Measured between 1.5 and 30°K; two frequencies	204	131
	$BaMoO_4$	1, 2, 3	Measured between 1.5 and 30°K	204	131
($4f^5$:Sm^{3+})	$Sm_2Mg_3(NO_3)_{12} \cdot 24H_2O$	1, 3	Bottlenecked direct process in this concentrated salt	190	13
	$La_2Mg_3(NO_3)_{12} \cdot 24H_2O$	1, 3	No bottleneck observed; limited temperature range of data, 1.4–2.3°K	190	13
		1, 2, 3	$\Delta = 55$°K, pulse data for 1.4–4°K	202	127
		1, 2, 3	Anisotropic direct rate; isotropic resonant and Raman process	202	118

		2	Linewidth data between 6 and 10°K in either resonant process ($\Delta = 66$°K) or Raman process (T^9)		
	$La_2(C_2H_5SO_4)_3 \cdot 9H_2O$	1, 2, 3	Two excited doublets ($\Delta = 46, 72$°K) contribute to resonant process	202	127
		1, 2, 3	Anisotropy factor of about 3 in direct process	202	118
		1, 3	Anisotropy factor of 6 in direct process, no resonant process identified	196	101
	$Y(C_2H_5SO_4)_3 \cdot 9H_2O$	1, 2, 3	2:1 anisotropy ratio of direct rate, isotropic resonant and Raman rates	202	118
		1, 2, 3	$\Delta = 51$°K	202	127
	LaF_3	1, 2	Six nonequivalent sites	204	136
	$LaCl_3$	1, 2, 3	Measured $T_1^D(\theta)$	197	121
		1, 2, 3	Measured Δ less than optical value	199	122
($4f^7$:Eu^{2+})	CaF_2	1, 3	T^5 Raman rate in this multilevel system	196	101
($4f^7$:Gd^{3+})	LaF_3	2	Six nonequivalent sites, $\Delta = 4.45$°K	204	136
	CaF_2	1, 3	T^5 Raman rate in this multilevel system; low-temperature rate is $T^{1/2}$	194	98
	Yttrium gallium garnet	—	Measured effects of strain on fine structure in orthorhombic symmetry	188	99
($4f^8$:Tb^{3+})	$Y(C_2H_5SO_4)_3 \cdot 9H_2O$	1, 3	Compared anomalous field dependence to theory	202	127
	$LaCl_3$	1, 2, 3	Measured $T_1^D(\theta)$	197	121
		1, 2, 3	Compared anomalous field dependence of all three processes to theory	202	128
	$CaWO_4$	1, 2, 3	$\Delta = 111$°K	204	132
	CaF_2	1, 2	Tetragonal sites studied	194	98
	SrF_2	1, 2	Tetragonal sites studied	195	111

Table II (*continued*)

Ion	Host	Mechanism	Comments	Text	Ref.
($4f^9$:Dy^{3+})	$Dy(C_2H_5SO_4)_3 \cdot 9H_2O$	1	Measured anisotropy in T_1^D with ac susceptibility techniques; $g_\perp = 0$	190	104
		2	Used transient Faraday rotation measurements	196	119
	$Y(C_2H_5SO_4)_3 \cdot 9H_2O$	2	Measurements on ground and first excited doublet using pulse and linewidth data	194	108
	$YCl_3 \cdot 6H_2O$	1, 2, 3	Monoclinic host, $T_1(\theta, \varphi)$ not symmetric with respect to principal axes at the g tensor	204	137
	CaF_2	—	$T_1 < 3$ μsec at $T = 2°K$; cubic, tetragonal, and trigonal sites	194	98
	CeO_2	1, 2	Trigonal sites	204	135
	LaF_3	1, 3	Six nonequivalent sites, cross-relaxation effects	204	136
	$La_2Mg_3(NO_3)_{12} \cdot 24H_2O$	2	No direct process observed as low as 1.3°K	202	127
($4f^{10}$:Ho^{3+})	$Ho(C_2H_5SO_4)_3 \cdot 9H_2O$	1	Transient Faraday rotation data imply harmonic cross-relaxation with excited state leading to oscillatory field dependence	196	119
	$LaCl_3$	—	Measured $T_1 = 3.5 \times 10^{-4}$ sec at $T = 1.14°K$	199	122
($4f^{11}$:Ho^{2+})	CaF_2	1, 2	Nearly degenerate Γ_7 and Γ_8 states may both contribute to the resonant process	196	101
($4f^{11}$:Er^{3+})	CaF_2	1, 2	Tetragonal, trigonal, and cubic sites	194	98
		2	Trigonal sites	195	114
	BaF_2	1, 2, 3	Trigonal sites	195	112
		1, 2	Tetragonal sites, anomalous low-temperature rates; possible bottleneck	195	114

	SrF_2	1, 2, 3	Trigonal sites	195	112
	CdF_2	1, 2	Cubic site, anomalous H dependence	195	114
		1, 2	Cubic site	195	117
	LaF_3	1, 2, 3	Six nonequivalent sites, concentration dependence at low temperature	204	136
	CeO_2	1, 2	Cubic and trigonal sites; required additional T^6 rate	204	134
	$CaWO_4$	1, 2	Two resonant processes, $\Delta = 26, 58°K$	204	132
	$YCl_3 \cdot 6H_2O$	1, 2	Monoclinic crystal, anisotropic resonant process	204	137
	Yttrium aluminum garnet	—	Measured effects of strain on g-tensor in orthorhombic symmetry	188	99
	$La(C_2H_5SO_4)_3 \cdot 9H_2O$	1, 2, 3	Isotropic resonant and Raman rates, anisotropic direct process which is not in agreement with estimate	202	118, 127
		1	Studied cross-relaxation by incorporating Ce^{3+}	202	118
	$Y(C_2H_5SO_4) \cdot 9H_2O$	1, 2, 3	Isotropic resonant and Raman rates, anisotropic direct process which is not in agreement with estimate	202	118, 127
		1	Hyperfine effects on T_1^D for Er^{167}	202, 203	118
	$LaCl_3$	1, 2, 3	Measured $T_1^D(\theta)$	121	121
		1, 2, 3	Measured Δ less than optical value	199	122
	$LaCl_3{:}LaBr_3$	2	Compared Δ measurements in mixed crystals	199, 200	80
$(4f^{13}{:}Tm^{2+})$	CaF_2	1, 3	T^9 Raman rate up to 20°K	196	101
$(4f^{13}{:}Yb^{3+})$	$Y(C_2H_5SO_4) \cdot 9H_2O$	—	Anisotropic T_1^D leads to use in a proton spin refrigerator, $g_\perp = 0$	190	105
	$CaWO_4$	1, 3	T_1 measured between 1.5 and 40°K, T_1^D measured at two frequencies	204	130
		1, 3	Spin echo techniques, Raman rate $\propto T^{10.1}$ and slightly anisotropic	203	129
	CaF_2	1, 2, 3	Cubic and tetragonal sites studied	194	98
		—	Static stress measurements	185	93
		1, 3	Cubic sites studied	195	113
	$PbMoO_4$	1, 3	T_1 measured between 1.5 and 30°K, T_1^D measured at two frequencies	204	130
	BaF_2	1, 3	Trigonal sites studied	195	113

Table II (*continued*)

Ion	Host	Mechanism	Comments	Text	Ref.
	SrF_2	1, 3	Trigonal sites studied	195	113
	CdF_2	1, 3	Cubic sites, anomalous frequency dependence of direct process	195	116
	ThO_2	—	Static stress data	185	93
		1, 3	Concentration-dependent T_1^D	186	94
	Yttrium gallium garnet	—	Measured stress effects on g-tensor in orthorhombic symmetry	188	99
		1, 3	Low-temperature rate is concentration-dependent	196	101
	Yttrium aluminum garnet	1, 3	Low-temperature rate is concentration-dependent	196	101
	Lutetium gallium garnet	1, 3	Low-temperature rate $\propto T^{1.7}$	196	101
	CeO_2	1, 3	Cubic sites; additional $T^{4.5}$ rate required	204	135
	LaF_3	1, 3	Six nonequivalent sites, cross-relaxation important at low temperatures	204	136
($5f^3$:U^{3+})	CaF_2	1	Compared T_1^D to calculations based on strain data	186, 187	95,97
		—	Uniaxial stress measurements	186	95

[a] The ion is listed in the first column, the host crystal in the second; column three contains a code indicating which relaxation mechanisms are observed: 1 implies a direct process, 2 a resonant process, and 3 a Raman mechanism. It should be noted that a designation of 1 may not necessarily imply a true direct relaxation mechanism such as $H^4 \coth(g\beta H/2kT)$ or $H^2 \coth(g\beta H/2kT)$, but rather some mechanism dominant at the lowest temperatures and different from both 2 and 3. The appropriate references (column six) should be consulted for any evidence of anomalous behavior, not specifically mentioned under the comments in column three, such as a concentration dependence, an unusual field dependence, or cross-relaxation effects. Column five cites the page in this review where the data are mentioned or discussed in detail. The complexity of relaxation data makes a more detailed description of such data too difficult to tabulate for the number of systems considered here, and the reader is referred to the original papers.

5.2. Relaxation Rates Predicted from the Dynamic Spin Hamiltonian Formalism

In Section 2, the use of the dynamic spin Hamiltonian to calculate relaxation rates was described. This formalism has the advantage of circumventing several difficult problems such as the validity of the point-charge model and the need for distant-neighbor interactions. Instead, the relaxation Hamiltonian is rewritten in an empirical form for which the only necessary assumption is the "long-wavelength approximation" for the phonons. Under this assumption, the elements of the dynamic spin Hamiltonian can be measured statically in a paramagnetically doped crystal. The question of bulk strain versus local strain at a paramagnetic site is then totally irrelevant to a relaxation calculation. All that is implied is that the phonons important in the relaxation mechanism are sufficiently long in wavelength that the local strain they produce at the paramagnetic site is the same as that produced by the corresponding static bulk strain. As mentioned earlier, this assumption is obviously best for the direct process which involves only phonons of microwave frequencies with $\lambda_{ph} \sim 10^3$ Å.

We begin, therefore, with systems in which a comparison between measurement and theory has been made for the direct relaxation process using static strain data. One such system which has been studied extensively is ruby. Hemphill *et al.*[27] have reported the effects of stress on the ESR of Cr^{3+} in Al_2O_3. Their measurements were done at room temperature near 10.1 GHz on a ruby crystal with a 0.05% Cr^{3+} concentration. Stress was applied to the sample with a screw and the strain was measured with several gauges. The static spin Hamiltonian for this system is axially symmetric with $S = \frac{3}{2}$ and contains a fine-structure term $D_0(S_z^2 - \frac{5}{4})$, reflecting the C_3 point symmetry at the Cr^{3+} site. The effects of a small uniaxial strain can be fully described by (2.16) and (2.17).

While all the Cr^{3+} sites in Al_2O_3 are magnetically equivalent in the absence of an externally applied strain, there are two nonequivalent sites under stress. A total of ten independent components of the strain coupling tensor G exist for each site. Six of these are identical for both sites, while the other four are opposite in sign. In a coordinate system where the x axis is along the crystallographic $+a$ axis and the z axis is along the crystallographic c axis, the measured values of the G tensor (in units of cm^{-1}/unit strain) are[27]

$$G_{11} = 4.57 \pm 0.3 \qquad G_{14} = -0.43 \pm 0.13$$
$$G_{12} = -1.94 \pm 0.3 \qquad G_{41} = -0.63 \pm 0.30$$
$$G_{33} = 6.40 \pm 0.13 \qquad G_{25} - 0.070G_{16} = \pm 1.50 \pm 0.2$$
$$G_{44} = 1.97 \pm 0.15 \qquad G_{52} + 0.159G_{45} = \pm 1.43 \pm 0.3$$

Donoho,[27] using preliminary G_{ij} values close to those listed above, computed the angular anisotropy of the direct relaxation rates for Cr^{3+} in Al_2O_3 in the long-wavelength limit. His thorough computer calculation took into account: effects due to an anisotropic phonon phase velocity; the Christoffel equations; phonon polarizations; the anisotropic effects of the magnetic field; and the effects of cross-relaxation within the multilevel spin system. For any given transition among the four spin levels (see discussion in Section 3.2), there are, in general, three time constants involved in the recovery to thermal equilibrium. Except for a scaling factor of one-half, Donoho's results correctly predict the angular anisotropy of the relaxation time of the high field (2–3) transition between 10° and 80°, as measured by Standley and Vaughan[83] at 4.2°K and 9.3 GHz. The comparison between theory and experiment is shown in Fig. 6. The relaxation of the 3–4 transition involves two time constants, and both were in good agreement with Donoho's calculations. Only samples which were magnetically dilute ($\leq$0.2 at.%) and magnetically pure gave results which were in good agreement with theory. The magnetic purity seemed to depend upon the crystal growing process, and of three methods tested, only the vapor-phase modification of the Verneuil process produced samples of the required purity.

The samples of Standley and Vaughan were used again by Lees *et al.*[84] to measure relaxation times at 35 GHz. As before, only samples grown with the vapor-phase modification showed any appreciable angular anisotropy in the relaxation time. The magnitudes were only about one-fourth of those predicted by Donoho, and the angular dependence was also in poor agreement with his calculations. The authors consider cross-relaxation to Cr^{2+} ions as a possible cause of this disagreement.

Considering the effort expended to make the relaxation calculations in ruby as accurate as feasible, it is discouraging to find that the excellent agreement which is present for the 9-GHz data fails to occur at 35 GHz. One might be a bit more comfortable if the comparisons with theory at the two frequencies were reversed.

Another system in which strain data were used to compute direct process relaxation rates is Co^{2+} in MgO. Here, Tucker[28] measured strain effects on both the Zeeman and hyperfine interactions, and found that, within experimental error, the strain-induced changes in the Zeeman and hyperfine tensors were proportional. In this cubic system, the effective spin is 1/2 and three nonzero elements of the G' tensor (2.23) are listed in Table I since G'_{11} and G'_{12} are now independent because the trace of $\mathscr{H}_{SP}$ is no longer required to be zero. From these data, Tucker estimated T_1

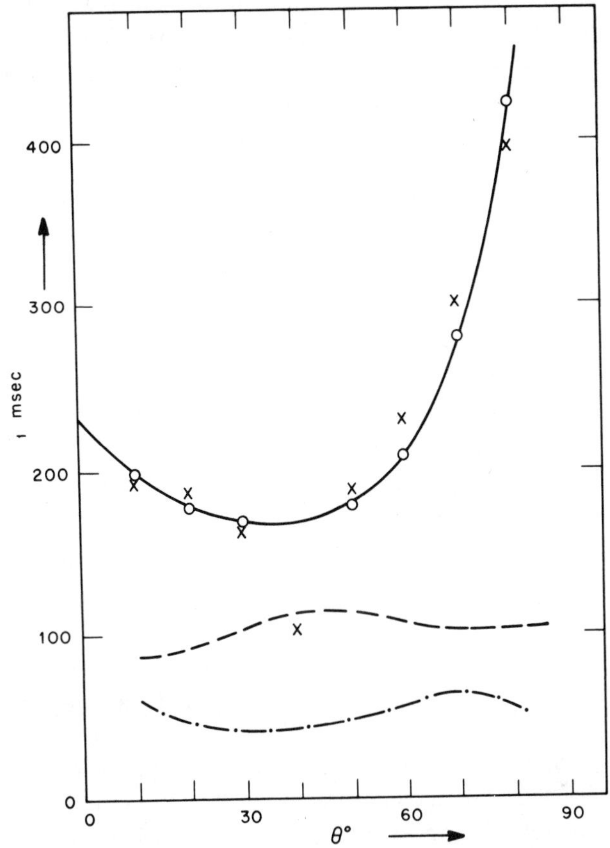

Fig. 6. Comparison of experimental and calculated relaxation data for the 2–3 transition of Cr^{3+} in Al_2O_3 at 4.2°K and 9.3 GHz. The open circles represent Donoho's calculated values (reduced by a factor of two) and the crosses are experimental points for a sample with low Cr^{3+} concentration and grown by the vapor-phase modification of the Verneiul process. The two lower curves give experimental results for samples grown by other methods. (From Standley and Vaughan.[83])

to be 60 sec, four to five orders of magnitude longer than the experimentally observed value. He concludes from this that the relaxation mechanism is not the direct process.

These results serve to illustrate the poor agreement between experiment and theory for direct-process relaxation rates when the dynamic spin-Hamiltonian approach is used along with experimental strain data. It seems that very often long direct-process T_1's are short-circuited by other processes such as cross-relaxation to other impurities. The theoretical approach appears rather invulnerable, so we shall retreat momentarily from relaxation rates and instead compare static strain data with theory.

5.3. Calculation of Strain Parameters and Comparison with Experiment

In the previous section, we have considered the strain parameters G and G' which appear in the dynamic spin Hamiltonian to be measured parameters obtained from static strain data. In this section, we turn our attention to the theoretical evaluation of these strain parameters and a comparison with experimental data. A fair amount of attention has been devoted to this subject, so we include it here with the *caveat* that such comparisons are more a test of assumptions which are extraneous to the theory of spin–lattice relaxation. For example, one may make a model calculation assuming point-charge ligands, only nearest-neighbor interactions, and local strain equal to bulk strain. All of these assumptions are neatly side-stepped in the long-wavelength limit of the dynamic spin-Hamiltonian approach.

We shall begin this selected review with experimental and theoretical data on strain effects in the cubic crystals MgO and CaO doped with iron group elements. Such data are presented in Table I in terms of the strain coefficients G_{11} and G_{44} defined by Eq. (2.17). It should be noted that some authors[28] define the off-diagonal elements of the lattice strain without the factor of one-half which appears in Eq. (2.18). Point-charge expressions for G_{11} and G_{44} have been listed by Tucker[28] using expressions derived by Al'tschuler *et al.*[32] For V^{2+} ($3d^3$), Cr^{3+} ($3d^3$), and Ni^{2+} ($3d^8$), the point-charge expressions are

$$G_{11} = \frac{100}{9} \frac{ee_{\text{eff}}}{R^5} \langle r^4 \rangle \frac{\lambda^2}{\Delta_1{}^2} \tag{5.1}$$

and

$$G_{44} = -\frac{2\lambda^2}{21\Delta_1} \frac{ee_{\text{eff}}}{R} \left\{ 15\left(\frac{2}{\Delta} + \frac{1}{\Delta_1}\right) \frac{\langle r^2 \rangle}{R^2} + 5\left(\frac{3}{\Delta} - \frac{2}{\Delta_1}\right) \frac{\langle r^4 \rangle}{R^4} \right\} \tag{5.2}$$

where Δ_1 and Δ are the crystal field splittings of the first and second excited states, respectively, relative to the ground level; λ is the spin–orbit coupling constant, e_{eff} is the effective ligand charge, $\langle r^n \rangle$ is the mean nth-power radius of the $3d$ electron, and R is the impurity–ligand separation.

For Fe^{2+} ($3d^6$), the coupling elements are

$$G_{11} = \frac{4\sqrt{3}}{10} \frac{ee_{\text{eff}}}{R} \left\{ \frac{3}{7} \frac{\langle r^2 \rangle}{R^2} + \frac{25}{63} \frac{\langle r^4 \rangle}{R^4} \right\} \tag{5.3}$$

and

$$G_{44} = \frac{1}{10\sqrt{2}} \frac{ee_{\text{eff}}}{R} \left\{ \frac{\langle r^2 \rangle}{R^2} + \frac{5}{33} \frac{\langle r^4 \rangle}{R^4} \right\} \tag{5.4}$$

These expressions are independent of λ, since the orbit–lattice matrix elements are allowed internal to the ground triplet.

Expressions for the S-state ions Mn^{2+} and Fe^{3+} are more involved, but the problem has been treated by Blume and Orbach,[17b] Leushin,[48a] and Sharma *et al.*[34]

Calvo *et al.*[29] have made a comparison of the experimental G_{11} values in MgO and CaO for d^3 (V^{2+}), d^8 (Ni^{2+}), and d^5 (Mn^{2+}, Fe^{3+}) configurations. They note that if the point-charge relation $1/R^5 \propto \Delta_1$ is incorporated into Eq. (5.1), then $G_{11}(\text{CaO})/G_{11}(\text{MgO})$ should change as $\Delta_1(\text{MgO})/\Delta_1(\text{CaO})$. Estimating this ratio to lie between 1.5 and 2.0, they compare with the measured G_{11} ratios of 3.0 ± 0.4 for V^{2+} and 1.8 ± 0.2 for Ni^{2+}. Closer agreement is found for the S-state ions using the expression[17a]

$$G_{11} = \frac{5\sqrt{10}}{9} \frac{ee_{\text{eff}}}{R} \frac{\langle r^4 \rangle}{R^4} \lambda^2 (2\Sigma' - \Sigma) \tag{5.5}$$

where the quantities Σ, Σ' refer to weighted inverse excited-state energies, and $(2\Sigma' - \Sigma)$ varies almost linearly with Δ_1 in the region of interest. The ratio $G_{11}(\text{CaO})/G_{11}(\text{MgO})$ would then be proportional to $[\Delta_1(\text{CaO})/\Delta_1(\text{MgO})]^2$ and thus lie between 0.44 and 0.25. Remarkably, the experimental ratios are 0.32 (Mn^{2+}) and 0.28 (Fe^{3+}) in spite of the fact that covalence effects in G_{11} for the latter are appreciable.

Feher and Sturge[85] have measured the change in the fine-structure splitting ($2D$) of ions with a $3d^3$ configuration in Al_2O_3 under uniaxial compression along the c axis. Their X-band results for V^{2+} and Cr^{3+} at 77°K and for Mn^{4+} at 1.4°K were $dD/dp = -20.3 \times 10^{-13}$ cm/dyn, and -31.8×10^{-13} cm/dyn, respectively. They also measured stress effects on the excited-state (^2E) splitting in these crystals by observing both the mean and the individual energy shifts of the R_1 and R_2 fluorescence lines under uniaxial stress applied parallel or perpendicular to the c axis. Both the unstressed ground-state splitting and the additional stress-induced splittings were found to vary only slightly among these isoelectronic ions.

However, the small variation was in the direction predicted from a point-charge model. The ^2E splitting and its change under stress varied greatly from ion to ion. The authors interpret these results in terms of the theory of Macfarlane[86] and conclude that v', the off-diagonal trigonal crystal field parameter, depends primarily on the long-range electrostatic interactions; while v, the diagonal trigonal crystal field parameter, depends on short-range effects. Geschwind *et al.*[81] have studied the spin–lattice relaxation rate within the metastable $\bar{E}(^2E)$ state in ruby using combined optical and microwave methods discussed earlier. The return of one of the Zeeman components of the R_1 fluorescence line to its steady-state intensity under optical pumping was followed after a microwave pulse saturation of the $\bar{E}$ magnetic sublevels. The relaxation rate was found to vary with temperature as $e^{-\Delta/kT}$, with $\Delta = 38.8 \pm 1.0\ \text{cm}^{-1}$, indicating a resonant relaxation mechanism involving the $2\bar{A}(^2\bar{E})$ levels which lie 29 cm^{-1} above the $\bar{E}$ states.

Blume *et al.*[87] have calculated the strength of the resonant relaxation process in the $\bar{E}$ states of ruby using data from static strain measurements. Their expression for the phonon-induced transition probability between the $2\bar{A}_-$ and the $\bar{E}_+$ states is

$$W_{2\bar{A}_- \to \bar{E}_+} = \frac{27}{4}\left(\frac{\Delta E}{T_{Z'Z'}}\,\frac{1}{S_{33}-S_{13}}\right)^2 \frac{\delta^3}{135\pi\varrho h^4} \times\left(\frac{1}{v_l^5}+\frac{3}{2}\,\frac{1}{v_t^5}\right)[n(\delta)+1] \qquad (5.6)$$

where ϱ is the mass density of the crystal, v_l and v_t are the longitudinal and transverse sound velocities, respectively, δ is the splitting between the $2\bar{A}$ and $\bar{E}$ states (29 cm^{-1} in ruby), and $n(\delta)$ is the thermal phonon occupation number at energy δ. The term $\Delta E/T_{Z'Z'}$ is known in ruby from the strain data of Schawlow[88] to be 5.5×10^{-11} cm/dyn. The difference between the elastic constants, $S_{33} - S_{13}$, is known to be 0.232×10^{-12} cm^2/dyn. With $\varrho = 4$ g/cm^3, $v_l \sim 10^6$ cm/sec, and $v_t \sim v_l/\sqrt{3}$, one obtains for ruby an estimated relaxation rate for the $\bar{E}$ states of $T_1^{-1} = 4W = 1.3\times10^8 e^{-\delta/kT}$ sec^{-1}. The coefficient, 1.3×10^8 sec^{-1}, should be compared with the experimental value of 2.6×10^8 sec^{-1}.

Recently, Adde and Geschwind[89] and Adde *et al.*[90] have noted that, while the temperature dependence of this relaxation rate is independent of the intensity of optical pumping, the rate itself depends upon it. Optical pumping determines the steady-state population of the $\bar{E}$ levels, so this experiment is equivalent to a series of ordinary pulse saturation ESR measurements with different effective Cr^{3+} concentrations. However, the same host crystal is being used. The authors suggest that this is evidence of a phonon bottleneck in the resonant relaxation process. Additional evidence and further discussion of this phenomenon will be given later.

We shall conclude our discussion of strain effects in iron group impurities in various host lattices with mention of the pulse saturation work of Orton and co-workers[91,92] to measure the relaxation times of Cr^{3+} in emerald and in $ZnWO_4$. They interpret their data in both hosts in terms of a cubic crystal field model, so that only G_{11} and G_{44} describe the spin–phonon interaction. The actual symmetry of the Cr^{3+} site in emerald (Cr^{3+}:$Be_3Al_2Si_6O_{18}$) is trigonal, like ruby. Relaxation measurements were performed at 9.3 GHz between 1.5 and 20°K, as a function of the angle between H and the trigonal crystal axis. They found that the relaxation rate was dominated by a resonant process which was nearly isotropic, except at angles where harmonic cross-relaxation occurred. Charge compensation effects were studied in the $ZnWO_4$ host by the incorporation of various amounts of Li^+. They found that T_1 became longer as the charge compensation was increased until full compensation was achieved. Excess Li^+ seemed to have no effect on T_1. They found no anisotropy in T_1 if charge compensation was complete and about 50% anisotropy otherwise.

5.4. Comparison of Strain and Relaxation Data on Crystals Containing Rare Earth Ions

We turn now to stress measurements on crystals containing rare earth ions. Tachiki *et al.*[23,93] have reported stress effects on Yb^{3+} in ThO_2 and Ho^{2+} in CaF_2. In both cases, the ions were in cubic sites with a Kramers doublet lying lowest. Measurements were made at 4° and 1.2°K in a 35-GHz spectrometer for stresses of up to 10^3 kg/cm² applied perpendicular to the static magnetic field. A single measurement of the resulting g-shift ($\Delta g_\perp$) permits all the stress-induced changes in the g tensor to be determined for a given stress axis in these crystals ($2\Delta g_\perp = -\Delta g_\parallel$). Those states which would produce a nonzero trace of Δg_{ij} are too far removed in energy to be effective. Hence, for these compounds, there exist only two independent elements of G'_{ij}. They may be determined from the measured g-shifts when the crystals are stressed along the [001] and [111] directions.

The results for Yb^{3+}:ThO_2 were $\Delta g_\perp[111] = 6.18\times10^{-6}$ kg^{-1} cm^2 and $\Delta g_\perp[001] = -(1.1 \pm 0.3)\times10^{-6}$ kg^{-1} cm^2, in close agreement with theoretical values based on a point-charge model. The spectrum of Ho^{2+}:CaF_2 is complicated by a hyperfine interaction with $I = 7/2$, but changes from the cubic hyperfine tensor are simply related to changes in the g-tensor. The experimental results were $\Delta g_{\perp}[001] = 1.43 \times 10^{-4}$ kg^{-1} cm^2 and $\Delta g_{\perp}[111] = 5.3 \times 10^{-5}$ kg^{-1} cm^2. Theoretical calculations yield 1.427×10^{-4} kg^{-1} cm^2 and 1.286×10^{-5} kg^{-1} cm^2. In both $Yb^{3+}:ThO_2$ and $Ho^{2+}:CaF_2$, the point-charge model gave theoretical values which agreed very well with the experimental results when the $l = 2$ term of the crystalline field potential dominated the contribution.

Relaxation measurements on Yb^{3+} in cubic sites of ThO_2 have been reported briefly by Wagner *et al.*[94] They measured T_1 between 1.3 and 22°K by the inversion-recovery method at a frequency near 9 GHz. They could fit their data to a sum of direct and Raman processes, but the strength of the former depended upon the Yb^{3+} concentration even at dopings as low as 0.001 at.%. Furthermore, they observed a different direct-process rate for Yb^{173} than for the even–even isotopes. Therefore, a comparison between the T_1^D calculated using the static stress measurements of Sroubek *et al.*[23,93] and experiment was not very informative.

Black and Donoho[95] have reported uniaxial stress measurements on the rare earth ion Nd^{3+} and its actinide analog U^{3+} in calcium fluoride. The ions occupied sites of tetragonal symmetry and had an isolated doublet lying lowest. Here again, one can introduce a perturbed g-tensor, Δg_{ij}, depending linearly upon the strain ε_{ij} through a fourth-rank tensor G'_{ijkl}. While ε_{ij} is symmetric, the authors point out that it is not necessary that Δg_{ij} be symmetric; hence there are, in general, 54 independent elements of G'_{ijkl} and Voigt notation breaks down unless symmetry considerations permit it. Unfortunately, the Black and Donoho experiments are appropriate to C_{4v} symmetry, so that there are eight independent nonzero elements, two of which are G'_{2323} and G'_{3232}. The effects of these two elements cannot be distinguished experimentally from each other. For simplicity, the authors took them equal, allowing Voigt notation. The seven independent elements were denoted by the subscripts 11, 12, 13, 31, 33, 44, and 66. Here, 3 denotes a tetragonal symmetry axis which lies along the equivalent [100] directions in the fcc CaF_2 lattice. The shifts in the EPR spectrum were measured at 4.2°K and at a frequency of 9.4 GHz. A single strain gauge measured strains up to 3×10^{-4}. The results are given in Table III (per unit strain).

From these data, the intensity of the ultrasonic paramagnetic resonance (UPR) absorption was calculated and found to be marginally detectable. Experimentally, this is in agreement with the negative UPR results of Wetsel and Donoho[96] for Nd^{3+} in CaF_2. The direct-process relaxation times T_1^D were then calculated for these ions, again using the G'_{ij} values, and the results were compared with experimental data. As shown in Fig. 7, the calculation

Table III

	U^{3+}	Nd^{3+}
G'_{11}	1.60 ± 0.5	4.60 ± 1.0
G'_{12}	1.30 ± 0.5	-3.10 ± 1.0
G'_{13}	1.00 ± 0.5	4.60 ± 1.5
G'_{31}	2.00 ± 1.0	1.80 ± 1.0
G'_{33}	3.20 ± 1.0	-0.70 ± 1.0
G'_{44}	-8.90 ± 1.0	-8.90 ± 0.2
G'_{66}	-5.85 ± 0.1	-0.06 ± 0.1

for both ions predicts some anisotropy in T_1^D with respect to both θ and φ, the polar angles of the magnetic field with respect to the symmetry axis. The predicted T_1^D values for U^{3+} at 4.2°K and 9.4 GHz start around 5.5 sec at $\theta = 0$°K, rise to 10 sec near 55°, and fall back to 2 sec at $\theta = 90°$. Berulava *et al.*[97] have measured relaxation times for U^{3+} in CaF_2 and find $T_1^D \approx 2.5$ sec at $\theta = 0°$ and ≈ 0.08 sec at $\theta = 90°$, but with evidence that a concentration dependence exists even at doping levels as low as 0.1 at.%. Black and Donoho's calculated T_1^D values for Nd^{3+}-doped CaF_2 under the same conditions of temperature and microwave frequency start at 8.5 sec at $\theta = 0$°K, rise to 15 sec near 65°, and fall to 1.5 sec at $\theta = 90°$. Bierig *et al.*[98] have measured the relaxation times of rare earths in CaF_2

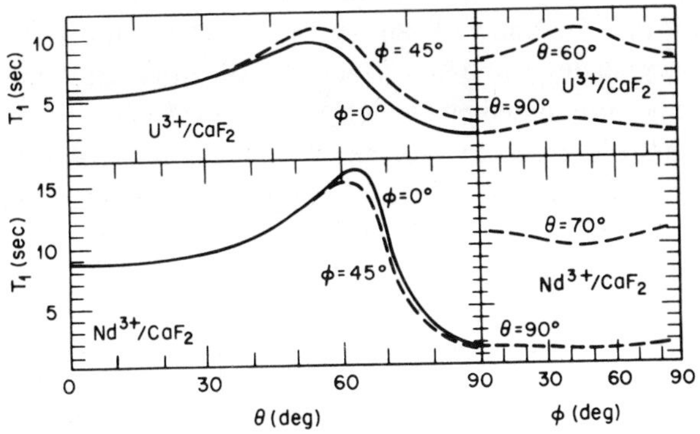

Fig. 7. Computed spin–lattice relaxation times of U^{3+}/CaF_2 and Nd^{3+}/CaF_2 for the direct relaxation process, using experimentally determined elements of the G' tensor and assuming $T = 4.2$°K and $\nu_0 = 9.4$ GHz. (From Black and Donoho.[95]

at 9.6 GHz and find for Nd^{3+} that $T_1^D \approx 180$ μsec at 4.2°K and $\theta = 90°$. For the parallel orientation, T_1^D is about twice as long. Further details on this work will be presented later, but for the moment, we simply note that these exceptionally short values of T_1^D are not understood and are in sharp disagreement with the estimates of Bierig *et al.*,[98] who used the static crystal field splittings to estimate the orbit–lattice interaction, and with the calculations of Black *et al.*[95]

Phillips and White[99] have measured the full orthorhombic magnetoelastic tensor for Gd^{3+} in yttrium gallium garnet (YGaG), for Yb^{3+} in YGaG, and for Er^{3+} in yttrium aluminum garnet (YAG). Since Yb^{3+} and Er^{3+} have an effective spin 1/2, these latter two tensors relate changes in the g-tensor to the strain, while the first involves changes in the fine structure. The measurements were made at 26 GHz. Phillips *et al.*[100] have also measured the magnetoelastic tensor for Fe^{3+} in Al_2O_3.

5.5. Other Relaxation Data on Rare Earth Ions

For rare earth ions, relaxation times are more frequently expressed using an approximate orbit–lattice interaction than a spin-Hamiltonian formalism. One reason is that many common rare earth crystals are not very amenable to stress (e.g., ethyl sulfates, double nitrates, trichlorides, and tribromides).

A common approach, due to Orbach,[54] is to expand the orbit–lattice interaction in terms of spherical harmonics and the strain. The magnitudes of the dynamic coupling coefficients are estimated from the static crystalline field values, if they are known from optical and ESR data. The methods of estimation vary slightly among various investigators, and three differing schemes have been proposed by Orbach,[54] Scott and Jeffries,[13] and Huang.[101] The basic argument, common to all three methods, is as follows. The static crystal field Hamiltonian for a nonoverlapping charge distribution can be written as

$$\mathscr{H}_0 = \sum_{i,m} \sum_{k=2,4,(6)} B_k^m r_i^k Y_{km}(\theta_i, \varphi_i) \tag{5.7}$$

where $(r_i, \theta_i, \varphi_i)$ are the unpaired electron coordinates on the paramagnetic ion and Y_{km} are normalized spherical harmonics. The values of $B_k^m\langle r^k\rangle$ are actually treated as empirical parameters to be determined experimentally, but a point-change model would yield

$$B_k^m = \sum_j \frac{q_j e^2}{\varrho_j^{k+1}} \left(\frac{4\pi}{2k+1}\right) Y_{km}^*(\alpha_j, \beta_j) \tag{5.8}$$

where $(\varrho_j, \alpha_j, \beta_j)$ are the coordinates of the jth external charge $(-q_j e)$. For the purposes of this discussion, we will assume that there is only one external charge so that the notation can be simplified. The effects of phonon-induced strain are then computed from changes in $B_k^m(\boldsymbol{\rho})$:

$$B_k^m(\boldsymbol{\rho}) \approx B_k^m(\boldsymbol{\rho}_0) + (\nabla B_k^m) \cdot \delta\boldsymbol{\rho} + \tfrac{1}{2}[\nabla\{\nabla B_k^m \cdot \delta\boldsymbol{\rho}\} \cdot \delta\boldsymbol{\rho}] \tag{5.9}$$

where $\delta\boldsymbol{\rho} = \boldsymbol{\varepsilon} \cdot \boldsymbol{\rho}_0$ and $\boldsymbol{\rho}_0$ is the static vector. We then simplify the situation even further by choosing to ignore the tensor properties of the strain $\boldsymbol{\varepsilon}$ and denote it simply as ε. Some authors then approximate

$$| \nabla B_k^m | \sim | B_k^m(\boldsymbol{\rho}_0) | / \varrho_0 \tag{5.10}$$

which is not unreasonable provided $B_k^m(\boldsymbol{\rho}_0) \neq 0$ and is inversely proportional to some power of ϱ_0. Unfortunately, the number of independent $B_k^m\langle r^k\rangle$ required to describe the various normal modes of vibration of the paramagnetic ion and its complex of nearest neighbors exceeds the number required to describe the static configuration, so additional relationships are required. Scott and Jeffries[13] used the empirical relationship

$$| B_k^m\langle r^k\rangle |_{\text{vibrational}} \approx | B_k^0\langle r^k\rangle |_{\text{static}} \tag{5.11}$$

and thus obtain the additional terms needed to write the perturbation Hamiltonian

$$\begin{aligned} \mathscr{H}' &= \sum_{ikm} \{B_k^m(\boldsymbol{\rho}) - B_k^m(\boldsymbol{\rho}_0)\} r_i^k Y_{km}(\theta_i, \varphi_i) \\ &\sim \sum_{ikm} (\varepsilon + \varepsilon\varepsilon') | B_k^0(\boldsymbol{\rho}_0) | r_i^k Y_{km}(\theta_i, \varphi_i) \end{aligned} \tag{5.12}$$

to second order in strain, neglecting phase factors. One further point is that one generally uses unnormalized tensor operators with coefficients $A_k^m\langle r^k\rangle$ rather than $B_k^m\langle r^k\rangle$ so that Eq. (5.11) is translated into

$$| A_k^m\langle r^k\rangle |_{\text{vib}} \approx g_n^{|m|} \, | A_k^0\langle r^k\rangle |_{\text{static}} \tag{5.13}$$

where the $g_n^{|m|}$ are ratios of normalizing factors and are listed by Scott and Jeffries.[13] The required k and m values occurring in Eq. (5.12) are determined by applying group-theoretical arguments to the normal modes of vibration of the complex.[54]

However crude this approach, the estimates of the various relaxation rates are usually within an order of magnitude of those observed, and generally any anisotropy is also correctly predicted. We will now review some experimental results in which calculations have been based on this orbit–lattice interaction model.

Finn *et al.*[12] first reported a relaxation time varying as $e^{\Delta/T}$ in their ac susceptibility measurements on undiluted cerium magnesium nitrate at liquid helium temperatures. They obtained data at frequencies of 175, 500, and 900 Hz with fields of 300, 500, and 1000 Oe and found $\Delta = 34°K$ They estimated the strength of the spin–phonon coupling from the known static crystal field parameters and obtained $T_1^{-1} \approx 10^{+10} e^{-34/T}$ sec, in good agreement with their data. Leifson and Jeffries,[102] using pulse saturation techniques, also reported the Ce^{3+} relaxation rate in Ce^{3+}-doped lanthanum magnesium nitrate (Ce:LaMN) could be fitted to $T_1^{-1} \propto e^{-32\pm2/T}$. Cowen and Kaplan[103] used spin echo methods to measure T_1 and T_2 in Ce:LaMN, also finding $T_1^{-1} \propto e^{-34/T}$.

Cooke *et al.*[104] reported a large anisotropy in the direct relaxation process in dysprosium ethyl sulfate (DyES). From ac magnetic susceptibility data, they obtained T_1 as a function of temperature for two orientations of the dc magnetic field with respect to the hexagonal crystal axis. The direct relaxation rate $(T_1{}^D)^{-1}$ of a Kramers transition depends upon magnetic admixtures from excited doublets. The ground doublet in DyES is characterized by $g_\perp = 0$, and the closest excited doublet is admixed only by the perpendicular component of H. Thus, the relaxation rate at low temperatures should be approximately proportional to $H_\perp{}^2 H_\parallel{}^2 = (\sin^2\theta \times \cos^2\theta) H^4$, if higher excited states can be ignored. Between 1.4 and 2.5°K, Cooke *et al.* could observe only a resonant relaxation mechanism ($\Delta = 23$ °K) if H was parallel to the crystal axis. With H at an angle of 45°, a direct process became dominant around 1.5°K. More recently, Langley and Jeffries[105] have used a similar anisotropy in $(T_1{}^D)^{-1}$ for Yb^{3+} in yttrium ethyl sulfate (Yb:YES) to polarize protons in a spin refrigerator. In this system, the theoretical values of $[T_1{}^D(\theta)]^{-1}$ follow $(\sin^2\theta \cos^2\theta) H^4$ even more rigorously.

A comprehensive review of the theory of relaxation in rare earth salts has been given by Scott and Jeffries[13] in their paper, which presented relaxation data on Nd:LaMN, Pr:LaMN, Sm:LaMN, SmMN, Ce:LaES, and Nd:LaES. They made pulse saturation measurements at 9.3 and 34 GHz between 1.4 and 5°K. Figure 8 shows their T_1^{-1} data for Nd:LaES in which Raman and direct processes clearly dominate and exhibit their characteristic temperature dependences. Figure 9 shows their relaxation data for Nd:LaMN in which the direct and resonant relaxation mechanisms contribute. In addition to the three relaxation mechanisms, direct, Raman,

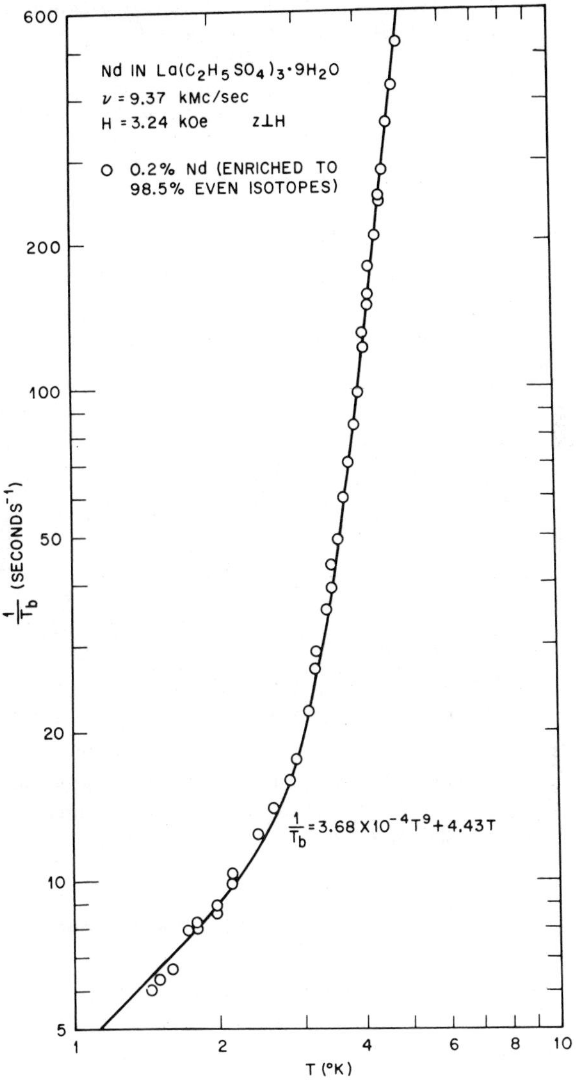

Fig. 8. Relaxation data for 0.2% Nd^{3+} in lanthanum ethyl sulfate, showing direct and Raman relaxation processes. (From Scott and Jeffries.[13])

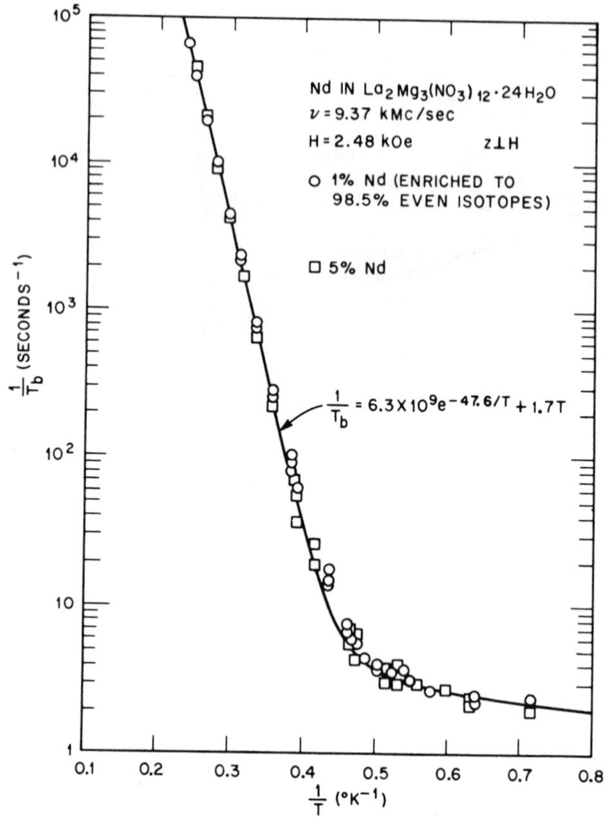

Fig. 9. Relaxation data for Nd^{3+} in lanthanum magnesium nitrate, showing direct and resonant relaxation processes. (From Scott and Jeffries.[13])

and resonant, they observed in several instances a phonon bottleneck in the direct process, leading to a concentration-dependent, T^2, low-temperature relaxation rate. They estimated the various relaxation rates as described above, using static crystal field parameters. In addition, they interpreted their phonon bottleneck results using the method of Faughnan and Strandberg.[106] The predicted bottleneck existed, phonon thermalization seemed to occur at the surface of the crystal. Data for Ce and Nd in LaMN were extended down to 0.25°K in a following paper by Ruby *et al.*[107] The measurements were made at 9.6 GHz in a cavity cooled by adiabatic demagnetization. Their Ce:LaMN data, shown in Fig. 10, displayed a resonant relaxation process with $\Delta = 34$°K, and a bottlenecked direct process.

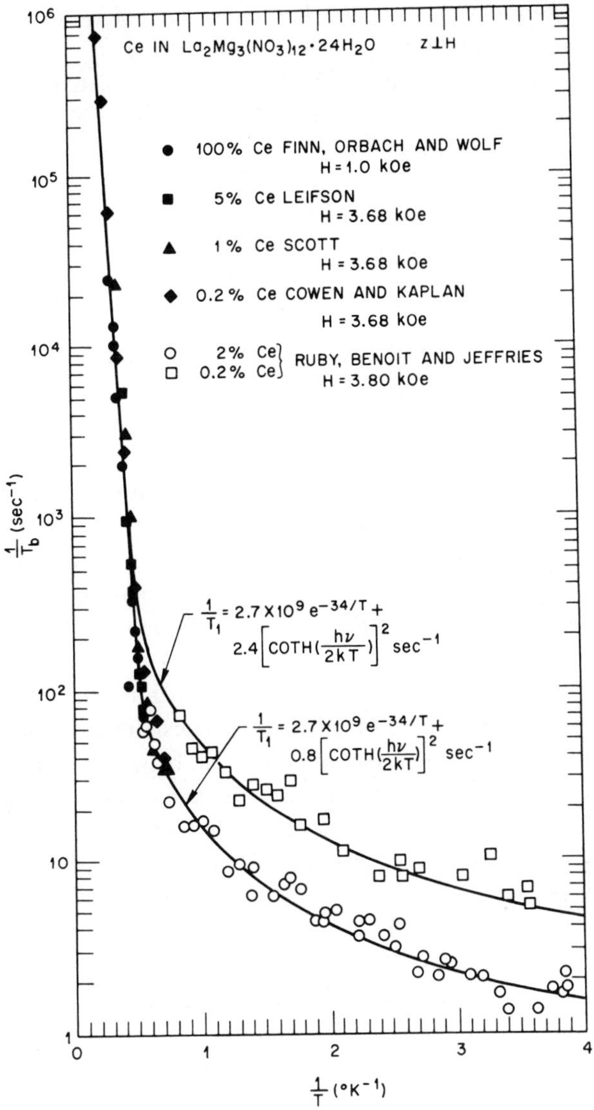

Fig. 10. Relaxation data for Ce^{3+} in lanthanum magnesium nitrate as measured by various investigators. The low-temperature data of Ruby *et al.* have a temperature dependence indicative of a phonon bottleneck and are also concentration-dependent, as expected for such a system. (From Ruby *et al.*[107])

The relaxation times for the lowest two doublets in 0.1% Dy:YE have been reported by Gill.[108] Small local distortions from C_{3h} apparentl introduce a nonzero value of $g_{\perp}$ in the ground doublet, and allow puls saturation methods to be used at 9.5 GHz to measure T_1. For 1.25–2°K the relaxation time of the ground doublet was the same for a magneti field orientation of 0° or 45° and was given by $T_1{}^0 = 3\times10^{-9}(e^{19.8/T} - 1$ sec. Between 4.2 and 20°K, T_1 for the first excited doublet was measure from its linewidth. The expected low-temperature limit for T_1 of 3×10^{-9} se was observed, as well as the influence of transitions to the third and fourt excited levels. These latter contributions to $T_1{}^0$ were calculated using th method described above and were in reasonable agreement with the line width measurements.

Bierig *et al.*[98] studied trivalent rare-earth relaxation rates in CaF_2 The trivalent ions enter the lattice at cubic, tetragonal, or trigonal loca symmetry sites. Pulse saturation techniques were used to study Ce, Nd Dy, Er, Yb, Tb, and Gd at 9.6 GHz. Below 7°K, the relaxation times o Ce^{3+} in tetragonal sites usually varied at T^{-1} but were very concentration dependent. All the transitions of Gd^{3+} in tetragonal sites showed equa relaxation rates at liquid helium temperatures. The overall temperatur dependence of T_1 for this ion was proportional to $T^{-1/2}$ at the lowest tem- perature, and to T^{-5} at somewhat higher temperatures. While the low- temperature $T^{-1/2}$ dependence suggests a mechanism other than the norma direct process, the T^{-5} dependence, observed between 25 and 40°K, is expected on the basis of the discussion in Section 4.6 concerning ions with multilevel ($S > 1/2$) spin systems. Indeed, the criteria for the dominance of the T^5 process, (4.30), is satisfied for λ and Δ equal to[109] 1600 cm^{-1} and 30,000 cm^{-1}, respectively.

The relaxation times for Dy^{3+} in all three sites were too fast (<3 μsec) to be observed. The cubic Yb^{3+} resonance could be seen up to 115°K, while the tetragonal resonance was not detected at 77°K. From linewidth data on the cubic resonance at temperatures between 50° and 105°K, a Raman process could be fitted using a Debye temperature of 680°K, rather than with the specific heat value of 480°K. Between 5 and 20°K, the cubic Yb^{3+} relaxation rate could be fitted by either a $T^{3.5}$ or an $e^{-38/T}$ dependence, in addition to a linear dependence on T which dominated at the lowest temperatures. Using the optical data of Weber and Bierig,[110] all the theo- retical relaxation rates were estimated and the resulting values for the direct rates were generally two orders of magnitude slower than the observed values. On the basis of this, together with the fact that the low-temperature signal recoveries were often multiexponential and concentration-dependent, e authors concluded that the normal direct relaxation process was not bserved in spite of the apparently correct temperature dependence.

Rare-earth relaxation in lattices similar to CaF_2 has been reported by everal Russian groups, some of whom combine pulse saturation data ith continuous saturation and linewidth measurements in order to extend e relaxation data up to 50°K or more. Antipin and co-workers report everal relaxation studies which include Tb^{3+} in SrF_2 from 1.4 to 30°K t 36 GHz,[111] Er^{3+} in trigonal sites in BaF_2 and SrF_2 at 36 and 9.4 GHz,[112] nd Yb^{3+} in trigonal sites in BaF_2 and SrF_2 and Yb^{3+} in cubic sites of aF$_2$.[113] At 36 GHz, T_1 for Yb^{3+} in cubic sites of CaF_2 was six times larger han at 9.4 GHz, and the high-temperature data (13–60°K) could be fit o a Raman process with $\theta_D = 420 \pm 20$°K, in disagreement with the esults of Bierig *et al.*,[98] but in closer agreement with specific heat data. elaxation times of Er^{3+} in CdF_2, BaF_2, and CaF_2 have been reported by verev and Smirnov.[114] Other related papers include those on Nd^{3+} in aF$_2$ by Kask *et al.*,[115] Yb^{3+} in CdF_2 by Pashinin and Prokhorov,[116] and r^{3+} in CdF_2 by Zverev *et al.*[117]

5.6. Field and Angular Dependences

The magnetic field dependence of the direct-process relaxation rate for Kramers transitions has been studied by several groups to determine if the expected H^4T [more precisely, $H^5 \coth(g\beta H/2kT)$] behavior is exhibited. Pashinin and Prokhorov[58] compared the relaxation rates of Fe^{3+} in $K_3(Co, Fe)(CN)_6$ at two frequencies, 8.750 and 3.170 GHz, and obtained $T_1^{-1} = 5.4T + 0.0054T^9$ and $T_1^{-1} = 0.093T + 0.0054T^9$, respectively. The ratio of the experimental direct process rates is 58, in agreement with the H^4 law, since the ratio of microwave frequencies is $(8.75/3.17)^4 = 58$.

Davids and Wagner[59] made similar measurements on the same salt, $K_3(Co, Fe)(CN)_6$, at several frequencies in the S and X microwave bands, corresponding to magnetic fields between 1100 and 3700 Oe. They found that the field dependence of the direct rate was between $H^{3.88}$ and $H^{4.06}$, further confirming the fourth-power dependence.

Baker and Ford[61] examined the magnetic field dependence of T_1^D for Nd^{3+} in LaMN and LaF_3. Operating at frequencies between 8.42 and 17.23 GHz, they observed a resonant process and a partially bottlenecked direct process in a LaMN sample doped with 98.5% even–even isotopes of Nd^{3+}. Correcting for the bottleneck at each frequency, under the assumption that the pertinent phonon bandwidth and thermalization time were frequency-independent, they found that the true direct relaxation rate went as $H^{3.8\pm0.3}$. In addition to the usual H^4 mechanism in LaF_3, the could fit their field dependence data to additional terms in H^2 and c^2H where c is the Nd^{3+} concentration. They interpreted the H^2 dependence a due to cross-relaxation to the 20% of the Nd^{3+} ions for which the nuclea spin was nonzero. The time-reversal symmetry in the wave functions c ions with hyperfine structure may be broken by the hyperfine interactior resulting in an H^2 field dependence. The theoretical treatment of hyperfin effects on the direct process relaxation rate, as given by Baker and Forc has also been used, with some modification, by Larson and Jeffries[1] to explain their relaxation data on ions with hyperfine structure. The c^2H term in Baker and Ford's data suggested cross-relaxation to pairs of couple Nd^{3+} ions.

A very elegant verification of the H^4T dependence for the direct proces has been given by Sabisky and Anderson[119] in their experiments on Tm^{2+} in CaF_2, SrF_2, and BaF_2. The field dependence of the direct process i measured over a large number of magnetic field values without the necessit for using any microwave excitation. The spin population is disturbed b optical excitation and its recovery is monitored by optical circular di- chroism. Many data points were taken over a range of fields from 2 to 8 kg and excellent agreement is found with the H^4 dependence. Further details of this technique are to be found in an accompanying article by S. Gesch- wind in this volume.

Kalbfleisch[119a] made transient Faraday rotation measurements, using pulsed magnetic fields, to study relaxation rates in concentrated holmium and dysprosium ethyl sulfate. He observed an oscillating behavior of T_1 in holmium ethyl sulfate as a function of magnetic field between 100 and 5300 Oe. This was explained by harmonic cross-relaxation between the Zeeman levels of the ground doublet and the singlet excited state at 6 cm^{-1}.

Huang[101] measured relaxation rates for Yb^{3+} in yttrium gallium garnet (YGaG), yttrium aluminum garnet (YAG), and lutetium gallium garnet (LGaG); Nd^{3+} in YGaG and YAG; Eu^{2+}, Ho^{2+}, and Tm^{2+} in CaF_2; and Sm^{3+} in LaES. Measurements were made at 8.9 GHz by pulse methods, and then compared to theoretical estimates using either the normal modes of vibration of an XY_8 molecular cluster, or an expansion in spherical harmonics. In general, agreement between experiment and theory was good. In some of the Yb-doped garnets, the low-temperature relaxation rates were concentration-dependent and were not proportional to T. In Sm:LaES, an angular anisotropy of 600% was observed in the direct relaxa- tion rate. Although the estimated magnitudes of $T_1{}^D$ in Sm:LaES were off, e anisotropy factor of six was correctly predicted using arguments similar those of Orbach.[120] A T^5 process was observed for Eu^{2+} in CaF.

Mikkelson and Stapleton[121] measured the point-by-point angular aniso-opy of the relaxation rates of trivalent Ce, Pr, Nd, Sm, Tb, and Er in aCl_3 at 9 GHz between 1.65 and 4.2°K. A comparison was made to com-uter calculations which used the static crystal field parameters to estimate e dynamic coefficients. Except for Ce, the observed angular dependence f T_1^D followed the theory almost exactly, and the overall magnitudes of $_1^D$ were generally predicted to within a factor of five. No anisotropy in e Raman or resonant relaxation rates was predicted or observed. The

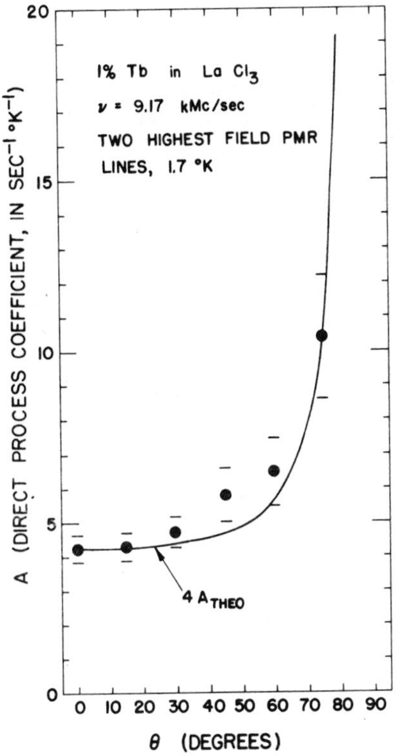

Fig. 11. Angular dependence of the direct process coefficient $A(\theta)$ ($sec^{-1}\ {}^\circ K^{-1}$) for $LaCl_3$ containing 1% Tb^{3+}. The solid curve represents four times the calculated value of $A(\theta)$, where θ is the angle between the magnetic field and the hexagonal crystal axis. (From Mikkelson and Stapleton.[121])

large anisotropy in T_1^D for non-Kramers transitions in Tb^{3+} shown Fig. 11 is predicted theoretically, due to anisotropic magnetic field a mixtures. A phonon bottleneck was predicted and observed for the dire process in the Pr sample. In addition to the expected concentration an temperature dependences, the bottleneck relaxation rate was found to proportional to the frequency bandwidth of the paramagnetic resonanc i.e., $\Delta\nu(\theta) \propto g\,\Delta H$. This is shown in Fig. 12. The Ce relaxation rate involve a direct and a resonant process ($\Delta = 46 \pm 3$°K), although no optic data were then available for the excited-state energies.

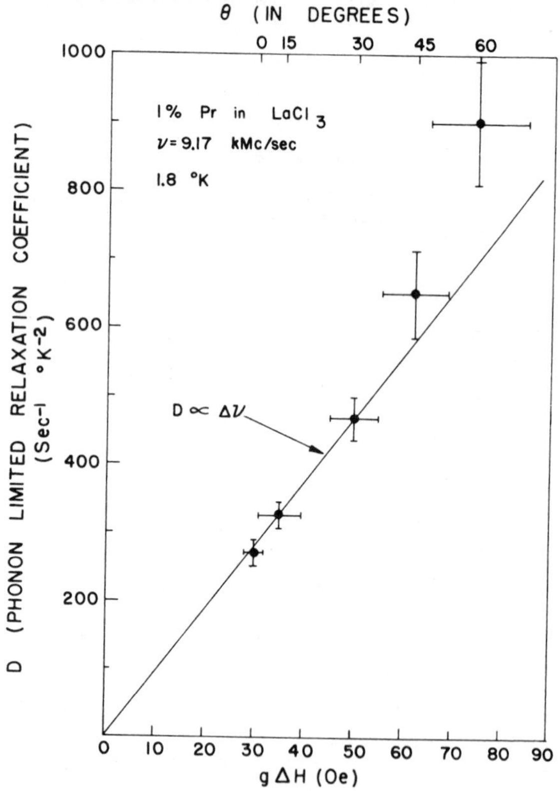

Fig. 12. The phonon bottleneck coefficient D (sec^{-1} °K^{-2}) for 1% Pr^{3+} in $LaCl_3$ at 9.17 GHz and 1.8°K plotted against $g(\theta)\,\Delta H(\theta)$, where ΔH is defined as the full-width at half maximum absorption. These data are collected by measuring the relaxation rate and the linewidth as a function of θ, the angle between the magnetic field and the hexagonal crystal axis. The frequency bandwidth of the ESR line is proportional to $g\,\Delta H$. (From Mikkelson and Stapleton.[121])

.7. Other Features of Relaxation Rates in Solids

Mangum and Hudson[122] reported an independent study of several are-earth relaxation rates in $LaCl_3$. They also found a resonant process or Ce with $\Delta = 46°K$. Using naturally abundant Nd^{3+} isotopes, they report an anisotropic Raman rate at 10°K. No resonant process was bserved in this salt. For Sm and Eu, their resonant relaxation data implied n excited state closer than that given by optical work. The values of Δ lso varied from sample to sample. They suggested, by way of explanation, he possibility of an inhomogeneously broadened excited state, and a esultant lowering of an effective Δ due to cross-relaxation.

Shortly after these two relaxation measurements[121,122] on Ce:$LaCl_3$ ndicating a value of 46°K for Δ, Hellwege *et al.*[123] reported optical data on the same dilute system which placed the first excited state at 53.9±4.5°K. ndependently, Varsanyi and Toth[124] reported Δ to be 55.1°K. Young and Stapleton[125] then examined further the suggestion of Mangum and Hudson.[122] They treated a model in which Δ varied, from one paramagnetic ite to another, according to a Gaussian distribution centered at the optical value Δ_{op} and with a standard deviation σ. If only Kramers transitions are considered, a rather wide distribution of Δ values will not necessarily broaden the ESR of the ground doublet excessively. By assuming that the ith ion relaxed at a rate given by $T_{1i}^{-1} = B' \Delta_i^3 e^{-\Delta_i/kT}$ and that a spin–spin interaction acted to bring the spin system into equilibrium as it relaxed to the lattice, they estimated that the observed relaxation rates near the end of each recovery would result in an effective Δ which is lower than Δ_{op}. If $\sigma^2/\Delta_{op}kT \ll 1$, then this effective Δ can be approximated by

$$\Delta \cong \Delta_{op} + (3\sigma^2/\Delta_{op}) - (\sigma^2/kT) \tag{5.14}$$

The authors point out that since measurements of the resonant relaxation process are frequently made over small ranges of $1/T$, the temperature dependence of Δ may not have been noticed in other measurements, especially in the presence of other relaxation mechanisms.

Young and Stapleton[80] later measured rare-earth relaxation in a series of mixed crystals of $LaCl_3$:$LaBr_3$ in order to test their calculations in a system over which they had some control of the inhomogeneous broadening. In both the Ce^{3+}- and Er^{3+}-doped samples, they observed a systematic lowering of Δ from Δ_{op} as the Br^- concentration was increased. Figure 13 shows their cerium data. Infrared absorption data were also exhibited which, verifying their relaxation data analysis, established the width (2σ) of the first excited state of Ce:$LaCl_3$ to be 9 cm^{-1}.

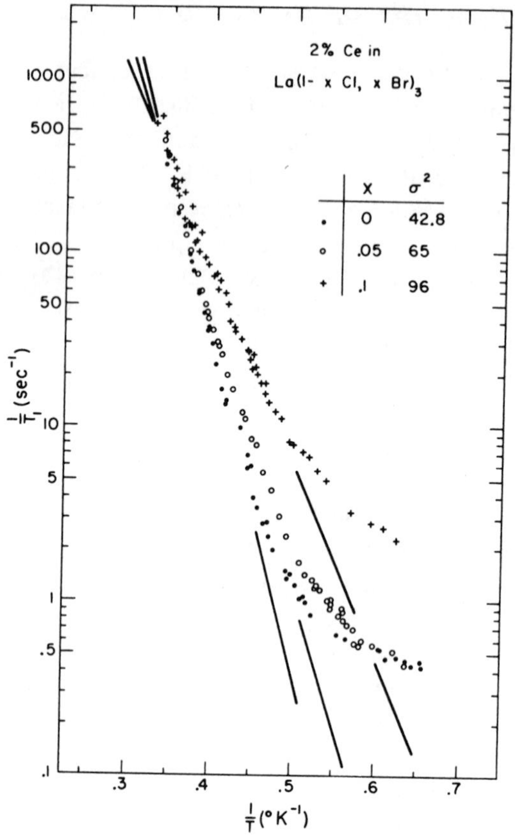

Fig. 13. Temperature dependence of the relaxation rate for 2% Ce^{3+} in mixed $La(Cl, Br)_3$ crystals containing 0, 5, and 10% Br^-. The decrease in the magnitude of the semilogarithmic slope with increased Br^- concentration is attributed to a broadened distribution of first excited states throughout the crystal. The listed values of σ^2 pertain to the inferred width of this distribution in units of K^2. (From Young and Stapleton.[80])

By studying the return of a spin system to thermal equilibrium after an inversion of the spin populations, rather than after saturation, Brya and Wagner[56] have observed the development of a "phonon avalanche" in Ce:LaMN, where the direct relaxation rate is phonon-bottlenecked. The temporal development of this avalanche, as shown in Fig. 14, is characterized by an inverted ESR signal which starts to decay at the unbottlenecked rate, then falls very rapidly to about zero with rapid emission of phonons,

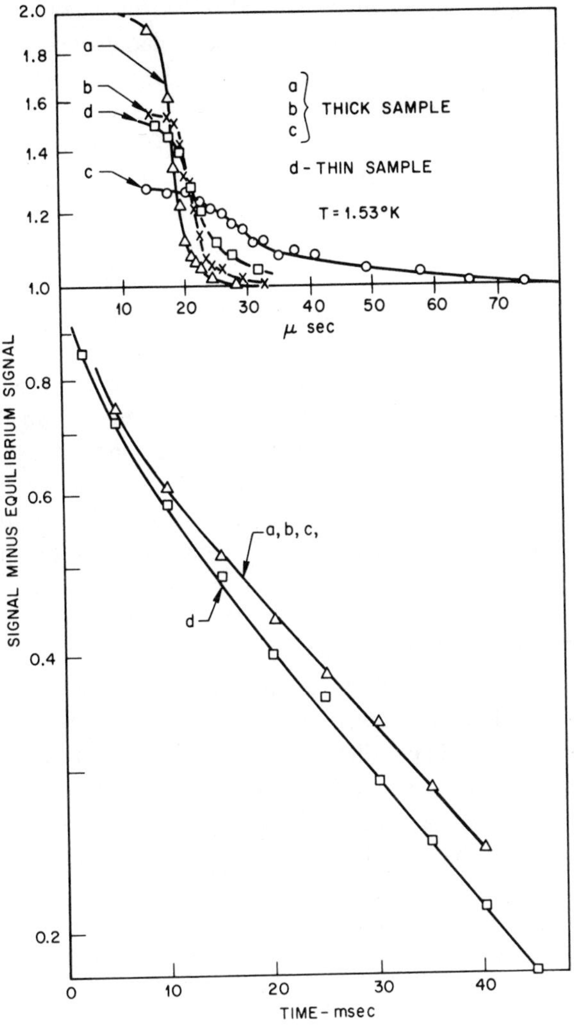

Fig. 14. Normalized plot of recovery after inversion, for different values of initial inversion and for two sample sizes of lanthanum magnesium nitrate containing 0.2% Ce^{3+}. Complete inversion is 2.0; saturation is 1.0. Fast passage occurs at $t = 0$. Lower curves are continuations of the upper curves with a change in time scale of 10^3. (From Brya and Wagner,[56] *Phys. Rev. Letters*.)

and finally returns much more slowly to equilibrium. The time scal for the two decays may differ by several orders of magnitude. Recentl Joseph *et al.*[57] have theoretically considered spatial effects in a phon avalanche.

Standley and Wright[126] have examined the entire nonexponenti recovery after saturation in two organic copper complexes in which phonon bottleneck was present. Their results were in agreement with t predictions of Faughnan and Strandberg[106] under the condition that t unbottlenecked direct relaxation time is much longer than the phonc thermalization lifetime.

Larson and Jeffries[118,127] have reported the effects of temperatur magnetic field direction, hyperfine interactions, and cross-relaxation c spin–lattice relaxation times. They studied several rare earth ions in LaM LaES, and YES. Within experimental error, the resonant and Raman rat were all isotropic except for Nd:LaMN, where the angular variation of t resonant rate could be explained on the basis of the anisotropy of $g'(\theta)/g(\theta$ where g' and g are the g values of the first excited- and ground-state doublet respectively. As predicted, the observed anisotropy in the bottlenecke relaxation of Pr:LaMN was found to be proportional to the product c $g(\theta)$ and $\Delta H(\theta)$. In both LaES and YES, the direct relaxation processes fc Pr^{3+} were bottlenecked. Both Er^{167} in YES and Nd^{143} in LaMN we studied to determine the effects of hyperfine interactions on the direc relaxation rate of Kramers transitions. At a constant microwave frequenc and with a magnetic field parallel to the hexagonal crystal axis, all hyperfin lines were expected to have, and indeed exhibited, the same relaxation rat As shown in Fig. 15, for a perpendicular field orientation, the relaxatio rates varied among the hyperfine states in accord with calculations whic assumed that "forbidden" transitions dominated the relaxation proces Larson and Jeffries also studied the effects of cross-relaxation by incor porating Ce^{3+} in Er:LaES.

The frequency dependence of the relaxation rate for an ion such a Tb^{3+} ($4f^8, {}^7F_6$) in a crystal of C_{3h} symmetry must be classified separatel from the usual behavior for a non-Kramers transition. This is because magnetic field *parallel* to the crystal axis mixes the lowest two singlets which have a well-defined zero-field splitting of the order of 0.1 cm^{-1} Thus the eigenstates in a parallel magnetic field are frequency-dependen and make all three relaxation processes frequency-dependent. Comparison between theory and experiment have been made for Tb:YES,[127] in whic a direct and Raman process exist, and for Tb:$LaCl_3$,[128] in which a resonan relaxation mechanism is also present.

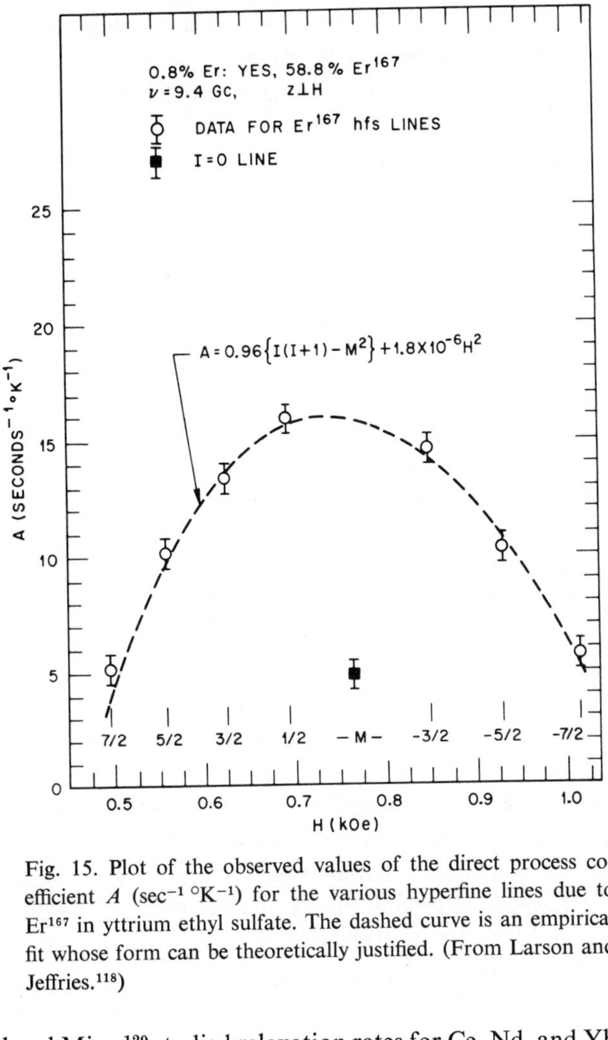

Fig. 15. Plot of the observed values of the direct process coefficient A (sec^{-1} $°K^{-1}$) for the various hyperfine lines due to Er^{167} in yttrium ethyl sulfate. The dashed curve is an empirical fit whose form can be theoretically justified. (From Larson and Jeffries.[118])

Kiel and Mims[129] studied relaxation rates for Ce, Nd, and Yb in $CaWO_4$ ısing electron spin-echo techniques at 9.4 GHz between 1.8 and 9°K. Their results exhibited direct and Raman-type processes. However, they found it necessary to fit their Raman data to a T^n law where n was between 10 and 11, rather than 9. They explained this temperature dependence on the basis of dispersion in the phonon spectrum and a more exact form of the Raman integral. Their values for n in the Ce-, Nd-, and Yb-doped salts were 10.9 ± 0.2, 10.4 ± 0.2, and 10.1 ± 0.2, respectively, for the field parallel to the c axis. These important results should serve to caution investigators against glibly assuming specific forms for the Raman process in the analysis of their data. From pulse saturation data alone, it seem quite unlikely that both the strength and the most accurate power-la dependence of a weak Raman process can be accurately established if strong resonant process is also present, in which the Δ value must also b determined. Two nonlinear parameters in such a fit are usually sufficie to send a computer fitting program into oscillation. Increasing the tempera ture range of the measurements by the use of linewidth data may hel establish the true Raman rate.

Antipin *et al.*[130] also studied the relaxation of Ce^{3+} and Yb^{3+} in tetrag onal sites of $CaWO_4$ and $PbMoO_4$ at 9.35 and 36 GHz. They used puls saturation, CW saturation, and linewidth measurements to obtain dat between 1.5 and 50°K. For Ce in both crystals, the Raman rates wer frequency-independent and the data could be fitted to $T^9 J_8(\theta/T)$, where was 250°K and 220°K for $CaWO_4$ and $PbMoO_4$, respectively. The basis fo the discrepancy in the Raman rates, as measured by Kiel and Mims[1 and by these investigators,[130] is not clear. Antipin *et al.*[130] also found th frequency dependence of the Ce direct relaxation rate in both CaWO and $PbMoO_4$ to be closer to ν^3 than to ν^4.

Rare-earth relaxation data in several crystals of the scheelite structur ($CaWO_4$) have been reported by Antipin and co-workers. Their data ar generally taken over a wide temperature range through the use of pulse CW, and linewidth techniques with spectrometers at 9 and 36 GHz. The have reported results for Nd in tetragonal sites of $CaWO_4$, $CdMoO_4$ $CaMoO_4$, $PbMoO_4$, and $BaMoO_4$;[131] for Er in $CaWO_4$,[132] where two reso nant relaxation mechanisms contribute; and for Tb in $CaWO_4$.[132] Result for Nd in $CaWO_4$ have also been reported by Kask *et al.*[133] Rare earth re laxation in CeO_2 has been reported by Antipin *et al.*[134] and Shekun *et al.*[13

Schultz and Jeffries[136] studied trivalent rare-earth relaxation rates ir LaF_3, where there are six inequivalent magnetic sites. They also studiec relaxation rates of rare-earth ions in monoclinic $YCl_3 \cdot 6H_2O$.[137] They observed that the angular dependence of the direct relaxation rate of Dy was not symmetric with respect to the principal axes of the g tensor. They explained that this is possible because the C_2 site symmetry of the rare- earth ion requires only that one principal axis of each g-tensor be along the C_2 crystal axis. The other two principal axes need not be common among the g tensors of the ground and excited doublets. Thus, Zeeman admixtures can be a very complex function of the magnetic field direction.

The body of experimental and theoretical work on the angular de- pendence of the direct process in iron group salts includes a calculation of $T_1^D(\theta)$ for Ti^{3+} in Al_2O_3 by Al'tshuler *et al.*[138] using a point-charge, normal-ode model. Kask *et al.*[139] measured these relaxation rates between 1.7 nd 3.5°K using pulse saturation techniques at 3.2 cm. Besides a resonant echanism ($\Delta = 30 \pm 3\ \text{cm}^{-1}$) which dominated the relaxation above °K, they found that the angular dependence of the direct process rate ent as $H^2 \sin^2 \theta$ for $\theta > 15°$, as predicted by Al'tshuler *et al.*[138] In this ılt, $g_{\parallel} = 1.067$ and $g_{\perp} \sim 0$.

David and Wagner[140] made point-charge, normal-mode calculations or $^2T_2$ states in rhombic symmetry, and applied these results to their data n $Fe:K_3Co(CN)_6$. They measured relaxation rates in the frequency anges 2–4 and 8–12 GHz by observing the recovery after adiabatic fast assage. They found that, aside from the angular variation in T_1^D intrinsic o the *g*-tensor anisotropy, the remaining factors which influence the relax- tion time were isotropic to within a factor of two, in agreement with their alculations.

Culvahouse and co-workers have reported theoretical[19] and experi- nental[19,141] work on Co^{2+} in lanthanum zinc nitrate (LaZN) and $ZnSiF_6 \cdot 6H_2O$. Olsen and Culvahouse[19] have made three model calcula- ions of the Co relaxation rate in LaZN using: (1) a point water dipole nodel with a homogeneous Debye spectrum; (2) a more realistic motion or the water dipoles, and (3) an estimation of the dynamic crystal field arameters using static crystalline field values.

We shall conclude this selected review with some further comments n the resonant relaxation process. From Eqs. (4.20), (4.21), and (4.24), ve see that the resonant relaxation rate is expected to be proportional to

$$B'\,\Delta^n \cosh(\hbar\omega_0/kT)\cosh(\hbar\omega_0'/kT)[e^{\Delta/kT} - 1]^{-1}$$

vhere B' is independent of Δ and T and $n = 3$ in the long-wavelength limit $\Delta \ll k\theta_D$), or $n = 1$ in the short-wavelength limit ($\Delta \approx k\theta_D$). Here, $\hbar\omega_0$ ınd $\hbar\omega_0'$ are the Zeeman splittings of the ground and excited doublets, espectively, and the two terms containing them can frequently be set equal o unity. If the resonant process becomes phonon-limited, the temperature lependence remains the same, but n becomes 2 and B' becomes concentra- ion-dependent, at least for the simplest models.[13]

It is therefore interesting to attempt a measurement of the Δ dependence of the resonant relaxation process. However, changes in Δ will also act in consort with changes in the orbit–lattice coupling. Thus, it is usually impossible to separate changes in B' from the Δ^n effects. A notable exception is the work of Imbusch *et al.*[142] in which they compared the resonant relax- ation rates of V^{2+}, Cr^{3+}, and Mn^{4+} (all $3d^3$) in the $\bar{E}(^2E)$ excited state in

Al_2O_3. The relaxation takes place via the $2\bar{A}(^2E)$ doublet with Δ values
12.3, 29, and 80 cm^{-1} for V^{2+}, Cr^{3+}, and Mn^{4+} respectively. In these pa
ticular systems, the use of Eq. (5.6) allows one to relate measuremen
from static strains along the c axis to changes in B'. Their results verifi
a Δ^3 dependence for these systems. (See also Section 2.3 of the chapt
by S. Geschwind in this volume.)

The question of a phonon bottleneck in a resonant relaxation proce
is also of interest. Since the temperature dependence of the resonant proce
is unchanged in the phonon-limited regime, more subtle effects constitu
the existing evidence. The change of the resonant relaxation rate with i
tensity of optical pumping in ruby, as reported by Adde *et al.*[89,90], ha
already been mentioned. Brya and Wagner[143] first suggested that such
mechanism might be responsible for the frequency dependence which the
observed for the resonant rate in Ce:LaMN. Their data were taken
frequencies between 1.70 and 11.14 GHz, and at temperatures betwee
1.2 and 2.6°K. They were able to fit the relaxation rates to a sum of direc
resonant, and Raman rates, where the direct rate went as ν^4 and the resonar
rate was characterized by $\Delta = 36.7 \pm 0.4$°K. It should be noted that thi
value of Δ is in closer agreement with the IR value[79] of 36.25 ± 0.4°K
than with the relaxation data of several others,[12,102,103,107] which yielde
approximately 34°K. Brya and Wagner[143] observed a frequency-independen
Raman process, but a resonant rate which decreased with increasing micro
wave frequency. Hoffman and Sapp[144] report a similar behavior in th
strength of the resonant process in CeMN using a self-induction bridg
to study the magnetic susceptibility.

Stapleton and Brower[145] have reported linewidth measurements o
Ce, Nd, and Sm in LaMN as a function of temperature. The temperatur
dependence of the linewidth of Ce:LaMN between 6 and 12°K fitted
resonant process with $\Delta = 33.9 \pm 1$°K, but its strength was an order o
magnitude larger than that reported by others using pulse saturation tech
niques below 4°K. Stapleton and Brower carried out both linewidth (u
to $T = 26$°K) and pulse saturation measurements ($2.5 < T < 3.7$°K) o
their Nd:LaMN sample, and found that resonant relaxation processe
dominated in both cases, but that the strengths of the rates were in the ratio
of 8.6:1, respectively. Culvahouse and Richards[68] have shown that this
effect cannot be due to the nature of the orbit–lattice interaction alone
They show that the contribution to the ESR linewidth from the resonan
relaxation process is $1/T_2 = (B_1 + B_2)n(\Delta)$, while $1/T_1 = 4B_1B_2n(\Delta)/(B_1 + B_2)$
where B_1 and B_2 are defined in (4.22). For a perpendicular orientation of
H with respect to the hexagonal crystal axis, B_1 equals B_2. Thus T_1 should

ual T_2 for this orientation, in disagreement with experiment.[145] They ggest that a phonon bottleneck, involving only certain phonon modes, a reasonable explanation of both the linewidth and pulse saturation data. , for example, the lowest-lying transverse acoustic phonon, at $\hbar\omega(\mathbf{k}, s)$ Δ, has an extraordinarily long lifetime because certain decay modes are rbidden,[146] then at some concentration of magnetic ions, that mode ill become ineffective in bringing about thermal equilibrium of the spin stem. A concentration dependence in the resonant relaxation rate would ot necessarily be observable by saturation techniques when this one ng-lived phonon mode became bottlenecked, since the remaining un-ottlenecked phonon modes at $\hbar\omega(\mathbf{k}, s') = \Delta$ would eventually control e relaxation. Thus, any concentration dependence of the resonant process ould be saturated at concentrations usually employed in pulse saturation SR studies. Culvahouse and Richards[68] also note that a phonon bottle-eck of the resonant relaxation process will not affect T_2 measurements. his is because T_2 depends on the sum of all the phonons with pertinent equencies, and that remains constant in a resonant process, which merely ansfers a phonon from one pertinent frequency to another. Linewidth easurements would thus yield the true magnitude of the resonant relaxa-on rate. Further study is necessary to clear up this intriguing question.

. SUMMARY

The gross features of electron spin–lattice relaxation appear to be well nderstood and the basic interaction is now recognized to be the phonon odulation of ion–ligand potentials. We have cited many gratifying ex-erimental results which, for example, exhibit the proper temperature ependences predicted by the theory for the different processes we have iscussed. However, there are numerous examples where quantitative agree-ent between theory and experiment is very poor indeed or completely bsent. We wish now to summarize these difficulties with a view to achieving reasonable perspective for evaluating the experimental results as well s to provide a guide for future work.

One of the more pressing theoretical problems is our general failure o predict theoretically the magnitude of the spin–phonon coupling con-tants from a "first principles calculation." However, this failure is the same which one encounters in static crystalline field problems. Covalent con-ributions to the static crystalline field can only be dealt with using a limited et of basis functions, limiting substantially the accuracy of the calculations. The lack of agreement between theory and experiment for the static crystalline field augers poorly for the dynamic spin–phonon coupling constant. It is suggested that only when this difficulty is removed will quantitativ estimates of the dynamic coupling constants become possible.

However, even starting with the phenomenological spin–phonon cou pling constants, determined from static strain measurements, we ofte find orders of magnitude disagreement between the calculated T_1's and th experimentally observed values. Moreover, rarely do Kramers transition exhibit the expected field dependence, i.e., $H^5 \coth(g\beta H/2kT)$ law. Eve more rarely does the two-phonon process obey the $(H^2+\beta H_{\text{int}}^2)/(H^2+H_{\text{int}}^2$ law, where β is bounded theoretically in terms of the parameters of th specific system (between two and three) and H_{int} represents the root mea square spin–spin field [see Eq. (4.26)].

The usual explanation for the failure to predict the correct T_1 is tha the spin system being studied is cross-relaxing to a rapidly relaxing defec center. The situation is most severe at small concentrations because of th widely varing strengths of coupling constants. For example, in the iro group ions, Fe^{2+} possesses a coupling constant 100 times that of Mn^{2+} o Fe^{3+} or other S-state ions such as Gd^{3+}. The spin–lattice relaxation rate i proportional to the square of the coupling constant, so that relative im purity concentrations of less than 10^{-4} are required to observe the intrinsi spin–lattice relaxation rate of the latter ions (allowing for a 1% cross-tal efficiency and allowable error). The example of ruby is perhaps the bes to illustrate this point. The spin–lattice relaxation rate reported for th Cr^{3+} ion seems to get longer every time purer crystals are grown. Hopefully the results exhibited in Table II represent a convergence of measurement and a generally accepted value. A similar remark holds for divalent rare earth ions in CaF_2. A hint of cross-relaxation can be obtained from the field dependence, for in general the more the field is increased, the poorer the cross-talk (because of differeing g factors) and hence the longer the observed spin–lattice relaxation rate. This is opposite of what one would expect from intrinsic spin–lattice relaxation alone.

The aforementioned difficulty is most marked in crystals with gemlike hardness such as MgO (see for example Tucker's work on Co^{2+} in MgO, Ref. 28), Al_2O_3, ThO_2, CaF_2, etc. In contrast, it is curious that measurements on many soft crystals, such as hydrated rare-earth salts, produce relaxation rates which are in reasonably good agreement with the crudest of theoretical estimates. This seems true for the direct and resonant relaxation processes, both in magnitude and angular anisotropy. (It is unlikely that the reported measurements of the Raman rates in these materials have much significance if a resonant process is present, because the Raman ate is usually considerably weaker than the resonant mechanism over the emperature ranged covered in, say, the region of the pulse saturation nethod. The most obvious distinction between these classes of host materials s that the sound velocities in the hard crystals are faster so that the direct elaxation rates are slower. The spin system can then relax more easily by ome mechanism which short-circuits the direct process. Such arguments eem convincing when accompanied by additional frequency, concentration, r temperature dependence data which are also at variance with the direct-process relaxation mechanism. Even the proper temperature dependence of the direct process over a narrow temperature range can be misleading, s cross-relaxation may exhibit a linear temperature dependence at low emperature. However, positive identification of the short-circuiting mech-nism is infrequently reported. A very elegant verification of the direct process via its field dependence is to be found in the work of Sabisky and Anderson.[119] (See also the discussion and Figs. 16 and 17 in Section 4.2 of the chapter by S. Geschwind.)

Another quantitative feature of the spin–lattice relaxation problem which has not been adequately dealt with as yet is the use of correct phonon states. Eigenvectors and dispersion laws have now been obtained for a number of the common host materials, principally through neutron dif-fraction measurements. It seems only logical that a projection of the local vibrational coordinates onto the correct running wave modes be carried out. A more satisfactory estimate of the strength and temperature depen-dence of the Raman relaxation process would obtain, in place of that found from a simple (in reality, oversimple) long-wavelength Debye mode. Low-lying optical branches, and the large weighting of high-frequency modes in the relaxation integral (ω^6 and ω^8 for non-Kramers and Kramers transitions, respectively) make this approach almost mandatory if one wishes to obtain a correct magnitude and temperature dependence for Raman rates.

The resonant relaxation mechanism still poses some interesting ques-tions with regard to the phonon bottleneck. As we have pointed out, such a bottleneck does not produce a change in the temperature dependence but will produce a linear dependence of T_1 upon concentration, as has been verified in the $\bar{E}(^2E)$ state of ruby.[90] The many results in the literature displaying concentration and magnetic field dependences, as well as the differences noted between pulse and linewidth data, could all be associated with this bottleneck problem. It would be most satisfying if some inde-pendent method of identifying the presence of this bottleneck were possible. Perhaps phonon sidebands in optical spectroscopy could be studied at the same time that the magnetic resonance was being performed. The "dip" and "peak" associated with the bottleneck of the resonance process might then be observable.

Studies of the Raman relaxation rate has been somewhat neglected in many rare-earth salts where low-lying crystal field states produce a dominant resonant process. Some ions, when incorporated into these lattices, possess a level structure which precludes the possibility of a resonant relaxation process, and such systems should be studied to determine the nature of the Raman temperature dependence to as high a value of kT/Θ_D as possible using both pulse and linewidth techniques.

The separation of the relaxation process into resonant and Raman components is itself an interesting question. In point of fact, there are two-phonon processes which can interfere with one another if the phonon lifetime broadening of the intermediate spin level is sufficiently great. This phenomenon is familiar in nuclear physics and optical spectroscopy and is shown by peculiarities in the lineshape (e.g., antiresonances). For the case at hand, it will alter the temperature dependence of the two-phonon process, especially in the temperature regime where the resonance and Raman processes are of comparable strength. This regime is often near the point at which the two-phonon processes begin to dominate the one-phonon, or direct, relaxation process. Preliminary calculations (Lys, to be published) exhibit marked departures from the simple exponential form for the resonance process, always in the direction of a smaller effective exponent or apparent position of the excited state. This interference can remain effective for from two to four orders of magnitude in $1/T_1$, so that the importance of measurements over wide dynamic ranges is clear if one wishes to extract a meaningful excited-state splitting from relaxation time measurements. It is not impossible that this interference could explain, in place of Stapleton and Young's static broadening mechanism, the reason why most spin–lattice relaxation rate studies obtain smaller values for excited state energies than those deduced from other measurements (e.g., optically).

Finally, we point out an obvious, but neglected, area of experimental investigation which is particularly appropriate to crystals which are too soft for static strain measurements. In spite of this restriction, certain symmetry predictions of the dynamic spin-Hamiltonian approach can be checked. Fourth-rank tensors are involved, so that a more complex angular anisotropy can result for T_1 values than for the ESR spectrum. The latter's angular dependence depends only on the transformation properties of second-rank tensors. Thus one might expect angular variations in T_1

when **H** is varied in a plane perpendicular to a C_n symmetry axis, where n is equal to or greater than three. The g-value would remain constant, but T_1 would vary. This might prove useful in aligning crystals in the basal plane with axially symmetric g-tensors for studies of pair spectra or other interactions with yet lower symmetries.

The theoretical and experimental suggestions above are merely indications of which areas of this rich field we believe to be ripe for further investigation. A part from its own intrinsic interest, knowledge of the spin–lattice relaxation rate, and an understanding of its microscopic origin, enables one to extract information not necessarily otherwise obtainable. Thus, the temperature dependence of the resonance relaxation process can be used to determine the energy splitting of ionic levels which might not be observable in spectroscopic experiments for a variety of reasons. By choosing an ion with a large splitting between the ground and first excited level, one can eliminate the resonance relaxation process and thereby use the temperature dependence of the two-phonon process to study local modes at the site of the resonating ion. This is not trivial, because the spin–lattice relaxation is sensitive to local modes which are even under inversion, whereas optical techniques more often than not are sensitive to only odd, localized vibrational modes. In a similar way, the temperature and field dependences of spin–lattice relaxation processes can shed considerable light on peculiarities of the phonon spectrum internal to the phonon Brillouin zone. This region is not accessible optically, and can be studied by neutrons only if one wishes to go to some trouble.

Knowledge of the spin–phonon coupling constant from ground-state measurements alone can also enable the investigator to predict the non-radiative relaxation rate of excited optical states and optical transition linewidths, knowledge crucial to the design and choice of laser materials. The possible presence, and an estimate of strength, of the Jahn–Teller effect can also obtain from such knowledge. Strengths of phonon sidebands in spectroscopic measurements can be estimated. Energy transfer rates via virtual phonon exchange can be determined. In short, knowledge of the spin–phonon coupling constants, most easily obtainable from the magnitude of the spin–lattice relaxation rate, enables one to treat a wide variety of physical phenomena. The theory of spin–lattice relaxation is almost forty years old (dating from Waller's original paper), yet its challenge and surprises continue.

ACKNOWLEDGMENTS

The authors are indebted to Mr. S. K. Lyo, who checked the equation in Section 4 and corrected some significant errors. Also, appreciation t Dr. W. B. Mims is expressed for his corrections to Section 3 covering th phenomenological theory of the spin–phonon interaction. We have als benefited from critical remarks from Drs. A. Abragam, B. Bleaney, M Baker, and K. W. H. Stevens.

REFERENCES

1. I. Waller, *Z. Physik* **79**, 370 (1932).
2. C. J. Gorter, *Paramagnetic Relaxation* (Elsevier, New York, 1947).
3. J. H. Van Vleck, *Phys. Rev.* **59**, 724 (1941).
4. W. Heitler and E. Teller, *Proc. Soc.* (*London*) **A155**, 629 (1936).
5. M. Fierz, *Physica* **5**, 433 (1938).
6. M. F. Deigen and V. Ya. Zevin, *Zh. Eksperim. i Teor. Fiz.* **39**, 1126 (1960) [Englisl transl.: *Soviet Phys.—JETP* **12**, 785 (1961)] and D. W. Feldman, R. W. Warren and J. G. Castle, Jr., *Phys. Rev.* **135**, A470 (1964).
7. S. A. Al'tshuler, *Izv. Akad. Nauk SSSR* **20**, 1207 (1956).
8. H. A. Kramers, *Proc. Roy. Acad. Amsterdam* **33**, 959 (1930).
9. R. de L. Kronig, *Physica* **6**, 33 (1939).
10. J. H. Van Vleck, *J. Chem. Phys.* **7**, 72 (1939); *Phys. Rev.* **57**, 426 (1940).
11. R. D. Mattuck and N. W. P. Strandberg, *Phys. Rev.* **119**, 1204 (1960).
12. C. B. P. Finn, R. Orbach, and W. P. Wolf, *Proc. Phys. Soc.* (*London*) **77**, 261 (1961).
13. P. L. Scott and C. D. Jeffries, *Phys. Rev.* **127**, 32 (1962).
14. R. Orbach and M. Blume, *Phys. Rev. Letters* **8**, 478 (1962).
15. R. L. Micher, *Phys. Rev.* **125**, 1537 (1962).
16. B. Bleaney and K. W. H. Stevens, *Rep. Progr. Phys.* **16**, 108 (1953).
17a. A. L. Schawlow, A. H. Piksis, and S. Sugano, *Phys. Rev.* **122**, 1469 (1961).
17b. M. Blume and R. Orbach, *Phys. Rev.* **127**, 1587 (1962).
18. C. Y. Huang and M. Inoue, *J. Phys. Chem. Solids* **25**, 889 (1964); T. J. Menne, D. P. Ames, and Sook Lee, *Phys. Rev.* **169**, 333 (1968).
19. J. W. Culvahouse, Wesley P. Unruh, and David K. Brice, *Phys. Rev.* **129**, 2430 (1963); A. G. Taylor, L. C. Olsen, D. K. Brice, and J. W. Culvahouse, *Phys. Rev.* **152**, 403 (1966); L. C. Olsen and J. W. Culvahouse, *Phys. Rev.* **152**, 409 (1966).
20. D. K. Ray, T. Ray, and P. Rudra, *Proc. Phys. Soc.* (*London*) **87**, 485 (1966); T. Ray and D. K. Ray, *Phys. Rev.* **164**, 420 (1967).
21. I. V. Ovchinnikov, *Fiz. Tverd. Tela* **4**, 1597 (1962) [English transl.: *Soviet Phys.—Solid State* **4**, 1170 (1962)].
22. A. K. Morocha, *Zh. Eksperim. i. Teor. Fiz.* **43**, 1804 (1962) [English transl.: *Soviet Phys.—JETP* **16**, 1275 (1963)]; N. G. Koloskova, *Fiz. Tverd. Tela* **5**, 61 (1963) [English transl.: *Soviet Phys.—Solid State* **5**, 40 (1963)].
23. M. Tachiki, Z. Sroubek, P. H. Zimmermann, and M. Abrahams, *J. Appl. Phys.* **39**, 977 (1968); Z. Sroubek, M. Tachiki, P. H. Zimmermann, and R. Orbach, *Phys. Rev.* **165**, 435 (1968).

24. J. Kondo, *Progr. Theor. Phys.* (*Japan*) **28**, 1026 (1962).
25. H. G. Koloskova, *Paramagnetic Resonance* (collected articles) (Kazan University, Kazan, USSR), p. 115.
26. G. Watkins and Elsa Feher, *Bull. Am. Phys. Soc.* **7**, 29 (1962); Elsa Rosenvasser Feher, *Phys. Rev.* **136**, A145 (1964).
27. P. L. Donoho, *Phys. Rev.* **133**, A1080 (1964); R. B. Hemphill, P. L. Donoho, and E. D. McDonald, *Phys. Rev.* **146**, 329 (1966).
28. E. B. Tucker, *Proc. IEEE* **53**, 1547 (1965); E. B. Tucker, *Phys. Rev.* **143**, 264 (1966).
29. R. Calvo, Z. Sroubek, R. S. Rubins, and P. H. Zimmerman, *Phys. Letters* **27A**, 143 (1968).
30. K. D. Bowers and J. Owen, *Rep. Progr. Phys.* **18**, 304 (1965); J. W. Orton, *Rep. Progr. Phys.* **22**, 204 (1959).
31. K. Zdansky, *Phys. Rev.* **159**, 201 (1967).
32. S. A. Al'tschuler, B. I. Kochalaev, and A. M. Leushin, *Usp. Fiz. Nauk* **75**, 459 (1961) [English transl.: *Soviet Phys.—Usp.* **4**, 880 (1962)].
33. J. Kondo, *Progr. Theoret. Phys.* (*Kyoto*) **23**, 106 (1960).
34. R. R. Sharma, T. P. Das, and R. Orbach, *Phys. Rev.* **149**, 257 (1966); *Phys. Rev.* **155**, 338 (1967); *Phys. Rev.* **171**, 378 (1968).
35. A. Abragam and M. H. L. Pryce, *Proc. Roy. Soc.* **A205**, 135 (1951); **A206**, 164 (1951); **A206**, 173 (1951).
36. A. R. Edmonds, *Angular Momentum in Quantum Mechanics* (Princeton University Press, Princeton, 1957); J. S. Griffith, *The Theory of Transition Metal Ions* (Cambridge University Press, Cambridge, 1961).
37. M. B. Walker, *Can. J. Phys.* **46**, 1347 (1968).
38. Y. Yafet, to be published.
39. R. Orbach and M. Tachiki, *Phys. Rev.* **158**, 524 (1967).
40. J. Kanamori, *Progr. Theoret. Phys.* (*Kyoto*) **17**, 197 (1956).
41. M. H. L. Pryce, *Proc. Phys. Soc.* (*London*) **A63**, 25 (1950).
42. J. M. Baker and B. Bleaney, *Proc. Roy. Soc.* **A245**, 156 (1958).
43. K. A. Muller, *Phys. Rev.* **171**, 350 (1968).
44. R. D. Mattuck and N. W. P. Strandberg, *Phys. Rev. Letters* **3**, 369 (1959).
45. P. G. Klemens, *Phys. Rev.* **125**, 1795 (1962); D. L. Mills, *Phys. Rev.* **146**, 336 (1966); A. A. Maradudin, *Solid State Physics* (III), **19**, 1 (1966).
46. M. B. Walker, *Phys. Rev.* **162**, 199 (1967); M. B. Walker, *Can. J. Phys.* **46**, 161 (1968).
47. D. W. Feldman, J. G. Castle, Jr., and J. Murphy, *Phys. Rev.* **138**, 1208 (1965).
48a. A. M. Leushin, *Fiz. Tverd. Tela* **5**, 605 (1963) [English transl.: *Soviet Phys.—Solid State* **5**, 440 (1963)].
48b. M. H. L. Pryce, *Phys. Rev.* **80**, 1107 (1950).
48c. H. Watanabe, *Progr. Theoret. Phys.* (*Kyoto*) **18**, 405 (1957).
49. R. Calvo, R. Isaacson, and Z. Sroubek, *Phys. Rev.*, accepted for publication.
50. E. R. Andrew and D. P. Tunstall, *Proc. Phys. Soc.* (*London*) **78**, 1 (1961).
51. F. Bridges, *Phys. Rev.* **164**, 299 (1967).
52. E. B. Tucker, in *Physical Acoustics*, Vol. IVA, ed. by W. Mason.
53. T. J. Menne, *Phys. Rev.*, to be published.
54. R. Orbach, *Proc. Roy. Soc.* **A264**, 458 (1961).
55. G. E. Pake, *Paramagnetic Resonance* (W. A. Benjamin, New York, 1963).
56. W. J. Brya and P. E. Wagner, *Phys. Rev. Letters* **14**, 431 (1965) and *Phys. Rev.* **157**, 400 (1967); N. S. Shiren, *Phys. Rev. Letters* **17**, 958 (1966).

57. C. H. Anderson and E. S. Sabisky, *Phys. Rev. Letters* **21**, 987 (1968); R. I. Joseph, David H. K. Liu, and Peter E. Wagner, *Phys. Rev. Letters* **21**, 1679 (1968); W. J. Brya, S. Geschwind, and G. E. Devlin, *Phys. Rev. Letters* **21**, 1800 (1968).
58. P. P. Pashinin and A. M. Prokhorov, *Fiz. Tverd. Tela* **5**, 2722 (1963) [English transl.: *Soviet Phys.—Solid State* **5**, 1990 (1963)].
59. Douglas A. Davids and Peter E. Wagner, *Phys. Rev. Letters* **12**, 141 (1964).
60. R. J. Elliott and K. W. H. Stevens, *Proc. Roy. Soc.* **A215**, 437 (1952); **A218**, 553 (1953); **A219**, 387 (1953).
61. J. M. Baker and N. C. Ford, Jr., *Phys. Rev.* **136**, A1692 (1964).
62. A. Abragam and B. Bleaney, *Electron Paramagnetic Resonance of Transition Ions* (Clarendon Press, Oxford, 1970), pp. 643–664.
63. L. C. Hebel and C. P. Slichter, *Phys. Rev.* **113**, 1504 (1959).
64. A. A. Manenkov and A. M. Prokhorov, *Zh. Eksperim. i Teor. Fiz.* **42**, 1371 (1962) [English Transl.: *Soviet Phys.—JETP* **15**, 951 (1962)].
65. R. Orbach, *Proc. Roy. Soc.* (*London*) **A264**, 485 (1961).
66. D. L. Huber, *Phys. Rev.* **131**, 190 (1963).
67. A. A. Manenkov and A. M. Prokhorov, *Zh. Esperim. i Teor. Fiz.* **42**, 75 (1962) [English transl.: *Soviet Phys.—JETP* **15**, 54 (1962)].
68. J. W. Culvahouse and Peter M. Richards, *Phys. Rev.* **178**, 485 (1969).
69. E. Abrahams, *Phys. Rev.* **107**, 491 (1957).
70. J. H. Van Vleck, *Magnetic and Electric Resonance and Relaxation*, ed. by J. Smit (North-Holland Publishing Co., Amsterdam, 1963), p. 1.
71. A. Abragam, *The Principles of Nuclear Magnetism* (Oxford University Press, Oxford, 1961), p. 408.
72. W. M. Rogers and R. L. Powell, *Natl. Bur. Std.* (*U.S.*) *Circ.* 595 (1958).
73. L. A. Vredevoe, *Phys. Rev.* **153**, 312 (1967).
74. Frank Bridges and W. Gilbert Clark, *Phys. Rev.* **164**, 288 (1967).
75. G. M. Zverev and N. G. Petelina, *Zh. Eksperim. i. Teor. Fiz.* **42**, 1186 (1962) [English transl.: *Soviet Phys.—JETP*, **15**, 820 (1962)].
76. W. Heitler, *Quantum Theory of Radiation* (Oxford University Press, Oxford, 1957), p. 196.
77. J. C. Phillips, *Phys. Rev. Letters* **12**, 447 (1964).
78. Theodore G. Castner, *Phys. Rev. Letters* **8**, 13 (1962); *Phys. Rev.* **155**, 816 (1967).
79. J. H. M. Thornby, *Phys. Rev.* **132**, 1492 (1963).
80. B. Arlen Young and H. J. Stapleton, *Phys. Rev.* **176**, 502 (1968).
81a. S. Geschwind, G. E. Devlin, R. L. Cohen, and S. R. Chinn, *Phys. Rev.* **137**, A1087 (1965).
81b. S. K. Lyo, *Phys. Rev.* **5**, 795 (1972).
81c. A. M. Stoneham, *Phys. Stat. Sol.* **19**, 787 (1967).
82a. R. J. Benzie and A. H. Cooke, *Proc. Phys. Soc.* (*London*) **A63**, 201 (1950).
82b. J. F. Broer and D. C. Schering, *Physica* **10**, 631 (1943).
82c. C. Starr, *Phys. Rev.* **60**, 241 (1941).
82d. A. H. Eschenfelder and R. T. Weidner, *Phys. Rev.* **92**, 869 (1953).
82e. A. A. Antipin, A. N. Katyshev, I. N. Kurkin, and L. Ya Shekun, *Fiz. Tverd. Tela* **9**, 813 (1967) [English transl.: *Soviet Physics—Solid State* **9**, 636 (1967)].
82f. P. L. Scott and C. D. Jeffries, *Phys. Rev.* **127**, 32 (1962).
82g. A. Kiel and W. B. Mims, *Phys. Rev.* **161**, 386 (1967).
82h. J. G. Castle, Jr., P. F. Chester, and P. E. Wagner, *Phys. Rev.* **119**, 953 (1960)

82i. Heinz Kalbfleisch, *Z. Physik* **181**, 13 (1964).
82j. S. Geschwind, G. E. Deviin, R. L. Cohen, and S. R. Chinn, *Phys. Rev.* **137**, A1087 (1965).
83. K. J. Standley and R. A. Vaughan, *Phys. Rev.* **139**, A1275 (1965).
84. R. A. Lees, W. S. Moore, and K. J. Standley, *Proc. Phys. Soc.* (*London*) **91**, 105 (1967).
85. Elsa Feher and M. D. Sturge, *Phys. Rev.* **172**, 244 (1968).
86. R. M. Macfarlane, *J. Chem. Phys.* **39**, 3118 (1963); **42**, 442 (1965); **47**, 2006 (1967).
87. M. Blume, R. Orbach, A. Kiel, and S. Geschwind, *Phys. Rev.* **139**, A314 (1965).
88. A. L. Schawlow, *Advances in Quantum Electronics*, ed. by J. R. Singer, (Columbia University Press, New York, 1961), p. 50.
89. R. Adde and S. Geschwind, *Bull. Am. Phys. Soc.* **13**, 457 (1968).
90. R. Adde, S. Geschwind, and L. R. Walker, in *Proc. XVth Colloque AMPERE* (North-Holland Publishing Co., Amsterdam, 1969).
91. P. T. Squire and J. W. Orton, *Proc. Phys. Soc.* (*London*) **88**, 649 (1966).
92. J. W. Orton, A. S. Fruin, and J. C. Walling, *Proc. Phys. Soc.* **87**, 703 (1966).
93. *Phys. Rev.* **170**, 606 (1968) [Erratum].
94. G. R. Wagner, J. G. Castle, and D. W. Feldman, *Bull. Am. Phys. Soc.* **13**, 458 (1968).
95. T. D. Black and P. L. Donoho, *Phys. Rev.* **170**, 462 (1968).
96. G. C. Wetsel, Jr. and P. L. Donoho, *Phys. Rev.* **139**, A334 (1965).
97. B. G. Berulava, T. L. Sanadge, and O. G. Khakhanashvili, *Fiz. Tverd. Tela* **7**, 640 (1965) [English transl.: *Soviet Phys.—Solid State* **7**, 509 (1965)].
98. R. W. Bierig, M. J. Weber, and S. I. Warshaw, *Phys. Rev.* **134**, A1504 (1964).
99. T. G. Phillips and R. L. White, *Phys. Rev.* **160**, 316 (1967).
100. T. G. Phillips, R. L. Towsend, Jr., and R. L. White, *Phys. Rev.* **162**, 382 (1968).
101. C. Y. Huang, *Phys. Rev.* **139**, A241 (1965).
102. O. S. Leifson and C. D. Jeffries, *Phys. Rev.* **122**, 1781 (1961).
103. J. A. Cowen and D. E. Kaplan, *Phys. Rev.* **124**, 1098 (1961).
104. A. H. Cooke, C. B. P. Finn, B. W. Mangum, and R. L. Orbach, *J. Phys. Soc. Japan* **17**, 462 (1962).
105. K. H. Langley and C. D. Jeffries, *Phys. Rev.* **152**, 358 (1966).
106. B. W. Faughnan and M. W. P. Strandberg, *J. Phys. Chem. Solids* **19**, 155 (1961).
107. R. H. Ruby, H. Benoit, and C. D. Jeffries, *Phys. Rev.* **127**, 51 (1962).
108. J. C. Gill, *Proc. Phys. Soc.* (*London*) **82**, 1066 (1963).
109. B. Bleaney, *Proc. Phys. Soc.* (*London*) **A68**, 937 (1955).
110. M. J. Weber and R. W. Bierig, *Phys. Rev.* **134**, A1492 (1964).
111. A. A. Antipin, L. D. Livanova, and L. Ya Shekun, *Fiz. Tverd. Tela* **10**, 1286 (1968) [English transl.: *Soviet Phys.—Solid State* **10**, 1025 (1968)].
112. A. A. Antipin, A. N. Katyshev, I. N. Kurkin, and L. Ya Shekun, *Fiz. Tverd. Tela* **9**, 1370 (1967) [English transl.: *Soviet Phys.—Solid State* **9**, 1070 (1967)].
113. A. A. Antipin, A. N. Katyshev, I. N. Kurkin, and L. Ya Shekun, *Fiz. Tverd. Tela* **9**, 3400 (1967) [English transl.: *Soviet Phys.—Solid State* **9**, 2684 (1968)].
114. G. M. Zverev and A. I. Smirnov, *Fiz. Tverd. Tela* **6**, 96 (1964) [English transl.: *Soviet Phys.—Solid State* **6**, 76 (1964)].
115. N. E. Kask, L. S. Kornienko, and M. Fakir, *Fiz. Tverd. Tela* **6**, 549 (1964) [English transl.: *Soviet Phys.—Solid State* **6**, 430 (1964)].
116. P. P. Pashinin and A. M. Prokhorov, *Fiz. Tverd. Tela* **5**, 359 (1963) [English transl.: *Soviet Phys.—Solid State* **5**, 261 (1963)].

117. G. M. Zverev, L. S. Kornienko, A. M. Prokhorov, and A. I. Smirnov, *Fiz. Tver. Tela* **4**, 392 (1962) [English transl.: *Soviet Phys.—Solid State* **4**, 284 (1962)].
118. G. H. Larson and C. D. Jeffries, *Phys. Rev.* **145**, 311 (1966).
119. E. S. Sabisky and C. H. Anderson, *Phys. Rev.* **1** (5), 2028 (1970).
119a. Heinz Kalbfleisch, *Z. Physik* **181**, 13 (1964).
120. R. Orbach, *Phys. Rev.* **126**, 1349 (1962).
121. R. C. Mikkelson and H. J. Stapleton, *Phys. Rev.* **140**, A1968 (1965).
122. B. W. Mangum and R. P. Hudson, *J. Chem. Phys.* **44**, 704 (1966).
123. K. H. Hellwege, E. Orlich, and G. Schaak, *Phys. Kondens. Materie* **4**, 196 (1965).
124. F. Varsanyi and B. Toth, *Bull. Am. Phys. Soc.* **11**, 242 (1966).
125. B. A. Young and H. J. Stapleton, *Phys. Letters* **21**, 498 (1966).
126. K. J. Standley and J. K. Wright, *Proc. Phys. Soc.* (*London*) **83**, 361 (1963).
127. G. H. Larson and C. D. Jeffries, *Phys. Rev.* **141**, 461 (1966).
128. Thomas L. Bohan and H. J. Stapleton, *Phys. Rev.* **182**, 385 (1969).
129. A. Kiel and W. B. Mims, *Phys. Rev.* **161**, 386 (1967).
130. A. A. Antipin, A. N. Katyshev, I. N. Kurkin and L. Ya Shekun, *Fiz. Tverd. Tela* **10**, 1433 (1968) [English transl.: *Soviet Phys.—Solid State* **10**, 1136 (1968)].
131. A. A. Antipin, A. N. Katyshev, I. N. Kurkin, and L. Ya Shekun, *Fiz. Tverd. Tela* **9**, 813 (1967) [English transl.: *Soviet Phys.—Solid State* **9**, 636 (1967)].
132. A. A. Antipin, A. N. Katyshev, I. N. Kurkin, and L. Ya Shekun, *Fiz. Tverd. Tela* **10**, 595 (1968) [English transl.: *Soviet Phys.—Solid State* **10**, 468 (1968)].
133. N. E. Kask, L. S. Kornienko, A. M. Prokhorov, and M. Fakir, *Fiz. Tverd. Tela* **5**, 2303 (1963) [English transl.: *Soviet Phys.—Solid State* **5**, 1675 (1964)].
134. A. A. Antipin, Z. N. Zonn, A. N. Katyshev, I. N. Kurkin, and L. Ya Shekun *Fiz. Tverd. Tela* **9**, 2646 (1967) [English transl.: *Soviet Phys.—Solid State* **9**, 2080 (1968)].
135. L. Ya Shekun, A. A. Antipin, Z. N. Zonn, A. N. Katyshev, and I. N. Kurkin *Fiz. Tverd. Tela* **10**, 1065 (1968) [English transl.: *Soviet Phys.—Solid State* **10**, 843 (1968)].
136. M. B. Schultz and C. D. Jeffries, *Phys. Rev.* **149**, 270 (1966).
137. M. B. Schultz and C. D. Jeffries, *Phys. Rev.* **159**, 277 (1967).
138. S. A. Al'tshuler, Sh. Sh. Bashkivov, and M. M. Zaripov, *Fiz. Tverd. Tela* **4**, 3367 (1962) [English transl.: *Soviet Phys.—Solid State* **4**, 2465 (1963)].
139. N. E. Kask, L. S. Kornienko, T. S. Mandel'shtam, and A. M. Prokhorov, *Fiz. Tverd. Tela* **5**, 2306 (1963) [English transl.: *Soviet Phys.—Solid State* **5**, 1677 (1964)].
140. Richard F. David and P. E. Wagner, *Phys. Rev.* **150**, 192 (1966).
141. Wesley P. Unruh and J. W. Culvahouse, *Phys. Rev.* **129**, 2441 (1963).
142. G. F. Imbusch, S. R. Chinn, and S. Geschwind, *Phys. Rev.* **161**, 295 (1967).
143. William J. Brya and P. E. Wagner, *Phys. Rev.* **147**, 239 (1966).
144. J. T. Hoffman and R. C. Sapp, *J. Appl. Phys.* **39**, 837 (1968).
145. H. J. Stapleton and K. L. Brower, *Bull. Am. Phys. Soc.* **13**, 459 (1968); *Phys. Rev.* **178**, 481 (1969).
146. R. Orbach and L. A. Vredevoe, *Physics* **1**, 91 (1964).

Chapter 3

Dynamic Polarization of Nuclei

C. D. Jeffries

Department of Physics
University of California, Berkeley

1. INTRODUCTION

Dynamic nuclear polarization (DNP) is a general method of enhancing by several orders the polarization of a nuclear spin system in interaction with an electron spin system by saturation of a suitable resonance of the combined system. Although the idea was first proposed by Overhauser[1] for nuclei in hfs interaction with the conduction electrons in a metal, similar ideas apply to many physical systems. The electron spins S may be: paramagnetic ions, atoms, or free radicals in solids, liquids, or gases; F-centers; trapped donors in semiconductors; conduction electrons in metals. The nuclear spins I may be: nuclei in the paramagnetic ions, atoms, radicals, or F-centers; neighboring nuclei in diamagnetic atoms; or the nuclei in metals and semiconductors. The coupling between I and S may be contact hfs or dipolar or both. This coupling affects the resonance spectrum of the system as well as the relaxation transitions. Besides the physical system and the type of I, S coupling, one can further characterize DNP according to whether one induces "allowed" transitions ($\Delta S_z = 1$, $\Delta I_z = 0$) or "forbidden" transitions ($\Delta S_z = 1$, $\Delta I_z = \pm 1$). It is not surprising that a large number of cases can be distinguished.[2]

As an example, we consider a case that has had rather full development and is important in polarizing targets for nuclear scattering: nuclei in dipolar coupling with paramagnetic ions diluted in a solid, the dynamic nuclear polarization being produced by saturation of forbidden transitions.[3,4] To fix ideas, consider a single crystal of $(Nd_{0.01}, La_{0.99})_2Mg_3(NO_3)_{12} \cdot 24H_2O$, denoted as Nd:LaMN. The Nd^{3+} ions form the $S = 1/2$ electron spin system, the protons in the waters the nuclear spin system, $I = 1/2$. Figures 1 and 2 show the typical experimental arrangement to keep in mind: The crystal is mounted in a microwave cavity in a field $H \sim 10^4$ Oe, and in a helium bath at temperature $T \sim 1°K$. The crystal is also coupled to an rf coil by which one measures the nuclear polarization from the NMR signal. In the simple Q-meter circuit shown, the polarization is proportional to the change in coil voltage ΔV at nuclear resonance, provided $(\Delta V/V) \ll 1$; large polarizations require corrections for radiation damping.[5] The energy levels of a typical I, S pair are shown in Fig. 3(a), labeled in zero order by the high-field spin states $|M, m\rangle^0$, where $M = \langle S_z \rangle$, $m = \langle I_z \rangle$. However, the I, S dipole–dipole interaction admixes these slightly, so that in addition to the usual allowed EPR transition W_1, one observes two weak, forbidden transitions, W_2 and W_3, with relative intensity of order (dipolar energy/nuclear Zeeman energy)$^2 \sim (g\beta/r^3H)^2$. If W_3 is

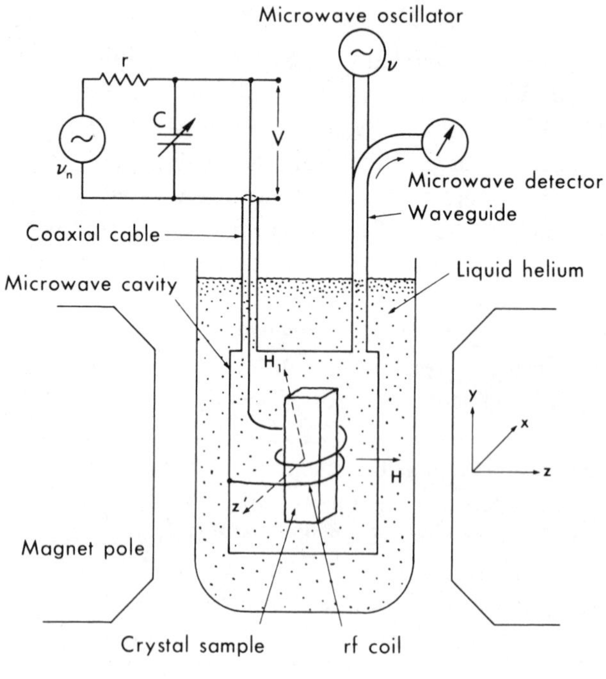

Fig. 1. Typical experimental arrangement for dynamic nuclear polarization (DNP).

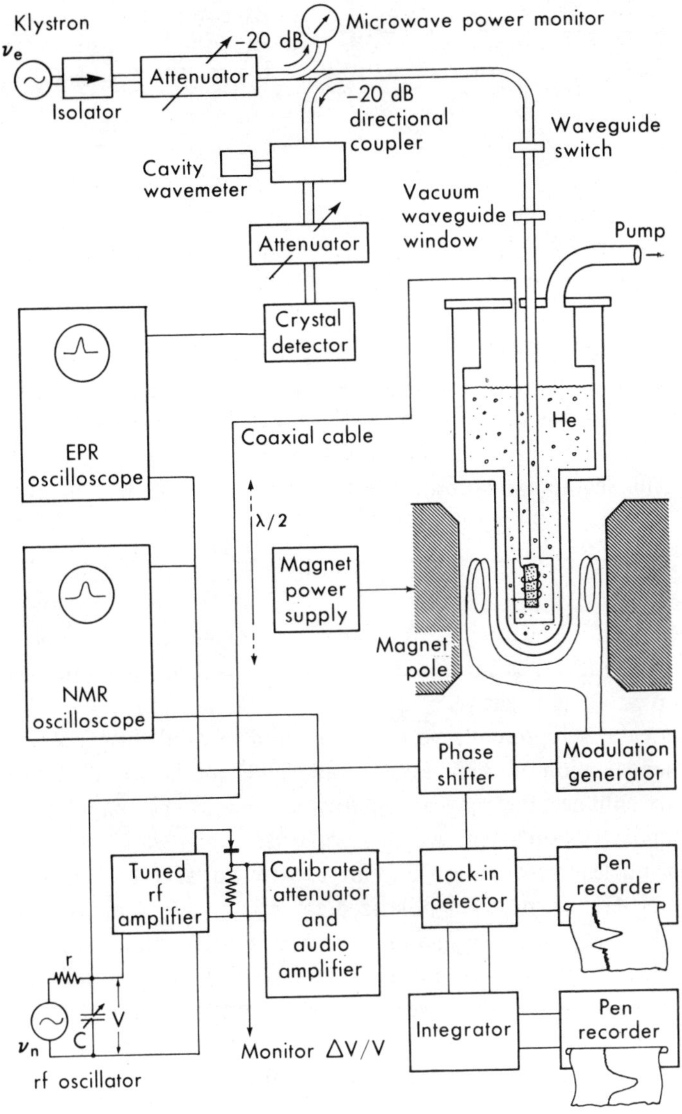

Fig. 2. Block diagram of DNP apparatus.

saturated, the relative populations of $|\ 1/2, 1/2\rangle^0$ and $|\ -1/2, -1/2\rangle^0$ become equal to, say, unity. Compared to the nuclei, the ions have a strong direct coupling to the thermal lattice vibrations, and the principal relaxations are of the type $\Delta M = \pm 1$, $\Delta m = 0$. These establish Boltzmann populations of the other two states as shown in Fig. 3(a), resulting in a

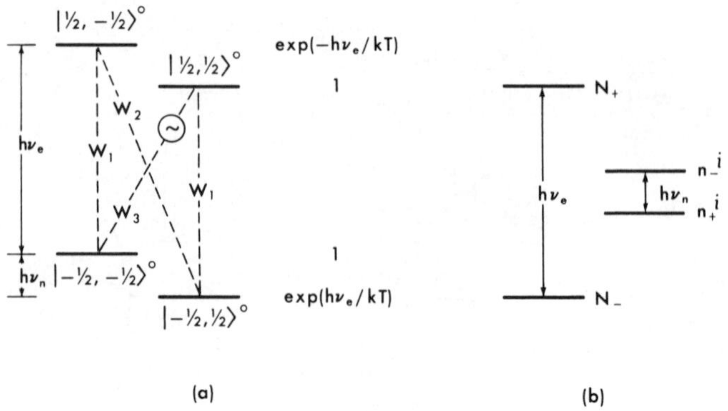

Fig. 3. (a) Energy levels and populations for I, S dipolar coupled pair in a field H. (b) Energy levels of S and I^i, no coupling.

dynamic nuclear polarization

$$p \equiv \frac{\overline{\langle I_z \rangle}}{I} = \frac{n(m = +\frac{1}{2}) - n(m = -\frac{1}{2})}{n(m = +\frac{1}{2}) + n(m = -\frac{1}{2})} = \tanh\left(\frac{h\nu_e}{2kT}\right) \tag{1}$$

which is enhanced by $\nu_e/\nu_n \approx 10^3$ over the thermal equilibrium polarization $p_0 = h\nu_n/2kT$. If W_2 is saturated, the same enhancement is obtained, but the polarization is negative.

This case is representative of the method of DNP in general: By the simultaneous action of applied microwave fields and lattice relaxations, one is able to enhance the nuclear polarization up to the magnitude of the electron polarization itself, which is of order unity at 10 kOe and 1°K. Dynamic nuclear polarization may be used to enhance the signal-to-noise ratio of NMR; to produce polarized targets; to align radio nuclei and observe the γ-ray anisotropy; and to study I, S interactions and the relaxation processes in diverse physical systems. Rather than survey all the various variations of DNP, we consider in Section 2 the above case in quantitative detail, as a representative example of the analytical approach used. In Section 3, we discuss briefly several other cases.

2. DIPOLAR CASE IN SOLIDS

2.1. Model and Rate Equations

Consider a unit volume of a magnetically dilute solid in a high field at low temperature, containing an electron spin system with N particles of effective spin $S = 1/2$ and numerous identical nuclei of interest with spin

$I = 1/2$ in diamagnetic atoms. As an example, the electron spins may be the Nd ions in Nd:LaMN, and the nuclei the protons in the waters of hydration. The system is described by the Hamiltonian

$$\mathscr{H} = \beta \sum_k \mathbf{H} \cdot \mathsf{g} \cdot \mathbf{S}_k - g_n\beta \sum_i \mathbf{H} \cdot \mathbf{I}_i + \sum_{i,k} \mathscr{H}_{I_i,S_k} + \sum_{i,j} U_{ij} \tag{2}$$

The terms represent electron Zeeman, nuclear Zeeman, electron–nuclear, and nuclear–nuclear dipole interactions. Electron–electron interactions are omitted because we assume the ions are highly diluted by the nuclei; $\mathscr{H}_{I_i,S_k}$ may be written as a contact hfs term,

$$\mathbf{S}_k \cdot \mathsf{A}_{ki} \cdot \mathbf{I}_i \tag{3a}$$

plus the dipole–dipole interaction

$$\mathscr{H}_{\text{dd}} = -\frac{g_n\beta^2}{r_{ik}^3}\left[\mathbf{I}_i \cdot \mathsf{g} \cdot \mathbf{S}_k - \frac{3(\mathbf{I}_i \cdot \mathbf{r}_{ik})(\mathbf{S}_k \cdot \mathsf{g} \cdot \mathbf{r}_{ik})}{r_{ik}^2}\right] \tag{3b}$$

where $\mathbf{r}_{ik}$ is the distance between $\mathbf{I}_i$ and $\mathbf{S}_k$. Now, the electron and the nucleus are each strongly coupled to the field, but only weakly coupled together by the third term. In zeroth order, one expects an EPR line at $\nu_e = g\beta H/h$ and an NMR line at $\nu_n = g_n\beta H/h$. However, the third term has several effects: for nuclei near a paramagnetic ion, the terms in $I_\pm S_z$ admix the states and give forbidden transitions at $\nu_e \pm \nu_n$; the NMR line is broadened and shifted by the dipole field of the ion; and the EPR linewidth is provided by the dipolar field of the nuclei. The fourth term in Eq. (2) provides the NMR linewidth and possibly a structure for the nuclei remote from an ion; it also provides for mutual nuclear spin flips through terms like $I_{+i}I_{-j}$, leading to a spatial diffusion of nuclear polarization.

To make the problem tractable, we introduce an approximate model: Imagine that all the nuclei in the sample may be grouped into shells about the ions, and consider a single shell as representative of the entire sample, a typical shell being $r_1 \leqq r \leqq r_2$, where r_1 is the minimum ion–nucleus spacing, determined by the crystal structure, and $r_2 = (4\pi N/3)^{-1/3}$ is half the average ion–ion spacing. We further divide the nuclei into two groups: (1) an inner shell $r_1 \leqq r \leqq d$, containing a number n' per unit volume of "near" nuclei, where d ($\approx$10 Å typically) is the diffusion barrier within which mutual nuclear flips are inhibited by the ion local field; the near nuclei interact more strongly with the ion than with each other and are relaxed directly by dipolar coupling with the ion at the average rate $T_{1n'}^{-1}$, to be calculated later. (2) An outer shell $d \leqq r \leqq r_2$ containing a large number n per unit volume of "distant" nuclei which rapidly come into internal equilibrium in a time t_d by diffusion. This time is of order $t_d \sim (r_2/a)^2 T_{2n}$, where a is the internuclear spacing and T_{2n} is the inverse NMR linewidth; the distant nuclei are not directly coupled to the ion, but rather to the nuclei at the interface $r \approx d$, with an effective cross-relaxation rate T_{12}^{-1}. It is reasonable to assume that $t_d^{-1} \gg T_{12}^{-1} \gg T_{1n'}^{-1}$. This model is summarized by the thermal (i.e., spin temperature) block diagram of Fig. 4, which shows an additional "leakage" relaxation rate T_{1nl}^{-1} of the distant nuclei directly to the lattice through, e.g., undesirable magnetic impurities or molecular rotation. In this figure, T_{1e}^{-1} is the spin–phonon relaxation rate of the ion and T_{ph}^{-1} is the phonon–bath relaxation rate. We will assume for the present that T_{ph}^{-1} is very large, that there is no phonon bottleneck, and thus T_{1e}^{-1} is the usual spin–lattice relaxation rate of the paramagnetic ion.

Our objective is to find the rate equation for the polarizations p, p', and P of the distant nuclei, near nuclei, and ions, respectively. From the above model, we see that, in the absence of microwave fields,

$$\left.\frac{dp}{dt}\right|_R = -\frac{p - p_0}{T_{1nl}} - \frac{n'}{n}\frac{p - p'}{T_{12}} \tag{4}$$

$$\left.\frac{dp'}{dt}\right|_R = -\frac{p' - p_0}{T_{1n'}} - \frac{p' - p}{T_{12}} \tag{5}$$

$$\left.\frac{dP}{dt}\right|_R = -\frac{P - P_0}{T_{1e}} \tag{6a}$$

Note that in these equations the nuclear polarizations p and p' are defined by Eq. (1), and $p_0 = h\nu_n/2kT$ is the (positive) thermal equilibrium value. The ion polarization is similarly defined by

$$P = (N_+ - N_-)/(N_+ + N_-) \tag{6b}$$

where N_+ is the probability to find the ion in the $M = +1/2$ state, and

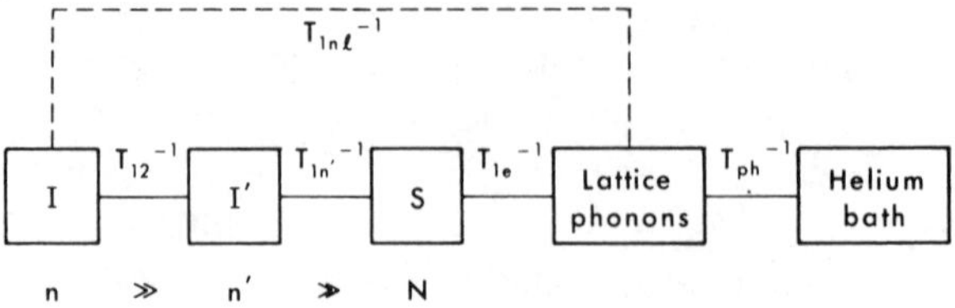

Fig. 4. Thermal block diagram of distant nuclei I, near nuclei I', electron spins S in a crystal, coupled by the indicated relaxation rates.

N_- to find it in the $M = -1/2$ state. Thus, in this convention, the thermal equilibrium value is negative: $P_0 = -\tanh(h\nu_e/2kT)$.

To find the effect of the microwaves, as well as the relation between $T_{1n'}$ and T_{1e}, we consider a system consisting of an isotropic ion and n' nuclei in the inner shell. Neglecting the dipolar interaction $\mathscr{H}_{\rm dd}$, the zero-order wave functions can be written as a product of one-particle functions $\psi^0 = \psi(M) \cdot \psi(m_1) \cdot \psi(m_2) \cdots \psi(m_i) \cdots \psi(m_{n'})$, where $m_i = \langle I_z^i \rangle$ for the ith proton at $(r_i, \theta_i, \varphi_i)$. Since the energy $E^0 = g\beta HM - g_n\beta H \sum_i m_i$ depends only on M and $\sum m_i$, the ψ^0 are highly degenerate. Dipolar terms like $I_+^i S_z$ will admix a certain $\psi_a{}^0(M, m_1, \ldots, m_i = 1/2, \ldots)$ with $\psi_b{}^0(M, m_1, \ldots, m_i = -1/2, \ldots)$ by the amount

$$\alpha_i = \tfrac{3}{2} M[(\sin\theta_i \cos\theta_i)/r_i^3](g\beta/H) \exp(\pm i\varphi_i) \tag{7}$$

where θ_i and φ_i are the polar angles of the $\mathbf{r}_{ik}$ in Eq. (3b). However, $\psi_a{}^0$ will also be admixed by $I_+^j S_z$ with $\psi_c{}^0(M, m_1, \ldots, m_j = -1/2, \ldots)$, etc., so that $\psi_a = \psi_a{}^0 + \alpha_i \psi_b{}^0 + \alpha_j \psi_c{}^0 + \cdots$. The normalizing factor $(1 + \sum_i \alpha_i^2)^{-1}$ is of order unity and we drop it for simplicity. The microwave transition probabilities induced by a microwave field $2H_1 \cos 2\pi\nu t$ are

$$W_1 = \tfrac{1}{4}(\gamma_e H_1)^2 g(\nu - \nu_e) \ \sec^{-1}, \qquad \Delta M = \pm 1, \ \Delta m_i = 0 \tag{8}$$

$$W_2{}^i = \sigma_i{}' \tfrac{1}{4}(\gamma_e H_1)^2 g(\nu - (\nu_e + \nu_n)), \qquad \Delta M = \pm 1, \ \Delta m_i = \mp 1 \tag{9}$$

$$W_3{}^i = \sigma_i{}' \tfrac{1}{4}(\gamma_e H_1)^2 g(\nu - (\nu_e - \nu_n)), \qquad \Delta M = \pm 1, \ \Delta m_i = \pm 1 \tag{10}$$

$$\sigma_i{}' \equiv 4\alpha_i^2 = (3/10)(g\beta/H)^2(1/r_i^6) \tag{11}$$

where we have used the angular average $\langle \cos^2\theta \sin^2\theta \rangle = 2/15$; the EPR lineshape function $g(\nu)$, with a maximum value $g(0) = 2T_{2e}$; T_{2e} is the inverse EPR linewidth; and $\gamma_e \equiv g\beta/h$. Note that $W_2{}^i$ and $W_3{}^i$ are the transition probabilities for the ion to flip simultaneously with nucleus I^i for a given set of other nuclear quantum numbers $m_1, \ldots, m_{n'}$ which do not change. Let $n_+{}^i$ be the probability to find I^i in the $m_i = +1/2$ state, and $n_-{}^i$ in the $m_i = -1/2$ state. Then, from Fig. 3(b), the rate of change of N_- due to W_1, and to all the $W_2{}^i$ and $W_3{}^i$ for the nuclei in the inner shell, is

$$\begin{aligned} dN_-/dt = {} & N_+ W_1{\downarrow} - N_- W_1{\uparrow} + N_+ \sum_i n_-{}^i W_2{}^i{\downarrow} - N_- \sum_i n_+{}^i W_2{}^i{\uparrow} \\ & + N_+ \sum_i n_+{}^i W_3{}^i{\downarrow} - N_- \sum_i n_-{}^i W_3{}^i{\uparrow} \end{aligned} \tag{12}$$

Similarly,

$$dn_+{}^i/dt = N_-(n_-{}^i W_3{}^i{\uparrow} - n_+{}^i W_2{}^i{\uparrow}) + N_+(n_-{}^i W_2{}^i{\downarrow} - n_+{}^i W_3{}^i{\downarrow}) \tag{13}$$

If we introduce the ith nuclear polarization $p^i = (n_+{}^i - n_-{}^i)/(n_+{}^i + n_-{}^i$
and the electron polarization P from Eq. (6b), and note that $W_1\uparrow = W_1\downarrow$
etc. for the microwave induced transitions, then Eqs. (12) and (13) can be rewritten as

$$\left.\frac{dP}{dt}\right|_{\text{rf}} = -2W_1P - \sum_i W_2{}^i(P - p^i) - \sum_i W_3{}^i(P + p^i) \qquad (14a)$$

$$\left.\frac{dp^i}{dt}\right|_{\text{rf}} = -W_2{}^i(p^i - P) - W_3{}^i(p^i + P) \qquad (14b)$$

In addition to these rf terms due to the microwaves, the thermal lattice vibrations induce relaxation transitions through the phenomenological Hamiltonian $g\beta\mathbf{S}\cdot\mathbf{H}'(t)$, where $\mathbf{H}'(t)$ is an effective field of thermal origin. Since the same operator $S_\pm$ is involved as in the rf case, the relative magnitudes of the relaxation transition probabilities are given by w_1, $w_2{}^i = \sigma_i{}'w_1$, and $w_3{}^i = \sigma_i{}'w_1$, by analogy to Eqs. (8)–(11). However, since these are of thermal origin, we cannot say that $w_1\uparrow = w_2\downarrow$, but rather $(w_1\downarrow/w_1\uparrow) = \exp(h\nu_e/kT)$ in order that $N_+w_1\downarrow = N_-w_1\uparrow$ in thermal equilibrium. Thus we weight symmetrically the relaxation rates by Boltzmann factors:

$$w_1\uparrow = w_1\exp(-h\nu_e/2kT) \qquad (15a)$$

$$w_1\downarrow = w_1\exp(h\nu_e/2kT) \qquad (15b)$$

$$w_2{}^i\uparrow = w_2{}^i\exp[-h(\nu_e + \nu_n)/2kT] \qquad (15c)$$

$$w_2{}^i\downarrow = w_2{}^i\exp[h(\nu_e + \nu_n)/2kT] \qquad (15d)$$

$$w_3{}^i\uparrow = w_3{}^i\exp[-h(\nu_e - \nu_n)/2kT] \qquad (15e)$$

$$w_3{}^i\downarrow = w_3{}^i\exp[h(\nu_e - \nu_n)/2kT] \qquad (15f)$$

$$1/T_{1e} \equiv w_1\uparrow + w_2\downarrow = 2w_1\cosh(h\nu_e/2kT) \qquad (15g)$$

The overall contribution of relaxation to the rate equation is obtained by replacing W_1, $W_2{}^i$, and $W_3{}^i$ in Eqs. (12) and (13) by Eqs. (15). The result is

$$\left.\frac{dP}{dt}\right|_R = -[(P - P_0)/T_{1e}]\left[1 - \sum_i \sigma_i{}'p^ip_0\right] \qquad (16a)$$

$$\left.\frac{dp^i}{dt}\right|_R = -[\sigma_i{}'(1 - PP_0)/T_{1e}](p^i - p_0) \qquad (16b)$$

The second term in Eq. (16a) is negligible. The total rate equation for P

is the sum of Eqs. (14a) and (16a); to reduce the p^i to an average nuclear polarization p' over the inner shell, we set

$$\sum_i W_2^i(P - p^i) \approx (P - p') \sum_i W_2^i = (P - p')(n'/N)\overline{W}_2$$

in Eq. (14a), to find

$$\frac{dP}{dt} = -\frac{P - P_0}{T_{1e}} - \overline{W}_2(P - p')\frac{n'}{N} - \overline{W}_3(P + p')\frac{n'}{N} - 2W_1P \qquad (17)$$

We add Eqs. (14b) and (16b), sum over i, and set

$$\sum_i \sigma_i'(p^i - p_0) \approx (p' - p_0) \sum_i \sigma_i' \approx (p' - p)(n'/N)\bar{\sigma}'$$

where $\bar{\sigma}'$ is the average value of Eq. (11) over the inner shell,

$$\bar{\sigma}' = (3/10)(g\beta/H)^2(1/r_1{}^3d^3) \qquad (18)$$

We obtain

$$\left.\frac{dp'}{dt}\right|_R + \left.\frac{dp'}{dt}\right|_{\text{rf}}$$

$$= -\frac{\bar{\sigma}'(1 - PP_0)}{T_{1e}}(p' - p_0) - \overline{W}_2(p' - P) - \overline{W}_3(p' + P) \qquad (19)$$

We identify the first term of Eq. (19) with that of Eq. (5) and define

$$1/T_{1n}' = \bar{\sigma}'(1 - PP_0)/T_{1e} \qquad (20)$$

Adding the cross-relaxation term to Eq. (19), we get the total rate equation

$$\frac{dp'}{dt} = -\frac{p' - p_0}{T_{1n}'} - \frac{p' - p}{T_{12}} - \overline{W}_2(p' - P) - \overline{W}_3(p' + P) \qquad (21)$$

Equations (17) and (21) together with

$$\frac{dp}{dt} = -\frac{p - p_0}{T_{1nl}} - \frac{n'}{n}\,\frac{p - p'}{T_{12}} \qquad (22)$$

are the desired rate equations for our system. In these equations, W_1, $\overline{W}_2$, and $\overline{W}_3$ have maxima at ν_e, $\nu_e + \nu_n$, and $\nu_e - \nu_n$, respectively, with the maximum values $W_1 = \frac{1}{2}T_{2e}(\gamma_e H_1)^2$, $\overline{W}_2 = \bar{\sigma}' W_1$, and $\overline{W}_3 = \bar{\sigma}' W_1$.

This calculation has assumed an isotropic g-factor for the electron spin system. However, if it is anisotropic, with $g(0)^2 = g_\parallel{}^2 \cos^2\theta + g_\perp{}^2 \sin^2\theta$, and the sample is a single crystal with orientation $\theta = \angle z, H$, then th above results are valid if in σ_i' and $\bar{\sigma}'$ we replace g by a new factor g_d defined by[6]

$$\left[\frac{3d_d^2}{10}\right] \equiv \frac{3}{10} g^2(\theta) + \frac{7}{20}\left[\frac{g_\perp^2 - g_\parallel^2}{g(\theta)}\right]^2 \sin^2\theta \cos^2\theta \tag{23a}$$

We must also replace $\frac{1}{4}(\gamma_e H_1)^2$ by the expression

$$\frac{\beta^2}{4\hbar^2}\left[\left(\frac{g_\parallel g_\perp}{g(\theta)}\right)^2 \frac{H_{1x}^2}{4} + g_\perp^2 \frac{H_{1y}}{4}\right] \tag{23b}$$

where H_{1y} is that component of the microwave field mutually perpendicular to H and the crystal axis z; and H_{1x} is the component mutually perpendicular to H and H_{1y}.

2.2. Nuclear Relaxation

As a test of this model, we note that the nuclear relaxation is determined, in the absence of microwaves, by Eqs. (4) and (5), with $T_{1n'}$ defined by Eq. (20), and $P \to P_0$, which reduces to $T_{1n'}^{-1} = \bar{\sigma}' T_1 . \operatorname{sech}^2 \chi$, where

$$\chi \equiv g\beta H/2kT \tag{24}$$

For $n \gg n'$, these equations have the time constants

$$\frac{1}{\tau_s} \approx \frac{1}{T_{1nl}} + \frac{n'}{n} \frac{1}{T_{12} + T_{1n'}} \tag{25}$$

$$\frac{1}{\tau_f} \approx \frac{1}{T_{12}} + \frac{1}{T_{1n'}} \tag{26}$$

The principal change in p occurs with the slow time constant τ_s, which we identify as the relaxation time of the distant nuclei. Actually, the near nuclei are not as abundant and do not contribute much to the observed NMR signal since their resonance is broadened and shifted by the local field of the ion.* If we further neglect T_{1nl} and take $T_{12} \ll T_{1n}'$, we obtain

$$1/\tau_s \to (n'/n)(1/T_{1n}') = 1/T_{1n} \tag{27}$$

as the nuclear relaxation rate expected for the whole sample. The model

* Recently the NMR resonances of the near nuclei have been resolved: A. R. King, J. P. Wolfe, and R. L. Ballard, *Phys. Rev. Letters* **28**:1099 (1972). These new experiments provide a site-selective probe of spin-lattice relaxation, spin diffusion, and dynamic polarization.

hus predicts an essentially unique relaxation rate, given by the average ear nuclear relaxation rate, reduced by the ratio of heat capacities (n'/n) $\approx (d^3/r_2{}^3)$. This yields, finally,

$$\frac{1}{T_{1n}} = \frac{3}{10}\left(\frac{g_d\beta}{H}\right)^2 \frac{1}{r_1{}^2 r_2{}^3}\,\frac{\operatorname{sech}^2\chi}{T_{1e}} \equiv \frac{\bar{\sigma}\operatorname{sech}^2\chi}{T_{1e}} \tag{28}$$

which defines $\bar{\sigma}$. This prediction is insensitive to the exact value of the diffusion barrier provided $r_1{}^3 \ll d^3 \ll r_2{}^3$. It also shows that although diffusion is important, and leads to a unique observed relaxation rate, it is too rapid to enter explicitly into the observed rate, which is limited rather by the direct relaxation of the near nuclei through the ion. Although Eq. (28) was derived from a phenomenological model, the same result can also be obtained by a more rigorous solution of the spin diffusion equation, assuming a small but finite diffusion constant in the inner shell and rapid diffusion in the outer shell.[6]

The factor $\operatorname{sech}^2\chi$ in Eq. (28) arises in the following way: For $\chi \gg 1$, it is proportional to $\exp(-g\beta H/kT)$, which is the probability that the ion is in its upper Zeeman state. The nuclear relaxation rate is proportional to this probability since it proceeds by a $M = 1/2 \to -1/2$, $m = -1/2 \to 1/2$ transition.

Equation (28) is not the most general expression for nuclear relaxation[7] but is particularly appropriate for dilute paramagnets, e.g., 1% Nd:LaMN, in high fields and low temperatures. More specifically, the validity conditions are $(T_{1e}\omega_n)^2 \gg 1$, $T_{1e} \gg T_{2n}$, and $d \gg b$, where b is the scattering length introduced by Khutshishvilli,[8] typically ~1 Å in the Nd:LaMN case. In the opposite limit of $d \ll b$, i.e., free diffusion, the theoretical prediction is $T_{1n}^{-1} = 4\pi NbD \propto (T_{1e})^{-1/4}$.

Although at low temperatures the approximation $T_{1e}\omega_n \gg 1$ assumed in Eq. (28) is usually valid, it is possible to generalize the treatment, and one finds that the nuclear relaxation rate at arbitrary values of $T_{1e}\omega_n$ is given by Eq. (28), multiplied by the factor

$$1/[1 + (\omega_n T_{1e})^{-2}] \tag{29}$$

The general experimental indications bear out this result at higher temperatures, where $\omega_n \sim T_{1e}^{-1}$.

We give two examples of an experimental test of Eq. (28), the first being a study of the proton relaxation time in Nd:LaMN,[9,10] for which $g_\parallel = 0.36$ and $g_\perp = 2.7$. The relaxation of Nd^{3+} in this crystal has been studied in detail;[11] for crystal orientation $c \perp H$, the observed spin–lattice relaxation rate is

$$T_{1e}^{-1} = 3\times 10^{-18} H^5 \coth(g_\perp\beta H/2kT) + 6\times 10^9 \exp(-47/T) \quad \text{sec}^{-1} \tag{30}$$

where H is in Oe. The first term represents the direct process, which dominates roughly below 2°K; the second is the Orbach process, dominating at higher temperatures. The Raman process is too weak to be observed in the helium range. Although the direct process becomes phonon-bottlenecked at fields above ~6 kOe, and the observed spin–bath relaxation rate is *not* the spin–lattice relaxation rate; we should use for T_{1e}^{-1} in Eq. (28) the true spin–lattice relaxation rate, except in cases of very extreme bottlenecking, as discussed in Section 2.5. The proton relaxation time T_{1n} was measured for several crystals of Nd:LaMN with $c \perp H$ over the wide ranges $0.5 \leqq T \leqq 4.2$°K, $1 \leqq H \leqq 50$ kOe, by dynamically enhancing the proton polarization through saturation of a forbidden transition at $\nu_e - \nu_n$, and then observing the decay of the NMR signal to thermal equilibrium. Although the crystals were grown from a 1% Nd:LaMN solution, there is evidence that the Nd concentration in the solid is only of the order of 0.2%. For this concentration, $r_2 \approx 41$ Å; $r_1 \approx 4.4$ Å from x-ray data; thus Eqs. (28) and (29) predict

$$T_{1n}^{-1} = [0.87\times 10^{-16} H^3 \coth(2.7\beta H/2kT) + (2\times 10^{11}/H^2)\exp(-47/T)]\,\mathrm{sech}^2(2.7\beta H/2kT) \quad \mathrm{sec}^{-1} \tag{31}$$

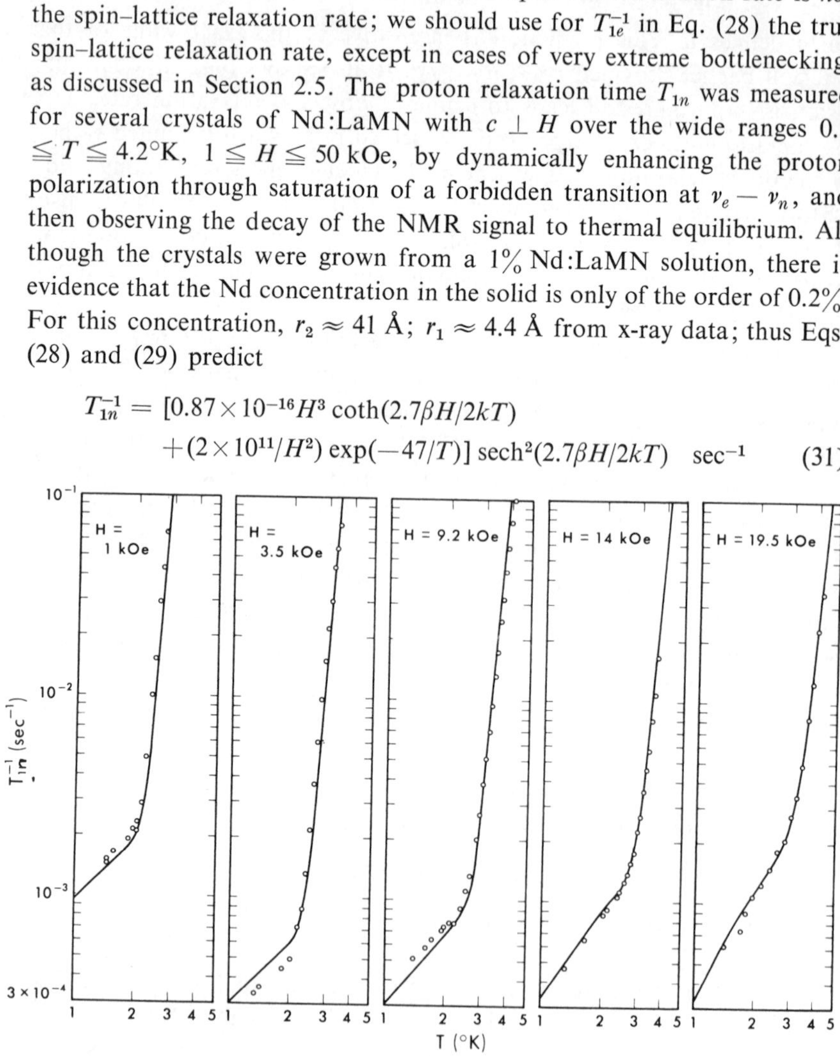

Fig. 5. Observed[9] proton relaxation rate in 1% Nd:LaMN with $c \perp H$. The solid line is the equation

$$T_{1p}^{-1} = [1.4\times 10^{-16} H^3 \coth(2.7\beta H/2kT + 1.25\times 10^{12} H^{-2}\exp(-47/T)] \times \mathrm{sech}^2(2.7\beta H/2kT) + 2.1\times 10^{-4} T + 7\times 10^2 H^{-2} T \quad \mathrm{sec}^{-2}$$

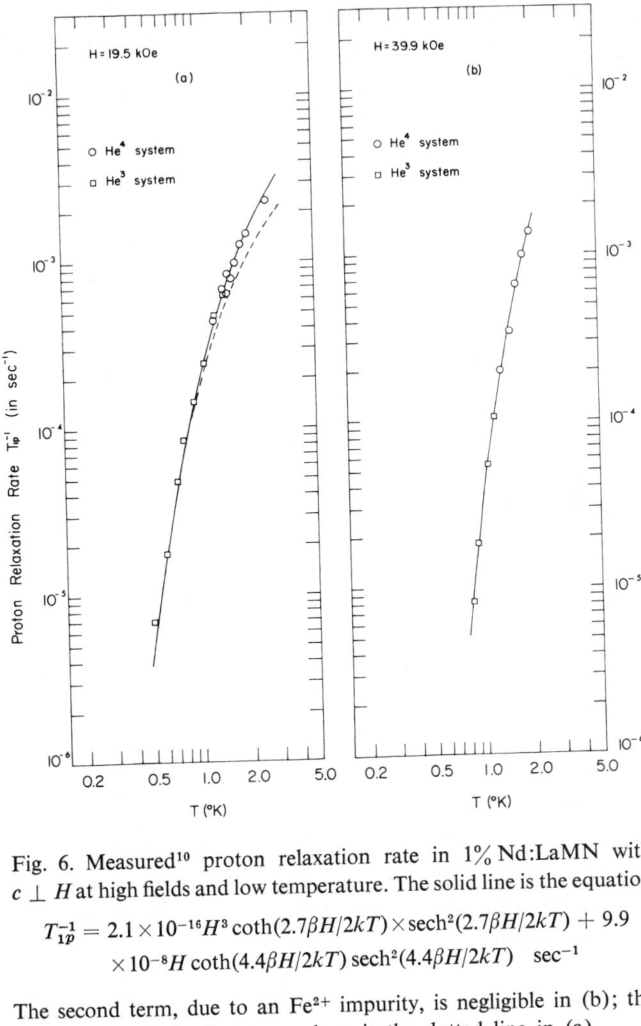

Fig. 6. Measured[10] proton relaxation rate in 1% Nd:LaMN with $c \perp H$ at high fields and low temperature. The solid line is the equation

$$T_{1p}^{-1} = 2.1 \times 10^{-16} H^3 \coth(2.7\beta H/2kT) \times \mathrm{sech}^2(2.7\beta H/2kT) + 9.9 \times 10^{-8} H \coth(4.4\beta H/2kT)\, \mathrm{sech}^2(4.4\beta H/2kT) \quad \mathrm{sec}^{-1}$$

The second term, due to an Fe^{2+} impurity, is negligible in (b); the contribution of the first term alone is the dotted line in (a).

Data for T_{1n}^{-1} are shown in Fig. 5 and clearly display the predicted dependence on H and T at these fields and temperatures, where the sech² factor is approximately unity. Figure 6 shows data at larger values of H/T where the $\mathrm{sech}^2(g\beta H/2kT)$ dependence is very well displayed. The consequence is that the relaxation time becomes exponentially larger, $T_{1n} \propto \exp(g\beta H/2kT)$, with important consequences for polarized target technology.

We take as a second example the proton relaxation[12] in $(Yb_{0.02}, Y_{0.98}) \times (C_2H_5SO_4)_3 \cdot 9H_2O$, denoted as Yb:YES, which is the crystal used in

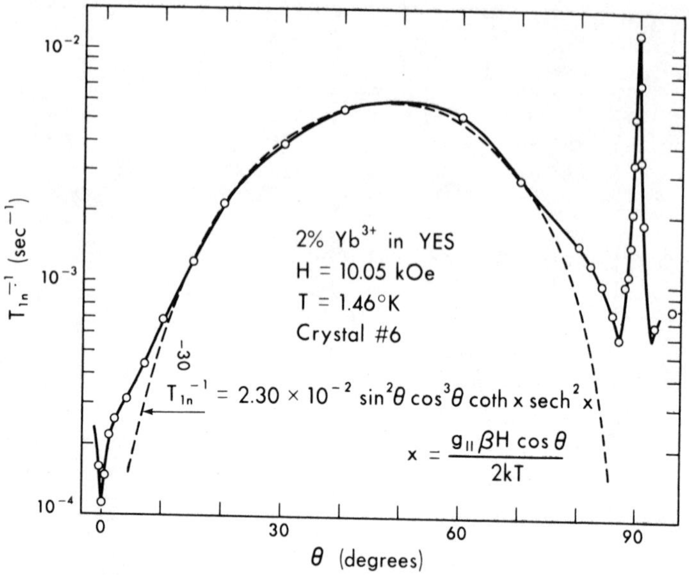

Fig. 7. Measured[12] proton relaxation rate in Yb:YES.

the spin refrigerator discussed in Section 4. The Yb^{3+} ions in YES have both an anisotropic g-factor ($g_{\|} = 3.35$, $g_{\perp} \approx 0$) and an anisotropic direct process, Eq. (69). The data for T_{1n}^{-1}, Fig. 7, display the expected angular dependence and also show a sharp spike at $\theta = 90°$ due to direct cross-relaxation of the protons with the Yb^{3+} ions.

To summarize, there is reasonably satisfactory agreement between observed nuclear relaxation rates and the predictions of the model, which provides some justification for extending its use to the dynamic case where microwaves are applied.

2.3. Dynamic Polarization

2.3.1. Ideal case

We take Eqs. (17), (21), and (22) as our starting point in predicting the enhanced proton polarization in the ideal case, where the resonance lines W_1, $\overline{W}_2$, and $\overline{W}_3$ are clearly resolved; let us induce only one, say $\overline{W}_2 = \bar{\sigma}' \cdot \frac{1}{2}T_{2e}(\gamma_e H_1)^2$. Assuming negligible leakage relaxation, i.e., $T_{1nl}=0$, these equations can be solved for the steady-state value of the polarization of the abundant distant protons, to find

$$p_{ss} \approx \frac{P_0}{1+f}\left[\frac{s}{s+s_{1/2}}\right] \tag{32}$$

here we have introduced the factor

$$f \equiv nT_{1e}/NT_{1n} \tag{33}$$

nd the usual saturation factor

$$s \equiv 2T_{1e}W_1 = (\gamma_e H_1)^2 T_{1e} T_{2e} \tag{34}$$

nd its particular value for half-saturation in this case,

$$s_{1/2} \approx 2Nf/\bar{\sigma}n(1+f) = 2Nf/\sigma' n'(1+f) \tag{35}$$

vhere $\bar{\sigma}$ and $\bar{\sigma}'$ are defined in Eqs. (28) and (18). The factor f is just the atio of the number of nuclei relaxing per second to the number of electrons elaxing per second, and this requires $f \ll 1$ for optimum enhancement. For 1% Nd:LaMN under typical conditions, $f < 10^{-2}$, and then, from Eqs. (35) and (28),

$$s_{1/2} \approx 2 \operatorname{sech}^2(g\beta H/2kT) \tag{36}$$

This usually is of order unity, but at large H/T values, it becomes very small, essentially because of the depopulation of the upper ion state and the increase in T_{1n}.

The transient solutions of Eqs. (17), (21), and (22) are complicated, but one can show that p builds up approximately at the rate $\tau_{on}^{-1} \approx \bar{\sigma} W_1$. At very high saturation, τ_{on}^{-1} goes over to the constant value N/nT_{1e}.

In the case of significant leakage relaxation, one can show that $s_{1/2}$ is increased roughly by the factor (T_{1n}/T_{1nl}) and that $p_{ss}|_{\max}$ is decreased.

As an example, we discuss briefly some DNP experiments on 1% Nd:LaMN.[9] Figure 8 shows the observed steady-state enhancement of the proton polarization versus magnetic field at $T = 4.2°$K and $\nu_e = 73.9$ GHz for several values of the microwave attenuator in Fig. 2. At fields 20 Oe below and above H_0, one observes large negative and positive enhancements corresponding, respectively, to saturation of the forbidden transitions W_2 and W_3. No enhancement is observed when the main line W_1 at H_0 is saturated, in agreement with predictions,[13] since the relaxation rates w_2 and w_3 are equal in a solid with I, S coupling of dipolar form.

At lower temperatures, $T \approx 1.5°$K, enhancements of 540, corresponding to proton polarizations of 72%, were observed in crystals weighing ~200 mg. The theoretical ideal under these conditions is $p = P_0 = 83\%$. This experiment showed for the first time that large proton polarizations could be obtained, with the subsequent application to polarized targets

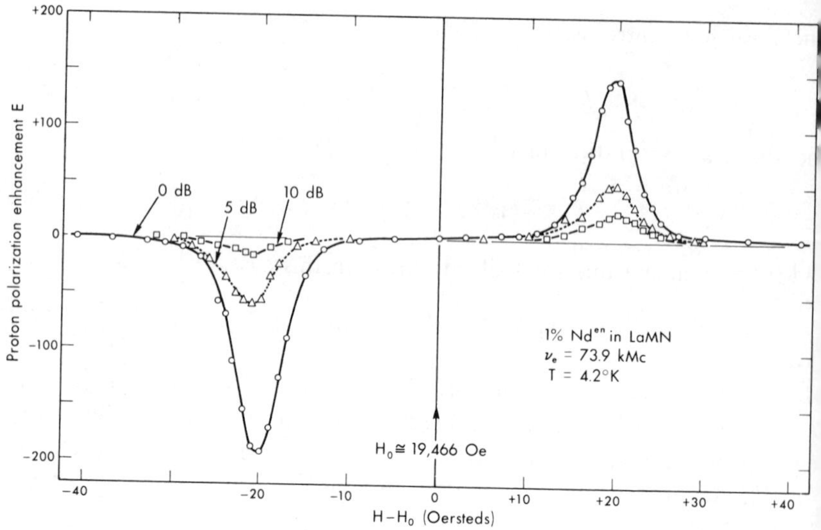

Fig. 8. Observed[9] DNP of protons in 1% ^{144}Nd:LaMN.

discussed in Section 2.6. The observed steady-state enhancement E_{meas} may be fit approximately by the expression

$$E_{\text{meas}} = E_{\text{sat}} \mathscr{P} / (\mathscr{P} + \mathscr{P}_{1/2}) \tag{37}$$

which is that expected from Eq. (32), where $\mathscr{P}$ is the cavity microwave power, and $\mathscr{P}_{1/2}$ is empirically determined for a best fit. The power lost in the cavity walls required to produce a peak microwave field H_{1y} in a high-mode cavity of volume V and quality factor Q is

$$\mathscr{P} \approx (3H_{1y} V \nu_e / 8Q) \times 10^{-7} \quad \text{W} \tag{38}$$

For Nd:LaMN with $c \perp H$, we have

$$W_1 \approx T_{2e} g_\perp{}^2 \beta^2 H_{1y}^2 / 8\hbar^2 \ \sec^{-1}, \qquad s = 2T_{1e} W_1, \qquad \text{and} \qquad T_2 = (\tfrac{1}{2} g_\perp \beta H_{1/2}/\hbar)^{-1},$$

where $H_{1/2}$ is the EPR linewidth at half-maximum. For the values $H_{1/2} = 6$ Oe, $V = 1$ cm^3, $T_{1e} = 10^{-4}$ sec from Eq. (30) at 1.5°K, $g_\perp = 2.7$, $Q = 10^3$, and $\nu_e = 75$ GHz, we find $\mathscr{P} \approx 15s$ mW, leading to $\mathscr{P}_{1/2} = 15 s_{1/2} \approx 10$ mW, as the predicted cavity power required to half-saturate the enhancement. This agrees approximately with the experimental result for small crystals. If the forbidden lines are not well resolved from the main line, then the cavity Q may be significantly lowered by the paramagnetic resonance losses, requiring much more power.

2.3.2. Unresolved Lines

If the EPR linewidth T_{2e}^{-1} is comparable or greater than ν_n, then the above analysis is invalid, since in attempting to saturate only $\overline{W}_2$, which yields a negative enhancement, one also partially saturates $\overline{W}_3$, which tends to produce positive enhancements, and also to partially saturate W_1, which tends to produce zero enhancement. The result is that there is only a differential net enhancement, considerably reduced from the ideal value. If the EPR line is inhomogeneously broadened, with a packet width $\delta\nu \ll T_{2e}^{-1}$, each thermally isolated, then at a microwave frequency ν, one saturates the transition $\overline{W}_2$ for a packet at $\nu - \nu_n$, giving a negative partial polarization of magnitude $p_{ss}g(\nu - \nu_n)\,\delta\nu$; and simultaneously, one saturates the $\overline{W}_3$ transition for a packet at $\nu + \nu_n$, giving a positive partial polarization of magnitude $p_{ss}g(\nu + \nu_n)\,\delta\nu$. The net polarization is thus

$$p = |\, p_{ss} \,| \, (\delta\nu)[g(\nu - \nu_n) - g(\nu + g_n)] = 2 \,|\, p_{ss} \,| \, (\delta\nu)\nu_n \, dg/d\nu \qquad (39)$$

which is proportional to the derivative of the EPR line shape. This analysis is also valid for a powdered sample with an anisotropic g-factor. The differential dependence $p \propto dg/d\nu$ has been observed for F-centers in KCl,[14] and for n-irradiated $(CH_2)_n$,[15] where the EPR line is much wider than ν_n and is believed to be inhomogeneous.

In the opposite limiting case of wide, homogeneous EPR linewidth, one can solve Eqs. (17), (21), and (22) including the explicit dependence $W_1(\nu)$, $\overline{W}_2(\nu)$, and $\overline{W}_3(\nu)$.[16,17] The result is that at low power, a differential effect $p \propto dg/d\nu$ is expected, but at increasing power, the peaks of the enhancement are pushed out and greatly reduced. [See also Eqs. (48) and (49).]

The true situation is intermediate between these two limiting cases and a proper treatment must include the effects of cross-relaxation, i.e., diffusion of spin temperature out of the packets. It has been shown[17a,17b] that for an inhomogeneously broadened EPR line, there is an interesting second-order effect: Microwave saturation of the center of the line will produce dynamic nuclear polarization if the NMR line itself has a structure, for example, due to dipole–dipole interactions, or to quadrupole interactions.

2.4. Spin Temperature Theory

Although the phenomenological rate equation theory of Sections 2.1–2.3 is approximately correct, and does explain the principal features observed in DNP, it is not strictly correct, since it uses time-dependent perturbation theory, valid only at low powers, and it does not proper take into account spin–spin interactions. A better approach is to use t density matrix and the concept of spin temperature in the rotatir frame.[18–20] We first consider an electron spin system S in a dc field H an a very strong microwave field $2H_1$, described by

$$\mathscr{H} = g\beta H \sum_k S_z^k + 2g\beta H_1 \cos \omega t \sum_k S_x^k + \mathscr{H}_{ss} \qquad (4($$

where the last term is the spin–spin interaction, represented by a gener bilinear form. To fix ideas, think of the spins as dilute paramagnetic io in a crystal. A canonical transformation $U\mathscr{H}U^{-1}$, to a rotating frame reference, by the operator $U = \exp(-i\hbar\omega S_z t)$, will yield an effective tim independent spin Hamiltonian in this rotating frame,

$$\mathscr{H}^* = \hbar(\omega_e - \omega)S_z + \hbar\omega_1 S_x + \mathscr{H}_{ss}^0 \qquad (41$$

where $\hbar\omega_e = g\beta H$, $\hbar\omega_1 = g\beta H_1$, $S_z = \sum_k S_z^k$, $S_x = \sum_k S_x^k$, and $\mathscr{H}_{ss}^0$ is th part of $\mathscr{H}_{ss}$ that commutes with U. We now make the assumption th $|\omega_1| \ll |\omega_e - \omega|$, but that H_1 is still strong enough to make transition between the Zeeman energy reservoir and the spin–spin energy reservoi and thus to bring both reservoirs to a common spin temperature T_s in th rotating frame. This means that one can compute the ensemble average o the Zeeman energy $\overline{\langle Z^* \rangle} = \hbar(\omega_e - \omega)\overline{\langle S_z \rangle}$ and the spin–spin energ $\overline{\langle \mathscr{H}_{ss}^0 \rangle}$ from a common density matrix

$$\varrho^* = \exp(-\mathscr{H}^*/kT_s)/\mathrm{Tr}[\exp(-\mathscr{H}^*/kT_s)]$$

One finds that

$$\overline{\langle Z^* \rangle} = \mathrm{Tr}(\varrho^* Z^*) \approx \frac{\mathrm{Tr}(\mathscr{H}^* Z^*)}{-kT_s \,\mathrm{Tr}\,(1)} = \frac{\hbar^2(\omega_e - \omega)^2 \,\mathrm{Tr}\,(S_z^2)}{-kT_s \,\mathrm{Tr}\,(1)} \qquad (42$$

$$\overline{\langle \mathscr{H}_{ss}^0 \rangle} \approx \frac{\mathrm{Tr}\,(\mathscr{H}_{ss}^0)}{-kT_s \,\mathrm{Tr}\,(1)} = \frac{\hbar^2 \omega_L^2 \,\mathrm{Tr}\,(S_z^2)}{-kT_s \,\mathrm{Tr}\,(1)} \qquad (43$$

under the approximation $\exp(-\mathscr{H}^*/kT_s) \approx 1 - \mathscr{H}^*/kT_s$. In contrast t the rate equations theory of Section 2.1, where the exponential factors ca be treated exactly, with predictions valid for large polarizations, one mus make this high-temperature approximation in order to make the proble tractable in the density matrix approach; the results will still be useful however. We define $\hbar^2\omega_L^2 = \mathrm{Tr}\,(\mathscr{H}_{ss}^0)/\mathrm{Tr}\,(S_z^2)$; it can be shown that ω_L^2 is one-third of the second moment of the EPR line, and thus $H_L = \hbar\omega_L/g\beta$ is roughly the local interaction field.

We assume that the H_1 field produces a common spin temperature and that equilibrium with the lattice is slowly approached according the relaxation equations

$$(\partial/\partial t)\overline{\langle Z^*\rangle} = -(1/T_{1e})\{\overline{\langle Z^*\rangle} - [(\omega_e - \omega)/\omega]\overline{\langle Z\rangle}\} \tag{44a}$$

$$(\partial/\partial t)\overline{\langle \mathscr{H}^0_{ss}\rangle} = -(2/T_{1e})\overline{\langle \mathscr{H}^0_{ss}\rangle} \tag{44b}$$

here

$$\overline{\langle Z\rangle} = \mathrm{Tr}(\varrho Z) \approx \hbar^2\omega_e^2 \mathrm{Tr}\,(S_z^2)/-kT\,\mathrm{Tr}\,(1) \tag{44c}$$

the thermal average value of the Zeeman energy in the laboratory frame : lattice temperature T. In the steady state, $\partial\overline{\langle Z^*\rangle}/\partial t + \partial\overline{\langle \mathscr{H}^0_{ss}\rangle}/\partial t = 0$, ading to

$$T/T_s = (\omega_e - \omega)\omega_e/[(\omega_e - \omega)^2 + 2\omega_L^2] \tag{45}$$

his equation shows that the electron spin temperature in the rotating ame is reduced by the large factor $\pm\omega_e/2\sqrt{2}\omega_L$ at $\omega = \omega_e \mp \sqrt{2}\omega_L$ a very strong rf field at ω is applied. The case for arbitrary values of H_1 as also been treated.[19]

To see how this low spin temperature may be transferred to nuclei, e now include the nuclear spins in the crystal as part of the system; this dds the term $\hbar\omega_n I_z$ to Eq. (41), along with spin–spin interaction terms ke $\mathscr{H}^0_{SI}$ and $\mathscr{H}_{II}$. If $|\omega_e - \omega| \sim \omega_L \sim \omega_n$, then we can expect that lectron spins in the rotating frame will be roughly on "speaking terms" vith the nuclei in the laboratory frame and thus will transfer their temerature T_s to the nuclei. This results in an enhancement of the nuclear olarization just given by Eq. (45), which represents in the ideal case a lirect nuclear spin cooling, which is distinct from the dynamic polarization y saturation of the forbidden transitions considered in Sections 2.1–2.3.

Of course, both direct spin cooling and saturation of forbidden transiions may occur simultaneously. A general treatment has been given by bragam and Borghini[20] for the system of Section 2.1: an I, S system under he influence of an rf field H_1 of arbitrary strength, and lattice relaxation ates T_{1e}^{-1} and $T_{1n}^{-1} = \bar{\sigma}T_{1e}^{-1}\,\mathrm{sech}^2\,\chi$, and assuming a homogeneous EPR inewidth. The result for the enhancement of the nuclear polarization is

$$E = \frac{(\omega_e/\omega_n)(\overline{W}_3 - \overline{W}_2)T_{1n} + [\omega_e(\omega_e - \omega)/2\omega_L^2](\overline{W}_3 + \overline{W}_2)W_1 T_{1n} T_{1e}}{[1 + (\overline{W}_3 + \overline{W}_2)T_{1n}]\{1 + (1/2\omega_L^2)[(\omega_e - \omega)^2 + 2\omega_L^2]W_1 T_{1e}\}} \tag{46}$$

The first term in the numerator represents the dynamic polarization by saturation of the forbidden transitions $\overline{W}_2$ or $\overline{W}_3$; the second term gives the polarization through the direct spin cooling. If the EPR line is ver narrow compared to ω_n, and $\omega = \omega_e - \omega_n$, then W_1 and $\overline{W}_2$ are negligibl and Eq. (46) reduces to

$$E \approx \frac{\omega_e}{\omega_n} \frac{\overline{W}_2 T_{1n}}{1 + \overline{W}_2 T_{1n}} = \frac{\omega_e}{\omega_n} \frac{s}{s + s_{1/2}} \quad (47$$

where $s_{1/2} = 2$. This is in agreement with the previous result, Eqs. (32 and (36), in the high-temperature limit with $f \ll 1$, thus verifying that th phenomenological approach is essentially correct in the ideal case.

If the EPR line is broad and homogeneous, for low rf fields,

$$E \approx \frac{\omega_e}{\omega_n} \frac{(\overline{W}_3 - \overline{W}_2)T_{1n}}{1 + (\overline{W}_3 + \overline{W}_2)T_{1n} + \{[(\omega_e - \omega)^2 + 2\omega_L{}^2]/2\omega_L{}^2\} W_1 T_{1e}} \quad (48$$

For very high rf fields,

$$E \approx \omega_e(\omega_e - \omega)/[(\omega_e + \omega)^2 + 2\omega_L{}^2] \quad (49$$

These predictions, Eqs. (46), (48), and (49), have not yet been well verifie experimentally. Recent DNP experiments[21,21a] on Ce:LaMN can be inter preted in terms of Eq. (46) with modifications to take into account th phonon bottleneck.

Borghini[21b] has proposed still another mechanism for dynamic polariza tion by saturation of the wings of an electron line broadened by spin–spin interactions, called the DONKEY effect, meaning dynamic orientation of nuclei by cooling of electron interactions. If the microwave frequency ν is slightly below resonance, the rf field produces an electron spin flip, and the extra energy $h(\nu_e - \nu)$ comes from the spin–spin reservoir, which becomes cooled thereby. The dipolar coupling of the nuclei with the spins allows for thermal contact with this reservoir, thereby cooling the nuclei. The overall effect of sweeping through the electron resonance is to give positive and negative enhancements below and above the center of the line, respectively, and is similar but distinguishable from the saturation of forbidden I_+S_- transitions. The two mechanisms may occur simultaneously.*

2.5. Phonon Bottleneck Effects

The above sections have assumed no phonon bottleneck, but at high fields where the direct relaxation process becomes strong, the limiting process in Fig. 4 becomes the phonon-bath rate T_{ph}^{-1}, and the phonon

* For a recent review of the theory of all mechanisms of dynamic polarization see the article by M. Borghini, Ref. 23b, page 1.

excitation number $q = [\exp(h\nu_e/kT_p) - 1]^{-1}$ must now be considered a dynamic variable. The number of phonons in the system is qN_p, where N_p is the number of phonon modes per unit volume,

$$N_p = 12\pi\nu_e^2\, d\nu_e/v^3 \tag{50}$$

where v is the velocity of sound, and $d\nu_e$ is the EPR linewidth. If $(N/T_{1e}) \gg (qN_p/T_{\mathrm{ph}})$, it is clear that the phonons will heat up during spin relaxation and also during microwave resonance saturation. In the absence of nuclear interactions, this phonon bottleneck results in a longer spin–bath relaxation time, given by[11]

$$T_b \approx (1 + \sigma^*)T_{1e} \tag{51}$$

where σ^* is the phonon bottleneck factor [see also Section 6.1 in Chapter 5]

$$\sigma^* = (NT_{\mathrm{ph}}/N_pT_{1e})P_0^2 \tag{52}$$

The influence of the phonon bottleneck on the steady-state nuclear polarization obtained by saturation of the forbidden transitions, as in Section 2.1, may be calculated by introducing as additional dynamical variables the phonon excitations $q_1(\nu_e)$, $q_2(\nu_e + \nu_n)$, and $q_3(\nu_e - \nu_n)$. If $\overline{W}_2$ is saturated (assuming $f \ll 1$), then it can be shown that the proton polarization p is reduced from its value P_0, in the absence of a bottleneck, to the value given in Fig. 9, as a function of the parameter

$$\alpha' = \frac{\sigma^* n\bar{\sigma}}{-2NP_0} = \frac{n/T_{1n}}{N_p/T_{\mathrm{ph}}}\left[\frac{-P_0}{2(1 - P_0^2)}\right] \tag{53}$$

where $\bar{\sigma} = (T_{1e}/T_{1n})\,\mathrm{sech}^2\,\chi = T_{1e}/T_{1n}(1 - P_0^2)$. Except for the P_0 dependence, α' is the ratio of the number of nuclei relaxing per second to the

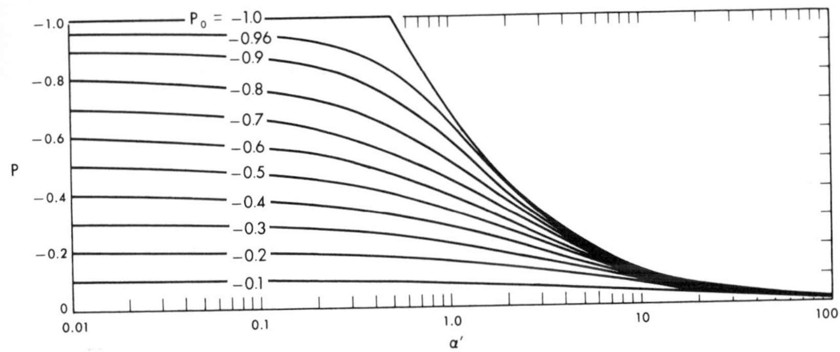

Fig. 9. The reduced polarization p versus the bottleneck parameter α', Eq. (53), with the ideal polarization P_0 as a parameter. Saturation of $\overline{W}_2$ is assumed.

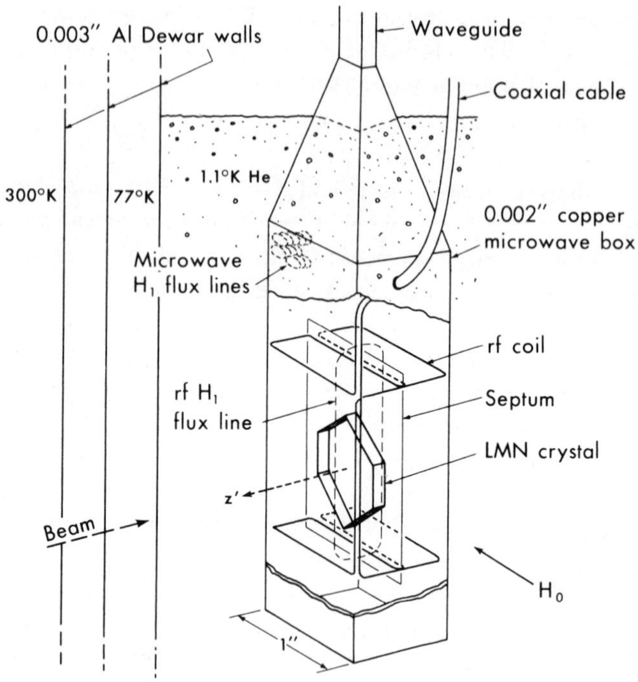

Fig. 10. Details of the Berkeley polarized target.[24] Only one of a stack of crystals is shown.

number of phonons relaxing per second. For 1% Nd:LaMN crystal 1 cm thick at 20 kOe and 1°K, one estimates $\sigma^* \approx 460$, a sizeable bottleneck factor. However, under these same conditions, $P_0 = -0.95$, and $\alpha' \approx 0.23$ the dynamic proton polarization is not reduced significantly. One can also show that proton relaxation rate is slightly reduced by the factor $(1 - P_0\alpha')^{-1}$.

For nuclear polarization by the direct spin cooling of Section 2.4, it is necessary to multiply $2\omega_L{}^2$ in Eq. (45) by the factor $(1 + \sigma^*)$ in order to take into account the bottleneck; this gives moderate agreement with the experiments on Ce:LaMN.[21,21a]

2.6. Applications*

An important application of this method of DNP has been the creation of polarized proton targets for use in nuclear scattering experiments.[23]

* For more recent results, see Ref. 23b.

igure 10 shows the essential part of the first high-energy target, made at erkeley.[24] It is a 1-in. cube of Nd:LaMN crystals in a copper box, which as many microwave resonance modes and thus a reasonably uniform H_1 eld throughout the large sample, required to induce uniformly the forbidden transitions. The proton NMR signal is obtained from a figure-of-ght coil designed to sample uniformly the polarization. The proton NMR gnal in Nd:LaMN has a dipole–dipole structure which changes conderably as the polarization is enhanced. To measure the actual polarizaon requires a careful comparison of signal areas, including corrections for etector nonlinearity for large signals. The coil Q and filling factor must be eld down, otherwise one gets maser oscillations at negative enhancements.

The structure in the NMR line is a superposition of a number of soalled Pake doublets arising from the fact that each proton in an H_2O olecule has a nearest-neighbor proton which can be either parallel or ntiparallel to H; this gives two NMR lines, separated by the dipole–ipole interaction. At thermal equilibrium, these have nearly equal innsities, but as the polarization builds up, the intensity ratio becomes $\sim\exp(h\nu_e/kT)$ for ideal enhancements. The ratio is reversed for negative nhancements. When one reverses the polarization by switching frequency rom $\nu_e - \nu_n$ to $\nu_e + \nu_n$, the weakest line reverses first and then becomes he strongest negatively enhanced line.

Although large polarizations are obtained in Nd:LaMN, unforunately it is only 3% hydrogen by weight; DNP experiments in more ydrogeneous materials have yielded polarization of $\sim$30% in waxes oped with the free-radical di-tertiary butyl nitroxide,[24a] frozen toluene or polystyrene doped with diphenyl picryl hydrazyl,[24b] neutron-irradiated olyethylene,[15] and frozen alcohol–water mixtures containing free radials.[23,23a] Quite recently polarizations of $\sim$80% have been obtained in naterials containing 15% hydrogen.*

3. OTHER CASES

In the above case of electron–nuclear dipolar coupling in solids, the sample is dilute in electron spins, but concentrated in nuclear spins belonging o diamagnetic atoms. The interaction $\mathscr{H}_{IS}$ is well represented by the static

* For more recent compilations of results, see C. D. Jeffries in *Polarization Phenomena in Nuclear Reactions,* edited by H. H. Barschall and W. Haeberli, University of Wisconsin Press, Madison (1971), page 351. See also Ref. 23b: A. Abragam, page 247; M. Borghini, page 1; K. Scheffler, page 271; H. Glattli, page 281; S. Mango, page 289.

dipolar interaction $\mathscr{H}_{\rm dd}$, Eq. (3b), especially at low temperatures, wher $T_{1e} \gg \nu_n^{-1}$; the principal effect of $\mathscr{H}_{IS}$ is a slight admixing of the wave functions $| M, m\rangle^0$, giving rise to the forbidden transitions. The effect of the lattice vibrations are represented by a phenomenological relaxatio operator $\mathscr{H}_R$, which in this case may take the form $g\beta \mathbf{S} \cdot \mathbf{H}'(t)$; its physica origin lies in the spin–orbit–crystal field interactions of the electron spins. We now discuss several other cases of DNP, differing principally by the form of $\mathscr{H}_{IS}$ and the form of $\mathscr{H}_R$.[13]

3.1. Strong HFS Coupling in Paramagnetic Ions

We consider a magnetically dilute crystal where the spin S is a paramagnetic ion and the nucleus I belongs to the ion, corresponding to the static spin Hamiltonian

$$\mathscr{H} = \beta \mathbf{S} \cdot \mathsf{g} \cdot \mathbf{H} + \mathbf{I} \cdot \mathsf{A} \cdot \mathbf{S} + \mathbf{S} \cdot \mathsf{D} \cdot \mathbf{S} + \mathbf{I} \cdot \mathsf{Q} \cdot \mathbf{I} - g_n \beta \mathbf{H} \cdot \mathbf{I} \tag{54}$$

Bound donors in lightly doped semiconductors also belong to this case; the sample is dilute both in electron spins and nuclear spins of interest. The energy levels are shown in Fig. 11 for $I = 1$, $S = 1/2$, assuming that the first two terms in Eq. (54) are dominant. The applied rf field $\mathbf{H}_1 \cos \omega t$

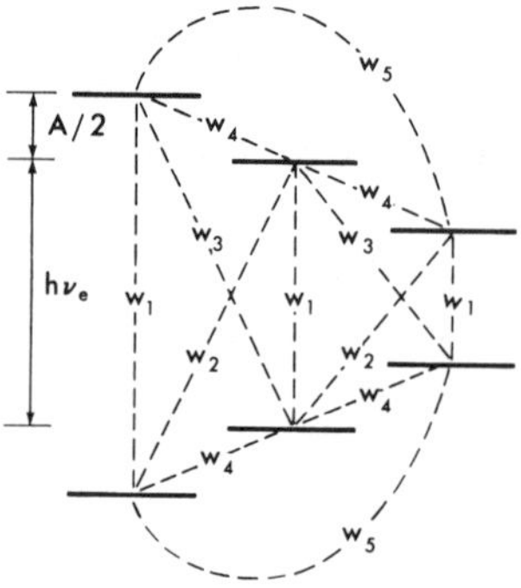

$\|M, m\rangle^\circ$	(a)	(b)
$\|\tfrac{1}{2}, 1\rangle^\circ$	$\exp(-h\nu_e/kT)$	1
$\|\tfrac{1}{2}, 0\rangle^\circ$	1	$\exp(-h\nu_e/kT)$
$\|\tfrac{1}{2}, -1\rangle^\circ$	1	$\exp(-h\nu_e/kT)$
$\|-\tfrac{1}{2}, -1\rangle^\circ$	$\exp(h\nu_e/kT)$	1
$\|-\tfrac{1}{2}, 0\rangle^\circ$	$\exp(h\nu_e/kT)$	1
$\|-\tfrac{1}{2}, 1\rangle^\circ$	1	1
Polarization:	$-\dfrac{h\nu_e}{3kT}$	$\dfrac{h\nu_e}{6kT}$
Alignment:	$-\dfrac{h\nu_e}{9kT}$	$\dfrac{h\nu_e}{18kT}$

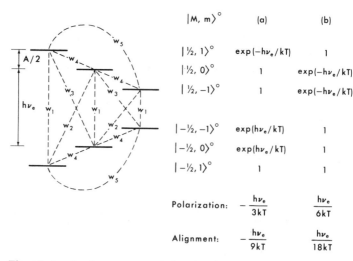

Fig. 11. Levels of a paramagnetic ion $S = \frac{1}{2}$ and nucleus $I = 1$ in strong hfs coupling and in a field H. (a) dynamic populations when forbidden transition $| \frac{1}{2}, 0\rangle \leftrightarrow | -\frac{1}{2}, 1\rangle$ is saturated; (b) populations when $| \frac{1}{2}, 1\rangle \leftrightarrow | -\frac{1}{2}, 1\rangle$ is saturated.

gives various allowed and forbidden transitions; tables have been given for the transition rates.[25] If the hyperfine tensor A is isotropic, the allowed transitions are of the type w_1 in Fig. 11, and correspond to the usual hfs lines in the EPR spectrum; the term AI_+S_- admixes the zero-order functions $| M, m\rangle^0$, however, so that forbidden transitions of the type $\Delta M = +1$, $\Delta m = -1$ are also observed, weaker by the factor $(A/g\beta H)^2 \sim 10^{-2}$–$10^{-4}$ typically. If A is anisotropic, or D or Q are important, the forbidden transitions $\Delta M = +1$, $\Delta m = +1$ are also observable.

The form of the static Hamiltonian Eq. (54) suggests the phenomenological relaxation Hamiltonian

$$\mathscr{H}_R = \beta \mathbf{S} \cdot \mathsf{g}'(t) \cdot \mathbf{H} + \mathbf{I} \cdot \mathsf{A}'(t) \cdot \mathbf{S} + \mathbf{S} \cdot \mathsf{D}'(t) \cdot \mathbf{S} + \mathbf{I} \cdot \mathsf{P}'(t) \cdot \mathbf{I} \quad (55)$$

where the terms represent the fluctuations induced by the lattice vibrations in the g, A, D, and Q tensors; in Section 2, we considered only the first term, rewritten as $g\beta\mathbf{S} \cdot \mathbf{H}'(t)$, which gives rise to relaxation transitions w_1, w_2, and w_4 if $\mathsf{A}'(t)$ is isotropic, and also to w_3 if anisotropic; $\mathsf{Q}'(t)$ additionally gives rise to w_5. If the relaxation rates w can be estimated or measured, one can set up $(2S + 1)(2I + 1)$ rate equations for the population N_i of the ith level of the form

$$dN_i/dt = \sum_{j \neq i} [N_j(W_{ji} + U_{ji}) - N_i(W_{ij} + U_{ij})] \quad (56)$$

where $W_{ji} = W_{ij}$ is the rf transition probability and $U_{ij} = w_{ij} \exp(-E_j/kT)$ is the total relaxation rate from N_i to N_j due to all thermal processes, and $U_{ji} = w_{ij} \exp(-E_i/kT)$ is the corresponding rate from N_j to N_i. As in Section 2.1, the relaxation rates are weighted symmetrically by Boltzmann factors to yield the correct thermal equilibrium populations. The steady-state solution of Eq. (56) will yield the dynamic populations and the dynamic nuclear polarization $p = \overline{\langle I_z \rangle}/I$ and alignment

$$p_2 = I^{-2}[\overline{\langle I_z^2 \rangle} - \tfrac{1}{3}I(I + 1)] \quad (57)$$

As the simplest example, column (a) of Fig. 11 shows the populations for saturation of the forbidden transition $| 1/2, 0\rangle \leftrightarrow | -1/2, 1\rangle$, assuming only relaxations w_1 and w_2 are important, and $g\beta H \gg A$. The polarization $p \approx -h\nu_e/3kT$ and alignment $p_2 \approx -h\nu_e/9kT$ are orders of magnitude larger than the thermal equilibrium values, and this method has been used to orient radionuclei and observe the anisotropy of the γ-radiation.[3,25,26]

Column (b) of Fig. 11 shows the populations and dynamic orientation achieved by saturation of the allowed transition $|1/2, 1\rangle \leftrightarrow |-1/2, 1\rangle$. When the allowed transition is saturated, the nuclear polarization arises from the forbidden relaxation w_2 of the flip-flop type, I_+S_-; a comparable relaxation rate w_3 of the flip-flip type I_+S_+ will cancel the overall effect. On the other hand, rf saturation of the forbidden transitions directly polarizes the nuclei, and is less susceptible to undesirable relaxation leakage of the polarization.

3.2. Overhauser Effect

The case originally considered by Overhauser is found in metals, highly doped semiconductors, and sometimes in concentrated free radicals; the sample is concentrated in both electron spins and nuclear spins. The electron spins are in strong exchange interaction with each other and in weaker contact hfs with the nuclei. The $\mathscr{H}_{IS}$ interaction takes the form of Eq. (3a), but it is very rapidly fluctuating, either because of the translational motion of the electrons or their rapid spin flips due to strong exchange forces, e.g., in concentrated free radicals. The secular part of the hfs is effectively averaged to zero and the EPR spectrum is typically a single narrow line. The static spin Hamiltonian for this system is simply

$$\mathscr{H} = g\beta \mathbf{S} \cdot \mathbf{H} - g_n\beta \mathbf{I} \cdot \mathbf{H} \tag{58}$$

and the relaxation Hamiltonian is the first two terms of Eq. (55). Figure 12 shows the energy levels for $S = 1/2$, $I = 1/2$. The rf field H_1 induces only

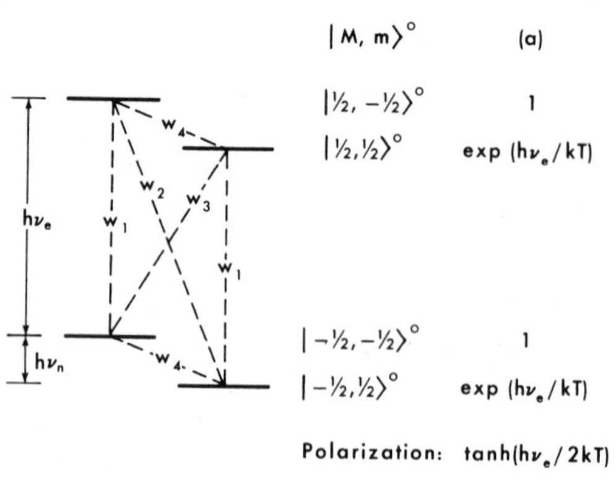

Fig. 12. Levels and populations of a conduction electron $S = \frac{1}{2}$ and a nucleus $I = \frac{1}{2}$ when electron spin resonance is saturated.

the allowed transitions $\Delta M = 1$, $\Delta m = 0$; the first relaxation term induces only allowed w_1 relaxations. If the hfs interaction is of the pure Fermi–Segrè contact type due to s electrons, then the $A'(t)I_+S_-$ terms give rise to the w_2 relaxations only. However, orbital and dipolar contributions will give w_3 relaxations, while impurities (or quadrupole relaxation for $I \geqq 1$) will contribute to the "extraneous" leakage relaxation w_4. In the simplest case, $w_1 \gg w_2 \gg w_3, w_4$, and complete saturation of the EPR line yields the populations of column (a), Fig. 12, and a dynamic nuclear polarization $p = \tanh(h\nu_e/2kT) = P_0$. This example has assumed a negative magnetic moment for S and a positive moment for I; p will be negative for nuclei with a negative magnetic moment. To take into account competing relaxation w_4 and a finite saturation factor s, the rate equations may be solved to obtain[1]

$$p \approx \frac{I+1}{3}\left(\frac{h\nu_n}{kT}\right)\left[1 + \left(\frac{w_2}{w_2 + w_4}\right)\frac{\nu_e}{\nu_n}\,\frac{s}{s+1}\right] \tag{59}$$

The Overhauser effect was first demonstrated in metallic lithium;[27] for metals, the principal experimental problem is to make the sample thickness smaller than the microwave skin depth. This is not so difficult in semiconductors, and sizable enhancements have been obtained in phosphorous doped silicon,[28] and in charcoal and graphite,[29] as well as in colloidal Li particles in irradiated LiH.[30]

3.3. Liquids

This subject has been well reviewed,[30a] and we consider it only briefly here. The system we consider consists of the nuclear spins in the solvent molecules in a liquid containing paramagnetic ions, or free radicals, or other magnetic species. The interaction $\mathscr{H}_{IS}$ may be a contact hfs form, Eq. (3a), or dipolar, Eq. (3b), or both. It will be rapidly fluctuating because of the translation motion of the solvent molecules as well as rapid relaxation flips of the electron spins. Actually, the EPR spectra of liquids display extensive hfs, but this is due to a fixed hfs interaction with nuclei incorporated into the structure of the magnetic species, and not to the solvent nuclei, which are of interest here.

If $\mathscr{H}_{IS}$ is purely scalar, of the form $A'(t)\mathbf{I}\cdot\mathbf{S}$, the situation is much like that of the previous section and Fig. 12. Saturation of the EPR line (assuming no hfs) leads to a dynamic nuclear polarization given by Eq. (59), which we rewrite for $I = 1/2$ as

$$p = p_0 - P_0 f'[s/(s+1)] \tag{60}$$

where $P_0 = -h\nu_e/2kT$ is the electron polarization and f' is a leakage factor equal to the fraction of the total nuclear relaxation rate due to the electron spins. Thus under ideal conditions, a positive enhancement of $E \approx g/g_n \approx 660$ can be obtained for solvent protons. The first experiments[27] were on Na–ammonia solutions, and yielded $E \simeq +400$.

On the other hand, if $\mathscr{H}_{IS}$ is predominantly dipolar with a short correlation time, the relaxation rates have the ratios $w_2:w_3:w_4 = 2:12:3$, yielding the dynamic polarization[13]

$$p = p_0 + \tfrac{1}{2}P_0 f'[s/(s+1)] \tag{61}$$

which is negative, with a maximum proton enhancement of 330. Thus from the sign and magnitude of E, one can deduce chemical information on the nature of the ion–solvent interactions. It is found, for example, that in a solution of tri-tertiary butyl phenoxy radical in difluorobenzene, the proton enhancement is negative, indicating a predominant dipolar interaction, while that for fluorine is positive, indicating a scalar interaction.[31] There is some evidence for a three-spin process.[32] A full treatment of relaxation and DNP in liquids has been given.[33]

4. SPIN REFRIGERATORS

4.1. Introduction

We discuss now a general method of cyclically transferring a large electron spin polarization to nuclear spins without the use of microwaves: the nuclear spin refrigerator.[34,35] In its simplest form, one polarizes the nuclei by simply rotating a suitable crystal in a magnetic field. To fix ideas, we immediately consider a suitable crystal, the ethyl sulfate (Yb, Y) $\times (C_2H_5SO_4)_3 \cdot 9H_2O$, doped with $\sim 1\%$ paramagnetic Yb^{3+} ions, which form the electron spin system. The nuclei of interest are the protons in the ethyl and water groups, which are in dipolar coupling with the Yb spins. Figure 13 shows the experimental arrangement: the crystal is mounted so that the angle $\theta = \angle c, H$ may take any value; c is the crystal symmetry axis, and $H \sim 10^4$ Oe is an applied dc field. The fixed rf coil is used to measure the proton NMR signal. The Yb spins have both an anisotropic g-factor ($g_\parallel = 3.35$, $g_\perp \approx 0.003$) and an anisotropic direct spin lattice relaxation time $T_{1e} \propto [\cos^2\theta \sin^2\theta]^{-1} \sim 10^{-3}$ sec at 45°, typically as discussed in greater detail in Section 4.2. If we hold $\theta \approx 45°$ for a short time, the Yb spins become highly polarized parallel to H, as shown by the popula-

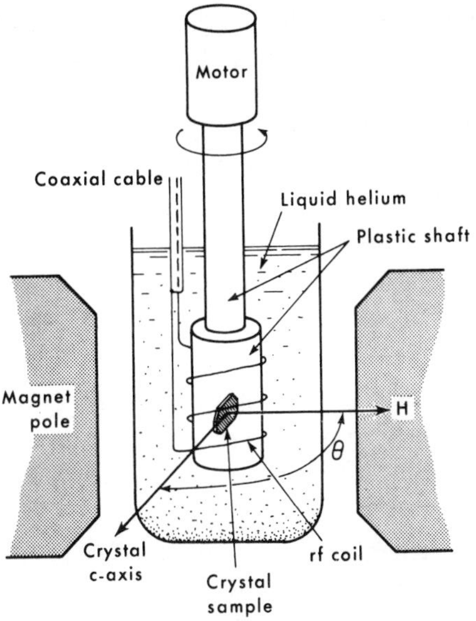

Fig. 13. Experimental arrangement for nuclear spin refrigerator.

tions in column (a) on the energy level diagram of Fig. 14, where $\Delta = g(45°)\beta H/kT$. If we now rotate the crystal to $\theta = 90°$ quickly compared to T_{1e} but slowly compared to the electron Larmor period, i.e., adiabatically in the Ehrenfest sense, the Yb spins remain polarized along H. The levels and populations are now shown in column (b); T_{1e} is here very long. A Yb spin now finds itself on speaking terms with a proton neighbor, i.e., because $g_{\perp} \approx g_n$, they engage in a mutual spin flip, which con-

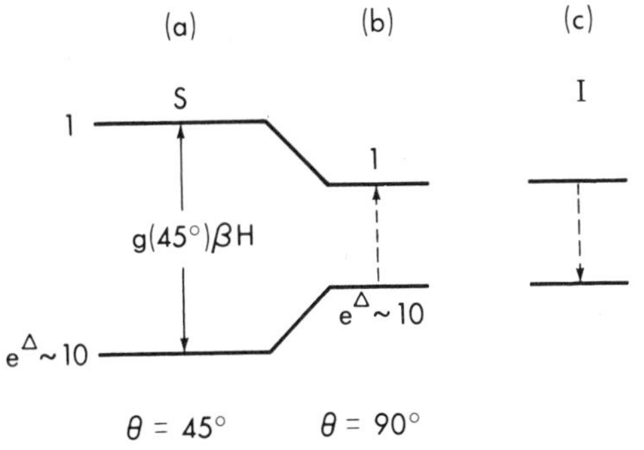

Fig. 14. (a) Energy levels of Yb^{3+} at 45°; (b) levels of Yb^{3+} at 90° showing mutual spin flips with protons as dashed lines; (c) proton energy levels.

serves energy approximately, the Yb spin flipping up, the proton down, in a time $\sim T_2$. The crystal is then quickly rotated to $\theta = 135°$, where the Yb spin gets repolarized; also during this time, the proton polarization diffuses out to the farther protons. Upon rotation to 270°, the Yb spin again polarizes a neighbor proton, etc., so that after n/N cycles, all the protons in the crystal become polarized, where n/N is the relative abundance of protons to Yb ions. More exactly, the proton polarization acquires the steady-state value

$$p = \tanh(\Delta/2) \approx g(45°)\beta H/2kT \tag{62}$$

which exceeds the thermal equilibrium value p_0 by $\sim 10^3$. The protons will be depolarized by relaxation, but at a much slower rate than the polarization process. This ideal spin refrigerator is thus potentially as effective as the dynamic microwave method in polarizing nuclei.

An alternative generalized description of the refrigerator using the concepts of spin temperature[36] and cross-relaxation[37] is illustrated in Fig. 15, a thermal block diagram of the weakly interacting systems: proton spins, Yb spins, crystal lattice phonons, and helium bath. The protons have a common spin temperature T_n defined by $p = g_n\beta H/2kT$; similarly, T_e is the Yb spin temperature; and T is the phonon temperature, here assumed to be that of the helium bath. Thermal switch S_1 schematically represents the Yb spin–lattice relaxation, and is closed at roughly 45°, 135°, . . . ; S_2 represents the cross-relaxation between protons and Yb spins when $g(\theta) \approx g_n$, and is thus closed only at $\theta = 90°, 270°, \ldots$. The Yb spins are an anisotropic working substance cyclically transferring heat from the protons to the bath as θ takes on the successive values 45°, 90°, etc. At 45°, $T_e \to T \sim 1°$K. Then, as $\theta \to 90°$, S_1 opens and the Yb spins are isentropically cooled according to the relation

$$g(45°)/T_e(45°) = g(90°)/T_e(90°) \tag{63}$$

yielding $T_e(90°) \sim 10^{-3}$ °K. At 90°, S_2 closes, putting the cold Yb spins into thermal contact with the proton spins, initially at $T_n \sim 1°$K. Con-

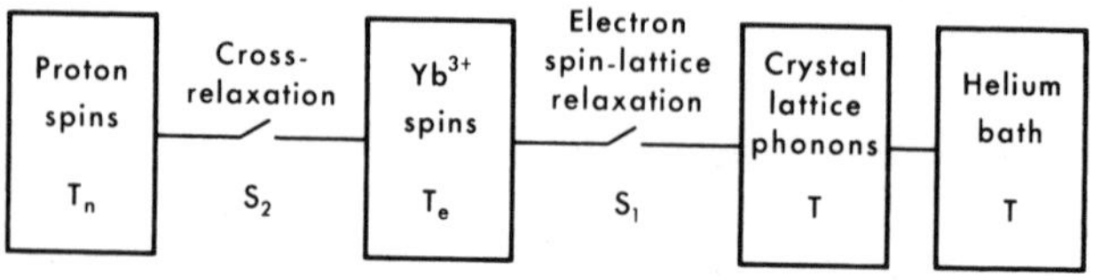

Fig. 15. Spin temperature block diagram of spin refrigerator.

servation of energy leads to the common temperature T_c after mixing

$$(n/T_n) + [N/T_e(90°)] = (n + N)/T_c \tag{64}$$

which, for $n/N \sim 10^3$, yields $T_c \sim 0.5°K$, i.e., the proton polarization is doubled. It is again doubled in the next cycle, etc., reaching after many cycles $T_n \approx T_e(90°)$, resulting in the enhanced proton polarization of Eq. (62). Continuous rotation instead of the discrete θ sequence leads to only slightly smaller polarizations.

To generalize, the method is a solid-state quantum mechanical spin refrigerator. One external parameter θ automatically operates the switches S_1 and S_2 and the isentropic cooling cycle, all in proper sequence. The switches are internal and microscopic in the sense that they operate by virtue of the dependence of the Yb^{3+} spin wave functions on θ. Energy is taken from the protons in quanta $g_n\beta H$ and exhausted as phonons of energy $\sim g(45°)\beta H$ which travel with the velocity of sound to the helium bath. The refrigerator can easily operate at 10^3 cycles per second, but only the spins, not the lattice, are cooled. These features very clearly distinguish it from classical adiabatic demagnetization. Other nuclei in the crystal besides the protons could be similarly polarized. There are many possible varieties of spin refrigerators: e.g., one could operate S_1 not only by θ, but by the magnitude of H, light, pressure, temperature, or electric fields; the electron spin splitting can be varied by $g(\theta)$, by crossing levels, or by magnitude of H in third-order Zeeman splitting. One can also cross the Yb spin levels with a hyperfine system, thus cooling the nuclei in paramagnetic atoms; another case of interest is to cross levels with a nuclear electric quadrupolar system.

Proton polarizations of 19% have been achieved[12] upon rotation of a Yb:YES crystal at $f_r = 60$ rps in 10 kOe at 1.42°K, the limitation being due to insufficient rotation speed. In the same crystal, polarizations of 35% at $T = 1.3°K$ are obtained by rotation of a magnetic field of 20 kOe, rather than the crystal, at $f_r \approx 10^3$ Hz.[38] It is clear that Yb:YES is a favorable substance, and in the following sections, we make a more quantitative analysis based on phenomenological rate equations, as representative of a generally useful approach. Small polarizations have also been obtained in Ce:LaMN[39–41] and $Cr:Al_2O_3$.[42]

4.2. Properties of Yb:YES

Magnetic properties. The Yb^{3+} free ion $4f^{13}$ has a $^2F_{7/2}$ ground state and a Landé g-factor $\Lambda = 8/7$. The spin–orbit interaction places the next

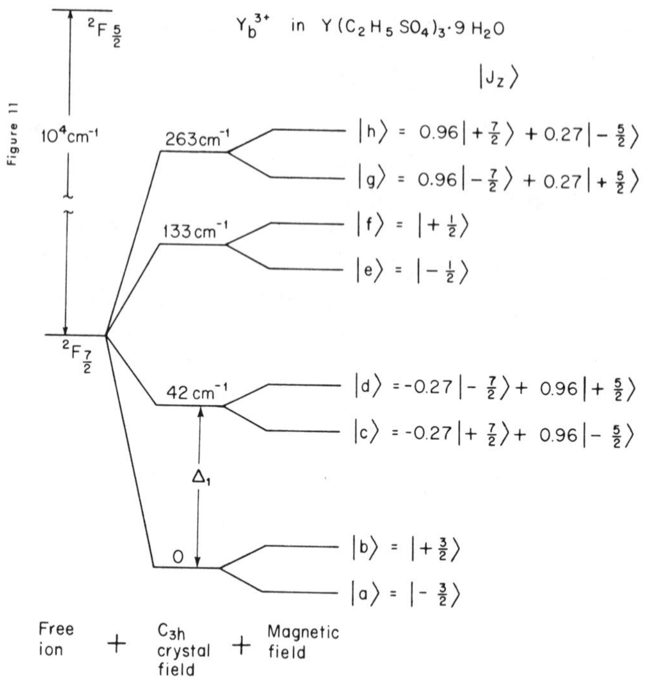

Fig. 16. Levels of Yb^{3+} in ethyl sulfate crystal.[12]

multiplet $^2F_{5/2}$ higher by 10,200 cm^{-1}. Figure 16 shows the further splitting produced by the crystal field of C_{3h} symmetry in the ethyl sulfate, and by the applied dc field H. Only the lowest Kramers doublet $|a\rangle, |b\rangle$ is significantly populated at helium temperatures and forms the effective $S = 1/2$ Yb "spin" system introduced above. The wave functions $|J = 7/2, J_z\rangle$ shown in Fig. 16 are the eigenfunctions of the crystal field interaction[43]

$$\mathscr{H}_c = A_2^0\langle r^2\rangle\alpha O_2^0 + A_4^0\langle r^4\rangle\beta O_4^0 + A_6^0\langle r^6\rangle\gamma O_6^0 + A_6^6\langle r^6\rangle\gamma O_6^6$$

with the values $A_2^0\langle r^2\rangle = 140$ cm^{-1}, $A_6^0\langle r^6\rangle = -29$ cm^{-1}, $A_6^6\langle r^6\rangle = 410$ cm^{-1}, obtained by an extrapolation procedure, and $A_4^0\langle r^4\rangle = -68$ cm^{-1}, which best yields the optically measured value $\Delta_1 = 42$ cm^{-1}. For the doublet $|a\rangle, |b\rangle$, these zero-order wave functions yield $g_\parallel = 3.43$ and $g_\perp = 0$, in agreement with susceptibility measurements[44,45] ($g_\parallel = 3.40 \pm 0.07$, $g_\perp < 0.05$) and paramagnetic resonance ($g_\parallel = 3.35$, $g_\perp \approx 0$). That $g_\perp$ is zero follows from the fact that the lowest doublet is $|\pm 3/2\rangle$, which has no matrix elements for $J_\pm$. That the lowest doublet is pure $|\pm 3/2\rangle$

is a consequence of the relative magnitudes of the crystal field parameters and also the symmetry: For C_{3h}, only states differing by $\Delta J_z = \pm 6$ are admixed by $\mathscr{H}_c$. Actually, $g_\perp$ does not entirely vanish (otherwise, paramagnetic resonance would not be observable), because the Zeeman perturbation $\mathscr{H}_z = \Lambda\beta\mathbf{H} \cdot \mathbf{J}$ does admix the zero-order states slightly, yielding at $\theta = 90°$ an estimated third-order Zeeman splitting $E_3 \approx 4.5H^3 \times 10^{-12}$ MHz, where H is in Oe; this does not exceed the proton splitting for $H < 30$ kOe. At angles near 90°, the total Yb spin splitting is $E = (E_3^2 + E_1^2)^{1/2}$, where $E_1 = g_\parallel \beta H \cos\theta$, so that proton–Yb cross-relaxation can always occur for $H < 30$ kOe. Although this theoretical prediction has not yet been experimentally confirmed, nevertheless, existing data indicate that Yb:YES is unusually favorable in this respect: It is exceedingly anisotropic, by a factor 10^3, even in sizable fields.

Yb^{3+} relaxation. Thermal vibrations of the crystal lattice add to the crystal field interaction a random, time-dependent term $\mathscr{H}_c' \sim \varepsilon \mathscr{H}_c$, where ε is the thermal strain. This induces relaxation transitions between $|a\rangle$ and $|b\rangle$ at the rate T_{1e}^{-1}. At the upper helium temperatures, the experimental results for concentrated YbES are[45]

$$T_{1e}^{-1}|_{\mathrm{O+R}} = 7 \times 10^{11} e^{-60/T} + 1.5 \times 10^{-2} T^9 \quad \mathrm{sec}^{-1} \tag{65}$$

due to the Orbach and Raman processes, which are independent of H and θ. At lower temperatures, $T < 1.5$°K, where the spin refrigerator is operated, the direct process T_{1d}^{-1} dominates. Since it is central to the operation of the refrigerator, we sketch its H, θ dependence, starting from the standard expression

$$T_{1d}^{-1} \approx 2\pi\hbar^{-1}\varrho(\nu)[|\langle a, q+1 | \mathscr{H}_c' | b, q|^2 + |\langle b, q | \mathscr{H}_c' | a, q+1\rangle|^2] \tag{66}$$

where $\varrho(\nu)$ is the density of states, proportional to the number of lattice oscillators per unit frequency, i.e., to ν^2, and q is the phonon occupation number. In Eq. (66), the bracketed term will yield a factor $\nu \coth(h\nu/2kT)$ from the strain ε, and a factor $|\langle a | O_n^m | b\rangle|^2$ from $\mathscr{H}_c$, which vanishes by Kramers theorem unless $|a\rangle$ and $|b\rangle$ are admixed by the Zeeman perturbation to higher doublets $|i\rangle$ at Δ by an amount of order $\langle a | \Lambda\beta H[(\cos\theta)J_z + (\sin\theta)J_x] | i\rangle/\Delta$. From Fig. 16, it is evident that only J_x will admix, the overall result being

$$T_{1d}^{-1} \propto \nu^3 \coth(h\nu/2kT) H^2 \sin^2\theta \tag{67}$$

Using $h\nu = g_\parallel \beta H \cos\theta$ and $\chi \equiv g_\parallel \beta H(\cos\theta)/2kT$, we get

$$T_{1d}^{-1} = A' H^5 \sin^2\theta \cos^3\theta \coth\chi \tag{68}$$

which takes the simpler form for $\chi \ll 1$

$$T_{1d}^{-1} = AH^4T \sin^2 \theta \cos^2 \theta \tag{69}$$

Theoretical estimates give $A' \approx 1.38 \times 10^{-16}$.*

Proton relaxation in Yb:YES. Neglecting other impurities, the relaxation of the abundant protons ($n/N \approx 1650$ for 2% Yb:YES) is determined entirely by the Yb^{3+} ions. We recognize two cases: (a) $\theta \neq 90°$, where the relaxation is predominantly through forbidden I_+S_- transitions, already discussed in Section 2.2; and (b) $\theta \approx 90°$, where relaxation is through energy-conserving I_+S_- cross-relaxation flips. For $\theta \neq 90°$, we expect T_{1n}^{-1} to be given by Eq. (28), using for T_{1e}^{-1} the sum of Eqs. (65) and (68). The best fit, Fig. 7, is obtained for $A' = 3.2 \times 10^{-17}$ in Eq. (68), which is somewhat smaller than the theoretical estimate but more in accord with tentative microwave saturation-recovery data.

In case (b) at $\theta \approx 90°$, $g \approx g_n$, and we assume all the protons are coupled together by rapid diffusion and relax at the cross-relaxation rate T_{21}^{-1} to the Yb spins, which themselves relax to the lattice at the rate T_{1e}^{-1}. Note that T_{21}^{-1} is the I, S cross-relaxation rate (obtained from Fig. 4 by $T_{1n'}^{-1} \to T_{21}^{-1}$), in contrast to the I, I' cross-relaxation rate T_{12}^{-1}. The proton spin-bath relaxation rate is expected to be, by analogy to Eq. (25),

$$\left.\frac{1}{T_{1n}}\right|_{90°} \approx \frac{N}{n} \frac{1}{T_{1e} + T_{21}} \tag{70}$$

which may be several orders greater than Eq. (28). The spike at 90° in Fig. 7 is due to cross-relaxation, and the observed magnitude of $T_{1n}^{-1}|_{90°}$ is within a factor of two of that predicted by Eq. (70) assuming $T_{21} \ll T_{1e} \approx 0.3$ sec from the Raman process, dominant at $\theta = 90°$ and 1.46°K. Actually, the proton relaxation is not quite exponential at $\theta = 90°$. The angular width 0.3° of the cross-relaxation spike is probably due to a finite Yb^{3+} linewidth. To summarize, the magnitude of the proton relaxation rate, and its dependence on T, H, and θ, are somewhat understood for Yb:YES.

4.3. Rate Equations[12]

Section 4.2 may be summarized as follows: At some given values of H, T, and $\theta \neq 90°$, the Yb spin polarization P and the abundant "distant"

* Recently the value $A' = 2 \times 10^{-17}$ sec^{-1} Oe^{-5} has been experimentally determined by J. P. Wolfe and C. D. Jeffries, *Phys. Rev.* **B4**:731 (1971).

proton polarization p obey the effective relaxation rate equations

$$\left.\frac{dP}{dt}\right|_R \approx -\frac{P-P_0}{T_{1e}} \tag{71a}$$

$$\left.\frac{dp}{dt}\right|_R = -\frac{p-p_0}{T_{1n}} \tag{71b}$$

At $\theta \approx 90°$, where the proton splitting Δ_n equals the Yb spin splitting Δ_e, we must add the cross-relaxation terms[46]

$$\left.\frac{dP}{dt}\right|_c = \frac{p-P}{T_{21}}(1-\gamma) \approx \frac{p-P}{T_{21}} \tag{72a}$$

$$\left.\frac{dp}{dt}\right|_c = \frac{P-p}{T_{21}}\gamma \approx \frac{P-p}{T_{21}}\,\frac{N}{n} \tag{72b}$$

where we have assumed $N \ll n$, and introduced the cross-relaxation rate T_{21}^{-1}, where

$$\gamma \equiv N/(N+n) \approx N/n \tag{73a}$$

$$T_{21}^{-1} = \left(\frac{N+n}{Nn}\right)\sum_i^n \sum_k^N w_{ik} \tag{73b}$$

where w_{ik} is the transition probability of a mutual energy-conserving spin flip between proton I_i and Yb spin S_k induced by the dipole terms $I_{i\pm}S_{k\mp}$. The overall behavior of P and p at $\theta = 90°$ is obtained by adding Eqs. (71a,b) and (72a,b). Suppose that initially $P \gg p$; the solutions show that P drops to nearly p in a short time constant $\tau_f \approx T_{21}$, and P and p then decay together to $P_0 = p_0$ with the longer time constant τ_s given by

$$\frac{1}{\tau_s} \cong \frac{1}{T_{1n}} + \frac{\gamma}{T_{1e}+T_{21}} \tag{74}$$

In a spin refrigerator, we can break a cycle of operation into two regions: I, of duration τ_1, in which p and P are not coupled by cross-relaxation, and P is built up to some large value; and II, of duration τ_2, during which P and p cross-relax, and p is built up. In region I of the next cycle, p decays slightly at the rate $R_n = \langle T_{1n}^{-1}\rangle_{\tau_1}$, while P is built up again. Figure 17 shows schematically the overall behavior of $P(t)$ and $p(t)$. After many cycles, the values of P and p at the beginning of any region II, denoted by $\bar{P}$ and $\bar{p}$, do not vary, but reach the steady-state values denoted by $\bar{P}_s$ and $\bar{p}_s$. Under the reasonable approximations $\tau_1 R_n \ll 1$, $\tau_2 \ll \tau_s$, and $(\tau_2/\tau_s) \ll \tau_1 R_n$, all valid for Yb:YES in the region of operation, it can be

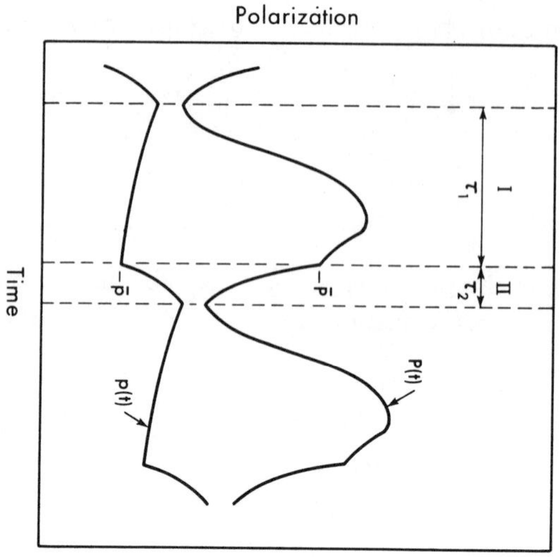

Fig. 17. Schematic of electron polarization $P(t)$ and nuclear polarization $p(t)$ in region I (no cross-relaxation) and region II (cross-relaxation) of spin refrigerator.

shown that the steady-state proton polarization is

$$\bar{p}_s \approx (\tau_1 R_n p_0 + \gamma f \bar{P}_s)/(\tau_1 R_n + \gamma f) \tag{75}$$

where

$$f \equiv 1 - \exp(-\tau_2/\tau_f) \tag{76}$$

is a measure of the completeness of cross-relaxation. With the same approximations, the build up of p should be exponential at the rate

$$1/\tau_{\text{on}} \cong R_n + \gamma(f/\tau_1) \tag{77}$$

At fast operation $\tau_1 \to 0$, and $\bar{p}_s \to \bar{P}_s$, $\tau_{\text{on}} \to \tau_1(n/Nf)$, as expected from elementary considerations.

Equations (75) and (77) assume that cross-relaxation occurs only if $\Delta_n = \Delta_e$, i.e., 1:1 proton–Yb spin flips. It is also energetically possible to have 2:1 flips if $2\Delta_n = \Delta_e$, etc., so we introduce the factor

$$\varepsilon \equiv \Delta_e/\Delta_n \tag{78}$$

to take into account multiple spin flips. We also introduce the factor

$$K \equiv 1 - \exp(-\tau_1/\langle T_{1e}^{-1}\rangle_{\tau_1}) \tag{79}$$

as a measure of the completeness of lattice relaxation of P in region I. In Eq. (75), $\bar{P}_s$ is the steady value obtained in the absence of the effect of the protons on P, i.e., assuming $K = 1$. With corrections for K and ε, Eqs. (75) and (77) become

$$\dot{p}_s = \frac{\tau_1 R_n p_0 + \{\varepsilon\gamma f K \bar{P}_s/[K + f(1-K)]\}}{\tau_1 R_n + \{\varepsilon^2 \gamma f K/[K + f(1-K)]\}} \tag{80}$$

$$\frac{1}{\tau_{\text{on}}} = \frac{1}{\tau_1}\left(\tau_1 R_n + \frac{\varepsilon^2\gamma f\{[K/(1-K)] - \tau_1 R_n\}}{[K/(1-K)] - \tau_1 R_n + f}\right) \tag{81}$$

We note that for fast operation, $\tau_1 \to 0$, and Eq. (80) predicts that $\bar{p}_s \to \bar{P}_s/\varepsilon$, showing that only multiple spin flips can prevent the proton polarization from reaching the value $\bar{P}_s$, the Yb polarization at $\theta = 90°$.

One can calculate $\bar{P}_s$ by integrating directly Eq. (71a). For simple rotation of the crystal $\theta = 2\pi f_r t$, and in the limit $f_r \to \infty$, one finds

$$\bar{P}_s\,|_\infty = (32/15\pi)g_{\parallel}\beta H/2kT \tag{82}$$

assuming $T_{1e}^{-1}(\theta) \propto \cos^2\theta \sin^2\theta$ as for Yb:YES. For finite rotation speeds, a computer calculation gives the results of Fig. 18, with the rotation speed as a parameter in the form f_r/A', where A' is the direct relaxation constant in Eq. (68). For given f_r, $\bar{P}_s \propto H$ up to a certain value and then decreases like $\bar{P}_s \propto H^{-1/3}$, as it turns out, because T_{1e} is becoming so short owing to the H^5 dependence in Eq. (68), that P tends to follow P_0 as $\theta \to 90°$.

Although we have implicitly assumed that the crystal c axis is oriented as in Fig. 13, this is not strictly required for Yb:YES, or any other material in which $g_\perp = 0$. That is, if c is at some angle $\Phi < 90°$ with the vertical, this only reduces the maximum g-factor to $g_\parallel \sin\Phi$, but still allows $\theta \to 90°$ at some time during the rotation, since $g_\perp = 0$ in the whole plane perpendicular to c. For a single crystal, the ideal proton polarization will be reduced by a factor $\sin\Phi$, and for a random powdered sample, by $\pi/4$. At very high speeds, the θ dependence of T_{1e} changes this slightly, and one predicts that a powdered sample will yield a polarization $10\pi/36 \approx 0.87$ of that of a single crystal, optimally oriented. Paramagnetic ions in dilute liquid solutions often experience a reasonably well-defined crystal field, and frozen solutions may exist in which $g_\perp \approx 0$, by virtue of the crystal field symmetry in the solid matrix. Nuclear polarization by a spin refrigerator is possible in such materials.

Our rate equation treatment of spin refrigerators should be valid if internal equilibrium is maintained in the proton and the Yb spin systems, respectively, and if changes in $\mathscr{H}$ are adiabatic, i.e., occur slowly compared

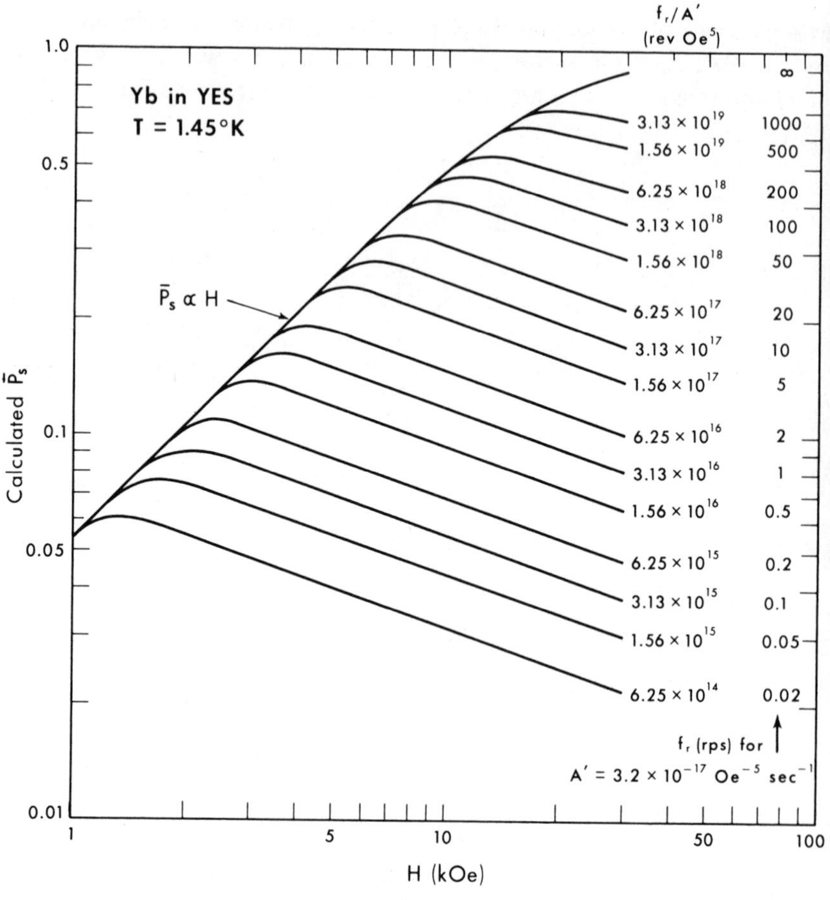

Fig. 18. Calculated $\bar{P}_s$ versus H for Yb:YES spin refrigerator for various rotation speeds f_r.[12]

to Larmor periods. The treatment seems to be adequate to give a semi-quantitative explanation of the data in Section 4.4 although a more rigorous analysis using the density matrix may be necessary at higher speeds.*

4.4. Experimental Results on Yb:YES[12]

Experiments were done over the ranges $1 < H < 20$ kOe, $1.2 < T < 2.7$°K, and $0.5 < f_r < 60$ rps. The proton NMR signal $p_0(H, T_0)$ was

* More recently, R. L. Ballard (Thesis, University of California, Berkeley, 1971, unpublished) has made a detailed analysis of nonadiabatic effects and harmonic cross-relaxation in spin refrigerators.

first measured at some given field H and temperature T_0; the crystal was then rotated at some constant speed f_r, while the signal was observed to build up at the rate τ_{on}^{-1} to the final steady-state enhanced proton signal $p_{\text{enh}}(H, T)$. Friction caused a slight rise in temperature to $T \approx 1.2T_0$, typically. The measured enhancement is $E = p_{\text{enh}}/p_0$, and the steady-state polarization defined by $p_{\text{ss}} \equiv Ep_0$ is essentially the polarization that would have obtained without heating. Figure 19 shows p_{ss} versus H in a 2% natural Yb crystal for various speeds and displays a behavior like that of $\bar{P}_s$ in Fig. 18 except that p_{ss} is smaller by a factor ~3. To eliminate the ^{173}Yb hfs and consequent line broadening at $\theta = 90°$, crystals were then grown using Yb enriched in ^{172}Yb ($I = 0$). The largest proton polarization achieved was 19%, in a crystal of 2% ^{172}Yb:YES at 10 kOe, 1.42°K, and $f_r = 60$ rps. The crystal was ground to a powder, and a polarization of 17% was observed under the same conditions.

The measured polarization buildup was exponential, and in general exhibited the behavior predicted by Eq. (77): $\tau_{\text{on}}^{-1} \approx R_n + 2\gamma f f_r \rightarrow 2\gamma f f_r$

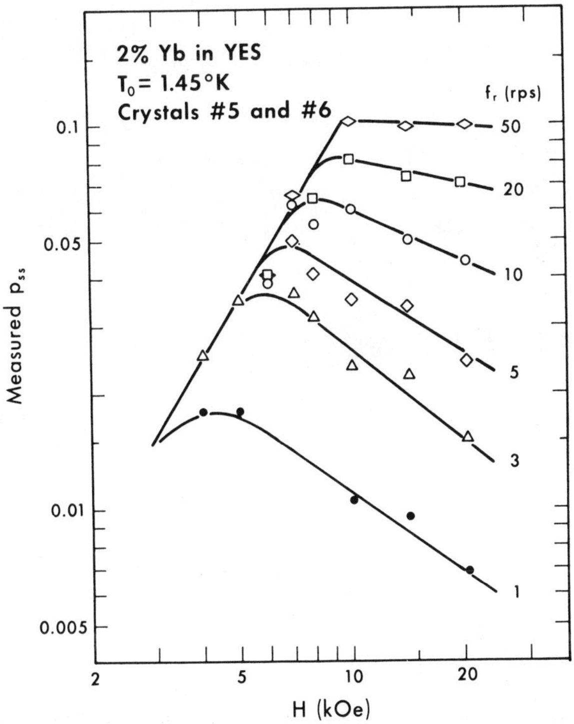

Fig. 19. Measured proton steady-state polarization versus H in Yb:YES spin refrigerator for various rotation speeds f_r.[12]

at high speeds. The data roughly showed τ_{on}^{-1} to be (1) proportional to γ, i.e., to Yb concentration; (2) independent of H and T; (3) proportional to f_r. This means that $f = 1 - \exp(-\tau_2/T_{21})$ must be nearly unity, even at $f_r \approx 60$ rps, i.e., $T_{21} \lesseqq 10^{-4}$ sec, roughly.

Failure to achieve ideal polarization is best explained by multiple spin flips. The data can be fit to Eq. (80) with $f = 1$; $\bar{P}_s$ from Fig. 18 with $A' = 3.2 \times 10^{-17}$; K calculated from Eq. (79); and $\varepsilon = \varepsilon_{\text{eff}}$, an adjustable parameter. The minimum value $\varepsilon_{\text{eff}} = 1.9$ is found at 50 cps and 10 kOe for the enriched crystal and increases with decreasing field and decreasing frequency f_r. This is not unexpected, since at low f_r, more time is spent at orientations where multiple flips can occur, allowing more Yb spin polarization to leak to the protons at higher spin temperatures; and at lower H, the fractional linewidth $\Delta H/H$ may be greater, allowing more overlap in the tails of the lines at a given θ.

These experiments clearly showed that higher rotation speeds were needed. This was achieved by rotating the field electrically, rather than the crystal, by subjecting the crystal to a dc field $H_{\text{dc}} \sim 15$ kOe, and a pulsed field $H_{\text{p}}(t) \sim 15$ kOe produced by discharging a capacitor through a copper solenoid.[38] The pulse had the approximate shape of a half sine wave of duration 0.2×10^{-3} sec. When the pulse is on, the net field is 22 kOe at $\theta = 45°$; this quickly polarizes the Yb spins. As the pulse turns off, the net field rotates down to $\theta \approx 91°$, passing through the cross-relaxation region in a time $\tau_2 \approx 10^{-5}$ sec. A proton polarization of 35% at $H_{\text{p}} = 20$ kOe and $H_{\text{dc}} = 15$ kOe was observed at 10 pulses per sec, with an exponential build up time of ~10 min.*

5. OPTICAL PUMPING

The suggestion of Kastler[47] that the population of Zeeman levels and hfs levels of gaseous atoms could be preferentially enhanced by pumping with circularly polarized resonance light has had ample verification. It is also possible to transfer the polarization of one atomic species to another through exchange collisions,[48] the most dramatic example being the polarization of ^{3}He nuclei in helium gas.[49] We consider briefly in this section methods for enhancement of nuclear polarization by optical pumping in solids

* More recently, W. H. Potter and H. J. Stapleton, *Phys. Rev.* **B5**:1729 (1972) have reported results on a spin refrigerator with Yb:YES rotated by a high-speed cryogenic gas turbine; polarizations of 35% were obtained. They also analyzed sample heating by nonresonant relaxation.

ENPOPS),[50] which are related both to optical pumping in gases and to the dynamic microwave methods discussed in Sections 2 and 3. The methods involve (1) production of an electron spin polarization by pumping with circularly polarized light; (2) transfer of this polarization to nuclear spins through hfs coupling, preferential relaxation processes, or saturation of microwave transitions; (3) transfer of the polarization to abundant nuclei through cross-relaxation. The production of significant electron spin polarization in solids has been achieved, e.g., in *F*-centers in alkali halides by Karlov *et al.*[51] and for Tm^{2+}:CaF_2 by Anderson *et al.*[52]

To fix ideas, consider such a magnetically dilute crystal, represented by the spin Hamiltonian $\mathscr{H} = g\beta\mathbf{H} \cdot \mathbf{J} + A\mathbf{J} \cdot \mathbf{I}$, the first term representing the Zeeman interaction of the electron spins in the ion or *F*-center with an external field H; the second term, assumed much smaller, represents the

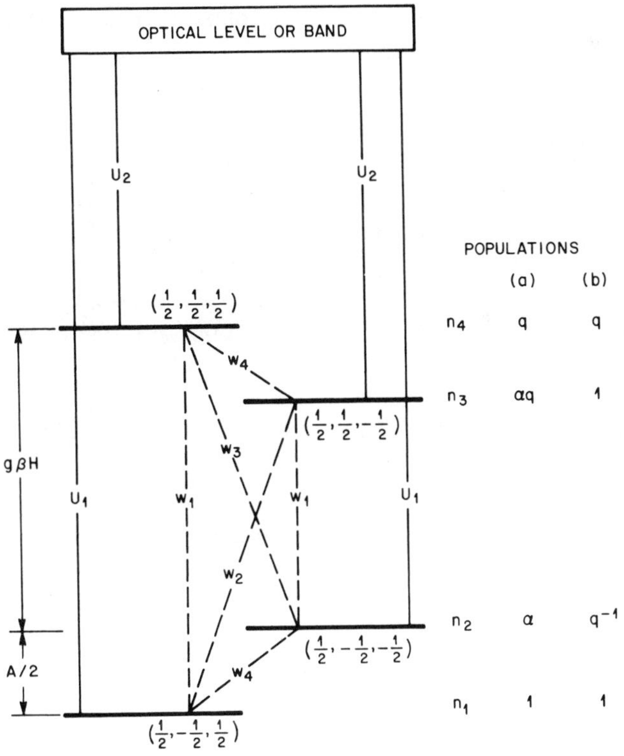

Fig. 20. Levels and transitions for a paramagnetic ion in high field with $S = \frac{1}{2}$ and $I = \frac{1}{2}$. The populations in columns (a) and (b) are obtained by enhancement of the nuclear polarization by optical pumping.

hfs interaction with the nucleus of the ion (or in the case of the *F*-cente with a near-neighbor nucleus). These energy levels and wave function (J, J_z, I_z) are shown in Fig. 20 for $J = 1/2$, $I = 1/2$, along with an optic level or band to which we induce transitions by illuminating the cryst with light. We assume that by pumping with, say, right-hand polarize light we induce the transition probabilities shown, where U_2 is significantl different from U_1. This comes about because the field decouples the electro and the nucleus, and the light wave is coupled only to the electron; th transitions obey the selection rule $\Delta J_z = +1$, $\Delta I_z = 0$. For example, i the ground state is $^2S_{1/2}$ and the excited state is $^2P_{1/2}$, the relative transitio probabilities are $U_1 = 2$ and $U_2 = 0$; if the excited state is $^2P_{3/2}$, the $U_1 = 1$ and $U_2 = 3$, since $\langle \frac{3}{2}, \frac{3}{2} \mid J_+ \mid \frac{1}{2}, \frac{1}{2} \rangle = \sqrt{3} \langle \frac{3}{2}, \frac{1}{2} \mid J_+ \mid \frac{1}{2}, -\frac{1}{2} \rangle$. If w pump both states of the *LS* multiplet, however, then $U_1 = U_2$; in solid where the optical lines or bands may be broad, we thus require a sufficientl large spin–orbit coupling to partially resolve the multiplets in order t selectively pump out of the ground state. For forther discussion of selectiv pumping of electronic states with light, see the article by S. Geschwind i this volume.

In Fig. 20, w_1 represents the paramagnetic spin–lattice relaxation arisin from the thermal modulation of the crystalline electric fields; w_2 and w represent relaxation arising, say, from modulation of the hfs interactio $A'(t)(J_+I_- + J_-I_+)$, which makes $w_2 \gg w_3$. It is just this preferential relax ation which makes the Overhauser effect possible as discussed in Section 3 In considering the downward relaxation from the optical level, we postulat two extreme cases.

(1) *Nuclear spin memory*. This means that ions optically pumped ou of the left-hand side of Fig. 20 ($I_z = +\frac{1}{2}$) will decay to the left-hand sid before thermalization can occur; and ions on the right-hand side retur to the right. The overall effect of pumping with circularly polarized ligh in competition with w_1 is to establish the relative populations shown i column (a), where $q \to U_1/U_2$ for strong pumping and α is to be determine by the relaxations w_2 and w_3. For $w_2 \gg w_3$, thermal equilibrium require $\alpha q = \exp(-g\beta H/kT) \equiv \exp(-\Delta)$. This ideal ENPOPS thus yields a nuclea polarization

$$p = \frac{n_1 + n_4 - n_2 - n_3}{n_1 + n_4 + n_2 + n_3} = \frac{q - \exp(-\Delta)}{q + \exp(-\Delta)} \quad (83)$$

We assume that the population of the optical level remains negligible.

olution of the rate equations for arbitrary light intensity yields

$$p = \frac{(U_1/w_1) - (U_2/w_1)\exp(-\Delta)}{4(\exp -\frac{1}{2}\Delta) + (U_1/w_1) + (U_2/w_1)\exp(-\Delta)} \quad (84)$$

alf-saturation occurs for $U \sim T_{1e}^{-1} \exp(-\frac{1}{2}\Delta)$, where T_{1e}^{-1} is the ground-ate relaxation rate. At very low temperatures, where $q \gg \exp(-\Delta)$, Eq. 33) shows that the nuclear polarization is essentially complete and obtains ven if $q = 1$, i.e., for unpolarized light and even if $U_1 = U_2$. At high mperatures, $\exp(-\Delta) \approx 1$, and Eq. (83) becomes

$$p = (q - 1)/(q + 1) \quad (85)$$

howing that a large polarization could be obtained even at room tempera-ures. Reversing the light polarization requires that $q \to 1/q$, which reverses he sign of p. For an oscillator strength of order unity and moderate pumping ntensities (~W/cm²), it is possible to achieve $U \sim 10^5\ \text{sec}^{-1}$, which is omparable with T_{1e}^{-1} for favorable substances at room temperature. If we annot be sure that $w_2 \gg w_3$, then one knows from dynamic nuclear olarization that it is feasible to saturate the forbidden microwave transi-ion $(\frac{1}{2}, \frac{1}{2}, -\frac{1}{2}) \leftrightarrow (\frac{1}{2}, -\frac{1}{2}, \frac{1}{2})$. This, together with optical pumping, will ead to the populations of column (b) in Fig. 20, and a polarization given y Eq. (85), again large and independent of temperature. This is es-entially because in ENPOPS, the enhanced polarization is determined y matrix element ratios rather than by Boltzmann factors as in dynamic olarization.

(2) *Randomized optical relaxation.* In this case, we postulate that ions n the optical band relax with equal probability to the four ground levels. Solution of the rate equations shows that very strong optical pumping yields no nuclear polarization, because the optical relaxation in effect short-circuits the relaxations w_2, w_3; however, at intermediate light in-tensities, a polarization is obtained if $w_1 \approx w_2 \gg w_3$, a requirement met in *F*-centers, for example. Or, instead, one could saturate the forbidden micro-wave transition, yielding again the polarization of Eq. (85).

It is possible to transfer the polarization of the rather few nuclei of the ions to the abundant nuclei I' at diamagnetic sites in the crystal by cross-relaxation, for example, by operating in a field such that $g_n'\beta H = \frac{1}{2}A$; it is well known that the polarization will diffuse throughout the sample by rapid mutual spin flips.

The first significant nuclear polarization achieved by ENPOPS has been for ^{169}Tm in Tm^{2+}:CaF_2.[53] The primary practical difficulty is in making the optical transition rate U comparable to the electron spin–lattice relaxation rate w_1. For this reason, it was necessary to operate at helium temperatures, where at $H = 800$ G and 1.6°K, a ^{169}Tm polarization of 9% is obtained by optical pumping alone. If, in addition, the I_+S_- forbidden transition is saturated, the polarization increases to 19%, showing that the relaxation w_2 is not quite strong enough itself to give the ideal ENPOPS. However there is good evidence for nuclear spin memory. The abundant ^{19}F nuclei are also polarized at fields near a crossing of the levels of a pair of Tm^{2+} ions.[53] Results have also been obtained for ^{29}Si in silicon,[55] and for the protons in anthracene by a spin-selective deexcitation of an optically pumped excited state.[56]

A new method of nuclear polarization has been developed,* in which the protons in Yb:YES are polarized to a value of $p \approx -2\%$ by pumping the infrared vibrational modes of the CH_2 and CH_3 complexes surrounding the Yb^{3+} ions. Maser oscillation at the proton frequency is observed.†

REFERENCES

1. A. Overhauser, *Phys. Rev.* **89**:689 (1953); **92**:411 (1953).
2. W. A. Barker, *Rev. Mod. Phys.* **34**:173 (1962).
3. C. D. Jeffries, *Phys. Rev.* **106**:164 (1957); **117**:1056 (1960).
4. A. Abragam and W. G. Proctor, *Compt. Rend.* **246**:2253 (1958).
5. C. H. Schultz, Report No. UCRL-11149, Lawrence Radiation Laboratory, Berkeley, California (1964); M. Borghini, P. Roubeau, and C. Ryter, *Nucl. Instr. Methods* **49**:248 (1966).
6. J. R. McColl, Thesis, University of California, Berkeley (unpublished, 1967).
7. H. E. Rorschach, *Physica* **30**:38 (1964).
8. G. R. Khutshishvilli, *Soviet Phys.—JETP* **15**:909 (1962).
9. T. J. Schmugge and C. D. Jeffries, *Phys. Rev.* **138**:A1785 (1965).
10. T. E. Gunter and C. D. Jeffries, *Phys. Rev.* **159**:290 (1967).
11. P. L. Scott and C. D. Jeffries, *Phys. Rev.* **127**:32 (1962).
12. K. H. Langley and C. D. Jeffries, *Phys. Rev.* **152**:358 (1966).
13. A. Abragam, *Phys. Rev.* **98**:1729 (1955).
14. M. Abraham, M. A. H. McCausland, and F. N. H. Robinson, *Phys. Rev. Letters* **2**:249 (1959).
15. Chester Hwang and T. M. Sanders, Jr., in *Proceedings of the VIIth International Conference on Low Temperature Physics*, p. 148, Univ. of Toronto Press, Toronto (1961).
16. J. L. Motchane and J. Uebersfeld, *J. Phys. Radium* **21**:194 (1960).
17. O. S. Leifson and C. D. Jeffries, *Phys. Rev.* **122**:1781 (1961).

17a. C. M. Verber and R. G. Leconder, *Phys. Letters* **25A**:179 (1967).

* J. P. Wolfe and C. D. Jeffries, Ref. 23b, page 386.

† J. P. Wolfe, A. R. King, and C. D. Jeffries, *Bull. Am. Phys. Soc.* **17**:148 (1972).

7b. Sook Lee, *Phys. Letters* **26A**:572 (1968).
3. A. Redfield, *Phys. Rev.* **98**:1787 (1955).
). B. N. Provotorov, *Soviet Phys.—JETP* **14**:1126 (1962).
). A. Abragam and M. Borghini, in *Progress in Low Temperature Physics*, Vol. IV, p. 384, North-Holland Publishing Co., Amsterdam (1964).
1. T. J. B. Swanenburg, G. M. van den Heuvel, and N. J. Poulis, in *Proceedings of the Colloque Ampere, Lubjana, 1966*; North-Holland Publishing Co., Amsterdam (1967).
1a. T. J. B. Swanenburg, G. M. van den Heuvel, and N. J. Poulis, *Commun. Kamerlingh Onnes Lab.* No. 354, a, Leiden; *Physica* **33**:707 (1967).
1b. M. Borghini, *Phys. Rev. Letters* **26A**, 242 (1968).
2. C. D. Jeffries, *AEC Technical Report UCB-34P20-T-1*, University of California, Berkeley (1966, unpublished).
3. *Proceedings of the Conference on Polarized Targets and Polarized Beams, Saclay, December 1966*, published by La Documentation Française, Paris (1967).
3a. S. Mango, O. Runolfsson, and M. Borghini, *Nucl. Instr. and Methods* **72**:45 (1969).
3b. Gilbert Shapiro, editor, *Proceedings of the II International Conference on Polarized Targets*, Lawrence Radiation Laboratory, Berkeley, California, University of California (Berkeley) Report LBL-500, UC-34 Physics (1971).
4. O. Chamberlain, C. D. Jeffries, C. H. Schultz, G. Shapiro, and L. Van Rossum, *Phys. Letters* **7**:293 (1963).
4a. A. Moretti, A. Yohosawa, F. W. Markley, and R. C. Miller, *Rev. Sci. Instr.* **38**:1335 (1967).
4b. R. J. Wagner and R. P. Haddock, *Phys. Rev. Letters* **16**:1116 (1966).
5. C. D. Jeffries, *Dynamic Nuclear Orientation*, John Wiley and Sons, New York (1963).
6. F. M. Pipkin and J. W. Culvahouse, *Phys. Rev.* **106**:1102 (1957); **109**:1423 (1958).
27. T. R. Carver and C. P. Slichter, *Phys. Rev.* **92**:212 (1953); **102**:975 (1956).
28. J. Combrisson and I. Solomon, *J. Phys. Radium* **20**:683 (1959).
29. A. Abragam, A. Landesman, and J. M. Winter, *Compt. Rend.* **247**:1852 (1958).
30. M. Gueron and C. Ryter, *Phys. Rev. Letters* **3**:338 (1959).
30a. K. H. Hausser and D. Stehlik, *Advances in Magnetic Resonance*, Vol. 3, Academic Press, New York (1968), p. 29.
31. R. E. Richards and J. W. White, *Proc. Roy. Soc. London* **A283**:459 (1965).
32. K. H. Hausser and F. Reinbold, *Phys. Letters* **2**:53 (1962).
33. J. Korringa, D. O. Seevers, and H. C. Torrey, *Phys. Rev.* **127**:1143 (1962).
34. C. D. Jeffries, *Cryogenics* **3**:41 (1963).
35. A. Abragam, *Cryogenics* **3**:42 (1963).
36. A. Abragam and W. G. Proctor, *Phys. Rev.* **109**:1441 (1958).
37. N. Bloembergen, in *Proceeding of the VII International Conference on Low Temperature Physics*, p. 36, Univ. of Toronto Press, Toronto (1961).
38. J. R. McColl and C. D. Jeffries, *Phys. Rev. Letters* **16**:316 (1966); *Phys. Rev.* **B1**:2917 (1970).
39. F. N. H. Robinson, *Physics Letters* **4**:180 (1963).
40. J. Combrisson, J. Ezratty, and A. Abragam, *Compt. Rend.* **257**:3860 (1963).
41. V. I. Luschikov, B. S. Neganov, L. B. Parfenov, and Yu V. Taran, *Soviet Phys.—JETP* **22**:285 (1966).
42. W. G. Clark, G. Feher, and M. Weger, *Bull. Am. Phys. Soc.* **8**:463 (1963).
43. R. J. Elliott and K. W. H. Stevens, *Proc. Roy. Soc. (London)* **A215**:437 (1962); **A218**:553 (1953); **A219**:387 (1953).

44. A. H. Cooke, F. R. McKim, H. Meyer, and W. P. Wolf, *Phil. Mag.* **2**:928 (1957
45. J. Van den Broek and L. C. Van der Marel, *Physica* **29**:948 (1963); **30**:565 (1964
46. N. Bloembergen, S. Shapiro, P. S. Pershan, and J. O. Artman, *Phys. Rev.* **114**:4
(1959).
47. A. Kastler, *J. Phys. Radium* **11**:255 (1950).
48. H. G. Dehmelt, *Phys. Rev.* **109**:38 (1958).
49. F. D. Colegrove, L. D. Schearer, and G. K. Walters, *Phys. Rev.* **132**:2561 (1963
50. C. D. Jeffries, *Phys. Rev. Letters* **19**:1221 (1967).
51. N. V. Karlov, J. Margerie, and V. Merle D'Aubigue, *J. Phys. Radium* **24**:717 (1963
52. C. H. Anderson, H. A. Weakliem, and E. S. Sabisky, *Phys. Rev.* **143**:223 (1966
53. L. F. Mollenauer, W. B. Grant, and C. D. Jeffries, *Phys. Rev. Letters* **20**:488 (1968
54. W. B. Grant, L. F. Mollenauer, and C. D. Jeffries, *Phys. Rev.* **B4**:1428 (1971).
55. G. Lampel, *Phys. Rev. Letters* **20**:491 (1968).
56. G. Maier, U. Haeberlen, H. C. Wolf, and K. H. Hausser, *Phys. Letters* **25A**:38
(1967).

Chapter 4

Electron Spin Echoes*

W. B. Mims

Bell Telephone Laboratories
Murray Hill, New Jersey

1. INTRODUCTION

1.1. General

The principles involved in the generation of electron spin echoes and nuclear spin echoes are essentially the same. The differences between the two kinds of experiment are related to the technical problems of scaling from the radiofrequency range to the microwave range, and to the physical circumstances which distinguish NMR and EPR studies.

The translation of spin echo methods into the microwave range leads to the use of pulsed magnetrons, klystrons, or traveling-wave tubes as signal sources, to the replacement of coil-condenser combinations by resonant cavities, to the use of heterodyne or homodyne detection systems, etc. It also involves the choice of a shorter time scale for the pulsing sequences. In rf fields of comparable strength, electron moments are turned over by 90° or 180° in $\sim 10^{-3}$ the time required for nuclear moments. Besides this, electron moments are more susceptible to time-dependent perturbations in the environment, and tend to be associated with shorter phase memory times. Scaling according to the ratio of magnetic moments, one might suppose that it would be necessary to go over to a nanosecond time scale in place of the microsecond time scale which is characteristic of nuclear spin-echo work. Fortunately, however, this is not essential under the experi-

* See Notation Section at back of this chapter.

mental conditions which obtain in many EPR studies, and it is usuall possible to compromise with an intermediate time scale in which the pulse are from 10 to 100 nsec in length.

A typical electron spin echo experiment involves liquid helium tem peratures, and magnetically dilute materials. The reasons for this choic are essentially the same as in conventional CW (continuous wave) EPR The spin–lattice relaxation rate T_1 is often too rapid for any resonance t be obtained at room temperature, and spin–spin interactions are liable t obscure many of the finer details of resonance behavior in magneticall concentrated systems. It is, in fact, often more important in echo experi ments than in CW experiments to choose the conditions so that spin–spi and spin–lattice interactions are relatively weak in order to minimiz homogeneous line broadening. This does not rule out the possibility o obtaining electron spin echoes at room temperature, nor does it mean tha echoes can never be found in crystals containing appreciable amounts o a paramagnetic impurity. T_1 itself is quite long for certain centers,[1] even a room temperature, and the effects of spin–spin interactions are sometime reduced by narrowing mechanisms (see Section 3). The discussion whic follows will, however, be primarily based on experiments which have bee made at low temperatures and at spin concentrations $\gtrsim 10^{18}$ cm^{-3}.

1.2. Two-Pulse Spin Echoes

The spin echo process may be described quantum mechanically by means of the density matrix formalism, or classically in terms of the model of precessing gyroscopic magnets. It will be more helpful in forming a clear physical picture of the mechanism if we begin by using the classical descrip-tion. We assume that the resonance line is inhomogeneous, i.e., that it consists of a distribution of spectral components or "spin packets" all independent of one another, and related only by their common interaction with the microwave field H_1. The dynamics of the system can most readily be described in reference to a coordinate system which rotates about the Zeeman field (the z axis) at ω_0, where ω_0 is the microwave frequency in radians/sec (see Fig. 1). In this rotating coordinate system,* the microwave driving field is represented by a fixed vector H_1 which we take as defining the direction of the y axis. The rotating coordinate transformation also has the effect of reducing the apparent Zeeman field,† giving a value $H_{0,j}$

* If the microwave field in the resonant cavity is linearly polarized, it will require a total field amplitude of $\sqrt{2}H_1$ to give the rotating component assumed here.

† A derivation of this result is given by Slichter.[2]

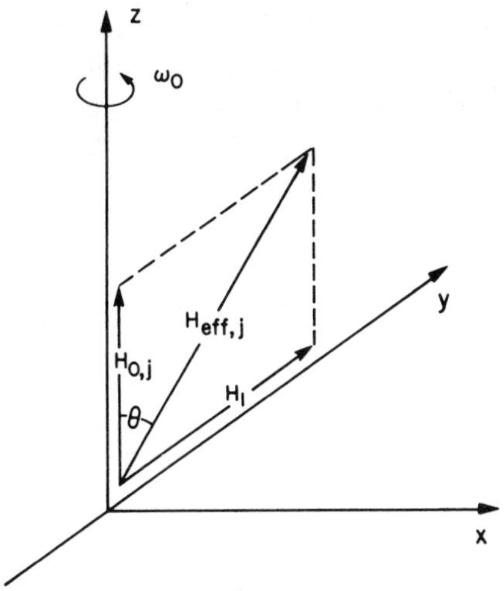

Fig. 1. Magnetic fields in a coordinate system rotating at the microwave frequency ω_0. In this system, the microwave field vector H_1 is stationary. $H_{0,j}$ corresponds to the discrepancy between the actual Zeeman field and the Zeeman field which would be required in order to make the jth spin packet resonate at ω_0. The magnetic moment of the jth spin packet precesses about $H_{\text{eff},j}$ with a frequency $\gamma_j H_{\text{eff},j}$.

for the jth spin packet, where $\gamma_j H_{0,j} = \gamma_j H_0 - \omega_0$, and where $\gamma_j = g_j\beta/\hbar$ is the appropriate gyromagnetic ratio. When no microwave pulses are applied, precession takes place about the z axis with an angular velocity* $\omega_j = \gamma H_{0,j}$. When the microwave field is present, precession occurs with an angular velocity $\omega_{i,j} = \gamma H_{\text{eff},j}$ about a field $H_{\text{eff},j} = (H_1{}^2 + H_{0,j}^2)^{1/2}$ which is oriented at an angle of $\theta_j = \arctan(H_1/H_{0,j})$ with respect to the z axis. Since each spin packet is characterized by a different $H_{\text{eff},j}$, each will have its own distinct dynamical history, the overall magnetization at any time in the spin echo cycle of events being obtained by summing the magnetic moments of individual spin packets. The computation for the general case is straightforward, but tedious. It is discussed in more detail in Section 5. The essential features of the echo generating process can, however, be described fairly easily by specializing to the case in which $H_1 \gg H_{0j}$ for all the spin packets in the resonance line (i.e., the case where

* We shall drop the subscripts in g_j and γ_j, since differences in these quantities only enter into the calculations to the second order of smallness.

$H_1 \gg$ linewidth). For that portion of the echo cyle that coincides with t application of microwave power, we can then make the simplifying assum tion that precession of all the spin packets takes place about the y ax with a uniform angular velocity $\omega_1 = \gamma H_1$. To begin with, we shall addition suppose that the two microwave pulses are applied for tim t_{pI} and t_{pII} such that $\omega_1 t_{\text{pI}} = \pi/2$ and $\omega_1 t_{\text{pIII}} = \pi$, i.e., we shall consider "90–180" pair of pulses.

The sequence of events is illustrated in Fig. 2. In the initial restir condition, the spin packets have a net moment $M = \sum_j M_{0,j}$ along t z axis. The first microwave pulse brings M into alignment with the —

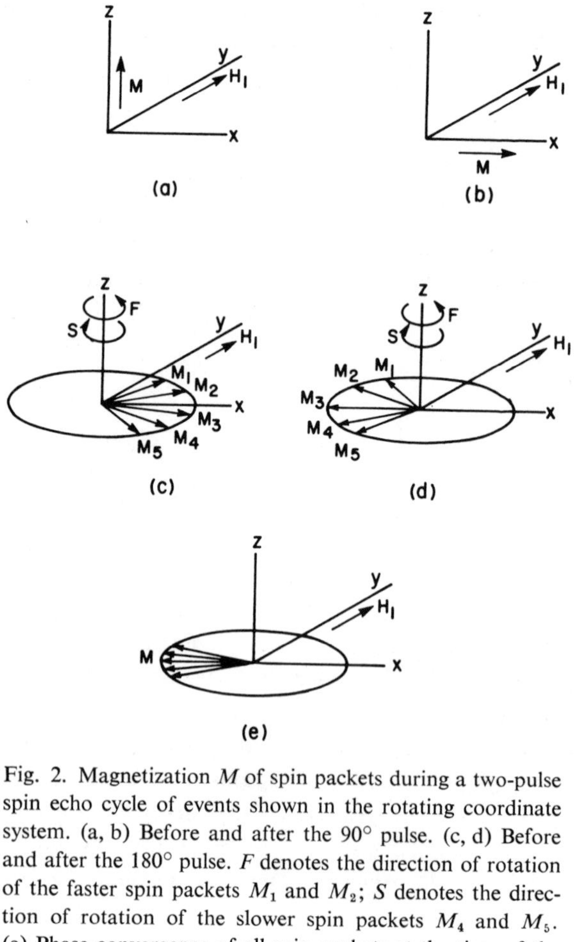

Fig. 2. Magnetization M of spin packets during a two-pulse spin echo cycle of events shown in the rotating coordinate system. (a, b) Before and after the 90° pulse. (c, d) Before and after the 180° pulse. F denotes the direction of rotation of the faster spin packets M_1 and M_2; S denotes the direction of rotation of the slower spin packets M_4 and M_5. (e) Phase convergence of all spin packets at the time of the echo.

is. As soon as microwave power is removed, the approximate uniformity the effective fields $H_{\text{eff},j}$ vanishes and the spin packets precess at different tes, giving a resultant moment at a time t after the end of the pulse,

$$M_{x+iy,j}(t_{\text{pI}} + t) = \sum_j M_{0,j} \exp(i\omega_j t) \tag{1}$$

learly, M_{x+iy} is simply the Fourier transform of the spin packet distribu- on $M_{0,j}$, which defines the line shape.* In the laboratory system of coor- nates, it corresponds to a moment precessing at ω_0, and it gives rise to a gnal termed the "free-induction signal." The effect of pulse II can be ppreciated by making a comparison of Figs. 2c and 2d. The moments M_j hose phases had advanced furthest during the interval τ following pulse I ow find themselves retarded in relation to the rest, and those that had gged behind are now advanced. The resultant moment immediately after ulse II is given by

$$M_{x+iy,j}(t_{\text{pI}} + \tau + t_{\text{pII}}) = -\sum_j M_{0,j} \exp(+i\omega_j\tau) \tag{2}$$

he M_j retain, of course, their former precession frequencies ω_j, so that the period following the end of pulse II the spin packets with more rapid recession frequencies begin to recover from their setback, and the more lowly precessing ones lose their temporary lead. At time τ after the end f pulse II, there is then once again a net resultant magnetization, this time long the $-x$ axis, and a signal will be generated. The resultant magnetiza- ion at a time $\tau + t$ after the end of the second pulse is

$$M_{x+iy,j}(t_{\text{pI}} + t_{\text{pII}} + 2\tau + t) = -\sum_j M_{0,j} \exp(i\omega_j t) \tag{3}$$

t may be noted that the right-hand side of (3) has the same form as the ex- ression in (1), and will similarly generate a signal proportional to the ourier transform of the line shape. In (3), t can of course take negative alues, thus giving an "echo" signal which consists of two free-induction ignals back to back, peaking at a time τ after pulse I.

Now that the essential picture is clear, we may begin to discuss the ffect of relaxing some of the conditions. Firstly, it has been implied, by aking H_1 along the same axis throughout, that pulses I and II are coherent. Coherence is often not an important consideration. A phase lag of φ in

* For the ideal inhomogeneous line assumed here, this result is trivial. It has been derived for the more difficult case of an NMR line broadened by interactions with like nuclear neighbors by Lowe and Norberg.[2a]

pulse II corresponds to a rotation of the y axis in Fig. 2c by an angle – and will introduce a common phase factor $e^{i\varphi}$ for all the precessing momen due with respect to this new axis system. Pulse II reverses all phases, ar the echo will therefore contain a phase factor $e^{-i\varphi}$ relative to the coordina system used to describe pulse II, or $e^{-2i\varphi}$ relative to the coordinate syste used to describe pulse I. Effects of this kind will not be observed, howeve unless the detection system is phase-coherent in relation to the applie pulses.

The 90°–180° sequence is likewise not essential. Let us assume th $\omega_1 t_{\text{pI}} = \theta_{\text{pI}}$ and $\omega_1 t_{\text{pII}} = \theta_{\text{pII}}$. Then, immediately after pulse I, the ma netization consists of components $M_{0,j} \cos \theta_{\text{pI}}$ along the z axis and M_0 $\times \sin \theta_{\text{pI}}$ along the x axis. Only the latter contributes to the free-inductio signal and echo. At the beginning of pulse II, the components in the x plane are

$$M_{x+iy,j}(t_{\text{pI}} + \tau) = M_{0,j}(\sin \theta_{\text{pI}}) \exp(i\omega_j \tau) \qquad (4$$

During pulse II, only the x component in (4) nutates, leaving

$$M_{y,j}(t_{\text{pI}} + \tau + t_{\text{pII}}) = M_{0,j}(\sin \theta_{\text{pI}}) \sin \omega_j \tau \qquad (5$$

unchanged, and giving new components

$$M_{x,j}(t_{\text{pI}} + \tau + t_{\text{pII}}) = M_{0,j} \sin \theta_{\text{pI}} \cos \theta_{\text{pII}} \cos \omega_j \tau \qquad (6$$

$$M_{z,j}(t_{\text{pI}} + \tau + t_{\text{pII}}) = -M_{0,j} \sin \theta_{\text{pI}} \sin \theta_{\text{pII}} \cos \omega_j \tau \qquad (7$$

along the other two axes of the rotating system. The z component make no further contribution in a two-pulse echo sequence, and we may combin the x and y components to give

$$M_{x+iy,j}(t_{\text{pI}} + \tau + t_{\text{pII}}) = M_{0,j} \sin \theta_{\text{pI}} \cos^2(\tfrac{1}{2}\theta_{\text{pII}}) \exp(i\omega_j \tau) - M_{0,j} \sin \theta_{\text{pI}} \sin^2(\tfrac{1}{2}\theta_{\text{pII}}) \exp(-i\omega_j \tau) \qquad (8$$

The accumulation of phase during the time interval $\tau + t$ following puls II phase results in the introduction of a further factor $\exp[i\omega_j(\tau + t)]$ The second term in (8) is therefore the only one giving rise to an echo signal and, summing over all spin packets, we obtain a precessing magnetizatio

$$M_{x+iy,j}(t_{\text{pI}} + t_{\text{pII}} + 2\tau + t) = -\sin \theta_{\text{pI}} \sin^2(\tfrac{1}{2}\theta_{\text{pII}}) \sum_j M_{0,j} \exp(i\omega_j t) \qquad (9$$

It will be apparent from this result that the spin echo phenomenon is intrinsically a nonlinear one. For small turn angles, the echo signal is near in θ_{pI} and quadratic in θ_{pII}. If, as assumed here, the same microwave urce is used to generate both pulses, the precessing magnetization will be roportional to H_1^3 in the limit of low pulse power.

.3. Three-Pulse "Stimulated" Echoes

The distribution $M_{z,j}$ in (7) can be made to generate a spin echo if a ird microwave pulse is applied. This echo is termed a "stimulated echo" nd is useful in some experiments since the pattern $M_{z,j}$ is not subject to ephasing effects, and may persist much longer than the pattern $M_{x+iy,j}$ hich gives rise to the simple two-pulse echo. Let us suppose that the third ulse lasts for a time t_{pIII}, where $\omega_1 t_{\text{pIII}} = \theta_{\text{pIII}}$. This pulse may be applied t an arbitrary time T after pulse II, the value of T being irrelevant within ne context of the present discussion. After pulse III, the magnetization attern derived from (7) consists of a z component, which we may ignore, nd an x component $-M_{j,0} \sin\theta_{\text{pI}} \sin\theta_{\text{pII}} \sin\theta_{\text{pIII}} \cos\omega_j\tau$. Evolution for time $\tau + t$ introduces the factor $\exp[i\omega_j(\tau + t)]$, and the stimulated cho is generated by the precessing moment

$$M_{x+iy,j}(t_{\text{pI}} + t_{\text{pII}} + t_{\text{pIII}} + T + 2\tau + t) = -\tfrac{1}{2} \sin\theta_{\text{pI}} \sin\theta_{\text{pII}} \sin\theta_{\text{pIII}} \times \sum_j M_{0,j} \exp(i\omega_j t) \qquad (10)$$

The stimulated echo amplitude may be maximized by taking $\theta_{\text{pI}} = \theta_{\text{pII}} = \theta_{\text{pIII}} = \pi/2$. When considering applications of spin echo techniques, it s sometimes useful to look at this result from another point of view. Let s take a 90°–180°, two-pulse echo sequence and divide the second pulse nto two equal portions, separating the two portions by a time T. Comparing the results obtained from (9) and (10), we see that the echo amplitude s simply that of the original two-pulse sequence divided by a factor of wo. It is more or less as if we had been able to suspend the process of phase lispersal by applying the "first half" of pulse II and then storing part of he pattern along the z axis. Then, after an arbitrary (and relatively long) period T, the cycle of events can be continued by applying the "second half" of pulse II. The factor of two appears because only the M_x portion of the pattern [Eq. (4)] which exists at the beginning of pulse II is stored along the z axis. The M_y portion gives a two-pulse echo at a time τ after pulse II.

1.4. Quantum Mechanical Treatment of Spin Echoes

We conclude this introductory section by indicating briefly how spin echoes may be treated from a quantum mechanical standpoint. We take the simple case of a 90°–180° pulse sequence applied under conditions su that $H_1 \gg$ linewidth. Let us suppose that the Hamiltonian for all sp systems which constitute the jth spin packet can be written in the for

$$\mathscr{H}_T = \mathscr{H}_s + \Delta\mathscr{H}_j \tag{1}$$

where $\Delta\mathscr{H}_j$ represents the time-independent perturbation which is respo sible for the inhomogeneous broadening of the line. The physical even occurring during the two pulses and during the periods of free precessic between them can be described by transformations of the density matrix The initial density matrix ϱ_0 is diagonal, with elements which are determine by the occupation probabilities for the eigenstates of $\mathscr{H}_T$ (i.e., the energ levels of the system). After pulse I, we have $\varrho(t_{\text{pI}}) = R_{\text{I}}\varrho_0 R_{\text{I}}^{-1}$, where is the transforming operator which corresponds to pulse I. Similarly, the end of the first free precession interval τ, we have

$$\varrho(t_{\text{pI}} + \tau) = R_\tau \varrho(t_{\text{pI}}) R_\tau^{-1} = R_\tau R_{\text{I}} \varrho_0 R_{\text{I}}^{-1} R_\tau^{-1}$$

and at the end of the complete two-pulse echo sequence

$$\varrho(t_{\text{pI}} + \tau + t_{\text{pII}} + \tau + t) = \varrho_E = R\varrho_0 R^{-1} \tag{12}$$

where R is the product

$$R = R_{\tau+t} R_{\text{II}} R_\tau R_{\text{I}} \tag{12a}$$

R_{II} is the 180° pulse operator, and R_τ the operator corresponding to th evolution of the system over a time interval τ in the absence of microwav fields. The precessing magnetization which gives rise to the echo is obtaine from the final density matrix ϱ_E by evaluating

$$M_p = \text{Tr}(\varrho_E M_+) \tag{13}$$

where $M_+ = M_x + iM_y$.

The problem of calculating M_p is essentially that of finding the matri elements of the transformation operators in some suitable representation and of forming the required matrix products. As in the case of the classica treatment, it is advantageous to perform the calculation in a rotating coordinate system. Here, this change is effected by transforming all the operators $\mathscr{O}$ (including the density operators) according to

$$\exp(i\mathscr{H}_s t/\hbar)\mathscr{O}\exp(-i\mathscr{H}_s t/\hbar) \tag{14}$$

ıis has the effect of removing $\mathscr{H}_s$ from the Hamiltonian, while leaving the uations of motion between the new operators the same as the equation motion between the old ones.* The initial density $\varrho(0)$ is unaltered, as ay be seen by setting $t = 0$ in (14).

To illustrate the details of the calculation, we may consider an ensemble systems each characterized by two energy levels, although in fact the ıantum mechanical treatment affords no new insight into the spin echo henomenon for this simple case. It has been pointed out by Feynman *al.*[3] that the equations of motion for any pair of levels interacting with radiation field can be represented in terms of classical spin precession, ppropriate parameters computed from the actual Hamiltonian taking the lace of γ and H_1. The simple classical treatment given in Sections 1.2 nd 1.3 therefore affords a satisfactory description for any spin echo exeriment involving the resonance of only two levels, whatever may be the verall level scheme associated with the spin Hamiltonian.

In order to evaluate the precessing magnetization (13), we must obtain ıe operators ϱ_0, R_{I}, R_{II}, R_τ, and M_+ corresponding to a two-level system. he initial density is given by $\varrho_0 = [\exp(-\mathscr{H}/kT)]/\mathrm{Tr}[\exp(-\mathscr{H}/kT)]$. For he purpose of calculating ϱ_0, we can set $\mathscr{H} \simeq \mathscr{H}_S = 1/\beta H_0 \sigma_z$, where σ_z s the Pauli spin matrix†

$$\begin{pmatrix} 1 & 0 \\ 0 & -1 \end{pmatrix}$$

nd thus write the numerator in the form

$$[\mathrm{I} \cosh(\tfrac{1}{2} g\beta H_0/kT) - \sigma_z \sinh(\tfrac{1}{2} g\beta H_0/kT)]$$

rom (12) and (13), it is easily seen that the unit matrix I makes no conribution to M_p and may therefore be dropped, leaving $\varrho_0 = \alpha\sigma_z$, where

$$\alpha = -[\sinh(\tfrac{1}{2} g\beta H_0/kT)]/\mathrm{Tr}[\exp(-\mathscr{H}/kT)] \tag{15}$$

To find R_{I} and R_{II}, we choose the microwave frequency ω_0 such that $\tfrac{1}{2}\hbar\omega_0\sigma_z = \mathscr{H}_S$, i.e., such that the spin packet with zero perturbation $\Delta\mathscr{H}_j$ s the one in exact resonance with the microwave field. Then, the interaction vith the microwave field can be described by adding a term

$$\tfrac{1}{2} g\beta H_1(\sigma_y \cos\omega_0 t - \sigma_x \sin\omega_0 t = \tfrac{1}{2} g\beta H_1 \exp(-\tfrac{1}{2} i\omega_0 t\sigma_z)\sigma_y \exp(\tfrac{1}{2} i\omega_0 t\sigma_z)$$

* See Ref. 2, p. 133.

† For a two-level system, the spin operator $\mathbf{S} = \tfrac{1}{2}\boldsymbol{\sigma}$. The magnetic moment operator $\mathbf{M} = \tfrac{1}{2} g\beta\boldsymbol{\sigma}$. Exponential operators may be expanded in the form $\exp(i\sigma_y\theta) = I\cos\theta + i\sigma_y \sin\theta$, etc. The commutation relations reduce to $\sigma_x\sigma_y = i\sigma_z$, etc.

to the Hamiltonian. In the laboratory coordinate system, this gives

$$\mathscr{H} = \mathscr{H}_T + \tfrac{1}{2}g\beta H_0 \exp(-\tfrac{1}{2}i\omega_0 t\sigma_z)\sigma_y \exp(\tfrac{1}{2}i\omega_0 t\sigma_z) \qquad (16$$

and in the transformed coordinate system

$$\mathscr{H} = \Delta\mathscr{H}_j + \tfrac{1}{2}g\beta H_1\sigma_y \qquad (17$$

Since the Hamiltonian (17) is time-invariant, the R matrices can be derive by substituting the appropriate Hamiltonians $\mathscr{H}$ and time intervals t_R and in expressions of the form $\exp(-i\mathscr{H}t/\hbar)$. During free precession, $\mathscr{H} = \Delta\mathscr{H}_j$ During the pulses, we approximate by setting $\mathscr{H} \simeq \tfrac{1}{2}g\beta H_1\sigma_y$ (this corre sponds to the $\omega_1 \gg$ linewidth assumption). We thus have

$$R_{\mathrm{I}} = \exp(-i\sigma_y\pi/4) = (1/\sqrt{2})(I - i\sigma_y) \qquad (18a$$

$$R_{\mathrm{II}} = \exp(-i\sigma_y\pi/2) = -i\sigma_y \qquad (18b$$

$$R_\tau = \exp(-i\,\Delta\mathscr{H}_j\tau/\hbar) \qquad (18c$$

The value of R_τ will of course depend on the particular spin packet con sidered. In a representation consisting of eigenstates of $\mathscr{H}_T$, we obtain th diagonal matrix

$$R_\tau = \begin{bmatrix} e^{-i\lambda_+\tau} & 0 \\ 0 & e^{-i\lambda_-\tau} \end{bmatrix} \qquad (19$$

where $\hbar(\omega_0 + \lambda_+)$ and $\hbar(\omega_0 + \lambda_-)$ are the eigenvalues of $\mathscr{H}_T$, and $\lambda_+ - \lambda_-$ is the frequency shift corresponding to the perturbation $\Delta\mathscr{H}_j$ (i.e., $\lambda_+ - \lambda_-$ is the frequency ω_j describing the precession of the jth spin packet in the rotating coordinate system). The matrix for M_+ is $\tfrac{1}{2}g\beta(\sigma_x + i\sigma_y)$.

The evolution of the density matrix ρ takes place according to the following pattern. The transformation $R_{\mathrm{I}}\varrho(0)R_{\mathrm{I}}^{-1}$ changes σ_z to the matrix

$$\begin{bmatrix} 0 & 1 \\ 1 & 0 \end{bmatrix}$$

(i.e., to σ_x). Transformation by R_τ attaches phase factors giving

$$\begin{bmatrix} 0 & e^{-i(\lambda_+-\lambda_-)\tau} \\ \mathrm{cc} & 0 \end{bmatrix}$$

(where cc stands for complex conjugate), R_{II} interchanges the two off-diagonal elements, $R_{\tau+t}$ attaches further phase factors, and finally

$$\varrho_E = \begin{bmatrix} 0 & e^{-i(\lambda_+-\lambda_-)t} \\ \mathrm{cc} & 0 \end{bmatrix}$$

he component of precessing magnetization due to the jth spin packet is ound by computing $\mathrm{Tr}(\varrho_E M_+)$ for the jth subensemble [i.e., by evaluating e constant α_j in (15) for the jth spin packet]. Summing over all such ibensembles, we obtain the precessing magnetization due to the whole homogeneous line. At a time $\tau + t$ after pulse II, this is given by

$$M_p = \sum_j \mathrm{Tr}(\varrho_E M_+)_j = -g\beta \sum_j \alpha_j \exp(i\omega_j t)$$

ince the initial magnetization of the jth spin packet is given by $M_{0,j}$ $= \frac{1}{2} g\beta\, \mathrm{Tr}(\varrho_0 \sigma_z) = g\beta\alpha_j$, we can rewrite this expression in the form

$$M_p = -\sum_j M_{0,j} \exp(i\omega_j t) \tag{20}$$

ius obtaining the classical result (3).

The calculation as given above for the two-level case is of course elatively trivial. We shall, however, have occasion to use the same general iethods where more than two levels are involved and where the classical escription would not suffice (see Section 4).

2. INHOMOGENEOUS LINE BROADENING

2.1. Line Broadening and Electron Spin Echoes

We begin this section with an attempt to clarify the meaning of the erms "inhomogeneous broadening" and "homogeneous spin packet." It vas assumed in the description of spin echo generation in Section 1 that a) the resonance line consisted of a distribution of spin packets having listinct Larmor frequencies, (b) the precession frequency of each spin acket remained constant for all time, and (c) there were no agencies, other han the microwave fields, tending to reorient spins. In this idealized case, he line could be described as exhibiting pure inhomogeneous broadening, he spin packets could be chosen to have an infinitesimally small spectral vidth, and echoes could be obtained for times τ of any length. Actual naterials do not, of course, show this behavior. Instead, the echo signal alls off according to some decay function $E(2\tau)$ which represents the ratio etween the signal strength at a time 2τ after pulse I and the unattenuated alue at $\tau = 0$. We can conveniently use the Fourier transform of $E(2\tau)$ o define the homogeneous line shape of an individual spin packet. The verall resonance line then has the structure shown schematically in Fig. 3.

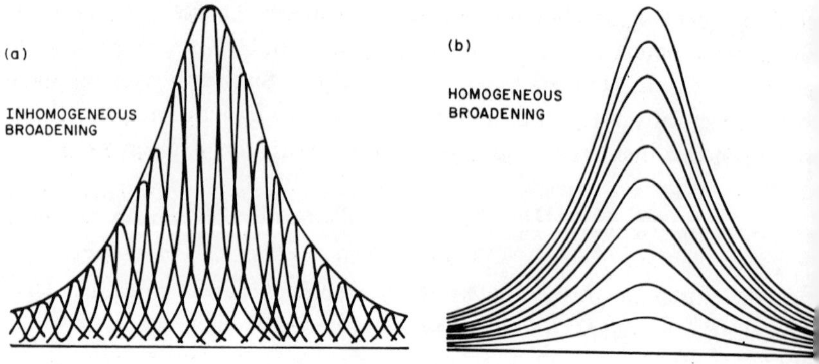

Fig. 3. (a) Schematic diagram showing an inhomogeneously broadened line. Individu component lines or "spin packets" correspond to the Fourier transform of the spin ech envelope decay function $E(2\tau)$. (b) A homogeneously broadened line shown as th superposition of a number of similar spin packets; the individual spin packet width here the same as the width of the overall line.

From Eq. (1) (which shows that the free-induction signal is the Fourie transform of the line shape), it will be apparent that the echo signal ca only be separated from the free-induction tail if the homogeneous spi packet width is appreciably less than the width of the whole line.* Thi inhomogeneous broadening criterion is satisfied for many dilute para magnetic materials at low temperatures. In these materials, the broadenin mechanisms are effectively static. This need not mean that the perturbation which give rise to the distribution of resonance frequencies are wholl time-independent, but only that they vary on a relatively long time scale (See, for example, the discussion in Section 3.)

Static line broadening can originate in two distinct ways. It can b caused by macroscopic spatial fluctuations in the physical state of th sample, such as wandering of the crystal axes, the presence of a mosai structure,[4,5] or strains induced during crystal growth. We shall not discus this macroscopic broadening any further here since it does not posses any general interest for spin echo studies, and is largely an accidenta property of particular samples. In contrast to this, we can classify micro

* This condition has to be modified if the microwave field is such that $\omega_1 <$ linewidth In this case, a distribution of spin packets $\sim 2\omega_1$ wide is selected from the line (se Section 5), and the free-induction tail is longer than that which would be deduce from the Fourier transform of the overall line shape. The condition for separabilit of the echo and free-induction tail is roughly that the homogeneous spin packet widt should be appreciably less than ω_1.

copic broadening mechanisms as those which correspond to a more or ess random variation in Larmor frequencies as we proceed from point to nearby point in the host crystal lattice. These mechanisms are typified by the line broadening caused by point defects,[6] and form the principal opic of discussion in the remainder of this section. Microscopic broadening has important consequences for spin echo experiments. By detuning nearby pins from one another, it reduces the mean spin–spin flip time and lengthens he phase memory which can be obtained in the limit of low temperatures when lattice relaxation effects become negligible. Because of the effects of uch detuning mechanisms, the homogeneous spin packet width is often very considerably less than the width which would be deduced from spin–spin interactions considered alone (Section 3.3).

2.2. The Statistical Theory of Line Broadening

We shall used the remainder of Section 2 to discuss the application of the "statistical theory" of Margenau[7] to the problem of broadening in EPR lines. This affords some useful insights into the nature of the microscopic inhomogeneous broadening which we have just mentioned, and it can, in some circumstances, be applied to calculate the phase memory, or homogeneous spin packet broadening (see Section 3.4). In this section, we shall use it to calculate broadening due to randomly distributed magnetic dipoles,* to point defect strains,[6] and to irregularities in the distribution of lattice point charges.[8,9] Since the argument follows the same general lines in all these cases, we present it in outline before introducing specific details.†

The following assumptions are made:

1. Only one type of defect or lattice irregularity is involved in the broadening.

2. Such a defect causes a shift $\varepsilon(\mathbf{z}_i)$ in the resonance frequency, where $\mathbf{z}_i$ denotes the coordinates of the defect relative to the paramagnetic center. $\varepsilon(\mathbf{z}_i)$ is linear in the perturbation, and the contributions from a number of defects are additive, i.e.,

$$\varepsilon(\mathbf{z}_1, \mathbf{z}_2, \ldots, \mathbf{z}_N) = \sum_i \varepsilon(\mathbf{z}_i)$$

* We assume in this section that the orientation of the dipoles remains constant. The effects of spin flips, which generate fluctuating local fields and cause some degree of homogeneous broadening, are discussed in Section 3.

† In this, we follow the procedure of a review of line broadening mechanisms by Stoneham.[10]

3. The positions at which defects are found are not correlated with each other, and the probability $p(\mathbf{z}_1, \mathbf{z}_2, \ldots, \mathbf{z}_N)$ of finding a given defect configuration can be expressed as a product of the individual probabilities $p(\mathbf{z}_i)$, i.e.,

$$p(\mathbf{z}_1, \mathbf{z}_2, \ldots, \mathbf{z}_N) = \prod_i p(\mathbf{z}_i)$$

4. The lattice can be approximated by a continuous medium.

Before proceeding with the argument, we comment in turn on these four assumptions.

1. It is quite easy to generalize the treatment to cover defects of more than one kind when the perturbation law $\varepsilon(\mathbf{z})$ has the same functional form for all defects. This will be illustrated shortly when we discuss magnetic dipolar broadening due to environmental spins in different magnetic substates. If $\varepsilon(\mathbf{z})$ has a different functional form for the various kinds of defects, the statistical method is still valid provided that the several kinds of defect constitute independent random arrays. In the latter case, however, the results may have to be computed numerically, the problem being essentially that of evaluating the appropriate convolution integrals.

2. Nonlinear shifts $\varepsilon(\mathbf{z}_i)$ are not usually important in practice unless, for symmetry or other reasons, there is no perturbation which produces a linear shift. One exception (i.e., small magnetic perturbations of a line characterized by a highly anisotropic g-tensor) is mentioned briefly later on.

3. Although correlations between the positions of the defects are discounted here, they may in actual fact be a common occurrence (e.g., in materials where charge compensation of the paramagnetic centers occurs). A general analysis of this case would involve complicated statistical arguments and will not be attempted. It may, however, be noted that in the particular case where the correlation merely concerns the positions of the defects in relation to the paramagnetic centers, the problem can be treated relatively easily by incorporating a suitable weighting factor in $p(\mathbf{z}_i)$ (see Section 2.6).

4. The approximation of the crystal lattice by a continuous medium is justifiable if the defect density is low (e.g., $<1\%$ of the available lattice sites). Errors arise from the fact that the continuum model admits the possibility of more than one defect being simultaneously present at the same lattice site, but these errors are generally small enough to be ignored. The satellite lines due to nearest-neighbor perturbations, which are observed in many paramagnetic materials, do of course correspond to a failure of the continuum model to predict the correct line shape. However, it will be

pparent during the course of the statistical calculation that large perturba-ons due to near neighbors are automatically excluded when finding the ne shape nearer the center, and that errors which arise in predicting the ne shape out in the wings are not associated with serious inaccuracies lsewhere, as they may be when the method of moments is used.

We now continue with the statistical analysis of the perturbations which result from a distribution of defects throughout the crystal lattice. et us denote the probability of finding a defect over the range of positions $_i$ to $\mathbf{z}_i + d\mathbf{z}_i$ by $p(\mathbf{z}_i)\, d\mathbf{z}_i$. Then, the probability of finding a configuration f N defects at coordinates $\mathbf{z}_i$ ($i = 1, \ldots, N$) is given by

$$P(\text{config}) = p(\mathbf{z}_1)p(\mathbf{z}_2) \cdots p(\mathbf{z}_N)\, d\mathbf{z}_1\, d\mathbf{z}_2 \cdots d\mathbf{z}_N$$

his configuration will be associated with a frequency shift of $\sum_{i=1}^{N} \varepsilon(\mathbf{z}_i)$. t is our task to sum P(config) over all possible configurations by integrating V times with respect to the variables $\mathbf{z}_i$, while selecting only those con-igurations which result in shifts $\Delta\omega$ of a specified value. This will give us he intensity $I(\Delta\omega)$ of the resonance line at a frequency $\Delta\omega$ away from the nperturbed value. The required operation is denoted by the multiple ntegral

$$I(\Delta\omega) = V^{-N} \int d\mathbf{z}_1\, p(\mathbf{z}_1) \int d\mathbf{z}_2\, p(\mathbf{z}_2) \cdots \int d\mathbf{z}_N\, p(\mathbf{z}_N)\, \delta\Big\{\Delta\omega - \sum_i \varepsilon(\mathbf{z}_i)\Big\} \quad (21)$$

where $\delta\{\cdots\}$ stands for the Dirac delta function, and the value $V = \int^V d\mathbf{z}\, p(\mathbf{z})$ normalizes $p(\mathbf{z})$. Writing the delta function in the integral orm

$$\delta\Big\{\Delta\omega - \sum_i \varepsilon(\mathbf{z}_i)\Big\} = (1/2\pi) \int_{-\infty}^{+\infty} d\varrho\, \exp\Big\{-i\varrho\Big[\Delta\omega - \sum_i \varepsilon(\mathbf{z}_i)\Big]\Big\}$$

substituting it in (21), dropping subscripts i, and rearranging, we have

$$I(\Delta\omega) = (1/2\pi) \int_{-\infty}^{+\infty} d\varrho [\exp(-i\varrho\, \Delta\omega)] \times \Big\{(1/V) \int^V p(\mathbf{z})[\exp i\varrho\varepsilon(\mathbf{z})]\, d\mathbf{z}\Big\}^N \quad (22)$$

This expression gives the lineshape function due to N point defects randomly distributed throughout a volume V. To facilitate further analysis, we take V and N to infinity, avoiding divergences by rewriting $I(\Delta\omega)$ in the form

$$I(\Delta\omega) = (1/2\pi) \int_{-\infty}^{+\infty} d\varrho\, e^{-i\varrho\Delta\omega}[1 - (J/V)]^N \quad (23)$$

where

$$J = \int^{V} [1 - \exp i\varrho\varepsilon(\mathbf{z})]p(\mathbf{z})\, d\mathbf{z} \tag{24}$$

Then, provided that J remains finite, the line shape is obtained from the Fourier transform

$$I(\Delta\omega) = (1/2\pi) \int_{-\infty}^{+\infty} d\varrho\, e^{-i\varrho\Delta\omega} e^{-nJ} \tag{25}$$

where $n = N/V$ is the number of defects per unit volume.

2.3. Line Broadening Due to Dipolar Fields

As a convenient illustration of the statistical method, let us calculate the line broadening arising from dipolar interactions between two spin groups, group A constituting an observed resonance line, and group B constituting the perturbing centers. It is assumed here that neither species is undergoing spin flips at any significant rate, and that groups A and B are nonresonant, so that terms other than those of the form $S_z(A)S_z(B)$ in the dipolar Hamiltonian can be discarded. For simplicity, we also take g to be isotropic, and assume that the spin populations are confined to two levels for which the effective spin $S' = 1/2$. The frequency shift is then time-independent and is given by the function

$$\varepsilon(\mathbf{z}) = \gamma_A \mu_B (1 - 3\cos^2\theta)/r^3 \tag{26}$$

where γ_A is the gyromagnetic ratio $g_A\beta/\hbar$ for the A spins, and μ_B is the matrix element of $g_B\beta S_z$ for the B spins; r is the distance between A and B spins, and θ is the angle between the Zeeman field and the line joining them. Assuming that there is no correlation between the sites selected by A and B spins, $p(\mathbf{z})\, d\mathbf{z} = 2\pi r^2\, dr\, d(\cos\theta)$, and therefore, by (24),

$$J_{\pm 1/2} = 2\pi \int_0^{\infty} r^2\, dr \int_{-1}^{+1} d(\cos\theta) \times \{1 - \exp[\pm i(\varrho\gamma_A g_B\beta/2)(1 - 3\cos^2\theta)/r^3]\} \tag{27}$$

Since there are two types of perturbing B centers (corresponding to $\mu_B = \pm\frac{1}{2}g_B\beta$), $\varepsilon(\mathbf{z})$ has two possible values

$$\varepsilon(\mathbf{z}, \pm\tfrac{1}{2}) = \pm\tfrac{1}{2}\gamma_A g_B \beta(1 - 3\cos^2\theta)/r^3,$$

and we have to consider two independent random arrays of magnetic defects." In order to find their resultant effect on $I(\Delta\omega)$, we must modify ome of the earlier expressions. Let there be $N_{+1/2}$ and $N_{-1/2}$ spins corresponding to concentrations $n_{+1/2}$ and $n_{-1/2}$, in the $M_s = \pm\frac{1}{2}$ states. Then, ach probability $p(\mathbf{z}_i)$ corresponding to the spatial coordinate $\mathbf{z}_i$ must be esolved into two factors $p(\mathbf{z}_i, \frac{1}{2})$ and $p(\mathbf{z}_i, -\frac{1}{2})$, factors of each kind being ounted $N_{1/2}$ and $N_{-1/2}$ times, respectively, in the product. In Eq. (22), ve therefore make the change

$$\left\{(1/V)\int^{V} p(\mathbf{z})[\exp i\varrho\varepsilon(\mathbf{z})]\,d\mathbf{z}\right\}^{N}$$

$$\to\left\{(1/V)\int p(\mathbf{z}, 1/2)[\exp i\varrho\varepsilon(\mathbf{z}, 1/2)]\,d\mathbf{z}\right\}^{N_{1/2}}$$

$$\times\left\{(1/V)\int p(\mathbf{z}, -1/2)[\exp i\varrho\varepsilon(\mathbf{z} - 1/2)]\,d\mathbf{z}\right\}^{N_{-1/2}}$$

The argument then continues as before, concluding with the result

$$I(\Delta\omega) = (1/2\pi)\int_{-\infty}^{+\infty} d\varrho \exp(i\varrho\,\Delta\omega)\exp -(n_{1/2}J_{1/2} + n_{-1/2}J_{-1/2}) \tag{28}$$

in place of Eq. (25). In the high-temperature limit, $n_{1/2} = n_{-1/2} = \frac{1}{2}n_B$ and the remainder of the calculation is fairly straightforward. The exponent of (28) becomes

$$\begin{aligned} &n_{1/2}J_{1/2} + n_{-1/2}J_{-1/2} \\ &\quad = 2\pi n_B \int_0^{\infty} r^2\,dr \int_{-1}^{+1} d(\cos\theta)\{1 - \cos[(\varrho\gamma_A g_B\beta/2)(1 - 3\cos^2\theta)/r^3]\} \\ &\quad = |\varrho|\,(4\pi^2/q\sqrt{3})\gamma_A g_B\beta \end{aligned} \tag{29}$$

Hence, from Eq. (28),

$$I(\Delta\omega) = \frac{\Delta\omega_{\text{dip}}/\pi}{(\Delta\omega)^2 + (\Delta\omega_{\text{dip}})^2} \tag{30}$$

where

$$\Delta\omega_{\text{dip}} = (4\pi^2/9\sqrt{3})\gamma_A g_B\beta n_B \simeq 2.53\gamma_A g_B\beta n_B \tag{31}$$

The local field broadening of a resonance line by spins of the same species can be derived as a special case by dropping subscripts A and B. This leads to a result which is two-thirds of the result normally quoted,* and is ap-

* See Ref. 11.

plicable when flip-flop processes among the spins have been suppresse (as, e.g., when the spins are detuned by some extraneous broadenir mechanism). For a concentration of 10^{18} cm^{-3} B spins having $g = 2$, th dipolar half-width is 47 mG. If $g \simeq 2$ for the A spins, $\Delta\omega_{\text{dip}}/2\pi = 132$ kH

If there are different numbers of spins in the two M_s states, the ar gument from (28) onward is somewhat more involved, although the lir broadening turns out to be the same. Substituting $\zeta = 1/r^3$ and $A = \gamma_A g_B \beta /$ we can rewrite (27) in the form

$$J_{\pm 1/2} = (2\pi/3) \int_{-1}^{+1} d(\cos\theta) \int_{\zeta_1}^{\zeta_2} (d\zeta/\zeta^2)\{1 - \exp[\pm i\varrho A(1 - 3\cos^2\theta)\zeta]\} \quad (32$$

where, eventually, ζ_1 and ζ_2 have to be taken to the limits of 0 and ∞ respectively. $J_{\pm 1/2}$ consists of a real portion $J_{\pm 1/2, r}$ and an imaginary portio $J_{\pm 1/2, i}$, thus giving in (28) an exponent

$$\begin{aligned} n_{1/2}J_{1/2} + n_{-1/2}J_{-1/2} &= (n_{1/2} + n_{-1/2})J_{+1/2,\mathrm{r}} + i(n_{1/2} - n_{-1/2})J_{+1/2,i} \\ &= n_B J_{+1/2,\mathrm{r}} + i\,\Delta n_B J_{+1/2,i} \end{aligned} \quad (33$$

where Δn_B is the difference between the concentrations of spins in th $M_s = \pm 1/2$ substates. The real portion is easily integrated. Taking $\zeta_1 \to$ and $\zeta_2 \to \infty$, we obtain the same integral as in (29). Thus the distributio $I(\Delta\omega)$ retains the Lorentzian form and has the same width as was derived i (31) for the special case of equal substate populations. Unfortunately, th imaginary portion $J_{+1/2,i}$, which determines the frequency shift of th center of this Lorentz distribution, is sensitive to the form of the boundar adopted when performing the integration. For a spherical boundary $J_{+1/2,i}$ is zero and the shift is zero.* For other boundaries, the integral i hard to evaluate and the most practical procedure is to assume the materia lying outside an arbitrary spherical boundary to be uniformly magnetize

* From (24)–(26) and from the assumption $p(z)\,dz = 2\pi r^2\,dr\,d(\cos\theta)$, it can be readily shown that the first moment $\int I(\Delta\omega)\,\Delta\omega\,d(\Delta\omega)$ is zero, provided that the integration is performed over spherical shells between limiting radii R_1 and R_2. This is a familiar result for a uniformly magnetized body and will be the same for the statistical mode employed here. Since the line is symmetrically broadened, this zero first moment implies a zero shift due to spins lying between R_1 and R_2, the shift, if any, being due to nearby spins inside R_1 and to distant spins outside R_2. The shifts due to individual nearby spins will be large, but they will only affect the wings of the resonance line, for which the statistical model is invalid anyhow. The shifts due to distant spins can be calculated without taking statistical fluctuations in the distribution of paramagnetic centers into account. (The author is indebted to Dr. L. R. Walker for clarifying this point.)

ith a magnetization $M = g_B\beta\, \Delta n_B$. The local field at the center of this phere can then be calculated by elementary magnetostatics and the shift ω inferred accordingly. For example, in the case of a thin crystal with arallel sides, the local field in $8\pi M/3$ and the frequency shift

$$\Delta\omega(\text{shift}) = 8\pi\gamma_A g_B \beta n_B/3 \tag{34}$$

he point to note here is that the randomness of the distribution of B spins ives rise to a spread of frequencies which is independent of the partition etween the two magnetic substates. The shift of this distribution as a unction of temperature depends on the macroscopic geometry of the ample and can be determined without reference to the statistical considerations discussed above.

As an example of the way in which the statistical calculation can be xtended so as to apply to multilevel schemes, we outline briefly the argument for the case in which $S = 5/2$ for the B spins. For simplicity, M_s s assumed to be a good quantum number, thus reducing the problem to hat of calculating the A-spin line shape $I(\Delta\omega)$ when the perturbation is ue to six independent random arrays, one for each of the quantized B-spin moments $\mu_B = M_s g_B \beta$. Modifying the basic statistical expressions, we esolve $p(\mathbf{z}_i)$ into six factors, and rewrite the curly-bracketed term in (22) s a product of six terms taken to the powers N_j ($j = 1, \ldots, 6$). These erms involve frequency shifts

$$\varepsilon_j(\mathbf{z}) = (M_s)_j \gamma_A g_B \beta (1 - 3\cos^2\theta)/r^3$$

rom which we obtain a J integral with real part

$$J_{M_s,\mathrm{r}} = |\, 2M_s \,|\, J_{1/2,\mathrm{r}} = 10.1 \,|\, M_s \,|\,|\, A \,|\,|\, \varrho \,| \tag{35}$$

The six factors in the modified form of Eq. (22) lead, in place of (25), to a lineshape function

$$I(\Delta\omega) = 1/2\pi \int d\varrho \exp(i\varrho\, \Delta\omega) \exp\Big(-\sum_j n_j J_j\Big) \tag{36}$$

vhere the n_j are B-spin concentrations in the M_s substates. It may be seen eadily that, as before, we have a Lorentzian line and that the half-width

$$\Delta\omega_{\text{dip}} = 5.06(\gamma_A g_B \beta/2) \sum_j n_j \,|\, M_{s,j} \,| \tag{37}$$

n the high-temperature limit, the half-width becomes

$$\Delta\omega_{\text{dip}}(S) \simeq 7.6\gamma_A g_B \beta n_B \tag{38}$$

where

$$n_B = \sum_j n_j$$

Substituting $S = 5/2$, and taking $g_B \simeq g_A \simeq 2$, $n_B = 10^{18}\,\text{cm}^{-3}$ as in th previous numerical example, we obtain $[\Delta\omega_{\text{dip}}(5/2)]/2\pi \simeq 400$ kHz, (o 140 mG). This result corresponds to the situation in which all B-spi states are equally populated and is, as pointed out earlier, only appropriate t orientations of the Zeeman field which leave M_s as a good quantum numbe

It is perhaps instructive to note that we could have approached thi problem without necessarily rehearsing the whole statistical argument fror first principles onward. The six uncorrelated random distributions of M states are each independently capable of broadening and of shifting th center of the A-spin line. The overall distribution could therefore be ob tained by summing the shifts algebraically, and by convoluting the si distributions with each other. In the above case where all of the defec arrays cause perturbations of the same magnetic dipolar type, the convolu tion is easy to perform, since the distributions are Lorentzian, and con volute to give further Lorentzians, the individual width parameters simpl being added together. If similar arguments were used to derive the lin broadening due to arrays of defects of different types, a numerical con volution of the component lineshape functions would normally be needed The self-convoluting property of the Lorentzian functions is implicit i the previous analysis, where it is found that the sum $\sum_j n_j J_j$ required i order to compute $I(\Delta\omega)$ by (36) is linear in $|\varrho|$. This linearity holds fo any broadening mechanisms which correspond to a perturbation la $\varepsilon(\mathbf{z}) \propto r^{-3}$.

2.4. Broadening Due to Strains and to Quadrupola Electric Fields

The linewidths experimentally observed in dilute paramagnetic ma terials are often many times greater than the value $\Delta\omega_{\text{dip}}$ derived from Eq (31), indicating that mechanisms other than dipolar broadening are dom inant. In dilute materials, two other types of broadening mechanisms ar commonly found. They are: (a) broadening due to lattice strains generate by point defects, and (b) broadening due to the quadrupolar electrostati fields which result from the presence of point-charge irregularities in th lattice.* The two are physically distinct, as one can see by imagining th

* Broadening due to the familiar "first degree" electrostatic field can only occur if th paramagnetic ion is at a site lacking inversion symmetry. This case is considered next

ubstitution in the lattice of an ion of identical size but different charge nd vice versa, but in practice they may be hard to separate from one nother. The frequency shifts caused in either of these ways can be described y an expression of the form

$$\varepsilon(\mathbf{z}) = (1/r^3) \sum_{ij} A_{ij}[\delta_{ij} - (3x_i x_j/r^2)] \tag{39}$$

vhere the coordinates $\mathbf{z}$, r, etc. have the meanings assigned earlier, x_i $i = 1, \ldots, 3$) are the Cartesian components of $\mathbf{z}$, and δ_{ij} is the Dirac delta unction. The calculation proceeds along the previous lines, with the linear ombination $\sum_{ij} A_{ij} f_{ij}(\Omega)$ of the angular dependence functions $f_{ij}(\Omega) =$ $\delta_{ij} - (3x_i x_j/r^2)$ replacing $A(1 - 3\cos^2\theta)$.* Thus, instead of (32), we have

$$J_s = \tfrac{1}{3} \int d\Omega \int_{\zeta_1}^{\zeta_2} (d\zeta/\zeta^2)\left\{1 - \exp\left[i\varrho\zeta \sum_{ij} A_{ij} f_{ij}(\Omega)\right]\right\} \tag{40}$$

J_s consists of real and imaginary components $J_{s,r}$ and $J_{s,i}$, the real part eing associated with line broadening and the imaginary part with a shift of the mean frequency as in the magnetic dipolar case. Taking $\zeta_1 \to 0$ nd $\zeta_2 \to \infty$, we find that

$$J_{s,r} = -(\pi/6) \,|\, \varrho \,|\, \int \left| \sum_{ij} A_{ij} f_{ij}(\Omega) \right| d\Omega \tag{41}$$

$J_{s,r}$ is linear in ϱ, and hence, by Eq. (25), the strain- (or quadrupole-) broadened line has a Lorentzian form, the half-width at half-height being given by

$$\Delta\omega_{1/2}(\text{strain}) = (\pi n_s/6) \int \left| \sum_{ij} A_{ij} f_{ij}(\Omega) \right| d\Omega \tag{42}$$

The integral for the imaginary portion will, like the corresponding integral in the case of magnetic dipolar broadening, depend on the boundary which is chosen. Physically, this means that the bulk strains induced in the sample by the point defects may be anisotropic and shape-dependent. The essential useful result here is thus the derivation of the Lorentzian line shape. Actual magnitudes are hard to predict since the coefficients A_{ij} cannot be directly determined by experiment.†

* The magnetic dipolar interaction can be regarded as a special case of this more general law.

† In the case of broadening by quadrupolar fields, the A_{ij} must be calculated (e.g., by means of crystal field theory). In the case of strain broadening, measurements of the response of the resonance line to macroscopic strains can be used, but the strain fields caused by a given type of point defect must be calculated (see Ref. 10).

Strain broadening of resonance lines can be associated with dislocation as well as with point defects. We shall not reproduce the calculation her (it is similar to the one outlined above); but refer the reader instead to th papers of Stoneham.[10,12] The predicted line shape is Gaussian. Since ex perimental line shapes are often more nearly Lorentzian than Gaussian,[*] it might at first seem that point-defect broadening is more important tha dislocation broadening in practice. This conclusion does not necessaril follow, however. There may be a tendency for paramagnetic centers t associate with the dislocations, thus giving a line shape which is wider i the wings, and hence more Lorentzian in appearance, than the line shap which is derived by assuming an uncorrelated random distribution of centers

A number of studies of strain-broadened line shapes have been made. In MgO (the host lattice which has been most frequently investigated) the strains are typically 10^{-4} and the corresponding stresses are of the orde of 100 kg/cm^2. Agreement with theory is, up to a certain point, reasonabl good. For example, the linewidths of Mn^{2+} and Fe^{3+} in MgO vary with th orientation of the Zeeman field in the manner to be expected from th symmetry of the fourth-rank tensor relating mechanical strains with th "D" term in the spin Hamiltonian.[14] [This tensor determines the coefficient A_{ij} in (39).] The line shape is often not correctly predicted, however, and quantitative interpretation in terms of independently determined defec concentrations is usually not available.

2.5. Broadening Due to First Degree Electric Fields

We mention here a source of line broadening which, although perhap less universal than strain broadening, has at least the advantage that it is fairly easy to correlate with experimental parameters. This is the line broadening which is caused by internal electric fields of the familiar "first degree" kind when the paramagnetic centers show a linear electric field shift (i.e., when they are situated at lattice points which lack inversion symmetry). If the lattice defects possess a charge Ze, they give rise to in- ternal electric field components $E_i = Ze(x_i/r)/K_i r^2$, K_i representing the principal dielectric constants, and x_i the Cartesian components of the position vector $\mathbf{z}$ along the principal axes of the dielectric tensor.[†] The

* See Refs. 13 (in particular, p. 281) and 14–16. Reference 15 concerns the broadening of the resonant mode in KBr:Li^+, but the considerations are essentially the same as for an EPR line.

† The experimental electric field shift parameters, which form a third-rank tensor, may be transformed into this axis system if necessary.

requency shift due to an individual defect is then given by

$$\varepsilon(\mathbf{z}) = Ze \sum a_i x_i / K_i r^3 \tag{43}$$

here the a_i are experimental parameters obtained by measuring the shift $\omega_E = \sum_i a_i E_i$ of the resonance line in applied electric fields.* The resulting ne shape, according to the statistical theory, can be calculated by sub- tituting (43) in Eqs. (23) and (24). It is readily verified that the imaginary omponent of J is zero and that the line has therefore no net shift.† The eal component of J thus becomes

$$J_r = (4\pi/15)\left\{\left| \varrho Ze\left[\sum_i (a_i/K_i)^2\right]^{1/2} \right|\right\}^{3/2} \tag{44}$$

nd the lineshape function $I(\Delta\omega)$ is the Fourier transform of $e^{-n_e J_r}$ as n Eq. (25). The final result cannot be expressed analytically as in the nagnetic dipolar case, but $I(\Delta\omega)$ can easily be obtained by making a umerical integration. For this purpose, it is convenient to rewrite the xpression for $I(\Delta\omega)$ [derived from (25) and (47)] in the form

$$I(\Delta\omega) = (1/\omega_0)(1/\pi) \int_0^\infty \cos(\varrho\omega/\omega_0) \exp - | \varrho^{1.5} | \, d\varrho \tag{45a}$$

vhere

$$\omega_0 = n_e^{2/3} Ze\left(\sum_i (a_i/K_i)^2\right)^{1/2} \tag{45b}$$

ω_0 is a width parameter for this distribution, which is sometimes termed a Ioltzmark distribution.‡

This line shape is shown in Figs. 4 and 5 and is intermediate in form etween a Gaussian and Lorentzian. The parameters a_i can be deduced rom the coefficients of the third-rank tensor which specifies the paramag-

The charge defects giving rise to line broadening (as distinct from "pair" or compensa- tion spectrum effects) will be situated a number of lattice sites away. The dielectric may therefore be treated macroscopically for the purpose of calculating internal fields. Correction of the Clausius–Mossotti type are not needed here, since the fields due to *Ze* are calculated in the same manner as the fields due to the electrodes used in the experimental determination of the electric field effect parameters a_i.

The same conclusion could be reached at once by noting that the charge defects are equally likely to be found at coordinates $\pm z$, where they would give rise to local electric fields (and hence linear frequency shifts) of opposite signs.

The distribution derived in Ref. 17 is not, however, the one discussed here, but is, instead, the distribution of the scalar magnitudes of the local electric field.

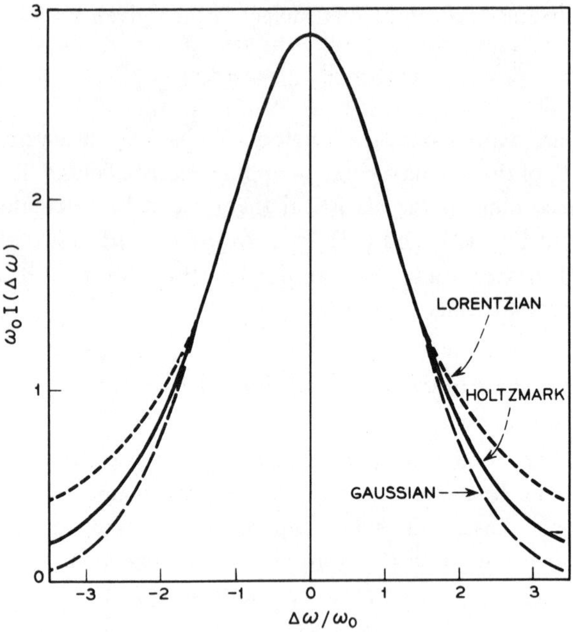

Fig. 4. Theoretical line shape for a paramagnetic resonance line broadened by internal electrostatic fields. The linewidth is proportional to $n^{2/3}$, where n is the number of charge defects per cm^3. Gaussian and Lorentzian curves are shown for comparison, all curves being matched to give the same height and full-width at the points of half-height.

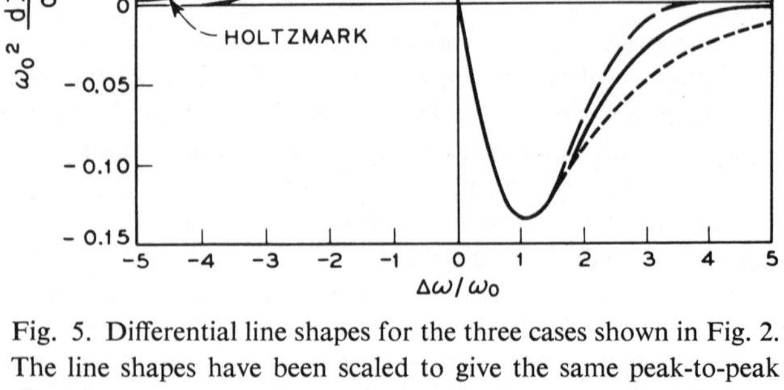

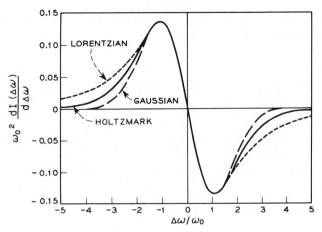

Fig. 5. Differential line shapes for the three cases shown in Fig. 2. The line shapes have been scaled to give the same peak-to-peak distance.

etic Stark effect for the material in question (see Ref. 8). Since the a_i ary in a characteristic manner with the orientation of the Zeeman field, t is sometimes possible to identify line broadening due to this source nambiguously and to estimate the magnitudes of the randomly oriented ocal electric fields which are responsible. In insulating crystals, these elds can be of the order of 50 kV/cm. Static electric fields lead, of course, o static inhomogeneous broadening of the resonance line. If, for any eason, moving charges are involved, effects on the phase memory of lectron spin echoes, and on the homogeneous spin packet width can be nticipated.

?.6. Further Comments on Line Broadening Mechanisms

The ideal situations considered above do not cover all the practically mportant cases, nor does the treatment which was given exhaust the vays in which magnetic dipoles, strains, and charge defects can induce line roadening. A single type of defect may contribute in several ways. For nstance, ionic vacancies may simultaneously give rise to broadening via irst-degree electrostatic fields, quadrupolar electrostatic fields, and me-hanical strains. The difficulty here is that the contributions are correlated ut cannot simply be added together because of differences in the perturba-ion law $\varepsilon(\mathbf{z})$. Nonlinear mechanisms may also be important in particular naterials, or at particular orientations of the crystal axes and the Zeeman ield. For example, in the vicinity of a maximum or minimum g, the res-nance frequency of a highly anisotropic center may depend quadratically n one or more components of the local magnetic field.* Assumption (b) hen breaks down and the perturbations $\varepsilon(\mathbf{z})$ are no longer additive. Holtz-nark[17] treats the analogous case of broadening via the quadratic Stark ffect by making a linear superposition of field amplitudes to obtain the listribution of local fields. The frequency distribution $I(\Delta\omega)$ can then be lerived from this result.† Another approach is suggested by Stoneham.[10]

Even in the domain in which the statistical theory as outlined here emains applicable, it is possible to proceed with quite an extensive catalog of possible interactions between paramagnetic ions and other crystal im-purities. For example, interest has recently focused on quadrupole–quadru-

* A quadratic dependence of resonance frequency on certain local field components is also frequently found when the resonance occurs between two levels of a multilevel scheme.

† $I(\Delta\omega)$ will generally be asymmetric in such a case.

pole interactions in paramagnetic resonance spectra,[18] the correspondin perturbation law being $\varepsilon(\mathbf{z}) = A_Q u(\Omega)/r^5$, where $u(\Omega)$ describes the angul dependence, and A_Q is a constant describing the strength of the interactio Without attempting to discuss this and other similar possibilities in an detail, we may here quote an interesting result obtained by Margenau which gives at once a prediction for the line shape. If the perturbatio law is of the form $u(\Omega)r^{-m}$, and the angular dependent portion $u(\Omega)$ has vanishing mean, then

$$I(\Delta\omega) = (1/\pi) \int_0^\infty \exp| -nk'\varrho^{3/m} | \cos(\varrho\, \Delta\omega)\, d\varrho \qquad (46$$

k' is a product of the interaction constant and of numerical factors whic arise in the integration over r and Ω. The shape of the line is, howeve determined solely by the power m.

Equation (46) may be useful when it is necessary to consider the clus tering of defects about the paramagnetic centers. Under some circumstances this clustering can be treated as equivalent to a change in the law of in teraction. For example, if the clustering probability is proportional to r^{-s} the probability factor $p(\mathbf{z})\, d\mathbf{z}$ in (22) becomes $r^{2-s}\, dr\, d\Omega$. By means of th substitution $r = u^{3/(3-s)}$, $p(\mathbf{z})\, d\mathbf{z}$ can, however, be restored to the form $u^2\, du\, d\Omega$, which it possesses in the case of an unweighted random distribu tion, the perturbation law being simultaneously changed from r^{-m} t $u^{-3m/(3-s)}$. Inserting $s = 1$ and $m = 3$, we see, for instance, that an r^{-} clustering function together with a magnetic dipolar interaction law would simulate an $r^{-9/2}$ interaction law in the lineshape formula (46).

In host lattices containing nuclei with large moments (such as ^{1}H ^{19}F, ^{27}Al) an appreciable degree of line broadening can originate in the electron nuclear interactions. The broadening of nuclear resonance line by other nuclei has been treated extensively in the literature by means o the method of moments* and we shall not treat it here. The results can be readily adapted to the case in hand. The mean square width is given by

$$\langle \Delta\omega^2(\text{nuclear}) \rangle = \tfrac{1}{3} I(I+1)(\hbar\gamma_n\gamma_e)^2 \sum_l (3\cos^2\theta_l - 1)^2/r_l^6 \qquad (47)$$

where γ_n and γ_e are the nuclear and electron gyromagnetic factors $g\beta/\hbar$ and θ_l and r_l are the coordinates of a given nucleus in relation to an electron spin. For a $g = 2$ electron in a CaF_2 lattice, $\Delta\omega(\text{nuclear})/2\pi$ ranges from ~4 to ~11 MHz according to the orientation of H_0. Actually, (47) may

* See Van Vleck.[19] Also see Ref. 11, p. 123.

underestimate the nuclear broadening. The perturbation due to the nearer nuclei is often increased by a contact interaction term. Taking this into account, Laurance *et al.*[20] estimate a linewidth of 24 MHz for Cr^{3+} ions in Al_2O_3, an amount which is four times larger than the width to be expected from (47) alone, and which represents 80% of the observed linewidth. The line shape due to nuclear dipolar and nuclear contact interactions tends to be approximately Gaussian in form. This follows from the central limit theorem of statistics, since the line shape represents the combined effect of a number of perturbations of random sign but of comparable magnitude due to the nearby nuclei.

It is hardly necessary to emphasize in concluding that the broadening of EPR lines can arise in a large variety of different ways, making it difficult to associate the observed linewidth or line shape with any one single cause. In view of this, it may appear that the whole subject offers little more than an occasion for making calculations, and that its usefulness in solid-state physics is minimal (although this situation could possibly be changed if the foundations were to be strengthened by a series of careful experimental studies). We need not, however, speculate further on this matter here, since our primary concern is with electron spin echo phenomena, and with the significance of line broadening for such properties as the phase memory time T_M. The models and mechanisms described above will, it is hoped, make it easier to form a picture of how the "detuning" of spins arises in actual materials, and will prepare the way for an understanding of the dynamical effects discussed in Section 3.

3. PHASE MEMORY AND SPECTRAL DIFFUSION

3.1. General

In this section, we shall focus attention on the factors which determine the echo envelope decay and the phase memory time T_M. This is a considerably more difficult problem than the one which we have discussed in the previous section. Inhomogeneous broadening is caused by static perturbations, phase memory decay by their dynamical counterparts. In the one case, solutions have been derived for a number of the more important practical situations, in the other, very little theoretical understanding exists at all. We must, however, attempt some discussion of the phase memory problem here, since it is fundamental to almost all electron spin echo experiments. One approach is illustrated by the analysis of echo envelope decay due to T_1 processes (Section 3.4). A physical model is set up, and its implications are studied in some detail using the statistical theory of Section 2. The case where T_M is limited by spin–spin flips is treated in a more cursory manner, although it corresponds to what is perhaps the most common practical situation. Here, we restrict ourselves to a more or less qualitative discussion of the mechanisms involved and try to derive some order-of-magnitude estimates for T_M, our primary aim being to show how the various physical parameters influence the result. The memory times may be of some value as a rough guide in planning experiments, but in view of the crudity of the arguments which are used, it should occasion little surprise if the estimates deviate widely from observations.

As was pointed out in Section 2, the decay of the echo envelope can arise from two distinct causes: spin flips of the echo-generating spins themselves, and disturbances in the environment which modify the Larmor frequencies, and thus randomize the phases of precession of each individual spin. We can classify the corresponding effects as direct or indirect. It is important to keep in mind that the effect in dilute materials is often predominantly an indirect one. Spin flips of the echo-generating spins, which we denote as the A spins may easily have less overall effect on the phase memory than the flips of environmental B spins, which cause fluctuations in the local magnetic field.* In such cases, T_M is not simply a measure of the spin-flip time, and caution should be exercised when using experimental values of T_M as a measure of the transverse relaxation time T_2.†

The simplest instance of the direct effect is that which is associated with lattice relaxation of the echo-generating spins (A spins). Here, $T_M = T_1$, and the echo decay function has the same form as the correlation function

$$\langle \mu_z(0)\mu_z(2\tau)\rangle = \langle [\mu_z(0)]^2\rangle \exp(-2\tau/T_1) \tag{48}$$

It would be convenient to describe the spin–spin effect in a similar manner, replacing T_1 by an effective spin-flip time T_f. Unfortunately, in a dilute material, the correlation function cannot be expected to have a simple

* A good indication of the extent to which the indirect effect normally determines T_M in a dilute paramagnetic material is afforded by the observations given in Ref. 1, p. 1212. The resonance lines due to exchange-coupled Fe^{3+} ions in TiO_2 gave unusually long phase memory times in zero Zeeman field, the reason being that here the Larmor frequencies are insensitive to changes in the local magnetic field.

† In order to avoid possible confusion on this point, we have used T_M rather than T_2 to denote the characteristic decay time of the echo envelope. $E(2\tau)$ may assume a variety of functional forms. Here, we define T_M as the time, measured from pulse I, which corresponds to an e^{-1} attenuation of the echo signal.

exponential form, and in the absence of any specific theory, T_f remains a somewhat imprecise parameter. (See, however, the footnote on p. 307.)

Indirect effects can arise in a wide variety of ways. Confining our attention to mechanisms which are associated with local magnetic field changes, we have the following kinds of motion among the B spins which can lead to echo envelope decay: (a) spin–lattice relaxation of electron spins; (b) spin–spin interactions among electron spins; (c) spin–spin interactions among nuclear spins in the host material (i.e., nuclear spins acting as B spins). In cases (a) and (b), it should be noted that the echo-generating A spins and the environmental B spins need not belong to different species, nor even to different resonance lines. The distinction applies to the role which is played. Relaxation of a given number of spins can remove spins from the precessing group (i.e., direct effect due to removal of A spins), and at the same time, change the local field seen by the rest of the spin system (indirect B-spin effect).

3.2. Phase Memory and Nuclear Spin Flip-Flops

One might hope that in case (c) at least, the memory time could be readily deduced by adapting results which have already been established by NMR studies. Let us, for example, take the case of CaF_2. It is known that the local field strength $(\overline{\Delta H_0^2})^{1/2}$ due to ^{19}F nuclei in CaF_2 ranges from $\sim$1.5 to $\sim$4 G (according to the orientation of H_0), and that the nuclear resonance line of ^{19}F is homogeneously broadened.* The nuclear phase memory time is therefore given by

$$T_M(^{19}F) \sim [\gamma(\overline{\Delta H_0^2})^{1/2}]^{-1} \sim 10\text{–}27 \quad \mu\text{sec} \tag{49}$$

If we were to assume that the fluctuating local fields seen by electron spins doped in CaF_2 remained the same as in the pure material, we should obtain

$$T_M \sim [\gamma_e(\overline{\Delta H_0^2})^{1/2}]^{-1} = 28\text{–}75 \quad \text{nsec} \tag{50}$$

for $g = 2$ electrons. Fortunately, this result is two orders of magnitude shorter than the experimentally observed times. The reason for this dis-

* A summary of linewidth data on CaF_2 is given in Ref. 11, p. 115. The statement that the line is homogeneous may be roughly justified by noting that the flip-flop term has about the same magnitude as the other terms in the dipolar Hamiltonian. There does not appear to be any rigorous deduction of this result. W. G. Clark (unpublished results) has verified it experimentally for ^{19}F nuclei in CaF_2 and for ^{27}Al nuclei in Al metal.

crepancy seems to be that the nuclear spins which are nearest to the electron are strongly detuned by the electron local field, and cannot take part in flip-flop processes with the remaining nuclei. In effect the electron creates around itself a volume in which the nuclear flip-flop processes are suppressed.

We can form a rough estimate of this volume in the following manner. Let us equate the electron–nuclear interaction at a radius r_s with the half-width

$$\Delta\omega_{nn} = \gamma_n \overline{(\Delta H_0{}^2)}^{1/2} \tag{51}$$

of the ^{19}F NMR line. The electron–nuclear interaction is a measure of the detuning effect, and $\Delta\omega_{nn}$ a measure of the interaction causing nuclear flip-flops (as in pure CaF_2). Then,

$$\hbar\gamma_n\gamma_e \,|\, 1 - 3\cos^2\theta \,|/r_s{}^3 = \Delta\omega_{nn} \tag{52}$$

By the moment method, we have*

$$\Delta\omega_{nn} \simeq 2n\gamma_n{}^2\hbar \tag{53}$$

Substitution of this value in (52) gives the result

$$r_s{}^3 = 0.5(\gamma_e/\gamma_n) \,|\, 1 - 3\cos^2\theta \,|/n \tag{54}$$

from which it can be inferred that there are $\sim 0.51\gamma_e/\gamma_n$ nuclei in the volume for which flip-flop motion is effectively suppressed. (For a $g = 2$ electron in CaF_2, this volume contains 360 ^{19}F nuclei.)

Only those ^{19}F nuclei that are outside this volume contribute to electron spin echo decay. We can obtain a rough estimate of their effect on T_M by making the following assumptions.

(a) The fluctuation in I_z is the same for all nuclei at radii greater than r_s, and its frequency spectrum can be characterized by an r.m.s. amplitude $\Delta\omega_{nn}$ determined as in Eq. (53).

(b) The effect of the motion of any one of these nuclei on the frequency spectrum of an electron spin packet is to contribute to the homogeneous broadening by an amount $\delta\omega_e$, where

$$\delta\omega_e = (\delta\omega_{ne})^2/\Delta\omega_{nn} \tag{55}$$

* The moment method gives $\Delta\omega_{nn} = 1.95n\gamma_n{}^2\hbar$ for a powder of crystallites, when each crystallite is characterized by a cubic lattice. See Ref. 11, page 112. Any approximate method of calculating $\Delta\omega_{nn}$ would suffice for the present purpose.

and the quantity

$$\hbar\,\delta\omega_{ne} = \hbar^2\gamma_e\gamma_n(1 - 3\cos^2\theta)/r^3 \tag{56}$$

is the electron–nuclear interaction energy.*

(c) The fluctuations are uncorrelated, and the resultant homogeneous broadening due to them is given by

$$(\Delta\omega_e)^2 = \sum (\delta\omega_e)^2 \tag{57}$$

If we approximate the lattice by a continuum, we have, for the summation (57), the integral

$$(\Delta\omega_e)^2 = \left\{2\pi n \int_{-1}^{1} d(\cos\theta) \int_{r_s}^{\infty} r^2\,dr[\hbar\gamma_e\gamma_n(1 - 3\cos^2\theta)/r^3]^4\right\} \Big/ (\Delta\omega_{nn})^2 \tag{58}$$

Integrating (58), combining the result with (52) and (53), and taking the square root, we obtain

$$\Delta\omega_e = 0.73\,\Delta\omega_{nn}(\gamma_e/\gamma_n)^{1/2} \tag{59}$$

For a $g = 2$ electron in CaF_2, (59) yields the result $\Delta\omega_e = 1.27\times 10^6$, which corresponds to a phase memory time $T_M \sim (\Delta\omega_e)^{-1} \sim 0.8$ μsec. This is still several times shorter than the experimental value, but it does at least approach the correct order of magnitude.

Equation (59) is based on such very rough arguments that it is hardly worth attempting to explain the remaining discrepancy. We may, however, note the following points. First, Eq. (53), which is derived by taking both the flip–flop and the $I_z(1)I_z(2)$ terms in the nuclear spin–spin Hamiltonian into account, probably overestimates the magnitude of the flip-flop interaction. Second, correlations between the nuclear flip-flop motions, which were neglected in taking the sum of squares (57), may tend to conserve the total moment in a given locality, and thus lead to a significant reduction in the perturbation experienced by an electron situated several lattice spacings away. We have, on the other hand, ignored the fact that the local field of the electron tends to shift the Larmor frequencies of nearby nuclei by similar amounts and have therefore somewhat overestimated the volume

* When $r > r_s$, we are in the "narrowing" regime, in which the frequency shifts $\delta\omega_{ne}$ caused by the fluctuating nuclear local fields (and hence the homogeneous broadening effects) are small compared with the rate $\sim\Delta\omega_{nn}$ at which the disturbance is taking place. "Narrowing" is a familiar phenomenon in resonance work and leads to expressions of the kind which we have adopted here in (56). [See, for example, Ref. 11, page 426, or Eq. (83).]

in which nuclear motion is frozen. The first two considerations suggest that $\Delta\omega_e$ should be smaller (and T_M longer) than our estimate here. The last consideration would have the opposite effect.

A more careful analysis of the above problem from a different standpoint has been made by Davies and Hurrell.[21] In order to understand the procedure adopted in their work, let us first suppose that we have a coupled system consisting of one electron spin ($S = 1/2$) and two ^{19}F nuclei. This system has eight eigenstates, from which we select for consideration the four states which define microwave intervals differing by an amount approximately equal to the ^{19}F nuclear spin–spin coupling. These four states may (assuming the appropriate state mixing) be excited in a coherent manner by the microwave pulses. If so, the resulting signals will interfere as shown in Section 4, causing a periodic modulation of the echo envelope.

Generalizing to the actual coupled system consisting of an electron spin with many ^{19}F neighbors, we find that coherences can be induced between many sets of closely spaced levels. Because of the variety of the level separations (representing the various nuclear spin–spin coupling energies as modified by the presence of the presence of the electron spin), the overall effect of the interference in this case is to lead to a net attenuation of the echo signal with time i.e., to an apparent phase memory decay. Good agreement with experimental data on the Ce^{3+}:CaF_2 system has been obtained by means of this model, which appears to be considerably more rigorous than the rudimentary model suggested earlier in this section.

3.3. Phase Memory and Electron Spin Flip-Flops

The situation with regard to electron spin–spin flips has some features in common with that discussed above. Detuning effects play an important role, and the memory is often much longer than the time given by the inverse dipolar linewidth. For example, in $CaWO_4$ doped with 0.83×10^{18} Ce^{3+} spins cm^{-1} $\Delta\omega_{\text{dip}} = 0.35 \times 10^6$ rad/sec [see Eq. (38)], and $(\Delta\omega_{\text{dip}})^{-1} = 2.9$ μsec, whereas the experimental value of T_M is* 125 μsec. Detuning in this instance is due to strains and to internal electric fields (see Sections 2.4 and 2.5), the experimentally measured half-width $\Delta\omega_{st}$ being 5.7×10^7 rad/sec, or $\sim 170 \times \Delta\omega_{\text{dip}}$. We can obtain a very rough estimate for the spin–spin flip time T_f by applying the Fermi "golden rule" to this situation, taking $\pi/\Delta\omega_{st}$ as the density of states in the center of the line and $\hbar\,\Delta\omega_{\text{dip}}$

* See Mims *et al.*,[22] Table 1. T_M was measured perpendicular to the *c* axis, where $g = 1.43$. $\Delta\omega_{1/2}$ has been calculated for this case in the example given.

as a mean value for the transition matrix element. We thus have

$$[T_f(\text{center})]^{-1} \simeq \pi(\Delta\omega_{\text{dip}})^2/\Delta\omega_{st} \tag{60}$$

Away from the center of the line, the density of states will follow the line-shape function. Assuming this to be approximately Lorentzian (as we did when taking the density of states to be $\pi/\Delta\omega_{st}$), we obtain a mean value for the spin–spin flip time given by

$$(T_f) \simeq \Delta\omega_{st}/[2\pi(\Delta\omega_{\text{dip}})^2] \tag{61}$$

In the instance just quoted, Eqs. (60) and (61), lead to a spin–spin flip time in the center of the line of 1.3 msec and a mean time T_f of 2.6 msec. Since this estimated value of T_f is an order of magnitude longer than the experimental value of T_M, we tentatively conclude that we are dealing here with an indirect effect. We can estimate a value for T_M due to the indirect effect by borrowing from the theory developed in Section 3.4 and treating T_f as it were a lattice relaxation time. Substituting $R = 1/T_f$ in the "Gauss Markov" formula for T_M [Eq. (82b)] we obtain $T_M = 57$ μsec. The "sudden jump" formula (86b) gives $T_M \simeq 170$ μsec. These estimates are surprisingly close to the experimental values, considering the qualitative nature of the argument. This is probably fortuitous, however. The physical models used in Section 3.4 to derive T_M for the indirect effect are based on the assumption that the motions of the B spins are uncorrelated. In fact, as we have pointed out in Section 3.2, the spin flip-flops will tend to conserve the magnetic moment in a given small volume, thus diminishing the fluctuation in the local magnetic fields. Besides this, we have no way of being sure that the measured strain width $\Delta\omega_{st}$ describes of the "microscopic" broadening associated with a random variation of the strain fields from point to point in the lattice and does not include macroscopic broadening effects. Probably the only valid conclusion to be drawn from the above argument is that in a dilute paramagnetic material, the inverse linewidths and the characteristic times will stand in the order

$$(\Delta\omega_{st})^{-1} < (\Delta\omega_{\text{dip}})^{-1} < T_M < T_f$$

The alternative procedure used by Davies and Hurrell[21] to calculate the effect of nuclear flip-flops[21] might in principle be applied to the case where electron flip-flops are the main source of phase memory decay. Better results could conceivably be obtained in this way if it were possible to devise a satisfactory description for the random array of strain-detuned electron spins.

3.4. Phase Memory and Lattice Relaxation

The direct effect of lattice relaxation processes on T_M is given by Eq. (48). The indirect effect is analyzed in this section by setting up physical models to represent the effects of the motion in a relaxing B-spin system on an echo-generating A-spin system.[23] This constitutes a relatively simple situation compared with the two which were discussed previously. We can reasonably assume that there is no correlation between spin flips, and that the spin-flip motion for all the B spins can be described by the same random function of time.*

The position of a given spin packet in the resonance line is determined by the static broadening mechanisms, which give a fixed perturbation ω_s, and by the A-spin, B-spin dipolar fields, which give a time-varying perturbation $\Delta\omega(t)$. The phases $\omega_s\tau$ accumulated as a result of the static perturbation during the two intervals τ cancel (as shown in Section 1) and are of no interest here. The phases accumulated due to $\Delta\omega(t)$ will, however, give a net resultant phase factor at the time† 2τ,

$$\exp[i\varphi(2\tau)] = \exp\left[i\int_0^{2\tau} s(t')\,\Delta\omega(t')\,dt'\right] \tag{62}$$

where

$$\Delta\omega(t) = \gamma_A \sum_j [\mu_j(t) - \mu_j(0)](1 - 3\cos^2\theta_j)/r_j^3 \tag{63}$$

The function $s(t)$ is the step defined by

$$\begin{aligned} s(t) &= +1 \qquad \text{if} \quad t < \tau \\ &= -1 \qquad \text{if} \quad t > \tau \end{aligned} \tag{64}$$

and takes account of the phase reversal caused by pulse II. The B-spin moment $\mu_j(t)$ is here assumed to be a Markovian stochastic variable characterized by the correlation function

$$\langle \mu_j(t')\mu_j(t + t')\rangle = \langle [\mu_j(0)]^2\rangle e^{-Rt} \tag{65}$$

where $R = 1/T_1$. The remaining variables in (63) are as defined in Section 2.

* The physical basis for this assumption lies in the fact that all B spins, whatever their position in the line, are driven in an equivalent manner by the lattice phonon spectrum. Coupling between B spins, whether due to dipolar forces or to any other mechanism, is discounted here.

† For simplicity, it will be assumed here that $\tau \gg t_p$, and that the echo signal appears at a time 2τ after the beginning of the pulsing cycle.

The echo signal will be proportional to the weighted sum of phase factors such as (62). Each element in the sum consists of a group of A spins specified by: (a) a particular value of the inhomogeneous strain perturbation, (b) a particular pattern of motion under the influence of the microwave pulses and (c) a particular geometric configuration of surrounding B spins.

Here, (a) and (b) are related to one another and determine the form of the echo signal in a manner which is discussed in detail in Section 5. These properties can be factored out in the sum over all spin packets, provided that ω_1 comfortably exceeds the intervals $\Delta\omega(2\tau)$ traversed as a result of spectral diffusion during the spin echo cycle of events. We are left therefore with a summation of phase factors due to various geometric configurations of B spins, each B spin being associated with a time-varying magnetic moment $\mu_j(t)$ which (in the first of the two models discussed here) will be assumed to be a Gaussian random variable. The resulting decay factor $E(2\tau)$ can be written as the double average

$$E(2\tau) = \left\langle\left\langle \exp i \int_0^{2\tau} s(t')\gamma_A\left\{\sum_j [\mu_j(t') - \mu_j(0)] \times (1 - 3\cos^2\theta_j)/r_j^3\right\} dt' \right\rangle_{\text{Av(i)}}\right\rangle_{\text{Av(ii)}} \tag{66}$$

Average (i) is taken over the distribution of values of $\mu_j(t)$ and average (ii) over the possible configurations of surrounding B spins. Strictly speaking, average (i), like average (ii), is a spatial average, and is taken over those A spins that see various possible B-spin moments $\mu_j(t)$ at a specific set of lattice points. However, by the ergodic hypothesis, average (i) can be replaced by a time average. Average (ii) cannot, of course, be treated in this way, since it is taken over a set of permanent configurations established when the crystal was grown and annealed. Writing

$$\xi_\gamma(t) = \int_0^t dt'\, s(t')[\mu_j(t') - \mu_j(0)] \tag{67a}$$

and

$$\alpha_j = \gamma_A(1 - 3\cos^2\theta_j)/r_j^3 \tag{67b}$$

Eq. (66) becomes

$$E(2\tau) = \left\langle\left\langle \exp i\left[\sum_j \alpha_j\xi_j(2\tau)\right]\right\rangle_{\text{time av}}\right\rangle_{\text{config. av}} \tag{68}$$

The first problem is that of finding the time average. Since $\xi_j(t)$ is a linear combination of the values of the Gaussian random variable $\mu_j(t)$, it is also itself a Gaussian random variable and therefore

$$\langle \exp[i\alpha_j \xi_j(2\tau)] \rangle = \exp[i\alpha_j \langle \xi_j(2\tau) \rangle] \exp\{-0.5\alpha_j \langle [\Delta\xi_j(2\tau)]^2 \rangle\} \qquad (69)$$

$\langle \xi(2\tau) \rangle$ is the mean, and $\langle [\Delta\xi_j(2\tau)]^2 \rangle^{1/2}$ the variance $\langle [\xi_j(2\tau) - \langle \xi_j(2\tau) \rangle]^2 \rangle^{1/2}$ of $\xi_j(2\tau)$. In the high-temperature limit, $\langle \mu_j(0) \rangle = 0$, therefore

$$\langle \xi_j(2\tau) \rangle = \int_0^{2\tau} dt'\, s(t') \langle \mu_j(t') \rangle = 0$$

and (69) reduces to

$$\langle \exp[i\alpha_j \xi_j(2\tau)] \rangle = \exp\{-0.5\alpha_j \langle [\xi_j(2\tau)]^2 \rangle\} \qquad (70)$$

Thus, remembering that the B-spin motions are uncorrelated, we have

$$E(2\tau) = \left\langle \exp - \left\{ 0.5 \sum_j \alpha_j \langle [\xi_j(2\tau)]^2 \rangle \right\} \right\rangle_{\text{config av}} \qquad (71)$$

where

$$\langle [\xi_j(2\tau)]^2 \rangle = \int_0^{2\tau} dt' \int_0^{2\tau} dt''\, s(t') s(t'') \times \langle \mu_j(t') \mu_j(t'') \rangle \qquad (72)$$

Substituting the Markovian correlation function (65) in (72), and setting $\langle [\mu_j(0)]^2 \rangle = \frac{1}{4} g_B{}^2 \beta^2$, we obtain

$$\langle [\xi_j(2\tau)]^2 \rangle = g_B{}^2 \beta^2 B(\tau)$$

where

$$B(\tau) = (1/R^2)[R\tau - (1 - e^{-R\tau}) - 0.5(1 - e^{-R\tau})^2] \qquad (73)$$

Thus

$$E(2\tau) = \left\langle \exp\left[-0.5 g_B{}^2 \beta^2 B(\tau) \sum_j \alpha_j{}^2 \right] \right\rangle_{\text{config av}}$$

$$= \left\langle \exp\left[-0.5 g_B{}^2 \beta^2 B(\tau) \sum_j (1 - 3\cos^2\theta_j)^2 / r_j{}^6 \right\rangle_{\text{config av}} \qquad (74)$$

The average over all B-spin configurations can be found by means of the statistical theory, using methods analogous to those employed in Section II. Each sum $\sum_j$ in the average corresponds to a particular configuration of B spins about an A spin. Let us suppose that this is arrived at by placing N B spins one in each of N specified volume elements $dV_1, dV_2, \ldots, dV_N$. If V is the volume of the sample, and if all points are equally likely as sites for the jth B spin, the probability of finding such a configuration is

$$\prod_{j}^{n} (dV_j/V) \tag{75}$$

The contribution to $E(2\tau)$ arising from this configuration is therefore given by the product of (75) and the exponential inside the brackets of (74), i.e., it is

$$\left[\prod_{j}^{N} (dV_j/V)\right] \exp\left[-0.5 g_B{}^2\beta^2\gamma_A{}^2 B(\tau) \sum_j (1 - 3\cos^2\theta_j)^2/r_j{}^6\right] \tag{76}$$

The average over all possible placements of N B spins in the neighborhood of an A spin is found by integrating over each of the volume elements dV_j. Following the same lines of argument as in Section 2, we have

$$E(2\tau) = \left\{(V^{-1}) \int^{\mathrm{vol}\,v} dV \exp[-0.5B(\tau) g_B{}^2\beta^2\gamma_A{}^2(1 - 3\cos^2\theta)^2/r^6]\right\}^N \tag{77}$$

whence

$$E(2\tau) = \exp(-n_B V') \tag{78a}$$

with

$$V' = 2\pi \int_0^\infty r^2\,dr \int_{-1}^{1} d(\cos\theta)$$
$$\times \{1 - \exp[-0.5B(\tau) g_B{}^2\beta^2\gamma_A{}^2(1 - 3\cos^2\theta)^2/r^6]\} \tag{78b}$$

Integrating (78b), and substituting numerical values where appropriate, we have

$$V' = 4.06\gamma_A g_B \beta [B(\tau)]^{1/2} \tag{79}$$

whence

$$E(2\tau) = \exp\{-4.06 n_B \gamma_A g_B \beta [B(\tau)]^{1/2}\} \tag{80}$$

Equation (80) can conveniently be reexpressed in terms of the dipolar half width $\Delta\omega_{\mathrm{dip}}$ [Eq. (31)], giving

$$E(2\tau) = \exp\{-1.88\,\Delta\omega_{\mathrm{dip}} [B(\tau)]^{1/2}\} \tag{81}$$

The required envelope decay function can be obtained by substituting $B(\tau)$ from (73). It assumes simple forms in the two limits $R\tau \gg 1$ and $R\tau \ll 1$. For $R\tau \ll 1$, $B(\tau) = R\tau^3$ and

$$E(2\tau) = \exp -(2\tau/T_M)^{3/2} \tag{82a}$$

where

$$T_M = 1.89[R(\Delta\omega_{\rm dip})^2]^{-1/3} \tag{82b}$$

For $R\tau \gg 1$,

$$E(2\tau) = \exp -(2\tau/T_M)^{1/2} \tag{83a}$$

where

$$T_M = 0.56R/(\Delta\omega_{\rm dip})^2 \tag{83b}$$

Provided that we are only concerned with the first one or two decay periods of $E(2\tau)$, these limiting conditions are equivalent to $R \ll \Delta\omega_{\rm dip}$ in (82), and $R \gg \Delta\omega_{\rm dip}$ in (83). The phase memory thus shortens with increasing B-spin relaxation rate as long as $(\Delta\omega_{\rm dip}/R) \gg 1$, and lengthens again when $(\Delta\omega_{\rm dip}/R) \ll 1$. The minimum can be found by differentiating the argument of (81) with respect to R, and is given by

$$T_M(\min) = 1.8/\Delta\omega_{\rm dip} = 3.8/R \tag{84}$$

[The form of $E(2\tau)$ at this minimum is shown in Ref. 23, Fig. 1.]

We see therefore that three different situations can arise due to B-spin relaxation. Equation (83) describes a result which is similar to motional narrowing results in NMR (e.g., Ref. 11, p. 426), the additional complications which enter here being a consequence of the random distribution of B spins in the magnetically dilute sample. At the minimum [Eq. (84)], the dipolar broadening is homogeneous. The inverse of this homogeneous width is of the order of the B-spin flip time and of the phase memory. Equation (82) corresponds to yet a third set of conditions in which the line is essentially inhomogeneous (under the influence of B-spin local fields as well as strain), but in which there is some degree of spectral diffusion.

The assumption that $\mu_j(t)$ can be treated as a Gaussian random variable constitutes an important element in the preceding calculation. This assumption represents an attempt to summarize the interaction of B spins and lattice phonons in a simple and manageable fashion. It may be justified by arguing that each relaxing B spin is in contact with many lattice modes, and is perturbed during one relaxation period by many small fluctuations occurring in the phonon spectrum. If these perturbations are all of comparable magnitude, then, according to the central limit theorem, one would expect their combined effect to result in a Gaussian distribution of values for the random variable $\mu_j(t)$. An argument of this kind is not always appropriate. It appears plausible enough where, as in the case of a Raman process, the interaction involves a large number of lattice modes. However, it is not clear that it can be applied in a case such as that of lattice relaxation via an Orbach process, which involves the interaction of the paramagnetic centers with a relatively low density of high-energy phonons.

The form of the random function $\mu_j(t)$ is not important in the "narrowing" regime, where $R\tau \gg 1$. In this case, $\mu_j(t)$ changes its orientation several times during the interval τ, and the integral in (67a) consists of a number of phase accumulations of comparable magnitude. By the central limit theorem, $\xi_j(t)$ will therefore tend to be a Gaussian random variable even if $\mu_j(t)$ is not one, and Eq. (70) will follow as before. In the "spectral diffusion" regime, however, where $R\tau \ll 1$, the form of the echo decay function differs according to the assumptions made regarding $\mu_j(t)$. To illustrate this, we take an extreme case, in which $\mu_j(t)$, instead of executing a random walk of many small steps, makes sudden jumps between the values $\pm\frac{1}{2}g_B\beta$. This we shall call the "sudden-jump" model to distinguish it from the "Gauss–Markov" model on which the previous calculation was based. The sudden-jump model has been described by Klauder and Anderson,[24] their argument being roughly as follows. In a very short time $\Delta t_1 \ll 1/R$, a small number of randomly selected B spins will reverse their orientation. This is equivalent to the insertion of $n_B R\,\Delta t_1$ moments, each of value $g_B\beta$, at random sites in the lattice. Therefore, according to the statistical theory of line broadening (Section 2), a spin packet with an initial frequency ω_i is broadened into a Lorentzian distribution:

$$K(\omega - \omega_i, \Delta t_1) = \frac{2R\,\Delta\omega_{\text{dip}}\,\Delta t_1/\pi}{(\omega - \omega_i)^2 + (2R\,\Delta\omega_{\text{dip}}\,\Delta t_1)^2} \tag{85}$$

where $\Delta\omega_{\text{dip}}$ is the inhomogeneous half-width corresponding to a concentration of N_B B spins per cm³ [Eq. (38)].* During the next time interval Δt_2, $n_B R\,\Delta t_2$ additional spin flips occur *at unrelated lattice sites*, and each frequency component in (85) gives rise to a new Lorentzian distribution of half-width $2R\,\Delta\omega_{\text{dip}}\,\Delta t_2$. The convolution of these two distributions yields a Lorentzian distribution whose width is given by the sum $2R \times \Delta\omega_{\text{dip}}(\Delta t_1 + \Delta t_2)$. As long as this picture holds good, therefore, the change in local fields seen by the full ensemble of A spins can be represented as a diffusion process characterized by a Lorentzian kernel, and having a width varying as Rt. Klauder and Anderson then show that the echo en-

* The factor 2 occurs here because each spin flip corresponds to the introduction of new moments $\pm g_B\beta$, whereas in calculating the line broadening in Section 2, we are dealing with a distribution of moments $\pm\frac{1}{2}g_B\beta$.

velope decay function is given (in our present notation) by

$$E(2\tau) = \exp -(2\tau/T_M)^2 \tag{86a}$$

where

$$T_M = 1.41(R\,\Delta\omega_{\text{dip}})^{-1/2} \tag{86b}$$

As in the Gauss–Markov case [Eq. (82)], the value of T_M lies somewhere between R^{-1} and $(\Delta\omega_{\text{dip}})^{-1}$, but $E(2\tau)$ has a different form here, and, if $R \ll \Delta\omega_{\text{dip}}$, then T_M is considerably longer.

The sudden-jump model as described here cannot be used to calculate $E(2\tau)$ outside the spectral diffusion regime. If $Rt \gtrsim 1$, it is not possible to make the assumption that the spin flips during successive time intervals correspond to the continual insertion of new moments $g_B\beta$ into the lattice at random and uncorrelated sites. As time progresses, spins which were assumed to be inserted earlier will have a significant chance of undergoing a return flip and thus being "removed." Removal cannot, of course, be regarded as taking place at purely random sites, and it does not affect all portions of a previously broadened spin packet (assumed to be a delta function at $t = 0$) in an equivalent way. On the contrary, a portion of the spin packet which has undergone more than the average shift of frequency in a time $t \sim 1/R$ will stand a more than average chance of making large frequency excursions in the future since it tends to represent configurations with more B spin near neighbors. Diffusion over times $\sim 1/R$ thus has the effect of sorting out the A spins according to their B-spin environment and cannot any longer be described by a "homogeneous Markovian" function of the type discussed by Klauder and Anderson in their derivation of Eq. (85).

We have devoted considerable space to the problem of T_1-limited phase memory in the hope that this will serve as a useful example of the way in which the echo decay function and the dynamics of local fields in crystals can be linked together. The effects of Jahn–Teller-type transitions in the environment (interacting via strain or magnetic dipole coupling) might be expected to yield to a similar analysis, as well as some of the situations mentioned in Section 6.4. The basic problem of spin–spin-limited phase memory is, as we have noted in Section 3.3 not easy to treat in this way, because of the correlations which exist between spin–spin flips.

3.5. Instantaneous Diffusion

At this point, we interrupt this discussion of phase memory and of the form of the echo envelope to draw attention to an effect which is related o the pulsing conditions rather than to the intrinsic properties of the sample. This is the "instantaneous diffusion" mentioned by Klauder and Anderon.[24,*] It results from the abrupt change in local fields caused by spin eorientation during pulse II and might perhaps be more appropriately lassified with the mechanisms described in Section 6.4. We introduce it ere, however, since it is primarily an experimental problem arising the tudy of envelope decay functions. It can, if overlooked, lead to a major rror in the determination of the phase memory.

Let us suppose that n' spins are reoriented by pulse II. These are A pins insofar as they are responsible for generating the spin echo signal, out must also be considered as a B-spin ensemble contributing to the local ields and to line broadening. If these B spins are randomly distributed throughout the lattice, then, by the arguments of Section 2.3, it can be shown that after their reorientation, each spin packet will have been replaced by the spectral distribution

$$I(\Delta\omega) = \frac{\Omega/\pi}{(\Delta\omega)^2 + \Omega^2}$$

where

$$\Omega \simeq 5.1\gamma g\beta n'$$

[This is twice the result given in (31) for the reason stated in the footnote on p. 279.] The spin packets whose phases would otherwise have converged at time τ after pulse II are thus each replaced by a Lorentzian distribution of precessing moments with phase factors $\exp(i\,\Delta\omega\tau)$. The resultant precessing magnetization is reduced by a factor which can be found by taking the Fourier transform of the Lorentzian, i.e., by a factor $\exp(-\Omega\tau)$.

Under ideal 90°–180° pulsing conditions (i.e., with $\omega_1 \gg$ linewidth), n' is equal to the total number of spins in the line. The effect of instantaneous diffusion is therefore to reduce the echo amplitude by $\exp(-2\,\Delta\omega_{\text{dip}}\tau)$, where $\Delta\omega_{\text{dip}}$ is the dipolar half-width given in (31). Typically, for a concentration of 10^{18} cm^3 of spins with $g = 2$, this would multiply the echo envelope by an exponential decay factor with time constant ~1.2 μsec. The decay rate due to instantaneous diffusion can, however, be reduced very considerably by working in a regime where $\omega_1 \ll$ linewidth, and where, as a consequence, fewer spins undergo reorientation. (Spin echo waveforms obtainable under the conditions of $\omega_1 \ll$ linewidth are discussed in Section 5.3.) The effective number n' corresponding to given values of ω_1 and the

* Effects of this type have been noted in nuclear spin echo experiments by B. G. Silbernagel *et al.*[25] See, in particular, p. 539.

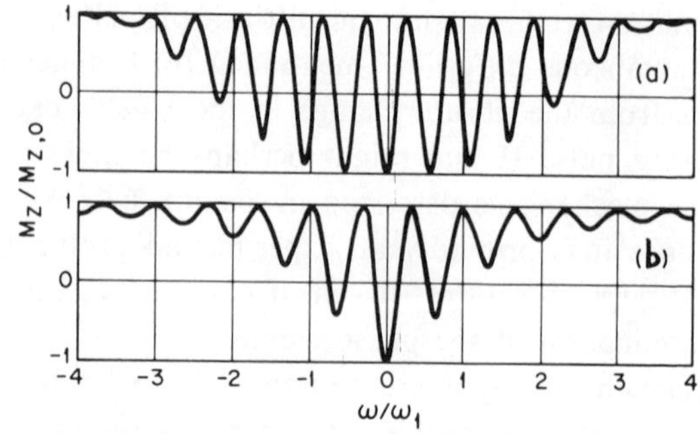

Fig. 6. Pattern of magnetization M_z existing after pulse II of a stimulated echo sequence for the case in which $\omega_1 \ll$ linewidth. (a) $\omega_1 t_{pI} = \omega_1 t_{pII} = \pi/2$, $\tau = 6t_{pI}$; (b) A Lorentzian envelope of half-width ω_1 containing a sinusoidal modulation pattern of period τ shown for comparison. From a comparison of (a) and (b), it is evident that, in the $\omega_1 \ll$ linewidth regime, the period of the modulation is approximately given by $\tau + t_p$, and not by τ, as it would be in the ideal case. The reason for this is suggested in the text shortly after Eq. (127).

pulse duration t_p may be obtained by computing spectral distribution patterns such as those shown in Fig. 6, and integrating the area under each curve. If pulse II is a 120° pulse and the linewidth is $\gg \omega_1$, then $n' \simeq \omega_1 N_{z,\omega}$, where $N_{z,\omega}$ is the spin concentration per unit spectral interval (in radians/sec). (The instantaneous diffusion effect can only be estimated in this simple way if the local dipolar fields due to the spins in a given portion of the line are merely proportional to the height of the line at that point, and are not correlated with spectral position in any other fashion. In the case of a typical strain-broadened line, this seems to be a reasonable assumption to make.)

3.6. Stimulated Echoes and Spectral Diffusion

In the last two subsections, we discussed briefly the connection between spin echo experiments, the concept of spectral diffusion, and the transfer of excitation throughout a resonance line. The spectral diffusion concept has already been invoked as a possible way of interpreting two-pulse echo decay. It is also useful as means of explaining the decay function associated with the three-pulse, or stimulated echoes. As was shown in Section 1.3, the information impressed on the spin packets by the microwave pulses exists s a spectral pattern of magnetization $M_x(\omega) + iM_y(\omega)$ during the two ntervals τ, and is stored as a pattern $M_z(\omega)$ during the interval T. We hall therefore tentatively write the overall echo amplitude decay factor $(2\tau, T)$ as a product $D_\tau(2\tau)D_T(\tau, T)$, where the subscripts refer to the time ntervals τ and T in the three-pulse echo cycle. $D(2\tau)$ describes the decay f phase memory during the two intervals τ, and is essentially the same as ne function $E(2\tau)$, while $D_T(\tau, T)$ describes the erasure of the pattern* $I_z(\omega) = M'(\omega)\cos(\omega\tau)$ [Eq. (7)]. This erasure can of course be a direct ffect of lattice relaxation of the A spins, in which case, $D_T(\tau, T)$ $= \exp(-T/T_1)$. We shall, however, be primarily interested here in indirect ffects, and in particular with the diffusion effect caused by B-spin lattice elaxation. We assume that diffusion behavior is independent of the position f a spin packet in the resonance line, and that the diffusion kernel can be ritten in the form $K(\omega_f - \omega_i, T)$, where ω_i and ω_f are initial and final pin packet frequencies, respectively. The M_z spectrum at the end of the nterval T is therefore given by

$$M_z(\omega_f) = M' \int \cos(\omega_i\tau) K(\omega_f - \omega_i, T)\, d\omega_i \tag{87}$$

etting $\omega' = \omega_f - \omega_i$, we obtain

$$M_z(\omega_f) = M'[\cos\omega_f(\tau)] \int [\cos(\omega'\tau)] K(\omega', T)\, d\omega' \tag{88}$$

hus demonstrating that the decay factor $D_T(\tau, T)$ is the Fourier transform f the diffusion kernel. This result is quite generally true provided that iffusion is a linear process [as is implied by the convolution integral (87)] nd depends on the difference $\omega_f - \omega_i$. The mathematical form of the ernel and the nature of the physical process giving rise to it are then rrelevant (e.g., it does not matter whether $K(\omega', T)$ satisfies the requirements or a homogeneous Markovian function[24] or not).

The diffusion kernel, according to the sudden-jump model for long imes T, is derived in the same manner as the kernel given in Eq. (85). The probability of a B spin being found in the opposite orientation (after ne or more flips) at the end of a time interval T is $\frac{1}{2}(1 - e^{-2RT})$. The hange in local fields is therefore equivalent to that which would be produced y the insertion of $\frac{1}{2}n_B(1 - e^{-2RT})$ new moments of $\pm g_B\beta$ at random

* If the first two pulses are incoherent, the magnetization pattern $M_z(\omega)$ is proportional to $\cos(\omega\tau + \varepsilon)$. This does not effect the diffusion arguments which follow.

points in the lattice,* giving the kernel

$$K(\omega', T) = \frac{\Delta\omega_{\mathrm{dip}}(1 - e^{-2RT})/\pi}{(\omega')^2 + [\Delta\omega_{\mathrm{dip}}(1 - e^{-2RT})]^2} \tag{89}$$

This results in the decay function

$$D_T(\tau, T) = \exp[-\Delta\omega_{\mathrm{dip}}\tau(1 - e^{-2RT})] \tag{90}$$

A diffusion kernel can also be derived according to the Gauss–Markov model, giving

$$K(\omega', T) = \frac{\Delta\omega_{\mathrm{dip}}(1 - e^{-2RT})^{1/2}/\pi}{(\omega')^2 + (\Delta\omega_{\mathrm{dip}})^2(1 - e^{-2RT})} \tag{91}$$

whence

$$D_T(\tau, T) = \exp[-\Delta\omega_{\mathrm{dip}}\tau(1 - e^{-2RT})^{1/2}] \tag{92}$$

The derivation of (71) is not reproduced here, since it follows an already familiar outline and introduces no new concepts.†

Several qualifications must be made regarding these results. First it is important that $t_p \ll \tau$ (under normal pulsing conditions such as those specified in Section 1 or 5). Otherwise, the period $2\pi/\tau$ of the pattern in M_z may be comparable with the spectral width $2\omega_1$ excited by the microwave pulses, and we cannot approximate the pattern by $M' \cos \omega\tau$. Second the A-spin resonance line which is being studied should be characterized by a strain width $\Delta\omega_{\mathrm{st}} \gg \Delta\omega_{\mathrm{dip}}$ in order that a full representative distribution of dipolar fields should be seen by each spin packet as we have assumed in the previous analysis. Third, some consideration should be given to the errors which can arise from spin sorting effects of the type discussed at the end of Section 3.4. Although the decay function $D_T(\tau, T)$ follows quite straightforwardly from the diffusion kernel, the kernels (89) and (91) themselves may not be strictly applicable in the physical situations which exist after pulse II. For example, if there is a considerable amount of spectral diffusion during the first interval τ, it will tend to have eliminated those A spins that have many near B-spin neighbors from the $\cos(\omega_i\tau)$ pattern

* The points at which a net change of orientation has occurred are selected at random from the randomly distributed sites occupied by B spins. Since we are only concerned here with the end result of the spectral diffusion process and not with the sequence of individual steps by which this is reached, we do not encounter the difficulty discussed in Section 3.3 in connection with the derivation of $E(2\tau)$.

† The result is derived by first using a Gaussian diffusion kernel to describe the probable motions of each B spin and then applying the statistical theory of Section 2 to take account of the random distribution of B spins in the lattice. See Ref. 23.

which is established at the beginning of the interval T. Presumably the result of eliminating contributions due to A spins situated in dense B-spin neighborhoods from the echo would be to cause the local field environments to appear more nearly uniform. The spin–spin interactions would then be more or less comparable in magnitude and (by the central limit theorem) we might expect diffusion kernel to approach the Gaussian form.

The diffusion model which we have assumed in the discussion of three-pulse echoes is largely meaningless in the regime where $R\tau \gg 1$. At the end of pulse II, we have, as before, a magnetization spectrum $M_z(\omega) \propto \cos \omega_i \tau$. But the constituent A spins will already have seen the full range of local fields, and during the time T, their Larmor frequencies will merely continue to fluctuate about a mean value, this mean value being determined by the strain perturbations only. Likewise, it does not greatly matter where, in their dipolar excursions, the local fields are overtaken by pulse III, since further fluctuations will occur before there has been time for a significant accumulation of random phase. Ideally, the echo signal intensity should therefore be independent of T in this regime [provided that $T \ll T_1(A)$].

3.7. *The Diffusion of Excitation through a Resonance Line*

Very little is understood regarding the transfer of excitation through an inhomogeneous EPR line and there is no theory enabling one to predict this behavior for many of the commonly studied materials.* In order to illustrate the nature of the problem (and to make it quite clear that our earlier discussion only applies to certain aspects of the process), we shall comment on this problem here and suggest mechanisms by which the transfer may take place in a typical resonance line. Let us consider the following hypothetical case. A sample contains 10^{18} spins cm^{-3} belonging to a species with $S = 1/2$, $g = 2$. The resonance line is strain-broadened with a half-width of 2 G, which corresponds to a frequency half-width $\Delta\omega_{st} = 3.5 \times 10^7$ rad/sec. The phase memory time $T_M = 50$ μsec, $T_1 = 50$ msec, and the spin-flip time $T_f \simeq 2$ msec.† In a CW resonance experiment, only those

* Some detailed studies of this kind have been made by Grant in a sequence of four papers.[26] These papers deal with various aspects of the cross-relaxation problem, but do not discuss the influence of strain broadening on the results.

† At low temperatures, T_M is often controlled by T_f. As the temperature is raised, T_1 begins to play a more important role. A measurement of T_1 at the point where both spin–spin and spin–lattice flips of the B spins appear to be equally important may be used to estimate T_f.

spins within a spectral interval defined by the homogeneous half-widt $\Delta\omega_h$ (or ω_1 if this is greater*) are directly driven by the microwave field Setting $\Delta\omega_h = 1/T_M$, we see that this accounts for only one spin in 2000 The remainder must acquire excitation by means of some diffusion o energy-transfer process.

The diffusion of dipolar fields discussed earlier in this section is quit inadequate for this purpose. In an infinite time, this would only result i burning out a Lorentzian hole with a half-width $\Delta\omega_{\text{dip}} = 8.3 \times 10^5$ rad/se [see Eq. (31)], accounting for only 2.4% of the linewidth. To achieve wider transfer of excitation, some additional mechanism, not previousl discussed here, would be required. The most likely process seems to be th mutual flip-flop of two spins characterized by different strain perturbation but temporarily brought into resonance with one another by the dipola fields. The second spin would then be able to transfer energy across further interval $\sim\Delta\omega_{\text{dip}}$, and in a series of steps of this kind, excitatio could diffuse across the line. The rate at which excitation is transferred wil in turn depend on how these successive steps are related to one another At first sight, it might seem that these steps could best be represented as typical random walk. The total diffusion distance after n_d steps would the be $\sim(n_d)^{1/2}\,\Delta\omega_{\text{dip}}$. We must remember, however, that in a randomly dilute paramagnetic material, these steps will vary greatly in size according to the variations in the dipolar fields seen by each spin. As we have notec earlier, a situation of this kind is sometimes better represented by a Lorentzian diffusion function with a width increasing linearly with time.† If this is an appropriate description, excitation should be pictured as spreading throughout the line more in the manner of a relay race than a random walk, and the diffusion distance will be $\sim n_d\,\Delta\omega_{\text{dip}}$ In the particular case discussed here, $T_1 = 25T_f$. This is therefore time for 25 steps (assuming that flip-flop transfers and local field diffusion takes place concurrently) before lattice relaxation erases the diffusion effect. The fraction of the line covered in T_1 would be 12% according to the random-walk model and 60% according to the relay-race model.

* The homogeneous width $\Delta\omega_h$ is usually greater than ω_1 under normal Cw resonance conditions. If, for example, we set $T_2 = T_M$ and assume that the spin packets on resonance are half saturated, we obtain $\omega_1 = 2/(T_1 T_2)^{1/2} = 0.6 \times 10^3$ rad/sec in the present case.

† This can be rationalized by assuming that the transfer of excitation from the first spin to the second spin occurs when the second sees a zero dipolar field. The diffusion kernel corresponding to a two-step transfer would then be given by the convolution of two Lorentzian functions of half-width $\Delta\omega_{\text{dip}}$ each. n similar steps would give a Lorentzian distribution $n\,\Delta\omega_{\text{dip}}$ wide.

Not enough is known about the behavior of actual resonance lines to say whether either one of the above models comes reasonably close to representing the facts. It has been known for some time that "holes" can be burned in certain resonance lines,[27] and that they will persist for times of the order of T_1, but the data are more or less qualitative in nature and such parameters as the spin-flip rate are not known. Detailed studies of spectral diffusion have been made,[22,28] but these studies concern a different group of materials and are of limited value since they cover a spectral range $< \Delta\omega_{\text{dip}}$. The experimental results probably relate to the local field diffusion process without flip-flop transfer. It is to be hoped that more complete studies will eventually be made and will lead to a better understanding of the mechanisms governing energy transfer through resonance lines.

4. THE NUCLEAR MODULATION EFFECT

4.1. Two-Pulse Echoes

In many materials, the envelope of echoes shows a periodic modulation at frequencies which are identifiably related to the nuclear resonance frequencies of nuclei present in the host crystal.[22,29,30] We shall call this phenomenon the "nuclear modulation effect." An illustration of it is shown in Fig. 7. The effect arises from terms of the form S_zI_x and S_zI_y in the Hamiltonian describing the interaction of the electron spins with these nuclei in their immediate neighborhood. We can understand the effect qualitatively in the following manner. The reorientation of the electron spins during pulses I and II is a sudden event on the time scale of nuclear precession. At a certain instant in time, the surrounding nuclei therefore experience an abrupt reorientation of the local magnetic field. Free nuclear precession ensues. This then leads to a periodic change in the local field seen by the electron and gives rise to a variation in intensity of the echo signal.

A picture of this kind is useful for presenting the general features of the situation, but it will not, in this case, serve as an adequate description. Fortunately, however, it is easy to give an account of the situation in quantum mechanical terms by means of the density matrix formalism (see Section 1.4). We consider the sample to consist of an ensemble of isolated electron–nucleus systems. At first, we restrict discussion to systems consisting of an electron with an effective spin $S' = 1/2$ interacting with one adjacent nucleus of spin 1/2. The Hamiltonian for the jth spin packet may

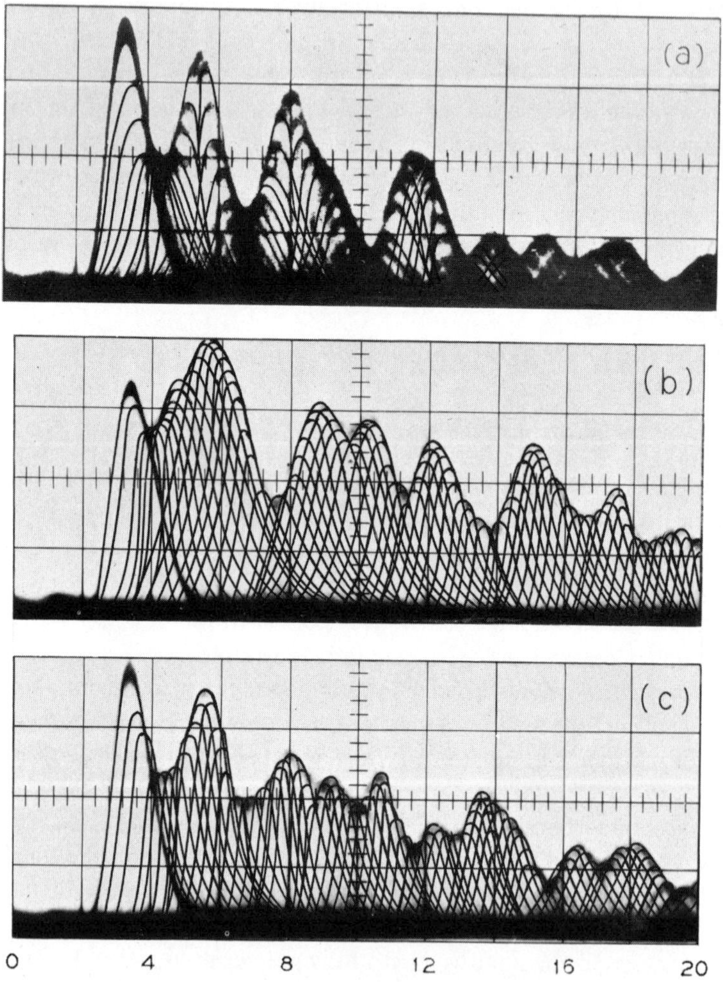

Fig. 7. Echo envelopes illustrating the nuclear modulation effect for Ce^{3+} in $CaWO_4$. The nucleus responsible for the modulation is ^{183}W. The oscilloscope is triggered at the end of pulse II. The resonance frequency is 6.7 GHz. Envelopes (a), (b), (c) were obtained with H_0 at 0°, 10°, 20° to the c axis, respectively.

e written in the form [compare Eq. (11)]

$$\mathscr{H}_T = \mathscr{H}_s + \Delta\mathscr{H}_j - \beta_I g_I \mathbf{H}_0 \cdot \mathbf{I} + \hbar \mathbf{I} \cdot \alpha \cdot \mathbf{S} \tag{93}$$

vhere $\mathscr{H}_s$ is the electron spin Hamiltonian*

$$\mathscr{H}_s = +\beta \mathbf{H}_0 \cdot \mathsf{g} \cdot \mathbf{S} \tag{94}$$

and β_I are the electron and nuclear Bohr magnetons, g_I is the nuclear g actor, g is the electron g tensor, and α describes the tensor coupling between he electron and nucleus.†

As in the simple two-level case treated in Section 1.4, we can use a epresentation consisting of eigenstates of‡ $\mathscr{H}_T$ to describe the evolution of he system during the spin echo cycle of events. These eigenstates are conected by the nutation operators R_{I} and R_{II} and accumulate phase as pecified by the free-precession operators R_τ. This is the essence of the roblem. The nutation operators simultaneously connect each of the $M_S = \pm 1/2$ states and thus establish a coherent relationship between all our eigenfunctions (see Fig. 8). The time evolution of these four eigenunctions then leads to interference effects which are manifested in the cho envelope. In order to preserve these general features of the situation luring the analysis (and incidentally to simplify the actual calculation), ve shall avoid specifying the exact constitution of the eigenstates of $\mathscr{H}_T$ n terms of M_s substates, etc. Instead, we merely write down parameters u and v which specify relationships between the transition matrix elements $\langle a | S_x | c \rangle$, $\langle a | S_x | d \rangle$, etc., which enter into the calculation of the nutation operators. We can then reduce the problem of finding the product of a

* The argument given here can readily be extended to include more general forms of $\mathscr{H}_s$ containing crystal field terms, etc. In many instances, this does not introduce any further complications, since only one interval belonging to the level scheme of $\mathscr{H}_s$ is in resonance with the microwave field, and only those nuclear hfs states that are associated with this transition need be considered in the density matrix calculation. If this is not the case, and if two intervals with a level in common are close enough to be simultaneously excited by the microwave pulses, further contributions to the envelope modulation function may be expected on this account.

† Coupling between electron spins and the nearest-neighbor nuclei are often predominantly due to forces other than the classical dipolar interaction. The general tensor interaction will have the form α = T · g, where T is a symmetric tensor.

‡ A representation in eigenstates of $\hbar\omega_s S_z + \hbar\omega_I I_z$ was used in making the calculation in Ref. 30. In this representation, a number of transformations are needed in order to simplify the operator expressions for R_τ.

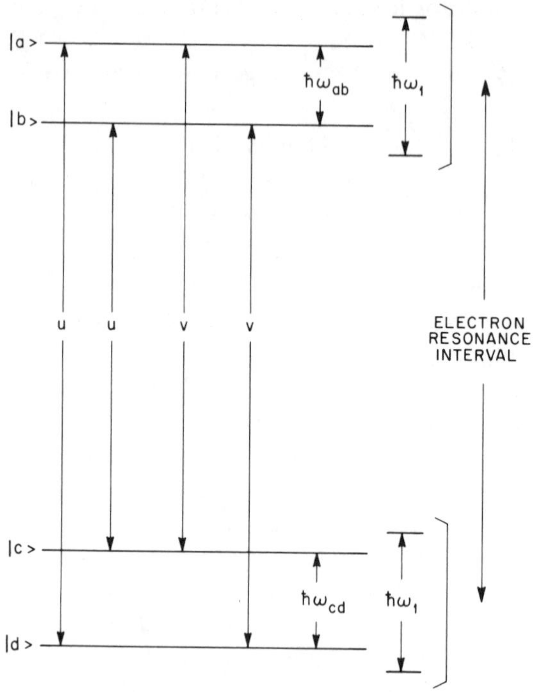

Fig. 8. Four-level scheme for coupled electron spin $S = \frac{1}{2}$ and nuclear spin $I = \frac{1}{2}$. The quantities u and v represent elements appearing in the unitary matrices which describe the evolution of the system during 90° and 180° pulses. It is important that the microwave field should be intense enough so that $\omega_1 \gg \omega_{ab}, \omega_{cd}$; otherwise, the coherent excitation of the four-level system which is implied by the nutation matrices (96) and (97) will not occur.

number of 4×4 matrices to a submatrix calculation involving the 2×2 unitary submatrix*

$$U = \begin{bmatrix} v & u \\ -u^* & v^* \end{bmatrix} \tag{95}$$

where U takes care of the branching of the transitions induced by the 90° and 180° pulses (see Fig. 8). For example, the 180° nutation matrix R_{I}

* U must be unitary in order to give unitary matrices for R_{I}, R_{II}. The choice of phase factor in the element u corresponds to the selection of an axis (in the xy plane) for the nutation which takes place during the pulse.

consists of a combination of U with the 180° operator $\exp(-i\sigma_y\pi/2) = -\sigma_y$ given in Eq. (18b). We have

$$R_{\text{II}} = \begin{bmatrix} 0 & U \\ -U^+ & 0 \end{bmatrix} = \begin{array}{c} \begin{matrix} |a\rangle & |b\rangle & |c\rangle & |d\rangle \end{matrix} \\ \begin{bmatrix} 0 & 0 & v & u \\ 0 & 0 & -u^* & v^* \\ -v^* & u & 0 & 0 \\ -u^* & -v & 0 & 0 \end{bmatrix} \end{array} \begin{matrix} |a\rangle \\ |b\rangle \\ |c\rangle \\ |d\rangle \end{matrix} \tag{96}$$

Elements such as $\langle a | R_{\text{II}} | a\rangle$ are zero, since we assume that the "180°" pulse will continue until systems initially in state $|a\rangle$ have completed a transition to both or either of states $|c\rangle$, $|d\rangle$ under the influence of the microwave field.* No elements of the type $\langle a | R_{\text{II}} | b\rangle$ appear, since this interval is not in resonance. We obtain matrices for the operator R_I and M_+ in a like manner by combining **U** with $(1/\sqrt{2})(I - i\sigma_y)$ and with $\frac{1}{2}g\beta(\sigma_x + i\sigma_y)$. Thus we have

$$R_{\text{I}} = 1/\sqrt{2} \begin{bmatrix} I & U \\ -U^+ & I \end{bmatrix} \tag{97}$$

$$M_+ = g\beta \begin{bmatrix} 0 & U \\ 0 & 0 \end{bmatrix} \tag{98}$$

The matrix R_τ consists, as in the two-level case, of diagonal elements $e^{-i\lambda_i\tau}$, where $\hbar\lambda_i$ denotes the eigenvalues of $\mathscr{H}_T - \mathscr{H}_s$. To conform with the other matrix operators, we write it in terms of submatrices as

$$R_\tau = \begin{bmatrix} P & 0 \\ 0 & Q \end{bmatrix} \tag{99}$$

where

$$P = \begin{bmatrix} e^{-i\lambda_a\tau} & 0 \\ 0 & e^{-i\lambda_b\tau} \end{bmatrix} e^{-i\lambda_+\tau} \tag{99a}$$

and

$$Q = \begin{bmatrix} e^{-i\lambda_c\tau} & 0 \\ 0 & e^{-i\lambda_d\tau} \end{bmatrix} e^{-i\lambda_-\tau} \tag{99b}$$

The quantities $\lambda_\pm$ in the factors $e^{-i\lambda_\pm\tau}$ are the matrix elements of $\Delta\mathscr{H}_j$. They represent the inhomogeneous broadening of the resonance line, $\lambda_+ - \lambda_-$ being the frequency ω_j which describes the precession of the jth

* This is not an essential condition, but it simplifies the working in much the same way as the assumption of 90°–180° pulses simplified the classical description in Section 1.

spin packet in the rotating coordinate system. Provided that $\Delta\mathscr{H}_j$ does not mix the hfs states, the $e^{-i\lambda_{\pm}\tau}$ appear as simple scalar factors in P and Q and enter the calculation in much the same way as in Section 1.4. Since we are assuming here that ideal 90°–180° pulsing conditions obtain ($\omega_1 \gg$ linewidth, etc.), these factors will not appear in the nutation operators $R_{\rm I}$, $R_{\rm II}$. The initial density ϱ_0 is obtained by expanding

$$\exp(-\mathscr{H}_T/kT)/\text{Tr}[\exp(-\mathscr{H}_T/kT)]$$

Dropping the unit matrix as in Section 1.4 and approximating $H_T \sim H_s$, we have

$$\varrho_0 = \alpha\begin{bmatrix} I & 0 \\ 0 & -I \end{bmatrix} \tag{100}$$

where α is the constant given in (15).

The 4×4 matrices (96)–(100), expressed in the submatrix form, may now be used to calculate the spin echo signal. Evaluating

$$\text{Tr}(\varrho_E M_+)_j = \text{Tr}_j(R_{\tau+t}R_{\rm II}R_\tau R_{\rm I}\varrho_0 R_{\rm I}^{-1}R_\tau^{-1}R_{\rm II}^{-1}R_{\tau+t}^{-1}M_+) \tag{101}$$

we find that the precessing moment due to the jth spin packet

$$M_{p,j} = M_{p,j}^{(0)} \exp(i\omega_j t)$$

where

$$M_{p,j}^{(0)} = -g\beta\alpha_j \times \text{Tr}(QU^+PUQ^+U^+P^+U) \tag{102}$$

and is the value of $\text{Tr}(\varrho_E M_+)_j$ obtained by ignoring $\Delta\mathscr{H}_j$. To obtain the precessing magnetization due to the whole spin ensemble, we take the sum

$$M_p = \sum_j M_{pj} = \sum_j M_{p,j}^{(0)} \exp(+i\omega_j t)$$

If the frequencies $\lambda_a - \lambda_b$, etc. contained in $M_{p,j}^{(0)}$ are small compared with the range of values ω_j (i.e., if the period of envelope modulation is long compared with the duration of the echo), we can express the result as the product of two terms, one describing the echo pulse waveform, and one describing the modulation of the envelope. Thus

$$M_p = \left[\sum_j (\text{Tr}\ \varrho_E M_+)_{0,j} \exp(i\omega_j t)\right] E(\text{mod}, \tau) \tag{103}$$

where $\text{Tr}(\varrho_E M_+)_{0,j}$ is the precessing magnetization of the jth spin packet calculated by eliminating branching transitions* in Fig. 8 [i.e., setting $U = I$ in (102)]. The first factor in (103) is, of course, the precessing magnetization calculated in Section 1.4 for a simple two-level system (i.e., the Fourier transform of the inhomogeneous line shape). The second factor

$$E(\text{mod}, \tau) = \text{Tr}(\varrho_E M_+)_j / \text{Tr}(\varrho_E M_+)_{0,j} \quad (104)$$

describes the effect of the branching transitions on the echo envelope† and is given by $\frac{1}{4}\text{Tr}(QU^+PUQ^+U^+P^+U)$, where $\Delta\mathscr{H}_j$ is omitted in finding the product of the submatrices.

The evaluation of the trace in (102) and the derivation of the echo envelope factor is straightforward, if somewhat tedious. We find that

$$E(\text{mod}, \tau) = |v|^4 + |u|^4 + |v|^2 |u|^2 [2\cos(\omega_{ab}\tau) + 2\cos(\omega_{cd}\tau) - \cos(\omega_{ab} - \omega_{cd})\tau - \cos(\omega_{ab} + \omega_{cd})\tau] \quad (105)$$

where ω_{ab} and ω_{cd} are the frequencies $(\lambda_a - \lambda_b)$ and $(\lambda_c - \lambda_d)$ (i.e., the ENDOR frequencies in the two nuclear intervals‡). Equation (105) may also be written in the form

$$E(\text{mod}, \tau) = 1 - 2\sin^2 2\psi \sin^2(\omega_{ab}\tau/2) \sin^2(\omega_{cd}\tau/2) \quad (106)$$

where $\tan^2\psi = |u|^2/|v|^2$ is the ratio of the two branching transition probabilities in the four-level scheme of Fig. 8.

* For want of a better term, we use the word "branching" here to denote the fact that each of the levels in Fig. 8 has transition matrix elements to two other levels.

† In practice, if the microwave pulses are not sufficiently narrow, the echo waveform will last long enough to show some of the structure belonging to $E(\text{mod}, \tau)$. Treatment of this case involves more than merely making a careful summation of the $M_{p,j}$ in (102). If $t_p \sim 1/\omega_1 \sim$ the envelope modulation period, then it is also likely that H_1 is too small to operate in an equivalent manner on all the intervals in the hfs level scheme, and the operators R_I and R_II do not have the simple forms which were assumed in deriving (102).

‡ The way in which the ENDOR frequencies enter the result may be understood by noting that, when branching transition such as those shown in Fig. 8 can occur, the wave function after the 90° pulse will consist of a coherent admixture of all four states. The 180° pulse produces a new admixture. However, because of branching, the phase differences accumulated in the first and second intervals will not cancel, as they do in the simple two-level case. Instead, the various possible combinations as in (105) appear. (If no branching takes place, the splitting of two levels into four by the nuclear interaction is equivalent to inhomogeneous broadening. The effects of this, which could in principle be observed in the echo waveform, will normally be masked by strains and other static broadening mechanisms.)

As an illustration of the general result above, let us consider the Hamiltonian $\mathscr{H}_T$ from Eq. (93). Eliminating the inhomogeneous broadening $\Delta\mathscr{H}_j$, we have

$$\mathscr{H}_T = g_\perp\beta(H_xS_x + H_yS_y) + g_\parallel\beta H_zS_z - g_n\beta_n(H_xI_x + H_yI_y + H_zI_z) + \hbar\sum_{ij}\alpha_{ij}S_iI_j \tag{107}$$

where $i, j = x, y, z$. A Hamiltonian of this form describes, for example, the interaction between a Ce^{3+} spin in $CaWO_4$ and a neighboring ^{183}W nuclear moment.* In order to calculate the transition probabilities and the ENDOR frequencies, we make two coordinate transformations. First, we rotate the reference axes of I and S, reducing (107) to the form

$$\mathscr{H}_T'/\hbar = \omega_sS_z + \omega_II_z + \sum_{ij}\alpha'_{ij}S_iI_j \tag{108}$$

where $\omega_s = g\beta H_0/\hbar$, $\omega_I = -g_n\beta_nH_0/\hbar$, $g = (g_\parallel{}^2 + g_\perp{}^2)^{1/2}$, and α'_{ij} is the transform of the tensor α_{ij}. Next, we note that all terms in $\alpha'_{ij}S_iI_j$ except $\alpha'_{zz}S_zI_z$, $\alpha'_{zx}S_zI_x$, and $\alpha'_{zy}S_zI_y$ can be discarded on account of the large disparity between the electron Zeeman energy and the electron–nuclear coupling. A further rotation then gives the Hamiltonian

$$\mathscr{H}_T''/\hbar = \omega_sS_z + \omega_II_z + AS_zI_z + BS_zI_x \tag{109}$$

Eigenfunctions of $\mathscr{H}_T''$ are derived as linear combinations of eigenfunctions of the Hamiltonian $\omega_sS_z + \omega_II_z$. The four states are products of pure S_z states and mixed I_z states, the mixing of the latter being brought about by the term BS_zI_x. They are constituted as follows (see Fig. 8 for labeling):

$$\begin{aligned}
|a\rangle &= [\cos(\eta/2)]\,|++\rangle + [\sin(\eta/2)]\,|+-\rangle \\
|b\rangle &= -[\sin(\eta/2)]\,|++\rangle + [\cos(\eta/2)]\,|+-\rangle \\
|c\rangle &= [\cos(\zeta/2)]\,|-+\rangle + [\sin(\zeta/2)]\,|--\rangle \\
|d\rangle &= -[\sin(\zeta/2)]\,|-+\rangle + [\cos(\zeta/2)]\,|--\rangle
\end{aligned} \tag{110}$$

$|++\rangle$ denotes $M_s = +1/2$, $M_I = +1/2$, etc.; $\tan\eta = B/(A + 2\omega_I)$; and $\tan\zeta = B/(A - 2\omega_I)$. The branching parameters u and v in the nutation matrices can be derived by evaluating matrix elements of $\exp(-i\sigma_y\pi/2) = -i\sigma_y$, where σ_y operates on the electron spin only. We thus have

* This case is discussed in detail in Ref. 30. The electron–nuclear interaction parameters α_{ij} have been measured for this system by an ENDOR method. See Mims.[31]

$u = \sin[\frac{1}{2}(\eta - \zeta)]$ and $v = \cos[\frac{1}{2}(\eta - \zeta)]$. The angle 2ψ in Eq. (106) is $(\eta - \zeta)$ and the two ENDOR frequencies are

$$\begin{aligned} \omega_{ab} &= \{(B/2)^2 + [(A/2) + \omega_I]^2\}^{1/2} \\ \omega_{cd} &= \{(B/2)^2 + [(A/2) - \omega_I]^2\}^{1/2} \end{aligned} \tag{111}$$

Substituting these values in (106), we obtain

$$E(\text{mod}, \tau) = 1 - (2\omega_I{}^2B^2/\omega_{ab}^2\omega_{cd}^2) \sin^2(\omega_{ab}\tau/2) \sin^2(\omega_{cd}\tau/2) \tag{112}$$

Some calculated envelopes for the Ce^{3+} in $CaWO_4$ system are shown in Fig. 9. The modulation depth decreases for very high and very low Zeeman fields. For the low-field case, this can be seen at once by setting $\omega_I = 0$ in (112). For the opposite case of $\omega_I \gg A, B$, Eq. (112) reduces to

$$E(\text{mod}, \tau) = 1 - 2(B/\omega_I)^2 \sin^4(\omega_I\tau/2) \tag{113}$$

The disappearance of the modulation effect in these limiting cases results, of course, from the fact that nutation in the microwave field no longer induces branching, since two out of the four microwave transitions are forbidden.

4.2. Comments on More Complex Situations

In the materials which are commonly studied, a number of circumstances may arise to complicate the analysis further. If the nuclear neighbors belong to two or more isotopic species, the resultant echo signal will be given by the weighted sum of contributions due to each type of electron–nucleus pair. For example, if there is only one isotope with a nuclear magnetic moment, and this is present in a proportion x, the precessing magnetization

$$M_p = xM_p(I = I_1) + (1 - x)M_p(I = I_2 = 0) \tag{114}$$

whence it can readily be inferred that

$$E(\text{mod}, \tau) = 1 - 2x \sin^2 2\psi \sin^2(\omega_{ab}\tau/2) \sin^2(\omega_{cd}\tau/2) \tag{115}$$

The case of $I > 1/2$ is more complicated and will not be discussed here. However, the calculation follows similar lines and the same type of behavior is to be expected. The submatrices U, P, Q will be of larger dimensions [e.g., 3×3 for $(I = 1)$] but the result (102) will still hold. The ENDOR

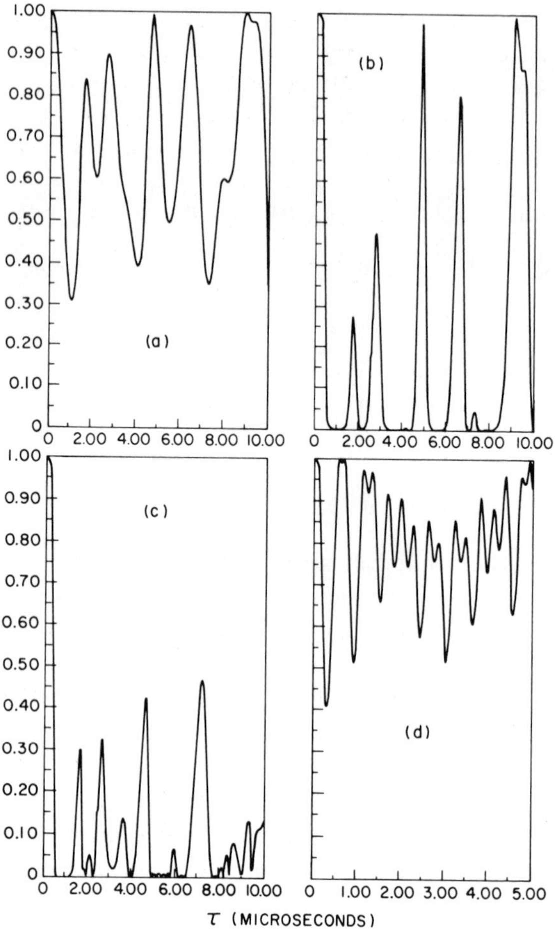

Fig. 9. Nuclear modulation envelopes calculated for Ce in $CaWO_4$ using the electron–nuclear coupling parameters given in Ref. 31. Absolute magnitudes of the echo signal are plotted. Times are measured from pulse II. (a) $\theta = 0$, $H_0 = 2.3$ kG (X-band resonance), normal 15% concentration of ^{183}W. (b) Effect of making 100% enrichment of the ^{183}W isotope in case (a). (c) $\theta = 20$, $H_0 = 2.43$ kG, with 100% enrichment of ^{183}W. Case (c) typifies the echo envelopes which are observed when a number of nonequivalent nuclei are responsible for the modulation and when mixing of the hfs states is appreciable. (The echo signal in this case undergoes phase reversals.) (d) $\theta = 0$, $H_0 = 9.2$ kG ($\simeq$38 GHz resonance) with 100% enrichment of ^{183}W. The effect of increasing the Zeeman field, and thus reducing the mixing of hfs states, may be seen by comparing this with case (b).

frequencies and their combinations will appear in the echo envelope, admixtures of different components depending on the branching probabilities between the $M_s = +1/2$ and $M_s = -1/2$ manifolds of hfs levels.

An increase in the complexity of the hfs level scheme and hence an increase in the dimension of R_{I}, R_{II}, etc. will also result if the electron has more than one near-neighbor nucleus. Fortunately, in this particular case, the matrix operators can usually be factored out as shown in the following argument. Let us write the Hamiltonian (93) (deleting $\Delta\mathscr{H}_j$) in the form $\mathscr{H}_T = \mathscr{H}_{s,I_1} = \mathscr{H}_s + \mathscr{H}_{I_1}$, and assume that the effect of adding a second nucleus to the system is to add a term

$$\mathscr{H}_{I_2} = -\beta_{I_2} g_{I_2} \mathbf{H}_0 \cdot \mathbf{I}_2 + \hbar \mathbf{I}_2 \cdot \alpha \cdot \mathbf{S}$$

giving a total Hamiltonian

$$\mathscr{H}_T = \mathscr{H}_{s,I_1,I_2} = \mathscr{H}_s + \mathscr{H}_{I_1} + \mathscr{H}_{I_2}$$

The precessing magnetization is derived from a matrix product as in (102), where the matrices P, Q, U have the dimension 4×4 (if $I_1 = I_2 = 1/2$). The eigenstates of $\mathscr{H}_{s,I_1,I_2}$ in terms of which the matrix elements must be evaluated bear a simple relation to the eigenstates of $\mathscr{H}_{s,I_1}$ which were used in the preceding calculation. Suppose we express them as linear combinations of the simultaneous eigenstates of $\mathscr{H}_s$, $g_{I_1}\beta_{I_1}\mathbf{H}_0 \cdot \mathbf{I}_1$ and $g_{I_2}\beta_{I_2}\mathbf{H}_0 \cdot \mathbf{I}_2$. Since $\mathscr{H}_s, \mathscr{H}_{I_1}, \mathscr{H}_{I_2}$ effectively commute with one another, the states of nucleus 1 are mixed by the term $\hbar\mathbf{I}_1 \cdot \alpha_1 \cdot \mathbf{S}$ only* and the states of nucleus 2 by $\hbar\mathbf{I}_2 \cdot \alpha_2 \cdot \mathbf{S}$ only. It is therefore easy to see that the matrices consist of the tensor products

$$P = P_1 \times P_2, \qquad Q = Q_1 \times Q_2, \qquad U = U_1 \times U_2$$

where P_1, P_2, etc. are derived from states involving nucleus 1 or nucleus 2 only. The matrix $W = QU^+PUQ^+U^+P^+U$ is likewise a tensor product $W_1 \times W_2$, and by the theorem[32] $\text{Tr}\, W = (\text{Tr}_1\, W_1) \times (\text{Tr}_2\, W_2)$, we infer that

$$E(\text{mod}, \tau)_{I_1,I_2} = E(\text{mod}, \tau)_{I_1} \times E(\text{mod}, \tau)_{I_2}$$

* Since $\mathscr{H}_s \gg \mathscr{H}_I$, terms in $\hbar\mathbf{I} \cdot \alpha \cdot \mathbf{S}$ which mix opposite electron spin states can be ignored. Hence, there is no second-order mixing of the I_1 and I_2 states via the electron coupling. Direct coupling of the two nuclei has been discounted on the grounds that the coupling energies would be small compared with the ENDOR interval separations $\hbar\omega_{ab}$, $\hbar\omega_{cd}$, etc. This assumption may not be justifiable if there are degeneracies in the hfs level scheme, as, e.g., when a number of like nuclei are arranged symmetrically about the electron.

Generalizing to N neighbors, this expression becomes

$$E(\text{mod}, \tau)_N = \prod_i E(\text{mod}, \tau)_i \tag{116}$$

From Eqs. (115) and (116), it follows that the echo envelope will contain appreciable components at comparatively high frequencies (corresponding to transition sequences linking widely separated hfs levels) if the nuclear Zeeman states are strongly mixed by the electronic interaction terms. However, some of these components are likely to be missing in an actual experimental situation. ω_1 may be too small to span the required intervals, and the microwave pulses may therefore be unable to excite all the transitions in accordance with the probabilities which are implicitly assumed when forming combinations such as (96)–(98) with the hfs submatrices U. The condition for full excitation is that $\omega_1 > \omega_{\text{hfs}}$. For practical purposes, this is more or less equivalent to saying that high ENDOR frequencies can be observed in the modulation envelope if these frequencies are present in the Fourier spectrum of the applied microwave pulses.* Arguments based on the Fourier spectrum of the pulses are not strictly correct, however, when strongly nonlinear interactions such as those required to generate spin echoes are involved.

In host lattices where there are a number of nuclei in the vicinity of the electron, all having interaction parameters of the same general magnitude, the modulation envelope may contain so many periods that it simulates an echo envelope decay function. This is liable to happen at intermediate Zeeman fields, where state mixing is greatest, the modulation is deep, and the individual factors in (116) are undergoing reversals of sign. The electron spin echo envelope may then appear to be the decay envelope associated with a homogeneous spin packet width of $\sim\omega_{\text{hfs}}$. There are, of course, a number of ways in which this situation differs from that which gives rise to true homogeneous broadening and to the associated echo envelope decay. Since the number of interacting spins is relatively small, the echo signal may reappear for certain times τ. It may also be possible to lengthen the decay time if it is due to this cause by using a smaller H_1 (thus exciting a smaller band of hfs transitions) or to change the decay function drastically by working at high Zeeman fields. Yet another property may sometimes help to distinguish echo envelope decay due to nuclear modulation effects from true homogeneous broadening. If the surrounding nuclei are equi-

* In a typical echo experiment, $\omega_1 t_p = 2\pi/3$ (see Section 5) and the requirement therefore becomes $\omega_{\text{hfs}} < 2/t_p$.

'alent to one another in pairs, it may be seen from Eq. (116) that $E(\text{mod}, \tau)_N$ s positive-definite, and is therefore not likely to result in cancellation of the ·cho signal for an extended period of time. For example, if the electron pins are at sites with a C_2 point symmetry and only one kind of neighboring ıucleus is involved, echoes may be seen with H_0 aligned along the c axis ·ven when they are invisible at other orientations.

4.3. Experimental Application of the Modulation Effect. The Effect in a Stimulated Echo Sequence

The nuclear modulation effect can in principle be used to measure :lectron–nuclear interactions. Yudanov *et al.*[33] have in this way investigated he proton–electron coupling in a single crystal containing the radical CH(COOH)_2. In many situations, however, the analysis of the echo en- ·elope is inconvenient and the results are too inaccurate to be very much ·alue. As may be seen from Eq. (105), the ENDOR frequencies and their ıums and differences all appear together in the envelope. Moreover, the :otal duration of the measurement is limited to a time of the order of T_M, and an uncertainty equivalent to the electron spin packet width is thus ntroduced into the nuclear frequency determinations. Better results can ısually be obtained by means of double resonance techniques.

Some of the more obvious disadvantages of the echo envelope method can be eliminated by using a stimulated echo sequence in place of the simple two-pulse sequence. The same submatrix formalism can be used to compute the result. Let us assume an ideal 90°–90°–90° pulse sequence. At the end of pulse II, the density is then given by

$$\begin{aligned}\varrho_{\text{II}} &= R_{\text{I}} R_\tau R_{\text{I}} \varrho(0) R_{\text{I}}^{-1} R_\tau^{-1} R_{\text{I}}^{-1} \\ &= \frac{1}{2}\begin{bmatrix} (PUQ^+U^+ + UQU^+P^+) & (PUQ^+ - UQU^+P^+U) \\ (QU^+P^+ - U^+PUQ^+U^+) & (-U^+PUQ^+ - QU^+P^+U) \end{bmatrix}\end{aligned} \tag{117}$$

It is not necessary to carry all of (117) through the remainder of the calculation. To see this, let us, as before, denote the shift off resonance of the jth spin packet due to inhomogeneous line broadening by λ_j. The off-diagonal elements then evolve with phase factors $\sim\exp(i\lambda_j T)$ during the interval T after pulse II. This phase factor cannot be cancelled during the interval τ which follows pulse III. Therefore, because of the spread in the λ_j, the off-diagonal elements of (117) make no resultant contribution to the stimulated echo and can be dropped. Of course, the submatrices appearing on the diagonal of (117) contain in themselves some off-diagonal elements.

These elements cause the precessing magnetization to evolve with phase factors of the form $\exp(i\omega_{ab}T)$ and $\exp(i\omega_{cd}T)$ which are not cancelled in the final interval τ and which are the source of the nuclear modulation effect. We have assumed hitherto that there will be no spread in the values of the ENDOR frequencies ω_{ab} and ω_{cd}. If there are significant spreads $\Delta\omega_{ab}$ and $\Delta\omega_{cd}$ in these intervals, the relevant modulation patterns in the stimulated echo will be erased in times $T \lesssim 1/\Delta\omega_{ab}$, $1/\Delta\omega_{cd}$.

Returning to our calculation, we find that the stimulated echo signal is given by $\mathrm{Tr}(\varrho_{\mathrm{stim}}M_+)$, where

$$\varrho_{\mathrm{stim}} = R_\tau R_{\mathrm{I}} R_T \varrho'_{\mathrm{II}} R_T^{-1} R_{\mathrm{I}}^{-1} R_\tau^{-1}$$

ϱ'_{II} is the matrix obtained by deleting the off-diagonal entries in (117), and R_T has the same form as (99) with T substituted for τ. $\mathrm{Trace}(\varrho_{\mathrm{stim}}M_+)$ is proportional to $\mathrm{Tr}(QMU^+P^+U + QU^+NP^+U)$, where $N = P_T(UQU^+P^+ + PUQ^+U^+)P_T^+$ and $M = Q_T(QU^+P^+U + U^+PUQ^+)Q_T^+$, and P_T and Q_T are analogs of (99a) and (99b). Substituting for the submatrices, and taking the trace, we find that the stimulated echo is proportional to

$$\begin{aligned} E(\mathrm{mod}, \tau, T) = |v|^4 + |u|^4 + |u|^2|v|^2\{&\cos(\omega_{ab}\tau) + \cos(\omega_{cd}\tau) \\ &+2\sin^2(\tfrac{1}{2}\omega_{cd}\tau)\cos[\omega_{ab}(\tau + T)] \\ &+2\sin^2(\tfrac{1}{2}\omega_{ab}\tau)\cos[\omega_{cd}(\tau + T)]\} \end{aligned} \tag{118}$$

Let us now consider an experiment in which τ is fixed and the echo envelope is obtained as a function of T. From (118), we see that the envelope contains only the ENDOR frequencies ω_{ab} and ω_{cd} themselves and not their sums and differences. Moreover, by making a suitable choice of τ, one or the other of these frequency components can be suppressed. The times T available in a stimulated echo experiment are normally much longer than T_M, and the accuracy of ω_{ab} or ω_{cd} is not directly limited by the homogeneous spin packet width of the electron transition. The widths (inhomogeneous or otherwise) of the intervals ω_{ab} and ω_{cd} will of course eventually lead to a randomization of the phase factors $\exp(i\omega_{ab}T)$ and $\exp(i\omega_{cd}T)$ which enter into the calculation, but this merely means that the ENDOR linewidths limit the accuracy with which we can determine the ENDOR frequencies.

4.4. Modulation Effects in Other Physical Systems

It should be emphasized that the envelope modulation effect is not peculiar to electron–nucleus systems, but is a general interference property, which may occur in any multilevel system when branching transitions are excited by the applied pulses. A phenomenon of this type is observed* in NMR when pairs of nuclei in organic molecules are coupled by a term $JI_1 \cdot I_2$, thus forming four-level systems with branching from the outer levels to the middle ones. Similar arguments could also be applied in the case of optical transitions in any material for which multilevel systems, characterized by branching and susceptible to coherent excitation, exist.†

5. EXPERIMENTAL CRITERIA: PULSE POWER, ECHO WAVEFORMS, DETECTION SENSITIVITY

At the end of this chapter, in Section 6, we shall suggest some ways in which electron spin echo methods can be applied for the purpose of making measurements in solid-state physics. As a preliminary, it may be useful to consider the experimental problems which arise in setting up an electron spin echo experiment. We shall therefore attempt here to define the essential criteria of performance, and indicate how the experimental parameters can be chosen in order to obtain the best results. Attention will be focused on the general specifications of an electron spin apparatus, such as power level, cavity Q, etc., rather than on the details of items of equipment or particular pulsing systems.[37,38]

5.1. Transmitter Pulse Power

In the design of a spin echo experiment, the primary factor to be considered is the phase memory of the material to be investigated. A short phase memory will necessitate short pulses; short pulses, in turn, necessitate large microwave fields H_1. As a rough rule, the pulse duration t_p should be at least ten times shorter than the phase memory T_M. In order to produce a 180° pulse, we therefore need a value of H_1 (for the circular component which interacts with the precessing spins) given by‡ $H_1 = \pi/\gamma t_p \geq 10\pi/\gamma T_M$. Although an ideal 90°–180° pulse sequence is not usually necessary, it is essential that fields of the appropriate magnitude should be available,

* See Ref. 34 and Ref. 11, p. 497.

† Observations made during the course of photon echo experiments by Kurnit *et al.*[35] have subsequently been interpreted in terms of an envelope modulation effect by Grischkowsky and Hartmann.[36]

‡ When $g = 2$, $\gamma/2\pi = 2.80$ MHz per gauss. $H_1 = 1$ G therefore corresponds to the pulse durations $t_p \sim 0.1$–0.2 μsec.

since the echo phenomenon is a nonlinear one, and, in the low pulse-ang limit, the signal falls off as H_1^{-3}.*

A microwave signal source can be made to provide a large H_1 by eith (a) using a high-Q resonant cavity or (b) confining microwave energy int a small space in the vicinity of the sample. The first approach is not usual practicable in spin echo experiments, since a smaller Q than that which electrically attainable may have to be chosen in order to admit the ban of frequencies present in the pulse, and the Q value must in any case b small enough to avoid excessively long ringing times. The bandwidth lim can be estimated by matching the cavity response and the Fourier transfor of a Gaussian pulse at their half-power points. This approximately give the condition†

$$Q = \omega_0 t_p/4 \tag{119}$$

where ω_0 is the microwave frequency ($\omega_0 = 2\pi \times$frequency in Hz/sec) In estimating the limit imposed by cavity ringing, we note that a decay o the order of 140 dBs‡ is needed to reduce the power level from the pea pulse level to noise. Since the power in the cavity decays as $\exp(-\omega_0 t/Q)$ the effective ringing time is therefore

$$t(\text{ringing}) \simeq 14(\log_e 10)Q/\omega_0 \simeq 32Q/\omega_0 \tag{120}$$

If Q has the value denoted by the bandwidth condition (119), this lead to the result $t(\text{ringing}) \simeq 8t_p$. It may in practice be desirable to use a somewhat smaller Q (e.g., in cases where the phase memory is short and the signal weak). In order to have some definite relation between Q and t_p for the purpose of discussing pulse power and sensitivity, we shall, however, assume in general that the apparatus has been designed so as to satisfy (119).

* See Eq. (9). In deriving this, it was assumed that $\omega_1 \gg$ linewidth. If $\omega_1 <$ linewidth, the number of spins contributing to the echo will fall off as the first power of H_1, thus giving a total precessing magnetization which is proportional to H_1^4 [i.e., to (pulse power)^2].

† See Ref. 38, Eq. (7a). Values of Q determined in this way will not, of course, admit microwave pulses with a rectangular envelope, such as the pulses which were assumed in Section 1. In practice, this is unimportant, however, since short pulses having this idealized form are not usually available from signal sources.

‡ A 500-W, 20-nsec transmitter pulse has an energy of 10^{-5} J. The noise energy with which the echo signal has to compete is $\sim kT_N$ [Eq. (133)]. Assuming a noise temperature $T_N = 5000$°K, we thus have $kT_N \simeq 7.10^{-20}$ J. The transmitter pulse should therefore undergo an attenuation ~140 dBs before che appearance of the echo.

The second approach to the problem of obtaining a large H_1, by onfinement of the field,* may be made in a variety of ways, as, for example, y loading the cavity with a high dielectric material, by loading with metal locks,[39] or by choosing the cavity geometry so that H_1 is concentrated n the vicinity of the sample. Special microwave structures such as the helix[40] ave also been employed. Although such solutions are not commonly mployed in CW spectrometry, since they usually degrade Q, they may be uite acceptable in spin echo experiments where Q is already restricted by andwidth considerations. The relation connecting power P_0, Q, and avity volume can be approximately inferred from the steady-state formula

$$P_0 = V_c H_1^2 \omega_0 / 8\pi Q \tag{121a}$$

P_0 refers to one circular component, and V_c is an effective cavity volume btained by writing

$$\overline{\{H_1(\text{sample})\}^2} V_c = \int^{\text{cavity vol}} H_1^2 \, dV \tag{121b}$$

he steady-state condition will not of course be reached during a short ulse, and actual H_1 values will be somewhat below those derived according o (121a).

It may help in forming a realistic idea of the power levels which are nvolved if we give specifications for a somewhat mediocre but easily built ystem, and indicate how the parameters must be scaled to improve perormance. Let us suppose that a signal source of 5 W power at X band 10 GHz) is available, and is used in conjunction with a conventional ectangular box cavity of $Q = 1000$. The effective volume $V_c = 4$ cm^3. When losses in the waveguide system, in couplers, and in the unwanted ircularly polarized component of H_1 in the cavity are allowed for, we are eft with $P_0 \simeq 1$ W. By (121a), this gives a field of 1 G, which will suffice o generate 120° pulses for $g = 2$ spins in a time $t_p = 0.12$ μsec. In this xample, the value of Q has deliberately been chosen to be less than that llowed by (119) in order to reduce t(ringing) to ~0.5 μsec, or ~$4t_p$. f we assume that echo signals cannot be seen unless they begin a time (ringing) after pulse II, we find that the shortest observable phase memory

* The extent to which this is desirable (or possible) will depend partly on the form in which samples are available. If unlimited quantities of material are available, the best signal-to-noise ratio may be obtained by maximizing the number of spins and by using a large sample in a large cavity. If, on the other hand, the sample to be studied is in the form of a small crystallite, it may be better to use a geometry, such as the quarter-wave coaxial geometry, which gives a high concentration of H_1 over a small volume.

is $\sim$1.2 μsec (i.e., the limiting phase memory time according to the roug rule $T_M \simeq 10t_p$ suggested earlier). Suppose now that it is desired to stud materials with shorter phase memories, down to $\sim$0.1 μsec, and to upgrad the apparatus by reducing the time scale 10:1. We then require that H should be ten times larger, and Q ten times smaller. According to (121a) P_0 must therefore be increased 1000 times to 5 kW unless it is possible t make some simultaneous reduction in the effective cavity volume V_c

5.2. Echo Waveforms

In Section 1, it was assumed that H_1 could be made large enough s that $\omega_1 \gg$ linewidth. It will be apparent from the discussion above tha this is not always a practical matter when dealing with the lines encountere in EPR. Fortunately, the linewidth condition is in no way a necessary one,* and was postulated earlier merely in order to make the description of th echo-generating process as simple as possible. This condition should not b considered as imposing any constraints on the design of a spin echo spec trometer. However, it is useful to understand what is involved in workin in the regime where $\omega_1 <$ linewidth, and we shall pause to analyze thi situation in some detail before proceeding further.

The motion of a single spin packet during a two-pulse echo sequenc has been calculated by Bloom[41] using a convenient spinor representation as indicated by Jaynes,[42] to describe the reorientations of the magnetizatior in the rotating frame. H_1 is taken to be along the x axis. If we denote the static magnetization of a spin packet by $M_{0,j}$ and the precessing moment along the y axis at a time later by $M_{p,j}$, then the ratio $E_{p,j} = M_{p,j}/M_{0,j}$ is given by

$$\begin{aligned} E_{p,j} = &-\sin^3\theta \sin(bt_{\mathrm{pI}}) \sin^2(\tfrac{1}{2}bt_{\mathrm{pII}}) \cos\omega_j t \\ &-(\sin 2\theta)(\sin^2\theta) \sin^2(\tfrac{1}{2}bt_{\mathrm{pI}}) \sin^2(\tfrac{1}{2}bt_{\mathrm{pII}}) \sin\omega_j t \end{aligned} \tag{122}$$

In Eq. (122), $b^2 = \omega_1{}^2 + \omega_j{}^2$, $\tan\theta = \omega_1/\omega_j$, $\omega_1 = \gamma H_1$, and ω_j is the difference γH_j between the microwave frequency and the Larmor frequency of the jth spin packet.† The total time measured from the beginning of

* It may actually be undesirable to fulfill this condition in some circumstances. Instantaneous diffusion, which is caused by the abrupt change in local dipolar fields occurring when the spins are rotated by pulse II, is liable to become a more serious problem as ω_1 is increased and as more of the spins become involved.

† Equation (122) is essentially Eq. (28) of Ref. 41 with some minor changes in the notation. The first factor on the right-hand side of this equation has been corrected to $\sin^3\theta$.

pulse I is $t_{pI} + t_{pII} + 2\tau + t$ as in Section 1.2. In the general case, it would be necessary to calculate an x-axis component $M_{x,j}$ of the precessing magnetization as well as the y-axis component. This step can be omitted, however, if the inhomogeneous line is assumed to be symmetric about the exact resonance point, since the contributions from the $M_{x,j}$ then cancel out* in the final summation over all spin packets. In order to keep down the number of independent parameters, it is also helpful to set $t_{pI} = t_{pII} = t_p$.† Equation (122) can then be written in the form

$$E_p(\omega, t) = -(\sin^2 \theta) \sin^2(\tfrac{1}{2}bt_p) \times \{(\sin \theta)(\sin bt_p)(\cos \omega t) + (\sin 2\theta)[\sin^2(\tfrac{1}{2}bt_p)] \sin \omega t\} \tag{123}$$

The subscripts j have been dropped so that E_p, ω, etc. can be treated as continuous variables. The total precessing magnetization $M_p(t)$ is obtained by making a suitably weighted summation of $E_p(\omega, t)$ over the inhomogeneous line. Let us denote the normalized lineshape function (referred to the microwave transmitter frequency as zero origin) by $S(\omega)$.‡ Then the static magnetization due to spin packets lying in the interval from ω to $\omega + d\omega$ is $M_0 S(\omega)\, d\omega$, where M_0 is the total static magnetic moment in the line, and we have

$$M_p(t) = M_0 \int_{-\infty}^{+\infty} S(\omega) E_p(\omega, t)\, d\omega \tag{124}$$

Simple results analogous to those obtained in Section 1 can be derived as a special case from Eqs. (122) and (124) by setting $\omega_1 \gg \omega$, thus reducing

* A more elaborate analysis, taking account of line asymmetry, would lead to some minor changes in the calculated echo waveforms. This analysis seems scarcely worth the effort, however, since larger errors could easily arise from the approximation of the actual microwave pulse envelopes by rectangular waveforms.

† Equal pulses often form a convenient experimental choice, and may also make it easier to optimize Q, if Q is limited by bandwidth considerations. If the pulses are equal, and $\omega_1 \gg$ linewidth, it can be seen from Eq. (9) that the signal is maximized when $\theta_p = 2\pi/3$. This signal is only 65% as large as that obtained by means of a 90°–180° sequence. Calculations for a $t_{pII} = 2t_{pI}$ sequence have not been made in the $\omega_1 <$ linewidth regime. Experiments suggest, however, that little is lost in this regime by equalizing the pulse durations.

‡ $S(\omega)$ must be symmetric if it is to satisfy the earlier assumption that $M_{x,j}$ can be ignored in calculating the echo signal. The function $I(\Delta\omega)$ used to define the lineshape in Section 2 has deliberately been changed to $S(\omega)$ here in order to avoid confusion regarding the different zero origins of $\Delta\omega$ and ω. In $S(\omega)$, ω is measured from the microwave transmitter frequency; in $I(\Delta\omega)$, $\Delta\omega$ is measured from the Larmor frequency of a center which is not perturbed in any way by local fields.

b to ω_1, and θ to $\pi/2$. Equation (124) then states that $M_p(t)$ is proportiona to the Fourier cosine transform of $S(\omega)$. In the present instance, howeve ω may exceed ω_1 and Eq. (124) must be evaluated by numerical integratio for selected values of the pulsing parameters. In making the necessary com putation, it is convenient to reexpress (123) and (124) in terms of dimension less parameters $\omega' = \omega/\omega_1$ and $t' = t/t_p$. Thus $bt_p = \omega_1 t_p(1 + \omega'^2)^{1/2}$ $\omega t = \omega' t' \times \omega_1 t_p$, $\theta = \cot^{-1}(\omega')$, and, from (124),

$$M_p'(t', \omega_1 t_p) = M_0 \int_{-\infty}^{+\infty} S'(\omega') E_p'(\omega', t', \omega_1 t_p)\, d\omega' \tag{125}$$

where M_p', S', and E_p' are functions obtained by reexpressing S and E_p in terms of the new arguments ω' and t'. By writing $\sin^2\theta$ in (123) in the form $\omega_1^2/(\omega^2 + \omega_1^2)$, it can be seen that the major contribution from E_p in the integral (124) is made by spin packets which lie within $\sim\pm\omega_1$ o exact resonance. If, then, $S(\omega)$ varies slowly over this range (i.e., if ω_1 $\ll$ linewidth), $S(\omega)$ can be taken outside the integral, and in place of (125), we have

$$M_p'(t', \omega_1 t_p) \simeq M_0 \omega_1 S(0) \int_{-\infty}^{+\infty} E_p'(\omega', t', \omega_1 t_p)\, d\omega' \tag{126}$$

where $S(0)$ is the height of the normalized lineshape function in the center. The numerical integration merely concerns E_p', and can be denoted by the dimensionless ratio

$$r_M(t') = M_p'(t', \omega_1 t_p)/M_0 \omega_1 S(0) \tag{127}$$

The modulus* $|r_M|$ is shown graphically for several values of $\omega_1 t_p$ in Fig. 10. (The results of computations made for the case in which $\omega_1 \simeq$ linewidth are shown in Fig. 7 of Ref. 31 and are essentially similar.)

The following general comments may be made on the results. For $\omega_1 t_p = 2\pi/3$ (120° pulsing of spin packets on exact resonance), the echo is approximately the same width as the driving pulses. It is timed so that the centers of pulse I, pulse II, and the echo are very nearly equally spaced, the

* The modulus $|r_M|$ gives the experimental waveform when spin echoes are detected by an incoherent superheterodyne system. If the transmitter pulses are coherent, and if a homodyne or coherent superheterodyne detection system is used, then the echo waveform will be proportional to r_M. Coherence of the two transmitter pulses was implicitly assumed in Section 1, and in the derivation of (122) by taking H_1 along the same axis in the rotating frame for both pulses. See the discussion in the paragraph after Eq. (3).

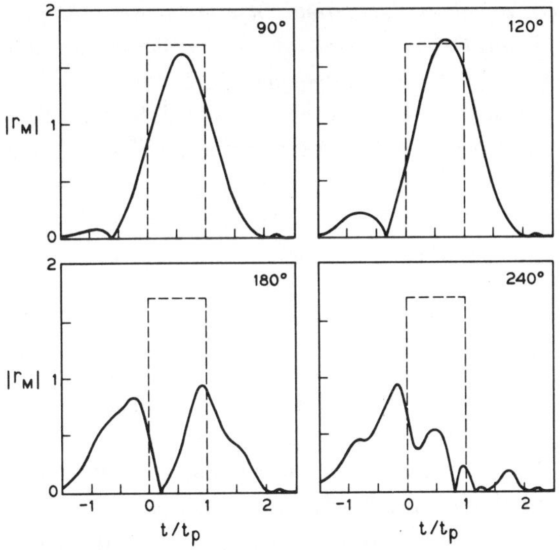

Fig. 10. The echo waveform function $|r_M|$ shown as a function of t/t_p for the case in which the applied microwave pulses are equal and $\omega_1 \ll$ linewidth. The angles denote the values of $\omega_1 t_p$ (i.e., the angle turned by those spins that are on exact resonance). The time measured from the beginning of pulse I is $2t_p + 2\tau + t$. The broken line shows the applied pulse I as it would be if mirrored in time about the center of pulse II.

echo thus occurring $\sim\frac{1}{2}t_p$ later than in the case of $\omega_1 \gg$ linewidth. This time shift is not surprising when one remembers that, in the case of $\omega_1 \ll$ linewidth, a certain amount of precession about the z axis occurs during the pulses, whereas in the $\omega_1 \gg$ linewidth case, the z-axis precession is suppressed in the presence of the microwave field. The magnetization in the peak of the echo is $\sim 1.7 M_0 \omega_1 S(0)$. In the previous paragraph, we commented that the major contribution to the echo was made by spin packets lying within $\pm\omega_1$ of exact resonance. It is perhaps interesting to note therefore that the peak precessing magnetization is not very different from that which would be calculated by assuming that the microwave field selects from the broad line a Lorentzian distribution of spin packets characterized by a height $M_0 S(0)$ and a half-width at half-height ω_1. If two 120° pulses are applied to such a selection of spin packets under ideal $\omega_1 \ll$ linewidth conditions, we obtain, by (9), a peak precessing magnetization of $2.0 M_0 \omega_1 S(0)$. The height of the echo calculated from (126) remains almost stationary as $\omega_1 t_p$ varies from 90° to just over 120°. *The echo signal is there-*

fore insensitive to minor fluctuations in pulse power and provides a fairl reliable and simple means of sampling the magnetization at various point in a broad line. The normal consequence of overdriving the spin system i to produce extreme distortions of the echo signal. It may be noted, however that this does not occur under all circumstances. For approximately equa pulses and very large turning angles, it is sometimes possible to obtain narrow and comparatively clean echo signal.[43] This effect depends on th inhomogeneity of H_1 throughout the sample. The width of such an over drive signal is $\sim 2.4/\omega_1$ and is unrelated to the pulse width t_p.

5.3. Detection Sensitivity

We conclude with a discussion of the signal strength in relation to thermal noise, when $\omega_1 \ll$ linewidth. This case is easier to define than the $\omega_1 \sim$ linewidth case and it leads to essentially similar results.* We start from the calculation of precessing magnetization M_p which was made in the last section. From M_p, we next find the microwave field H_E generated in the cavity, and the energy $H_E{}^2 V_c/8\pi$ delivered to the cavity mode during the echo pulse [V_c is the effective cavity volume defined in (121b)]. This echo pulse energy is then compared with the noise energy which would be sampled by a gating pulse $\sim t_p$ long. Finally, consideration is given to the improvement in signal-to-noise ratio which can be obtained by accumulating a number of spin echo pulses, and to the limitations imposed by the lattice relaxation time of the spin system. Although such calculations are inherently of an order-of-magnitude kind, we shall retain approximate numerical factors in the intermediate steps to avoid any unnecessary pileup of errors.

A circuit model, such as that given by Bloembergen and Pound,[44] can be used to derive H_E from M_p. Assuming that the change in amplitude during one microwave cycle is small, we obtain†

$$H_E = (2\pi\omega_0/V_c)\exp(-\omega_0 t/2Q)\int_0^t M_p \exp(\omega_0 t''/2Q)\,dt'' \qquad (128)$$

* If $\omega_1 \gg$ linewidth, M_p may be derived as in Section 1. The sensitivity will, however, tend to be lower than that obtained here. See the discussion which follows Eq. (138).

† It is assumed in this circuit model that the spins and the cavity field are weakly coupled, i.e., that M_p is not significantly perturbed by the field H_E to which it gives rise. Breakdown of this condition is indicated experimentally by the appearance of a satellite echo occurring at a time τ after the primary echo, which under these circumstances assumes the role of an additional transmitter pulse. Strong coupling of this kind is unimportant in a discussion of limiting sensitivity, since it will not be found at very low spin concentrations.

ince, from (127),

$$M_p(t, \omega_1 t_p) = M_p'(t', \omega_1 t_p) = M_0 S(0) \omega_1 r_M$$

ve therefore have

$$H_E = 2\pi\omega_0\omega_1 t_p M_0 S(0) r_H(t')/V_c \tag{129}$$

vhere $r_H(t')$ is a dimensionless quantity

$$r_H(t') = \exp(-\omega_0 t'/2Q) \int_{t'''=-\infty}^{t'''=t'} r_M(t''') \exp(\omega_0 t''' t_p/2Q)\, dt''' \tag{130}$$

lescribing the response of the cavity to r_M, and t''' is the dimensionless parameter t''/t_p. The form and the peak amplitude of r_H depend on the value of $\omega_0 t_p/Q$. The quantity $\omega_0 t_p/Q$ might, from the point of view of detection sensitivity alone, be treated as an independent parameter. It is, however, already restricted by the requirement that the bandwidth of the cavity should be sufficient to preserve the short duration of the transmitter pulses. As was pointed out earlier, bandwidth arguments do not necessarily determine the value of $\omega_0 t_p/Q$ which will be best from an experimental standpoint. It is nevertheless convenient to settle on some number and we therefore adopt the value $\omega_0 t_p/Q = 4$ suggested in (119), using this in conjunction with the 120° pulse condition

$$\omega_1 t_p = 2\pi/3 \tag{131}$$

to define a pair of optimizing conditions on which the sensitivity criteria can be based. Using the function r_M shown in Fig. 10, we find that the peak value of r_H under these conditions is 0.45.* Thus, by (129) and (131), the peak energy delivered to the cavity

$$U_{\text{peak}} = H_E^2 V_c/8\pi = 1.4[\omega_0 M_0 S(0)]^2/V_c \tag{132}$$

This energy should be transferred as efficiently as possible from the cavity to the detection system. Here, we shall make the *ad hoc* assumption that the transfer can be effected in such a way that half of this peak energy actually reaches the detector, the other half being dissipated in the cavity walls, waveguide, etc. The energy received at the detector due to a single echo signal is then

$$U_E = 0.70\omega_0^2[M_0 S(0)]^2/V_c \tag{132a}$$

* See Ref. 38, Fig. 9.

It should perhaps be pointed out that no advantage is to be gained by introducing lossy material into the cavity either in order to reduce Q to th value Q_{opt} prescribed by (119), or to reach a critically coupled conditio This will merely reduce the echo signal intensity and impair signal-to-nois performance. If Q must be reduced, this should be done by increasing th coupling to the detection system.

It may be useful to pause here momentarily and consider what woul be the effect on signal strength if Q were made either larger or smaller tha the "optimum" value Q_{opt} defined by Eq. (119). Computed curves given i Ref. 38, Fig. 9 show that r_H rises from 0.45 to 0.54 when $\omega_0 t_p/Q$ is change from 4 to 3. Thus, supposing that Q could be increased without alterin t_p or ω_1, the echo signal could initially be increased in energy. Only limited improvement could be obtained in this way, however, even if al the practical problems involved in lengthening the cavity ringing time wer solved. If $Q \gg Q_{\text{opt}}$, then $r_H(t')$ becomes a smooth step with height r_H(max $\simeq 1.3$, followed by a decay proportional to $\exp -[\omega_0 t'/2Q]$. The energy in the cavity $U_c(t') \propto [r_H(t')]^2$. In the absence of dissipative loss, the energ delivered from the cavity to the detector is given by $U_E = \int P_{\text{det}}\, dt$ $= (\omega_0/Q) \int U_c(t')\, dt'$, and tends therefore, in the limit $Q \gg Q_{\text{opt}}$, to b independent of Q. If, on the other hand, $Q \ll Q_{\text{opt}}$, r_H has the same functional form as r_M and a peak height proportional to Q. The peak valu of $U_c(t')$ is proportional to Q^2, the duration of $U_c(t')$ is independent of Q and the energy received by the detector is thus proportional to Q. It wil be apparent from the analysis which follows in the next few paragraph that in this second limiting case, the sensitivity is proportional to $Q^{1/2}$.

In calculating the noise signal with which U_E has to compete, we assume that the detection system has the same bandwidth ω_0/Q as the cavity. The noise power is then $kT_N(\omega_0/2\pi Q)$, where T_N is the effective noise temperature of the detection system. The noise energy emitted during the time interval $\sim t_p$ which contains the echo signal is $kT_N t_p(\omega_0/2\pi Q)$. Introducing the optimizing conditions (119) and (131), we thus have a noise energy

$$U_N = (2/\pi)kT_N \tag{133}$$

An expression for the minimum number of spins detectable by means of a single spin echo cycle can be derived by equating U_E and U_N. In this way, we obtain

$$M_0(\text{min}, 1E)S(0) = (0.95/\omega_0)(kT_N V_c)^{1/2} \tag{134}$$

where

$$M_0(\text{min}, 1E)S(0) = (d/d\omega)[M_0(\text{min}, 1E)] \tag{135}$$

'he quantity on the left-hand side of (134) and (135) denotes the static nagnetization per unit spectral interval corresponding to that part of the ne which is taking part in the generation of the echo. The limit (134) an also be expressed in terms of a minimum spectral density of spins by neans of the relation

$$M_0 = \tfrac{1}{3} N_0 \mu H_0 S(S+1)/kT \tag{136}$$

iving the result

$$(d/d\omega)[N_0(\text{min}, 1E)] = 2.9(kT_N V_c)^{1/2}/[\omega_0 \mu^2 S(S+1) H_0] \tag{137}$$

t will be noted that the large detection bandwidth which is required in lectron spin echo experiments does not enter into the expressions for ensitivity. This is essentially because we confine our observations to a articular interval t_p, during which time we compare the echo energy with he equivalent noise energy $\sim kT_N$ which characterizes the fundamental node of the resonant cavity.

A "boxcar" circuit can be used to integrate a succession of echo oulses and thus obtain an improvement in $(d/d\omega)[N_0(\text{min})]$ proportional to he square root of the number of pulses integrated. This number is $n_E \,\Delta\nu$, vhere $\Delta\nu$ is the bandwidth of the output circuit of the boxcar, and n_E s the recurrence rate for the echo cycle of events. n_E may of course be letermined by purely instrumental considerations, in which case we merely livide the right-hand side of (137) by $(n_E \,\Delta\nu)^{1/2}$ to obtain the final result of the sensitivity calculation. At low temperatures, however, n_E is often imited by the lattice relaxation time T_1. Under these circumstances, fast ecurrence rates yield diminishing returns, since saturation increases linarly with n_E, whereas the gain in sensitivity via boxcar integration is only oroportional to $(n_E)^{1/2}$. The optimum rate can be estimated as follows. f relaxation recovery is exponential, then the magnetization M' at the oeginning of each cycle is related to its ideal maximum value by $M'/M_0 = 1 - \exp(-1/n_E T_1)$. Differentiating the quantity $(n_E)^{1/2} M'$, we find that ensitivity is maximized when $n_E T_1 = 0.79$ and $M'/M_0 = 0.72$. When these oarameters are combined with (137), we obtain a limiting sensitivity under he conditions of boxcar operation, and partial saturation of the resonance, given by

$$d/d\omega)[N_0(\text{min, boxcar}, T_1)] = 4.5(kT_N V_c T_1 \,\Delta\nu)^{1/2} kT/[\omega\mu^2 H_0 S(S+1)] \tag{138}$$

This sensitivity specifies the spectral density of spins in a wide line which can be observed with unity signal-to-noise under the given experimental conditions.

It may seem surprising that Q does not appear in the final result. Thi is a consequence of assuming that $\omega_1 <$ linewidth and of setting $Q = Q_{op}$ by Eq. (119). The effect of taking $Q < Q_{opt}$ has already considered It leads to a sensitivity proportional to $Q^{1/2}$. The situation with regard t sensitivity outside the regime where $\omega_1 <$ linewidth is closely related t this, as we now show. For, as long as $\omega_1 <$ linewidth, a reduction in t will be associated with an increase in ω_1, an increase in the number of spi packets involved in echo signal generation, and a reduction in echo signa duration. We thus have a larger precessing magnetization for a shorte time. Q must be reduced in order to accommodate the shorter transmitte pulses, but as long as we retain the relationship $Q = Q_{opt}$, the same pea field will be generated in the cavity and the same energy will be passed o to the detector. If t_p is reduced to such a degree that $\omega_1 >$ linewidth, n new spin packets are taken in. The echo does not shorten any further (i becomes the Fourier transform of the line shape), and it would yield a constant amount of energy to the detector if Q could be left unchange during further reductions of t_p. But in order to admit the transmitte pulses, Q must also be reduced as in (119). From (128), we can deduc that, when the time $2Q/\omega_0$ is much less than the duration of $M_p(t)$, the $H_E \mathrel{\widetilde{\propto}} 1/Q$ and the energy U_c in the cavity falls off as Q^2. As we showe earlier, this implies that the energy delivered to the detector is proportiona to Q and the sensitivity to $Q^{1/2}$. In the regime $\omega_1 >$ linewidth, we se therefore that the sensitivity is proportional to $(t_p)^{1/2}$.

An EPR spectrum obtained by using a spin echo spectrometer is shown in Fig. 11. In practice, the sensitivity obtainable in this way is comparable

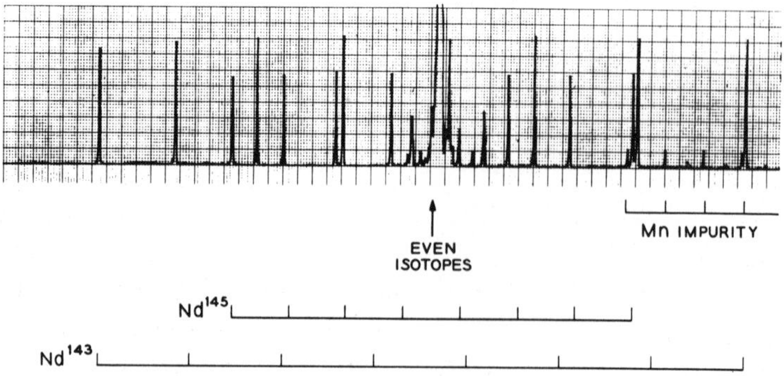

Fig. 11. EPR spectrum of (Ca, Nd)WO_4 at 4.2°K and 90° to the c axis, showing hyperfine lines, etc. There are 5.4×10^{14} spins in each ^{143}Nd line. The spectrum was obtained by generating electron spin echoes and accumulating echo signals in a "boxcar unit," with a time constant of 1 sec.

) that which can be obtained by CW resonance methods, but a point-by-oint comparison is difficult to make without entering into a discussion 'hich lies outside the scope of the present chapter. The relative sensitivity f CW versus pulse spectrometry may in any case be decided by experi-iental considerations which cannot easily be taken into account in a iathematical analysis. For instance, the power level in a CW experiment ; often limited to a few milliwatts by the difficulty of balancing out noise :om the signal source. This problem is avoided in the pulsed mode of peration, since the two functions of driving and observing the signals mitted by the spin system are separated in time. On the other hand, certain dvantages can be gained in CW spectroscopy by the use of field modulation echniques, especially where a limit is set by spin saturation. Of course, CW methods are more universally applicable for the purpose of obtaining esonance spectra since they do not require the existence of appreciably ong phase memory times.

6. THE APPLICATION OF ECHO TECHNIQUES IN EPR EXPERIMENTS

6.1. General

Although spin echo techniques have been used extensively in nuclear nagnetic resonance, comparatively little has been done by these methods n the companion field of EPR. This may be partly due to technical diffi-culties. However, with the increasing availability of high-performance microwave tubes and fast-pulse circuitry, it seems likely that more studies of this type will be attempted in the future. One cannot compile a catalog of possible applications at this stage. The discussion and the practical illustrations will therefore have the purpose of drawing attention to the following three considerations: (i) In some circumstances, it is advantageous to adopt a pulse method rather than a CW method so that one can separate the function of driving the spin system from the function of observing the signal emitted by the precessing magnetization. (ii) Some phenomena are most conveniently observed in the time domain by the use of echoes or other pulsed methods. (iii) It is often much easier to observe perturbations of the homogeneous spin packet than to observe perturbations of the whole resonance line. The homogeneous spin packet in many dilute materials has a width corresponding to a high Q value ($\sim 10^6$) and yet is a compara-tively sensitive detector of changes induced in the crystalline environment.

6.2. EPR Spectra by Pulsed Methods

We begin by elaborating further on the advantages and disadvantage of using a pulsed system for the purpose of making straightforward measure ments of EPR spectra. As we have seen in the previous section, the ech signal is proportional to the magnetization M_z present in a spectral interva $\sim 2\omega_1$ wide. A broad line can therefore be traced out by repeating the spi echo cycle of events while gradually varying the Zeeman field. In this way it is possible, by accumulating echo signals in a boxcar integrator, to obtai a degree of sensitivity which is comparable with that obtained in CW reso nance experiments.

For studying certain samples, the pulse method may be preferabl to the CW method. For instance, if a resonance line is very broad and ha a long, relatively flat portion in the middle,* the signal obtained by ac cumulating spin echoes is likely to be stronger than that which can b obtained by χ' detection or by field modulation methods, and ambiguitie are eliminated by the fact that no echo signal can appear where there ar no spins. The advantages of echo spectrometry are somewhat offset by certain difficulties peculiar to the method. For example, if the broad line corresponds to sizeable variations in g, or if there are a number o lines having different strengths of interaction with the microwave field the sampling width $2\omega_1$ will vary and the pulsing conditions will deviate sufficiently from the optimum to cause distortion of the signal or reductior in its size† at some points in the spectrum. An even larger error in the plotted spectrum may arise if the envelope is deeply modulated as a result of nuclear interactions (see Section 4). This effect may depress or remove signals for certain values of the pulse spacing τ and for certain components of the spectrum. Signals will of course also be eliminated if they correspond to a species having $T_M \ll \tau$.

Although no relevant experiments have yet been reported, it seems that in one type of experimental situation the pulse method would possess considerable advantages over CW methods. That is, when it is required to

* Echo methods have also been used for measuring broad lines in NMR.[45]

† The difficulty is not a serious one if single-crystal samples are being studied. The drive conditions can be tested by varying the pulse power, and, if necessary, they can be separately optimized in different portions of the spectrum. If this is done, there should be no differences in the sampling width $2\omega_1$. A check can be made by finding the power level required to produce the characteristic 180°–180° echo waveform (Fig. 10). These procedures may not be satisfactory when examining powder samples, or samples which induce an inhomogeneity of H_1, since each point in the spectrum will then represent a set of spin packets characterized by a distribution of ω_1 values.

tudy the properties of transient states such as those produced by optical xcitation, by electron beams,* or by chemical reactions.† Continuous-wave esonance methods in conjunction with a steady-state excitation of the ystem can be used in some situations of this kind, but the CW approach s not always practicable, either because the sources of excitation themelves operate in the pulsed mode, or because CW operation would lead o excessive heating of the sample. Pulse methods possess, of course, the urther advantage of making it possible to study the dynamics of the roduction and decay of the resonant centers.

.3. *The Measurement of* T_1

One of the principal applications of spin echo methods in the field f nuclear resonance has been in the measurement of lattice relaxation imes T_1. A typical experiment involves the use of three pulses as follows. The first is a 180° pulse, which brings about an inversion of the spin system. variable time interval is then permitted to elapse, and the magnetization M_z is measured by using a standard 90°–180° pulse sequence to generate spin echo (Section 1.2). The time interval τ adopted for the spin echo equence is unimportant, provided that it is short compared with the phase nemory time, and with the lattice time which is being determined.

The same procedure can, in principle, be used in EPR but some nodification is desirable, since it is often difficult to achieve true 90°–180° pulse conditions (or even to excite the line uniformly during the first pulse) ecause of the width of the lines involved. Unless $\omega_1 \gg$ linewidth, the quantity $M_{z,t}(\omega)/M_{z,0}(\omega)$, which denotes the ratio between the z component f magnetization of a spin packet after and before the initial pulse, will vary according to the position in the line (see Fig. 12) and errors due to ross-relaxation or spectral diffusion are liable to enter into the result. The problem of cross-relaxation is of course not peculiar to the spin echo echnique, but is one which must be faced in any measurement of T_1 made by pulsed or CW methods. In pulse experiments, it can be solved fairly asily by using a field sweep to achieve adiabatic fast passage inversion of the spin system. The timing sequence used in one such T_1 experiment is shown[47] in Fig. 13. Many variations are possible, not all of them involving

* The writer is indebted to D. K. Wilson for the suggestion that an electron spin echo spectrometer should be used in studies of excitation by pulsed electron beams.

† An electron spin echo spectrometer is being used at the Institute of Chemical Kinetics, Novosibirsk, USSR, for the study of radicals produced in chemical reactions.[46]

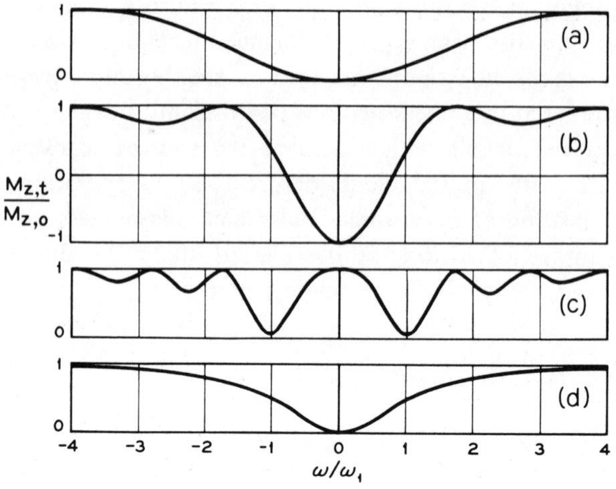

Fig. 12. Effect of applying a single microwave pulse to an inhomogeneous line when $\omega_1 \ll$ linewidth. (a) $\omega_1 t_p = \pi/2$, (b) $\omega_1 t_p = \pi$, (c) $\omega_1 t_p = 2\pi$ (where t_p is the pulse duration). A Lorentzian "hole" (d) is shown for comparison. ω is the difference between the microwave frequency and the frequency of a given spin packet.

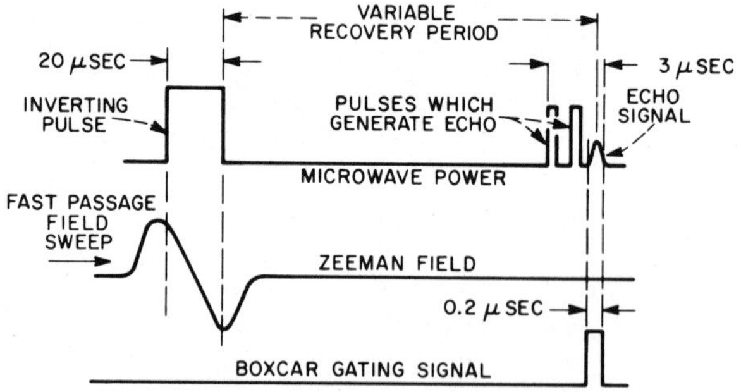

Fig. 13. Example of timing diagram for pulses and waveforms which may be used in a lattice relaxation experiment. The field sweep and the longer microwave pulse invert the spin system by means of an adiabatic fast passage. The two short pulses generate an electron spin echo signal which measures the magnetization of the spin system at some point during its return to Boltzmann equilibrium. A boxcar unit makes it possible to accumulate echoes and to translate echo amplitude into a voltage which can be applied to a pen recorder.

lectron spin echoes. Relaxation recovery may be monitored by applying weak CW signal or by making a second passage through the line.[48] rändle *et al.*[49] have described a relaxation experiment in which the narrow ne due to a free radical is monitored by causing it to generate a free-nduction signal.

.4. *The Paramagnetic Stark Effect and Similar Measurements*

In the instances given so far, the echo phenomenon has been used nerely as a means of making a rapid measurement of the magnetization M_z. The free-induction signal could in principle have been used instead, lthough for broad, inhomogeneous lines, a spin echo method may be more onvenient since the free-induction signal is short and is hard to separate rom the driving pulse. We next turn to a type of experiment which is based n the intrinsic property of the echo envelope to yield information regarding he spin packets themselves. In dilute paramagnetic materials, the spin ackets are often remarkably narrow. At X-band, values of T_M greater han 100 μsec have sometimes been observed,* corresponding to spin acket widths of ~1 kHz. This width is often considerably less than the lipolar width [Eq. (38)], and may be several orders of magnitude less than he inhomogeneous width (~10 MHz) of the resonance lines which are eing observed. Perturbations which do not readily show up in the normal EPR spectrum but which cause shifts or changes in the spectral constitution of the individual spin packets may thus be detected by their effects on the cho envelope.

We can illustrate this by describing a particular experiment in which pin echo methods were used to measure the paramagnetic Stark effect† (i.e., to measure the small shifts in resonance frequency which occur when lectric fields are applied to samples containing paramagnetic ions occuping sites without inversion symmetry). Figure 14 denotes a two-pulse echo cycle of events in which a voltage step is applied to the sample at the end of pulse II. Let us suppose that this voltage step changes the precession frequencies of all the spins by a small and uniform amount ω_v. Thus, at the end of pulse II, the magnetization is still the same as in Eq. (8), but the precession frequencies have just undergone a change from ω_j to $\omega_j + \omega_V$.

* Dyment[28] obtains $T_M \sim 90$ μsec for 2.6×10^{17} Cr^{3+} ions cm^{-3} in $K_3Co(CN)_6$ at 9.4 GHz. Mims *et al.*[22] obtained $T_M \sim 125$ μsec for 8.3×10^{17} Ce^{3+} ions cm^{-3} in $CaWO_4$, at 6.7 GHz, with H_0 parallel to the c axis.

† Further experimental details are given by Mims.[58]

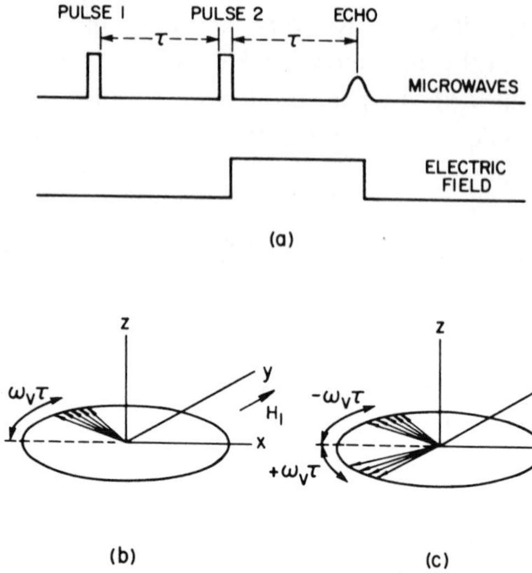

Fig. 14. (a) Timing of the electric field and the microwave pulses. (b) Phase convergence of spin packets at the time of the echo ($t = 0$) if all Larmor frequencies are shifted uniformly by ω_V at the end of pulse II (compare Fig. 2e). (c) Phase convergence for an inversion image crystal.

In the succeeding period $\tau + t$, phase will accumulate according to the factor $\exp[i(\omega_j + \omega_V)(\tau + t)]$, giving, in place of Eq. (9), the result

$$M_{x+iy}(t_{\text{pI}} + t_{\text{pII}} + 2\tau + t) = -i(\sin \theta_{\text{pI}})[\sin^2(\tfrac{1}{2}\theta_{\text{pII}})] \times \exp[i\omega_V(\tau+t)] \sum_j M_{0j} \exp(i\omega_j t) \quad (139)$$

If $\omega_V \ll \omega_1$, the factor $\exp[i\omega_V(\tau + t)]$, which embodies the effect of the electric field step, can be approximated by $\exp(i\omega_V\tau)$, since the change in this factor is small during the time ($\approx 1/\omega_1$) taken by the echo pulse itself.

The phase factor $\exp(i\omega_V\tau)$ could be detected by comparing the phase of the echo with the phase of the primary signal source in a coherent pulsing system. Alternatively, in an incoherent system, the phase comparison might be made by placing two samples in the cavity and applying the field to only one of them. Actually, for the study of a large class of materials, neither of these methods need be used. The spins in these materials are equally divided between two kinds of lattice site, the point symmetry at one site being the inversion image of the point symmetry at the other. In the absence of applied fields, spins in both types of site have the same resonance frequency, but when the voltage step is applied, they undergo frequency shifts in opposite directions, so that the total precessing magnetization at the time of the echo proportional to a factor $[\exp(i\omega_V\tau) + \exp(-i\omega_V\tau)] \sim \cos\omega_V\tau$ (see Fig. 14c). This factor modulates the echo decay envelope. The electric field shift can then be inferred from the relation $\omega_V = (2n+1)\pi/2\tau_n$, where τ_n is the time corresponding to the nth null in the echo envelope. The null times are easy to measure unless the shift is very inhomogeneous. In this case, it can be shown, by an extension of the above argument, that the echo envelope is proportional to the Fourier transform $\int g_V(\omega_V)\exp(i\omega_V\tau)\,d\omega_V$, where $g_V(\omega_V)$ denotes the distribution of the shifts.

An example may be useful here. Let us suppose that a voltage pulse of 100 V is applied across a sample of Ce^{3+}-doped $CaWO_4$ that is 2 mm thick. Let the resulting electric field of 500 V/cm lie along the c axis, and let the Zeeman field H_0 be oriented in the ab plane in the direction where the electric field effect is at a maximum. In this case, the electric field changes g by an amount $\delta_g \sim 1.5\times10^{-5}$ (Ref. 50, Fig. 4). Since $g_\perp = 1.43$, this would correspond to a shift $\omega_v/2\pi \simeq 0.1$ MHz/sec in an overall spin resonance frequency of 10 GHz/sec. The first null would appear at a time $\tau_1 \sim 2.4$ μsec after pulse II of the echo sequence, the second at $\tau_2 \sim 7.2$ μsec, etc. The phase memory of a Ca, $CeWO_4$ sample containing $\sim10^{18}$ spins cm^{-3} with H_0 in the ab plane is ~50 μsec. This time corresponds to a homogeneous spin packet width (full-width) ~0.006 MHz/sec, and is, of course, more than adequate for the observation of the electric field shift caused by our 100-V pulse. The full inhomogeneous linewidth under the specified conditions is ~30 MHz/sec, i.e., 300 times larger than the shift we are considering. To shift the spin resonance frequency across an interval of this size would require an electric field of 150 kV/cm.

The essential feature of the above experiment is the sudden change of the precession frequencies in the middle of the spin echo cycle of events,* which leads to an accumulation of phase $\omega(\tau)$ during the second half of the cycle which fails to cancel the accumulation of phase during the first half. Formally, we may express the net phase accumulation over the whole

* The end of pulse two was chosen earlier to coincide with the change in order to keep the explanation as simple as possible. If $\omega_1 \gg$ linewidth, it will not matter whether the voltage step occurs during pulse II or at the beginning of pulse II. If $\omega_1 \gtrsim$ linewidth, the maximum effect will be achieved by synchronizing the voltage step with the center of pulse II. The motion of the spins in the simultaneous presence of H_1 and the varying voltage need not be analyzed, since these effects are absorbed in an experimentally determined correction factor as noted in Ref. 50.

echo cycle by the factor

$$\exp\left[-i\int_0^{2\tau} s(t')\omega_V(t')\,dt'\right] \qquad (140$$

where $s(t)$ is the step function defined in (64), and ω_V is the frequency per turbation as a function of time. The adoption of a step function for $\omega_V(t$ is convenient, and easy to realize experimentally in the case just given but is not in any way essential. As an example of a different choice, let u consider a hypothetical experiment designed to measure resonance fre quency shifts under an applied mechanical stress. The stress is applie periodically by means of a piezoelectric resonant transducer operating i the frequency range ~100 kHz, and the time τ is adjusted so that 2τ i equal to the period of the transducer (see Fig. 15). For the phase facto (140), we then have

$$\exp\left[-i\omega_s\int_0^{2\tau} s(t')\sin(2\pi t'/2\tau)\,dt'\right] = \exp -i(4\omega_s\tau/\pi) \qquad (141$$

where ω_s is the frequency shift at the peak of the stress, and the phase relation between the piezoelectric stress and the echo cycle has been chosen

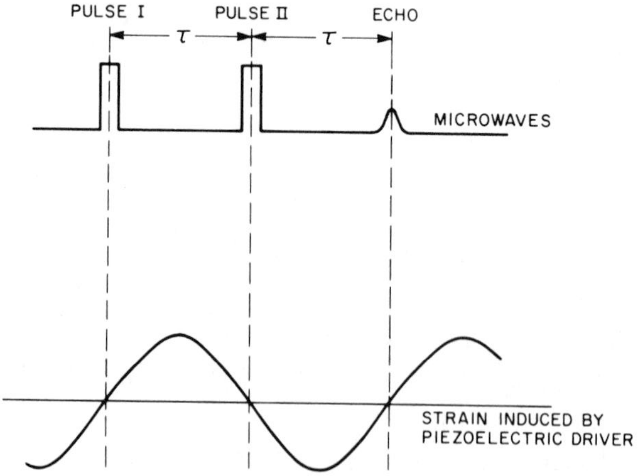

Fig. 15. Timing sequence for an experiment in which electron spin echoes are used to measure the shift of Larmor frequency associated with a lattice strain. The variation in the strain shown above causes there to be a difference between the precessional phase accumulated during the two halves of the echo cycle of events. The phase of the echo signal is therefore shifted as shown in Eq. (141) and may be used to provide a sensitive measure of the strain effect.

ɔ give the maximum effect. It should be noted that, for materials having hase memory times of a few microseconds or longer, very small shifts ould be measured in this way,* thus obviating the need to apply large tresses to the sample. The accumulated phase factor (141) could be meaured, as was suggested earlier, either by placing stressed and unstressed amples in the same cavity, and observing interference of the two echo ignals, or by setting up a phase-coherent pulsing and detection system.

.5. The Measurement of Local Field Dynamics

The perturbations discussed above were brought about by means of nacroscopic laboratory fields. But shifts in the precession frequencies are lso caused, on the microscopic scale, by changes taking place in the internal ields in the sample. Mechanisms of this kind have already been discussed n Section 3, the changes in question being the slow, random fluctuations n local magnetic fields associated with the lattice relaxation of impurities.

It appears that no work along these lines has been attempted in EPR, lthough such effects have been extensively investigated in NMR, where hey are usually observed by studying changes in the line shape. The lineshape approach is not always practical in EPR because of the masking of the effect by inhomogeneous line broadening.† However, as we have seen before, the difficulty can be overcome by measuring the echo decay envelope, since this displays the Fourier transform of the line shape corresponding to individual spin packets.

6.6. Free-Induction Spectroscopy

A spectrometer designed to generate pulses suitable for spin echo experiments will, *a fortiori*, be capable of being used to generate free-induction signals. Comparatively few applications of this type have been reported, however. This is because EPR lines are often several gauss wide,

* Some experimental tests have been made in which echo envelope interference effects have been observed, but no measurements of the strain shift in EPR have yet been made by this method.

† The nuclear spin echo method for measuring self-diffusion coefficients forms a closer analogy to the cases we are interested in here. In such experiments, the (artificially induced) inhomogeneous broadening is large, and the echo envelope is used as a means of observing the effects of diffusion on the "line" which would correspond to an individual spin packet. Line narrowing can of course be used to detect exchange effects in EPR. But the relevant interaction here is much larger than there, which is needed to modify the echo envelope, and the fluctuation rates are much more rapid.

and the resulting free-induction signals (the Fourier transforms of the lines would decay in times $\lesssim$25 nsec. Unless the time resolution of the equipment is considerably better than this, the free-induction signal will be overwhelmed by cavity ringing or by amplifier overload effects. (If $\omega_1 <$ line width, the decay time $\simeq 1/\omega_1 \simeq t_p$.) The lines due to free radicals constitute an exception since they are often only a fraction of a gauss wide.[49] Lines due to conduction electron spin resonance in high-purity metals have also been observed to give reasonably long free-induction signals.[51]

Where the experimental conditions result in the burning out of a narrow hole in a broad line, free-induction methods may be adequate to display the Fourier transform of this hole, if not to display the Fourier transform of the overall line.* The stimulated echo signals described in Section 1.3 are in fact free-induction signals of this type and represent the transform of the serrated pattern $M_z \propto \cos \omega\tau$ induced in the line by pulses I and II. In some special instances, a free-induction method may also be used to reveal structure which is hard to resolve by CW resonance.[53]

6.7. An Application of the Stimulated Echo Sequence. ENDOR Spin Echo Methods

In concluding, we describe an experiment which illustrates how a stimulated echo sequence may be used in a double resonance experiment. The purpose of the experiment was to measure the ENDOR† frequencies of the system constituted by Ce^{3+} spins and the surrounding ^{183}W nuclei in $CaWO_4$. The attempt was made to obtain these frequencies by a method involving a perturbation of the echo envelope, because the frequencies were <1 MHz and could not be detected by the well-known saturation method.‡ A two-pulse echo sequence, used in conjunction with an rf field to induce the ENDOR transitions, proved to be unsatisfactory for two reasons. First, because it proved impossible to avoid generating small (and very inhomogeneous) components of the ENDOR rf field parallel to the Zeeman field, thus directly causing interference effects such as those denoted by the integral (140), and second, because the time T_M was not long enough to bring about the required transitions. A three-pulse stimulated echo sequence was therefore adopted, with the ENDOR signal applied between

* See Ref. 52. The holes in this case are burnt out by a phonon avalanche.

† For a discussion of ENDOR (electron nuclear double resonance) processes, see Ref. 54.

‡ The probable reason for this difficulty is that cross-relaxation across intervals ~1 MHz occurs in times shorter than the T_1 of the electron spin.

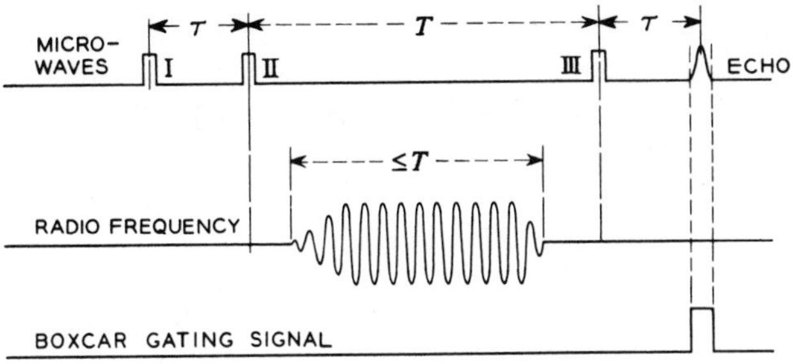

Fig. 16. Timing sequence for an experiment in which the stimulated echo signal is used to detect electron nuclear double resonance. The echo signal amplitude is converted into a dc voltage by means of a boxcar unit. Reasons for using a three-pulse sequence rather than the simpler two-pulse sequence are discussed in the text.

pulses II and III. This eliminated both of the above-mentioned difficulties. No low-frequency rf fields were present during the periods of free precession, and longer times, of the order of a few milliseconds were made available for inducing ENDOR transitions. The timing sequence is shown in Fig. 16.

The effects on the electron–nucleus system of this sequence of signals can be understood as follows. The first two microwave pulses produce a pattern $M_z(\omega) \propto \cos \omega t$ in the resonance line [Fig. 17b, Eq. (7)]; the ENDOR resonance shifts this pattern up in frequency for some components, down for others (Fig. 17c), thus causing a general smoothing of its features and a reduction in the stimulated echo intensity. To express this process in density matrix terms, we refer to Fig. 18, assuming for simplicity that the nuclear modulation effect is negligible, and that the microwave field induces only the transitions $|a\rangle \rightarrow |d\rangle$ and $|b\rangle \rightarrow |c\rangle$. If ω_0 is the microwave frequency, it will drive spin packets with frequencies ranging from $\sim\omega_0 - \omega_1$ to $\sim\omega_0 + \omega_1$ (or all spin packets if $\omega_1 >$ linewidth). A spin packet at $\omega_0 + \omega$ in this band will be composed of two different sets of four-level systems. For one set, $\omega_0 + \omega = \omega_{ad}$ and for the other, $\omega_0 + \omega = \omega_{bc}$. Assuming that these two sets everywhere contain equal numbers of spins, the diagonal density matrix elements after pulse II are as follows:

$$\begin{aligned}
&\varrho_{aa}(1) = \alpha \cos \omega\tau, \qquad \varrho_{bb}(1) = \alpha \cos(\omega - \omega_{ab} - \omega_{cd})\tau \\
&\varrho_{cc}(1) = \alpha \cos(\omega - \omega_{ab} - \omega_{cd})\tau, \qquad \varrho_{dd}(1) = -\alpha \cos \omega\tau \\
&\varrho_{aa}(2) = \alpha \cos(\omega + \omega_{ab} + \omega_{cd})\tau, \qquad \varrho_{bb}(2) = \alpha \cos \omega\tau \\
&\varrho_{cc}(2) = -\alpha \cos \omega\tau, \qquad \varrho_{dd}(2) = -\alpha \cos(\omega + \omega_{ab} + \omega_{cd})\tau
\end{aligned} \tag{142}$$

The arguments 1,2 denote four-level systems belonging to the first and second sets. $\alpha = \frac{1}{4}(g\beta H_0/kT) \sin \theta_{pI} \sin \theta_{pII}$, is the product of the Boltz mann factor corresponding to the initial state of the system and a facto defining the pulsing conditions [compare Eq. (7)]. The z component o magnetization is given by

$$M_z(\omega) = \mathrm{Tr}_\omega(\varrho M_z) \tag{143}$$

where the trace is taken only over those elements in systems 1 and 2 whic are associated with the states separated by an interval $\hbar(\omega + \omega_0)$. In th absence of ENDOR transitions, we have, from (142),

$$\mathrm{Tr}_{\omega,0}(\varrho M_z) = M_0 \cos \omega\tau \tag{144}$$

ENDOR resonance in the $| a\rangle + | b\rangle$ interval will equalize ϱ_{aa} and ϱ_{b}

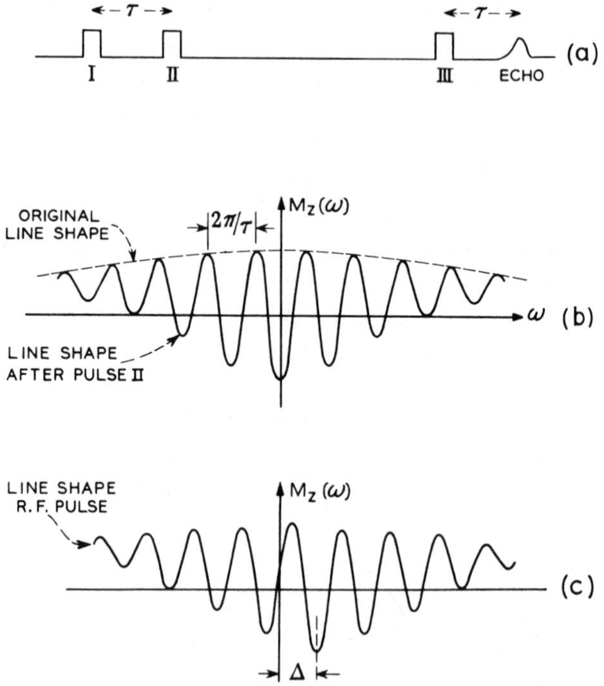

Fig. 17. (a) Stimulated echo sequence. (b) Artificial structure imposed on the resonance line by pulses I and II. (c) Displacement of this structure caused by resonating nuclei for which $h(\omega_{ab} - \omega_{cd}) = h\Delta$ corresponds to the difference between the interaction energies in the two nuclear orientations.

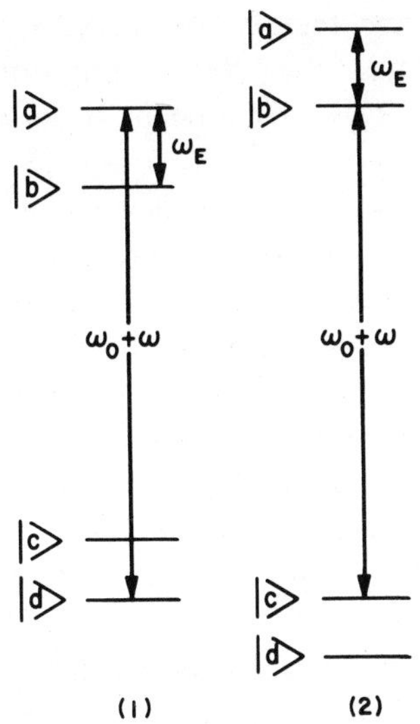

Fig. 18. Two types of four-level electron–nuclear systems which contribute to the spin packet at $\omega_0+\omega$. The microwave transitions $|a\rangle \to |d\rangle$, $|b\rangle \to |c\rangle$ are allowed. Inhomogeneous broadening accounts for the difference between the intervals $|a\rangle \to |d\rangle$, etc. in the two cases shown. ω_0 is the frequency of the microwave signal source; ω_E is the frequency of the rf ENDOR signal.

For both sets of systems,* leaving the relevant elements as follows:

$$\begin{aligned} \varrho_{aa}(1) &= \tfrac{1}{2}\alpha\{\cos\omega\tau+\cos[(\omega-\omega_{ab}-\omega_{cd})\tau]\}, \qquad \varrho_{dd}(1) = -\alpha\cos\omega\tau \\ \varrho_{bb}(2) &= \tfrac{1}{2}\alpha\{\cos\omega\tau+\cos[(\omega+\omega_{ab}+\omega_{cd})\tau]\}, \qquad \varrho_{cc}(2) = -\alpha\cos\omega\tau \end{aligned} \tag{145}$$

Then,

$$\mathrm{Tr}_\omega(\varrho M_z) = \tfrac{1}{4}M_0(\cos\omega\tau)\{3+\cos[(\omega_{ab}+\omega_{cd})\tau]\} \tag{146}$$

The ratio between (144) and (146) thus gives the factor

$$F(\text{ENDOR}) = \tfrac{1}{4}\{3+\cos[(\omega_{ab}+\omega_{cd})\tau]\} \tag{147}$$

This is the factor by which the magnetization pattern, and hence the stimulated echo signal, is reduced as a result of the ENDOR resonance. It will

* A 180° pulse in the interval $|a\rangle \to |b\rangle$ would interchange the two densities and produce twice as large an effect on the stimulated echo signal.

be noticed that $F = 1$, and the effect disappears, if $(\omega_{ab} + \omega_{cd})\tau = 2n\pi$ The reason is that, in this case, the ENDOR transition shifts the spin packets in the sinusoidal pattern $M_z(\omega) \propto \cos \omega\tau$ by an integral number of periods $2\pi/\tau$, thus leaving the pattern essentially unaltered. This problem does not arise if the shift is large enough to take the spin packets out of the portion of the resonance line that is excited by microwave pulses, i.e., in the above instance, if $\omega_{ab} + \omega_{cd} \gg \omega_1$. In this case, it can be shown, by arguments similar to those used above, that the ratio F(ENDOR) is simply 3/4.

An alternative method of performing an ENDOR experiment in which a π pulse is used to invert (or to burn a hole in) the microwave resonance line and a two-pulse sequence is applied a time T later to monitor the changes induced by rf resonance is described by Davies.[55] It is suggested that this method should be preferable to the one described above when a longer time interval is required in order to induce ENDOR transitions. As we have remarked earlier, a pulsed microwave apparatus, designed primarily for the generation of spin echoes, will be capable of being used in a variety of ways not necessarily involving spin echo mechanisms at each step and in every set of experimental circumstances.

ACKNOWLEDGMENTS

Figures 1, 4–6, and 12–14 are reproduced by permission of *The Physical Review*, Figs. 10 and 11 by permission of *The Review of Scientific Instruments*, and Figs. 16 and 17 by permission of *The Proceedings of the Royal Society of London*. Figure 3 is reproduced by permission of A. M. Stoneham and of *The Review of Modern Physics*.

The author would like to thank numerous colleagues at UCLA and at Bell Telephone Laboratories who have helped to clarify various topics discussed here. Particular thanks are due to Professor W. G. Clark for making an experimental study verifying the point mentioned in the footnote on p. 291, and to Alvin Kiel for making a detailed criticism of the manuscript.

NOTATION

A spins	spins which contribute directly to the precessing magnetization which generates the spin echo signal
B spins	spins in the environment which perturb the precession of the *A* spins; motion among the *B* spins gives rise to fluctuating local fields which destroy the phase memory of the *A* spins

$E(2\tau)$	echo envelope as a function of the time 2τ between pulse I and the spin echo, $E(0) = 1$
$E(\text{mod}, \tau)$	function specifying the modulation of the echo envelope due to the "nuclear modulation effect"; see Section 4
H_0	Zeeman field
H_1	microwave magnetic field amplitude
M_0	magnetic moment due to spin system in Boltzmann equilibrium
M_p	precessing moment which generates the echo signal
pulse I	the first transmitter pulse of a spin echo sequence; ideally, a "90°" pulse.
pulse II	the second transmitter pulse of a spin echo sequence; ideally, a "180°" pulse in a simple two-pulse sequence or a "90°" pulse in a stimulated echo sequence
pulse III	the third transmitter pulse in a stimulated echo sequence; ideally, a "90°" pulse
T_M	phase memory time, i.e., time interval between pulse I and the echo which corresponds to e^{-1} attenuation of echo signal
T	time between pulse I and pulse II in a stimulated echo sequence
t_p	microwave pulse duration
t	time coordinate for the echo signal; the total time measured from the beginning of pulse I is $t_{pI} + t_{pII} + 2\tau + t$
γ	gyromagnetic ratio $g\beta/\hbar$
$\Delta\omega_{dip}$	half-width at half-height of the theoretical Lorentzian line due to dipolar broadening [see Eq. (31)]
τ	time between pulse I and pulse II in a spin echo sequence
ω, ω_j	Larmor frequency of a spin packet in the rotating coordinate system; in the laboratory coordinate system, the frequency is $\omega_0 + \omega$, or $\omega_0 + \omega_j$
ω_0	signal frequency of microwave transmitter
ω_1	spin nutation frequency; for an $S = 1/2$ system, $\omega_1 = \gamma H_1$

REFERENCES

1. D. A. Bozanic, D. Mergerian, and R. W. Minarik, *Phys. Rev. Letters* **21**, 541 (1968).
2. C. P. Slichter, *Principles of Magnetic Resonance*, Harper and Row, New York (1963), p. 11.

2a. I. J. Lowe and R. E. Norberg, *Phys. Rev.* **107**, 46 (1957).
3. R. P. Feynman, F. L. Vernon, and R. W. Hellwarth, *J. Appl. Phys.* **28**, 49 (1957
4. D. Shaltiel and W. Low, *Phys. Rev.* **124**, 1062 (1961).
5. R. F. Wenzel and Y. W. Kim, *Phys. Rev.* **140**, A1592 (1965).
6. A. M. Stoneham, *Proc. XIVth Colloque Ampere, Ljubljana (1966)*, North Hollan Publishing Co., Amsterdam (1967).
7. H. Margenau, *Phys. Rev.* **82**, 156 (1951).
8. W. B. Mims and R. Gillen, *Phys. Rev.* **148**, 438 (1966).
9. Y. Ayant and E. Belorizsky, *J. Phys. (France)* **27**, 24 (1966).
10. A. M. Stoneham, *Rev. Mod. Phys.* **41**, 82 (1969).
11. A. Abragam, *The Principles of Nuclear Magnetism*, Oxford University Press (1961 p. 126.
12. A. M. Stoneham, *Proc. Phys. Soc.* **89**, 909 (1966).
13. M. F. Lewis and A. M. Stoneham, *Phys. Rev.* **164**, 271 (1967).
14. E. Feher, *Phys. Rev.* **136**, A145 (1964).
15. A. E. Hughes, *Solid State Comm.* **6**, 61 (1968).
16. D. H. McMahon, *Phys. Rev.* **134**, A128 (1964).
17. J. Holtzmark, *Physik. Z.* **20**, 162 (1919).
18. J. M. Baker and A. E. Mau, *Can. J. Phys.* **45**, 403 (1966).
19. J. H. Van Vleck, *Phys. Rev.* **74**, 1168 (1948).
20. N. Laurance, E. C. McIrvine, and J. Lambe, *J. Phys. Chem. Sol.* **23**, 515 (1962
21. E. R. Davies and J. P. Hurrell, *Solid State Comm.* (1971), in press.
22. W. B. Mims, K. Nassau, and J. D. McGee, *Phys. Rev.* **123**, 2059 (1961).
23. W. B. Mims, *Phys. Rev.* **168**, 370 (1968).
24. J. R. Klauder and P. W. Anderson, *Phys. Rev.* **125**, 912 (1962).
25. Silbernagel, M. Weger, W. G. Clark, and J. H. Wernick, *Phys. Rev.* **153**, 535 (1967)
26. W. J. C. Grant, *Phys. Rev.* **135**, 1554, 1565, 1574; *Phys. Rev.* **135**, 1265 (1965).
27. G. Feher, *Phys. Rev.* **114**, 1219 (1959).
28. John C. Dyment, *Can. J. Phys.* **44**, 637 (1966).
29. J. A. Cowen and D. E. Kaplan, *Phys. Rev.* **124**, 1098 (1961).
30. L. G. Rowan, E. L. Hahn, and W. B. Mims, *Phys. Rev.* **137**, A61 (1965).
31. W. B. Mims, *Proc. Roy. Soc. London* **A283**, 452 (1965).
32. A. Messiah, *Quantum Mechanics*, Interscience, New York (1961), p. 279.
33. V. F. Yudanov, A. M. Raitsimirov, and Yu. D. Tsvetkov, *Teor. i Eksperim. Khin* Vol. 4, 520 (1968).
34. E. L. Hahn and D. E. Maxwell, *Phys. Rev.* **88**, 1070 (1952).
35. N. A. Kurnit, I. D. Abella, and S. R. Hartmann, *Phys. Rev. Letters* **13**, 567 (1964)
36. D. Grischkowsky and S. R. Hartmann, *Phys. Rev. Letters* **20**, 41 (1968).
37. D. E. Kaplan, M. E. Browne, and J. A. Cowen, *Rev. Sci. Instr.* **32**, 1182 (1961)
38. W. B. Mims, *Rev. Sci. Instr.* **36**, 1472 (1965).
39. J. D. McGee, *IRE Trans. on Microwave Theory and Techniques* **MTT-8** (1960)
40. F. Volino, F. Csakvary, and P. Servoz-Gavin, in *Proc. XIV Colloque Ampere, Lju bljana* (1966), North-Holland Publishing Co., Amsterdam (1967), p. 1010.
41. A. L. Bloom, *Phys. Rev.* **98**, 1105 (1955).
42. E. T. Jaynes, *Phys. Rev.* **98**, 1099 (1955).
43. W. B. Mims, *Phys. Rev.* **141**, 499 (1966).
44. N. Bloembergen and R. V. Pound, *Phys. Rev.* **95**, 8 (1954).
45. J. Itoh, K. Asayama, and S. Kobayashi, *J. Phys. Soc. Japan* **18**, 455, 458 (1963)

46. V. V. Voevodsky, in *Eighth International Symposium on Free Radicals, Novosibirsk (1967)*.
47. A. Kiel and W. B. Mims, *Phys. Rev.* **161**, 386 (1967).
48. J. G. Castle, P. F. Chester, and P. G. Wagner, *Phys. Rev.* **119**, 935 (1960).
49. R. Brändle, G. J. Kruger, and W. Müller-Warmuth, *Z. Naturforsch.* **25a**, 1 (1970).
50. W. B. Mims, *Phys. Rev.* **133**, A835 (1964).
51. D. R. Taylor, R. P. Gillen, and P. H. Schmidt, *Phys. Rev.* **180**, 427 (1969).
52. W. B. Mims and D. R. Taylor, *Phys. Rev.* **3B**, 2103 (1971).
53. W. B. Mims, *Phys. Rev.* **3B**, 2840 (1971).
54. A. J. Freeman and R. B. Frankel, *Hyperfine Interactions*, Academic Press, New York (1965), pp. 126ff.
55. E. R. Davies, in *International Conference on RF Spectroscopy, Karl Marx University, Leipzig, 1969*.

Chapter 5

Optical Techniques in EPR in Solids

S. Geschwind

Bell Telephone Laboratories
Murray Hill, New Jersey

1. INTRODUCTION

During the past several years, there has been an increasing application of different optical techniques to a variety of EPR problems. The common thread that runs through most of these applications is that the population redistribution among magnetic substates, in passage through magnetic resonance in the ground or excited state of a paramagnetic center, produces a change in some aspect of either emitted or absorbed light associated with the center. The change in light, in effect, acts as the indicator of magnetic resonance rather than the direct observation of microwave power absorption by the paramagnetic specimen. The scaling up of the EPR detection from the microwave to the optical region renders these optical EPR methods extremely sensitive, as will be seen below. Therefore, a strong motivation for using these techniques has been to study the EPR spectra in sparsely populated excited states, which would be more difficult or impossible to do by the more conventional methods. Thus, the observation of excited-state EPR by optical detection comprises a good portion of this review. While major emphasis is placed on this topic in Section 2, different aspects of it will be treated throughout this chapter.

A very general result has emerged in the study of excited paramagnetic states which indicates that the magnetic sublevels of these states are selectively populated during the radiationless transitions from the higher-lying absorption bands. This selectivity is connected with ground-state polarization for ions with paramagnetic ground states and has been dubbed spin memory. The selective feeding of magnetic sublevels is also operative in nonradiative decay from singlet to triplet levels in a wide variety of systems. At the root of this selective feeding are spin selection rules for the optical transitions and equally strong selection rules for the radiationless decay. Even in the case of excited triplet states of aromatic compounds, where sufficient population is easily attainable to use standard EPR detection, optical EPR techniques have been applied to obtain information on the filling and emptying rates of the states that were not otherwise accessible. Excited-state spin memory is also an important factor which influences the ground-state polarization in the optical pumping of ground-state spins. In Section 3, these varied topics related to selective feeding of magnetic substates will be discussed.

Optical magnetic circular dichroism (MCD) has also been used as the probe of the recovery of ground-state spin temperature to measure ground-state spin–lattice relaxation T_1. Rather than being simply another way of doing the same thing, the optical detection of ground-state T_1 has often several very distinct advantages, among which may be the complete discarding of the microwaves (even as a saturating source for the spins) in the measurement of T_1. In addition, it has often proven to be a more reliable method of measuring T_1. This application of MCD will be treated in section 4. The combined use of magnetic circular dichroic absorption in the ground state and spin memory in the excited state to study excited-state EPR of *F*-centers will also be reviewed.

Excited-state EPR, in addition to producing a change in light due to population redistribution, also influences the light via the off-diagonal components or transverse magnetization. This results in a microwave modulation of the light which will be discussed in Section 5.

One of the topics in spin–lattice relaxation that has received a good deal of attention is the microwave phonon bottleneck, i.e., the failure of the phonons to act as a thermostat for relaxing spins because of the low phonon heat capacity. The behavior of the phonons in a bottlenecked situation has been invariably deduced by its back-influence upon the spins, i.e., the spins acted as a probe for phonon temperature. The advances made in the field of Brillouin light scattering in the last few years have made the direct and selective observation of these "hot" phonons possible for the first time. In Section 6, the study of these nonequilibrium phonons by Brillouin light scattering will be presented.

2. OPTICAL DETECTION OF EPR IN EXCITED STATES USING FLUORESCENCE

2.1. Use of Fluorescence As Monitor of Excited-State EPR

The vast array of data and understanding that characterizes EPR as a mature field is primarily confined to ground-state EPR. Increased interest has been shown in recent years in the magnetic structure of excited states in solids, whose Zeeman effects in some cases have been examined by conventional optical spectroscopic techniques. It would be desirable, however, to study these excited states with the high resolution characteristic of microwave EPR and ENDOR. In attempting to do EPR in optically excited states in crystals by the conventional method of observing the direct absorption of microwave power, one is usually hampered by the small number of ions that can be maintained in these states by optical pumping. The combination of optical pump power, quantum efficiency of absorption, and lifetime of excited states results in an excited-state population that is below the minimum number that can be detected by direct microwave absorption. Exceptions to this situation are found in the case of triplet excited states of organic molecules, which are generally so long-lived (of the order of 1 sec) that one can achieve the necessary population by optical excitation.* The study of excited-state triplets by standard microwave EPR has thus been and continues to be a rich field. However, even in this case, there has been a good deal of activity involving optical detection techniques which has provided particular information not otherwise available, and which will be briefly discussed in Section 3. Even in those other, exceptional cases where direct EPR absorption may be observed in optically excited states, optical techniques would usually prove more sensitive.

If the excited-state population is, however, too small to allow direct EPR, but is a fluorescent state, then one can look to change some aspect of this fluorescence in passing through magnetic resonance in the excited state. Optical detection of excited-state EPR is thus a "trigger detection" in that the absorption of a microwave photon triggers a change in emission of an optical photon. The detection of photons is thus displaced from the microwave range to the far more sensitive optical range. This is the exact analog in solids of the optical–rf double resonance techniques suggested by Brossel and Kastler[2] and first demonstrated by Brossel and Bitter[3]

* See Ref. 1. This was the first EPR in the triplet excited state.

and which have been so extensively applied to the study of excited state in gases.[4]

As an illustration of how the method would be applied, consider an io in a solid which has a ground-state magnetic doublet and an excited-stat magnetic doublet in which a steady-state population is maintained by som means of optical pumping. In solids, this pumping is invariably done b initial excitation to a higher-lying, broad absorption band and subsequen rapid, radiationless decay to the excited state in question. Assume that th excited state fluoresces to the ground state with selection rules given i Fig. 1, emitting two Zeeman components as shown. These lines will be circularly polarized along the field direction and linearly polarized (σ light) when viewed at right angles to the magnetic field. The π components corresponding to the $\Delta m_s = 0$ transitions are not shown for the sake of simplicity. We also assume that the required difference of populations for magnetic resonance exists between the Zeeman levels a and b, for example, $n_a < n_b$. This may have been brought about by thermalization in the excited state or by some selective means of feeding level b as compared to a, as will be described in Section 3. Several methods appropriate to differing experi-

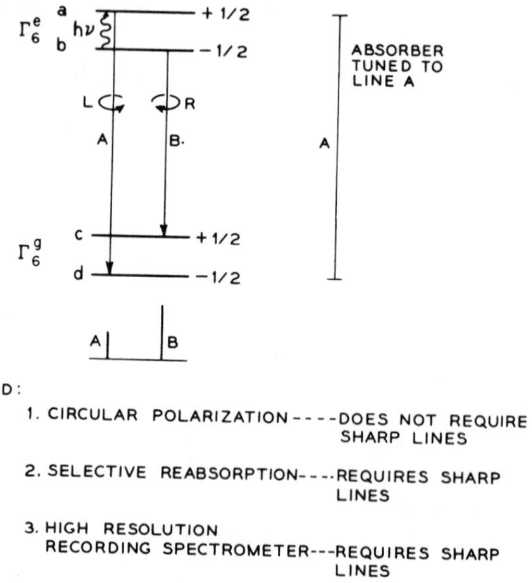

Fig. 1. Optical detection of EPR in an excited state. $h\nu$ is an induced microwave transition which changes some aspects of the fluorescence from Γ_6^e. The π components are omitted for sake of simplicity. It is assumed that the population of level b is greater than that of a either as a result of thermalization or selective feeding.

ental situations suggest themselves at this point, some of which have een reviewed earlier.[5]

.1.1. *Circular Polarization**

Since we assume the matrix elements for emission of lines A and B are qual, their relative intensities will be proportional to n_a and n_b, respectively. hus if one looks along the field direction with an appropriate circular olarization analyzer the light will be found to have a net right circular olarization, as we have assumed $n_a < n_b$. If the analyzer is sensitive to ight circularly polarized light only, then, in effect, one is monitoring the opulation n_b. If microwave transitions are induced between a and b, n_b s decreased, while n_a is increased, and one will observe a decrease in right ircularly polarized light. The optical linewidths may be much greater than he Zeeman splitting with this technique. The method has general applica-ility in cubic crystals and in axial crystals with the dc field applied in the irection of the unique axis, as it is only in this direction that the light is ompletely circularly polarized.

2.1.2. *Selective Reabsorption*

Let the *total* fluorescent light be viewed after passing through some absorption filter which is sufficiently sharp to discriminate between lines A and B and is tuned to line A, as shown in Fig. 1. A microwave transition induced between levels a and b increases the number of ions in level a at the expense of b and so increases emission in line A compared to B. As the absorber is tuned to line A, however, the total amount of light passing through the absorber is decreased. The absorber need not be external, but may be the fluorescent ions themselves. If lines A and B have sufficient oscillator strength and are narrow enough, and if a sufficient number of the paramagnetic centers are present in the specimen, then lines A and B will be internally reabsorbed in the sample. Reabsorption will decrease the intensities of the lines, as there is also present nonradiative decay to the ground state which competes with the fluorescence. However, if one is at low temperatures such that the population of level d, n_d, is greater than that of c, n_c, then line A will be more strongly reabsorbed than B. Thus a micro-wave transition in the excited state which transfers excitation from line B to A will decrease the total amount of emitted light from the sample due to this selective reabsorption. This technique has been used on a number of

* C. Hermann and G. Lampel have recently used the circularly polarized properties of the recombination radiation to detect spin resonance of photoexcited carriers in p-type GaSb [*Phys. Rev. Letters* **27**, 373 (1971)].

occasions in gases.[6] It was this method that was employed in the first optical detection of EPR in a solid in the study of excited $\bar{E}(^2E)$ state of Cr^{3} in Al_2O_3 by Geschwind *et al.*[7] following the study of this reabsorption by Varsanyi *et al.*[8]

Selective reabsorption can be applied only to very sharp emission line and to the lowest temperatures, as at higher temperatures, it becomes les selective as n_c approaches n_d. If, however, one encounters systems where A and B have different oscillator strengths, so that the selectivity of the reabsorption is not dependent upon $n_c \neq n_d$, then this method may be applied at higher temperatures as well.[6] For many paramagnetic center in solids, the combination of reduced oscillator strength, low concentration of impurity ions, and broad absorption lines makes this method inapplicable and it is presented here primarily for historical interest.

2.1.3. Use of High-Resolution Optical Spectrometer

If the widths of lines *A* and *B* are less than the Zeeman splitting, so that they can be resolved with a high-resolution spectrometer, then the spectrometer may be set at a fixed wavelength to coincide with line *A*, for example, and an increase in its intensity will be observed in passing through resonance in the excited state as ions are transferred from *b* to *a*. Of course, in sweeping through the excited-state EPR by varying the field and keeping the microwave frequency fixed, line *A* will not remain stationary and with spectrometer wavelength fixed, an intensity change will be recorded even off resonance. This unwanted background changes gradually as compared to the true EPR signal and can be eliminated by chopping the microwaves on and off and using lock-in detection. In cases where it is not feasible to use lock-in detection, it may be eliminated by taking one magnetic field sweep with microwave power on and the next with microwaves off and subtracting the two from each other electronically.[9]

2.1.4. Spin-Dependent Nonradiative Transitions

Imagine that a degree of nonradiative decay from the excited state exists in addition to the fluorescence and that it involves spin-dependent interactions, such that, for example, the nonradiative decay is stronger when the ion is in level *a* compared to *b*. Then, a microwave transition induced between *b* and *a* will *increase* the total nonradiative decay and *decrease* the total amount of emitted light from the excited state. Such a spin-dependent nonradiative decay plays a partial role in the optical detection of EPR in the excited triplet state of naphthalene, first done by Sharnoff[10].

Similar detection has been used to observe the triplet state of a number of other aromatic molecules.[11,12]* This spin-dependent nonradiative decay will be referred to again in Section 3. While, as was indicated earlier, enough population can be obtained in excited triplet states to do ordinary EPR, optical detection allows one to study the dynamics of the radiationless transitions, as will be briefly discussed in Section 3.

A spin-dependent nonradiative decay has recently been elegantly demonstrated by Porret and Lüty[13] for the relaxed excited state of the F-center in KCl, and they have indicated how this effect could be used to detect excited-state EPR of F-centers and have also suggested that this may have been what was seen in earlier experiments.[14]

2.2. Sensitivity and Saturation in Optical Detection

2.2.1. Signal-to-Noise Ratio of Optical Detection of EPR

It is fairly straightforward to derive a simple expression for the signal-to-noise ratio in the optical detection of excited-state EPR. Referring to Fig. 1, let n_a be the number of centers in the crystal in level a whose fluorescence is being monitored. This is being done either by circular polarized light for either sharp or broad emission lines, or a high-resolution spectrometer for sharp lines only. The total number of photons emitted per unit time in line A is n_a/τ_R and a fraction η is incident on the phototube; η takes account of solid angle and other geometric factors such as the illuminated region of the sample from which light is brought to the photodetector. Let E be the photodetector efficiency and α the fractional change in total population of n_a on magnetic resonance; α depends upon the population difference between a and b, the degree of saturation of levels a and b, hole burning effects if the line is inhomogeneous, and hyperfine structure. The latter two factors are important, as the fluorescent light comes from all spin packets and hfs components and not just those that are saturated. Then, the change is detected fluorescent light due to magnetic resonance in the excited state, i.e., the signal, is given by

$$S = (\alpha n_a/\tau_R)\eta E \tag{2.1}$$

This signal is to be detected against the background of shot noise given by

$$N = [(n_a/\tau_R)\eta E]^{1/2} \tag{2.2}$$

* For a review of this technique applied to molecular triplet states, see the article by A. L. Kwiram in *MTP International Review of Science (Series One), Physical Chemistry*, Vol. 4, ed. C. A. McDowell, Butterworths, London (1972).

The resultant S/N ratio is then

$$S/N = (\alpha^2 n_a \eta E/\tau_R)^{1/2}$$ (2.3

We have implicitly assumed a time constant of 1 sec in considering th number of photons emitted per unit n_a/τ_R. For a S/N of unity and th 1-sec time constant, one finds for the minimum number of centers in stat a which may be detected

$$n_a = \tau_R/\alpha^2 \eta E$$ (2.4

The parameters appearing in the above equation may vary by wide margin for different systems. For a typical example such as one finds in the excite state of ruby, i.e., $\alpha = 0.1$, $\eta = 10^{-3}$, $E = 10^{-2}$, $\tau_R = 3\times 10^{-3}$, one find $n_a \sim 3\times 10^4$, which is far smaller than one can detect by the more standar techniques. It is interesting to note the absence of kT in our expression fo minimum detectable sensitivity, as in the optical region one is limited b shot noise rather than thermal noise.

2.2.2. Line Shapes and Saturation Effects

The normal microwave magnetic resonance absorption *signal* is propor- tional to $\chi'' \sqrt{P}$, where χ'' is the imaginary part of the susceptibility an P the microwave power.[15] χ'' is proportional to the transverse componen of M, which decreases at high power levels as $1/P$, so that the EPR signa decreases as $1/P$. In contrast, the optical signal is proportional to th change in $(n_a - n_b)$, or to the change in the excited state magnetizatio ΔM_0^z produced by microwave saturation. Thus the optical signal approache asymptotically a maximum value $\sim\Delta M_0^z = M_0^z/2$ as P increases.

2.3. Experimental Procedure for Optical Detection of Excited-State EPR

Figure 2 shows a block diagram of a system that has been extensively used to observe excited-state EPR by optical detection techniques.[5,9] The sample is placed in a microwave cavity which has a hole on the bottom for admitting the broadband exciting light from a mercury lamp and slots cut in its side to allow the fluorescent light to escape from the cavity for analysis. The slots are cut in a direction parallel to the microwave current flow so as not to degrade the Q of the cavity. A view of such a cavity is shown in Fig. 3. For low-temperature work, the cavity is placed in liquid helium in a glass Dewar which has flat windows at the bottom and is com- pletely unsilvered and transparent in a 6-in. region at the bottom, so as to

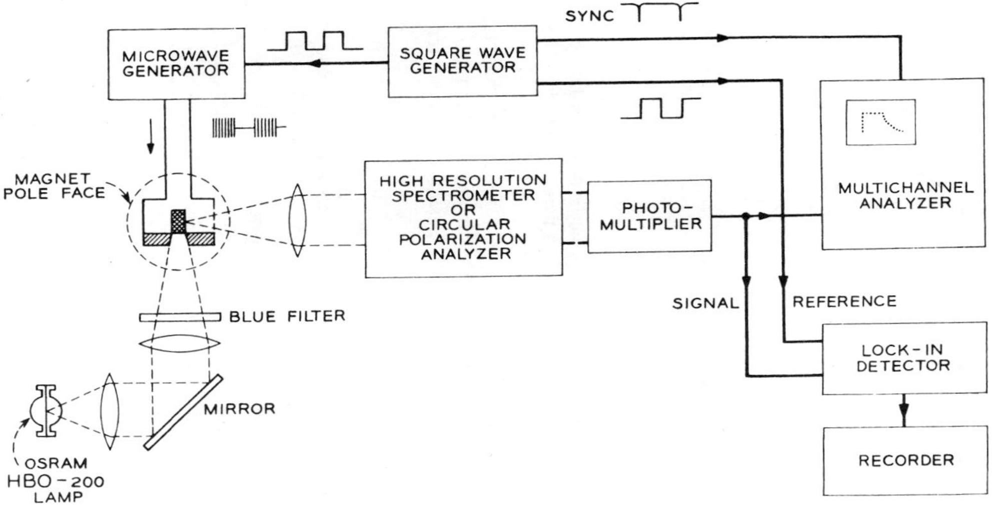

Fig. 2. Block diagram of system for optical detection of EPR in excited states. The lock-in detector is used for recording of spectra, while the multichannel analyzer is used for relaxation studies.[9]

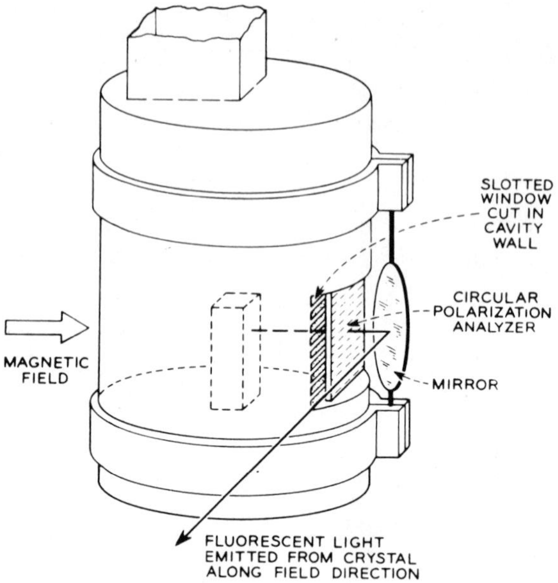

Fig. 3. Cavity used to detect circularly polarized fluorescence, of a given sense, from crystal. Paramagnetic ions are excited by light entering the cavity in a vertical direction from a hole cut in its base. A small mirror deflects the analyzed light out of the Dewar and magnet region into the more convenient direction perpendicular to the field.[9]

allow unimpeded entry and exit for the pump and fluorescent light. Thi Dewar is then surrounded by a similar outer Dewar containing liquid nitrogen

When the circular polarization technique is used, the light must be collected for analysis in a direction parallel to the magnetic field. If the cavity is in a Dewar between the magnet pole faces, then this light should be turned through 90° to be brought out to the photomultiplier at the more convenient direction at right angles to the pole faces. Nonetheless, the circular polarization analysis is best done in the liquid helium itself to reduce depolarization by the Dewar windows. This is done by a combination quarter-wave plate made from a piece of mica and a piece of sheet Polaroid placed at the cavity slots as shown in Fig. 3.[9] The analyzed light is then brought out of the Dewar by a small mirror mounted on the cavity. Attempts should also be made to reduce light reflected off the walls of the cavity from reaching the detector,[9] as it will be depolarized. An alternate method of bringing the analyzed circular polarized light out of the magnet region with a light-pipe has been used by Chase.[16] If one is dealing with harp lines and method 2.1.3 above, i.e., a high-resolution optical spectro-neter is being used, then the light may be brought out directly at right angles o the field into the spectrometer.

The microwaves are chopped at some convenient audio frequency. n slowly sweeping the magnetic field through resonance, this frequency omponent appears in the light signal and is detected with standard lock-in etection techniques. This chopping frequency must obviously be slower han the excited-state relaxation time. If the spin–lattice relaxation time in he excited state, T_1, is to be measured, the magnetic field is kept fixed and he microwaves modulated on and off as before, and the modulated light ignal, which contains T_1 information, is observed directly on an oscilloscope r, in the case of low S/N, is fed to an averaging computer.[7,9]

?.4. Illustrative Results

2.4.1. $\bar{E}(^2E)$ state of V^{2+}, Cr^{3+}, and Mn^{4+} in Al_2O_3

Extensive work has been done on the $\bar{E}(^2E)$ excited state of the isoelec-ronic sequence of d^3 ions, V^{2+}, Cr^{3+}, and Mn^{4+} in Al_2O_3 [5,9] This system herefore serves as a good illustration of the more general discussion that as preceded. These ions enter substitutionally for Al in the Al_2O_3 lattice n such a way as to preserve the local site symmetry C_3, the necessary charge ompensation for V^{2+} and Mn^{4+} apparently occurring at a sufficient distance from the site. The energy level diagram appropriate to these ions in Al_2O_3, shown in Fig. 4, was derived by Sugano and Tanabe[17] and early studies of the optical spectra of Cr^{3+} were made by Sugano and Tsujikawa.[18] A steady-state population in the $\bar{E}(^2E)$ level is maintained by means of continuous pumping light from a high-pressure mercury lamp which first raises the ions to the broad $^4T_2$ and $^4T_1$ absorption bands, from which they rapidly decay by radiationless transitions to the $\bar{E}$ level. This is followed by radiative decay to the ground state, i.e., the R-line fluorescence, for which only one Zeeman component, labeled α, is shown. To observe EPR in the excited state, use is made of methods described in subsections 2.1.1 and 2.1.3 to distinguish the fluorescence between the $-1/2$ and $+1/2$ levels of $\bar{E}(^2E)$. For V^{2+} and Cr^{3+}, which have sharp lines, one monitors the intensity of α with a high-resolution spectrometer and records a diminution of its intensity on microwave resonance. In the case of Mn^{4+}, the emission lines are many cm^{-1} wide, and so one must utilize the circular polarization technique.[9] The polarization properties of all the fluorescent Zeeman lines between the $\bar{E}(^2E)$ levels and the $^4A_2$ ground state have been studied experimentally.[9,19] While there are six emission lines, each excited-state level

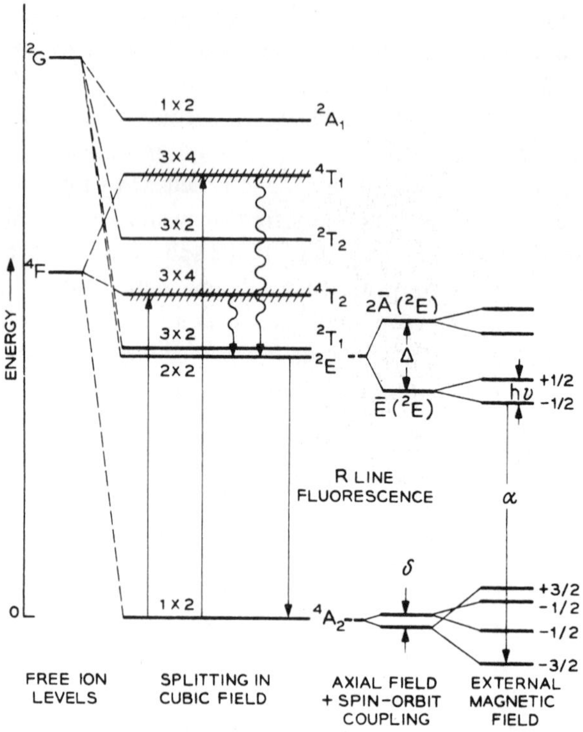

Fig. 4. Partial energy-level scheme of $(3d)^3$ ions in Al_2O_3 after Sugano and Tanabe.[17] Wavy lines indicate radiationless transitions from absorption bands to ^2E level.

emits preferentially one sense of circular polarization and this may b used to detect its population change in passing through resonance. Figure shows the EPR spectrum for Mn^{4+} with the expected six line hfs for Mn^{5} $(I = 5/2)$.[5,9] Similar g-value and hfs studies for V^{2+} and Cr^{3+} in Al_2O have led to as detailed an understanding of the hfs in the excited states o this isoelectronic sequence of three ions as one is accustomed to in ground state EPR.[9] Note that the EPR lines are only 10 G wide, while the optica lines being monitored are many cm^{-1} wide or 10^5 electron-gauss units i width. The optical line is inhomogeneously broadened so that each impurity center sees a different crystal field, but each ion has the same Zeeman splitting in the excited state throughout the spread of inhomogeneou crystal fields. This is very much akin to ENDOR in this regard. This point will be far more striking in the example of Eu^{2+}:CaF_2 to follow. The signa in Fig. 5 was obtained with approximately 5×10^{10} ions in the excited stat nd corresponds to a sensitivity of approximately 10^8 spins in a 1-G line-idth and integration time of 1 sec. While this S/N ratio is quite good by ormal EPR standards, it is actually lower here than might be hoped for, ue to a number of factors discussed in Ref. 9.

.4.2. Γ_8 *Excited State of* Eu^{2+} *in* CaF_2

The EPR spectrum of the excited-state Γ_8 ($4f^6 5d$) level of Eu^{2+} in the uorite lattice (CaF_2, SrF_2, and BaF_2) was studied by Chase,[20] who used n optical detection method which monitored the circular polarization of he intense blue fluorescence from the Γ_8 level to the ground state. Figure 6 hows the spectrum observed with the external magnetic field along a [111] irection of the crystal. The six hfs lines due to ^{153}Eu are marked and the

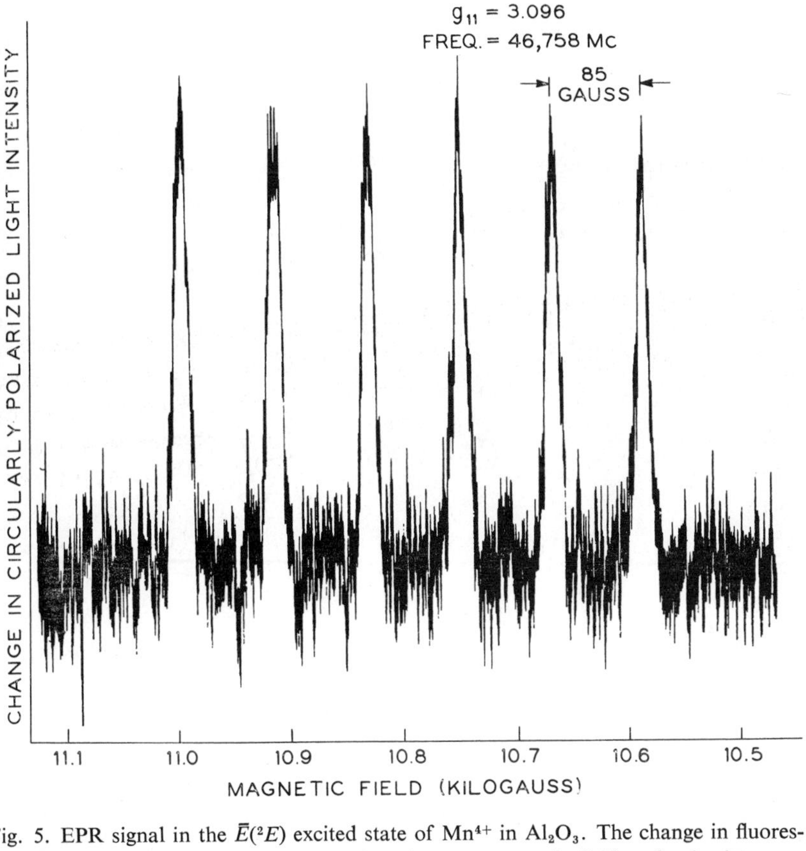

Fig. 5. EPR signal in the $\bar{E}(^2E)$ excited state of Mn^{4+} in Al_2O_3. The change in fluores-cent right circularly polarized light is plotted versus magnetic field at fixed microwave requency. [From Ref. 9.]

remaining five lines and the line overlapping with an Eu^{153} componen
are associated with ^{151}Eu. The spectrum shown in Fig. 6 resulted from
steady-state population of 10^9 ions in the Γ_8 excited state of Eu^{2+}.

This spectrum is remarkable on several counts; one is the good S/N
while another is the broad emission band that was monitored. The $4f^6 5d\,(\Gamma_8$

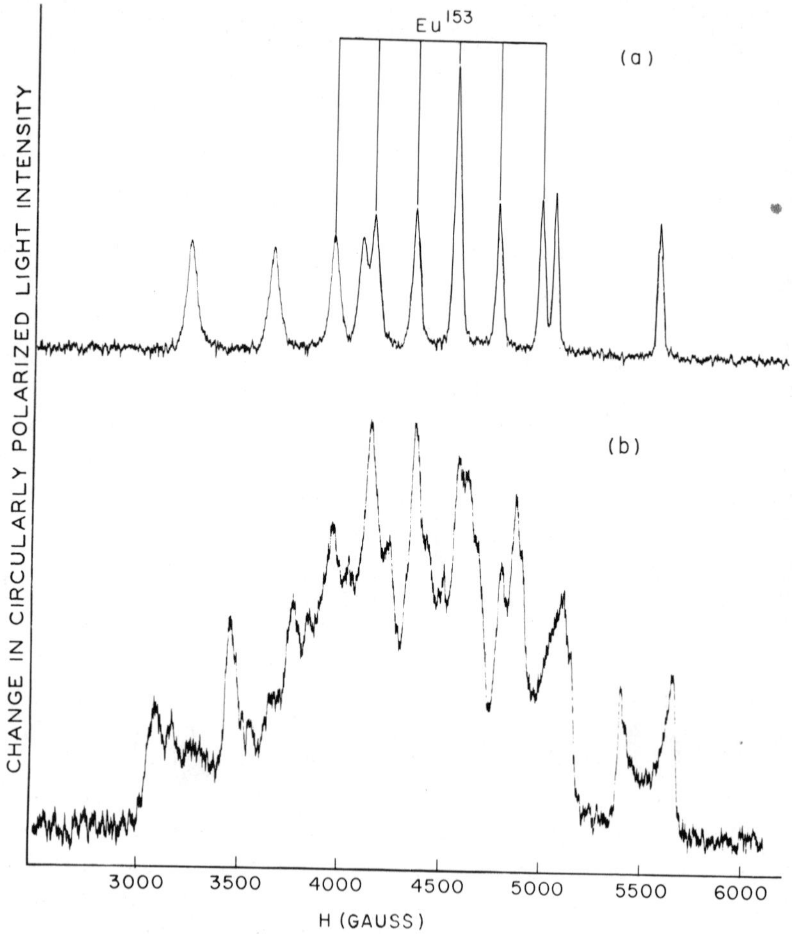

Fig. 6. Optically detected EPR signals for the Γ_8 excited state quartet of Eu^{2+} in CaF_2 (after Chase, Ref. 20). (a) Field orientation along [111] direction. The six hyperfine lines of ^{153}Eu are indicated, while the remaining five lines and the line overlapping an ^{153}Eu component are those of ^{151}Eu. (b) Spectrum observed 15° from [111] direction showing effects of random strain (see text and Ref. 20).

$\rightarrow 4f^7$ ($^8S_{1/2}$) transition had a sharp zero-phonon line at 4130 Å. However, nly 2% of the emission is in this line, the rest appearing in a broad vibronic pectrum extending over hundreds of angstroms. The fluorescence from the ntire band was monitored for the EPR detection as the signal increased in roportion to the width of that portion of the fluorescence spectrum that as used. Thus, this EPR spectrum emphasizes, in a far more striking way, ne remarks made above in connection with $Mn^{4+}:Al_2O_3$ about observing igh-resolution EPR in a broad optical spectrum.

The hyperfine splittings of a ^2E level or Γ_8 quartet, including the effects f strain, have been given by Chase[16] and Ham[21] as

$$\begin{aligned} E &= g\beta H + Am_I \\ g &= g_1 + qg_2[1 - 3(l^2m^2 + l^2n^2 + m^2n^2)]^{1/2}\cos(\varphi - \alpha) \qquad (2.5) \\ A &= A_1 + qA_2[1 - 3(l^2m^2 + l^2n^2 + m^2n^2)]^{1/2}\cos(\varphi - \alpha) \end{aligned}$$

I is the applied magnetic field, whose orientation relative to the cube xes is given by the direction cosines (l, m, n), and m_I is the component of uclear spin along H; q is an orbital reduction factor associated with oupling to the E_g vibrational modes of the fluorine ligands; φ is an angle vhose tangent measures the ratio of the $e_\theta \sim (3z^2 - r^2)$ and $e_\varepsilon \sim (x^2 - y^2)$ trains into which the random strains at a site can be decomposed; and α s a function of the magnetic field direction cosines. It is assumed that the train splitting of the orbital components at each lattice site is much greater han $g_2\beta H + A_2m_I$. [See Eq. (2.3.25) and the discussion in Section 2 of he chapter on the Jahn–Teller effect by F. Ham in this volume].

Equation (2.5) predicts two lines for each hfs component symmetrically plit about their average value g_1, which, however, coalesce in the [111] irection. Note the absence of this splitting in Fig. 6(a). For random distribution of strains φ, and of course away from the $\langle 111 \rangle$ direction, the pectrum resembles a powder pattern, with sharp boundaries at those oints of the spectrum corresponding to $\cos(\varphi - \alpha) = 0$ in (2.5). The plitting and characteristic sharp shoulder of such a strain-influenced Γ_8 uartet is shown in Fig. 6(b). Very similar strain-broadened spectra were bserved earlier by Chase[16] in the excited ^2E state of V^{2+} and Cr^{3+} in MgO. Iere, there is no axial field to help lift the orbital degeneracy of the ^2E tate as there is in Al_2O_3. Further analysis of the $Eu^{2+}:CaF_2$ spectrum by Chase[20] in field directions far from [111] revealed an asymmetric broadening or the different lines as one moved along the spectrum. This was identified s due to a Jahn–Teller singlet tunneling level some 10 cm^{-1} above the Γ_8 evel. This interpretation and analysis of Jahn–Teller EPR line shapes, done in such complete fashion for the first time actually in an excited sta of an ion, has clarified similar results seen in ground-state spectra whic had heretofore remained unexplained. For further details, see Ref. 20 an the chapter in this volume by F. Ham.

2.4.3. Measurement of Spin–Lattice Relaxation in the Excite State

Spin–lattice relaxation in the excited state may be measured by ob serving the recovery of the light signal on resonance after the microwave are switched off with the continuous pumping light kept on as shown i Fig. 7. The light signal may be derived by sitting on a resolved Zeema component with a high-resolution spectrometer or via circular polarizatio in either case, one is monitoring the recovery of the population of one c the Zeeman components. The rate equations for the populations of th two levels of $\bar{E}(^2E)$ labeled 1 and 2 may be written as

$$\dot{n}_1 = +U_1 - n_1 w_{12} + n_2 w_{21} - (n_1/\tau_R) \quad (2.6a)$$

$$\dot{n}_2 = +U_2 - n_2 w_{21} + n_1 w_{12} - (n_2/\tau_R) \quad (2.6b)$$

U_1 and U_2 are the feeding rates into levels 1 and 2 due to the exciting ligh

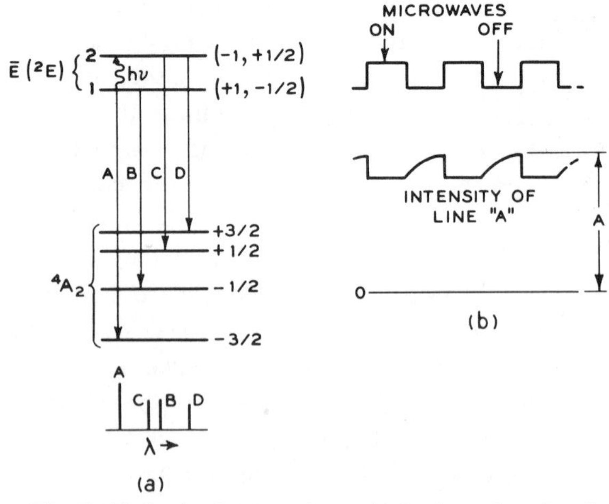

Fig. 7. Method of measuring spin–lattice relaxation in excited states. Population of lower level of excited-state doublet is monitored by observing intensity of line A for sharp line spectra or circularly polarized light for broad lines. Recovery rate of light signal is given by sum of rates of radiative decay to ground state and spin–lattice relaxation rate.

$_{12}$ and w_{21} are the phonon-induced flipping rates between 1 and 2, with $T_1 = w_{12} + w_{21}$, and τ_R is the radiative lifetime; $(n_1 + n_2)\tau_R = U_1 + U_2$ constant and $\dot{n}_1 = -\dot{n}_2$. We implicitly assume in the above equations at the pumping light does not disturb the ground-state equilibrium* (see ection 3.3). The solution to Eq. (2.6) corresponds to a recovery of n_1 or n_2 ward its equilibrium value with a time constant τ given by

$$1/\tau = (1/\tau_R) + (1/T_1) \tag{2.7}$$

The time τ_R may be measured in an independent experiment by ob- rving the decay of fluorescence after excitation by a fast light pulse, so at T_1 may be determined from a measurement of τ. We thus see that the easurement of T_1 in excited states is restricted to times that cannot be o much longer than τ_R.

In the case of sufficiently large signals at low temperatures, the recovery me τ may be seen directly on an oscilloscope screen. However, in many ases, especially above the λ-point of the He, where bubbling introduces vere noise, modern signal-averaging techniques must be used.[5,9] In Fig. 8a, e recovery time τ for the $\bar{E}(^2E)$ level of $Cr^{3+}:Al_2O_3$ is plotted on a semilog ale versus the reciprocal temperature $1/T$. It is apparent that above ap- roximately 2.5°K, τ is being limited by $\tau_R \ll T_1$, in accord with Eq. (2.7). he radiative lifetime of 4.3 msec in this sample is slightly longer than the ccepted value of 3.6 msec in very dilute ruby because of trapping.[22] If the oints in Fig. 8a are corrected for the 4.3-msec value of τ_R, values of T_1 re obtained which are plotted in Fig. 8b along with similar data from other rystals and at different frequencies.[5] It is seen that T_1 in this temperature ange is governed by an Orbach process (see Section 4.5 in the chapter n this volume by Orbach and Stapleton) involving the $2\bar{A}$ level, which is distance Δ above the $\bar{E}$ level,[5] as shown in Fig. 4, i.e.,

$$T_1 = \tfrac{1}{4}T(2\bar{A}_\pm \rightarrow \bar{E}_\mp)e^{\Delta/kT} \tag{2.8}$$

he direct process in the $\bar{E}$ level is apparently so much longer than τ_R hat it cannot be observed; $T(2\bar{A}_\pm \rightarrow \bar{E}_\mp)$ is the lifetime of the $2\bar{A}$ level with espect to a spontaneous phonon transition with a spin flip to the $\bar{E}$ level. he direct process rate from $2\bar{A}$ to $\bar{E}$ goes as $(1/\Delta^3)\tanh(\Delta/kT)$ (since $\bar{A}$ and $\bar{E}$ are non-time-reversed states) and for $\Delta \gg kT$ goes as $1/\Delta^3$, i.e., s the spontaneous emission rate. As Δ varies from 12 to 80 cm^{-1} in going cross the isoelectronic series V^{2+}, Cr^{3+}, and Mn^{4+}, a measurement of the oefficient $T(2\bar{A}_\pm \rightarrow \bar{E}_\mp)$ in the Orbach process for these ions provides a

We also neglect nonradiative decay to the ground state.

test of the Δ^3 dependence of the direct process for non-time-reversed stat in the regime $\Delta \gg kT$. While there have been previous experiments on io in ground states which showed a Δ^3 dependence by using the same io hosts with different Δ's,[23,24] they were not accompanied by a measureme of the orbit–lattice coupling parameter,[25] which was assumed to be th same. This orbit–lattice coupling parameter can be measured by stat strain and incorporated into the theoretical estimate of $T(2\bar{A}_+ \rightarrow \bar{E}_-$ The results of these measurements[9] are listed in Table I.

The reasonably good agreement between the experimental and theo retical values essentially verifies the $1/\Delta^3$ dependence of the direct proce for this case. The slight discrepancy for Mn^{4+} can be explained by phono dispersion.[9]

There is an advantage enjoyed by this optical technique of measurin T, besides the obvious one of sensitivity for excited states. This is connecte with the fact that the fluorescence from a given spin component level

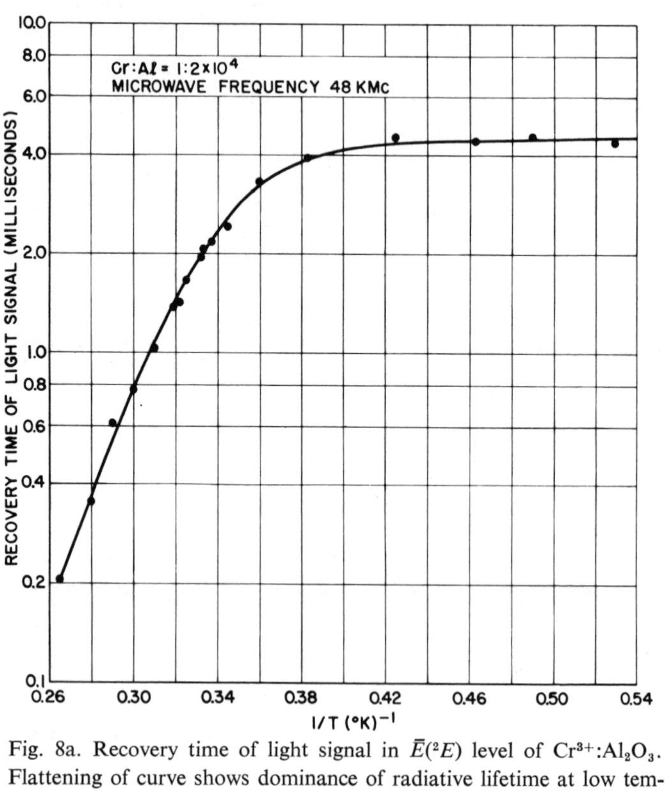

Fig. 8a. Recovery time of light signal in $\bar{E}(^2E)$ level of $Cr^{3+}:Al_2O_3$. Flattening of curve shows dominance of radiative lifetime at low temperatures.

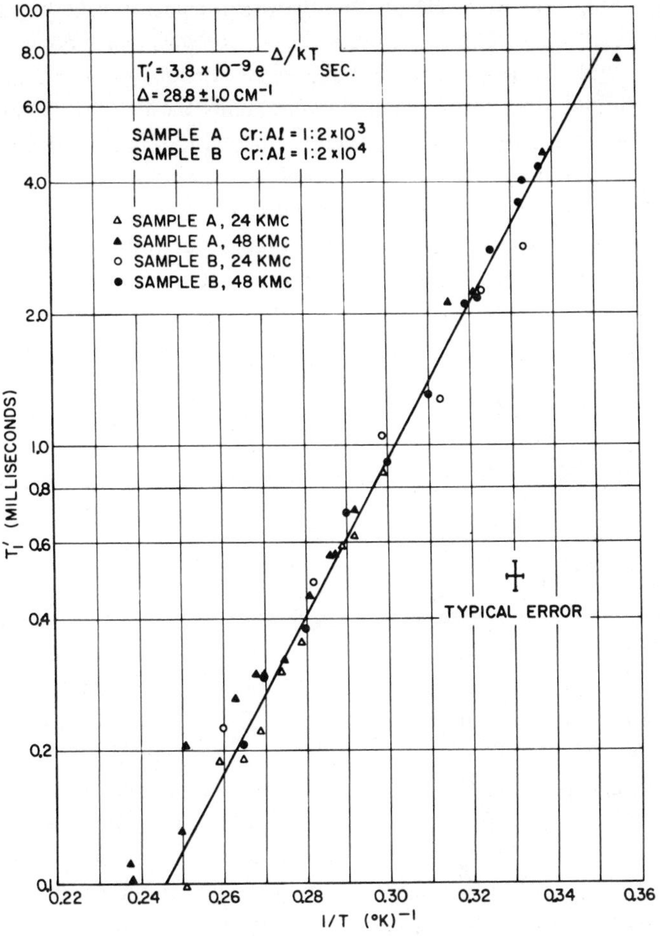

Fig. 8b. Plot of T_1' in $\bar{E}(^2E)$ of $Cr^{3+}:Al_2O_3$, after correction for radiative lifetime, showing Orbach relaxation involving $2\bar{A}(^2E)$ level 29 cm^{-1} above.

Table I

	Δ(cm^{-1})	$T(2\bar{A}_+ \to \bar{E}_-)$(theor)	$T(2\bar{A}_+ \to \bar{E}_-)$(expt)
V^{2+}	12.3	1.7×10^{-7}	2.1×10^{-7}
Cr^{3+}	29	3.0×10^{-8}	1.5×10^{-8}
Mn^{4+}	80	2×10^{-8}	6.4×10^{-10}

"coarse-grained" on the scale of microwave spin packet inhomogeneitie i.e., it looks at all the spin packets in the inhomogeneous EPR line. Thu if there is "hole burning" followed by spin diffusion within the line, th optical technique does not see this, but measures the true T_1. This is als true for cross-relaxation to other hfs components. In contrast, the standar microwave technique of pulsed saturation for measuring T_1 is often plague by spin diffusion. (See Section 3.7 in the chapter in this volume by W. Mim for a detailed discussion of spin diffusion through an inhomogeneous line Of course, the optical technique is still sensitive to cross-relaxation to foreig impurities.

3. SPIN MEMORY IN THE OPTICAL PUMPING CYCL

3.1. Spin Polarization Memory

3.1.1. Selective Feeding of Levels and Definition of Spin Memory

One of the essential requirements for performing excited-state EPR is the existence of a population difference among the magnetic sublevels This population difference may result from spin–lattice relaxation, in whic case there will be an equilibrium or Boltzmann distribution in the excite state. However, if the spin–lattice relaxation time in the excited state T_1^{ex}, is long compared to the radiative or nonradiative time out of the state the spins do not have time to thermalize. Such is the case in the excite $\bar{E}(^2E)$ state of Cr^{3+}:Al_2O_3 at very low temperatures, as may be seen b reference to Fig. 8b, where one observes that T_1 for the Orbach proces at very low temperatures extrapolates to seconds, whereas the radiativ lifetime to the ground state is 3.5 msec. Yet, even at the lowest temperatures excited-state EPR signals are easily observed, indicating therefore a selective feeding of the magnetic sublevels of the $\bar{E}(^2E)$ state. Evidence for thi selective feeding was recognized earlier, in the observation that the in tensities of the α and δ lines in Fig. 10 (Section 3.1.4) did not reflect a Boltzmann distribution at the lowest temperatures.[26] This selective feeding is present even when pumping with unpolarized light. A definitive demon stration that the feeding of the substates of $\bar{E}(^2E)$ was related to the popula tion distribution among the levels of the $^4A_2$ ground state was first given by Imbusch and Geschwind[27] and will be presented below. In effect, the excited-state magnetization retains some memory of the ground-state mag netization. This phenomenon of spin memory has been found to be generally revalent and to play an important role in many diverse systems and there-ore merits some general comments before proceeding to specific illustra-ons. In the most general fashion, spin memory in the optical pumping ycle refers to a connection between the magnetization M_0 in some initial et of magnetic substates (usually the ground state) and the magnetization $I' = f(M_0)$ in a later state to which the ion is carried by optical pumping adiationless decay, where M' may be positive or negative relative to M_0. 'his can be extended to the complete optical pumping cycle in which the aramagnetic species returns to its initial state. There are two parts to the ptical pumping cycle, i.e., the electromagnetic transitions and the phonon ecays. Spin selection rules (this would include fictitious spin as well) are perative for both types of transitions in such a way as to produce the onnection between M' and M_0.

3.1.2. Spin Selection Rules in Electromagnetic Transitions

First, let us consider spin memory associated with optical transitions. Group-theoretical and time-reversal symmetry arguments may be used to stablish optical selection rules. We will use the notation of Koster *et al.*'s ook[28] in the following discussion. For example, consider an optical transi-ion between two Γ_6 doublets in cubic symmetry as shown in Fig. 9(a) this, for example, could refer to Tm^{2+} in CaF_2, in which case they would e Γ_7 doublets).[29] In cubic O_h symmetry, the electric dipole operator trans-orms as Γ_4^- and the magnetic dipole operator as Γ_4^+. Depending upon the elative parity of the Γ_6's, they are connected by Γ_4^+ or Γ_4^- or both if either tate is of mixed parity. If one pumps with circularly polarized light rep-esented by the dipole operator $(x + iy)$, *along* the field direction Z, then

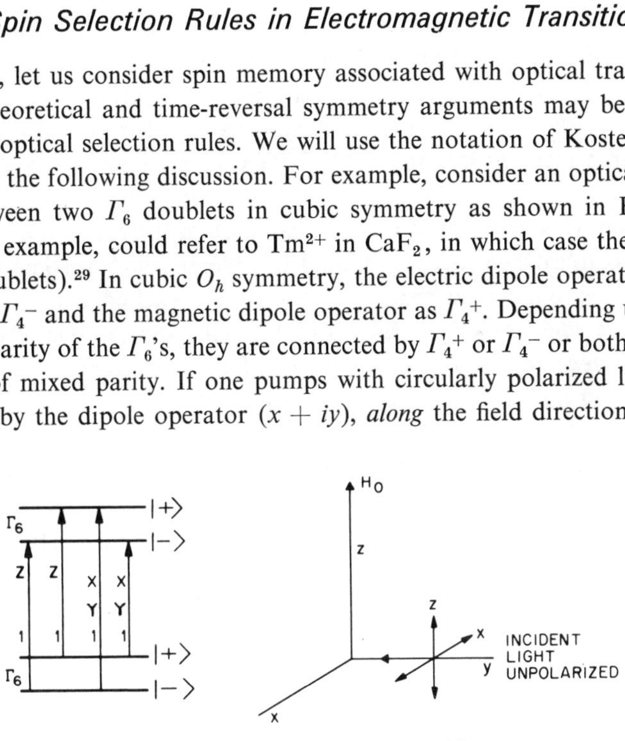

(a) (b)

Fig. 9. (a) Optical selection rules for Γ_6 (or Γ_7) doublets. (b) Pumping with unpolarized light at right angles to field populates both levels of excited state equally, independent of ground-state polarization, resulting in zero spin memory.

one finds from the tables of the vector coupling coefficients[28] that

$$\langle \Gamma_6^e + | x + iy | \Gamma_6^g - \rangle = 1 \tag{3.1}$$

and for the opposite sense of polarization $X - iy$

$$\langle \Gamma_6^e - | x - iy | \Gamma_6^g + \rangle = 1 \tag{3.2}$$

aside from the same reduced matrix element common to both transitions. Pumping with circularly polarized light therefore selectively connects one of the excited states to its opposite member in the ground state. This is perfect spin memory with a sign reversal.* To achieve selective pumping with unpolarized light along this direction, one must have a difference of population in the ground state. Now, if one pumps the same system with *unpolarized* light traveling at right angles to the magnetic field as shown in Fig. 9(b), then the matrix elements of x and z are again as shown in Fig. 9(a). It is seen that the transition probability from either of the ground states to both excited states is the same, so that, independently of any population difference in the ground state, the excited states will be equally populated and all spin memory is lost with this type of pumping between these Γ states.

However, similar group-theoretical arguments show that pumping with unpolarized light in ruby should in general result in some degree of transfer of ground-state magnetization to the excited state. The point symmetry at the Cr^{3+} site in ruby is C_3 and all states are one of two types of Kramers doublets. The components of one transform as Γ_4 and Γ_5 representations of C_3 and the components of the other as Γ_6 and $K\Gamma_6$ of C_3 (notation of Koster *et al.*[28] p. 51). The Γ_4 and Γ_5 transform respectively as $|+1/2\rangle$ and $|-1/2\rangle$, while Γ_6 and $K\Gamma_6$ as either $|-3/2\rangle$ or $|+3/2\rangle$. One cannot specify the relative intensities of an optical transition from $\Gamma_6 |-3/2\rangle$ to Γ_4 and Γ_5 since, even though Γ_4 and Γ_5 are time-reversed states of each other, they are different representations and as such will have different reduced matrix elements for the electric dipole operator with $\Gamma_6 |-3/2\rangle$. Similarly, there will be different matrix elements from the ground state $\Gamma_6 |-3/2\rangle$ to excited-state Γ_6 and $K\Gamma_6$ levels. Moreover, the different components of the electric dipole operator transform according to different representations. Thus, in general, the matrix elements for components of the electric dipole operator from the $\Gamma_6(-3/2)$ levels of the

* The term spin is used in its most general sense of fictitious spin. Real spin, on the other hand, will be explicitly referred to as such where the distinction is important.

;round state to all excited states are expected to be different, so that some legree of spin memory is anticipated in pumping with unpolarized light rom $\Gamma_6(-3/2)$. While time-reversal symmetry requires that the matrix elements from $K\Gamma_6(+3/2)$ and $\Gamma_6(-3/2)$ be the same to the corresponding et of conjugate excited states, a difference of ground-state population will esult in different net pumping rates from these levels, so that there will e magnetization transfer from ground to excited states.

Group-theoretical arguments as sketched above for a complicated ystem such as ruby cannot provide an answer to the question of the degree of spin memory. For this, one needs a knowledge of the reduced matrix elements, which can only be had in proportion to the accuracy with which excited-state wave functions are known.

3.1.3. Spin Selection Rules during Radiationless Transitions

Having seen how in general ground-state spin memory is operative in the optical transition, one must next examine the selection rules for the radiationless transitions involving phonon emissions. It is perhaps in connection with this part of the optical pumping cycle that spin memory came as a surprise to some. In going from the absorption band in ruby to the metastable level, the Cr^{3+} ion has to lose many thousands of cm^{-1} of energy. Since there are no phonons of this energy available in the lattice, the decay has to proceed via a cascade down many vibrational levels and "system crossings" to other electronic states and it was deemed improbable that spin memory could be preserved through so many complicated steps. The point is, however, that electronic states in a vibrational chain are essentially the same (to the extent that electronic–vibrational coupling may be neglected) so that phonon emission down a ladder of vibrational states proceeds without a change of electronic state (by definition), including m_s quantizations. System crossings or direct phonon transitions between different electronic states occur via an orbit–lattice operator[30,31] (see also Section 2.1 of the chapter by Orbach and Stapleton)

$$\mathscr{H}_{\mathrm{OL}} = \mathsf{V} : \boldsymbol{\epsilon} \tag{3.3}$$

where V is an electronic operator and $\boldsymbol{\epsilon}$ is a phonon operator. V and $\boldsymbol{\epsilon}$ are generally second-rank irreducible tensors whose components transform as the bases of the irreducible representation of the point group at the ion site and $\mathsf{V} : \boldsymbol{\epsilon}$ is a tensor dot product which transforms as the identity representation of the point group at the ion site. $\mathsf{V} : \boldsymbol{\epsilon}$ is analogous to $\mathbf{p} \cdot \mathbf{A}$ in the electromagnetic dipole radiation case, with $\mathbf{p}$ the electronic and **A** the photon operator. In principle, in a manner similar to that il lustrated above for the dipole operator, one can use the transformatio properties of the V's and group theory to arrive at selection rules fo phonon emission or absorption between electronic states. An illustratio of this procedure including detailed calculation of rates has been give for the transition between $2\bar{A}(^2E)$ and $\bar{E}(^2E)$ in ruby.[32] In practice, howeve one is often impeded in this because of lack of detailed knowledge of th excited-state absorption bands. Nonetheless, some general observation are possible. The orbit–lattice interaction does not operate on spins, bu only on orbital coordinates and as such can not *directly* flip real spin but may do so via the intermediary of spin–orbit coupling. $\mathscr{H}_{\text{OL}}$ can connec one component of a Kramers doublet, which we may label $A \mid -1/2\rangle$ to both components of another Kramers doublet $B \mid -1/2\rangle$ and $B \mid +1/2\rangle$ with, in general, different strengths only if they both have an $\mid S, m_s\rangle$ component of real spin in common with $A \mid -1/2\rangle$.

In those cases where the crystal field is much stronger than s–o cou pling so that spin and orbit are completely decoupled, the wave function may be described by unique m_s and m_l values. The fractional contaminatio of these states by other m_s values will then be given by something of th order of $\beta = \mathscr{H}_{\text{SO}}/E$, where $\mathscr{H}_{\text{SO}}$ is the s–o coupling and E is the energ separation of the nearest level which is coupled to the state by $\mathscr{H}_{\text{SO}}$. Thi situation is often encountered in transition metal ions, color centers i strong axial fields, and the triplet state of aromatic compounds. In th rare earth ions, s–o coupling is stronger than the crystal field and J is a good quantum number. The s–o coupling is now a strong handle for easily flipping real spins in transitions. It is then not very useful in the case of rare earths to think of spin memory in terms of real spin, but only in terms of fictitious spin with the selection rules given by the symmetry properties of the components of V and the symmetry of the states.

The smaller the quantity $\beta = \mathscr{H}_{\text{SO}}/E$, the more pronounced will be selective feeding to the magnetic substates and spin memory. However, there is a partial loss of spin memory in each phonon decay down a series of Kramers doublets. For phonon decay down a series of n Kramers doublets, the normalized difference of feeding rates into the two components of a final Kramers doublet starting from a given component of a higher doublet will be given by

$$\prod_{i=1}^{n} (1 - 2\alpha_j) \tag{3.4}$$

where α_i is the probability of a spin flip transition in going from the ith to he $(i+1)$th doublet down the chain. Since $|1-2\alpha_i|<1$. This product vill approach zero for large n and all spin memory will be lost. However, nost systems encountered involve both small enough α_i and a sufficiently mall number of different electronic states in the phonon decay so that pin memory seems to be quite a general phenomenon in optical pumping n solids. The case of an ion passing from an $S=0$ to $S=1$ state in the ptical pumping cycle, such as in non-Kramers systems as Ni^{2+} or the inglet–triplet transition, will be treated below.

3.1.4. Experimental Demonstrations of Ground-State Spin Memory

(a) *Ruby*. As indicated earlier, a nonthermal distribution was found etween Zeeman components of the $\bar{E}(^2E)$ level of ruby at very low temperatures, suggesting therefore a selective feeding of these levels. Referring o the inset of Fig. 10, the $\bar{E}$ level was populated by continuous pumping with unpolarized light into $^4T_2$ and subsequent radiationless decay to $\bar{E}(^2E)$. The relative populations of the $|-1/2\rangle$ and $|+1/2\rangle$ components of $\bar{E}$ were monitored by comparing the intensity of the α and δ lines as shown, recalling that the matrix elements for these transitions are equal. Pumping from the $^4A_2$ level to the quartet bands in this electric dipole transition[18] should follow a $\Delta m_s=0$ selection rule. To go from the quartet band to the doublet states, s–o interaction must be invoked and one would therefore anticipate a $\Delta m_s=0, \pm 1$ selection rule for this downward transition which would be the selection rule for the full pumping cycle. Of course, the $\Delta m_s=0$ and $\Delta m_s=\pm 1$ could have different intensities. To go from $^4A_2|-3/2\rangle$ (the most heavily populated state at low temperatures) to $^2E|+1/2\rangle$ corresponds to $\Delta m_s=+2$, which would require using s–o coupling twice and consequently be exceedingly weak. Thus ions pumped out of $^4A_2|-3/2\rangle$ will mainly connect to $\bar{E}|-1/2\rangle$.

Imbusch and Geschwind[27] provided direct experimental evidence for this link by saturating the ground state $|-3/2\rangle \rightarrow |-1/2\rangle$ transition with 48-GHz microwaves and observing a change in the relative populations of $\bar{E}$ via the intensities of the α and δ lines as shown in Fig. 10. Resonant trapping of the α radiation must be corrected for, but can be avoided by working in exceedingly dilute ruby, i.e., 8 ppm of Cr^{3+}. Figure 10 shows the fractional change in populations of $\bar{E}(\pm 1/2)$, measured as a fractional change in α and δ fluorescence, when the $(-3/2 \leftrightarrow -1/2)\ ^4A_2$ transition is saturated with microwaves. At the lowest temperatures, as a result of the $\Delta m_s=0, \pm 1$ selection rule, the $\bar{E}(-1/2)$ level is pumped from the

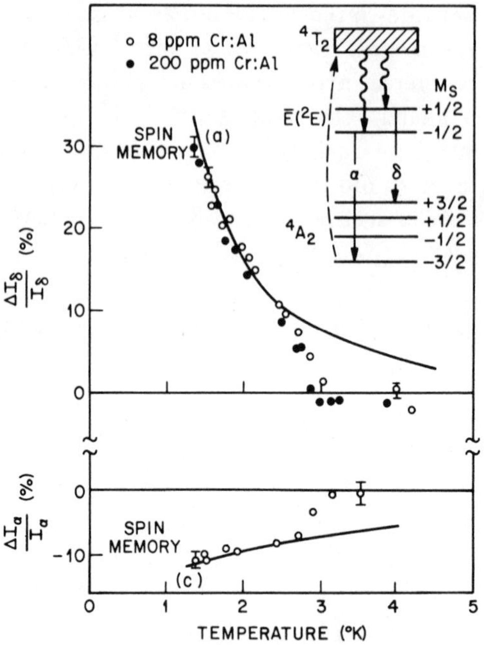

Fig. 10. Fractional change in α and δ fluorescence from $\bar{E}(^2E)$ level in ruby when $-3/2 \rightarrow -1/2$ level of ground state is saturated with microwaves. Spin memory is operative up to approximately 3°K, at which point, T_1 in excited state is faster than radiative time and population in $\bar{E}$ is no longer connected with groundstate polarization.

$^4A_2 \mid -3/2\rangle$ level, so that reducing the population of $^4A_2 \mid -3/2\rangle$ by microwave saturation should decrease the intensity of the α line and increase the intensity of the δ line, as is observed. These changes of intensity should be proportional to the difference in population ΔN between the $\mid -3/2\rangle$ and $\mid -1/2\rangle$ ground states, and to the relative efficiency of the $\Delta m_s = 0$ and $\Delta m_s = \pm 1$ selection rules. Solid curves were drawn which passed through a single low-temperature point for both ΔI_α and ΔI_δ and which were proportional to the Boltzmann variation of ΔN with temperature, as shown in Fig. 10. The remaining data points fit nicely on this curve until approximately 3°K. Above 3°K, microwave saturation of the ground-state transition produces no change in I_α or I_δ, i.e., the populations in $\bar{E}$ are independent of ground-state population above this temperature. This is as anticipated, since above 3°K, the spin–lattice relaxation time in $\bar{E}$ (see Fig. 8) is faster than the radiative decay rate to the ground state, so that thermalization occurs before radiation and spin memory is destroyed before fluorescence. Similar results[27] were obtained on more concentrated ruby, where corrections for trapping were necessary. Recent extension and analysis of such experiments by Adde and De Wijn[33] have enabled them to obtain the spin memory coefficients between all four ground-state levels and the components of $\bar{E}(^2E)$. A finite value is found for $^4A_2|-3/2\rangle \rightarrow \bar{E}(+1/2)$, but it is significantly smaller than the others, as anticipated.

(b) *Spin Memory in* $^2\Gamma_7$ *Metastable Level of* $Tm^{2+}:CaF_2$ Anderson and Sabisky[29] have done a very detailed experimental study and analysis of non-equilibrium population distributions in the excited $^2\Gamma_7$ level of $Tm^{2+}:CaF_2$. By saturating the ground-state $^2\Gamma_7$ resonances with microwaves, they have shown, in a fashion analogous to that described above for ruby, that there is spin polarization memory between the ground-state $^2\Gamma_7$ level in pumping with unpolarized light to higher-lying bands. The degree of this spin memory is strongly dependent upon the pump light wavelength, which enables one, by analysis of selection rules, to identify the symmetry character of the absorption bands. The spin memory selection rules are such in this case that an inverted population is obtained in the excited state $^2\Gamma_7$, as observed by the microwave lines being emissive.

In addition to spin memory, another process, optical cross-relaxation, is effective in establishing anomalous population distributions among the excited-state hyperfine levels. At low enough concentrations, however, such optical cross-relaxation disappears as expected. A high degree of nuclear spin memory is also demonstrated in the optical pumping cycle. This has been used by Jeffries and co-workers to polarize the Tm nuclei in the ground state as discussed by Jeffries in Section 5 of his chapter in this volume.

(c) *Spin Polarization Memory in Other Systems.* Spin memory is unquestionably operative in achieving the required population difference for many other excited-state EPR experiments at low temperatures. We cite as examples the analogs of the excited $\bar{E}(^2E)$ state of ruby, i.e., $V^{2+}:Al_2O_3$ and $Mn^4:Al_2O_3$,[9] and the relaxed excited state of the *F*-center, which will be discussed below.

3.2. Spin Selection Rules in Singlet–Triplet Transitions

3.2.1. Singlet–Triplet Selection Rules in Axial Fields

The first actual demonstration of spin selection rules in radiationless transitions in an optical pumping cycle was for a system with an $S = 0$ ground state of a coupled spin pair by Tanimoto *et al.*[34]

While there can obviously be no spin polarization memory in this case, strong spin selection rules are still operative. Spin–orbit coupling will mix the spin singlet and triplet states so that they may be connected by the electric dipole operator **p** and the orbital operator V in $\mathscr{H}_{OL}$. In low-symmetry crystalline fields, the $\Delta S_z = 0, +1, -1$ mixing between singlet and triplet may all be different from each other and thereby give rise to spin selection rules in going from singlet to triplet via photons or phonons. In axial fields, the mixings of $\Delta S = +1$ and $\Delta S_z = -1$ are equal so that the selective feeding produces a spin alignment rather than a spin polarization and one may refer to this as spin alignment memory. However, in terms of their origin, the distinction between spin polarization memory and spin alignment is academic, as both result from the same type of spin selection rules involving the electric dipole operator and the orbital operator V which were discussed earlier.

Tanimoto *et al.*[34] studied the F_t-center in CaO, which is approximately an exchange-coupled F-center pair in which the two electrons occupy the opposing O_2^- sites of a linear trivacancy $O^{2-} \cdots Ca^{2+} \cdots O^{2-}$. The ground state is a $^1A_1$ and the spin triplet $^3A_2$ lies a few hundred degrees above the ^1A as shown in Fig. 11. There is a zero field splitting of the triplet due to dipolar coupling. The excited optical states involve an *sp* configuration. Pumping with unpolarized light into the *sp* bands is followed by a cascade of primarily radiationless transitions down to the $^3A_2$ state. With H_0 either parallel or perpendicular to the axis of the center, the low-field

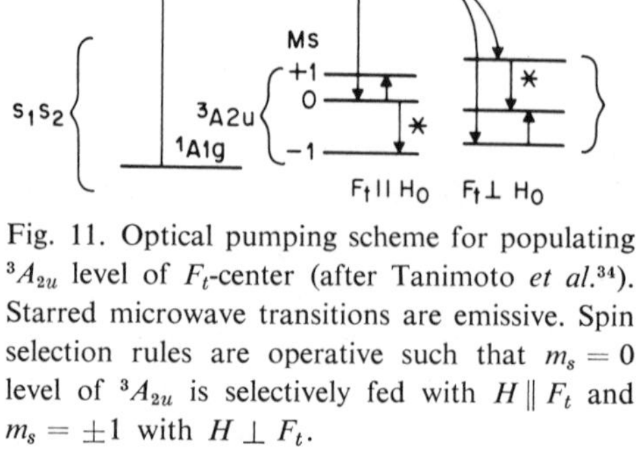

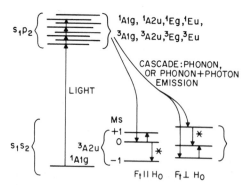

Fig. 11. Optical pumping scheme for populating $^3A_{2u}$ level of F_t-center (after Tanimoto *et al.*[34]). Starred microwave transitions are emissive. Spin selection rules are operative such that $m_s = 0$ level of $^3A_{2u}$ is selectively fed with $H \parallel F_t$ and $m_s = \pm 1$ with $H \perp F_t$.

EPR lines were found to be emissive, indicating the preferential feeding of the $M_s = 0$ level with $H \parallel F_t$ and the $M_s = \pm 1$ levels with $H \perp F_t$. These selection rules may be understood as follows.

If one neglects the crystal field and s–o coupling, then the ground states are ${}^1S$ and ${}^3S$ and the excited states are ${}^1P$ and ${}^3P$. In an axial field which is much stronger than the s–o coupling, the P states split into $P_{\pm 1}(E)$ and $P_0(A_2)$. If one now considers the s–o coupling, $\zeta(\mathbf{l}_1 \cdot \mathbf{s}_1 + \mathbf{l}_2 \cdot \mathbf{s}_2)$, it is clear that it will be strong along the Z axis (crystal field axis) and quenched at right angles to it, so that mixing of triplet and singlet is much stronger via the $l_z s_z$ matrix elements than via the $(l_+ s_- + l_- s_+)$. Another way of stating this is that $l_z s_z$ operator mixes the orbital components of ${}^3P$ and ${}^1P$ which are very close to each other, i.e., the ${}^3P_{\pm 1}$ with ${}^1P_{\pm 1}$, whereas the $l_+ s_-$ mixes the ${}^3P_{\pm 1}$ and ${}^1P_0$, which are split far apart by the crystal field. Thus, since it is $l_z s_z$ which strongly connects ${}^3P$ and ${}^1P$, the $\Delta S_z = 0$ mixing is strongest and the singlet states are contaminated chiefly with the $S_z = 0$ component of the triplet state, so that transitions into and out of the triplet state will involve the $\Delta S_z = 0$ selection rule. We have chosen the crystal field axis as the quantization direction for the spins and this quantization direction is maintained in ${}^3A_2$ with $H_0 \parallel F_t$. If H_0 is applied in a direction perpendicular to F_t, it is only the spin states that are labeled in H_0 as $|\pm 1\rangle$ which have the $M_s = 0$ spin state referred to the crystal field direction, so that they will now be preferentially fed as shown in Fig. 11.

A similar selective feeding into the very similar triplet state of an aluminum vacancy pair in silicon was also observed by Watkins.[35]

3.2.2. Singlet–Triplet Selection Rules in Lower Symmetries

From the discussion of the last paragraph, it is clear that in fields of orthorhombic symmetry, the different spin components may all have different selection rules for phonon and photon transitions. This was first demonstrated by Schwoerer and Wolf[36] for the excited triplet state of naphthalene, whose levels are shown in Fig. 12. Sufficient population can be easily maintained in the excited state by optical pumping, so that direct ESR is done. The zero-field states are labeled τ_x, τ_y, and τ_z as shown in Fig. 12 and correspond to the spins lying in planes perpendicular to the x, y, and z axes of the molecule. The high-field states labeled $|\pm 1, 0\rangle$ have the combinations of the τ's shown in the figure.[36,37]

They found both emission and absorption lines in the microwave spectrum as shown in Fig. 12, with the emission lines indicated by an arrow

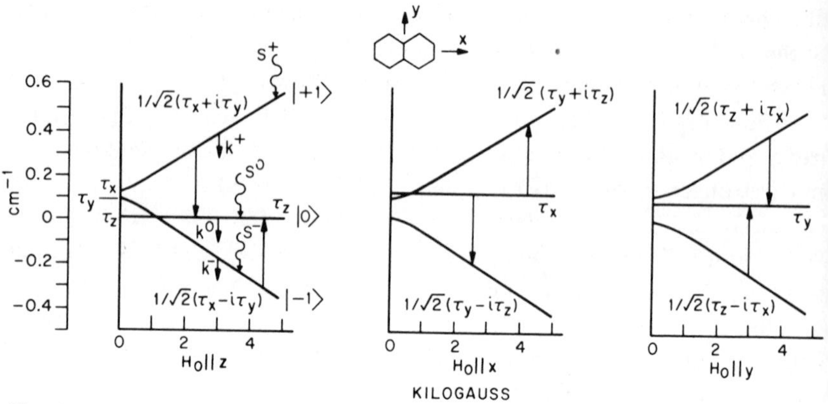

Fig. 12. Excited triplet levels in aromatic molecules (such as naphthalene diluted in benzene). S^+, S^0, and S^- refer to feeding rates into levels at high field and k^+, k^0, k^- refer to rates for leaving these levels. Arrows pointing down indicate emissive microwave transitions, while those pointing up are absorptive. Spin selection rules are such that high-field levels containing τ_x are selectively populated (after Schwoerer and Sixl.[39]).

pointing down. This implies that the selection rules are such that the feeding is strongest to those levels which in high-field contain the component τ_x. Transient experiments in naphthalene and similar molecules in which the pumping light is switched on and off[38–40] also verified that the molecule leaves the triplet state as well preferentially via τ_x. The origin of these selection rules is exactly the same as outlined in the preceding section except for the complication of lower symmetry. Transitions into and out of the triplet state become allowed by a weak s–o coupling into only those Zeeman sublevels of the triplet state that show the proper symmetry with respect to the singlet molecular states from which they are populated, as determined by the molecular axes.[37]

Since the emission intensities from the different levels are different (as well as the radiationless transitions), any redistribution of population among the levels, as, for example, by microwave transitions, will result in a change in the intensity of the total phosphorescence. This is the basis of optical detection in these triplet states mentioned in Section 2. Another variant of this idea is the measurement of the radiationless decay time out of some of the states, alluded to earlier, by "microwave-induced phosphorescence."[38] Suppose that in zero magnetic field, the triplet state is populated by a transient pulse of pumping light. After the light is shut off, the phosphorescent signal observed will be primarily due to emission from the rapidly decaying τ_x state. At a time t long compared to the decay time of the τ_x state, τ_x is practically empty and molecules are left in the slowly decaying and weakly emitting τ_y and τ_z states. By applying a saturating radio frequency between the τ_x and τ_y levels, for example, a sudden increase in light is seen due to those molecules transferred from the τ_y level into the strongly emitting τ_x level. Thus the intensity of this induced phosphorescence measures the molecules remaining in τ_y after the light has been shut off. By applying the radio frequency at different times after the light pulse, one can thus measure the decay time out of the τ_y level. By this elegant technique, the dynamics of filling and emptying of the triplet state may be studied.[38]

3.3. Optical Pumping of Ground-State Spin Polarization

Optical pumping of the ground-state spins refers to a change in the polarization of the ground state, P_e, with its total population remaining essentially constant, which is brought about by the optical pumping cycle. This well-known and widely applied method of polarizing atoms was first suggested by Kastler,[41] and Brossel[42] indicated its application to solids. As an illustration, if one pumps the Γ_6 doublets shown in Fig. 9(a) with right circularly polarized light, then one pumps out of the lower level of the ground-state doublet only and transfers the ions to the $|+\rangle$ level of the upper Γ_6 doublet. The ions radiate from this level back to both $|+\rangle$ and $|-\rangle$ levels of the ground state with equal probability, emitting right circularly polarized and π-polarized radiation with equal intensities. If all relaxation processes are neglected, and as there is no pumping out of $\Gamma_6^g|+\rangle$, one has a continuous transfer from $\Gamma_6^g|-\rangle$ to $\Gamma_6^g|+\rangle$, with the final equilibrium situation being attained under steady-state illumination in which all the ions are in $\Gamma_6^g|+\rangle$. If one had pumped with light of the opposite sense of circular polarization, then $\Gamma_6^g|-\rangle$ would in turn have been completely populated. We assume that the excited-state lifetime is much shorter than the reciprocal of the pumping rate u out of the ground state, so that the excited-state population is negligible. The time constant with which this equilibrium is reached is given by $T_p = 1/u$. Now, if the ground-state relaxation time T_1 is introduced, then it is obvious that one must have $T_p \lesssim T_1$ for effective optical pumping of the ground-state polarization P_e. If $T_1 \ll T_p$, the ground-state population will retain an essentially Boltzmann distribution, as was assumed earlier in Section 2.3.3 in the discussion of the measurement of T_1 in excited states.

* We refer to the electronic polarization. For a discussion of nuclear polarization by optical pumping, see Section 5 in Chapter 3 by C. D. Jeffries.

3.3.1. *Influence of Spin Memory*

In terms of our spin memory vocabulary, the example above is one in which there is a complete loss of spin memory in the full pumping cycle as the ions returned with equal probability to $\Gamma_6^g \mid +\rangle$ and $\Gamma_6^g \mid -\rangle$. If one considers, by contrast, a system of doublets whose properties are such that the spin retains perfect memory of its ground-state polarization during the complete pumping cycle of excitation, radiationless decay, and radiative decay back to the ground state, then obviously the pumping light, no matter how intense, will have no effect upon the ground-state polarization. This latter statement is a tautology, but becomes more meaningful when one considers the effect of rf resonance in the excited state. If saturating microwaves induce spin flips in the excited state at a rate which is faster than its decay rate, then the microwaves are destroying the spin memory, as the ions will tend to radiate to both ground-state levels equally from the equally populated excited states and the pumping of the ground-state spins will result. In this fashion, the almost perfect spin memory in the optical pumping cycle of *F*-centers has been used to observe EPR in the relaxed excited state of *F*-centers by monitoring the subsequent change in P_e as will be described in Section 4.3.

3.3.2. *Generalized Rate Equations for Optical Pumping*

We consider a fairly general cycle of optical pumping of ground-state electronic spin polarization as shown in Fig. 13, in which spin memory, rf perturbation of excited states, and relaxation are taken into account. Our

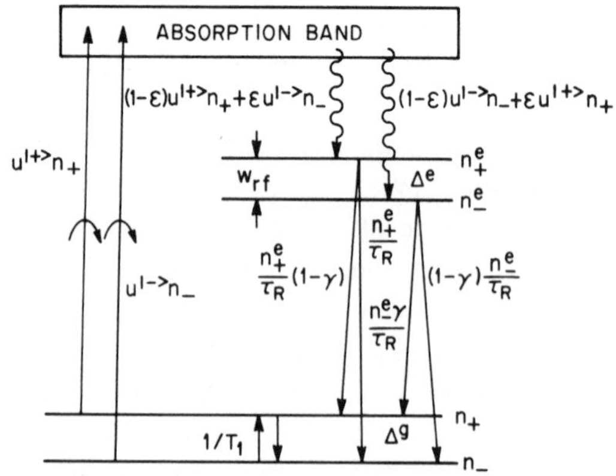

Fig. 13. Generalized optical pumping cycle in solids. $u^{|+\rangle}$ and $u^{|-\rangle}$ refer to pumping rates from $\mid +\rangle$ and $\mid -\rangle$ levels, respectively, with a given sense of circular polarization. ε and γ measure loss of spin memory. See text.

model system consists of a ground-state doublet which is pumped to an absorption band, rapid decay from the band to an excited doublet, and final radiative decay to the ground state. To treat higher-multiplicity ground states, such as ruby with $S = 3/2$ or Eu^{2+} with $S = 7/2$, would introduce much greater algebraic complexity and tend to obscure the basic physical elements. Our chosen model system also actually corresponds to many important cases, such as *F*-centers in alkali halides and Tm^{2+} in CaF_2.

Referring to Fig. 13, let $u^{|+\rangle}$ and $u^{|-\rangle}$ be the optical pumping rates respectively from the $|+\rangle$ and $|-\rangle$ levels of the ground state with populations n_+ and n_- for a *given sense* of circularly polarized light, for example, right circularly polarized light. $u^{|+\rangle}$ and $u^{|-\rangle}$ are proportional to light intensity and matrix elements. Recall that if the ground state is a simple Γ_7 doublet and the excited state another single Γ_7 doublet (which may extend over a broad vibrational band) then one of the u's would be zero. In general, however, because of more complex and overlapping bands, they are both finite, but $u^{|-\rangle} \neq u^{|+\rangle}$. We assume that the nonradiative decay from the absorption band to the excited-state doublet is much more rapid than the radiative decay back to the ground state, as is almost universally true, so that radiation back to the ground state from the band may be neglected. Let ε be a parameter which measures loss of spin memory in going from the band to the excited state, i.e., $\varepsilon = 0$ and $+1$ imply perfect spin memory without and with change of sign, respectively, while $\varepsilon = 1/2$ corresponds to complete loss of spin memory. Let γ be equal to the probability of a spin flip in a radiative transition, with lifetime τ_R, from the excited state to the ground state. w^e_{+-} and w^e_{-+} are the transition rates between excited-state levels due to spin–lattice relaxation, i.e., $1/T_1^{\text{ex}} = w^e_{+-} + w^e_{-+}$. w_{rf} is the transition rate in the excited state in the presence of resonant microwaves. T_1 is the ground-state relaxation time, and Δ^g and Δ^e are the respective Zeeman splittings in ground and excited states. The rate equations for the four levels of the ground and excited states are

$$\dot{n}_+ = -u^{|+\rangle}n_+ + (1-\gamma)\frac{n_+{}^e}{\tau_R} - \frac{n_+}{T_1} + \frac{1}{T_1}\frac{n_+ + n_-}{1+\exp \Delta^g} + \gamma\frac{n_-{}^e}{\tau_R} \tag{3.5a}$$

$$\dot{n}_- = -u^{|-\rangle}n_- + (1-\gamma)\frac{n_-{}^e}{\tau_R} - \frac{n_-}{T_1} + \frac{1}{T_1}\frac{n_+ + n_-}{1+\exp -\Delta^g} + \gamma\frac{n_+{}^e}{\tau_R} \tag{3.5b}$$

$$\dot{n}_+{}^e = (1-\varepsilon)u^{|+\rangle}n_+ + \varepsilon u^{|-\rangle}n_- + w_{\text{rf}}(n_-{}^e - n_+{}^e) - w^e_{+-}n_+{}^e + w^e_{-+}n_-{}^e - (n_+{}^e/\tau_R) \tag{3.5c}$$

$$\dot{n}_-{}^e = (1-\varepsilon)u^{|-\rangle}n_- + \varepsilon u^{|+\rangle}n_+ + w_{\text{rf}}(n_+{}^e - n_-{}^e) - w^e_{-+}n_-{}^e + w^e_{+-}n_+{}^e - (n_-{}^e/\tau_R) \tag{3.5d}$$

In dynamic equilibrium, with the $\dot{n}$'s $= 0$, one finds that the *ground*-state polarization $P_e = (n_+ - n_-)/(n_+ + n_-)$ is given by

$$P_e = \frac{P_{es} - (T_p/T_1)\tanh(\Delta^g/2kT)}{1 + (T_p/T_1)} \tag{3.6a}$$

where

$$P_{es} = \left[\frac{u^{|-\rangle} - u^{|+\rangle}}{u^{|-\rangle} + u^{|+\rangle}} + \frac{\sigma}{\alpha}\right] \Big/ \left[1 + \frac{\sigma}{\alpha}\,\frac{u^{|-\rangle} - u^{|+\rangle}}{u^{|-\rangle} + u^{|+\rangle}}\right] \tag{3.6b}$$

$$1/T_p = -\tfrac{1}{2}[(\alpha + \sigma)(u^{|-\rangle}) + (u^{|+\rangle})(\alpha - \sigma)] \tag{3.6c}$$

$$\alpha = \{(1 - 2\gamma)(1 - 2\varepsilon)/[2w_{\mathrm{rf}}\tau_R + 1 + (\tau_R/T_1^{\mathrm{ex}})]\} - 1 \tag{3.6d}$$

$$\sigma = \{(1 - 2\gamma)/[2w_{\mathrm{rf}}T_1^{\mathrm{ex}} + 1 + (T_1^{\mathrm{ex}}/\tau_R)]\}\tanh(\Delta^e/2kT) \tag{3.6e}$$

Eq. (3.6a) is the same as that given by Mollenauer *et al.*[43] except that T_p is now generalized to include T_1^{ex} in the excited state and loss of spin memory in the radiative decay as expressed by $\gamma \neq 0$. Complete spin memory in the optical pumping cycle would require $\varepsilon = 0$, $\gamma = 0$, $w_{\mathrm{rf}} = 0$, $T_1^{\mathrm{ex}} = \infty$, in which case $\alpha = \sigma = 0$ and $T_p \to \infty$, so that $P_e \to -\tanh(\Delta^g/2kT)$, i.e., its equilibrium value, and we have no optical pumping, as expected. One can also easily see how, if there is some loss of spin memory, so that T_p is finite, that if the pumping light is sufficiently intense to make $T_p \ll T_1$, the saturated polarization P_{es} is reached, which may be positive or negative depending upon the relative magnitudes of $u^{|+\rangle}$ and $u^{|-\rangle}$.

Inversion of ground-state spins by optical pumping has been demonstrated in a number of systems, including *F*-centers in KCl,[44] Tm^{2+} in CaF_2,[45] and Cr^{3+} in Al_2O_3.[46] Such reversal of ground-state polarization by optical pumping has been actually used to achieve maser action in Tm^{2+} in CaF_2[45] and Cr^{3+}:Al_2O_3.[46]

Optical pumping may also be used to drive the ground-state P_e from equilibrium, and then follow its recovery back to equilibrium after removal of the light, to measure T_1, as will be noted in Section 4.

4. MAGNETIC CIRCULAR DICHROIC TECHNIQUES IN EPR

4.1. General Background

4.1.1. Microscopic Origin of Magnetic Circular Dichroism

Magnetic circular dichroic absorption (MCDA) refers to the differential absorption for left and right circularly polarized light by a crystal in the presence of a static magnetic field H_0 with the light propagating along the field. The microscopic origin of MCDA is in the Zeeman splitting of levels and is most simply illustrated by reference to the ground- and excited-state doublets of Fig. 9. In the absence of the Zeeman splitting, the populations of the $|+\rangle$ and $|-\rangle$ levels of the ground state are equal, and since by time-reversal symmetry, the absorption of σ_+ and σ_- light would be equal, there would be no MCDA. The presence of H_0 gives rise to MCDA via three routes:

(i) *Paramagnetic effect*: the difference of populations of the magnetic sublevels of the ground state, i.e., $N_- - N_+ = N \tanh(g\beta H_0/2kT)$, where N is the total ground-state population.

(ii) *Mixing of states* due to H_0, so that time-reversal symmetry is broken and matrix elements for σ_+ and σ_- are no longer equal.

(iii) *Different energy denominators* appearing in the Kramers–Heisenberg dispersion and absorption formulas for σ^+ and σ^- due to Zeeman splitting.

These three factors are expressed quantitatively by the Kramers–Heisenberg dispersion formula for the complex electric susceptibility χ in an optical field for electric dipole transitions[47,48]

$$\chi_\pm(\omega) = \frac{(\langle n\rangle^2 + 2)^2}{9\langle n\rangle} \frac{Ne^2}{m} \sum_{ab} \frac{\varrho_a(\omega + \frac{1}{2}i\Gamma_{ba})/\omega_{ba}}{(\omega_{ba}^2 - \omega^2 + \frac{1}{4}\Gamma_{ba}^2) - i\omega\Gamma_{ab}} \times |\langle b \,|\, x \pm iy \,|\, a\rangle|^2 \tag{4.1}$$

where ϱ_a is the Boltzmann factor for the level; $\hbar/\Gamma_{ab}$ is the reciprocal linewidth for the transition; N is the number of centers per cm^3; ω is the angular frequency of light, $\omega_{ab} = (E_a - E_b)/\hbar$; m is the electron mass; and $\langle b \,|\, x + iy \,|\, a\rangle$ is the electric dipole moment matrix element between the states $|a\rangle$ and $|b\rangle$; $\langle n\rangle$ is the average index of refraction for left and right circularly polarized light. The factor $[(\langle n\rangle^2 + 2)^2/9\langle n\rangle]$ is the Lorentz–Lorenz local field correction to the free-atom χ. The real and imaginary parts of χ give rise, respectively, to Faraday rotation and MCDA.

The first of the three contributions to MCDA cited above, i.e., the paramagnetic effect, appears in ϱ_a. The mixing of states due to H_0 is not shown explicitly, but is implicit in modified wave functions $|a\rangle$ and $|b\rangle$. The final factor, i.e., the modified level positions due to the Zeeman splitting, is reflected in ω_{ba}.

The latter two origins of MCDA are temperature-*independent* and are customarily referred to in the literature as the "diamagnetic effect" and are proportional to H_0. They are generally negligible compared to the paramagnetic term, especially at low temperatures, and even for very low concentrations of paramagnetic impurities (one part in 10^4), the paramagnetic effect will dominate. Of course, for magnetic singlet ground states, as often occurs for ions with an even number of electrons, the paramagnetic effect will be absent at low temperatures and the MCDA is diamagnetic. We will be concerned, however, exclusively with the paramagnetic MCDA. Thus if ϱ_{a-} and ϱ_{a+} are the Boltzmann factors of the plus and minus magnetic sublevels of a ground-state Kramers doublet which is the only occupied level, for example, then the MCDA arises from the difference between ϱ_{a-} and ϱ_{a+} or ground-state magnetization as will be shown more explicitly below. It is in this regard that we are interested in MCDA, i.e., as a monitor of transient ground-state magnetization.

4.1.2. Classical Macroscopic Description of MCDA and Faraday Rotation

The macroscopic phenomenological description of MCDA is given by the complex index of refraction

$$\mathsf{n}_{\pm} = n_{\pm} - ik_{\pm} \tag{4.2}$$

with $\mathsf{n}_{\pm}$ appearing as phase factors in the $\sigma_{\pm}$ waves. In terms of the complex susceptibility $\chi_{\pm} = \chi_{\pm}' - i\chi_{\pm}''$ given by Eq. (4.1),

$$n_{\pm} \doteq 1 + 4\pi\chi_{\pm}' \tag{4.3}$$

$$k_{\pm} = 4\pi\chi_{\pm}'' \tag{4.4}$$

The more general tensor properties of χ do not appear, as we restrict ourselves to cubic crystals and to axial crystals with the magnetic field H_0 along the unique axis, for it is only in these cases that $\sigma_{\pm}$ light waves propagating along H_0 maintain their circular polarizations. Thus if a plane polarized wave traveling in the z direction with angular frequency ω is decomposed into two circularly polarized waves of equal amplitude and opposite sense of rotation, we write

$$E(\omega) = E_{+}(\omega) + E_{-}(\omega) \tag{4.5}$$

with

$$E_{\pm}(\omega) = \mathrm{Re}\{(\hat{\mathbf{i}} \pm i\hat{\mathbf{j}})E_0(\omega) \exp i\omega[t - (\mathsf{n}_{\pm}z/c)]\} \tag{4.6}$$

$\hat{\mathbf{i}}$ and $\hat{\mathbf{j}}$ are unit vectors at right angles to each other in the plane perpendicular to z. The real part of n gives rise to a differential phase retardation between E_+ and E_- which rotates the plane of polarization of E through an angle Φ per unit length, i.e., the well-known Faraday rotation, or magneto-optical rotation (MOR)

$$\Phi = (\omega/2c)(n_+ - n_-) \tag{4.7}$$

The imaginary part of n gives rise to the MCDA, which will be proportional to $(k_+ - k_-)$. If $\gamma_\pm$ are the absorption coefficients per unit length for $\sigma_\pm$ light, then

$$\gamma_\pm = 2\omega k_\pm/c \tag{4.8}$$

One refers to the complex Faraday rotation

$$\Phi = \Phi + i\Theta \tag{4.9}$$

where Φ is given by Eq. (4.7), and

$$\Theta = (\omega/2c)(k_+ - k_-) \tag{4.10}$$

Just as χ' and χ'' are related by a Kramers–Kronig relationship, so are Φ and Θ,[48,49] so that they each contain the same information and may be used interchangeably. Shen[48] has generalized the expression for the complex Faraday rotation to include magnetic dipole and electric quadrupole transitions as well.

MCDA is obviously a function of wavelength λ, depending upon, among other parameters, the degree of overlap of absorption bands, and its sign may change as a function of λ. The wavelength dependence in such cases may be used to identify the structure of excited-state bands as has been done, for example, in determining the s–o coupling in the ^2P F-center bands[50] and the structure of excited states of rare earth ions. Many examples are given in the review article by Starostin and Feofilov,[51] where many different cases of MCDA dispersion are examined and detailed references are given. In the application of primary concern to us in this review, i.e., *MCDA or MOR as monitors of transient ground-state magnetization*, one obviously selects a wavelength which maximizes the MCDA or MOR.

4.1.3. MCDA as a Monitor of Ground-State Magnetization

Many early and elegant experiments were performed using MOR to probe ground-state magnetization and in particular spin–lattice relaxation.[52–55] As we have emphasized, MOR and MCDA are equivalent in principle. However, experimental techniques in MCDA have been very highly developed recently. In addition, the particular advantage of the magnetooptical technique of being able to do away with microwaves and so measure T_1 over a continuous range of values of magnetic fields has had its most significant applications recently using MCDA. We will therefore restrict the following discussion to MCDA, but the interested reader should profit from an examination of the MOR work given in Refs. 52–55. Let $I_{0\pm}$ be the intensity of the incident $\sigma_\pm$ light and $I^\pm$ the intensity of the transmitted light through a paramagnetic crystal, i.e.,

$$I^\pm = I_0{}^\pm e^{-l\gamma_\pm} \tag{4.11}$$

where $\gamma_\pm$ is the absorption coefficient per unit length and l is the path length. We assume that the $|\,a\rangle$'s in Eq. (4.1) correspond to a ground-state Kramers doublet $|\,-\rangle$ and $|\,+\rangle$, which are the only occupied levels, with fractional populations ϱ_- and ϱ_+, respectively. Then $\gamma_\pm$ may be written

$$\gamma_\pm = \varrho_-(\alpha_\pm, |\,-\rangle) + \varrho_+(\alpha_\pm, |\,+\rangle) \tag{4.12}$$

where $(\alpha_\pm, |\,-\rangle)$ is the fractional absorption coefficient for $\sigma_\pm$ light for the $|\,-\rangle$ level, and similarly $(\alpha_\pm, |\,+\rangle$ corresponds to the fractional absorption with all the ions in the $|\,+\rangle$ state. We assume that the absorption band is much wider than the Zeeman splitting, so that for any $|\,b\rangle$ in $\sum_b$ in Eq. (4.1), its time-reversed state is included. Then, by time-reversal symmetry, we must have $\alpha_+(|-\rangle) = \alpha_-(|+\rangle) \equiv \alpha_-$ and $\alpha_+(|\,+\rangle) = \alpha_-(|-\rangle) \equiv \alpha_+$, so that

$$\gamma_\pm = \varrho_-\alpha_\mp + \varrho_+\alpha_\pm \tag{4.13}$$

Following Panepucci and Mollenauer,[56] an MCDA signal is defined by

$$S = 2(I^+ - I^-)/(I^+ + I^-) = \Delta I/I_{dc}$$

Then, under the assumptions that $l(\gamma_+ - \gamma_-) \ll 1$ and that $S \ll 1$, we have

$$S = 2l\gamma_0[(\alpha_+ - \alpha_-)/(\alpha_+ + \alpha_-)](\varrho_+ - \varrho_-) \tag{4.14}$$

where $(\varrho_- - \varrho_+)_{eq} = \tanh(g\beta H/kT)$ is the equilibrium ground-state polarization. Equation (4.14) can be extended to multiplicities greater than two as well, and one may show that in general, under similar conditions, MCDA is proportional to ground-state magnetization.* In the case of MOR, this is true even for large angles of rotation.[48]

* Reference 48 discusses general conditions under which MOR is proportional to ground-state magnetization for higher multiplicities.

4.1.4. Experimental Techniques for Measuring MCDA

There have been many refinements in the experimental technique for measuring MOR over the years.[51] However, one of the most significant advances recently in MCDA measurement has been the development of the stress-modulated quarter-wave plate by Billardon and Badoz,[57] Mollenauer *et al.*,[58] and Kemp,[59] whose principle is shown schematically in Fig. 14. An isotropic, transparent medium, such as fused quartz, is shaped into a bar of the proper length and resonantly vibrated along the X direction, which is the bar axis, by an acoustical transducer. Plane polarized light traveling in the Z direction and polarized at 45° to the X axis is incident on the bar. The stressed bar acts as a birefringent medium with the X axis being switched every half-period from being a slow to a fast axis. Sufficient extension can be easily achieved (a strain of the order of 10^{-5} is needed) with the element resonantly driven so that alternate $\pm\lambda/4$ phase retardations can be obtained between the X and Y components of the light beam and the bar transmits circularly polarized light whose sense of polarization is modulated from left to right at the vibration frequency of the bar. Frequencies in the 10–100-kHz range are quite feasible. The MCDA signal can thus be detected using lock-in techniques, resulting in very high sensitivity. MCDA signals as small as 10^{-6} can be detected with this technique. The high frequency is also necessary if one wishes to observe transient recovery times of M as short as 10 msec and still use lock-in detection.

One practical realization of the stressed modulator is due to Mollenauer.[60] The fused quartz bar is driven by a stack of ceramic transducers

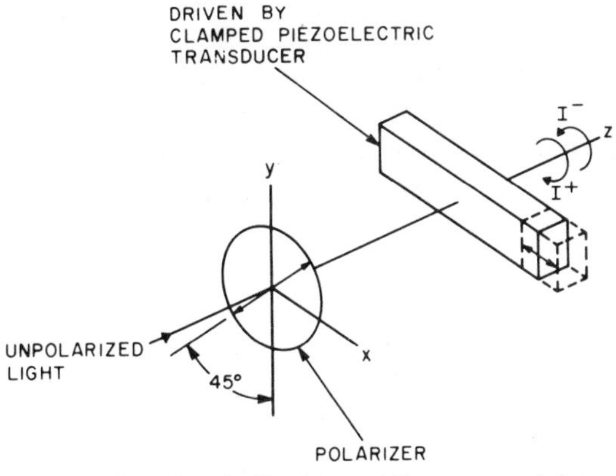

Fig. 14. Piezoelectrically driven $\lambda/4$ stress modulator to produce alternating sense of circular polarization (after Baldacchini and Mollenauer[60]).

which are cemented to the quartz and held fast at one end to a heavy mass The other end is freely suspended and the combination of transducers and quartz act as a quarter-wave acoustic line at the appropriate frequency so that the strain is maximum at the clamped end. The ceramic transducers are excited by a single transistor oscillator circuit. Further details regarding the stressed $\lambda/4$ modulator may be found in Ref. 58–60.

4.2. Measurement of Spin–Lattice Relaxation by MCDA

Since MCDA is a probe of ground-state magnetization M, it may be used to measure T_1 by observing the recovery of the MCDA signal after a perturbing influence which has driven M from equilibrium is quickly removed. Just as in the more conventional technique of measuring T_1 by microwave pulse saturation, a saturating microwave pulse may be used to drive M from equilibrium. The recovery, however, instead of being probed with low-level CW microwaves, is examined with MCDA. The advantage of MCDA and MOR over the conventional probe in this application is that in the case of inhomogeneously broadened lines, they look at the entire ground-state magnetization, and not just at a single spin packet if "hole burning" is present, so that they are not sensitive to spin diffusion though the line. This is analogous to the optical fluorescent techniques discussed in Section 2, used to measure T_1 in excited states. It is interesting to note how quite often considerably longer T_1's have been measured using MOR rather than the conventional microwave pulse saturation, indicating that some of these latter experiments were troubled by spin diffusion.

However, the most significant advantage of using MCDA to measure T_1 is that it allows one to do away completely with microwaves and thereby measure T_1 continuously over a wide range of magnetic fields. This is done by driving M from equilibrium by some disturbance other than microwave saturation and quickly removing it. This perturbation may be a sudden change of magnetic field,[61] optical pumping of the ground-state spins,[56,62] as outlined in Section 3, or a heat pulse.[62] The shortest T_1 which may be measured is of course governed by the speed with which these disturbances can be removed; nonetheless, T_1's as short as 10 msec have been measured using pulsed heaters.[62]

Before citing specific results, we illustrate in Fig. 15 the experimental arrangement used by Pannepucci and Mollenauer[56] to measure the ground-state spin–lattice relaxation of F-centers in KBr and KI. A burst of optical

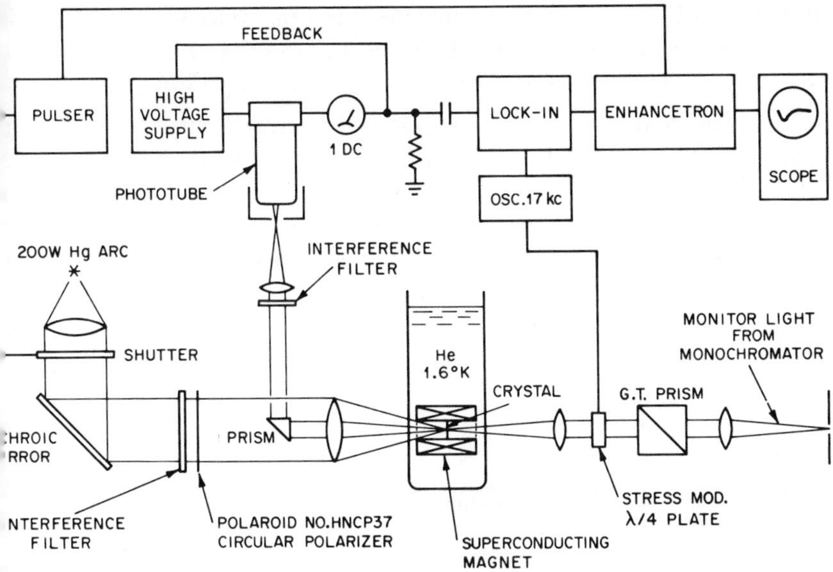

Fig. 15. Block diagram for measuring spin–lattice relaxation by magnetic circular dichroic absorption (MCDA). No microwaves are used. (After Panepucci and Mollenauer.[55])

pumping light is provided by a pulsed Hg arc lamp which shifts the ground-state F-center polarization P_e away from equilibrium. Light from a monochromator (this could be an interference filter) is switched alternately between left and right circular polarization at a 17-kHz rate by means of a stressed quarter-wave plate as discussed above, and monitors the recovery of P_e. Alternatively, switching of the magnetic field in a fraction of a second was also used for the longer T_1's, i.e., at lower field. The experimental results are shown in Fig. 16, where the points below 2 kG have been omitted, as they are connected with relaxation to clusters of F-centers at low fields. The relaxation rate is given by $T_1^{-1} = (AH^3 + BH^5)\coth(g\beta H/2kT)$, where A and B are constants. The $H^3\coth(g\beta H/2kT)$ corresponds to relaxation due to the modulation of the hyperfine interaction by the phonons[56,63] and is important up to 25 kG, while the $H^5\coth(g\beta H/2kT)$ term is the familiar Kronig–Van Vleck direct process[64] connected with the modulation of the crystal field by the phonons.* This term is very weak for F-centers as they are S-statelike and enters through higher orders of spin–orbit coupling; it is therefore only important above 25 kG, which for one thing would require microwave frequencies as high as 150 GHz at fields of 50 kG

* See Eq. (3.12) and discussion in the chapter by Orbach and Stapleton.

if measured by the microwave technique. The significance of these results is that the *field dependence* of the relaxation processes is a far more meaningful and sensitive test than the relatively weak linear temperature dependence.

Sabisky and Anderson[62] have performed a very similar experiment in using MCDA to measure the magnetic field and temperature dependences of Tm^{2+} in CaF_2, BaF_2, and SrF_2. In Figure 17 shows their results for Tm^{2+} in SrF_2 in three different crystals. These results have subtracted a constant rate (as a function of *H*) which appears at low field and which varies from crystal to crystal. This constant rate is ascribed to cross-relaxation to some other impurity and varies by more than an order of magnitude

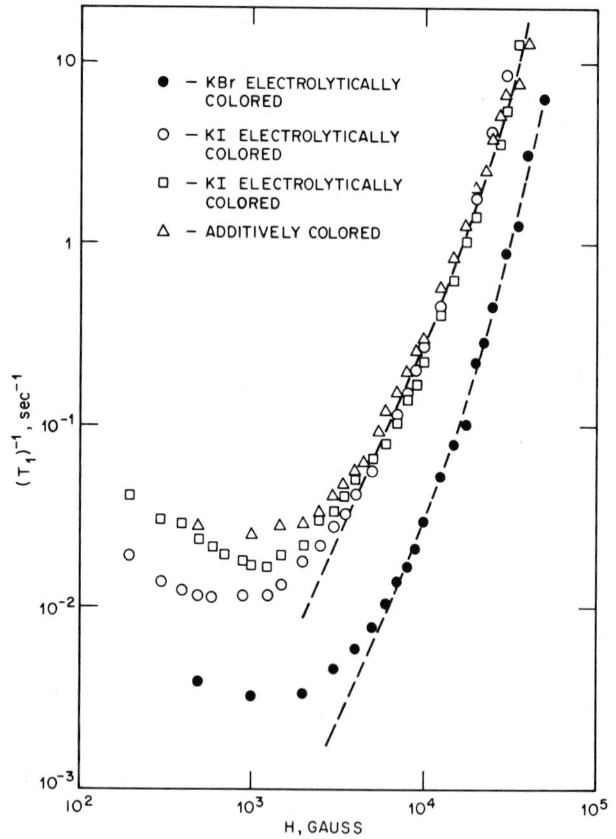

Fig. 16. Experimentally measured spin–lattice relaxation rates for *F*-centers in alkali halides. Departure of points from fitted curves at low field is due to clusters of *F*-centers. (After Panepucci and Mollenauer.[55])

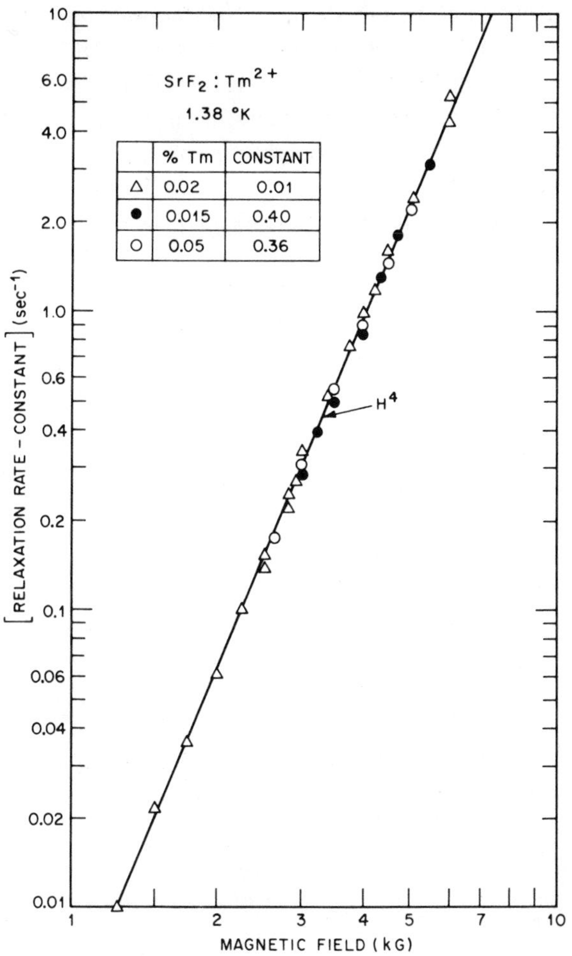

Fig. 17. Field dependence of relaxation rate minus a constant value for Tm^{2+} in SrF_2 (after Sabisky and Anderson[62]).

in different crystals. Note, however, that once this background rate is removed, all three crystals give the same results and that $1/T_1$ very accurately follows an H^4 curve* (since $g\beta H < kT$); a similar fit[62] was made for Tm^{2+}:CaF_2 using a heat pulse to drive the spins from equilibrium, and MCDA measurements were made from 1–12 kG at 1.27°K. Their results gave $1/T_1 = 6.9 \pm 0.5 \times 10^{-5} H^5 \coth(g\beta H/2kT)$. An earlier measurement[65] done by the standard microwave technique at 10 GHz gave a rate constant

* See Section 6 in the chapter by Orbach and Stapleton.

that was three orders of magnitude larger, yet showed a linear temperature dependence. This points up the pitfalls of relying solely on temperature measurements to properly identify a direct process, as there may be present strong cross-relaxation to a fast relaxing impurity and this can be temperature-dependent at very low temperatures.*

4.3. Excited-State EPR via MCDA

In Section 2, it was seen how an EPR transition in an excited state may produce a change in some aspect of the fluorescent light, for example its circular polarization, and how this may be used to detect excited-state EPR. However, one may encounter systems, such as the relaxed excited state of the F-center, where the polarization and intensity of the light emitted from the $|+1/2\rangle$ and $|-1/2\rangle$ spin states are nearly identical[6] so that there would be essentially no change in emitted light in going through resonance. This difficulty can be circumvented by a very elegant application of MCDA and spin memory which has been made by Mollenauer *et al.*[67] to observe the EPR in the relaxed excited state of the F-center in alkali halides.

Referring to Eqs. (3.6a) and (3.6c–e), we see that the ground-state spin polarization P_e depends upon the optical pumping time T_p, which in turn can be influenced by population changes in the excited state brought about by microwave resonance in the excited state as indicated by the w_{rf} term. Thus, excited-state EPR is detected by its change on the ground-state P_e, which in turn is monitored by MCDA. The variation of ΔP_e with ΔT_p is obtained by differentiating Eq. (3.5a):

$$\Delta P_e = \frac{\partial P_e}{\partial T_p}\,\Delta T_p = \frac{-1/T_1}{[1+(T_p/T_1)]^2}\left(P_{es} + \tanh\frac{\Delta^g}{2kT}\right)\Delta T_p \qquad (4.15)$$

which has its maximum value for $T_p = T_1$, and the pumping light intensity is adjusted to achieve this condition. The time T_1 is of the order of seconds and this value of T_p is very easily achieved with conventional light sources. At this point, making the approximations $T_1^{\text{ex}} \gg \tau_R$ and $\gamma \simeq 0$ appropriate to the excited-state F-center at low temperatures,[68] one has

$$\Delta T_p \simeq [w_{rf}\tau_R/(\varepsilon + q_{rf}\tau_R)]T_p \qquad (4.16)$$

One should recall that $\tau_R \lesssim 10^{-6}$ sec for the alkali halides, so that even with values of ε as small as 0.01 (almost complete spin memory), the available microwave power was such that $w_{rf}\tau_R < \varepsilon$ and $\Delta T_p \simeq w_{rf}\tau_R/\varepsilon$.

* See Section 6 in the chapter by Orbach and Stapleton.

Moreover, these expressions for ΔT_p tacitly assume a homogeneously broadened line, whereas invariably the EPR lines are generally inhomogeneously broadened due to hfs (as in the F-centers) or strains, so that ΔP_e is further reduced by the order of the width of the hole burned to the full inhomogeneous linewidth. The type of apparatus used for such an experiment is identical to that shown in Fig. 15, with the exception of the introduction of microwaves. Figure 18 shows the MCDA signal versus magnetic field for a fixed microwave frequency of 52 GHz. The low-field

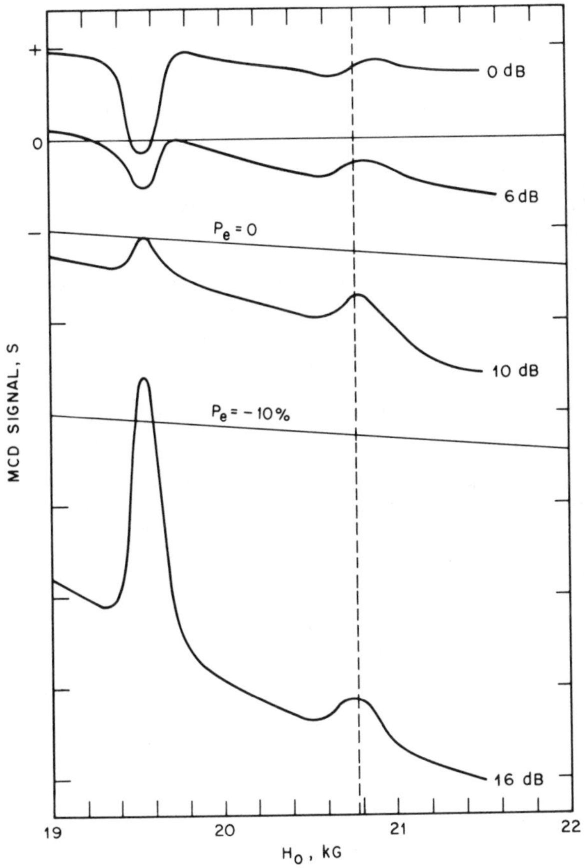

Fig. 18. MCDA signal versus magnetic field in F-center for fixed microwave frequency. The low-field line corresponds to a change in ground-state polarization P_e due to ground-state resonance, while the high-field line corresponds to change in P_e due to resonance in relaxed excited state of F-center. (After Mollenauer *et al.*[43])

line corresponds to a change in P_e due to the ground-state resonance, whil the high-field line is for the excited-state resonance. Similar experiment were reported on KBr and KI. The excited-state g-values for KCl, KBr and KI were found to be 1.976, 1.862, and 1.62, respectively. The significanc of these values in terms of the models for the excited state are discusse in Ref. 68.

As an indication of the sensitivity of the above technique, the smalles measurable excited-state population was 3×10^5 excess spins, or 10^3 *spin per gauss of linewidth for a 1-sec time constant.* This should not be too sur prising, as in principle these MCDA techniques should have similar sen sitivity to that discussed in Section 2.2 for fluorescent detection of excited state EPR, as basically a change in a microwave quantum of energy i being detected by a change in a much larger optical quantum. This sam consideration applies to the T_1 measurements with MCDA where no micro waves are used.

The rate at which P_e can change is determined by T_p, which wa adjusted to be equal to T_1 for maximum sensitivity, as indicated above Since T_1 is of the order of seconds, the microwaves could be switched n faster than T_1 for P_e to respond. Thus lock-in detection on a microwav switching frequency is ruled out. To circumvent this difficulty, Mollenaue *et al.*[68] adopted the following scheme. Instead of making $T_p \simeq T_1$, the used saturated optical pumping, i.e., $T_p \ll T_1$ (T_p was made equal t 10^{-2} sec). The optical pumping was switched at a 50-MHz rate betwee σ_+ and σ_- to give an average $P_e = 0$, since P_e cannot follow this rapidly. The microwaves were also switched *synchronously* with the light at this same rate, so that their "on" cycle coincided with σ_+ pumping. Then, o excited-state resonance, the σ_+ pumping effect on the ground state would be faster and a new value of $P_e \neq 0$ is reached. Now, the phase of the 50-kHz microwave modulation is reversed 180° after 0.1 sec, so that the microwave power is now on during σ_- pumping. P_e can now follow this 10-Hz switching of the phase of the microwaves, and one can use a second lock-in detector at 10 Hz. This, for example, gives far greater sensitivity in addition to removing the baseline in Fig. 18. To perform excited-state ENDOR, exactly the same technique is used except that the microwave power is kept constant and the ENDOR or rf power is switched at 50 kHz and phase-shifted every 0.1 sec.[68]

Until recently, optical decay times were all that were known about the relaxed excited state of F-centers. Now, attempts at constructing a model of the excited state will have far more precise information to fit, such as g-values and hfs.

5. COHERENCE EFFECTS IN OPTICAL–EPR EXPERIMENTS

5.1. General Discussion

5.1.1. Quantum Mechanical Description

The modification of some aspect of fluorescent light from excited tates on passage through magnetic resonance has been discussed until ıow solely in terms of population changes in the excited state, i.e., solely n terms of the diagonal elements of the density matrix. However, EPR n the excited states produces a coherent superposition of states, i.e., off-liagonal components as well, corresponding to a precessing transverse :omponent of magnetization which will modulate the intensity of fluorescent ight, when properly viewed, at the microwave frequency. This effect is vell known in gases and was first demonstrated at much lower frequencies.[69] Excellent accounts of this effect and related aspects of rf coherence in atoms nay be found in Dodd and Series[70] and the elegant thesis of Cohen-Tanıoudji.[71] The transposing of these effects to the microwave range and to nagnetic ions in solids has some interesting features and problems not encountered in gases, which are emphasized in the successful experiment of Chase[72] on the Γ_8 quartet excited state of Eu^{2+}:CaF_2. Before proceeding to this specific experiment, we will display the origin of this effect.

For purposes of illustration, let the fluorescent and ground states be Γ_6 doublets as shown in Fig. 19(a). Resonant microwaves of frequency $\omega_0(E_+^e - E_-^e)/\hbar$, besides tending to equalize the populations of $| e_+\rangle$ and

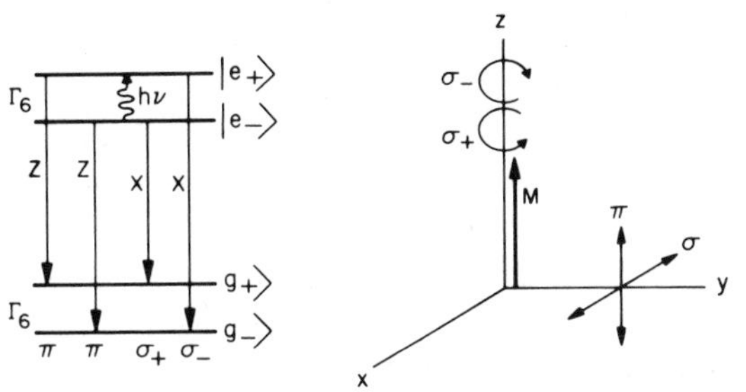

Fig. 19. (a) Excited- and ground-state doublets. Coherence is established in the excited state by microwave resonance. (b) When viewed along the y axis, coherence is manifested as interference between π and σ components giving rise to modulation of fluorescent light at the microwave frequency.

$| e_- \rangle$, will create a coherent superposition of states

$$\psi_e = a_+[\exp(-iE_+{}^e t/\hbar)] \,|\, e_+\rangle + a_-[\exp(-iE_-{}^e t/\hbar)] \,|\, e_-\rangle \qquad (5.1)$$

The ground states are labeled $\psi_g = | g_\pm \rangle$. If $P(x + iz)$ is the probability of emitting right circularly polarized light along the y direction, then it will be given by

$$P(x + iz) = |\langle \psi_e \,|\, x + iz \,|\, \psi_g \rangle|^2 \qquad (5.2a)$$

$$= \sum_g \langle \psi_e{}^e \,|\, x + iz \,|\, \psi_g{}^g \rangle \langle \psi_g{}^g \,|\, x - iz \,|\, \psi_e{}^e \rangle \qquad (5.2b)$$

$$= a_+{}^*[\exp(iE_+{}^e t/\hbar)] \langle e_+ \,|\, x + iz \,|\, g_- \rangle \langle g_- \,|\, x - iz \,|\, e_- \rangle$$

$$\times a_- \exp(-iE_-{}^e t/\hbar) + \text{dc terms} \qquad (5.2c)$$

$$= a_+{}^* a_-[\exp(i\omega_0 t)] \langle e_+ | x | g_- \rangle \langle g_- | -iz | e_- \rangle + \text{dc terms} \qquad (5.2d)$$

or

$$\overline{P(x + iz)} = \text{Re}[a_+{}^* a_- \exp(i\omega_0 t)] + a_+{}^* a_+ + a_-{}^* a_-$$

Thus the light is modulated at ω_0 and the depth of modulation is proportional to $a_+{}^* a_-$, which in turn is proportional to the perpendicular component of the precessing magnetization. Referring to Eq. (5.2d), the effective matrix elements which produce the light modulation appear as the product $\langle e_+ \,|\, x \,|\, g_- \rangle \langle g_- \,|\, z \,|\, e_- \rangle$. As the first term corresponds to σ radiation polarized along x and the second to π radiation polarized along z, we see that the modulation is due to an interference effect of σ and π light, propagating in the y direction as shown in Fig. 19(b), from the same atom.

5.1.2. Classical Picture

The modulation of the light by the transverse magnetization can be viewed classically in the following fashion. Assume that only the $| e_- \rangle$ component of the excited-state doublet is populated and let this correspond to a magnetization M_0 pointing in the z direction. In decaying from the excited state, the ions will emit only σ_+ light with intensity I along the z axis and σ and π light along the y axis as shown in Fig. 20(a). Both σ and π light will of course be emitted along the x axis as well, but are not shown in Fig. 20(a) for simplicity's sake. The application of a strong resonant microwave field H_1 to produce steady-state saturation of the excited state tips the magnetization 90° into the xy plane, where it precesses about the z axis at the resonant frequency, and under the condition $T_2 = T_1$,

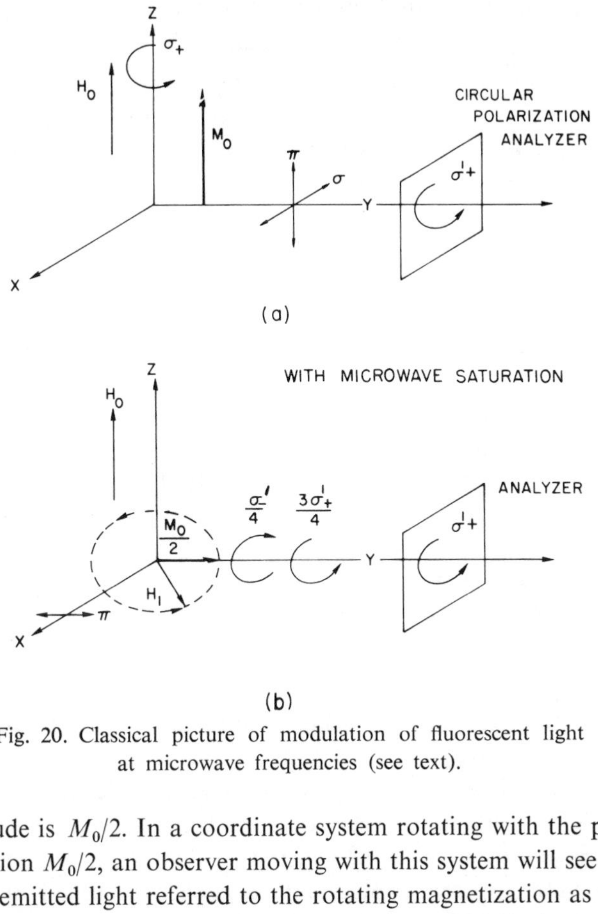

Fig. 20. Classical picture of modulation of fluorescent light at microwave frequencies (see text).

:s magnitude is $M_0/2$. In a coordinate system rotating with the precessing ıagnetization $M_0/2$, an observer moving with this system will see the same attern of emitted light referred to the rotating magnetization as would be een in the laboratory frame for a magnetization of $M_0/2$ along the z axis efore the application of resonant microwaves. For a magnetization of $I_0/2$, this would have corresponded to $\frac{3}{4}$ of the ions in $|e_-\rangle$ and $\frac{1}{4}$ in $e_+\rangle$, or to $\frac{3}{4}\sigma_+$ and $\frac{1}{4}\sigma_-$ emitted now along the direction of rotating mag-etization as shown in Fig. 20(b). A stationary observer looking along the direction of the laboratory coordinate system with a σ_+' analyzer will ee this pattern of light from the rotating frame sweep by him at the pre-ession frequency having a maximum intensity $3I_+/4$ when $M_0/2$ is pointing oward him and a value $I_+/4$ when it is pointing away from him one-half eriod later as, relative to his fixed viewing direction, the sense of circular olarization is then reversed. Thus he will see the light modulated at the ıicrowave frequency with a depth of modulation of 50% if all the ions vere originally in $|e_-\rangle$. If both levels are occupied, the signal will be smaller in proportion to the reduced initial magnetization. The discussio obviously can be generalized to Zeeman levels with higher multiplicit than two.

5.2. Experimental Observation in Γ_8 Level of $Eu^{2+}:CaF$

Chase[72] observed such a microwave modulation of the emitted ligh from the Γ_8 excited state of $Eu^{2+}:CaF_2$, discussed earlier in Section 2.4. at a 9-GHz EPR frequency. The microwave modulation of the light wa detected by a high-frequency photomultiplier* (using crossed electric an magnetic fields for focusing[73]). The microwave output of the photomultipli was detected by a superheterodyne receiver with phase-sensitive detectio at 60 MHz, which could be adjusted for phase sensitivity relative to th driving microwave field. Figure 21 shows the magnetic field derivative c the microwave-modulated fluorescence signals. Figure 21(a) correspond to the phase of the microwave receiver adjusted to detect the absorptiv component of the transverse moment, while in Fig. 21(b), the referenc microwave phase is shifted 90° to detect the dispersive component.

The actual signal seen in Chase's experiment was very much smalle than the maximum value of 50% predicted for the simple two-level schem in Section 5.1. This was due to a reduced circular polarization from eac level of the more complicated Γ_8 state, the presence of 12 hfs component insufficient microwave power to saturate the levels, and the fact tha $T_2 \ll T_1$. Nonetheless, Chase's experiment is remarkable for several reason The modulated light was detected over the entire vibronic band of man hundreds of angstroms, thus demonstrating that the rf modulation tech niques of atomic spectroscopy could be applied to broad optical ban transitions in solids. Second, the modulation frequency was in the 9-GH microwave range, instead of the lower rf range characteristic of hfs splitting used in similar experiments on atoms, which are listed below.

5.3. Suggested Applications

5.3.1. Crystals of Lower Symmetry

As we saw earlier, the modulated fluorescence is due to interferenc of the σ and π light from the *same* atom. To observe this interference in crystal, the σ and π light from the same ion must remain in phase over th

* The tube used by Chase was made at Bell Laboratories. A commercial version of thi tube is available from the Bendix Corp.

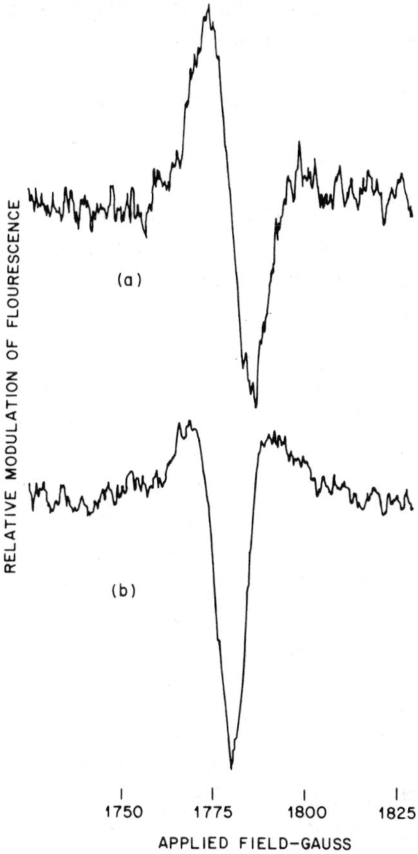

Fig. 21. Phase-sensitive detection of magnetic field derivative of microwave-modulated light in Γ_8 excited state of $Eu^{2+}:CaF_2$. (a) absorptive component, (b) dispersive component. (After Chase.[72])

ath length that the light travels in the crystal. In a cubic crystal, the velocity f light for σ radiation, v_σ, is the same as the velocity for π radiation, $_\pi$, so that this poses no problem. Nor will this present any difficulty if the direction shown in Fig. 20 is an optical axis of a lower-symmetry crystal. Iowever, for viewing along other directions in crystals with lower than ubic symmetry, the indices of refraction n_σ and n_π will be unequal, so that he interfering σ and π radiation will get out of phase in a distance d such hat $d \sim \lambda/(n_\sigma - n_\pi)$, where λ is the wavelength of the light. This is equivlent to saying that circular polarization is destroyed in a distance d. Thus, for example, for $(n_\sigma - n_\pi) \sim 10^{-3}$, one would have to use a crystal les than 0.05 cm thick.

5.3.2. Microwave Resonance with Modulated Light

As Chase[72] has indicated, one may reverse the procedure and us circularly polarized light modulated at the Larmor frequency[74] (or pulse rapidly in comparison with the Larmor period[75]) to induce precessin transverse magnetization in both ground and excited states, as has bee done in atomic spectroscopy. Similarly, partially absorbed light will b modulated by a precessing transverse magnetization and such an experimen has been proposed for ruby.[76]

5.3.3. Spin Echo Experiments in Excited States

An EPR spin echo pulse sequence (see Section 1 of the chapter b Mims in this volume) may be applied in the excited state and detected vi the pulses of microwave-modulated light. Such an experiment woul measure dephasing times T_2 in excited states, which has more than the usua interest if T_2 were dominated by rapid resonant transfer of the optica excitation between the impurity ions in solids. However, spin echo ex periments in excited states may be easier to do by conventional technique (especially at higher microwave frequencies which are above the frequenc response of the best phototubes) by using high-powered pulsed lasers t populate the excited states, as they need only be populated during th echo sequence, which is a time of order T_2.

5.3.4. Transfer of Coherence in Solids (Coherence Memory

In analogy to spin memory discussed in Section 3 which correspond to partial transfer of the diagonal components of the density matrix in radiationless transition from one electronic level to another, one may as whether the off-diagonal components or coherence may be transferre as well.

Such coherence transfer has been observed for spin-exchange collision in He gas in which coherence is transferred from one level to another o different precession frequency.[77] More recently, a transfer of coherence o transverse magnetization has been observed in a collision between metastabl He ions and Sr atoms, in which a metastable He atom and a Sr^+ ion pro duced.[78] Partridge and Series[77] have analyzed the criterion for such co herence transfer and shown that the collision time must be shorter than dephasing time corresponding to the reciprocal of the difference of pre-ession frequencies in the two states $1/(\omega_i - \omega_f)$; ω_i and ω_f are precession requencies in initial and final states, respectively. In a solid, τ_c would orrespond to the time for radiationless decay from one level to the other. The level to which the coherence is transferred is then initially driven at he precessional frequency ω_i of the originating state. For this coherence o persist in level f, it must decay out of this level again in a time less than $/(\omega_i - \omega_f)$. The fluorescent light from level f will then be modulated at ω_i. This transfer of coherence is connected with a spin selection rule, i.e., hat it does not change its orientation during a transition (even if precessing) inless coupled to orbital motion, which in turn is the origin of different precessional frequencies. Such experiments may be feasible in solids and nay give information on otherwise inaccessible states.

6. OBSERVATION OF THE MICROWAVE PHONON BOTTLENECK BY BRILLOUIN LIGHT SCATTERING

6.1. General Background

The conventional theory of paramagnetic relaxation assumes that spins driven from equilibrium relax via phonon interaction with the phonons erving as a constant-temperature heat reservoir at the temperature of the urrounding bath. However, many years ago, Van Vleck[79] pointed out that n the direct process of spin–lattice relaxation at low temperatures, the attice oscillators on "speaking terms" with the spins are too few in number o effectively conduct away surplus spin energy so that they may not "serve as a thermostat as ordinarily proposed." Thus the limited number of phonons on speaking terms with the spins present a "bottleneck" to the ransport of heat between the spin system and the helium bath. In this ection, we will present a new method of observing these heated phonons by Brillouin light scattering. First, however, a review will be given of the bottleneck, including the coupled spin–phonon equations, as well as a variety of other experiments which have been performed to observe this bottleneck using the more conventional spin temperature probe as an ndicator of the nonequilibrium phonons.

6.1.1. Thermodynamic Picture

Figure 22 shows a block diagram for the transport of energy from a two-level spin system to a bath with infinite heat capacity. The spins have been driven from equilibrium to a temperature T_s and have a heat capacity

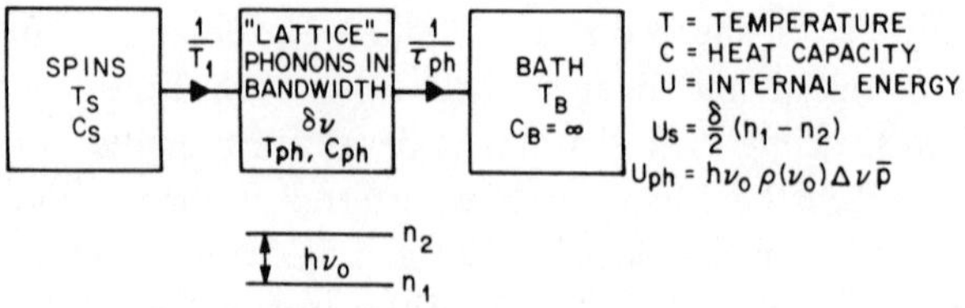

Fig. 22. Thermodynamic description of spin–lattice relaxation and microwave phonon bottleneck. $C_s \gg C_{ph}$.

C_s. The lattice consists of phonons on speaking terms with the spins, i.e. phonons in a bandwidth $\delta\nu$ centered at ν, where $\delta\nu$ is the EPR linewidth The number of such phonons is given by $\bar{p}\varrho(\nu)\,\delta\nu$, where the density o phonon states $\varrho(\nu)$ is

$$\varrho(\nu) \simeq 12\pi\nu^2/v^2 \qquad (6.1$$

and v is an average phonon velocity. At equilibrium, $\bar{p}$, the average occupa tion number per phonon mode, is given by

$$\bar{p}_0 = (e^{h\nu_0/kT_B} - 1)^{-1} \qquad (6.2$$

$1/T_1$ is the normal relaxation rate of the spins to equilibrium phonons, i.e. in the absence of a bottleneck, and $1/\tau_{ph}$ is the rate at which the phonon decay to the bath (the bath here may include decay by anharmonic pro cesses to other phonons not on speaking terms with the spins). Let n_0 be th equilibrium difference of population between the two levels, i.e., $n_0 = N$ $\times \tanh(h\nu/2kT)$. Then, the heat capacities of the spin and phonon system are given by

$$C_s = \partial U_s/\partial T = -\tfrac{1}{2}h\nu_0\, dn_0/dT \qquad (6.3$$

$$C_{ph} = \partial U_{ph}/\partial T = n\nu_0\varrho(\nu)\,\delta\nu\, d\bar{p}_0/dT \qquad (6.4$$

or

$$C_s/C_{ph} = [N/\varrho(\nu)h\nu]\tanh^2(h\nu/2kT_B) \qquad (6.5$$

Equation (6.5) may be written in the form given only under the assumptio that nothing is driven too far from equilibrium i.e., $T_s \simeq T_{ph} \simeq T_B$, es sentially the linearized description of the heat flow for a spin system.[80,8] As a typical illustration of the size of C_s/C_{ph}, consider the following situatior appropriate to 1% Ni^{2+} in MgO: $N \sim 3\times10^{20}$ cm^{-3}, $T \sim 1.4$°K, $\nu = 3\times10^{1}$ Hz, $\delta\nu = 5\times10^8$ Hz, and $v_{av}^3 \sim 5\times10^{17}$ gives $C_s/C_{ph} \simeq 2\times10^6$! Thus unless $1/\tau_{ph} > 2\times10^6/T_1$, the lattice (phonons in bandwidth $\delta\nu$) cannot transfe the energy fast enough to avoid being heated. At these low temperatures,

It will be seen that $\tau_{ph} \sim 3\times10^{-6}$ sec and $T_1 \sim 3\times10^{-3}$ sec, so that phonon heating should result. The thermodynamic picture can be used to write a set of rate equations for the energy flow, from which an effective bottlenecked T_1^{eff} may be derived.[82,83] However, in the Brillouin scattering experiments to be described, the spin system is driven to infinite temperature, so that the features of the spin and phonon behavior must be extracted from the more general nonlinear coupled rate equations.

6.1.2. Coupled Spin–Phonon Rate Equations for Direct Process

The rate equations for the two-level spin system shown in Fig. 22 may be written as

$$\dot{n}_1 = -n_1 w_{12} + n_2 w_{21} = -\dot{n}_2 \tag{6.6}$$

Let $1/K$ be the lifetime of level 2 against spontaneous emission of a phonon [see Eq. (3.11) and the discussion in the chapter by Orbach and Stapleton for evaluation of K], then (6.6) may be written

$$\dot{n}_1 = -n_1 K\bar{p} + n_2 K(\bar{p} + 1) \tag{6.7}$$

Letting $n = n_1 - n_2$ and $N = n_1 + n_2$,

$$\dot{n} = -nK(2\bar{p} + 1) + NK \tag{6.8}$$

If $\bar{p}$ does not depart from its thermal equilibrium value $\bar{p}_0$ given by (6.2), then we have

$$(d/dt)(n - n_0) = -(n - n_0)K \coth(h\nu/2kT) \tag{6.9}$$

Equation (6.9) describes the recovery of n toward its equilibrium value n_0 with the familiar time constant T_1 given by

$$1/T_1 = K \coth(h\nu/2kT) \tag{6.10}$$

However, in the presence of a phonon bottleneck, $p \neq p_0$, and we need an additional equation for the time dependence of the phonons, i.e.,

$$\frac{d}{dt}\bar{p} = \frac{-nK(2\bar{p} + 1) + NK}{2\varrho(\nu)\,\Delta\nu} - \frac{\bar{p} - \bar{p}_0}{\tau_{ph}} \tag{6.11}$$

The first term on the right of (6.11) describes the generation of phonons by the relaxing spins and the second term the decay of the phonons to the bath.

Equations (6.8) and (6.11) are two coupled nonlinear rate equation in n and p. It is customary to linearize these equations by assuming weal saturation of the spin system,[84] i.e., $n \simeq n_0$, in which case one finds tw time constants

$$T_1' = \tau_{ph}/\sigma \tag{6.12}$$

and

$$T_1^{eff} = T_1(1 + \sigma) \tag{6.13}$$

where

$$\sigma = \frac{\tau_{ph}}{T_1} \frac{n_0/2}{\varrho(\nu)(\delta\nu)(p_0 + \frac{1}{2})} = \frac{\tau_{ph}}{T_1} \frac{C_s}{C_{ph}} \tag{6.14}$$

σ is an important parameter in the theory and is called the bottleneck factor. In our earlier example, if we have $T_1 \sim 3\times 10^{-3}$ sec and $\tau_{ph} \sim 2\times 10^{-6}$ sec then $\sigma \simeq 100$. Thus T_1' is very fast and corresponds to an initial rapid deca of the spin system to a common spin–phonon temperature, while T_1^e measures the much slower decay of the combined system to the bath an is what is observed in experiments. It must be emphasized that this linearize description is only valid for small departures from equilibrium and so i completely inadequate for a saturated spin system in describing the recover of either the spins or the phonons except when both have decayed sufficientl to values near equilibrium. Faughnan and Strandberg[85] have given compute solutions to Eqs. (6.8) and (6.11) for the full nonlinear regime.

In the Brillouin scattering experiment, one operates in a steady stat of continuous, full microwave saturation of the spin system in order t produce the maximum amount of nonequilibrium resonant phonons from which to scatter light to test their presence. In the presence of microwave producing transitions between levels 1 and 2 at a rate w_{rf}, a term w_{rf} must be added to the right-hand side of Eq. (6.8) and Eq. (6.11) remain the same. Under conditions of steady-state rf saturation, i.e., $\dot{n} = \dot{\bar{p}} = 0$ $n = 0$, Eq. (6.11) becomes, using the value of K from (6.10) and $N = n$ $\times \coth(h\nu/2kT)$,

$$(\bar{p} - \bar{p}_0)_{max} = \frac{\tau_{ph}}{T_1} \frac{n_0/2}{\varrho(\nu)\,\delta\nu} \tag{6.15}$$

or in terms of σ,

$$(\bar{p} - \bar{p}_0)_{max} = \sigma(\bar{p}_0 + \tfrac{1}{2}) \tag{6.16}$$

For $kT \gg \hbar\omega$, it is customary to express $\bar{p}$ in terms of an effective tempera ture T_{eff} of the phonons, i.e., the temperature of a Planck distributior

* See Section 2.5 in Chapter 3 by C. D. Jeffries for effects of phonon bottleneck or dynamic nuclear polarization.

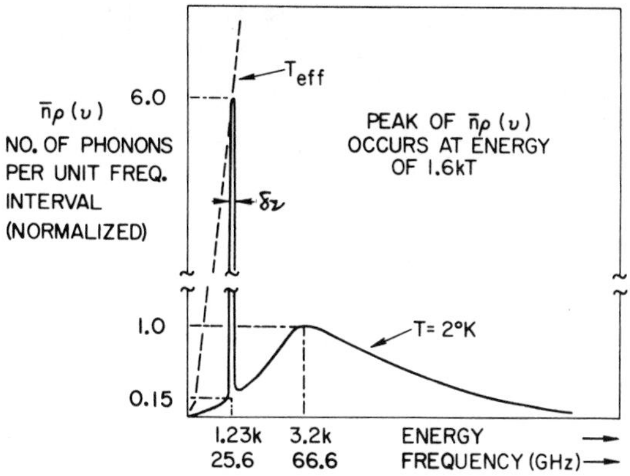

Fig. 23. Narrow band of resonant hot phonons produced by saturated spins relaxing via direct process, shown against background of thermal phonons at bath temperature of 2°K.

vhich would have the same value of $\bar{p}$. Equation (6.16) may then be rewritten n terms of T_{eff} as

$$T_{\mathrm{eff}} \sim (\sigma + 1)T_B \tag{6.17}$$

Thus even when the spin system is completely saturated and $T = \infty$, the phonons never rise above an effective temperature $\sim\sigma T_B$. Figure 23 illustrates the phonon bottleneck in terms of the production of these resonant phonons in a bandwidth $\delta\nu$, at an effective temperature considerably above ambient, to be observed by Brillouin scattering.

6.2. Observation of Bottleneck Using Spins as a Probe

6.2.1. Concentration Dependence

Equation (6.13) shows that for $\sigma \gg 1$, the effective relaxation time $T_1^{\mathrm{eff}} \sim T_1\sigma$. From (6.14), it is seen that $\sigma \sim n_0$, which in turn is proportional to concentration N, so that T_1^{eff} should increase with N in the bottlenecked regime. Such concentration dependences have been observed,[84,86] but the precise linear variation with N is hard to observe because of the difficulty of precisely controlling N and other factors which influence σ and which may change from sample to sample.

6.2.2. Size Effects

σ also varies linearly with τ_{ph}, so that to the extent that τ_{ph} is not due to anharmonic decay within the crystal but rather to escape at the boundaries, one would expect that, all else remaining the same, τ_{ph} would increase with increasing sample size. T_1^{eff} has been found to increase with sample size in a qualitative way,[84,86] except that here again it is hard to do controlled experiments, as even cutting the same crystal in half may change the surface and thereby influence τ_{ph}.

6.2.3. Temperature Dependence of T_1

In Eq. (6.14), $T_1 \sim \tanh(h\nu/2kT)$, $(p_0 + \frac{1}{2}) \sim \coth(h\nu/2kT)$, and $n \sim \tanh(n\nu/2kT)$, so that

$$\sigma \sim \tanh(h\nu/2kT) \quad (6.18)$$

so that $T_1^{eff} \sim 1/T^2$ for $h\nu > kT$. This inverse quadratic dependence upon temperature has also been seen quite often and cited as evidence for a phonon bottleneck.[84]

6.2.4. Phonon Avalanche

One of the most striking experiments to reveal the phonon bottleneck is the phonon avalanche first seen by Brya and Wagner.[88,89] A spin system (Ce-doped LaMN) was inverted by adiabatic fast passage and orders of magnitude *speedup* was observed in its T_1.* This was due to stimulated emission from the inverted spin system by the bottlenecked phonons from the spin relaxation. Further detailed experiments of this nature were performed by Mims and Taylor.[89]

6.2.5. Phonon Propagation Experiments Using Spins as Probe

Shiren[90] detected the bottlenecked phonons by generating them at one end of an Fe^{2+}-doped MgO rod with a phonon avalanche as described above. They were then detected at the other and by a double-quantum phonon–photon transition in Fe^{2+}. Rifman and Wagner[91] generated phonons by relaxing Ce^{3+} spins in one end of the crystal and observed their resonant absorption by Pr^{3+} spins in a different region of a LaMN crystal.

Anderson and Sabisky[92] have generated phonons at one end of a $Tm^{2+}:CaF_2$ crystal by CW microwave saturation and propagated them down the same crystal where they were detected by their heating of the Tm^{2+} spins in due farther region of the crystal which was well isolated from the saturating microwave power. The heated spins are detected by monitoring their spin temperature by the optical technique of MCDA described

* See Fig. 14 in Chapter 2 by Orbach and Stapleton.

in Section 3. In this way, the change in steady-state phonon temperature down the crystal could be mapped out. As we saw earlier, MCDA is extremely sensitive and can measure temperature changes as small as 10^{-3} deg K. One cannot obtain phonon mode selectivity with this technique by time-resolved experiments because of the long T_1 of Tm^{2+} in this lattice. Nonetheless, many elegant experiments have been done by Anderson and Sabisky using this detector and this optical technique is described in great detail in their review article,[93] so it will not be discussed any further here.

All the aforementioned methods in this section have in common the fact that information about the phonons is deduced via the behavior of the spins. Spin temperature, however, samples the phonons in an average way. It would be desirable to observe the different bottlenecked phonons (different modes, direction of travel, etc.) more directly and selectively. This may be done with Brillouin scattering as described in the next section.

6.3. Brillouin Scattering

6.3.1. Background and Techniques

The use of light scattering as a probe of lattice waves was first proposed by Brillouin in 1922.[94] However, it is only in recent years with the development of highly monochromatic laser sources and advances in photon counting techniques that Brillouin scattering has become such an active field of research. Excellent accounts of Brillouin scattering may be found in the paper by Benedek and Fritsch[95] and in Fabelinski's book,[96] so that only a brief account will be given here.

(a) *Picture in Terms of Classical Acoustic Waves.* The thermal fluctuations in a medium give rise to a local fluctuating dielectric constant to which the light wave couples and from which it scatters in all directions. The scattered light has imposed upon it a frequency modulation which reflects the frequency spectrum of the fluctuations. Light scattered from the *non*propagating isobaric or entropy fluctuations has a spectrum whose frequency is centered at that of the incident beam and whose width is a measure of the decay rate of the fluctuations. On the other hand, density fluctuations propagate as acoustic waves which set up a spatially varying dielectric constant which moves with the velocity of sound. Referring to Fig. 24(a), consider that spectral component of the lattice fluctuations that corresponds to a compressional wave of frequency ν_q moving with velocity v_q as shown. This wave sets up periodic regions of rarefaction and compression so that the dielectric constant of the medium corresponding to this

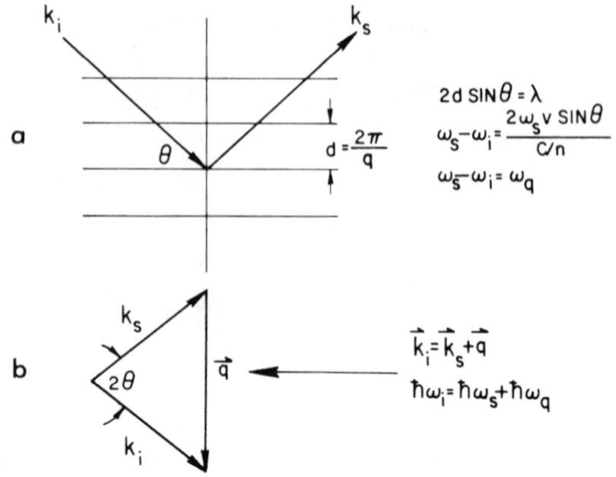

Fig. 24. (a) Brillouin scattering viewed classically as Doppler-shifted Bragg reflection from dielectric grating moving with phonon velocity. (b) quantum mechanical picture in which incident photon is scattered with absorption or emission of a phonon. Energy and quasi-momentum conservation give Brillouin formula.

wave will have a periodic spatial variation whose wavelength is $\lambda_q = v_q/\nu_q$ Incident light is therefore diffracted according to the Bragg condition

$$2\lambda_q \sin\theta = \lambda_i \quad (6.18)$$

However, the diffraction grating is also moving with velocity v_q, so tha the scattered light undergoes a Doppler shift given by

$$\omega_i - \omega_s = \pm[2v/(c/n)](\sin\theta)\omega_i \quad (6.19)$$

where ω_s is the frequency of the scattered light, c/n the velocity of ligh in the medium, and the $\pm$ correspond to phonons moving in opposite directions. Elimination of $\sin\theta$ from both equations gives

$$\omega_i - \omega_s = \pm\omega_q \quad (6.20)$$

Thus the scattered light is shifted to lower and higher frequencies by an amount corresponding to the frequency of the sound wave.

(b) *Simple Quantum Mechanical Picture*. Brillouin scattering can be viewed quantum mechanically as an inelastic scattering event, portrayed in Fig. 24(b), in which a photon of angular frequency ω_i and wave vector $\mathbf{k}_i$ is scattered into one of frequency ω_s and wave vector $\mathbf{k}_s$ with emission or absorption of a phonon of frequency ω_q and wave vector $\mathbf{q}$. Conservation of energy and quasimomentum, or wave vector, requires

$$\hbar\omega_i - \hbar\omega_s = \pm\hbar\omega_q \tag{6.21a}$$

and

$$\mathbf{k}_i - \mathbf{k}_s = \pm\mathbf{q} \tag{6.21b}$$

Since $v_q/c \sim 10^{-5}$ and $|q| \sim |\mathbf{k}|$, $\omega_q/\omega_i \sim 10^{-5}$, so that $\omega_i \simeq \omega_s$ and $|\mathbf{k}_i| \simeq |\mathbf{k}_s|$. Thus the wavevector diagram in Fig. 24(b) is an isosceles triangle with

$$|q|/2 = |k| \sin\theta$$

or

$$\omega_q = \pm[2v_q/(c/n)](\sin\theta)\omega_i$$

which is the same as Eqs. (6.19) and (6.20).

Thus out of the totality of acoustic waves into which the random motions of the lattice atoms may be decomposed, two particular Fourier components of frequency ω_q and traveling in opposite directions are selected by the direction of the incident beam and the direction chosen to view the light. As the scattering angle 2θ is varied from $2\theta = 0$ in the forward direction, the phonon frequency probed increases and has a maximum value of $\omega_{q_{\max}} = (2v/c)\omega_i$ for back-scattering. With $v_q/c \sim 10^{-5}$, $\omega_{q_{\max}}$ falls in the microwave range and thus is particularly suitable for examination of the nonthermal microwave phonons in the phonon bottleneck.

In a solid, corresponding to a given $\mathbf{q}$ vector selected by the scattering geometry, there will be in general three different frequencies corresponding to three different velocities on the three acoustic branches as shown in Fig. 25(a). Thus for fixed $\mathbf{q}$, if we scan in frequency about the incident frequency, we anticipate the spectrum shown in Fig. 25(b). The central component of unshifted frequency (sometimes called the Rayleigh line) is due to stray light scattered from crystal imperfections, with the triplet below and above the Rayleigh line corresponding to emission and absorption of phonons and called the Stokes and anti-Stokes lines, respectively.

The intensity of the scattered light[95,96] depends on the piezooptic constants of the material, which describe the variation of dielectric constant with strain. The strain in turn is determined by the phonon intensity, which for thermal phonons, is proportional to the crystal temperature and can be expressed in terms of the average occupation number $\bar{p}_0$ per phonon mode. More precisely, the Stokes line is proportional to $(\bar{p}_0 + 1)$, the

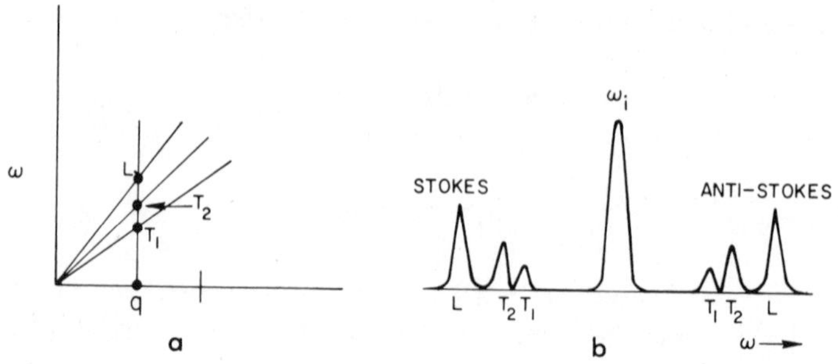

Fig. 25. (a) Three-phonon frequencies for a given **q** vector corresponding to three acoustic modes in a solid. (b) Stokes and anti-Stokes components of scattered light corresponding respectively to emission and absorption of phonons in (a). Also shown is the unshifted laser light scattered from imperfections.

factor one taking account of scattering with emission of a phonon, while the anti-Stokes line is proportional to $\bar{p}_0$.

To observe the phonon bottleneck, one observes the scattered light from thermal phonons whose frequency ν corresponds to the EPR transition of a paramagnetic impurity in the crystal. One then saturates the spin system, and the phonons generated by the relaxing spins should selectively increase the intensity of light scattering at frequency ν and not at other frequencies.

(c) *Experimental Techniques.* Since the frequency shifts of the Brillouin scattered light are of the order of 1 cm^{-1}, high resolution is required and this requirement is best met with a combination of single-mode laser as a source and a scanning Fabry–Perot spectrometer to frequency-analyze the scattered light. The efficiency of the scattering is determined by a number of factors discussed in Refs. 95 and 96 and is approximately 10^{-8} $\sim 10^{-12}$, providing another reason for intense light sources. The pass frequency of the interferometer is swept by varying the separation between the plates by means of piezoelectric transducers mounted on one of the plates to displace it electrically relative to the fixed plate at some low audio frequency. This mode of sweep seems preferable to the slower pressure scan, as it eliminates noise from drifts in light source intensity and electronics whose frequency is lower than the sweep rate. Further information of this mode of operation of a Fabry–Perot may be found in an article by Durand and Pine.[97]

The remaining details of the apparatus for viewing the phonon bottleneck are almost exactly the same as shown in Fig. 2 with the laser source replacing the Hg arc lamp and the Fabry–Perot used to analyze the scattered light. The multichannel analyzer is swept synchronously with the Fabry–Perot sweep and plots photon counts versus scattered light frequency. Additional details may be found in the articles by Brya *et al.*[98,99]

(d) *Factors Which Dictate Selection of Paramagnetic System for Observation of the Bottleneck.* One tries to select a sample to give as intense light scattering for a given $\bar{p}$ with minimum stray scattering and at the same time one must be able to incorporate a sufficient number of paramagnetic impurities, so chosen as to produce as large a bottleneck factor σ as possible. The Ni^{2+}:MgO and Nd^{3+}:$CaWO_4$ reported on below represent different compromises with these two requirements.

6.3.2. Experimental Results

(a) *Steady-State Phonon Heating in* Ni^{2+}:MgO *and* Nd^{3+}:$CaWO_4$. The first observation of a microwave phonon bottleneck by Brillouin scattering was reported by Brya *et al.*[99] A 1% Ni:MgO crystal was oriented as shown in Fig. 26(a) with the magnetic field perpendicular to the scattering plane, which was a [110] plane. A dielectric mirror was placed on one end of the sample to reflect the laser beam back on itself. There are thus two scattering triangles corresponding to the incident and reflected beams, respectively, shown in Figs. 26(b) and 26(c). The scattering geometry in Fig. 26(b) selects phonons of wave vector $\mathbf{q}_A$ to look at, while that in Fig. 26(c) probes phonons $\mathbf{q}_B$. While their magnitudes are equal, they are along different crystallographic directions and so the velocities of all three phonon branches are different along $\mathbf{q}_A$ and $\mathbf{q}_B$, so that the phonon frequencies for $\mathbf{q}_A$ and $\mathbf{q}_B$ on a given acoustic branch are different. Figure 26(d) shows the room-temperature Brillouin spectrum obtained with this geometry. This spectrum displays only the phonons from the lowest acoustic branch labeled T_1 in Fig. 25(a). T_2 is forbidden in this scattering geometry and the longitudinal phonon falls outside the range of scan shown. The Stokes and anti-Stokes Brillouin components (originating from adjacent interference orders of the Fabry–Perot) for the lowest acoustic branch for $\mathbf{q}_A$ and $\mathbf{q}_B$ are shown. $\mathbf{q}_A$ falls at approximately 26 GHz, the EPR transition frequency of the Ni^{2+} in the external magnetic field, while $\mathbf{q}_B$ falls at approximately 32 GHz. The 32-GHz phonons are to be used as a crystal temperature monitor.

Upon going to 2°K, the intensity of the thermal spectrum shown in Fig. 26(d) is reduced by a factor of 150 and is not visible for the length of integration time used to display Fig. 26(e). Upon application of the saturating microwaves to the spins as shown in Fig. 26(e), the Stokes and anti-Stokes phonons at 26 GHz are heated and nothing is observed at 32 GHz. By comparing integration times in Figs. 26(d) and 26(e), the selectively heated phonons at 26 GHz are found to have an equivalent temperature of approximately 800°K, while the crystal is at 2°K, a bottleneck factor of well over 400 for the anti-Stokes phonons. This asymmetry will be commented upon later. This degree of phonon heating is roughly in accord with theoretical expectations, as shown in Ref. 98.

The details of the collection optics are such that the Brillouin scattering probes a very small, almost pointlike region of the crystal. In certain regions of the crystal, the asymmetry between heated Stokes and anti-Stokes phonons described above disappears, whereas in other regions, it becomes even more severe, reaching asymmetries as large as 50:1. This is remarkable in that the only difference between Stokes and anti-Stokes phonons is that the phonons are traveling in opposite directions. These effects are discussed at greater length in Ref. 98 and are ascribed to parametric phonon oscillation connected with the three-level Ni^{2+} system and the selection of a phonon resonator mode in the sample which enhances a mode traveling in one direction relative to the opposite direction at a given point.

A similar experiment using Nd-doped $CaWO_4$ is shown in Fig. 27.[100] Again, 90° geometry is being used with a dielectric mirror at the end of the sample and the symmetry of the viewing direction is such that phonons $\mathbf{q}_A$ and $\mathbf{q}_B$ are of the same frequency and overlap in Fig. 27. The mirror is

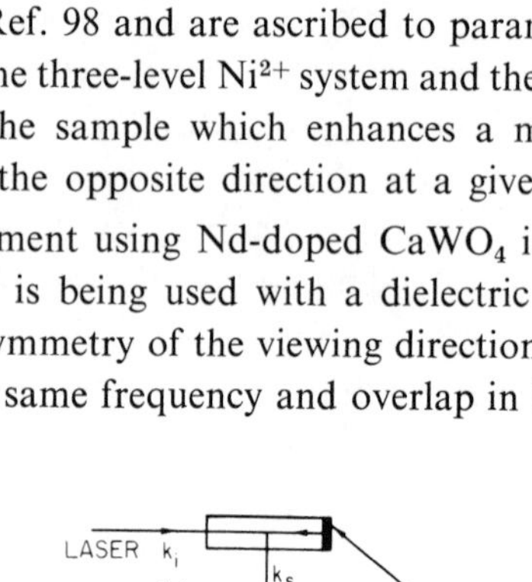

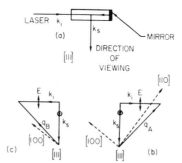

Fig. 26. (a) 90° Brillouin scattering geometry for Ni^{2+}:MgO. Dielectric mirror on sample reflects incident beam. (b) Scattering triangle for incident beam corresponding to scattering from phonon labeled q_A. (c) Scattering triangle for reflected beam associated with scattering from phonons q_B.

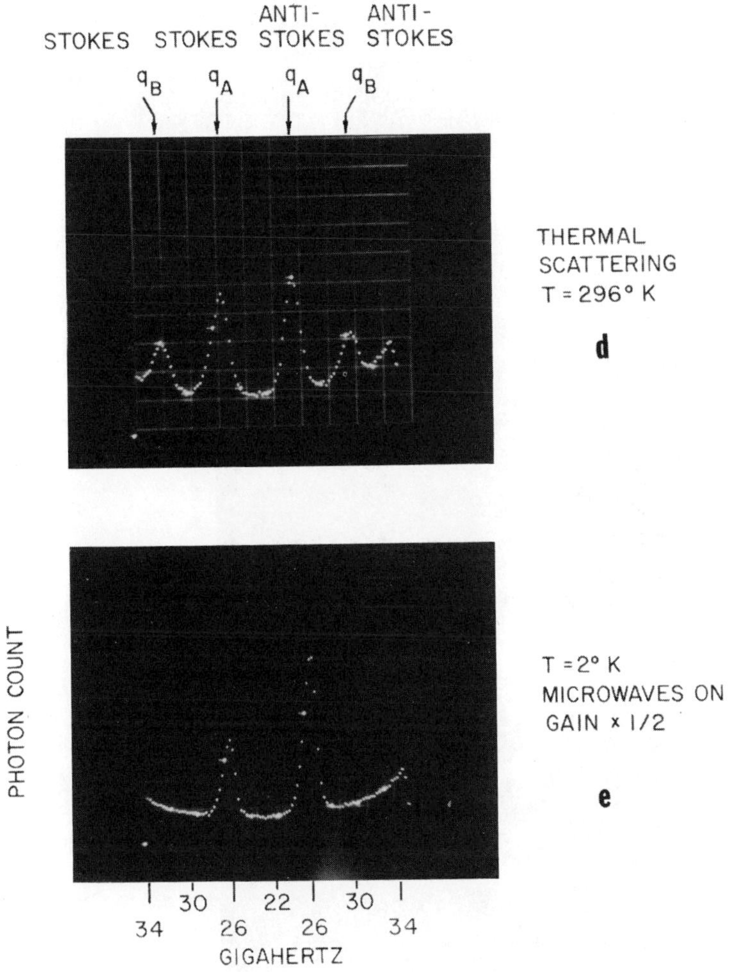

Fig. 26. (d) Observed room-temperature Brillouin spectrum. Scattering from only the lowest acoustic branches corresponding to q_A and q_B appears on this presentation. $\omega_A \sim 26$ GHz, $\omega_B \sim 32$ GHz. (e) Spectrum observed at 2°K with saturating microwaves at 26 GHz. Thermal scattering is not visible for this integration time. Note that only phonons at 26 GHz are heated by microwaves and not those at 32 GHz.

being used to double the intensity of the spectrum, but mainly to get the intense laser beam out of the sample and Dewar to avoid stray scattering. Longitudinal phonons now appear at 27.5 GHz and transverse phonons at approximately 15.4 GHz. $CaWO_4$ has very large piezooptic constants, so that now on going to 2°K, the thermal Brillouin spectrum is easily seen

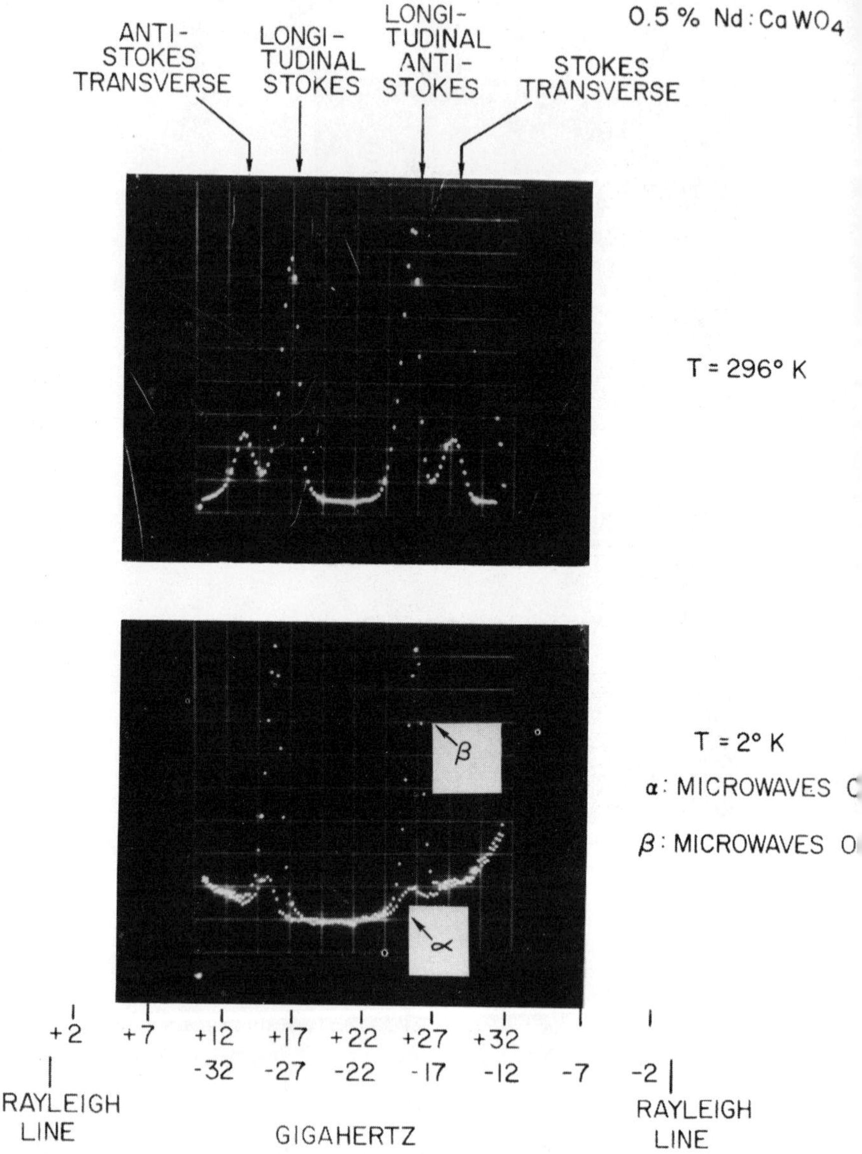

Fig. 27. (a) 90°, room-temperature scattering from Nd^{3+}:$CaWO_4$. (b) 2°K thermal Brillouin spectrum is labeled α, while spectrum labeled β corresponds to spectrum observed with 27-GHz microwave saturation. Note heating of only 27-GHz longitudinal phonons and no heating of lower-frequency transverse phonons.

and is labeled α in Fig. 27(b). Note the difference in intensity between the Stokes and anti-Stokes lines in the thermal spectrum, corresponding to $(\bar{p}_0 + 1)$ and $\bar{p}_0$, respectively. When the resonant microwaves at 27 GHz, corresponding to the Nd^{3+} EPR transition, are turned on, the spectrum shown in Fig. 27 β is obtained. Again, we see the selective heating only at 27 GHz. The bottleneck factor is considerably smaller here than for Ni^{2+}:MgO for reasons discussed in Ref. 100.

Similar observations of a microwave phonon bottleneck have been reported on concentrated cerium magnesium nitrate.[101,102]

(b) *Measurements of Phonon Bandwidth.* The free spectral range of 50 GHz and finesse of 50 of the Fabry–Perot resulted in an instrumental width that was approximately 1 GHz. If the heated phonon bandwidth were to be significantly narrower than this value, therefore, it could not be discerned by spectral scanning. This difficulty may be circumvented by making use of the unique relationship between the frequency shift of the Brillouin scattered light and the angle of scattering. From Eqs. (6.19) and (6.20), it is seen that if the phonons at frequency ν_q have a bandwidth $\delta\nu_q$, then the scattered light will have an angular aperture $\delta\theta$ given by

$$\delta\nu_q = [2v/(c/n)]\nu_i(\cos\theta)\,\delta\theta \qquad (6.22)$$

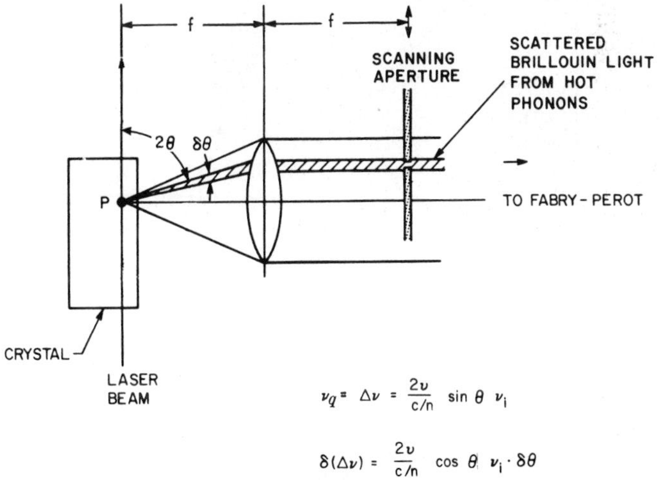

Fig. 28. Measurement of bandwidth of heated phonons by measuring angular aperture of Brillouin-scattered light from these phonons.

where again θ is half the scattering angle 2θ. Thus by measuring the angular aperture of the scattered light, one can measure the phonon bandwidth. This is most easily done, as illustrated in Fig. 28, by placing a narrow horizontal aperture behind the collecting lens as shown and in front of the Fabry–Perot. Figure 29 shows the intensity of Brillouin light scattered from the heated phonons in $Nd^{3+}:CaWO_4$ as a function of aperture position along the vertical.[100] The 1-mm scan in the vertical direction is related to an angular aperture of received light which corresponds to 30 MHz of bandwidth. The slight asymmetry observed is instrumental in origin. The bandwidth is seen to be 4 mm, or 120 MHz. This agrees quite well with the observed EPR linewidth in the same sample. Similar measurements have been made in CeMN[103] and in $Ni^{2+}:MgO$,[98] with an observed phonon bandwidth in the latter case which was significantly narrower than the EPR linewidth, the significance of which is discussed by Brya *et al.*[98]

(c) *Phonon Bottleneck at Elevated Temperatures.* If one examines the temperature of dependence of each term in the expression for $(\bar{p} - \bar{p}_0)_{max}$

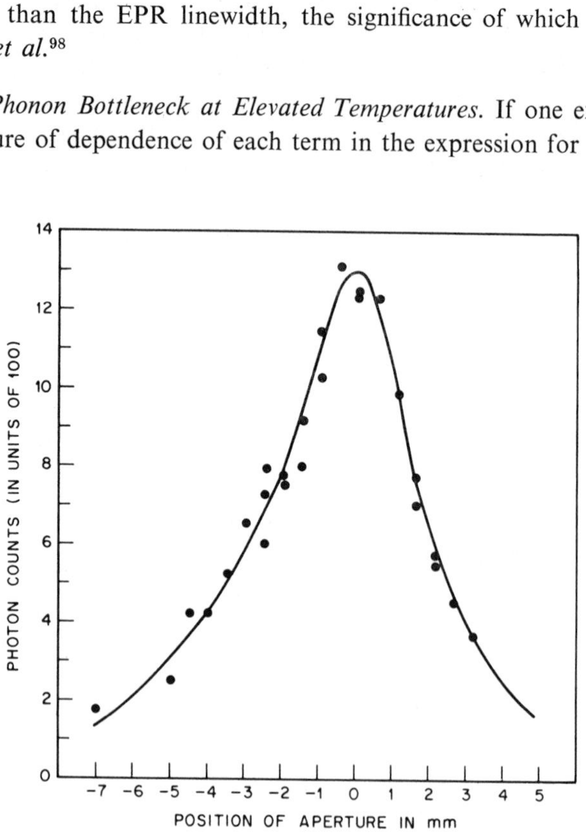

Fig. 29. Bandwidth of hot phonons in $Nd^{3+}:CaWO_4$ determined by method shown in Fig. 28. Here, 1 mm of aperture scan corresponds to 30 MHz bandwidth.

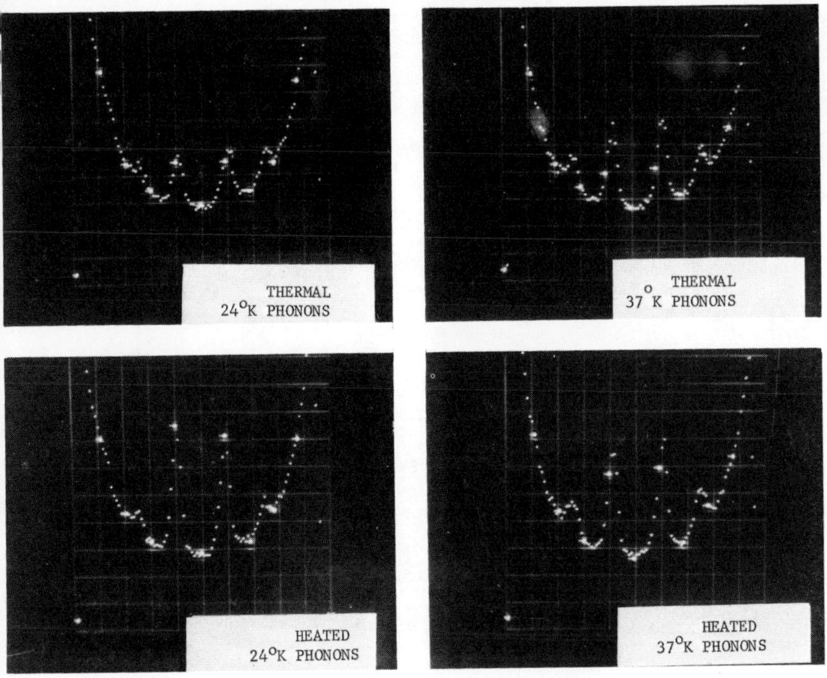

Fig. 30. Microwave phonon bottleneck observed in Ni^{2+}:MgO at higher crystal temperatures. (a, c) 24°K heated thermal phonons; (b, d) 37°K thermal phonons (from Ref. 98).

given by Eq. (6.15), then, since $n_0 \sim \tanh(h\nu/2kT)$ and $T_1 \sim \tanh(h\nu/2kT)$, $(\bar{p} - \bar{p}_0)_{\max}$ *is independent of temperatures as long as* τ_{ph} *is temperature-independent*. In this connection, it should be stressed that even at higher temperatures where the Raman relaxation rate may begin to exceed the direct process rate, the appropriate T_1 to use in Eq. (6.15) is still the unbottlenecked *direct* process T_1 as long as the spin system is saturated. The presence of the parallel Raman process does not influence the rate at which phonons in the direct process are generated as long as the faster Raman process does not make it difficult to saturate the spins. The constancy of τ_{ph} up to temperatures of 40°K in MgO is discussed in Ref. 98.

Figures 30(c, d) show the thermal Brillouin scattering corresponding to the same conditions as Fig. 26d except at the elevated temperatures of 24°K and 37°K, respectively. Figures 30(a, b) show the corresponding spectra with resonant saturation of the Ni^{2+} spins by microwaves. It is apparent that there is still significant heating of the phonons at 26 GHz, but not at 32 GHz.

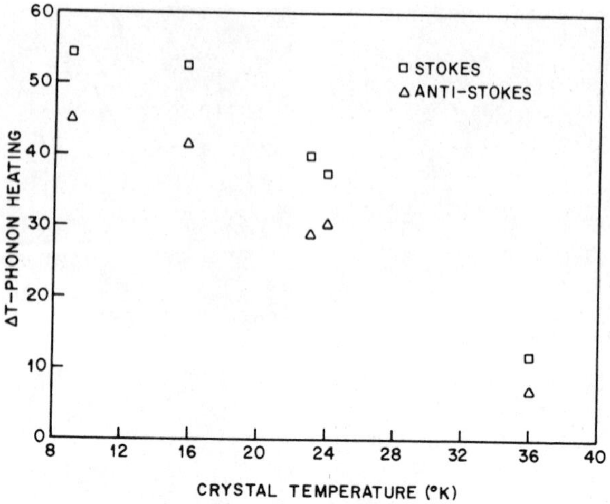

Fig. 31. Excess heating of resonant phonons ΔT versus crystal ambient temperature in Ni^{2+}:MgO. Near-constancy of heating is in accord with theory. Decrease of heating at higher temperatures is discussed in text.

The excess heating above thermal, i.e., $\bar{p} - \bar{p}_0$ or $T_{ph} - T_B$ taken from such figures as Figs. 30(a) and 30(b) at several different temperature points are plotted in Fig. 31. The two data points at each temperature correspond to Stokes and anti-Stokes lines whose intensities were different at this particular point in the crystal as described earlier. While there is a decrease of excess phonon heating ΔT_{ph} with increasing ambient temperature T, emphasis should be placed on its very slow variation with T, i.e., ΔT_{ph} is seen to be almost constant from 8°K to 16°K, and generally over the temperature range shown, varies less than $1/T$. This verifies the constancy of $(\bar{p} - \bar{p}_0)_{max}$ as given in Eq. (6.15). The actual observed decrease in ΔT_{ph} at higher temperatures is attributed to inability to saturate the spins completely at higher T because of a speedup in T_1 as well as τ_{ph} becoming temperature-dependent.

Measurements were also made on phonon decay times[98,100] which are controlled by escape time from the crystal at low temperatures, and experiments showing mode conversion of the phonons have also been performed.[98,100]

By consulting the references cited, the reader will undoubtedly sense the different possibilities this new technique presents for the study of the spin–phonon interaction.

ACKNOWLEDGMENTS

The author is indebted to many of his colleagues whose stimulation he has benefited from in discussing many of the topics treated in this review. These include W. J. Brya, L. Chase, G. E. Devlin, G. F. Imbusch, W. Mims, and L. R. Walker. He would also like to thank L. Mollenauer, C. H. Anderson, J. C. Kemp, and M. Schwoerer for permission to use figures from previous publications.

REFERENCES

1. C. A. Hutchison, Jr. and B. W. Nagnum, *J. Chem. Phys.* **34**, 908 (1961).
2. J. Brossel and A. Kastler, *Compt. Rend.* **229**, 1213 (1949).
3. J. Brossel and F. Bitter, *Phys. Rev.* **86**, 308 (1952).
4. G. W. Series, *Repts. Progr. Phys.* **22**, 280 (1959); B. Budick, in *Advances in Atomic and Molecular Physics*, Academic Press, New York (1967).
5. S. Geschwind, G. E. Devlin, R. L. Cohen, and S. R. Chinn, *Phys. Rev.* **137**, 1087 (1967).
6. H. Bucka, *Naturwiss.* **43**, 371 (1956); H. Bucka, *Z. Phys.* **151**, 328 (1958); R. H. Kohler, *Phys. Rev.* **121**, 1104 (1961).
7. S. Geschwind, R. J. Collins, and A. L. Schawlow, *Phys. Rev. Letters* **3**, 544 (1959).
8. F. Varsanyi, D. L. Wood, and A. L. Schawlow, *Phys. Rev.* **3**, 545 (1959).
9. G. F. Imbusch, S. R. Chinn, and S. Geschwind, *Phys. Rev.* **161**, 295 (1967); G. F. Imbusch and S. Geschwind, *Phys. Letters*, **18**, 109 (1965).
10. M. Sharnoff, *J. Chem. Phys.* **46**, 3263 (1967).
11. A. L. Kwiram, *Chem. Phys. Letters* **1**, 272 (1967).
12. J. Schmidt, I. A. M. Hesselmann, M. S. DeGroet, and J. H. Van Der Waals, *Chem. Phys. Letters* **1**, 434–436 (1967).
13. F. Porret and F. Lüty, *Phys. Rev. Letters* **26**, 843 (1971).
14. Y. Ruedin and F. Porret, *Helv. Phys. Acta* **41**, 1294 (1968).
15. A. Abragam, *Nuclear Magnetic Resonance*, Oxford University Press, p. 76.
16. L. L. Chase, *Phys. Rev.* **168**, 341 (1968).
17. S. Sugano and Y. Tanabe, *J. Phys. Soc. Japan* **13**, 880 (1958).
18. S. Sugano and I. Tsujikawa, *J. Phys. Soc. Japan* **13**, 899 (1958).
19. M. D. Sturge, *Phys. Rev.* **130**, 639 (1963).
20. L. L. Chase, *Phys. Rev.* **B2** (7), 2308 (1970).
21. F. Ham, *Phys. Rev.* **166**, 307 (1968).
22. D. F. Nelson and M. D. Sturge, *Phys. Rev.* **137**, A1117 (1965).
23. I. I. Eru, S. A. Peskovatskii, and A. N. Cherneto; *Soviet Phys.—Solid State* **7**, 263 (1965).
24. G. H. Larson, *Phys. Rev.* **150**, A264 (1966).
25. M. Blume, R. Orbach, A. Kiel, and S. Geschwind, *Phys. Rev.* **139**, A314 (1965).
26. F. Varsanyi and P. Pershan, private communication.
27. G. F. Imbusch and S. Geschwind, *Phys. Rev. Letters* **17**, 522 (1966).
28. G. F. Koster, J. O. Dimmock, R. C. Wheeler, and H. Statz, *Properties of the Thirty Two Point Groups*, MIT Press (1963).

29. C. H. Anderson and E. S. Sabisky, *Phys. Rev.* **178**, No. 2 (1969).
30. M. Blume and R. Orbach, *Phys. Rev.* **127**, 1587 (1962).
31. A. L. Schawlow, A. H. Piksis, and S. Sugano, *Phys. Rev.* **122**, 1469 (1961).
32. M. Blume, R. Orbach, A. Kiel, and S. Geschwind, *Phys. Rev.* **139**, A314 (1965).
33. R. Adde and H. De Wijn (to be published).
34. D. H. Tanimoto, W. H. Ziniker, and J. C. Kemp, *Phys. Rev. Letters* **14**, 645 (1965).
35. G. D. Watkins, *Phys. Rev.* **155** (1967) 802.
36. M. Schwoerer and H. C. Wolf, in *Proc. of the XIVth Colloque Ampère 1966*, ed. R. Blinc, North-Holland Publ. Co., Amsterdam (1967).
37. J. H. van der Waals and M. S. de Groot, *The Triplet State*, Cambridge University Press, London (1967).
38. J. Schmidt, W. S. Veeman, and J. H. van der Waals; *Chem. Phys. Letters* **4**, 341 (1969); D. A. Antheunis, J. Schmidt, and J. H. van der Waals, *Chem. Phys. Letters* **6**, 255 (1970).
39. M. Schwoerer and H. Sixl, *Chem. Phys. Letters* **2**, 14 (1968); *Z. Naturforsch.* **24a**, 952 (1969); *Chem. Phys. Letters* **6**, 21 (1970).
40. R. H. Clarke, *J. Chem. Phys.* **6**, 413 (1970).
41. A. Kastler, *J. Phys. Rad.* **11**, 255 (1950).
42. J. Brossel, *Quantum Electronics*, Columbia University Press, New York (1960), p. 81.
43. L. F. Mollenauer, S. Pan, and S. Y. Yngvesson, *Phys. Rev. Letters* **23**, 683 (1969).
44. N. V. Karlov, J. Margerie, and V. Merle D'Aubigne, *J. Phys. Rad.* **24**, 717 (1963).
45. E. Sabisky and C. H. Anderson, *IEEE J. Quant. Electron.* **QE-3**, 287 (1967).
46. D. P. Devor, I. J. D'Haenens, and C. K. Asawa, *Phys. Rev. Letters* **8**, 432 (1962).
47. A. S. Davydov, *Quantum Mechanics*, Pergamon Press (1965), Paragraph 81.
48. Y. R. Shen, *Phys. Rev.* **133**, A511 (1964).
49. F. C. Brown and G. Laramore, *Appl. Opt.* **6**, 669 (1967).
50. F. Luty and J. Mort, *Phys. Rev. Letters* **12**, 45 (1964); J. Mort, F. Luty, and F. C. Brown, *Phys. Rev.* **137**, A566 (1965).
51. N. V. Starostin and P. P. Feofilov, *Soviet Phys.—Uspekhi*, **12** (2), 252 (1969).
52. J. M. Daniels and M. Wesemeyer, *Can. J. Phys.* **36**, 405 (1958).
53. J. M. Daniels and K. E. Reickhoff, *Can. J. Phys.* **38**, 604 (1960).
54. C. K. Asawa and R. A. Satten, *Phys. Rev.* **127**, 1542 (1962).
55. Y. Hayashi, M. Fukui, H. Yoshioka, *J. Phys. Soc. Japan* **23**, 312 (1967); Y. Hayashi and M. Fukui, *J. Phys. Soc. Japan* **25**, 1043 (1968).
56. H. Pannepucci and L. F. Mollenauer, *Phys. Rev.* **178**, 589 (1969).
57. M. Billardon and J. Badoz, *Compt. Rend.* **262**, 1672 (1966); **263**, 139 (1966).
58. L. F. Mollenauer, D. Downie, H. Engstrom, and W. B. Grant, *Appl. Opt.* **8**, 661 (1969).
59. J. C. Kemp, *J. Opt. Soc. Am.* **59**, 915 (1960).
60. L. F. Mollenauer (private communication).
61. H. Kalbfleisch, Z. Physik **181**, 13 (1964).
62. E. S. Sabisky and C. H. Anderson, *Phys. Rev.* **B1** (5), 2028 (1970).
63. M. F. Deigen and V. Ya. Zevin, *Soviet Phys.—JETP* **12**, 785 (1961).
64. J. H. Van Vleck, *Phys. Rev.* **57**, 426 (1940); R. de L. Kronig, *Physica* **6**, 33 (1939).
65. C. Y. Huang, *Phys. Rev.* **139**, A241 (1965); **169**, 470 (1968).
66. M. P. Fontana and D. B. Fitchen, *Phys. Rev. Letters* **23**, 1497 (1969).
67. L. F. Mollenauer, S. Pan, and A. Winnacker, *Phys. Rev. Letters* **26**, 1643 (1971).
68. L. F. Mollenauer, S. Pan, and A. Winnacker, *Phys. Rev. Letters* **26**, 1643 (1971).

69. J. N. Dodd, W. N. Fox, G. W. Series, and M. J. Taylor, *Proc. Phys. Soc.* **74**, 789 (1959).
70. J. N. Dodd and G. W. Series, *Proc. Roy. Soc.* **A263**, 353 (1961).
71. C. Cohen-Tannoudji, These, Univ. de Paris (1963).
72. L. L. Chase, *Phys. Rev. Letters* **21**, 888 (1968).
73. R. C. Miller and N. C. Wittwer, *IEEE J. Quantum Electronics* **QE-1**, 47 (1965).
74. A. Corney and G. W. Series, *Proc. Phys. Soc. (London)* **83**, 331 (1961), and references cited therein.
75. E. B. Aleksandrov, *Opt. i Spectroskopiya* **14**, 436 (1963); **17**, 957 (1964) [English transl.: *Opt. Spectry. (USSR)* **14**, 233 (1963); **17**, 522 (1964)].
76. N. Bloembergen, P. S. Pershan, and L. R. Wilcox, *Phys. Rev.* **120**, 2014 (1960).
77. R. B. Partridge and G. W. Series, *Proc. Phys. Soc.* **88**, 983 (1966).
78. L. D. Schearer and L. A. Riseberg, *Phys. Rev. Letters* **26**, 599 (1971), and references cited therein.
79. J. H. Van Vleck, *Phys. Rev.* **59**, 724 (1941).
80. C. J. Gorter, *Paramagnetic Relaxation*, Elsevier (1947).
81. A. H. Cooke, *Rep. Prog. Phys.* **13**, 276 (1950).
82. A. M. Stoneham, *Proc. Phys. Soc.* **86**, 1163 (1965).
83. A. A. Abragam and B. Bleaney, *Electron Paramagnetic Resonance of Transition Metal Ions*, Clarendon Press, Oxford (1970), Sec. 10.6.
84. P. L. Scott and C. D. Jeffries, *Phys. Rev.* **127**, 32 (1962).
85. B. W. Faughnan and M. W. P. Strandberg, *Phys. Chem. Solids* **19**, 155 (1961).
86. J. A. Giordmaine, L. E. Alsop, F. R. Nash, and C. H. Townes, *Phys. Rev.* **109**, 302 (1958).
87. J. A. Giordmaine and F. R. Nash, *Phys. Rev.* **138**, A1510 (1965).
88. W. J. Brya and P. E. Wagner, *Phys. Rev. Letters* **14**, 431 (1965); *Phys. Rev.* **157**, 400 (1967).
89. W. B. Mims and D. R. Taylor, *Phys. Rev. Letters* **22**, 1430 (1969); *Phys. Rev.* **B3**, 2013 (1971).
90. N. S. Shiren, *Phys. Rev. Letters* **17**, 958 (1966).
91. S. S. Rifman and P. E. Wagner, *Solid State Commun.* **7**, 453 (1969).
92. C. H. Anderson and E. S. Sabisky, *Phys. Rev. Letters* **21**, 987 (1968).
93. C. H. Anderson and E. S. Sabisky, Chapter 1, in *Physical Acoustics*, Vol. 8, ed. by W. P. Mason and R. N. Thurston, Academic Press, New York (1971).
94. L. Brillouin, *Ann. Phys. (Paris)* **17**, 88 (1922).
95. G. Benedek and K. Fritsch, *Phys. Rev.* **149**, 647 (1966).
96. I. L. Fabelinskii, *Molecular Scattering of Light*, Plenum Press, New York (1968).
97. G. Durand and A. S. Pine, *IEEE J. Quantum Electronics* **4**, 523 (1968).
98. W. J. Brya, S. Geschwind, and G. E. Devlin, *Phys. Rev.* **6** (5), 1924 (1972).
99. W. J. Brya, S. Geschwind, and G. E. Devlin, *Phys. Rev. Letters* **21**, 1800 (1968).
100. S. Geschwind and G. E. Devlin, *Phys. Rev.* (to be published).
101. S. A. Altshuler, R. M. Valishev, and A. Kh. Khasanov, *Zh. Eksperim. i Teor. Fiz. Pisma* **10**, 179 (1969) [English transl.: *Soviet Phys.—JETP Letters* **10**, 113 (1969)].
102. S. A. Altshuler, R. M. Valishev, B. I. Kochalaev, and A. Kh. Khasanov, *Zh. Eksperim. i Teor. Fiz. Pisma* **13** (10), 535 (1971) [English transl.: *Soviet Phys.—JETP Letters* **13** (10), 382 (1971)].
103. R. M. Valishev and A. Kh. Khasanov, *Fiz. Tverd. Tela* **12**, 352 (1970) [English transl.: *Soviet Phys.—Solid State* **12**, 2859 (1971)].

Chapter 6

Pair Spectra and Exchange Interactions

J. Owen

Clarendon Laboratory, Oxford

and

E. A. Harris

Department of Physics, University of Sheffield

1. INTRODUCTION

This chapter will be concerned with EPR spectra from isolated pairs and clusters of magnetic ions in insulators. The study of these pair spectra gives detailed information about the exchange and other interactions between the magnetic ions, and by now is a standard technique for investigating such interactions. The first clear example of this type of investigation was for Cu^{2+}–Cu^{2+} pairs in copper acetate,[1,2] where all of the copper ions occur in pairs in the natural crystal structure.[3] The most common application, however, has been to magnetic salts which are diluted with an isomorphous diamagnetic one, so that in the mixed crystal, there can be isolated clusters of two or three or more magnetic neighbors.[4] This has proved to be particularly useful for investigating interactions in antiferromagnets and an early application of this kind was to some antiferromagnetic salts of iridium.[5–9] There have been many subsequent examples, most of which are collected together in Table VIII at the end of the chapter.

The main advantages of measuring exchange interactions in this way are that the method is direct and, especially for anisotropic interactions, very precise. Also, the symmetry of the pair structure is reflected in the symmetry of the spectrum, so it is fairly easy to distinguish between different types of pairs, e.g., first and second neighbors. Disadvantages are that the method is strictly limited to those pairs where it is possible to see an EPR spectrum at all—for example, the zero-field splittings of the energy levels may be so large that no microwave resonance is possible. Also, the the measurements usually refer to magnetically dilute crystals where the ion separations and therefore the interactions between them may be slightly different from those in the corresponding concentrated salt.

There are of course many other methods of estimating exchange interactions, the most traditional being from measurements of bulk properties such as the Curie–Weiss θ and the magnetic ordering temperature.[10] This macroscopic approach has the advantage of being applicable to almost any salt, but it is often difficult to distinguish the separate interactions in this way. Other methods, which have been developed more recently, include neutron scattering experiments which determine the dispersion in the spin wave spectrum of magnetically ordered systems;[11,12] ultrasonic resonance from pairs,[13] which is similar in many respects to EPR from pairs; and various optical absorption techniques,[14,158*] including those involving fluorescence[16] and those where infrared absorption by pair energy levels is detected directly.[17,18]

In what follows, we shall first discuss the various interactions which can occur between a pair of magnetic ions, giving a simple account of their origins. Then we consider the spin Hamiltonian, energy levels, and the resulting EPR spectrum from a pair. Next, some special topics are discussed, including exchange striction effects, hyperfine interactions, and spin–lattice relaxation. Finally, some experimental results on pairs are described and tabulated.†

2. MECHANISMS OF INTERACTIONS BETWEEN A PAIR OF IONS

In this section, we shall consider briefly the origin of some of the interactions which can occur between a neighboring pair of magnetic ions in a typical transition group salt such as MnO. In such salts, the magnetic ions are usually well separated, often with intervening ligands, so that effects

* For a review and further references, see Wickersheim.[15]

† Recent reviews of EPR pair spectra have been made by Kokoszka and Gordon[68] and Baker.[144]

of direct overlap tend to be small and the principal interaction arises from superexchange via intervening ligands. Following the early work of Kramers[19] and Anderson,[20] there have been many theoretical papers on this subject (for reviews, see Anderson[21] and Herring[22] and for recent calculations see Refs. 23–25). It is found that there are usually several contributions to the total isotropic interaction, some antiferromagnetic and some ferromagnetic, of which the antiferromagnetic ones usually predominate. In addition, there are anisotropic terms, some arising from the combined effect of isotropic exchange and spin–orbit coupling[26] and others which involve classical effects such as magnetic dipolar interaction and electric quadrupolar interaction. The latter are particularly important for rare-earth salts, where superexchange tends to be small.

We will first give a simple account of the superexchange mechanism and then go on to consider in turn the various anisotropic terms in the total interaction.

2.1. *Antiferromagnetic Contributions*

The structure often discussed in the literature is Mn^{2+}–O^{2-}–Mn^{2+} which occurs in MnO. Most of the proposed superexchange mechanisms involve the admixture of excited states in which an electron has hopped from one ion to another. Some examples are

(i) $(Mn^{2+}\text{–}O^{2-}\text{–}Mn^{2+}) + \alpha_1(Mn^{+}\text{–}O^{-}\text{–}Mn^{2+})$

(ii) $(Mn^{2+}\text{–}O^{2-}\text{–}Mn^{2+}) + \alpha_2(Mn^{+}\text{–}O\text{–}Mn^{+})$

(iii) $(Mn^{2+}\text{–}O^{2-}\text{–}Mn^{2+}) + \alpha_3(Mn^{+}\text{–}O^{2-}\text{–}Mn^{3+})$

where α_1, α_2, and α_3 are small admixture coefficients. In his review article, Anderson[21] concludes that mechanism (iii) proposed by Kondo[27] and independently by Anderson[28] gives much the largest contribution to the antiferromagnetic part of the exchange. This conclusion was supported by subsequent calculations.[24,25] In order to illustrate the superexchange mechanism, we will therefore discuss only mechanism (iii).

The two Mn^{2+} ions, with configuration $3d^5$, have un unpaired electron in each of the five d orbitals. If one of these d electrons is transferred from the Mn^{2+} ion to a d orbit on the other, its spin must be antiparallel to that of the electron already in this orbit by Pauli's principle. We can then picture the states given in Table I.

The energy U is that required to transfer the electron from Mn(2) to Mn(1) and can be estimated from ionization potentials (typically a few

Table I

		Mn(1)	Mn(2)	Energy
Mn^{+}–O^{2-}–Mn^{3+}	Excited state (spins antiparallel)	↑↓	—	U
Mn^{2+}–O^{2-}–Mn^{2+}	Ground state (spins parallel)	↑	↑	0
Mn^{2+}–O^{2-}–Mn^{2+}	Ground state (spins antiparallel)	↑	↓	0

eV). The self-consistent field Hamiltonian for the hopping electron (which is spin-independent) can only mix states of the same total spin. Thus there can be a matrix element, say b, between the excited state and the "spins antiparallel" ground-state so that in second-order perturbation theory the energy of the latter is depressed by approximately b^2/U relative to the "spins parallel" ground state. This is equivalent to saying that there is an antiferromagnetic exchange interaction of approximate energy b^2/U.

To find the magnitude of b, let us consider the interaction between an electron in a $d_{3z^2-r^2}$ orbit on Mn(1) on the left and an electron in a $d_{3z^2-r^2}$ orbit on Mn(2) on the right, where z is along the line joining the atoms (Fig. 1). These magnetic orbits each contain a small covalent and overlap admixture of $2p_z$ or p_σ orbit from the intervening O^{2-} ligand (configuration $1s^2 2s^2 2p^6$) and can be written as antibonding molecular orbitals

$$\psi_1 = -d_1 + A_\sigma p_\sigma, \qquad \psi_2 = d_2 + A_\sigma p_\sigma \tag{2.1}$$

where A_σ is the admixture coefficient. It can be seen that the fraction of unpaired spin on the ligand due to p_σ-bonding with one of the metal ions is

$$f_\sigma = A_\sigma^2 \tag{2.2}$$

which is usually called the spin transfer coefficient. The magnitude of A_σ

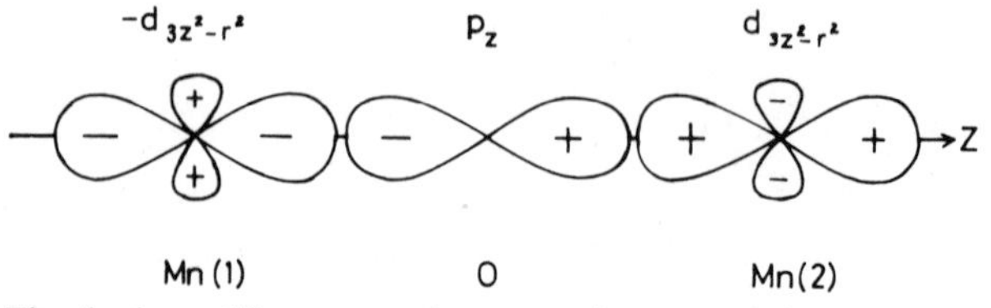

Fig. 1. An antiferromagnetic superexchange path between two Mn^{2+} ions via an O^{2-} ligand. Here, the process is shown in which $d_{3z^2-r^2}$ orbits are linked through an intermediate p_z orbit.

is expected to be approximately

$$A_\sigma \approx \langle d \mid h \mid p\rangle/(E_d - E_p) \tag{2.3}$$

where $\langle d \mid h \mid p\rangle$ is the matrix element of the one-electron Hamiltonian between the d and p orbits and $E_d - E_p$ is the energy separation between these orbits. (For a fuller discussion of these covalent effects see, for example, the review by Owen and Thornley[29]). It follows that the matrix element we are interested in, that connecting d_1 and d_2, is of approximate magnitude

$$\begin{aligned} b_\sigma &= \langle \psi_1 \mid h \mid \psi_2\rangle \\ &= \langle d_1 \mid h \mid p\rangle\langle p \mid h \mid d_2\rangle/(E_d - E_p) \\ &= f_\sigma(E_d - E_p) \end{aligned} \tag{2.4}$$

The proper states of the pair system of the molecular orbitals ψ_1 and ψ_2 of Eq. (2.1) are given in Table II.

Table II

State	Symmetry	Energy
$\mid D\rangle = (1/\sqrt{2})[\{\psi_1^+(1)\psi_1^-(2)\} - \{\psi_2^+(1)\psi_2^-(2)\}]$	Odd	U
$\mid C\rangle = (1/\sqrt{2})[\{\psi_1^+(1)\psi_1^-(2)\} + \{\psi_2^+(1)\psi_2^-(2)\}]$	Even	U
$\mid B\rangle = \begin{cases} \{\psi_1^+(1)\psi_2^+(2)\} \\ (1/\sqrt{2})[\{\psi_1^+(1)\psi_2^-(2)\} + \{\psi_1^-(1)\psi_2^+(2)\}] \\ \{\psi_1^-(1)\psi_2^-(2)\} \end{cases}$	Odd	0
$\mid A\rangle = (1/\sqrt{2})[\{\psi_1^+(1)\psi_2^-(2)\} - \{\psi_1^-(1)\psi_2^+(2)\}]$	Even	0

The curly brackets denote normalized determinantal functions, the plus and minus refer to spin up and spin down, and the odd and even refer to properties of the total wave function with respect to interchange of the subscripts 1 and 2. The only nonzero matrix elements of h are between states $\mid A\rangle$ and $\mid C\rangle$. The singlet ground state $\mid A\rangle$ is then found, to second order, to be depressed by

$$\mid \langle A \mid h \mid C\rangle \mid^2/U = 4b_\sigma^2/U \tag{2.5}$$

where b_σ is given by Eq. (2.4). Hence, for the ground states $|A\rangle$ and $|B$

$$E(\text{triplet–singlet}) = 4b_\sigma^2/U$$

which is equivalent to saying that there is an antiferromagnetic interactio between the two spins ($s_1 = s_2 = \frac{1}{2}$)

$$J_\sigma \mathbf{s}_1 \cdot \mathbf{s}_2 = (4b_\sigma^2/U)\mathbf{s}_1 \cdot \mathbf{s}_2 \quad (2.6$$

This is the result obtained by Anderson.[28] There are similar contribution from the two possible paths through the π-bonding orbits, i.e., d_{yz}–p_y–d_{yz} and d_{zx}–p_x–d_{zx} as indicated in Fig. 2. The total contribution from σ an π bonding is then approximately

$$J_\sigma + J_\pi = 4(f_\sigma^2 + 2f_\pi^2)(E_d - E_p)^2/U \quad (2.7$$

where the small σ bonding to ligand s orbits has been neglected throughout In terms of the total spins ($S_1 = S_2 = \frac{5}{2}$ for Mn^{2+}), this antiferromagneti contribution to the exchange interaction can finally be written

$$J\mathbf{S}_1 \cdot \mathbf{S}_2 = (\mathbf{s}^2/S^2)(f_\sigma^2 + 2f_\pi^2)[4(E_d - E_p)^2/U]\mathbf{S}_1 \cdot \mathbf{S}_2 \quad (2.8$$

To give an idea of the order of magnitude of the parameters, we pu $(E_d - E_p) \sim U \sim 10$ eV and $f_\sigma \sim f_\pi \sim 1.2\%$, giving $J/k \sim 10$ °K, which i roughly the observed value for Mn–O–Mn.

In Eq. (2.8), the term in square brackets does not vary very muc for ions in the same valence state and with the same ligands and in thes circumstances, J depends mainly on the relevant spin transfer coefficients

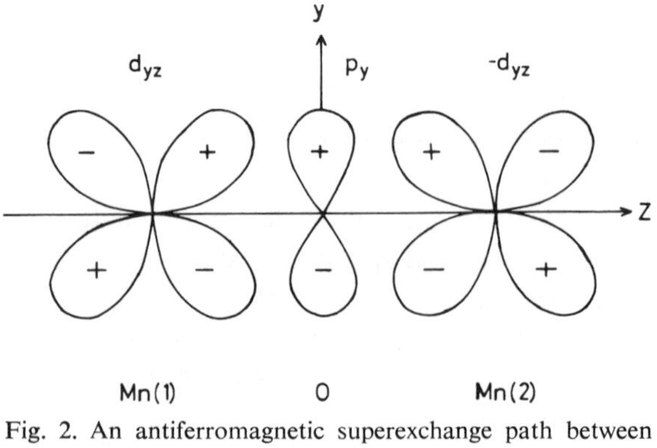

Fig. 2. An antiferromagnetic superexchange path between two Mn^{2+} ions via an O^{2-} ligand in which d_{yz} or d_{zx} orbits are linked through an intermediate p_y or p_x orbit.

Table III. The Approximate Dependence of the Contributions to the Superexchange Interaction on the Spin Transfer Coefficients f_σ and f_π for Some Octahedrally Coordinated Ions[a]

Ion		180°	90°
V^{2+} (d^3)	J(af)	$2f_\pi^2$	f_π^2
	J(f)	$2f_\pi^2$	$3f_\pi^2$
Mn^{2+} (d^5)	J(af)	$(2f_\pi^2 + f_\sigma^2)$	$(f_\pi^2 + 2f_\pi f_\sigma)$
	J(f)	$(2f_\pi^2 + 4f_\pi f_\sigma)$	$(3f_\pi^2 + f_\sigma^2 + 2f_\pi f_\sigma)$
Ni^{2+} (d^8)	J(af)	f_σ^2	(f_s^2)
	J(f)	$(2f_s f_\sigma)$	f_σ^2

[a] 180° means a structure such as Mn–O–Mn and 90° means one such as

Mn–O
|
Mn

Transfer to ligand s orbits is included only for d^8, where some of the terms would otherwise be zero. $J(af)$ and $J(f)$ are the antiferromagnetic and ferromagnetic contributions, respectively. (From Ref. 29.)

i.e., on $f_\sigma^2 + 2f_\pi^2$ for the 180° structure Mn–O–Mn. Some further examples for other configurations and also for 90° superexchange paths are given in Table III.

Almost all superexchange mechanisms which have been proposed in the literature lead to antiferromagnetic interactions which depend on the spin transfer coefficients in the way shown in Table III. Neglecting the smaller ferromagnetic terms for the moment, one can then predict the relative size of the interactions for different salts where the f's are known. This gives results in reasonable agreement with experiment (for example, see Refs. 21, 28, 29). However, attempts to calculate the absolute value of the antiferromagnetic exchange by putting calculated or measured parameters into Equations of type (2.8) often lead to a result higher, by a factor of about two, than the experimental value.[21,24] This may be partly due to the neglect or underestimate of the ferromagnetic contributions. Recently, Rimmer[25] has made a careful calculation, including ferromagnetic terms, for Ni^{2+}–F^-–Ni^{2+} which is the nearest-neighbor pair structure in $KNiF_3$. He finds $J = 125$°K, compared with the experimental $J = 90$°K.[10] This would appear to be the best agreement between theory and experiment yet obtained and is certainly very satisfactory.

2.2. Ferromagnetic Contributions

We will now indicate briefly one mechanism (see Fig. 3) by which the ferromagnetic contributions in Table III may arise in the Mn–O–Mn structure. There is a probability f_σ that an electron can hop from p_z on the oxygen to $d_{3z^2-r^2}$ on Mn(2) on the right and a probability f_π that an electron can hop from p_y on the oxygen to d_{yz} on Mn(1) on the left. There is thus an approximate probability $f_\sigma f_\pi$ that there are unpaired electrons in p_σ and p_y orbits on the ligand. These will try to couple parallel by the ordinary Heisenberg direct exchange integral which gives rise to Hund's rule and a ground-state spin of 1 in neutral oxygen. The antiparallel arrangement is higher by the exchange energy J(oxygen). The ferromagnetic contribution to the Mn–Mn interaction will then be roughly

$$J(\text{ferro}) \sim f_\sigma f_\pi J(\text{oxygen}) \tag{2.9}$$

though this may be an overestimate.[21] By adding up all such contributions, we obtain the results in Table III. It is usually thought that these ferromagnetic terms are much smaller than the antiferromagnetic ones, but there are a lot of them, especially if bonding to ligand s orbits is included, and they certainly cannot be neglected. In some cases where the antiferromagnetic terms are small or forbidden, the ferromagnetic terms will predominate. For example, in the structure d^3–oxygen–d^8, no antiferromagnetic interactions are allowed and so the exchange is expected to be ferromagnetic.[30]

Finally, it should be mentioned that there are other mechanisms for giving ferromagnetic contributions. For example, in Ni^{2+}–F^-–Ni^{2+}, Rim-

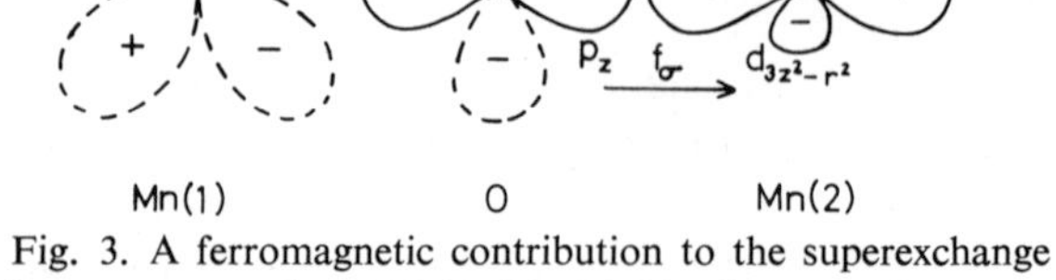

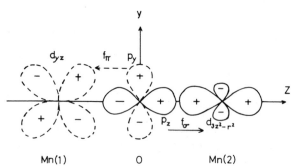

Fig. 3. A ferromagnetic contribution to the superexchange in a system such as Mn^{2+}–O^{2-}–Mn^{2+}. There are four processes like that shown here in which each contribution is proportional to $f_\sigma f_\pi$, and two involving only p_x and p_y orbits on the oxygen, each proportional to f_π^2.

ner[25] finds that the total ferromagnetic term is more than 10% of the sum of the antiferromagnetic terms even though the mechanism described above s forbidden in this case because there are no π bonding magnetic electrons o a first approximation.

?.3. *Biquadratic Exchange*

For magnetic ions which have more than one unpaired electron, further sotropic exchange terms can arise when perturbation theory is carried to igher orders. These include biquadratic terms of the form

$$-j(\mathbf{S}_1 \cdot \mathbf{S}_2)^2 \sim -(b^4/U^3)(\mathbf{S}_1 \cdot \mathbf{S}_2)^2 \qquad (2.10)$$

These terms are smaller than the bilinear isotropic exchange J by $/J \sim b^2/U^2$, which Anderson[21] estimates may be of order 1%, though j s probably smaller than this in most cases. It may be noted from Eq. 2.4) that $b^2/U^2 \propto f^2$, so one may expect j/J to increase as the covalent onding increases.

Physically, it may be pictured that such terms arise when there is a louble transfer in the superexchange process. For example, there is a robability b_σ^2/U^2 of transfer of a d_σ electron from one metal ion to the ext and of b_π^2/U^2 for a d_π electron, and therefore a probability $b_\sigma^2 b_\pi^2/U^4$ of both these electrons being transferred at the same time. This gives rise o terms of the form $(\mathbf{s}_1 \cdot \mathbf{s}_2)_\sigma(\mathbf{s}_1' \cdot \mathbf{s}_2')_\pi$ in the exchange interaction which ook like $(\mathbf{S}_1 \cdot \mathbf{S}_2)^2$ when expressed in terms of the total spin.

An explicit calculation for Mn–O–Mn has been carried out by Huang nd Orbach,[31] who find $j/J \sim 0.02$. However, the available experimental lata suggest that j/J is smaller than this and that exchange striction effects, which give rise to terms formally equivalent to Eq. (2-10), are probably nore important. A further discussion of these effects is given in Section 5.

This discussion refers only to magnetic ions with an orbital singlet round state. If there is orbital degeneracy, j/J may be much larger, as is hown, for example, by Elliott and Thorpe[32] in their theory of UO_2.

?.4. *Anisotropic Contributions*

The isotropic exchange discussed so far can lead to anisotropic interctions when spin–orbit coupling is included, as was first pointed out by /an Vleck in 1937.[26] Physically this is because the orbits of the magnetic ons have directional properties which reflect the symmetry of the pair and, if the spins are coupled to the orbits, they also will reflect the same directional properties. Most of the interactions which can arise in this way have been stated by Stevens,[33] some early analyses of experimental examples have been made by Bleaney and Bowers[1] and Judd,[7] and a good review is given by Kanamori.[34] We will now discuss briefly some of these interactions together with other anisotropic effects, and will include (i) magnetic dipolar interaction, (ii) pseudodipolar exchange, (iii) antisymmetric exchange, (iv) quadrupole–quadrupole interaction, and (v) virtual phonon interaction

2.4.1. Magnetic Dipolar Interaction

The classical interaction energy of two point magnetic dipoles $\boldsymbol{\mu}_1$ and $\boldsymbol{\mu}_2$ is

$$W = (1/r^3)[\boldsymbol{\mu}_1 \cdot \boldsymbol{\mu}_2 - (3/r^2)(\boldsymbol{\mu}_1 \cdot \mathbf{r})(\boldsymbol{\mu}_2 \cdot \mathbf{r})]$$

where $\mathbf{r}$ is the vector joining them. This can be expressed as a spin-Hamiltonian for a pair of magnetic ions by expressing the moment of each ion in the form

$$\boldsymbol{\mu} = \mathbf{i}\mu_x + \mathbf{j}\mu_y + \mathbf{k}\mu_z$$

where $\mu_x = -\beta(g_{xx}S_x + g_{xy}S_y + g_{xz}S_z)$, etc., and where S is the effective spin, g is the g-tensor, and β the Bohr magneton. In general, this gives quite a complicated expression, but in the simplest case when g_1 and g_2 are isotropic and equal, it reduces to

$$\mathscr{H} = D_d(3S_{1z}S_{2z} - \mathbf{S}_1 \cdot \mathbf{S}_2) \qquad (2.11)$$

where $D_d = -g^2\beta^2/r^3$, and the z axis is along the line joining the ions.

This value of D_d assumes point dipoles on ions 1 and 2. In a real crystal the magnetic moments are spread out over the neighboring ligands due to covalent transfer. Thus for a M_1^{2+}–F^-–M_2^{2+} pair in $KMgF_3$, M_1 transfers a fraction f of its moment to each of six F^- ligands, including the one shared with M_2, leaving a fraction $1 - 6f$ on M_1. Similarly for M_2. The net dipolar interaction between $(M_1F_6)^{4-}$ and $(M_2F_6)^{4-}$ must then include ligand–ligand and metal–ligand interactions. The result, found by Smith,[35] is

$$D_d' = -(g^2\beta^2/r_{12}^3)(1 + 8.6f - 71.6f^2) \qquad (2.12)$$

In addition, there is a correction due to transferred spin from M_1 and M being simultaneously on the shared ligand. Smith finds that this adds $-42f^2$ to the terms in Eq. (2.12) for the pair V^{2+}–F^-–V^{2+} in $KMgF_3$. Using the value $f = \frac{2}{3}f_\pi \approx 2\%$ for the $(VF_6)^{4-}$ complex, the final result is that D_d' is about 12% bigger* than the point-dipole value D_d. A similar increase in D_d is to be expected for other 180° pairs. It arises mainly because the dominant effect of the covalent transfer is to put magnetic moment from one metal ion on to the intervening ligand which is very close to the other metal ion.

For a 90° structure, such as a nearest-neighbor pair of Mn^{2+} ions in MgO, there is no intervening ligand, and then the dominant effect is expected to be the covalent decrease (by a factor $1 - 6f$) of the magnetic moment on each Mn^{2+} ion. This results in a decrease of about 10% in D_d. Such a decrease has been found experimentally by Harris[36] from analysis of the EPR pair spectrum in Mn:MgO.

Finally, it may be mentioned that for a concentrated salt of the MnO type which is antiferromagnetically ordered, the magnetic moments on the ligands do not contribute to the dipolar anisotropy energy. The only effect is the reduction $(1 - 6f)^2$ in D_d due to the reduced moments on the metal ions. This effect can be very large. For example, it has been estimated[29] that in NiO there is a 34% reduction in dipolar anisotropy energy due to this covalent transfer effect and this estimate appears to be confirmed by antiferromagnetic resonance measurements.[37]

2.4.2. Pseudodipolar Exchange

This is an interaction of the same dipolar form as Eq. (2.11) and arises from the combined effect of isotropic exchange and the spin–orbit coupling $\lambda\mathbf{L}\cdot\mathbf{S}$. Following Kanamori,[34] if the two ions have orbital singlet ground states g_1, g_2 (these are not to be confused with g-values) and excited states e_1, e_2, the interaction comes from a third-order perturbation process with terms such as

$$\frac{\langle g_1g_2 \mid \lambda\mathbf{L}_1\cdot\mathbf{S}_1 \mid e_1g_2\rangle\langle e_1g_2 \mid H_{ex} \mid e_1g_2\rangle\langle e_1g_2 \mid \lambda\mathbf{L}_1\cdot\mathbf{S}_1 \mid g_1g_2\rangle}{(E_e - E_g)^2} \tag{2.13}$$

where the term in H_{ex} is essentially the isotropic exchange interaction between e_1 and g_2. It can be seen that the magnitude of the effect is of order $J\lambda^2/\Delta^2$, where J is an exchange interaction and Δ the splitting between g_1 and e_1. This is also of order $(\Delta g)^2 J$, where Δg is the g-shift.

To illustrate (2.13), consider the simple case where each ion has only one unpaired electron so that $S_1 = S_2 = \frac{1}{2}$. The components of $\lambda\mathbf{L}\cdot\mathbf{S}$, i.e., λL_xS_x, λL_yS_y, and λL_zS_z, will in general admix different excited states,

* A more recent calculation,[35a] which allows for distortion in the pair, shows that D_d' may be up to 23% bigger than D_d.

say e_x, e_y, and e_z. The isotropic exchange interaction between these excited states on one ion and the ground state of the other ion may be written

$$\langle e_{1x}g_2 \mid H_{ex} \mid e_{1x}g_2 \rangle = J_x \mathbf{S}_1 \cdot \mathbf{S}_2 \tag{2.14}$$

with corresponding expressions for J_y and J_z.

The contributions to the spin-Hamiltonian from (2.13) are thus of the form

$$\alpha_x J_x S_{1x}(\mathbf{S}_1 \cdot \mathbf{S}_2) S_{1x} \tag{2.15}$$

where

$$\alpha_x = \lambda^2 \mid \langle g \mid L_x \mid e_x \rangle \mid^2 / (E_x - E_g)^2 \tag{2.16}$$

and there are similar terms with subscripts y and z. Using the relations $S_x S_x S_x = \frac{1}{4} S_x$, $S_x S_y S_x = -\frac{1}{4} S_y$ and $S_x S_z S_x = -\frac{1}{4} S_z$ which apply for $S = \frac{1}{2}$, (2.15) reduces to

$$(\alpha_x J_x / 4)(S_{1x} S_{2x} - S_{1y} S_{2y} - S_{1z} S_{2z}) \tag{2.17}$$

Collecting all such terms in x, y, and z, together with those where the suffixes 1 and 2 are interchanged, the total interaction is then

$$\mathscr{H} = \alpha_x J_x S_{1x} S_{2x} + \alpha_y J_y S_{1y} S_{2y} + \alpha_z J_z S_{1z} S_{2z} \tag{2.18}$$

plus an isotropic term $-\frac{1}{2}(\alpha_x J_x + \alpha_y J_y + \alpha_z J_z)\mathbf{S}_1 \cdot \mathbf{S}_2$.

It can be seen from (2.18) that there are two possible origins for the anisotropy. First, the isotropic exchange interactions. J_x, J_y, and J_z, between different pairs of orbits may be different. Second, the amounts α_x, α_y, and α_z of these different orbits which are present in the total wave function of the ground state of the magnetic ion may also be different.

In practice, the pair often has axial symmetry about the line joing the ions (the z axis) and then $J_x = J_y$ and $\alpha_x = \alpha_y$, so that apart from some isotropic terms, (2.18) can be expressed in the usual dipolar form

$$\mathscr{H} = D_E(3S_{1z}S_{2z} - \mathbf{S}_1 \cdot \mathbf{S}_2) \tag{2.19}$$

where

$$D_E = -\tfrac{1}{3}(\alpha_z J_z - \alpha_x J_x) \tag{2.20}$$

We will now consider some examples. Suppose that there is axial symmetry and that the ground state of each ion is $d_{3z^2-r^2}$ such as could occur for an octahedrally coordinated Cu^{2+} ion with a small tetragonal distortion. Then L_z admixes no excited states, giving $\alpha_z = 0$, while L_x and L_y admix

d_{zx} and d_{yz} equally, giving $\alpha_x = 3\lambda^2/\Delta_x^2$ and hence $D_E = J_x\lambda^2/\Delta_x^2$. Also, since $g_z = 2$ and $g_x = g_y = 2 - 6(\lambda/\Delta_x)$, we can express D_E in terms of the g-shift, $D_E = J_x(g_\perp - 2)^2/36$. In these expressions, J_x is the exchange interaction between $d(3z^2 - r^2)_1$ on one ion and $d(zx)_2$ on the other, which, if the pair has the 180° metal–ligand–metal structure, is expected (Section 2.2) to be a small ferromagnetic term perhaps only 10% to 20% of the main antiferromagnetic interaction J between the unperturbed ground states $d(3z^2 - r^2)_1$ and $d(3z^2 - r^2)_2$.

As a second example, consider the case where there is axial symmetry and $d_{x^2-y^2}$ is the ground state of each ion. Then L_z admixes the excited state d_{xy}, giving $\alpha_z = 4\lambda^2/\Delta_z^2$ and $g_z = 2 - 8\lambda/\Delta_z$; L_x admixes d_{zx}, giving $\alpha_x = \lambda^2/\Delta_x^2$ and $g_x = 2 - 2\lambda/\Delta_x$; also, $\alpha_x = \alpha_y$ and $g_x = g_y$. Hence, from Eq. (2.20)

$$D_E = -(1/12)[(\tfrac{1}{4})(g_\parallel - 2)^2 J_z - (g_\perp - 2)^2 J_x| \tag{2.21}$$

which is equivalent to the expression first given by Bleaney and Bowers[1] for the pseudodipolar exchange between pairs of Cu^{2+} ions in copper acetate. As a rough estimate, these authors assumed that $J_z = J_x = J_0$, where J_0 is the superexchange interaction between the ground states with known value $J_0 \approx 315$°K. Using the measured values of $g_\parallel$ and $g_\perp$, Eq. (2.21) then gives $D_E = +0.93\ \text{cm}^{-1}$. This compares with the experimental value $|D| = 0.23\ \text{cm}^{-1}$, which includes a dipolar contribution $D_d = -0.12\ \text{cm}^{-1}$. Hence, the experimental D_E is much smaller than the "calculated" value, suggesting that in this example, the exchange interactions J_x and J_z between the ground state of one ion and the excited states of the other may be an order of magnitude smaller than J_0.

For ions with an orbital singlet ground state but more than one unpaired electron, the calculation proceeds along similar lines. Kanamori[34] has treated the case of a Ni^{2+}–O^{2-}–Ni^{2+} pair in MgO*, for which $S_1 = S_2 = 1$. He finds after allowing for the differences between J_x, J_y, J_z, and J_0, that $D_E = 2J_0\lambda^2/\Delta^2 = J_0(\Delta g)^2/32$ and hence predicts that $D_E = 0.15\ \text{cm}^{-1}$ using $J = 91\ \text{cm}^{-1}$ and $g = 2.227$. A test of this theory has been made by May[38] on Ni^{2+}–F^-–Ni^{2+} pairs in $KMgF_3$. He finds that the experimental D_E is probably less than half of $2J_0\lambda^2/\Delta^2$, though there are some uncertainties about corrections which involve the unknown magnitude of the crystal field splitting for the Ni^{2+} ions in the pair. Also, ferromagnetic terms are neglected in the theory.

* For V^{2+}–O^{2-}–V^{2+} in MgO, it is found that $D_E = (1/96)(\Delta g)^2 J$, which gives an unobservably small anisotropy.[35a]

When the ground state is orbitally degenerate, the spin–orbit coupling is a first-order effect and then the pseudo dipolar exchange can be of comparable magnitude to the isotropic exchange. The first detailed analysis was by Judd,[7] who treated a pair of Ir^{4+} ions in $(NH_4)_2IrCl_6$. For this case, $D_E \approx J_0/10$ (Table VIII). There have been several recent papers[32,39,40,144] giving a general discussion of the case when there is orbital degeneracy to which reference should be made for further details.

2.4.3. Antisymmetric Exchange

This type of interaction was first pointed out by Stevens[33] and its importance, especially in weak ferromagnetism, has been emphasised by Dzialoshinski.[41] Moriya[42] has given a detailed analysis of its origin and we follow his review article[43] to give a simple derivation of this interaction.

For a pair of ions with orbital singlet ground states g_1, g_2 and excited states e_1, e_2, the interaction arises from a second-order perturbation which includes terms of the form

$$[\langle g_1g_2 \mid \lambda\mathbf{L}_1 \cdot \mathbf{S}_1 \mid e_1g_2\rangle\langle e_1g_2 \mid H_{ex} \mid g_1g_2\rangle + \langle g_1g_2 \mid H_{ex} \mid e_1g_2\rangle\langle e_1g_2 \mid \lambda\mathbf{L}_1 \cdot \mathbf{S}_1 \mid g_1g_2\rangle]/\Delta_1 \tag{2.22}$$

where Δ_1 is the energy separation g_1 to e_1 and H_{ex} is the exchange interaction between e_1 and g_2 as before [see Eq. (2.14)]. Since the states can be chosen real and $\mathbf{L}$ is pure imaginary, $\langle g_1 \mid \mathbf{L}_1 \mid e_1\rangle = -\langle e_1 \mid \mathbf{L}_1 \mid g_1\rangle$. The expression (2.22) can then be written

$$(\lambda J_1/\Delta_1)\langle g_1 \mid \mathbf{L}_1 \mid e_1\rangle[\mathbf{S}_1(\mathbf{S}_1 \cdot \mathbf{S}_2) - (\mathbf{S}_1 \cdot \mathbf{S}_2)\mathbf{S}_1]$$

There are corresponding terms with the subscripts 1 and 2 interchanged, so the total interaction is of the form

$$\mathscr{H} = \mathbf{C} \cdot (\mathbf{S}_1 \times \mathbf{S}_2) \tag{2.23}$$

with

$$C = (\lambda J_1/\Delta_1)\langle g_1 \mid \mathbf{L}_1 \mid e_1\rangle - (\lambda J_2/\Delta_2)\langle g_2 \mid \mathbf{L}_2 \mid e_2\rangle \tag{2.23a}$$

The first point to notice is that if the two ions are identical, then the two two terms in (2.23a) are also identical and therefore cancel, giving $C = 0$. Moriya gives the general rule that $C = 0$ if a center of inversion symmetry is located midway between the two ions. In the case $C \neq 0$, e.g., because each ion has a different site symmetry, the effect can be quite large since $C \sim (\lambda/\Delta)J \sim (\Delta g/g)J$ and is thus an order of magnitude greater than the pseudodipolar exchange. The effect of the vector coupling in (2.23) is that the spins try to orient perpendicular to each other; this can lead to canting of the spins and hence to weak ferromagnetism in a crystal which is predominantly antiferromagnetic.

This antisymmetric exchange interaction does not yet appear to have been identified unambiguously in the EPR spectrum of a pair of transition group ions mainly because the most detailed investigations so far have been of pairs with inversion symmetry.* However, it has recently been clearly demonstrated for pairs of O_2^- ions in potassium iodide.[157]

When there is orbital degeneracy and a large orbital contribution to the magnetic moment, the antisymmetric interaction can be a first-order effect, as can the pseudodipolar exchange interaction. Discussion of this case, with rare-earth and actinide ions particularly in mind, have been given by several authors.[32,39,40]

2.4.4. Electric Quadrupole–Quadrupole Interaction

This is the electrostatic interaction which results from the charge distribution on one ion of the pair contributing to the electric field gradient at the other. The electric dipole moment on each ion is zero if the ion site has inversion symmetry, as is normally the case, because then the charge cloud also has inversion symmetry. The lowest moment of interest is then the electric quadrupole moment, and there can also be higher moments.

The interaction between a pair of electric quadrupoles has been discussed by Finkelstein and Mencher[45] and the results have been given in operator equivalent form by Bleaney.[46] For a pair i and j with axial symmetry and with the z axis along the line joining the ions, the result is

$$\mathscr{H}_{QQ} = A_Q[4O_{2i}^0O_{2j}^0 - 16(O_{2i}^1O_{2j}^{-1} + O_{2i}^{-1}O_{2j}^1) + (O_{2i}^2O_{2j}^{-2} + O_{2i}^{-2}O_{2j}^2)] \quad (2.24)$$

where

$$O_2^0 = 3S_z^2 - S(S+1), \qquad O_2^{\pm 1} = \tfrac{1}{2}(S_zS_\pm + S_\pm S_z), \qquad O_2^{\pm 2} = S_\pm^2$$

and S is the effective spin of the ground state of each ion. We have

$$A_Q = 3e^2\langle r_i^2\rangle\langle r_j^2\rangle\alpha_i\alpha_j/8\varepsilon R^5hc \quad (2.25)$$

where R is the distance between the ions, $\langle r^2\rangle$ is the mean square radius of the magnetic electrons, and ε is the effective dielectric constant; ε is difficult to estimate in practice and may be very anisotropic.[47] α_i and α_j are numerical factors whose value depends on the ions concerned; for the rare earths, they

* Recently an antisymmetric exchange interaction has been reported[118] between Cr^{3+} ions in $SrTiO_3$.

are given by Elliott and Stevens[48]—for example, $\alpha = -2/35$ for Ce^{3+}, $4f^1$.

This quadrupole–quadrupole interaction is generally rather a small term, but its presence has been demonstrated by EPR measurements on pairs of rare-earth ions, notably pairs of Ce^{3+} ions in $LaCl_3$.[50] For Ce^{3+}, the ground state is a Kramers doublet with effective spin $S = \frac{1}{2}$, so the second degree terms of Eq. (2.24) have no first-order effect; they only contribute as a second-order perturbation via an excited state and this takes the form of an apparent contribution to the pseudodipolar interaction between the two effective spins in the ground state.

For S-state ions, there is spherical symmetry and therefore no quadrupole moment. For ions which have an orbital singlet ground state, the effect is also very small because the spins are only weakly coupled to the orbits (and therefore to the quadrupole moment) via spin–orbit coupling to excited states. Kanamori[34] has discussed the quadrupole–quadrupole interaction for a pair of Ni^{2+} ions ($S_i = S_j = 1$) in MgO; he finds that it is a fifth-order perturbation, so that the term $\alpha_i \alpha_j$ in Eq. (2.25) contains a factor $(\lambda/\Delta)^4$, which is of order $(\Delta g/g)^4 \sim 10^{-4}$.

Finally, it should be mentioned that Wolf and Birgeneau[51] have argued that the interactions between higher-order multipoles may give terms of comparable magnitude to those of the quadrupolar interaction and that in general it will be difficult to distinguish between them. A further difficulty is that the virtual phonon interaction gives rise to terms of similar form, as will now be seen.

2.4.5. Virtual Phonon Exchange Interaction

This type of interaction was first suggested by Sugihara.[52] The strain field $\mathscr{H}_{st}$, which is just the perturbation of the crystal field due to lattice waves, can interact with both spin i and spin j, and this leads to a coupling between them which can be regarded as emission by one spin of a phonon which is immediately reabsorbed by the other. $\mathscr{H}_{st}$ acting on spin i has matrix elements

$$\langle \psi_i \psi_j n \mid \mathscr{H}_{st}(i) \mid \psi_i' \psi_j n' \rangle$$

representing a transition of spin i form state ψ_i to ψ_i' accompanied by a change in the phonon excitation from n to n'. Interaction with spin j will yield a matrix element

$$\langle \psi_i' \psi_j n' \mid \mathscr{H}_{st}(j) \mid \psi_i' \psi_j' n \rangle$$

which represents a change of spin j from ψ_j to ψ_j' accompanied by a reversion of the phonon excitation from n' back to n. Thus, the matrix element between $\psi_i\psi_j$ and $\psi_i'\psi_j'$ is

$$\langle\psi_i\psi_j n \,|\, \mathscr{H}_{\rm st}(i) \,|\, \psi_i'\psi_j n'\rangle\langle\psi_i'\psi_j n' \,|\, \mathscr{H}_{\rm st}(j) \,|\, \psi_i'\psi_j' n\rangle/\Delta E$$

where ΔE is the energy of the intermediate state. The above expression encompasses diagonal elements as well, where initial and final states are the same ψ_i and ψ_j.

To illustrate the form and magnitude of this interaction, we will consider the simplest term in the theory developed by McMahon and Silsbee[53] for the interaction between Fe^{2+} ions in Fe^{2+}:MgO. For an isolated Fe^{2+} ion ($S = 1$) the spin-Hamiltonian describing the spin–strain interaction contains a term

$$\mathscr{H} = -G_{11}Q[3S_z^2 - S(S+1)] \tag{2.26}$$

and there are other second-order terms in S_+^2, S_-^2, S_+S_z, etc. Q represents the strain introduced by the appropriate noncubic symmetric mode of vibration of the surrounding octahedron of oxygen ligands, G_{11} is the corresponding energy per unit strain, which depends on the coupling between the spin and the crystal field (for example, $G_{11} = 720\ \text{cm}^{-1}$ per unit strain for Fe^{2+} in MgO). McMahon and Silsbee find that this spin–strain term leads to a virtual phonon coupling between two Fe^{2+} ions of the form

$$\mathscr{H} = A(G_{11}^2/R^3)[3S_{iz}^2 - S_i(S_i+1)][3S_{jz}^2 - S_j(S_j+1)] \tag{2.27}$$

where R is the distance between the ions and $A = 3(1 - 3\cos^2\theta)/64\pi\varrho v^2)$, ϱ being the crystal density, v the velocity of sound, and θ the angle between the line joining the ions and the [100] direction. There are also other second-order terms with different coefficients, so that in general the interaction is of the form

$$\mathscr{H}_{\rm VPE} = \sum a_{nm} O_{2i}^n O_{2j}^m \tag{2.28}$$

giving terms similar to those of the electric quadrupole–quadrupole interaction [Eq. (2.24)].

The difficulty of distinguishing between the EQQ and VPE interactions has been emphasised by Baker and Mau.[47] However, there are some examples, e.g., Ce^{3+} in $LaCl_3$, where it seems to have been established[50] EQQ predominates and VPE is negligible; while there are others, e.g., UO_2, where VPE is expected to make a substantial contribution.[54] One of the most interesting features of the VPE interaction is its long range—part of it is proportional to $1/R^3$ according to Eq. (2.27). Sugihara[52] estimates that it may be of comparable magnitude to the magnetic dipolar interaction (also with $1/R^3$ dependence) in the nickel Tutton salts and McMahon and Silsbee[53] estimate that the same is true for Fe^{2+}:MgO. The latter authors find that some of the coefficients a_{nm} in Eq. (2.28) even have a $1/R$ dependence. The VPE interaction would then certainly predominate over all others at large distances and would lead to a dependence of EPR linewidth on crystal size. This prediction has not yet been verified experimentally.

3. *SPIN-HAMILTONIAN AND* EPR *SPECTRUM FOR PAIRS AND CLUSTERS*

We consider how the interactions discussed above affect the EPR spectrum for a pair of neighboring ions. The largest interaction is often the isotropic exchange $J\mathbf{S}_i \cdot \mathbf{S}_j$ and the most important anisotropic terms are often of dipolar form $D(3S_{iz}S_{jz} - \mathbf{S}_i \cdot \mathbf{S}_j)$. Here, D includes the true dipolar interaction [Eq. (2.11)] and the pseudodipolar exchange [Eq. (2.18)]. In addition, if each ion has a Kramers doublet ground state, D may also include the only observable effects of quadrupole–quadrupole interaction [Eq. (2.24)] and virtual phonon exchange [Eq. (2.28)] as explained in Section 2.4.4. In practice, these higher-degree interactions tend to be relatively very small in the iron group and can be neglected to a first approximation; while in the rare-earth group most pair measurements have been made on ions with a Kramers doublet ground state so that the higher-degree terms are included in D.

It thus turns out that the EPR spectrum for a pair of similar ions with axial symmetry can be simply described in most cases by just two parameters, J and D. If the symmetry is lower than axial, an additional second-degree anisotropic term $E(S_{ix}S_{jx} - S_{iy}S_{jy})$ is added. Most of the available EPR results on pairs are tabulated at the end of the chapter (Table VIII) and are expressed in terms of J, D, and E. Though this formal description is very simple, it is often very difficult to distinguish the different contributions to each parameter and this remains a real problem in work on pair spectra.

The general arrangement of this section is as follows. First, the Hamiltonian for the simplest case of a pair of similar ions is discussed, and it is shown how the interaction parameters are derived from the spectrum in typical cases. Then, some of the new features which are introduced if the ions are dissimilar are considered briefly. Finally, a short account is given of larger clusters of interacting ions.

3.1. The Hamiltonian for a Pair of Similar Ions

The pair is assumed to have real or effective electronic spins $S_i = S_j$. The Hamiltonian for S_i, omitting interactions with S_j, is taken to be of the typical form

$$\mathscr{H}_i = \beta \mathbf{H} \cdot \mathsf{g} \cdot \mathbf{S}_i + D_c[S_{iz}^2 - \tfrac{1}{3}S(S+1)] + E_c(S_{ix}^2 - S_{iy}^2) + \mathbf{S}_i \cdot \mathsf{A} \cdot \mathbf{I}_i \quad (3.1)$$

There is a similar Hamiltonian $\mathscr{H}_j$ for S_j. The first term represents the energy in the applied magnetic field H. The terms in D_c and E_c are the usual second-degree crystal field terms with axial and rhombic symmetry respectively. These can be appreciable even in nominally cubic crystals, due to the distortions which are inevitably present round an impurity pair. Higher-degree crystal field terms are omitted for simplicity and in any case do not have a very important effect on most of the pair spectra discussed here. The last term in A is the hyperfine interaction; hyperfine structure effects for a pair of ions will be discussed separately in Section 4.

The two ions are now allowed to interact, so that the Hamiltonian for the pair system is

$$\mathscr{H}' = \mathscr{H}_i + \mathscr{H}_j + J\mathbf{S}_i \cdot \mathbf{S}_j + D_e(3S_{iz}S_{jz} - \mathbf{S}_i \cdot \mathbf{S}_j) + E_e(S_{ix}S_{jx} - S_{iy}S_{jy}) \quad (3.2)$$

The term in J is the isotropic exchange. Higher-degree isotropic terms, e.g., biquadratic exchange of the form $-j(\mathbf{S}_i \cdot \mathbf{S}_j)^2$ and exchange striction effects, are discussed separately in Section 5. The terms in D_e and E_e represent the anisotropic interactions with axial and rhombic symmetry, respectively. As already mentioned, these include the true dipolar interaction and pseudodipolar exchange, and may also include second-order effects of higher-degree terms such as quadrupolar interaction and virtual phonon exchange. It is assumed in the notation that the axes x, y, z of these interactions are the same as those of the crystal field. If z is along the line joining the ions, we can write

$$D_e = D_d(\text{dipolar}) + D_E(\text{pseudodipolar exchange}) + (\text{effects of higher degree terms}) \quad (3.3)$$

where $D_d = -g^2\beta^2/r_{ij}^3$ if g is isotropic* and a point-dipole model is assumed [Eq. (2.11)]. Deviations from a point-dipole model are discussed in Section 2.4.1.

If the term in J is much greater than the other terms in Eq. (3.2), the system is most conveniently described in terms of the total spin $\mathbf{S} = \mathbf{S}_i + \mathbf{S}_j$, where S can take values $S = S_i + S_j,\ S_i + S_j - 1, \ldots, 0$. The energies

* If g is not isotropic, the point-dipole model gives contributions to D_e, E_e, and J

$$D_d = (-\beta^2/3r_{ij}^3)[2g_z^2 + \tfrac{1}{2}(g_x^2 + g_y^2)], \qquad E_d = (\beta^2/2r_{ij}^3)(g_x^2 - g_y^2)$$

$$J_d = (\beta^2/3r_{ij}^3)(g_x^2 + g_y^2 - 2g_z^2)$$

of these total spin multiplets follow a Landé interval rule and are given by

$$W_S = (J/2)[S(S+1) - S_i(S_i+1) - S_j(S_j+1)] \quad (3.4)$$

Equation (3.2) can now be expressed in terms of the total spin S,

$$\mathscr{H} = W_S + \beta \mathbf{H} \cdot \mathsf{g} \cdot \mathbf{S} + D_S[S_z^2 - \tfrac{1}{3}S(S+1)] + E_S(S_x^2 - S_y^2) \quad (3.5)$$

where

$$\begin{aligned} D_S &= 3\alpha_S D_e + \beta_S D_c \\ E_S &= \alpha_S E_e + \beta_S E_c \end{aligned} \quad (3.6)$$

and

$$\begin{aligned} \alpha_S &= \tfrac{1}{2}[S(S+1) + 4S_i(S_i+1)]/(2S-1)(2S+3) \\ \beta_S &= [3S(S+1) - 3 - 4S_i(S_i+1)]/(2S-1)(2S+3) \end{aligned} \quad (3.7)$$

Equations (3.6) and (3.7) were first derived by Judd[55,56] Values of the coefficients α_S and β_S are listed for convenience in Table IV.

Table IV. Values of α_S and β_S for $S_i = S_j = \frac{1}{2}$ to $\frac{7}{2}$

$S_i = S_j =$	$\frac{1}{2}$	1	$\frac{3}{2}$	2	$\frac{5}{2}$	3	$\frac{7}{2}$
α_S							
$S = 1$	1/2	1	17/10	13/5	37/10	5	13/2
2		1/3	1/2	5/7	41/42	9/7	23/14
3			3/10	2/5	47/90	2/3	5/6
4				2/7	5/14	34/77	83/154
5					5/18	1/3	31/78
6						3/11	7/22
7							7/26
β_S							
$S = 1$	0	−1	−12/5	−21/5	−32/5	−9	−12
2		+1/3	0	−3/7	−20/21	−11/7	−16/7
3			+2/5	+1/5	−2/45	−1/3	−2/3
4				+3/7	+2/7	+9/77	−6/77
5					+4/9	+1/3	+8/39
6						+5/11	+4/11
7							+6/13

Higher-degree terms in the Hamiltonians (3.1) and (3.2) may be treated in a similar way and be represented by corresponding terms in (3.5). Also, if J is not very much larger than the other terms in (3.2), Eq. (3.4) and (3.5) are not applicable. In this case, the best procedure is to write down the substates of total spin $| M\rangle_S$ in terms of the individual spin magnetic quantum numbers

$$| M\rangle_S = a | M_i, M_j\rangle + b | M_i - 1, M_j + 1\rangle + \cdots$$

and calculate the matrix elements of (3.2). Such states for a pair of ions with $S_i = S_j = \frac{1}{2}$ and with $S_i = S_j = \frac{5}{2}$ are given as examples in Table V and the whole subject of the addition of angular momenta is given in a number of standard textbooks.[56]

3.2. EPR *Spectrum and Derivation of Parameters*

The form of the energy levels and the observed EPR spectrum which result from the pair Hamiltonian (3.2) depend on the relative magnitude of J and the other terms. It is therefore convenient to consider in turn the three cases $J \gg D_e$, $J > D_e$ and $J \sim D_e$.

3.2.1. $J \gg D_e$

This is one of the most commonly investigated cases, especially in the iron group. An example is nearest-neighbor Mn^{2+} ions in MgO,[57] for which $J \approx 10$ cm^{-1} and $D_e \approx 0.1$ cm^{-1}, so that $J/D_e \sim 100$.

The Hamiltonian (3.5) is then applicable, so that for a Mn^{2+}–Mn^{2+} pair with $S_i = S_j = 5/2$, for example, the total spin states are $S = 0, 1, 2, 3, 4$, and 5 with relative energies 0, J, $3J$, $6J$, $10J$, and $15J$, respectively. Since the perturbation due to the microwave magnetic field $\mathbf{h}$, $\beta\mathbf{h} \cdot \mathrm{g} \cdot (\mathbf{S}_i + \mathbf{S}_j)$, is diagonal in total spin, the allowed transitions occur only between the Zeeman components within each total spin state, so the selection rule $\Delta S = 0$ is obeyed. [Forbidden transitions $\Delta S = \pm 2$ are of order $(D_e/J)^2$ lower in intensity.]

The usual method of determining J is to use the temperature dependence of the signal intensity. Thus the population of an energy level belonging to S has a temperature variation

$$I_S = (1/Z) \exp(-W_S/kT) \tag{3.8}$$

where Z is the partition function, $Z = \sum_s (2S + 1) \exp(-W_S/k_T)$, and W_S is given by Eq. (3.4) in the approximation that $J \gg D$, $g\beta H$. The intensity

Table V. Substates for a Pair of Coupled Spins Written in the For $|M\rangle_S = \sum a\,|M_i, M_j\rangle$, (a) for $S_i = S_j = \frac{1}{2}$ and (b) for $S_i = S_j = \frac{5}{2}$

(a) $S_i = S_j = \frac{1}{2}$

$$|1\rangle_1 = |\tfrac{1}{2}, \tfrac{1}{2}\rangle$$

$$|0\rangle_1 = (1/\sqrt{2})(|\tfrac{1}{2}, -\tfrac{1}{2}\rangle + |-\tfrac{1}{2}, \tfrac{1}{2}\rangle)$$

$$|-1\rangle_1 = |-\tfrac{1}{2}, -\tfrac{1}{2}\rangle$$

$$|0\rangle_0 = (1/\sqrt{2})(|\tfrac{1}{2}, -\tfrac{1}{2}\rangle - |-\tfrac{1}{2}, \tfrac{1}{2}\rangle)$$

(b) $S_i = S_j = \frac{5}{2}$

$$|5\rangle_5 = |\tfrac{5}{2}, \tfrac{5}{2}\rangle$$

$$|4\rangle_5 = (1/\sqrt{2})(|\tfrac{3}{2}, \tfrac{5}{2}\rangle + |\tfrac{5}{2}, \tfrac{3}{2}\rangle)$$

$$|3\rangle_5 = \tfrac{1}{3}(\sqrt{2}\,|\tfrac{1}{2}, \tfrac{5}{2}\rangle + \sqrt{2}\,|\tfrac{5}{2}, \tfrac{1}{2}\rangle + (5)^{1/2}\,|\tfrac{3}{2}, \tfrac{3}{2}\rangle)$$

$$|2\rangle_5 = (1/2\sqrt{3})(|-\tfrac{1}{2}, \tfrac{5}{2}\rangle + |\tfrac{5}{2}, -\tfrac{1}{2}\rangle + (5)^{1/2}\,|\tfrac{1}{2}, \tfrac{3}{2}\rangle + (5)^{1/2}\,|\tfrac{3}{2}, \tfrac{1}{2}\rangle)$$

$$\begin{aligned}|1\rangle_5 = {} & [1/2(21)^{1/2}][\sqrt{2}(|-\tfrac{3}{2}, \tfrac{5}{2}\rangle + |\tfrac{5}{2}, -\tfrac{3}{2}\rangle) + 2(5)^{1/2}(|-\tfrac{1}{2}, \tfrac{3}{2}\rangle + |\tfrac{3}{2}, -\tfrac{1}{2}\rangle) \\ & +2(10)^{1/2}\,|\tfrac{1}{2}, \tfrac{1}{2}\rangle]\end{aligned}$$

$$\begin{aligned}|0\rangle_5 = {} & [1/6(7)^{1/2}][(|-\tfrac{5}{2}, \tfrac{5}{2}\rangle + |\tfrac{5}{2}, -\tfrac{5}{2}\rangle) + 5(|-\tfrac{3}{2}, \tfrac{3}{2}\rangle + |\tfrac{3}{2}, -\tfrac{3}{2}\rangle) \\ & +10(|-\tfrac{1}{2}, \tfrac{1}{2}\rangle + |\tfrac{1}{2}, -\tfrac{1}{2}\rangle)]\end{aligned}$$

$$|4\rangle_4 = (1/\sqrt{2})(|\tfrac{3}{2}, \tfrac{5}{2}\rangle - |\tfrac{5}{2}, \tfrac{3}{2}\rangle)$$

$$|3\rangle_4 = (1/\sqrt{2})(|\tfrac{1}{2}, \tfrac{5}{2}\rangle - |\tfrac{5}{2}, \tfrac{1}{2}\rangle)$$

$$|2\rangle_4 = [1/2(7)^{1/2}][3(|-\tfrac{1}{2}, \tfrac{5}{2}\rangle - |\tfrac{5}{2}, -\tfrac{1}{2}\rangle) + (5)^{1/2}(|\tfrac{1}{2}, \tfrac{3}{2}\rangle - |\tfrac{3}{2}, \tfrac{1}{2}\rangle)]$$

$$|1\rangle_4 = [1/(7)^{1/2}][(|-\tfrac{3}{2}, \tfrac{5}{2}\rangle - |\tfrac{5}{2}, -\tfrac{3}{2}\rangle) + (\tfrac{5}{2})^{1/2}(|-\tfrac{1}{2}, \tfrac{3}{2}\rangle - |\tfrac{3}{2}, -\tfrac{1}{2}\rangle)]$$

$$\begin{aligned}|0\rangle_4 = {} & [1/2(7)^{1/2}][(|-\tfrac{5}{2}, \tfrac{5}{2}\rangle - |\tfrac{5}{2}, -\tfrac{5}{2}\rangle) + 3(|-\tfrac{3}{2}, \tfrac{3}{2}\rangle - |\tfrac{3}{2}, -\tfrac{3}{2}\rangle) \\ & +2(|-\tfrac{1}{2}, \tfrac{1}{2}\rangle - |\tfrac{1}{2}, -\tfrac{1}{2}\rangle)]\end{aligned}$$

$$|3\rangle_3 = \tfrac{1}{3}[2\,|\tfrac{3}{2}, \tfrac{3}{2}\rangle - (\tfrac{5}{2})^{1/2}(|\tfrac{1}{2}, \tfrac{5}{2}\rangle + |\tfrac{5}{2}, \tfrac{1}{2}\rangle)]$$

$$|2\rangle_3 = (1/2\sqrt{3})[|\tfrac{1}{2}, \tfrac{3}{2}\rangle + |\tfrac{3}{2}, \tfrac{1}{2}\rangle - (5)^{1/2}(|-\tfrac{1}{2}, \tfrac{5}{2}\rangle + |\tfrac{5}{2}, -\tfrac{1}{2}\rangle)]$$

$$|1\rangle_3 = [1/(30)^{1/2}][(8)^{1/2}\,|\tfrac{1}{2}, \tfrac{1}{2}\rangle - (|-\tfrac{1}{2}, \tfrac{3}{2}\rangle + |\tfrac{3}{2}, -\tfrac{1}{2}\rangle) - (10)^{1/2}(|-\tfrac{3}{2}, \tfrac{5}{2}\rangle + |\tfrac{5}{2}, -\tfrac{3}{2}\rangle)$$

$$\begin{aligned}|0\rangle_3 = {} & [1/6(5)^{1/2}][4(|-\tfrac{1}{2}, \tfrac{1}{2}\rangle + |\tfrac{1}{2}, -\tfrac{1}{2}\rangle) - 7(|-\tfrac{3}{2}, \tfrac{3}{2}\rangle + |\tfrac{3}{2}, -\tfrac{3}{2}\rangle) - 5(|-\tfrac{5}{2}, \tfrac{5}{2}\rangle \\ & + |\tfrac{5}{2}, -\tfrac{5}{2}\rangle)]\end{aligned}$$

$$|2\rangle_2 = [1/2(7)^{1/2}][3(|\tfrac{1}{2}, \tfrac{3}{2}\rangle - |\tfrac{3}{2}, \tfrac{1}{2}\rangle) - (5)^{1/2}(|-\tfrac{1}{2}, \tfrac{5}{2}\rangle - |\tfrac{5}{2}, -\tfrac{1}{2}\rangle)]$$

$$|1\rangle_2 = [1/2(7)^{1/2}][2(|-\tfrac{1}{2}, \tfrac{3}{2}\rangle - |\tfrac{3}{2}, -\tfrac{1}{2}\rangle) - (10)^{1/2}(|-\tfrac{3}{2}, \tfrac{5}{2}\rangle - |\tfrac{5}{2}, -\tfrac{3}{2}\rangle)]$$

$$\begin{aligned}|0\rangle_2 = {} & [1/2(21)^{1/2}][4(|-\tfrac{1}{2}, \tfrac{1}{2}\rangle - |\tfrac{1}{2}, -\tfrac{1}{2}\rangle) - (|-\tfrac{3}{2}, \tfrac{3}{2}\rangle - |\tfrac{3}{2}, -\tfrac{3}{2}\rangle) - 5(|-\tfrac{5}{2}, \tfrac{5}{2}\rangle \\ & - |\tfrac{5}{2}, -\tfrac{5}{2}\rangle)]\end{aligned}$$

$$\begin{aligned}|1\rangle_1 = {} & [1/(105)^{1/2}][(27)^{1/2}\,|\tfrac{1}{2}, \tfrac{1}{2}\rangle + (15)^{1/2}(|-\tfrac{3}{2}, \tfrac{5}{2}\rangle + |\tfrac{5}{2}, -\tfrac{3}{2}\rangle) \\ & -(24)^{1/2}(|-\tfrac{1}{2}, \tfrac{3}{2}\rangle + |\tfrac{3}{2}, -\tfrac{1}{2}\rangle)]\end{aligned}$$

$$\begin{aligned}|0\rangle_1 = {} & [1/(70)^{1/2}][|-\tfrac{1}{2}, \tfrac{1}{2}\rangle + |\tfrac{1}{2}, -\tfrac{1}{2}\rangle - 3(|-\tfrac{3}{2}, \tfrac{3}{2}\rangle + |\tfrac{3}{2}, -\tfrac{3}{2}\rangle) + 5(|-\tfrac{5}{2}, \tfrac{5}{2}\rangle \\ & + |\tfrac{5}{2}, -\tfrac{5}{2}\rangle)]\end{aligned}$$

$$|0\rangle_0 = [1/(6)^{1/2}][(|-\tfrac{5}{2}, \tfrac{5}{2}\rangle - |\tfrac{5}{2}, -\tfrac{5}{2}\rangle) - (|-\tfrac{3}{2}, \tfrac{3}{2}\rangle - |\tfrac{3}{2}, -\tfrac{3}{2}\rangle) + (|-\tfrac{1}{2}, \tfrac{1}{2}\rangle - |\tfrac{1}{2}, -\tfrac{1}{2}\rangle)$$

[a] To obtain $|-M\rangle_S$ in (b), the signs of M_i and M_j are reversed.

f a transition within a state S is then proportional to I_S relative to the intensity from a normal paramagnetic ion which has no excited states. Equation (3.8) is plotted in Fig. 4 and is applicable to Mn^{2+}–Mn^{2+} pairs in CaO,[36] for example, where transitions within all five total spin states ($S = 1$ o $S = 5$) have been identified as shown in Table VIII.

Since it is difficult to measure intensities accurately, values of J determined in this way are rarely more accurate than $\pm 10\%$. However, such measurements in Mn:MgO gave the first direct demonstration[57] that the first- and second-neighbor interactions in this case are of comparable magnitude.

The anisotropic parameters D_S and E_S of the Hamiltonian (3.5) can be determined accurately in the usual way from the positions of the lines in the spectrum (see, for example, Fig. 5). In practice, this may not be easy because the spectra from the different total spin states overlap and also some of them may be obscured by the much stronger spectrum from isolated ions which are necessarily present in a mixed crystal. However, this has been successfully achieved in a number of instances, as can be seen in Table VIII.

Before leaving this case of the isotropic interaction J being much greater than the anisotropic interactions D, E, it should be mentioned that even if all the experimental parameters are measured, it may still be very

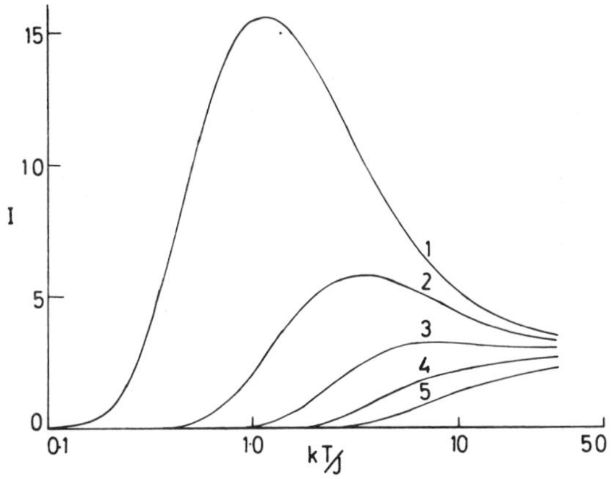

Fig. 4. The relative intensities I [Eq. (3.8)] of the transitions from a pair of spins $S_i = S_j = \frac{5}{2}$ coupled by an interaction $J\mathbf{S}_i \cdot \mathbf{S}_j$, plotted against the reduced temperature kT/J. Transitions within different total spin states $S = 1, 2, 3, 4$, and 5 are labeled. This type of behavior is observed for Mn^{2+} pairs in MgO, but there are deviations due to exchange striction (see Section 5).

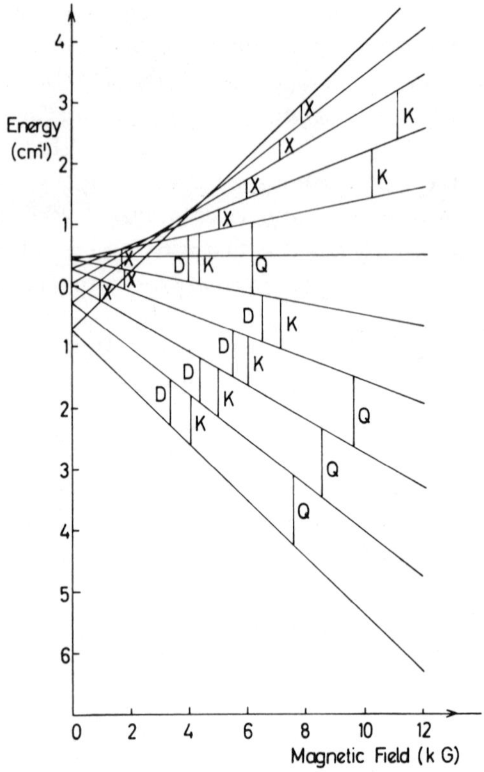

Fig. 5. Calculated energy levels and observed transitions within the $S = 5$ total spin state for a nearest-neighbor Mn^{2+} pair in ZnF_2 with H along the line joining the ions. In this example, the exchange is ferromagnetic and the $S = 5$ state lies lowest. The wavelengths used are in the region 3 cm (X), 1.2 cm (K and D), and 0.8 cm (Q). Similar good agreement was obtained for the $S = 4$ state. (From Brown *et al.*[123])

difficult to reduce them to the basic interactions J, D_d, and D_E etc. At first sight, it would appear that if D_S and E_S are known for each total spin state, they could be reduced by using Eqs. (3.3), (3.6), and (3.7). However, if exchange striction is present, as is likely when the exchange interaction is large, the separation of the ions is different for each total spin state. This means that the basic parameters are also different and the pair system cannot be described simply by three parameters J, D_e, and E_e as in (3.2). These effects are discussed in Section 5.

3.2.2. $J > D_e$

The terms in D_e and E_e in the Hamiltonian (3.2) have nonzero matrix elements between states differing by $\Delta S = 2$. This means that when D_e is not much smaller than J, the energy levels given by (3.4) and (3.5) are shifted by amounts of order D_e^2/J and that there can be microwave transitions between the Zeeman components of states differing by $\Delta S = 2$ with intensity of order D_e^2/J^2 relative to those within a total spin state. The measurement of these "forbidden" transitions allows a very accurate determination of the energy gap between total spin states and hence of J. This is only applicable, of course, if the microwave quantum is big enough to bridge the energy gap.

An example of this situation, studied by Windsor,[58] is that of second-neighbor pairs of Mn^{2+} ions in $KMgF_3$ where $J/D_e \sim 4$. Transitions between total spin states 0–2, 1–3, 2–4, and 3–5 were observed, giving almost isotropic lines in the spectrum separated from the single ion line by $\pm 3J$, $\pm 5J$, $\pm 7J$, and $\pm 9J$, respectively. Hence, a very accurate value for the second-neighbor isotropic exchange, $J/k = 0.043 \pm 0.005$°K, was obtained. This was the first direct demonstration that the second-neighbor interaction in this salt is much smaller (about 100 times) than the first.

3.2.3. $J \sim D_e$

For this situation, the energy levels are completely mixed and the only accurate treatment is to diagonalize the energy matrix exactly. An example is Gd^{3+} pairs ($S_i = S_j = 7/2$) in $LaCl_3$, where $J \sim 0.01$ cm^{-1} and $|D_e| \sim 0.02$ cm^{-1}. The diagonalization for this example has been carried out by Hutchings *et al.*[59] and hence they were able to find accurate values for the parameters (Table VIII) from their EPR results.

3.3. Pairs of Dissimilar Ions

So far, it has been assumed that the two ions in the pair are equivalent. We now describe briefly examples where the spins are the same ($S_i = S_j$), but the ions are dissimilar due respectively to different site symmetries, different hyperfine interactions, and different interactions with further neighbors.

3.3.1. Difference in Site Symmetry ($g_i \neq g_j$)

Here, it is assumed that the two ions have different g-values, and to illustrate the effect, we take the spin-Hamiltonian to be

$$\begin{aligned}\mathcal{H} &= g_i\beta\mathbf{H}\cdot\mathbf{S}_i + g_j\beta\mathbf{H}\cdot\mathbf{S}_j + J\mathbf{S}_i\cdot\mathbf{S}_j \\ &= \tfrac{1}{2}(g_i+g_j)\beta\mathbf{H}\cdot(\mathbf{S}_i+\mathbf{S}_j) + \tfrac{1}{2}(g_i-g_j)\beta\mathbf{H}\cdot(\mathbf{S}_i-\mathbf{S}_j) + J\mathbf{S}_i\cdot\mathbf{S}_j \qquad (3.9)\end{aligned}$$

If H is large, so that $|g_i - g_j|\beta H \gg |J|$, the term in J can be neglecte and then two separate resonances corresponding to g_i and g_j are observed On the other hand, if H is small, so that $|g_i - g_j|\beta H \ll |J|$, the off diagonal terms given by $\mathbf{S}_i - \mathbf{S}_j$ in the Hamiltonian can be neglected, an then only one resonance is seen at an average g-value, $(g_i + g_j)/2$. For th intermediate case, one might expect to find four transition frequencies fron which J could be directly determined.

Experimentally, these effects have not been observed for an isolate pair, but similar effects have been seen for the two types of Cu^{2+} io $(S_i = S_j = \frac{1}{2})$ in copper sulfate. In fact, this was one of the first direc observations of exchange effects in EPR spectra and was studied by Bagguley and Griffiths[60] and Pryce.[61] These authors found for a particular directio of H, at a frequency of 1.2 cm^{-1}, two well-resolved resonances with g_i=2.0 and $g_j = 2.46$. But at 0.3 cm^{-1}, they found only one resonance with $g = 2.27$ Observations were also made at intermediate frequencies and it was deduced from all the results that $|J| \sim 0.15$ cm^{-1} for nearest-neighbor Cu^{2+} ions in this salt.

3.3.2. Difference in Hyperfine Interaction

An example of this case is provided by the work of Baker[62] on Nd^{3+} pairs in lanthanum ethyl sulfate. Neodymium has several isotopes, an Baker observed transitions from pairs in which only one of the electroni spins was coupled to a Nd nuclear spin. The hyperfine interaction give large mixing between the total spin states $S = 1$ and $S = 0$, thus allowing transitions between them from which all interaction parameters could be determined directly. The results are given in Table VIII.

3.3.3. Difference in Further-Neighbor Interactions

An example here is given by the EPR spectra of pairs and triads of Ir^{4+} ions in diluted crystals of ammonium chloroiridate.[9] An isolated nearest-neighbor pair gives a well-defined spectrum, but if one of the two Ir^{4+} ions interacts with a second-nearest neighbor, the isolated pair lines are shifted and additional transitions are also allowed. From the results, an accurate value for the second-neighbor interaction J_2 was deduced showing that it is about 20 times smaller than J_1 (see Table VIII).

There are, of course, many other possible examples of dissimilar ions in a pair. If the site symmetry is different, then antisymmetric exchange, $\mathbf{C} \cdot (\mathbf{S}_i \times \mathbf{S}_j)$, could occur [see Eq. (2.23) and discussion] giving rise to additional anisotropy and also the possibility of transitions between total spin states differing by $\Delta S = \pm 1$. However, it would seem that EPR measurements on pairs have not yet identified this type of exchange unambiguously except for the work of Kanzig *et al.*[157] on O_2^- ions in KI. Another possibility is that the two ions of the pair are different, so that $S_i \neq S_j$. An example[63] is Cu^{2+}–Ni^{2+} pairs, where $S_i = \frac{1}{2}$ and $S_j = 1$. In general, it may be said that when the two members of the pair are dissimilar, the spectrum tends to be more complicated, but in principle at least, more information about exchange interactions can be obtained from it.

3.4. Larger Clusters of Ions

The Hamiltonian for a cluster of three or more ions is an obvious extension of that for a pair and can be written

$$\mathscr{H} = \sum_i \mathscr{H}_i + \sum_{i \neq j} \mathscr{H}_{i,j} \tag{3.10}$$

Here, $\mathscr{H}_i$ is the isolated ion Hamiltonian [Eq. (3.1)] and $\mathscr{H}_{ij}$ the interaction Hamiltonian [see Eq. (3.2)]. Assuming that the isotropic exchange terms, $J\mathbf{S}_i \cdot \mathbf{S}_j$, are much bigger than the other terms, the energy level scheme for the cluster consists, as for a simple pair, of a number of multiplets each of which can be described by a total spin quantum number $\mathbf{S}$, where $\mathbf{S} = \sum_i \mathbf{S}_i$.

The number of states is easily derived by noting that the spin states of S_i form a basis for the irreducible representation D_s. For example, the states of a cluster of three spins, each with $S = 1$, must form a basis for the representation

$$\begin{aligned} D_1 \times D_1 \times D_1 &= D_1 \times (D_0 + D_1 + D_2) \\ &= D_1 + (D_0 + D_1 + D_2) + (D_1 + D_2 + D_3) \\ &= D_0 + 3D_1 + 2D_2 + D_3 \end{aligned}$$

Thus we expect one singlet $S = 0$ state, three multiplets (triplets) with $S = 1$, two multiplets with $S = 2$, and one with $S = 3$. Similarly, for three spins, $S_i = S_j = S_k = \frac{1}{2}$, we expect two doublets $S = \frac{1}{2}$ and one quadruplet $S = \frac{3}{2}$.

For larger clusters, the number of levels increases as $(2S_i + 1)^n$, where n is the number of coupled spins. Thus for four exchange-coupled spins each with $S_i = \frac{5}{2}$, the 1296 energy levels are divided between 146 different total spin multiplets. For this reason, an exact treatment of the Hamiltonian for larger clusters rapidly becomes impracticable and the approximate methods used for treating concentrated magnetic materials start to become more useful.

The energies of the different multiplets depend on the magnitudes of the various interactions. One fairly simple case,[64] first considered by Kambe,[65] is that of three spins i, j, k, where the i–j and j–k coupling is the same but k–i is different. The Hamiltonian is

$$\mathscr{H} = J_1(\mathbf{S}_i \cdot \mathbf{S}_j + \mathbf{S}_j \cdot \mathbf{S}_k) + J_2\mathbf{S}_k \cdot \mathbf{S}_i \quad (3.11)$$

giving energy levels

$$W = \tfrac{1}{2}J_1[S(S+1) - S_i(S_i+1) - S_j(S_j+1) - S_k(S_k+1)] + \tfrac{1}{2}(J_2 - J_1)[S'(S'+1) - S_i(S_i+1) - S_k(S_k+1)] \quad (3.12)$$

where S is the total spin ($\mathbf{S} = \mathbf{S}_i + \mathbf{S}_j + \mathbf{S}_k$) and S' refers only to the combination of S_i and S_k ($\mathbf{S}' = \mathbf{S}_i + \mathbf{S}_k$). This energy level scheme was originally used to interpret the apparently low susceptibility of some salts containing Fe^{3+} ions ($S_i = S_j = S_k = \frac{5}{2}$) where it was thought that the Fe^{3+}–Fe^{3+} interactions were of this form. The energy levels for three spins coupled in this way are shown in Fig. 6(a) for $S_i = S_j = S_k = \frac{1}{2}$ and in Fig. 6(b) for $S_i = S_j = S_k = 1$.

The most detailed EPR investigation of a cluster would appear to be that of Harris and Owen[9] for three exchange-coupled Ir^{4+} ions ($S_i = S_j = S_k = \frac{1}{2}$) in mixed crystals of $(NH_4)_2(Ir, Pt)Cl_6$, where Eq. (3.12) and Fig. 3(a) are applicable. These authors identified all the allowed spectra from nearest-neighbor triads in the face-centered cubic lattice and also, as mentioned above, triads involving second nearest neighbors. Hence, it was possible to find all the interactions between first and second neighbors (Table VIII). To interpret the EPR spectrum exactly, it is, of course, necessary to include the anisotropic terms as well as the isotropic ones; these are added for a cluster in a similar way as for a pair and for a typical example of how this is done, the reader is referred to this paper on triads of Ir^{4+} ions.

Another very interesting EPR study of clusters is that of Clad and Wucher,[66] who investigated the isotropic part of the exchange interaction between the three coupled Cr^{3+} ions ($S_i = S_j = S_k = \frac{3}{2}$) in the compound $Cr_3(CH_3COO)_6(OH)_2Cl \cdot 8H_2O$. They assumed $J_1 \approx J_2$, and hence from Eq. (3.12), the two ground-state $S = \frac{1}{2}$ doublets are about $3J/2$ lower in energy than the first excited state, which consists of four almost degenerate $S = \frac{3}{2}$ quadruplets. From the temperature dependence of the signal intensity, they found this energy gap to be about 45°K and hence $J/k \approx 30$°K.

These authors also investigated[67] the similar salt where one of the three Cr^{3+} ions is replaced by an Fe^{3+} ion. Thus, in Eqs. (3.11) and (3.12), $S_i = S_k = \frac{3}{2}$, $S_j = \frac{5}{2}$, J_1 is the Fe^{3+}–Cr^{3+} interaction and J_2 is the Cr^{3+}–Cr^{3+}

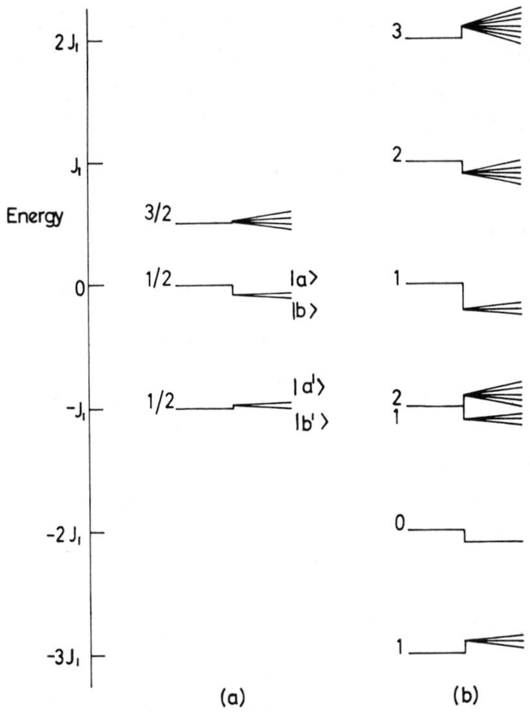

Fig. 6. The energy levels for three spins with coupling of the form $J_1(\mathbf{S}_i \cdot \mathbf{S}_j + \mathbf{S}_j \cdot \mathbf{S}_k) + J_2\mathbf{S}_k \cdot \mathbf{S}_i$ as in Eq. (3.11). (a) $S_i = S_j = S_k = 1/2$, giving total spin states $S = 1/2$ and $3/2$ as shown. The energies on the left correspond to $J_2 = 0$ and on the right to $J_2/J_1 = 0.1$, and the Zeeman splittings are also indicated schematically. Detailed measurements of this energy level scheme have been made for triads of Ir^{4+} ions.[9] The levels labeled $|a\rangle$, $|b\rangle$, etc. are referred to in Section 6. (b) The corresponding energy level diagram for $S_i = S_j = S_k = 1$.

interaction. Also, it is evident from the discussion above that if J_1 and J_2 are positive (antiferromagnetic), the lowest states consist of two $S = \frac{1}{2}$ doublets with energies $E(S = \frac{1}{2},\ S' = 0) = -4J_1 - 3.75J_2$, $E(S = \frac{1}{2},\ S' = 1) = -5J_1 - 2.75J_2$, so the separation is $J_1 - J_2$. The next group of levels consists of three close $S = \frac{3}{2}$ quadruplets which are higher in energy by about $3J/2$, where J is the average value of the exchange interactions J_1, J_2. The EPR results gave $J/k \approx 40°\mathrm{K}$ and $|\,(J_1 - J_2)/k\,| \approx 4°\mathrm{K}$.

There are other examples of clusters described in the literature which result mainly from three or four transition metal ions being coupled together in the natural crystal structure. The magnetic properties of these have been studied largely by bulk susceptibility methods. A good review of these studies and of the EPR work has been given by Kokoszka and Gordon.[6] Finally, it should be mentioned that a significant effect of clustering in the EPR spectrum is the marked change in the appearance of the hyperfine structure. We now go on to discuss this in some detail.

4. HYPERFINE INTERACTIONS

In this section, we consider the hyperfine interactions for a pair of ions i and j coupled by an exchange interaction $J\mathbf{S}_i \cdot \mathbf{S}_j$. It is assumed that S_i has the usual form of isotropic hyperfine interaction, $A_i\mathbf{S}_i \cdot \mathbf{I}_i$, with its own nucleus and that there is a similar term for ion j. It is also assumed that in the pair there is a small hyperfine interaction $a_j\mathbf{S}_j \cdot \mathbf{I}_i$ between nucleus i and electronic spin S_j, and that there is a corresponding term for nucleus j. The hyperfine Hamiltonian for the pair is then

$$\mathscr{H} = J\mathbf{S}_i \cdot \mathbf{S}_j + A_i\mathbf{S}_i \cdot \mathbf{I}_i + A_j\mathbf{S}_j \cdot \mathbf{I}_j + a_i\mathbf{S}_i \cdot \mathbf{I}_j + a_j\mathbf{S}_j \cdot \mathbf{I}_i \quad (4.1)$$

There may also be anisotropic terms, but these are often small and the principal effects in the EPR pair spectrum can be described most easily by omitting them in the first instance.

We first consider a pair of similar ions, such as a nearest-neighbor pair of Mn^{2+} ions in MgO, which is the type of example which has been most investigated experimentally; then, the transferred hyperfine interaction represented by the terms in a_i and a_j are discussed in more detail; and finally, some brief remarks are made about larger clusters.

4.1. A Pair of Similar Ions

In this case, $S_i = S_j$, $I_i = I_j$, $A_i = A_j$, and $a_i = a_j$. The Hamiltonian (4.1) can then be written

$$\begin{aligned}\mathscr{H} = \tfrac{1}{2}J[S(S+1) - S_i(S_i+1) - S_j(S_j+1)] + \tfrac{1}{2}(A+a)\mathbf{S} \cdot (\mathbf{I}_i + \mathbf{I}_i) \\ + \tfrac{1}{2}(A-a)(\mathbf{S}_i - \mathbf{S}_j) \cdot (\mathbf{I}_i - \mathbf{I}_j) \quad (4.2)\end{aligned}$$

where $\mathbf{S} = \mathbf{S}_i + \mathbf{S}_j$ is the total spin of the system which, as discussed earlier, can take values $S = S_i + S_j,\ S_i + S_j - 1, \ldots, 0$. If $J \gg A$, as is often the case in practice, the last term in (4.2) can be neglected since it only gives off-diagonal matrix elements of order A between different total spin

Fig. 7. Hyperfine structure from a nearest neighbor pair of Mn^{2+} ions in MgO. The spacing between adjacent lines is close to 43 gauss which is half the value for isolated Mn^{2+} ions in MgO (from reference 57).

states with separation of order J. The main features of the pair hyperfine structure are then determined by $\frac{1}{2}(A + a)\mathbf{S} \cdot (\mathbf{I}_i + \mathbf{I}_j)$. This gives a spacing $(A + a)/2$, which is close to half the isolated ion value A, and there are $4I_i + 1$ hyperfine lines with a characteristic intensity distribution.

One of the first examples of this type of hyperfine structure was found by Bleaney and Bowers[1] for pairs of Cu^{2+} ions in copper acetate. A similar effect was found[69] for As in Si and the theory was discussed by Slichter.[70] A good illustration (Fig. 7) is provided by nearest-neighbor pairs of Mn^{2+} ions in MgO.[57] Here, $I_i = I_j = \frac{5}{2}$ for ^{55}Mn, so the allowed values of the total nuclear magnetic quantum number are $M_I = 5, 4, \ldots, -5$. There are thus eleven hyperfine lines with relative intensities 1:2:3:4:5:6:5:4:3:2:1 corresponding to the number of ways of building up each M_I value. The spacing is close to 40×10^{-4} cm^{-1} (though it has not yet been measured accurately by ENDOR techniques), which is about half the value of A for an isolated Mn^{2+} ion in MgO. This characteristic structure provides an unambiguous way of identifying pair lines in the EPR spectrum. Similar structures have been found in other pair measurements, e.g., for nearest-neighbor pairs of V^{2+} ions in V^{2+}:$KMgF_3$.[35]

The presence of the transferred hyperfine term in a has not yet been detected.* It would show up as a small difference between the pair splitting parameter $(A + a)/2$ and half the isolated ion value, $A/2$. For the only case where ENDOR measurements have been made—that of a pair of Ir^{4+} ions—it is found[71] that this difference is less than the experimental error of 0.1%. This is not surprising because the Ir^{4+} ions are well separated and the transfer effect is expected to be small. For a pair like Mn^{2+}–O^{2-}–Mn^{2+}, on the other hand, a/A is estimated to be between 0.5 and 1%, which should be easily detectable by ENDOR methods. In spite of its small magnitude, this transferred hyperfine interaction has important consequences which we now discuss in more detail.

* See footnote on page 459.

4.2. Transferred Hyperfine Interaction

The most direct way of demonstrating this effect experimentally is b investigating a pair in which one ion is paramagnetic and the other diamag netic. If the transfer effect were zero, the diamagnetic ion would have n isotropic hyperfine interaction. Laurance *et al.*[72] first showed, using ENDO techniques on Cr^{3+} ions in Al_2O_3, that there is in fact an isotropic hyperfin field at the ^{27}Al nuclei due to the neighboring Cr^{3+} ions. Heeger an Houston[73] have also shown, using NMR methods on the salts $MnCr_2O$ and $MnFe_2O_4$, that the hyperfine field at the ^{55}Mn nuclei is 7% smalle when the Mn^{2+} ion has Cr^{3+} neighbors than when it has Fe^{3+} neighbors which again demonstrates this transfer effect.

We will illustrate this effect by describing ENDOR measurements[74] o the hyperfine field at ^{27}Al nuclei in Fe^{3+}:$LaAlO_3$. The nearest-neighbo metal ions of interest have the (approximately) 180° structure Fe^{3+}–O^{2-}– Al^{3+}. The hyperfine interaction with the ^{27}Al nucleus is found to be of th form

$$\mathscr{H} = a_s \mathbf{S} \cdot \mathbf{I} + a_p(3S_z I_z - \mathbf{S} \cdot \mathbf{I}) \qquad (4.3$$

where $S = \frac{5}{2}$ for Fe^{3+} and $I = \frac{5}{2}$ for ^{27}Al and z is along the line joinin the ions.

The measured values of the parameters are $a_s = 1.09 \times 10^{-4}\ \text{cm}^{-1}$ an $a_p = 0.21 \times 10^{-4}\ \text{cm}^{-1}$. Most of a_p results from the magnetic dipolar inter action between the magnetic moment localized on Fe^{3+} and the ^{27}Al nucleu (calculated to be $0.14 \times 10^{-4}\ \text{cm}^{-1}$).

The origin of a_s is ascribed to transfer of unpaired electron spin fro one metal ion to the next.* This is a similar mechanism to that alread discussed in Section 2 for the origin of superexchange, though there is difference because s orbits are particularly important for hyperfine inter actions but not so important for superexchange. For a structure like Fe^{3+}– O^{2-}–Al^{3+}, there is a fraction f_σ of unpaired spin in $2p_\sigma$ orbits on O^{2-} due t covalent transfer from O^{2-} to Fe^{3+} (see Section 2). This can be passed on t $2s$ orbits on Al^{3+} because of the overlap, $\mathscr{S} = \langle O^{2-}2p \mid Al^{3+}2s \rangle$, giving a unpaired spin fraction $f_\sigma \mathscr{S}^2$ in $2s$ orbits on Al^{3+} which is parallel to th spin on Fe^{3+}. The hyperfine interaction with the ^{27}Al nucleus is then $A_{2s} f_\sigma \mathscr{S}^2$ where A_{2s} is the interaction for a single unpaired electron in a $2s$ orbit Calculations[75] based on this mechanism give $a_s(\text{calc}) = 1.19 \times 10^{-4}\ \text{cm}^{-1}$ which is in good agreement with $a_s(\text{expt}) = 1.09 \times 10^{-4}\ \text{cm}^{-1}$.

* See Sec. 5.1.3 in Chapter 8 by Šimanek and Šroubek as well as ref. 74 for further details

Table VI. The Isotropic Part a_s of the Transferred Hyperfine Interaction for Some Pairs of Metal Ions

Pair	Salt	a_s (10^{-4} cm^{-1})	$6a_s/A$	Ref.
Fe^{3+}–O^{2-}–Al^{3+}	Fe^{3+}:$LaAlO_3$	1.09(expt) 1.19(calc)	7.85%	74, 75
Mn^{2+}–F^-–Mn^{2+}	$KMnF_3$	0.26(calc)	1.7%	76
Mn^{2+}–O^{2-}–Mn^{2+}	MnO	0.46(calc)	3.8%	76

Similar calculations have been made[76] for the transfer of electron from d orbits on a Mn^{2+} ion to 1s, 2s, 3s, and 4s orbits on a neighboring Mn^{2+} ion. The predicted transferred hyperfine interactions are summarized in Table VI.

In the last column of Table VI is given the magnitude of the transfer effect when six magnetic neighbors are contributing, as they do in the concentrated salt, relative to the hyperfine constant A for the isolated magnetic ion. In other words, for the examples given, this is the expected change in hyperfine interaction when going from a very dilute salt to a concentrated salt. Since A is negative, the total interaction is increased when the six neighbors are antiparallel as they are in the antiferromagnetic state, and is decreased in the paramagnetic state. This predicted change in hyperfine interaction has helped to explain some outstanding anomalies in experimental estimates of zero-point spin deviation in antiferromagnets.[74,76] However, an important problem in EPR measurements of pairs is to demonstrate that the expected change in hyperfine interaction for a pair such as Mn^{2+}–O^{2-}–Mn^{2+}, compared with an isolated Mn^{2+} ion in MgO, really does agree with the theoretical prediction.*

4.3. *Larger Clusters*

The first EPR measurements of the hyperfine structure for groups of three and four exchange-coupled ions were made by Feher *et al.*[77] on As and P donors in Si. The hyperfine interactions for large clusters of Mn^{2+} ions have also been discussed more recently by Ishikawa.[78] The general result is that the overall hyperfine spacing is $2IA$ (the isolated ion value), but the spacing between adjacent hyperfine lines in the spectrum is A/n,

* Recently May[38] has found that for V^{2+}–O^{2-}–V^{2+} pairs in MgO, the pair hyperfine interaction is about 0.5% greater than $A/2$ for the isolated V^{2+} ion and that this is approximately consistent with spin transfer theory.

where n is the number of ions in the cluster and the intensity of the central lines is very large relative to that for lines at the outside.

More specifically, if n similar ions are interacting in pairs, the hyperfine terms (neglecting the small transfer effect) can be written

$$\sum A\mathbf{S}_i \cdot \mathbf{I}_i = (A/n)[\sum \mathbf{S}_i \cdot \sum \mathbf{I}_i + \sum_i \sum_{\substack{j \\ i \neq j}} (\mathbf{S}_i - \mathbf{S}_j) \cdot (\mathbf{I}_i - \mathbf{I}_j)] \quad (4.4)$$

As for a simple pair, the second term in the square bracket of (4.4) can be neglected if $J \gg A$, so the effective hyperfine interaction is then $(A/n)\mathbf{S} \cdot \sum \mathbf{I}_i$ where S is the total spin $\mathbf{S} = \sum \mathbf{S}_i$. Thus, for transitions $\Delta S_z = \pm 1$ within any total spin state, the EPR spectrum contains $2nI_i + 1$ hyperfine lines, the spacing between adjacent lines is A/n, so the overall spacing is $2AI_i$ as for an isolated noninteracting ion. The intensity of each transition depends on the number of ways of building up the total nuclear magnetic quantum number $M_I = \sum I_{iz}$. For large n, this leads to an envelope of intensities which is very peaked at the center, giving a result similar to exchange narrowing (see Fig. 8).

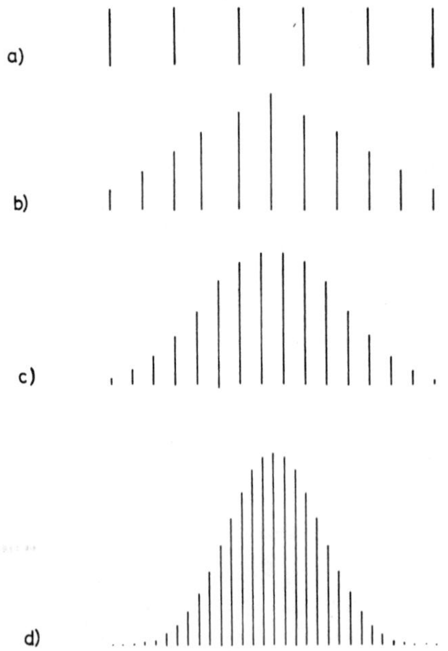

Fig. 8. Hyperfine structure from clusters of coupled magnetic ions each with nuclear spin $I = 5/2$. (a) single ion; (b) pair; (c) three ions; (d) six ions.

These effects have been observed by the authors mentioned above for P donors in Si[77] and for Mn^{2+} ions in ZnS.[78] The line shapes observed in other cluster measurements, e.g., triads of Ir^{4+} ions,[9] are also consistent with this analysis. Finally, it is of interest to note that EPR measurements[63] on a Cu^{2+}–Ni^{2+} pair give a hyperfine interaction with the copper nucleus which is about three times smaller than that for the isolated Cu^{2+} ion. In this case, $S_i = \frac{1}{2}$ and $I_i = \frac{3}{2}$ for Cu^{2+} and $S_j = 1$ and $I_j = 0$ for Ni^{2+}. There are thus three exchange-coupled electrons interacting with one nuclear spin and the result is analogous to that for a cluster of three exchange-coupled ions.

5. EXCHANGE STRICTION EFFECTS

So far, we have assumed that the magnetic ions remain fixed in the crystal. By allowing them to change their positions as a result of interactions between the ions, the energy levels can sometimes be significantly altered.[79] For example, anomalies in the EPR spectra of Mn^{2+} pairs in MgO[80] and also certain anomalous properties of the antiferromagnetic oxide MnO[81] can be explained using an isotropic interaction of the form

$$J\mathbf{S}_i \cdot \mathbf{S}_j - j(\mathbf{S}_i \cdot \mathbf{S}_j)^2 \tag{5.1}$$

The biquadratic term in j can arise as a high-order intrinsic exchange effect (as mentioned in Section 2.3), but in many cases, it seems more likely that it is due to exchange striction, which can give an effective term of the same form.

Exchange striction is a well-known phenomenon in concentrated magnetic materials. The total exchange energy of a ferromagnet or antiferromagnet depends not only on the total spin state, but also varies very rapidly with the distances r_{ij} between the neighboring ions. This means that the lattice tends to distort so as to lower the exchange energy until this is balanced by elastic forces. For example, MnO is an antiferromagnetic face-centered cubic material which orders at 117°K in such a way that all spins in a given [111] plane are parallel to each other but antiparallel to the spins in the adjacent [111] planes. Energy can therefore be gained below the transition temperature by a contraction along the (111) direction perpendicular to the ordered planes, and such a contraction is actually observed.[82] A number of other effects accompany this exchange striction.[83,84]

Considering the case of an isolated ion pair, we assume just an isotropic exchange interaction $J\mathbf{S}_i \cdot \mathbf{S}_j$, where J is a function of r_{ij} only.

In practice, J will be a function of all the coordinates of both ions, so this is a simplification. We assume also that for a distortion along the line joining the ions, there is a simple Hooke's law elastic restoring force of the form Kr/r_0, where K is an elastic constant, r_0 is the mean value of r_{ij} and the distortion is $r = r_{ij} - r_0$. The total exchange and elastic energy is then

$$E = J\mathbf{S}_i \cdot \mathbf{S}_j + (Kr^2/2r_0) \tag{5.2}$$

For equilibrium

$$\partial E/\partial r = 0 = (\partial J/\partial r)\mathbf{S}_i \cdot \mathbf{S}_j + (Kr/r_0) \tag{5.3}$$

so it is evident that the equilibrium distortion r depends on the total spin S and on $\partial J/\partial r$.

Assuming that $\partial J/\partial r$ does not differ appreciably from its value $(\partial J/\partial r)_0$ at $r_{ij} = r_0$, we can put

$$J = J_0 + r(\partial J/\partial r)_0 \tag{5.4}$$

and hence, using Eq. (5.3), Eq. (5.2) can be written

$$E = J_0\mathbf{S}_i \cdot \mathbf{S}_j - j(\mathbf{S}_i \cdot \mathbf{S}_j)^2 \tag{5.5}$$

where

$$j = (r_0/2K)(\partial J/\partial r)_0^{\,2} \tag{5.6}$$

Thus, in this approximation, the effect of exchange striction on the energy is identical to that of intrinsic biquadratic exchange.

The energy intervals given by (5.5) between successive total spin multiplets are

$$E_S - E_{S-1} = J_0S - jS[S^2 - S_i(S_i + 1) - S_j(S_j + 1)] \tag{5.7}$$

An example is shown in Table VII, where the EPR results of Harris and Owen[80] on Mn^{2+} pairs of MgO are compared with Eq.(5.7) using $J_0/k = 14.6°K$ and $j/J_0 = 0.05$. It may be seen that the observed deviations from the Landé interval rule predicted by conventional theory where $j = 0$ are very large; in particular, the splitting between $S = 0$ and $S = 1$ is nearly doubled by including the term in j. The agreement between Eq. (5.5) and experiment is quite good, being well inside the experimental errors.

We think that the origin of j for this particular example is mainly exchange striction rather than intrinsic biquadratic exchange, for several

Table VII. Energy Intervals $\Delta_{S-1,S} = E_S - E_{S-1}$ for Nearest-Neighbor Mn^{2+} Pairs in MgO in Units of °K

Energy interval (°K)	$\Delta_{0,1}$	$\frac{1}{2}\Delta_{1,2}$	$\frac{1}{3}\Delta_{2,3}$	$\frac{1}{4}\Delta_{3,4}$
Experiment	28.5 ± 2	24.5 ± 2	21.0 ± 2	17.5 ± 3
Biquadratic or simple exchange striction model ($J_0/k = 14.6$°K, $j/J_0 = 0.05$)	27.4	25.1	21.4	16.1
Exchange striction with $J = J_0 \exp(-br)$ ($r_0 J_0 b^2/2k = 0.0185$ and $J_0/k = 17$°K)	28.6	24.4	20.7	17.5

[a] These intervals would all be equal if $j = 0$. The second row of figures is calculated using the simple theory [Eq. (5.7)]; the third row is given by the more exact model[36] where an exponential dependence of J on ion separation is allowed for. (From Refs. 80, 85.)

reasons. First, the value $j/J = 0.05$ is considerably larger than that which could arise from an intrinsic effect according to the calculations of Huang and Orbach.[31] Second, the EPR results for the anisotropic interactions are consistent with a model in which r_{ij} is different for each total spin state. Third, if an improved exchange striction model[36] is used in which allowance is made for an exponential dependence of J on r_{ij}, the agreement with experiment is so good (see the third row of Table VII) that one is bound to think that the experimental errors quoted in Table VII are much too generous! In this improved model, it is assumed that $J = J_0 \exp(-br)$, so the energy is given by

$$E = J_0[\exp(-br)]\mathbf{S}_i \cdot \mathbf{S}_j + (Kr^2/2r_0) \tag{5.8}$$

with $\partial E/\partial r = 0$. This can be solved exactly in terms of the parameter $c = r_0 b^2 J_0/2K$ and the best fit (Table VII) is found using $c = 0.0185$ and $J_0 = 17$°K. A similar conclusion about the importance of exchange striction relative to that of intrinsic biquadratic exchange has been reached independently by Lines and Jones[84] from a careful analysis of the magnetic properties of MnO.

It is also of interest to note that the actual dependence of J on r_{ij} for Mn^{2+} pairs in oxides can be described empirically by $J = J_0 \exp(-15r/r_0)$, where, as before, $r = r_{ij} - r_0$ and r_0 is the mean separation. This is illustrated

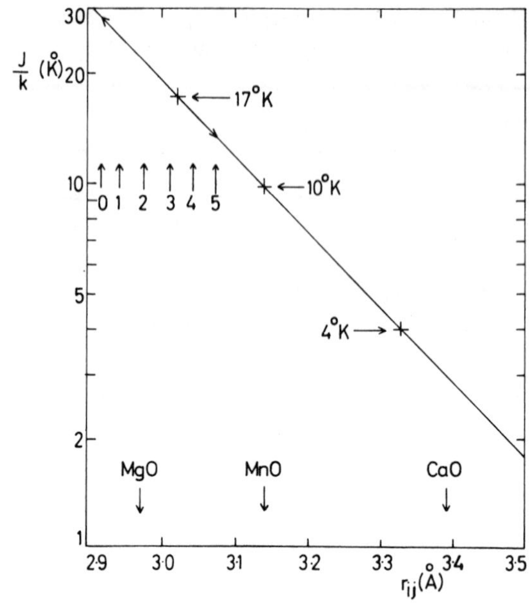

Fig. 9. The variation of J with ionic separation for nearest-neighbor Mn^{2+} ions in the MnO-type lattice. It can be described approximately by $J = J_0 \exp(-15r/r_0)$, where $r = r_{ij} - r_0$ (see text).* For Mn:MgO,[36,80] the ionic separations in different total spin states implied by an exchange striction model are indicated. This variation of J is consistent with the values derived[84] from the susceptibility of MnO ($J = 10$°K) and measurements[36] on Mn^{2+} pairs in CaO ($J = 4$°K).

in Fig. 9, which shows EPR results for nearest-neighbor pairs of Mn^{2+} ions in MgO[80] and CaO[36] and also the value of J deduced from susceptibility results for MnO.[84]

It is clear from this discussion that exchange striction effects are very important for Mn^{2+} pairs in MgO and that they are also likely to be for other pairs where the exchange is big and varies rapidly with distance. The evidence suggests that these effects may sometimes be larger for pairs than for the concentrated salt because of the difference in the effective elastic constant K, and this is one reason why it is dangerous to apply pair parameters to concentrated salts unless a complete analysis of the pair system is made. Further experimental evidence is required before the problem of the relative importance of intrinsic and striction effects can be fully understood.

* A more recent analysis[36] suggests $J = J_0 \exp(-19r/r_0)$.

One interesting experiment in this connection is the direct measurement of optical transitions between the total spin states, because it is then possible that the ions in the pair would not have time to adjust to their equilibrium positions and striction effects may be eliminated. This is a difficult experiment technically because the relevant transition probabilities are small.

6. RELAXATION EFFECTS IN CLUSTERS

It was first suggested by Van Vleck[86] that an important mechanism for spin–lattice relaxation might be through fast-relaxing "exchange-pockets," i.e., pairs or small exchange-coupled clusters of ions. Rather than relaxing directly to the lattice, isolated spins could relax much more effectively by cross-relaxation with these clusters, which then relax rapidly to the lattice. Since then many observations have been reported of strongly concentration-dependent spin–lattice relaxation times and the important role played by exchange-coupled clusters has become widely acknowledged.[44,87–91]

In this section, we discuss the mechanisms that sometimes lead to rapid spin–lattice relaxation for clusters, and the effects resulting from cross–relaxation between single ions and clusters.

6.1. Basic Ideas

Following the treatments of Kronig,[92], Van Vleck,[93] and others,[94,95] thermal vibrations of the lattice are represented by a phonon system at temperature T. These vibrations produce perturbations $\mathscr{H}_k$ of the crystal field acting on the spin of a magnetic ion in the lattice, and can induce transitions from one spin state $|a\rangle$ to another $|b\rangle$. The relaxation process which is usually dominant at low temperatures is the "Direct process," in which transitions are accompanied by absorption or emission of a single phonon of energy $\delta = E_a - E_b$, so as to conserve energy. The characteristic time T_1 in which these transitions bring the spin system into equilibrium with the lattice is called the spin–lattice relaxation time. For the direct process

$$1/T_1 \propto \delta^3 \coth(\delta/2kT) \sum_k |\langle a | \mathscr{H}_k | b\rangle|^2 \tag{6.1}$$

the summation being taken over the various modes of vibration.*

There are really two reasons why the spin–lattice relaxation rate for an exchange-coupled cluster of ions should be faster than for a single ion.

* See Eq. (3.11) in Chapter 2 by Orbach and Stapleton.

First, in addition to crystal field perturbations, the interactions between ion are modulated by the lattice vibrations. Exchange interactions are generall rather sensitive to small changes in the positions of the ions, and so th additional perturbations arising in this way can be considerable. Second the energy levels of a cluster often include exchange splittings $\Delta(\approx J)$ whic are much larger than the EPR splittings δ. In this situation, another re laxation process can occur in which the EPR levels $|a\rangle$ and $|b\rangle$ relax b two consecutive direct processes through a third level $|c\rangle$ of a differen total spin multiplet. This two-phonon process which was first invoked b Finn *et al.*,[96] can be much more rapid than the simple direct proces because it involves phonons of energy $\Delta \approx J \gg \delta$, which come from much more densely populated part of the phonon spectrum. There is Δ^3 dependence of the relaxation rate on phonon energy for this process which is given by*

$$\frac{1}{T_1} \propto \frac{\Delta^3}{e^{\Delta/kT} - 1} \frac{2P_1P_2}{P_1 + P_2} \tag{6.2}$$

where $P_1 = \sum_k |\langle a | \mathscr{H}_k | c\rangle|^2$ and $P_2 = \sum_k |\langle c | \mathscr{H}_k | b\rangle|^2$.

For a pair, all the terms in the Hamiltonian of Eq. (3.2) can be mod ulated by lattice vibrations and we now briefly discuss the contribution of the various terms to spin–lattice relaxation.

6.2. Modulation of Crystal Field

For phonons well below the Debye frequency, the wavelength λ i much greater than the interionic spacing r_{ij}. This means that neighborin ions will vibrate very nearly in phase and the main crystal field perturbation acting on each member of a pair will be of the same sign and similar to thos acting on a single ion, containing terms such as

$$\mathscr{H}_{ks} = D'[S_{iz}^2 - \tfrac{1}{3}S_i(S_i + 1) + S_{jz}^2 - \tfrac{1}{3}S_j(S_j + 1)] \tag{6.3}$$

For the direct process, the matrix elements in Eq. (6.1) are calculated i exactly the same way as in the static case (see Section 3) and the relaxatio rate is approximately

$$(1/T_1)_{\text{pair}} \approx \beta_S{}^2(1/T_1)_{\text{single}} \tag{6.4}$$

where β_S is the coefficient defined in Eq. (3.7).

* See Section 4.5 in Chapter 2 by Orbach and Stapleton for further details.

Much faster relaxation is possible by two-phonon processes between different multiplets. The Hamiltonian of Eq. (6.3) links states for which $\Delta S = \pm 2$, and so a two-phonon process can occur between these states.

A special case for which this is not possible is a pair of spins with $S_i = S_j = \frac{1}{2}$, and as an example of this, we discuss the case of nearest-neighbor Ir^{4+} pairs in $(NH_4)_2PtCl_6$,[90,97] for which $J \approx 7°K$, antiferromagnetic. It should be mentioned that the crystal field can have no matrix elements within the ground state in first order by the Kramers theorem, and the direct process for single ions occurs only as a result of second-order admixtures to highly excited states through the Zeeman interaction. The same applies to pairs, and so perturbations of the form of Eq. (6.3) lead to a direct-process relaxation rate approximately equal to that for single ions. The only possible two-phonon processes involve transitions between the $S = 1$ states and the $S = 0$ ground state, and these can only be induced by perturbations antisymmetric in S_i and S_j, such as

$$\mathscr{H}_{ka} = D'' \{[S_{iz}^2 - \tfrac{1}{3}S_i(S_i + 1)] - [S_{jz}^2 - \tfrac{1}{3}S_j(S_j + 1)]\} \tag{6.5}$$

Perturbations of this kind must result from antisymmetric modes in which the ions vibrate out of phase, and the amplitudes are about a factor $\pi r_{ij}/\lambda$ smaller than the amplitudes of the symmetric modes,[98] so that

$$D'' \approx (\pi r_{ij}/\lambda)D' \tag{6.6}$$

Thus the two-phonon relaxation rate for the pair can be related to the single-ion direct-process relaxation rate:

$$\left(\frac{1}{T_1}\right)_{\text{pair}} \approx \left(\frac{\pi r_{ij}}{\lambda}\right)^2 \frac{J^3}{\delta^3 \coth(\delta/2kT)} \left(\frac{1}{T_1}\right)_{\text{single}} \tag{6.7}$$

The product of the first two factors in this expression is of order unity and hence $T_1(\text{pair}) \approx T_1(\text{single})$. Experimentally, the relaxation rate measured for these Ir^{4+} pairs was found to have the temperature dependence expected for this two-phonon process and was very close to, though slightly faster than, that for single ions at 4.2°K.

6.3. *Modulation of Isotropic Exchange*

Since exchange interactions vary widely in magnitude, so do their effects on spin–lattice relaxation, but the effects are often very large. Modulation of the isotropic exchange can give contributions to $\mathscr{H}_k$ of the form

$$\mathscr{H}_{kJ} = J'(\mathbf{S}_i \cdot \mathbf{S}_j) \tag{6.8}$$

where $J' = r_0(\partial J/\partial r_{ij})$.

Since $\mathbf{S}_i \cdot \mathbf{S}_j$ is a scalar operator, the selection rules to first order are $\Delta S = 0$, $\Delta M_S = 0$. In the case of a pair, there is only one multiplet with any given value of S, and so this perturbation can not give any first-order transition probability. However, in clusters of three or more ions, these selection rules can give rise to transitions because there can occur more than one state having the same S and M values. In the $S_i = S_j = S_k = \frac{1}{2}$ triad of Fig. 6(a), for example, rapid relaxation is probable between states $|a\rangle$ and $|a'\rangle$ and between $|b\rangle$ and $|b'\rangle$.

For pairs, modulation of isotropic exchange can contribute to relaxation between states of different S and M as a second-order effect because the states are usually slightly mixed by other static terms in the Hamiltonian such as crystal field terms, Zeeman interaction, or anisotropic exchange. The relaxation rate depends on the amount of mixing, but J' is often so great that the rate can still be very rapid. The relaxation of Cr^{3+} pairs ($S_i = S_j = \frac{3}{2}$) in ruby is a case where this mechanism is probably effective, and has been discussed by Gill.[88] There is a comparatively large axial crystal field term D_c, as given by Eq. (3.1) and this introduces mixing of the order D_c/J between states of total spin $S = 1$ and $S = 3$. Thus the perturbation of Eq. (6.8) has matrix elements of order $J'D_c/J$ between these spin states, leading to a two-phonon relaxation process. Measurements of the spin-lattice relaxation of $S = 3$ transitions were found to be consistent with this process.

6.4. Modulation of Anisotropic Exchange

Although anisotropic exchange perturbations are often much smaller than the isotropic perturbations, the selection rules for spin–phonon transitions are much less stringent, and so they can lead to relaxation between quite a number of energy levels as a first-order effect. Waller[99] treated thermal modulation of magnetic dipolar interactions in a classic paper as long ago as 1932, predicting very weak relaxation in dilute crystals. For closely neighboring ions with large spins, however, this term can be comparable to other terms, and pseudodipolar exchange can sometimes have even stronger effects. For example, in the case of Ir^{4+} pairs in $(NH_4)_2PtCl_6$,[9] this type of exchange could contribute appreciably to the direct process within the $S = 1$ states. Modulation of pseudodipolar exchange can also lead to two-phonon processes between states for which $\Delta S = \pm 2$.

Antisymmetric perturbations of the form $\mathbf{C} \cdot (\mathbf{S}_i \times \mathbf{S}_j)$ can also lead to two-phonon processes through transitions between states for which $\Delta S = \pm 1$. Even when static interactions of this form are forbidden by symmetry, dynamic terms can still arise from the antisymmetric normal modes of vibration which destroy the inversion symmetry of a pair. It has been suggested that antisymmetric exchange is important in the relaxation of certain Cr^{3+} pairs in ruby.[44]

6.5. Cross-Relaxation Effects

When cross-relaxation can occur between single ions and exchange-coupled clusters, we may treat the energy flow between the spin systems and the lattice using the model indicated in Fig. 10. There are N_S single ions, which we will assume to each have a probability w_S per unit time of relaxing directly to the lattice, and N_C clusters, each with a similar spin–lattice relaxation probability w_C. If w_x is defined as the probability that any single ion will cross-relax with a cluster, then each cluster must have a probability $w_x N_S/N_C$ of cross-relaxing with a single ion. This cross-relaxation rate w_x is independent of temperature and is directly proportional to the concentration of clusters. It also depends on the degree to which energy can be conserved in the cross-relaxation process, and is greatest when the energy splittings for single ions and clusters are exactly equal.[100] We shall assume here that this is the case. The net relaxation of the single ions to the lattice has been considered in detail by Rannestad and Wagner[101] and normally is characterized by two time constants, but in two extreme cases, a single time constant appears. In the first case, the relaxation path via the clusters is limited by the cross-relaxation rate, and in the second case, it is limited by the cluster–lattice relaxation rate. The final rates in the two cases are

$$\text{If} \quad w_C \gg w_x N_S/N_C, \quad \text{then} \quad 1/T_1 = w_S + w_x \tag{6.9a}$$

$$\text{If} \quad w_C \ll w_x N_S/N_C, \quad \text{then} \quad 1/T_1 = [1/(N_S+N_C)](N_S w_S + N_C w_C) \tag{6.9b}$$

which in the limit $N_S \gg N_C$ becomes $w_S + w_C N_C/N_S$. Which of these alternatives occurs in any particular case depends on both temperature and

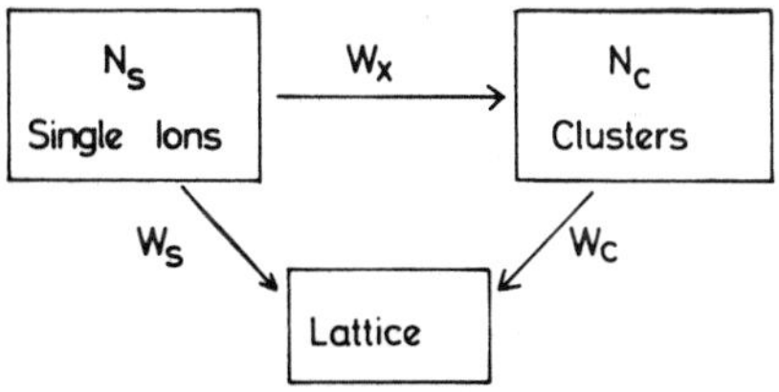

Fig. 10. Model for spin–lattice relaxation by cross-relaxation to clusters.

concentration. The second alternative is more probable at lower tempera tures, because w_C becomes smaller, and also at higher concentrations because w_x, being proportional to N_C, is likely to increase more rapidl than N_C/N_S.

An example where relaxation appears to be limited by cross-relax ation (case 6.9a) has been reported by Atsarkin[91] for Cr^{3+} in $ZnWO_4$ Under certain conditions, the relaxation rate of single ions varies wit concentration as $w \propto c^3$, and it is comparatively independent of tempera ture. Atsarkin suggests that the centers to which the single ions cross relax are exchange-coupled pairs with a third loosely coupled single io nearby. Cross-relaxation to any other arrangement of three ions would als give this concentration dependence.

An example of case (6.9b) was reported by Harris and Yngvesson[9] in which relaxation is limited by the cluster–lattice relaxation rate. I measurements on Ir^{4+} ions in $(NH_4)_2PtCl_6$ and K_2PtCl_6, it was found tha at concentrations $c < 0.1\%$, the relaxation rate for single ions was inde pendent of concentration, and could be interpreted in terms of the con ventional Kronig–Van Vleck theory. At higher concentrations, however a new relation process becomes important with a rate $1/T_1 \propto c^2$. It i known by direct measurement[97] that nearest-neighbor pairs relax too slowl to affect the relaxation of single ions, and cross-relaxation to these pair would also be rather slow. The suggested process indicated in Fig. 11 is on in which a single ion cross-relaxes to a triad consisting of a nearest-neighbo

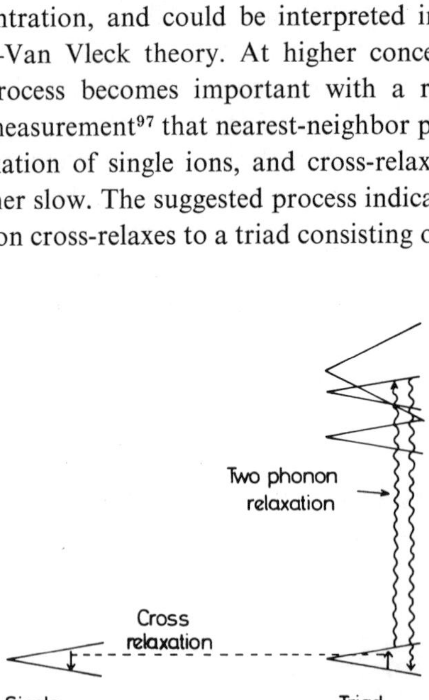

Fig. 11. Suggested relaxation process for Ir^{4+} ions in $(NH_4)_2PtCl_6$.[90] Single ions cross-relax rapidly to a triad, which then relaxes to the lattice by a two-phonon process.

pair with one more loosely coupled third ion. The observed temperature dependence in the region $J \gg kT$ can be fitted satisfactorily by $1/T_1 \propto \exp(-J/kT)$. This is consistent with a two-phonon process for the triad, and theoretical estimates of the relaxation rate give order-of-magnitude agreement with the experimental results. Recently, more direct evidence has been reported supporting this model.[102]

It is of interest to note that in both of these cases, triads appear to be more effective than pairs in providing a relaxation mechanism for single ions. This may prove to be a fairly general rule, occurring because modulation of isotropic exchange interactions can only act directly in relaxing clusters of three or more ions.

7. TABULATED RESULTS ON PAIRS

Experimental results have been published using a number of different notations, and for ease of reference, we have expressed them all here in terms of the parameters of the spin-Hamiltonian

$$\mathscr{H} = \mathscr{H}_i + \mathscr{H}_j + J\mathbf{S}_i \cdot \mathbf{S}_j - j(\mathbf{S}_i \cdot \mathbf{S}_j)^2 + D_e(3S_{iz}S_{jz} - \mathbf{S}_i \cdot \mathbf{S}_j) + E_e(S_{ix}S_{jx} - S_{iy}S_{jy})$$

where

$$\mathscr{H}_i = \mathbf{H} \cdot \mathbf{g} \cdot \mathbf{S}_i + D_c[S_{iz}^2 - \tfrac{1}{3}S_i(S_i + 1) + E_c(S_{ix}^2 - S_{iy}^2)]$$

For example, in much of the work on pairs for which $S_i = S_j = \frac{1}{2}$ and with axial symmetry, it has been found convenient to use a Hamiltonian of the form

$$\mathscr{H} = aS_{iz}S_{jz} + b(S_{ix}S_{jx} + S_{iy}S_{jy})$$

We have transformed the parameters using the relations

$$J = \tfrac{1}{3}(a + 2b), \qquad D_e = \tfrac{1}{3}(a - b)$$

The axes x, y, z in the spin-Hamiltonian are always the principal axes for the pair. Unless otherwise stated, the z axis is along the line joining the two ions. In some cases, where interpretation of the results is not straightforward, the parameters D_S and E_S of Eq. (3.5) have been given.

In the first column of Table VIII we list the various magnetic ions, their electron configurations, and their spins in the ground state. We adopt the usual notation of writing S_i for real spins, and S_i' for effective spins.

Tab

Magnetic ion and configuration	Host lattice and approximate structure of cluster	Isotropic interaction (cm^{-1}) and method of measurement
$V^{2+}(3d^3)$ $S_i = \frac{3}{2}$	$KMgF_3$ V–F–V $r_{ij} \approx 4.06$ Å	$J = 4.5 \pm 1.0$ from $S = 2$ intensities
	MgO *nnn*: V–O–V $r_{ij} \approx 4.13$ Å	$J^{nnn} = 50.2 \pm 3$ $j^{nnn} = 1.9 \pm 0.9$ $J^{nn} = -7.5 \pm 3$ from $S = 2$ intensities
$Cr^{3+}(3d^3)$ $S_i = \frac{3}{2}$	Al_2O_3 (ruby) O / \\ *nn*: Cr—O—Cr \\ / O $r_{ij} \approx 2.73$ Å	$J^{nn} = 240$ $J^{nnn} = 83.6$ $j^{nnn} = 9.7$ $J^{3n} = 11.59$ $j^{3n} = -0.06$ $J^{4n} = -6.99$ $j^{4n} = -0.14$
	BiI_3 I / \\ Cr Cr \\ / I $r_{ij} \approx 4.5$ Å	($J = -14$ from susceptibility CrI_3)
	$SrTiO_3$	$J = 40 \pm 8$ $J^{nnn} \leq 5$
	$ZnGa_2O_4$ Cr—O *nn*: | | O—Cr	$J = 11.5 \pm 0.7$ $j = 1.7 \pm 0.7$ $J^{nnn} = 0.83 \pm 0.15$ $J^{3n} = -1.0 \pm 0.15$ $J^{4n} = -1.0 \pm 0.15$ $J^{5n} = +0.76 \pm 0.15$ from intensities

ental Results

Anisotropic interaction parameters (cm⁻¹)	Calculated dipolar anisotropy (cm⁻¹)	Remarks	Ref.
$= -0.0339 \pm 0.0002$	$D_d = -0.027$	D_d can be increased by about 20% by allowing for spin transfer to the ligands, but D_e is still anomalously large; a pair hfs was observed	35, 35a
$= -0.0479$	$D_d = -0.0246$	The anomalously high value of D_e can only be partly explained by spin transfer to the ligands; a pair hfs was observed (see footnote, page 459)	103, 103a
— — — — — $^{4n}_{e} = -0.058 \pm 0.005$	— — — — — $D_d^{4n} = -0.037$	The quoted results[108] are representative of measurements on ruby pairs reported by several different authors, some with widely varying results; in addition to EPR, methods have included infrared[109,110] and optical[111–115] spectroscopy; Imbusch and Graifman[116] have reported EPR in an optically excited state of a pair	44, 104–108
$= 1.6 \pm 0.1$ $= 0.10 \pm 0.01$	$D_d = -0.017$	EPR was observed from the ground states ($S = 3$) only; the large anisotropic exchange is related to the high degree of covalency; the z axis is taken normal to the plane containing the pair and its intervening ligands	117
$_e = -0.0225 \pm 0.005$ $= 0.014 \pm 0.002$ $^{nnn}_{e} = 0.011 \pm 0.001$	$D_d = -0.029$ $D_d^{nnn} = -0.0102$	An antisymmetric exchange interaction was also observed	118
$_e = 0.016$	$D_d = -0.0672$	Anisotropic interactions for pairs other than nearest neighbor are consistent with dipolar values; the biquadratic term in the nn interaction probably results from exchange striction	119, 120

Tabl

Magnetic ion and configuration	Host lattice and approximate structure of cluster	Isotropic interaction (cm^{-1}) and method of measurement
Cr^{3+} (*cont.*)	$Cr_3(CH_3COO)_6(OH)_2Cl \cdot 8H_2O$ Cr Cr \ / O | Cr $r_{ij} \approx 3.28$ Å[121]	$J \approx 21$ $J' = 2.5 \pm 0.4$ from intensity measuremen where $\mathscr{H} = J(\mathbf{S}_1 \cdot \mathbf{S}_2 + \mathbf{S}_2 \cdot \mathbf{S}_3 + \mathbf{S}_3 \cdot \mathbf{S}_1) + J'\mathbf{S}_1 \cdot \mathbf{S}_2$
$Cr^{3+}(3d^3)$ $S_i = \frac{3}{2}$ $Fe^{3+}(3d^5)$ $S_i = \frac{5}{2}$	$Cr_2Fe(CH_3COO)_6(OH)_2Cl \cdot 6H_2O$	$J \approx 28,\ J' \approx 3$ from intensity measuremen where $\mathscr{H} = J(\mathbf{S}_1 \cdot \mathbf{S}_2 + \mathbf{S}_2 \cdot \mathbf{S}_3 + \mathbf{S}_3 \cdot \mathbf{S}_1) + J'\mathbf{S}_1 \cdot \mathbf{S}_2$
$Mn^{2+}(3d^5)$ $S_i = \frac{5}{2}$	MgO O / \\ *nn*: Mn Mn \ / O $r_{ij} \approx 3.06$ Å *nnn*: Mn–O–Mn $r_{ij} \approx 4.20$ Å	Energy intervals for nn from li intensities: $\Delta_{01} = 20.0 \pm 1.5$ $\frac{1}{2}\Delta_{12} = 17.0 \pm 1.5$ $\frac{1}{3}\Delta_{23} = 14.5 \pm 1.5$ $\frac{1}{4}\Delta_{34} = 12.0 \pm 2.0$ $J^{nnn} \approx \Delta_{01} = 19.5 \pm 3.0$
	CaO O / \\ *nn*: Mn Mn \ / O $r_{ij} \approx 3.32$ Å *nnn*: Mn–O–Mn $r_{ij} = 4.445$ Å	$J \approx 3$ from intensities

ued)

Anisotropic interaction parameters (cm^{-1})	Calculated dipolar anisotropy (cm^{-1})	Remarks	Ref.
—	—	The three Cr^{3+} ions are exchange-coupled so as to give two $S = 1/2$ doublets lowest; EPR was observed from these doublets, and also from an excited $S = 3/2$ state; magnetic susceptibility measurements support this model[122] (see Section 3.4)	66
—	—	As above, the two Cr^{3+} ions and the Fe^{3+} ion are again antiferromagnetically coupled; EPR was observed from the two ground-state doublets, and from excited states (see Section 3.4)	67
$= -0.776 \pm 0.010$ $= -0.182 \pm 0.002$ $= -0.084 \pm 0.001$ $= -0.050 \pm 0.001$ $= -0.149 \pm 0.005$ $= -0.024 \pm 0.003$ — $^{nnn} = 0.261 \pm 0.010$	$D_d = -0.061$ — — — — — — $3\alpha_1 D_d^{nnn} = -0.258$	For nn pairs, multiplets do not follow a Landé interval rule; results can be explained with a biquadratic term $j/J = 0.05$ with $J = 10.5$ cm^{-1}, but a more likely explanation is exchange striction, in which J varies from about 9 to 20 cm^{-1} (see Section 5); consistent with exchange striction is the fact that D_S cannot be expressed as $3\alpha_S D_e + \beta_S D_c$ unless D_e and D_c vary with S; D_e probably varies from -0.049 to -0.057 cm^{-1}; D_d(calc) can be reduced by about 10% by allowing for spin transfer to the ligands; a pair hfs was observed	36, 57, 80, 85
$= -0.433 \pm 0.002$ $= -0.126 \pm 0.001$ $= -0.0728 \pm 0.0003$ $= -0.0508 \pm 0.0002$ $= -0.0375 \pm 0.0002$ $= -0.030 \pm 0.001$ $^{nnn}_{1} = -0.316 \pm 0.002$	$D_d = -0.051$ $3\alpha_1 D_d^{nnn} = -0.175$	Like Mn:MgO, D_S cannot be expressed by $3\alpha_S D_e + \beta_S D_c$ unless D_e and D_c vary, probably indicating exchange striction; D_e probably varies from -0.0445 to -0.0485 cm^{-1}; D_d(calc) can be reduced by about 10% by allowing for spin transfer; a pair hfs was observed	36

Tab

Magnetic ion and configuration	Host lattice and approximate structure of cluster	Isotropic interaction (cm^{-1}) and method of measurement
Mn^{2+} (*cont.*)	ZnF_2	$J^{nn} = -0.30 \pm 0.14$ from $S = 4, 5$ intensities
	nn: 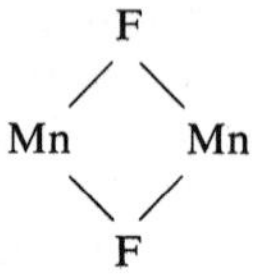 $r_{ij} \approx 3.13$ Å	
	nnn: 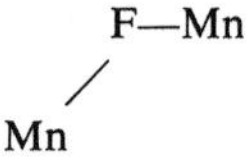 $r_{ij} \approx 3.69$ Å	$J^{nnn} = +2.8 \pm 1.0$ from $S = 1$ intensities
	$KMgF_3$ *nn*: Mn–F–Mn $r_{ij} \approx 3.99$ Å	$J = 3.0 \pm 1.5$ $S = 1, 2, 3, 4,$ intensities
	nnn: 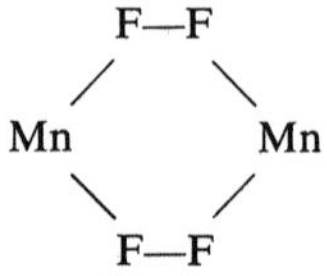 $r_{ij} \approx 5.64$ Å	$J^{nnn} = 0.030 \pm 0.003$ from $\Delta S = \pm 2$ line positior
	$KZnF_3$ Mn–F–Mn $r_{ij} \approx 4.06$ Å	$J = 6.9 \pm 0.4$ from $S = 1$ intensities
$Fe^{3+}(3d^5)$ $S_i = \frac{5}{2}$	Al_2O_3 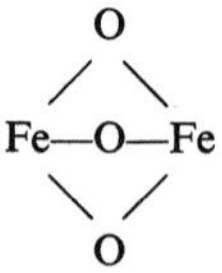 $r_{ij} \approx 2.73$ Å	$J = 250 \pm 50$ from $S = 1$ and $S = 2$ intensit
Fe^{3+} $S_i' = \frac{1}{2}$ (strong octahedral field)	$K_3CO(CN)_6$ 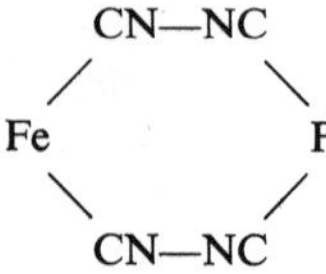 $r_{ij} \approx 7.00$ Å	Not measured

ed)

Anisotropic interaction parameters (cm^{-1})	Calculated dipolar anisotropy (cm^{-1})	Remarks	Ref.
$= -0.0135 \pm 0.004$ $= 0.00 \pm 0.006$ $= -0.051 \pm 0.009$	$D_d = -0.052$	D_c and E_c are similar to the values observed for single ions, $D_c = -0.0186$, $E_c = -0.0041$. Ferromagnetic J^{nn} confirmed by neutron measurements.[12]	123
—	—		
$= -0.17 \pm 0.005$ $= 0.011 \pm 0.005$ $D_e^{nnn} \mid = 0.0073 \pm 0.0010$ $D_c^{nnn} \mid = 0.0013 \pm 0.0010$	$D_d = -0.027$ $D_d^{nnn} = -0.0096$		58
$= -0.279 \pm 0.0009$	$3\alpha_1 D_d = -0.290$		124
$D_e \mid = 0.0910$	$D_d = -0.0855$		125
$^{I} = 0.026$ $^{I} = -0.004$ $^{II} = 0.035$ $^{II} = -0.004$	$D_d = -0.001$ $E_d = 0.007$	Two different pair spectra (I and II) were seen, both believed to arise from different types of nn pair; the line joining the ions is taken as the x axis with z along the crystal c axis; x and y are not principal axes of the g-tensor; in $K_3Fe(CN)_6$ J is believed to be about 0.25 cm^{-1}	126

Tab

Magnetic ion and configuration	Host lattice and approximate structure of cluster	Isotropic interaction (cm^{-1}) and method of measurement
$Co^{2+}(3d^7)$ $S_i' = \frac{1}{2}$ (octahedral site)	MgO 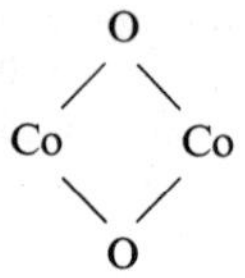 $r_{ij} \approx 2.97$ Å	$J = 18 \pm 3$ from $S = 1$ intensities
$S_i = \frac{3}{2}$ (tetrahedral site)	Cs_3ZnCl_5 *nn*: 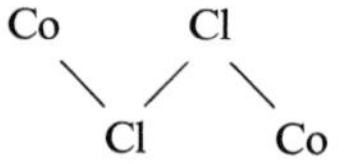 $r_{ij} \approx 6.54$ Å *nnn*: 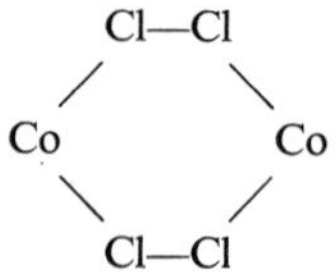 $r_{ij} \approx 7.25$ Å	$J = 0.0284 \pm 0.0003$ $J^{nnn} = -0.0214 \pm 0.0003$ both from line positions
	Cs_3ZnBr_5	$J = 0.0068 \pm 0.0007$ $J^{nnn} = -0.0147 \pm 0.0007$ from line positions
$Ni^{2+}(3d^8)$ $S_i = 1$	$KMgF_3$ Ni–F–Ni $r_{ij} \approx 3.99$ Å	$J = 64 \pm 10$ from $S = 1$ intensities
	K_2MgF_4 *nnn*: 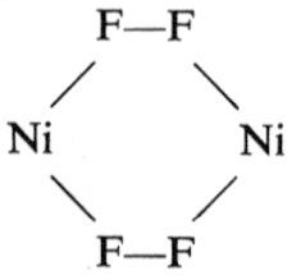 $r_{ij}^{nnn} \approx 5.61$ Å	$J^{nnn} = 0.400 \pm 0.005$ $J^{4n} \approx -0.035$ from line positions
	$ZnSiF_6 \cdot 6H_2O$	$J = -0.025 \pm 0.003$ $J^{nnn} = -0.014 \pm 0.003$ $J^{3n} \approx -0.008$ from line positions

ed)

Anisotropic interaction parameters (cm^{-1})	Calculated dipolar anisotropy (cm^{-1})	Remarks	Ref.
$_e \mid = 0.031 \pm 0.002$ $_e \mid = 0.060 \pm 0.004$	$D_d = -0.30$	There is expected to be a large contribution to D_e due to anisotropic exchange; a pair hfs was observed	35
—	—	No anisotropy detected	127
—	—		128
$_1 \mid = 0.277 \pm 0.003$	$D_d = -0.035$ $3\alpha_1 D_d = -0.105$	D_e is expected to include a large contribution from anisotropic exchange; second- and fourth-neighbor pairs have been observed using acoustic paramagnetic resonance[13]	38
—	—	D_e consistent with dipolar interactions	129
—	—	Anisotropy well described by same D_c as for single ions	131

Ta

Magnetic ion and configuration	Host lattice and approximate structure of cluster	Isotropic interaction (cm^{-1}) and method of measurement
$Cu^{2+}(3d^9)$ $S_i = \frac{1}{2}$	$CuSO_4 \cdot 5H_2O$ Cu–O–O–Cu $r_{ij} \approx 5.2$ Å	$\lvert J \rvert \approx 0.15$ from g-values at various quencies
	$Cu(CH_3COO)_2 \cdot H_2O$ (copper acetate) 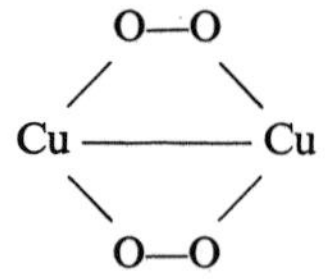 $r_{ij} \approx 2.64$ Å	$J = 260 \pm 50$ from $S = 1$ intensities
	$Cu(CH_2ClCOO)_2 \cdot 2.5H_2O$	$J \approx 274$[140] $J = 230 \pm 50$[138]
	$(C_5H_5NO)_2CuCl_2$ 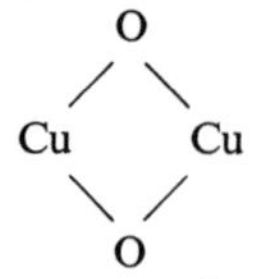 $r_{ij} \approx 3.3$ Å	$J = 550 \pm 100$
	$(C_5H_5NO)CuCl_2 \cdot H_2O$	$J = 885 \pm 180$

ed)

nisotropic interaction parameters (cm^{-1})	Calculated dipolar anisotropy (cm^{-1})	Remarks	Ref.
—	—	Average *g*-value effect seen in concentrated salt (see Section 3.3.1); similar effects have been observed in $K_2CUCl_4 \cdot 2H_2O$ and $(NH_4)_2CuCl_4 \cdot 2H_2O$[133,134]	60, 132
$\vert = 0.23 \pm 0.02$ $\vert = 0.02 \pm 0.01$	$D_a = -0.12$	Pairs of Cu^{2+} ions occur naturally in the concentrated salt; a large contribution to D_e is predicted due to anisotropic exchange (see Section 2.4); the *g*-values and hyperfine constants are consistent with those of "single" Cu^{2+} ions obtained by replacing one member of the pair by a Zn^{2+}[135] or Cu^{+}[136] ion; similar pairs have also been observed in copper propionate, $Cu(C_2H_5COO)_2 \cdot H_2O$,[137,138] copper butyrate, $Cu(C_3H_7COO)_2 \cdot H_2O$,[139] and copper monochloroacetate $Cu(CH_2ClCOO)_2 \cdot H_2O$[139]	1, 2
$\vert = 0.228 \pm 0.005$[141] $\vert \approx 0$ $\vert = 0.235 \pm 0.003$[140]		Naturally occurring pairs in the concentrated salt	140, 141
$= 0.140 \pm 0.006$ $= -0.04 \pm 0.02$	$D_a = -0.048$	Pairs occur naturally in both of these salts; in each case, *g*-values and hyperfine constants are consistent with those measured by Kokoszka *et al.*[63] for "single" Cu^{2+} ions obtained by using Zn-doped crystals; the same authors observed EPR from a ground-state doublet of Cu^{2+}–Ni^{2+} pairs in Ni-doped crystals of each sort; unfortunately, no direct information could be obtained concerning the exchange interaction for this mixed pair	142
$= 0.100 \pm 0.006$ $= -0.70 \pm 0.016$			

Ta

Magnetic ion and configuration	Host lattice and approximate structure of cluster	Isotropic interaction (cm^{-1}) and method of measuremen
Cu^{2+} (*cont.*)	$Zn(HCOO)_2 \cdot 2H_2O$ 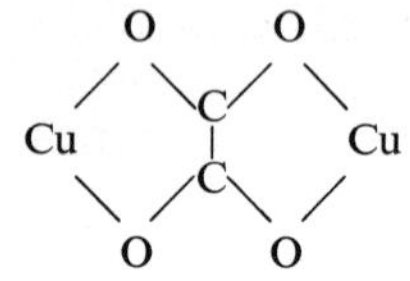 $r_{ij} \approx 4.62$ Å	$J = 33 \pm 1$ from $S = 1$ intensities
$Ir^{4+}(5d^5)$ $S_i' = \frac{1}{2}$	$(NH_4)_2PtCl_6$ *nn*: 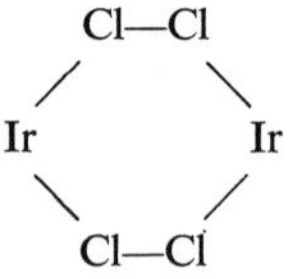 $r_{ij} \approx 6.95$ Å	$J = 5.2 \pm 0.8$ from pair $S = 1$, and t intensities
	nnn: 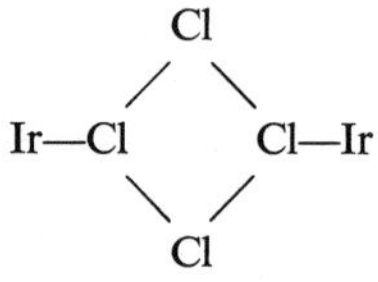 $r_{ij} \approx 9.82$ Å	$J^{nnn} = 0.28 \pm 0.03$ from triad line positions
	K_2PtCl_6 structure as for $(NH_4)_2PtCl_6$ $r_{ij}^{nn} \approx 6.88$ Å $r_{ij}^{nnn} \approx 9.73$ Å	$J = 8.0 \pm 0.8$ from pair $S = 1$, and tr intensities $J^{nnn} = 0.38 \pm 0.03$ from triad line positions
$Ce^{3+}(4f^1)$ $S_i' = \frac{1}{2}$	LaES[a] $r_{ij} \approx 7.1$ Å	—

ed)

nisotropic interaction parameters (cm^{-1})	Calculated dipolar anisotropy (cm^{-1})	Remarks	Ref.
—		No anisotropy was measured; a pair hfs was observed	143
$= +0.42 \pm 0.01$ $= -0.22 \pm 0.01$	$D_d = 0.002$ $E_d = -0.006$		5, 9, 71
$_e^{nnn} \mid < 0.015$	—	In both of these materials, anisotropy is almost entirely due to anisotropic exchange. The line joining the ions is defined as the x axis (a [110] crystal direction); y and z are the perpendicular [110] and [001] directions respectively	
$= +0.45 \pm 0.01$ $= -0.18 \pm 0.01$	$D_d = 0.002$ $E_d = -0.008$		5, 9
$_e^{nnn} \mid < 0.015$			
$= 0.0047$	$D_d = -0.0027$	In CeES and Ce:YES, a different doublet lies lowest, so the results are not directly comparable. Possibly VPE makes an important contribution to D_e[144]	49

Ta

Magnetic ion and configuration	Host lattice and approximate structure of cluster	Isotropic interaction (cm^{-1}) and method of measuremen
Ce^{3+} (*cont.*)	$LaCl_3$ *nn*: 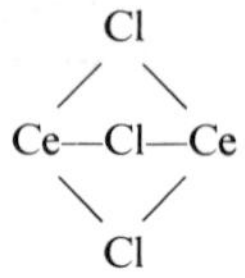 $r_{ij} \approx 4.375$ Å $r_{ij}^{nnn} \approx 4.84$ Å	—
	$LaBr_3$ *nn*: 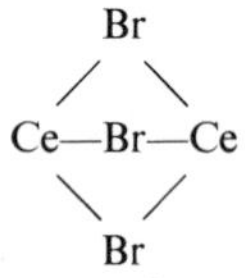 $r_{ij} \approx 4.510$ Å $r_{ij}^{nnn} \approx 5.13$ Å	
$Nd^{3+}(4f^3)$ $S_i' = \frac{1}{2}$	LaES[a] $r_{ij} \approx 7.1$ Å	$J = -0.0060 \pm 0.0003$ from line positions
	$LaCl_3$ $r_{ij} \approx 4.37$ Å	—
	$LaBr_3$ $r_{ij} \approx 4.51$ Å	—
	CaF_2 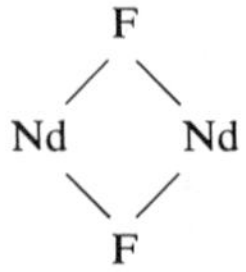 $r_{ij} \approx 3.85$ Å	—

ed)

nisotropic interaction parameters (cm^{-1})	Calculated dipolar anisotropy (cm^{-1})	Remarks	Ref.
$= 0.0710 \pm 0.0003$	$D_d = -0.056$		50, 145, 146
$^{n} = -0.0013 \pm 0.0010$	$D_d^{nnn} = -0.008$		
$= 0.1090 \pm 0.0005$	$D_d = -0.052$	There is probably a large contribution to D_e due to EQQ in these salts, as this is required to explain a small observed g-shift in g_{xz} for nnn; anisotropic superexchange is probably also important, and there is evidence that this is of high order	146, 147
$^{n} = -0.001 \pm 0.001$	$D_d^{nnn} = -0.0077$		
$= -0.0109 \pm 0.0002$	$D_d = -0.0120$ $J_d = -0.0068$	Measurement of J was possible because hyperfine interaction rendered the ions dissimilar (see Section 3.3)	62
$= 0.2082 \pm 0.0001$[148]	$D_d = -0.0163$	There is believed to be a major contribution to D_e from anisotropic exchange, including high-order terms; EQQ is probably negligible; parameters appear to vary between different samples; pairs out to 7th neighbor have also been measured, still showing anisotropic exchange effects.[151]	148–150
$= 0.212 \pm 0.004$[150]			
$= 0.2874 \pm 0.0001$[148]	$D_d = -0.0173$		148, 150
$= 0.2860 \pm 0.0005$[150]			
$= -0.114$ $= -0.048$	$D_d = -0.037$ $E_d = -0.026$	Anisotropic exchange is probably a major contributor to D_e	152

Ta

Magnetic ion and configuration	Host lattice and approximate structure of cluster	Isotropic interaction (cm^{-1}) and method of measuremen
Nd^{3+} (*cont.*)	SrF_2 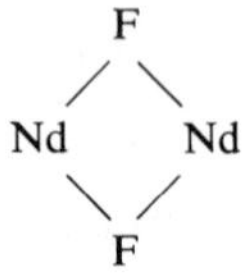 $r_{ij} \approx 4.14$ Å	—
$Eu^{2+}(4f^7)$ $S_i = \frac{7}{2}$	CaO and SrO 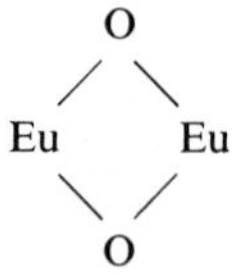	$J = -1.4 \pm 0.7$ from $S = 7,6$ intensities
	$r_{ij} \approx 4.79$ Å (CaO) ≈ 5.14 Å (SrO)	$\lvert J^{nnn} \rvert < 0.1$
$Gd^{3+}(4f^7)$ $S_i = \frac{7}{2}$	LaES[a] $r_{ij} \approx 7.1$ Å	$J = 0.0002 \pm 0.00005$ from line positions at 77°K
	$LaCl_3$ $r_{ij} \approx 4.375$ Å $r_{ij}^{nnn} \approx 4.84$ Å	$J = 0.0133 \pm 0.0005$ (77°K) $J^{nnn} = -0.0595 \pm 0.0020$ (77°K from line positions
	$EuCl_3$ $r_{ij} \approx 4.133$ Å $r_{ij}^{nnn} \approx 4.730$ Å	$J = 0.0488 \pm 0.003$ $J^{nnn} = -0.0637 \pm 0.0004$ from line positions
$Tb^{3+}(4f^8)$ $S_i' = \frac{1}{2}$	YES[a]	—
$Dy^{3+}(4f^9)$ $S_i' = \frac{1}{2}$	LaES[a]	—
$Ho^{3+}(4f^{10})$ $S_i' = \frac{1}{2}$	YES[a]	—

[a] ES = ethyl sulfate.

d)

isotropic interaction parameters (cm^{-1})	Calculated dipolar anisotropy (cm^{-1})	Remarks	Ref.
= -0.023 ± 0.001 = -0.018 ± 0.002	$D_d = -0.029$ $E_d = -0.024$		153
—	—	Measurements of anisotropy not given but reported to be consistent with a pure dipole–dipole interaction. Ferromagnetic J consistent with magnetic susceptibility.[159]	154
—	—		
= -0.0047 ± 0.0002	$D_d = -0.0047$		155
= -0.0219 ± 0.0003 ^{n} = -0.0159 ± 0.0005	$D_d = -0.0226$ $D_d^{nnn} = -0.0157$	J and D_e were accurately measured over a wide range of temperature (20–361°K), showing significant variations	59, 156
= -0.0246 ± 0.0003 ^{n} = 0.0165 ± 0.0005	$D_d = -0.0246$ $D_d^{nnn} = -0.0163$		156
= -0.257 ± 0.002	$D_d = -0.263$		47
= -0.097 ± 0.003	$D_d = -0.096$	The interaction was also measured between a pair of Dy^{3+} ions, one of which was in an excited state	47
= -0.201 ± 0.002	$D_d = -0.200$		47

Next, the host lattice is given, together with the approximate structure of the pair where possible. For further details of crystal structure, references should be made to the original papers. Interactions between nearest neighbor, next-nearest-neighbor, and third-nearest-neighbor pairs are indicated by superscripts, e.g., J^{nn}, J^{nnn}, and J^{3n}. The superscript *nn* has often been omitted, however, and so when there is no superscript the datum applies to the nearest-neighbor interaction.

The interaction parameters will generally contain some contribution due to magnetic dipole–dipole interaction. An estimate of this contribution is given in the fifth column and it is based on a model in which each ion is treated as a point dipole. The contributions to D_e, E_e, and J are given by

$$D_d = (-\beta^2/3r^3)[2g_z^2 + \tfrac{1}{2}(g_x^2 + g_y^2)]$$
$$E_d = (\beta^2/2r^3)(g_x^2 - g_y^2)$$
$$J_d = (\beta^2/3r^3)(g_x^2 + g_y^2 - 2g_z^2)$$

when the z-axis is along the line joining the ions.

REFERENCES

1. B. Bleaney and K. D. Bowers, *Proc. Roy. Soc.* **A214**:451 (1952).
2. H. Abe and J. Shimada, *Phys. Rev.* **90**:316 (1953).
3. J. N. van Niekirk and F. R. L. Schoening, *Acta. Crysta. Camb.* **6**:227 (1953).
4. J. M. Baker and B. Bleaney, in *Conference on the Physics of Low Temperature* Institut International du Froid, Paris (1953).
5. J. H. E. Griffiths, J. Owen, J. G. Park, and M. F. Partridge, *Proc. Roy. Soc.* **A250**:8 (1959).
6. A. H. Cooke, R. Lazenby, F. R. McKim, J. Owen, and W. P. Wolf, *Proc. Roy. Soc.* **A250**:97 (1959).
7. B. R. Judd, *Proc. Roy. Soc.* **A250**:110 (1959).
8. M. E. Lines, *Proc. Roy. Soc.* **A271**:105 (1963).
9. E. A. Harris and J. Owen, *Proc. Roy. Soc.* **A289**:122 (1965).
10. J. S. Smart, in *Magnetism*, ed. by G. T. Rado and H. Suhl, Vol. 1, Academic Press New York (1963).
11. B. N. Brockhouse, *Phys. Rev.* **106**:859 (1957).
12. G. G. Low, A. Okazaki, R. W. H. Stevenson, and K. C. Turberfield, *J. Appl. Phys.* **35**:998 (1964).
13. D. K. Garrod, H. M. Rosenberg, and J. K. Wigmore, *Proc. Roy. Soc.* **A315**:53 (1970).
14. J. Ferguson, H. J. Guggenheim, and Y. Tanabe, *J. Phys. Soc. Japan* **21**:692 (1966)
15. K. A. Wickersheim, in *Magnetism*, ed. by G. T. Rado and H. Suhl, Vol. 3, Academi Press, New York (1963).
16. K. W. Blazey and G. Burns, *Phys. Letters* **15**:117 (1965).

17. C. M. R. Platt and D. H. Martin, *Chemical Phys. Letters* **1**:659 (1968).
18. E. Belorizky, S. C. Ng, and T. G. Phillips, *Phys. Letters* **27A**:489 (1968).
19. H. A. Kramers, *Physica* **1**:182 (1934).
20. P. W. Anderson, *Phys. Rev.* **79**:350 (1950).
21. P. W. Anderson, in *Solid State Physics*, ed. by F. Seitz and D. Turnbull, Vol. 14, p. 99, Academic Press, New York (1963).
22. C. Herring, in *Magnetism*, ed. by G. T. Rado and H. Suhl, Vol. 4B, p. 1, Academic Press, New York (1963).
23. K. Gondaira and Y. Tanabe, *J. Phys. Soc. Japan* **21**:1527 (1966).
24. N. L. Huang and R. Orbach, *Phys. Rev.* **154**:487 (1967).
25. D. E. Rimmer, *J. Phys. C* **1**:329 (1969).
26. J. H. Van Vleck, *Phys. Rev.* **52**:1178 (1937).
27. J. Kondo, *Prog. Theor. Phys. Japan* **18**:541 (1957).
28. P. W. Anderson, *Phys. Rev.* **115**:2 (1959).
29. J. Owen and J. H. M. Thornley, *Rep. Prog. Phys.* **29**:675 (1966).
30. N. L. Huang, *Phys. Rev.* **164**:636 (1967).
31. N. L. Huang and R. Orbach, *Phys. Rev. Letters* **12**:275 (1964).
32. R. J. Elliott and M. F. Thorpe, *J. Appl. Phys.* **39**:802 (1968).
33. K. W. H. Stevens, *Rev. Mod. Phys.* **25**:166 (1953).
34. J. Kanamori, in *Magnetism*, ed. by G. T. Rado and H. Suhl, Vol. 1, p. 161, Academic Press, New York (1963).
35. S. R. P. Smith, Thesis, Oxford (1966).
35a. S. R. P. Smith and J. Owen, *J. Phys. C* **4**:1399 (1971).
36. E. A. Harris *J. Phys. C* **5**:338 (1972).
37. A. J. Sievers, III and M. Tinkham, *Phys. Rev.* **129**:1566 (1963).
38. C. E. C. May, Thesis, Oxford (1971).
39. R. M. White and R. L. White, *Phys. Rev. Letters* **20**:62 (1968).
40. P. M. Levy, *Phys. Rev. Letters* **20**:1366 (1968).
41. I. Dzialoshinsky, *J. Phys. Chem. Solids* **4**:241 (1958).
42. T. Moriya, *Phys. Rev.* **120**:91 (1960).
43. T. Moriya, in *Magnetism*, ed. by G. T. Rado and H. Suhl, Vol. 1, Academic Press, New York (1963).
44. H. Statz, M. J. Weber, L. Rimai, G. A. DeMars, and G. F. Koster, *J. Phys. Soc. Japan* **17**:B430 (1962).
45. R. Finkelstein and A. Mencher, *J. Chem. Phys.* **21**:472 (1953).
46. B. Bleaney, *Proc. Roy. Soc.* **77**:113 (1961).
47. J. M. Baker and A. E. Mau, *Can. J. Phys.* **45**:403 (1967).
48. R. J. Elliott and K. W. H. Stevens, *Proc. Roy. Soc.* **A218**:553 (1953).
49. J. M. Baker, *Phys. Rev.* **136**:A1633 (1964).
50. R. J. Birgeneau, M. T. Hutchings, and R. N. Rogers, *Phys. Rev.* **175**:1116 (1968).
51. W. P. Wolf and R. J. Birgeneau, *Phys. Rev.* **166**:376 (1968).
52. K. Sugihara, *J. Phys. Soc. Japan* **14**:1231 (1959).
53. D. H. McMahon and R. H. Silsbee, *Phys. Rev.* **135**:A91 (1964).
54. S. J. Allen, Jr., *Phys. Rev.* **166**:530 (1968).
55. J. Owen, *J. Appl. Phys.* **32**:2135 (1961).
56. B. R. Judd, *Operator Techniques in Atomic Spectroscopy*, McGraw-Hill, New York (1963).
57. B. A. Coles, J. W. Orton, and J. Owen, *Phys. Rev. Letters* **4**:116 (1960).

58. C. G. Windsor, Thesis, Oxford (1963).
59. M. T. Hutchings, R. J. Birgeneau, and W. P. Wolf, *Phys. Rev.* **168**:1026 (1968).
60. D. M. S. Bagguley and J. H. E. Griffiths, *Nature* **162**:538 (1948).
61. M. H. L. Pryce, *Nature* **162**:539 (1948).
62. J. M. Baker, *Phys. Rev.* **136**:1341 (1964).
63. G. F. Kokoszka, H. C. Allen Jr., and G. Gordon, *J. Chem. Phys.* **46**:3020 (1967).
64. J. Yvon, J. Horowitz, and A. Abragam, *Rev. Mod. Phys.* **25**:165 (1963).
65. K. Kambe, *J. Phys. Soc. Japan* **5**:48 (1950).
66. R. Clad and J. Wucher, *Compt. Rend.* **260**:4318 (1965).
67. R. Clad and J. Wucher, *Compt. Rend.* **262B**:795 (1966).
68. G. F. Kokoszka and G. Gordon, in *Transition Metal Chemistry*, ed. by R. I. Carlin, Vol. 5, p. 181, Dekker, New York (1969).
69. R. C. Fletcher, W. A. Yager, G. L. Pearson, and F. R. Merritt, *Phys. Rev.* **95**:844 (1954).
70. C. P. Slichter, *Phys. Rev.* **99**:479 (1955).
71. J. J. Davies and J. Owen, *J. Phys. C.* **2**:1405 (1969).
72. N. Laurance, E. C. McIrvine and J. Lambe, *J. Phys. Chem. Solids* **23**:515 (1962).
73. A. J. Heeger and T. W. Houston, in *Proc. Int. Conf. on Magnetism*, p. 395, Institute of Physics and the Physical Society, London (1965).
74. J. Owen and D. R. Taylor, *J. Appl. Phys.* **39**:791 (1968).
75. D. R. Taylor, J. Owen, and B. Wanklyn (to be published).
76. N. L. Huang, R. Orbach, E. Simanek, J. Owen, and D. R. Taylor, *Phys. Rev.* **156**:383 (1967).
77. G. Feher, R. C. Fletcher, and E. A. Gere, *Phys. Rev.* **100**:1784 (1955).
78. Y. Ishikawa, *J. Phys. Soc. Japan* **21**:1473 (1966).
79. J. Kanamori, *Prog. Theor. Phys.* (*Japan*) **17**:197 (1957).
80. E. A. Harris and J. Owen, *Phys. Rev. Letters* **11**:9 (1963).
81. D. S. Rodbell, I. S. Jacobs, J. Owen, and E. A. Harris, *Phys. Rev. Letters* **11**:10, 104 (1963).
82. N. C. Tombs and H. P. Rooksby, *Nature* **165**:442 (1950).
83. D. S. Rodbell and J. Owen, *J. Appl. Phys.* **35**:1002 (1964).
84. M. E. Lines and E. D. Jones, *Phys. Rev.* **139**:A1313 (1965).
85. E. A. Harris, Thesis, Oxford (1963).
86. J. H. Van Vleck, in *Quantum Electronics*, p. 392, Columbia University Press (1960).
87. J. C. Gill and R. J. Elliott, in *Advances in Quantum Electronics*, p. 399, Columbia University Press (1961).
88. J. C. Gill, *Proc. Phys. Soc.* **79**:58 (1962).
89. A. M. Prokhorov and V. B. Fedorov, *Soviet Phys.—JETP* **19**:1305 (1964).
90. E. A. Harris and K. S. Yngvesson, *J. Phys. C* **1**:990 (1968).
91. V. A. Atsarkin, *Soviet Phys.—JETP* **22**:106 (1966).
92. R. de L. Kronig, *Physica* **6**:33 (1939).
93. J. H. Van Vleck, *Phys. Rev.* **57**:426 (1940).
94. R. Orbach, *Proc. Roy. Soc.* **A264**:458 (1961).
95. P. L. Scott and C. D. Jeffries, *Phys. Rev.* **127**:32 (1962).
96. C. B. P. Finn, R. Orbach, and W. P. Wolf, *Proc. Phys. Soc.* **77**:261 (1961).
97. E. A. Harris and K. S. Yngvesson, *J. Phys. C.* **1**:1011 (1968).
98. A. M. Stoneham, *Phys. Letters* **18**:22 (1965).
99. I. Waller, *Z. Physik* **79**:370 (1932).

00. N. Bloembergen, S. Shapiro, P. S. Pershan, and J. O. Artman, *Phys. Rev.* **114**:445 (1959).
01. A. Rannestad and P. E. Wagner, *Phys. Rev.* **131**:1953 (1963).
02. E. A. Harris, *J. Phys. C* **2**:1413 (1969).
03. A. J. B. Codling, Thesis, University of Keele (1970).
03a. A. J. B. Codling and B. Henderson, *J. Phys. C* **4**:1409 (1971).
04. H. Statz, L. Rimai, M. J. Weber, G. A. de Mars, and G. F. Koster, *J. Appl. Phys.* **32**:2185 (1961).
05. Y. L. Shelekhim, M. P. Votinov, and B. P. Berkovskii, *Soviet Phys.—Solid State* **8**:469 (1966).
06. A. Jelenski, H. Szymczak, and J. Twarowski, in *Proc. of the Int. Conf. on Magnetic Resonance and Relaxation, 1966*, ed. by R. Blinc, p. 1205, North-Holland Publishing Co., Amsterdam (1967).
07. W. Gunsser, W. Hille, and A. Knappwost, *Z. Phys. Chem.* **58**:316 (1968).
08. M. J. Berggren, G. F. Imbusch, and P. L. Scott, *Phys. Rev.* **188**:675 (1969).
09. A. Hadni, *Phys. Rev.* **136**:A754 (1964).
10. J.-F. Moser, W. Zingg, H. Steffen, and F. K. Kneubühl, *Physics Letters* **24A**:411 (1967).
11. A. L. Schawlow, D. L. Wood, and A. M. Clogston, *Phys. Rev. Letters* **3**:271 (1959).
12. P. Kisliuk and W. F. Krupke, *Appl. Phys. Letters* **3**:215 (1963).
13. P. Kisliuk and W. F. Krupke, *J. Appl. Phys.* **36**:1025 (1965).
14. L. F. Mollenauer and A. L. Schawlow, *Phys. Rev.* **168**:309 (1968).
15. P. Kisliuk, N. C. Chang, P. L. Scott, and M. H. L. Pryce. *Phys. Rev.* **184**, 367 (1969).
16. G. F. Imbusch and M. B. Graifman, *J. Appl. Phys.* **39**:981 (1968).
17. R. W. Bene, *Phys. Rev.* **178**:497 (1969).
18. B. Elschner and D. Meierling, *J. de Physique* (1971).
19. J. C. M. Henning, *Phys. Letters* **34A**:215 (1971).
20. J. C. M. Henning and J. P. M. Damen. *Phys. Rev. B* **3**:3852 (1971).
21. B. N. Figgis and G. B. Robertson, *Nature* **205**:694 (1965).
22. N. Uryu and S. A. Friedburg, *Phys. Rev.* **140**:A1803 (1965).
23. M. R. Brown, B. A. Coles, J. Owen, and R. W. H. Stevenson, *Phys. Rev. Letters*, **7**:246 (1961).
24. J. J. Krebs, *J. Appl. Phys.* **40**:1137 (1969).
25. R. L. Gorifullina, M. M. Zaripov, and V. G. Stepanov, *Soviet Phys.—Solid State* **12**:43 (1970).
6. T. Ohtsuka, *J. Phys. Soc. Japan* **16**:1549 (1961).
7. R. P. van Stapele, J. C. M. Henning, G. E. G. Hardeman, and P. F. Bongers, *Phys. Rev.* **150**:310 (1966).
8. J. C. M. Henning, R. P. van Stapele, G. E. G. Hardeman, and P. F. Bongers, in *Proc. of the Int. Conf. on Magnetic Resonance and Relaxation 1966*, ed. by R. Blinc, p. 1203, North-Holland Publishing Co., Amsterdam (1967).
9. Y. Yamaguchi, *J. Phys. Soc. Japan* **29**, 1163 (1970).
0. K. W. Mess, E. Lagendijk, N. J. Zimmerman, A. J. van Duyneveldt, J. J. Giesen, and W. J. Huiskamp, *Physica* **43**:165 (1969).
1. S. A. Al'tshuler and K. M. Valishev, *Soviet Phys.—JETP* **21**:309 (1965).
2. D. M. S. Bagguley and J. H. E. Griffiths, *Proc. Roy. Soc.* **201A**:366 (1950).
3. H. Abe, K. Ono, I. Hayashi, J. Shimada, and K. Iwanaga, *J. Phys. Soc. Japan* **9**:814 (1954).

134. T. A. Kennedy, S. H. Choh, and G. Seidel, *Phys. Rev.* **B2**:3645 (1970).
135. G. F. Kokoszka, H. C. Allen, Jr., and G. Gordon, *J. Chem. Phys.* **42**:3693 (1965).
136. F. Apaydin and S. Clough, *J. Phys. C* **2**:1533 (1969).
137. H. Abe and H. Shirai, *J. Phys. Soc. Japan* **13**:987 (1958).
138. G. F. Kokoszka, M. Linzer, and G. Gordon, *Inorg. Chem.* **7**:1730 (1968).
139. H. Abe and H. Shirai, *J. Phys. Soc. Japan* **16**:118 (1961).
140. A. Dall'Olio, G. Dascola, and V. Varacca, *Nuovo Cimento* **43**:192 (1966).
141. G. F. Kokoszka, H. C. Allen, Jr., and G. Gordon, *J. Chem. Phys.* **47**:10 (1967).
142. G. F. Kokoszka, H. C. Allen, Jr., and G. Gordon, *J. Chem. Phys.* **46**:3013 (1967).
143. G. R. Wagner, R. T. Schumacker, and S. A. Friedberg, *Phys. Rev.* **150**:226 (1966).
144. J. M. Baker, *Rept. Progr. Phys.* (1971).
145. R. J. Birgeneau, M. T. Hutchings, and R. M. Rogers, *Phys. Rev. Letters* **16**:584 (1966).
146. J. M. Baker, R. J. Birgeneau, M. T. Hutchings, and J. D. Riley, *Phys. Rev. Letters* **21**:620 (1968).
147. J. D. Riley, Thesis, Oxford (1968).
148. K. L. Brower, H. J. Stapleton, and E. O. Brower, *Phys. Rev.* **146**:233 (1966).
149. J. M. Baker, J. D. Riley, and R. G. Shore, *Phys. Rev.* **150**:198 (1966).
150. J. D. Riley, J. M. Baker, and R. S. Birgeneau. *Proc. Roy. Soc. London* **A320**:369 (1970).
151. D. Marsh and J. M. Baker, in *Proc. of Magnetism Conf. (Grenoble) 1970, J. de Physique* **32**:948 (1971).
152. N. E. Kask, L. S. Kornienko, and E. G. Laviontsev, *Soviet Phys.—Solid State* **8**; 2058 (1967).
153. N. E. Kask and L. S. Kornienko, *Soviet Phys.—Solid State* **9**:1795 (1968).
154. B. A. Calhoun and J. Overmeyer, *J. Appl. Phys.* **35**:989 (1964).
155. R. J. Richardson and S. Lee, *Phys. Rev.* **1B**:108 (1970).
156. R. J. Birgeneau, M. T. Hutchings, and W. P. Wolf, *J. Appl. Phys.* **38**:957 (1967).
157. W. Kanzig, R. Baumann, H. U. Beyeler, and J. Muggli, in *Proc. of the XVI Colloque Ampère (1971)*, p. 127.
158. J. P. van der Ziel, *Phys. Rev.* **4**:2888 (1971).
159. T. R. McGuire, B. E. Argyle, M. W. Shafer, and J. S. Smart, *J. Appl. Phys.* **34**:1345 (1963).

Chapter 7

Electron Paramagnetic Resonance of Color Centers

R. H. Silsbee

Department of Physics and
Laboratory of Atomic and Solid State Physics
Cornell University
Ithaca, New York

1. INTRODUCTION

The application of the EPR technique to problems involving color centers has provided a rich harvest both for the physicist concerned with the deeper understanding of color-center problems and for the "resonator" whose primary concern is in the investigation of the intricacies of the magnetic resonance phenomenon itself. For the color-center physicist, EPR provides a more direct means of establishing defect models than other techniques, and typically a set of experimental parameters associated with the defect which provide a critical test for any detailed theoretical model. EPR has clearly demonstrated the striking lack of symmetry between the rather diffuse electron centers and the molecular hole centers in the alkali halides which was only hinted at before the use of EPR. EPR is used to measure the influence of external stress and electric fields upon the properties of color centers, to study color-center interactions, to measure concentrations, and to observe diffusion of centers. For the resonance physicist, the study of color-center problems has led to the introduction of concepts such as inhomogeneous broadening, "superhyperfine" interactions, and spectral diffusion and has provided examples of motional narrowing, various T_1 mechanisms, and the influence of the dynamic Jahn–Teller effect upon resonance properties.

There is no very clear line of demarcation between the study of color centers and the investigation of molecular and atomic impurities in solids. This is partly for historical reasons, the *U*–center in alkali halides not being recognized as an atomic hydrogen impurity until it had been studied in some detail by the Göttingen group, whose sphere of interest defined to a large extent the area of color-center physics; and partly for physical reasons, the problem of the molecular impurity O_2^-, for instance, being so closely related to that of the self-trapped hole or V_K-center. This review will perpetuate that lack of distinction and will draw freely on examples which some may claim lie outside the realm of color-center physics.

Further, this review will not attempt to summarize all EPR investigations of color-center problems, but rather will indicate the variety of ways in which the technique has been applied to color-center problems and in which color-center systems have provided models for the study of resonance physics, drawing on specific examples by way of illustration. A number of comprehensive reviews of the general problems of color centers are listed in the references,[1,2] as well as summaries of the applications of the EPR technique to these problems.[2–4]

2. DEFECT IDENTIFICATION

2.1. Introduction

The most dramatic introduction to the power of the EPR technique is the solid-state seminar which runs roughly as follows. "Slide 1" is an optical absorption spectrum, showing several broad bands, which is the basis of a 10-min summary of the last 20 years' work studying such and such a center, with rather indefinite conclusions. "Slide 2" then gives a 59-line EPR spectrum, which is systematically picked apart for the next 45 min to lead finally and unequivocally to a microscopic model for the defect in question.

Never discussed in the seminar is the EPR spectrum which has been analyzed in the same detail and which leads just as unequivocally to a model which does *not* fit, in any sensible way, into what one knows about the host crystal. Similarly, those spectra which are so complex that the overlap of one line onto another prevents detailed analysis or which are too weak to allow analysis do not provide attractive subjects for seminars. Also too rarely mentioned is the extent to which the "direct logical" analysis is in fact importantly influenced by the perhaps indefinite but still suggestive conclusions of earlier work. Nevertheless, the most important application of

EPR to color-center problems is in the construction of microscopic defect models.

The general philosophy here is rather different than in the analysis of the EPR spectra of transition element or rare-earth ions as substitutional impurities. In the case of magnetic ion impurities, one ideally knows the impurity under investigation, the principal ambiguity being its charge state. By considering the perturbation of the ground state of the "free ion" by the crystal field of the host crystal, one can deduce the important features of the spin-Hamiltonian for the ion and then confirm or deny this picture by comparison with observed spectra. In the color-center problem, one frequently has little or no starting information; and certainly in the case of many systems, it is nonsense to talk of the perturbation of the energy levels of the "free defect" by the host crystal. For the F-center, an electron trapped at an anion vacancy, for instance, what would be meant by the "free defect"?

The analysis of color-center spectra involves first the construction of a spin-Hamiltonian from which can be derived the observed spectrum and second the development of one or more models which are qualitatively consistent with this spin-Hamiltonian. The example of the V_K-center in LiF[5], for which the [110] spectrum is shown in Fig. 1(a), will serve as a basis of a discussion of this procedure since it represents close to an ideal center in terms of the complexity of the spectrum. The corresponding spectrum in KCl,[6] Fig. 1(b), was in fact analyzed earlier than the LiF spectrum, but the multiplicity of lines makes the analysis much more tedious and in fact gives no additional useful information. By contrast, the F-center spectrum in KCl, a single broad line 45 G wide which shows no structure and no anisotropy, is so uninteresting that at first sight one might expect to learn nothing from its analysis.

2.2. From Spectra to the Spin-Hamiltonian

A typical, or perhaps one should say an ideal, color center will yield EPR spectra which show a number of well-resolved lines whose positions depend upon the orientation of the dc magnetic field relative to the crystal axes. Such experimental results are conveniently summarized in a diagram such as Fig. 2 in which are plotted the positions of the lines in magnetic field as a function of the orientation of the field for the V_K-center in LiF. Also shown explicitly are the line positions for the spectrum of Fig. 1(a). The splitting of the resonance into many lines may arise from one or a combination of several mechanisms.

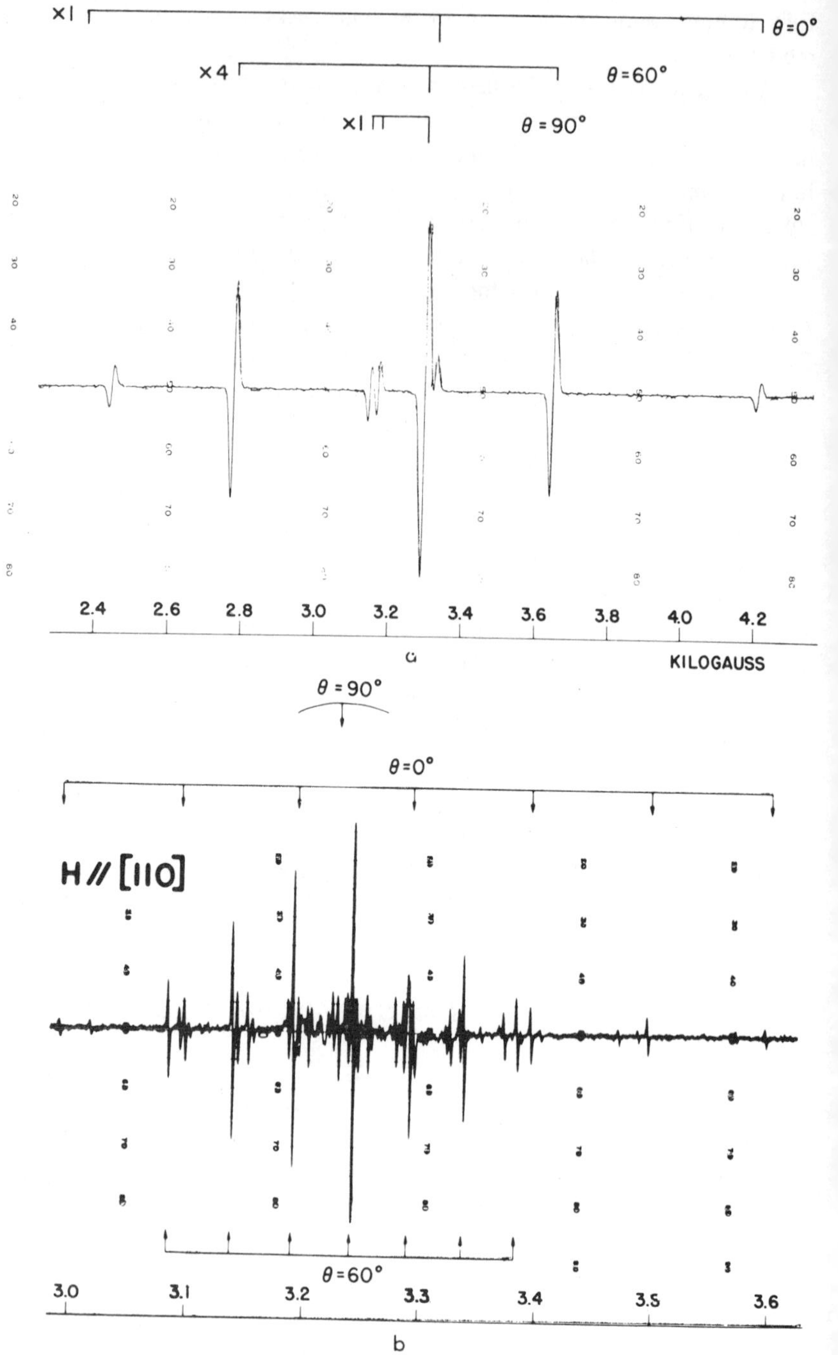
×1
θ=0°
×4
θ=60°
×1
θ = 90°
2.4
2.6
2.8
3.0
3.2
3.4
3.6
3.8
4.0
4.2
a
KILOGAUSS
θ = 90°
θ=0°
H // [110]
θ=60°
3.0
3.1
3.2
3.3
3.4
3.5
3.6
b

Fig. 1. (a) Spectrum of the V_K-center in LiF with the applied field parallel to the [110] direction (from Ref. 5). (b) Spectrum of the V_K-center in KCl with the applied field parallel to the [110] direction (from Ref. 6). The brackets indicate the various components of the spectrum associated with those centers whose molecular axis is at an angle θ with respect to the applied magnetic field.

In crystals of high symmetry, the paramagnetic species will frequently have a lower symmetry than that of the host crystal; i.e. the symmetry operations of the crystal point group may take a particular defect not into itself, but into an equivalent defect with different orientation. If the only significant term in the spin-Hamiltonian for a defect of orientation denoted by the superscript a is an anisotropic Zeeman interaction,

$$\mathscr{H}_g^{\,a} = \beta \mathbf{S} \cdot \mathrm{g}^a \cdot \mathbf{H} \tag{1}$$

then there will be in the crystal comparable numbers of defects with g-tensors related to g^a by the symmetry operations of the host crystal. For a general orientation of the field $\mathbf{H}$, there will be as many lines in the spectrum as there are distinct values of the g-tensors generated from g^a by the point

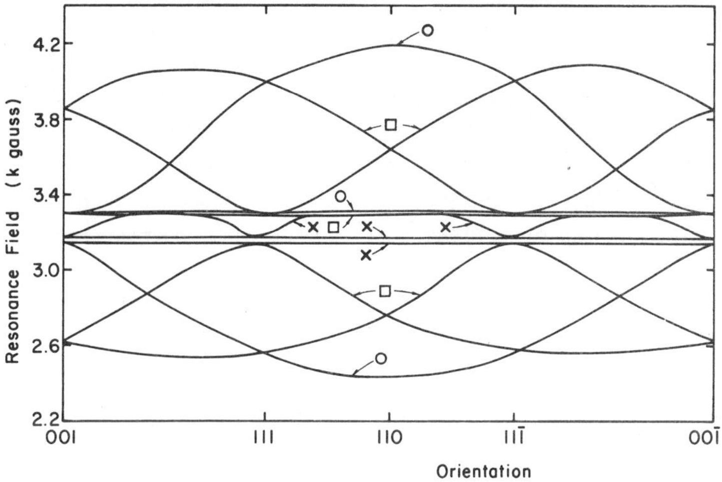

Fig. 2. Line splitting pattern for the V_K-center in LiF (from data of Ref. 5). The various groups of lines of Fig. 1(a) are indicated here, for the 110 direction, by the following symbols: (○) $\theta = 0°$; (□) $\theta = 60°$; (×) $\theta = 90°$.

symmetry operations of the crystal.[7] For special symmetry directions of the field, some of these lines will, of course, become degenerate.

If the defect spin is greater than $\frac{1}{2}$, there may be a zero-field splitting term for the defect of orientation a of the form

$$\mathscr{H}_D{}^a = \mathbf{S} \cdot \mathsf{D}^a \cdot \mathbf{S} \tag{2}$$

which can split the EPR line of defect a into two or more components. Again the symmetry of the tensor D^a will reflect the symmetry of the defect a and if this is less than the point symmetry of the host crystal, additional multiplicity will result, just as in the case of the g-anisotropy.

In crystals with nuclear species with magnetic moments, there will be additional splittings or line broadening resulting from the hyperfine interaction of the electrons with these species, described by a term in the Hamiltonian,

$$\mathscr{H}_A{}^a = \sum_i \mathbf{S} \cdot \mathsf{A}_i{}^a \cdot \mathbf{I}_i \tag{3}$$

The sum runs over those nuclei i for which there is significant hyperfine interaction with the electrons. The individual hyperfine interaction tensors $\mathsf{A}_i{}^a$ in general will have symmetry lower than that of the defect if the nuclei i are taken into one another by the symmetry operations of the defect. On the other hand, for those individual nuclei i which are taken into themselves by all the symmetry operations of the defect a, the coupling tensor $\mathsf{A}_i{}^a$ will show the same symmetry as the g^a and D^a tensors.

Although these three terms are usually the dominant terms in the spin-Hamiltonian, and the ones that most often determine the gross features of the spectrum, a number of qualifications should be noted. The most important is that if the ENDOR technique is applied for the determination of the hyperfine constants A_i or if partially forbidden lines involving simultaneous nuclear and electron spin flips are observed, then one must include the nuclear quadrupole

$$\mathscr{H}_Q{}^a = \sum_i \mathbf{I}_i \cdot \mathsf{Q}_i{}^a \cdot \mathbf{I}_i \tag{4}$$

and nuclear Zeeman

$$\mathscr{H}_N = \sum_i g_i \beta_N \mathbf{I}_i \cdot \mathbf{H} \tag{5}$$

interaction terms, since these importantly influence the positions of lines involving a nuclear spin transition.

If one considers all terms allowed by symmetry in the spin-Hamiltonian, one should consider terms of higher order in **S**, **H**, and $\mathbf{I}_i$ than those noted above. The allowed terms of this nature and the forms of the appropriate coupling tensors are discussed by various authors,[7–9] but are rarely noted in the analysis of color-center spectra. A possible exception is the *R*-center, in which the situation is complicated by the presence of a twofold orbital degeneracy, in addition to a twofold spin degeneracy. An appropriate spin-Hamiltonian is most naturally written by inclusion of an orbital Zeeman term and a spin–orbit term in addition to those noted above. On the other hand, an equivalent Hamiltonian could be expressed by representing the center as an $S = \frac{3}{2}$ system and including terms such as S^3H.

With the qualification of occasional exceptions for systems of high symmetry which allow orbital degeneracy, the spin-Hamiltonian

$$\mathscr{H}_S = \mathscr{H}_g + \mathscr{H}_D + \mathscr{H}_A + \mathscr{H}_Q + \mathscr{H}_N \tag{6}$$

will give a satisfactory description of the color-center spectra. The first three terms determine the positions of the freely allowed EPR transitions, the last three dominate the ENDOR spectrum of the center. Skill in the association of specific features in the spectra with appropriate terms in $\mathscr{H}_S$ is a matter of experience, but for well-resolved spectra, the problem is usually straightforward. Figure 3 illustrates the qualitatively different nature of the splitting patterns associated with the first three terms in the Hamiltonian. The three parts of the figure show the splitting patterns that would arise from each of the terms acting alone in the special case of tensors with axial symmetry about [110] in a cubic crystal. If two, or all three, are comparable in magnitude, the pattern is, of course, correspondingly more complex.

The relevant qualitative arguments in the analysis of the data for the V_K-center in LiF shown in Fig. 2 might go as follows.

1. The most apparent feature of the spectra is the presence of pairs of lines roughly symmetrically displaced with respect to the line or group of lines near the center. A comparison with Figs. 3a–3c suggests hyperfine interaction as the principal source of line splitting. Since the hyperfine multiplets seem to be triplets rather than the doublets of Fig. 3a, which assumes interaction with nuclear spin $\frac{1}{2}$, interaction with either a single spin-1 nuclear species or a pair of equivalent spin-$\frac{1}{2}$ nuclear species is indicated. Observation that the most abundant nuclear species are Li^7, with $I = \frac{3}{2}$, and F^{19}, with $I = \frac{1}{2}$, gives the first important conclusion, namely that the dominant splitting results from hyperfine interaction with two fluorine nuclei. The near-degeneracy of the central component suggests

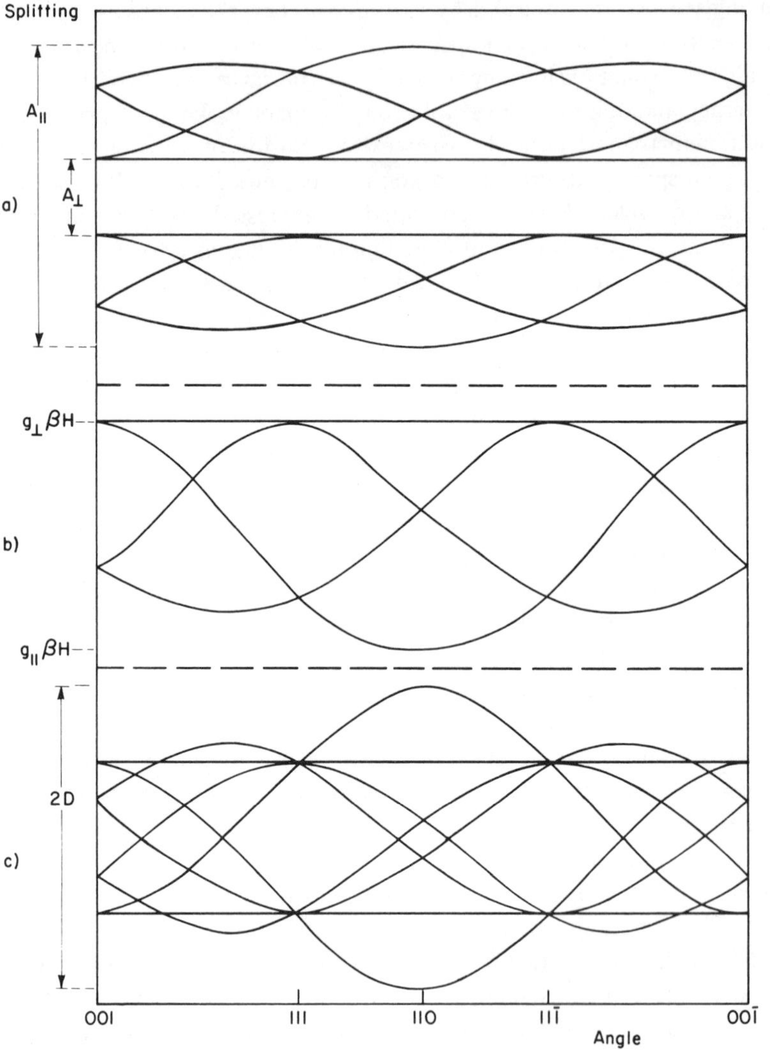

Fig. 3. EPR anisotropy splittings for a defect with a Hamiltonian showing axial symmetry about a [110] axis. The splittings are illustrated schematically for the applied magnetic field being rotated in a (1$\bar{1}$0) plane for the following cases: (a) Splitting due entirely to hyperfine interaction with a nucleus with spin $I = 1/2$; (b) splitting due entirely to g-anisotropy; (c) splitting due entirely to an axial crystal field splitting of a spin-one electronic defect, in the limit $D \ll g\beta H$.

that the interaction with the two fluorine is equivalent or nearly so, otherwise one would expect four well-resolved hyperfine components.

2. The superposition of lines when the field is parallel to high-symmetry directions implies that the splittings as one moves away from these directions are a consequence of the anisotropy of the hyperfine splittings. A quick comparison of Figs. 2 and 3a shows consistency of the pattern of Fig. 2 with a hyperfine tensor with nearly axial symmetry with large principal values in [110] directions and small principal values in the perpendicular directions. A small g-anisotropy gives observable splittings of the central components of the hyperfine triplets too small to be easily seen in Fig. 2 and the angular dependence of these splittings is suggestive of the pattern of Fig. 3c, implying again a defect of [110] symmetry. For this system, in which two components of the hyperfine coupling tensor are small compared with the third, the hyperfine splittings depart from the simple pattern for the magnetic field nearly perpendicular to the axis of symmetry, complicating the interpretation of the g-splittings.

3. There is no need to assume a fine-structure splitting in the interpretation of the gross features of the spectrum. Since the defect symmetry, [110], is low, such a splitting would be allowed by symmetry for spin $>\frac{1}{2}$ and so, in absence of evidence to the contrary, the spin of the defect is assumed to be one-half.

If ENDOR data are also available, or if one has analyzed the spectra which show forbidden transitions, then the additional nuclear Zeeman and quadrupolar interaction terms are relevant to the analysis. Typically, one finds the hyperfine and quadrupolar coupling constants for those nuclei with which the defect electron is in rather weak interaction, the symmetry of these coupling tensors giving a unique description of the nature of the defect surroundings.

The next step in the procedure is to verify that the constants of the proposed spin-Hamiltonian can be adjusted to give a quantitative fit to the data and that any ambiguities in interpretation of the data are removed. The quantitative comparison is straightforward but may be tedious if more than first-order perturbation theory is required to determine the eigenvalues of the spin-Hamiltonian with a precision comparable to that of the experimental results. Measurements at two microwave frequencies, say X and K band, are useful in providing unambiguous discrimination between splittings due to g-anisotropy and those due to crystal field interaction or hyperfine interaction. Occasional uncertainties in identification of the nuclear species responsible for hyperfine splittings may be resolved by using crystals in which a suspected nuclear species is replaced by an isotope of different magnetic moment. This isotopic substitution of nuclear species with magnetic moment is also useful, for systems with naturally occurring isotopes without spin, in elucidating the defect structure by giving the added information contained in the hyperfine constants. In many instances in which the naturally occurring isotopes have distinguishable moments, the relative intensities of the spectral lines produced by each species give an immediate identification of the chemical species by comparison of these intensities with the natural abundances. The analysis of the ENDOR spectra also gives unambiguous identification of nuclear species through the nuclear gyromagnetic ratios. Ambiguity in the assignment of the magnitude of the electronic spin is usually resolved, and correctly resolved, by the simple expedient of assuming a spin of $\frac{1}{2}$ unless the spectrum clearly indicates otherwise. This conclusion may be made more positive, if there is substantial hyperfine splitting, by establishing experimentally the absence of second-order hyperfine splitting which would be predicted for $S > \frac{1}{2}$.

2.3. From Spin-Hamiltonian to Defect

The spin-Hamiltonian does not imply a specific model, but simply provides a set of conditions which must be satisfied by any satisfactory model. The most important of these is the symmetry of the coupling tensors. The symmetry operations of the defect must leave the g and D tensors invariant and must take the hyperfine tensors either into themselves or into the tensors of equivalent nuclei. Thus, the symmetry of the coupling tensors, apart from accidental equality of some of the components, implies to a large extent the symmetry elements of the defect. The most important qualification is that the presence or absence of inversion symmetry cannot be deduced from the spin-Hamiltonian in the absence of experimental determination of the effects of electric fields. This is because time-reversal symmetry implies that the sum of the powers of **I**, **S**, and **H** which appear in any term in $\mathscr{H}_S$ must be even, which implies in turn that the coupling tensors will always be even under inversion, irrespective of the presence or absence of inversion symmetry for the defect.

In the example of the V_K-center, the data imply a model which is taken into itself by those operations of the cubic host crystal which take the [110] direction into itself. Although the data are analyzed here in terms of a model with full cylindrical symmetry, the published spectra show, as would be expected, that there are small deviations from axial symmetry, consistent with the twofold symmetry axis of the [110] direction in cubic crystals. As noted later, the near axial symmetry of the coupling tensors is a consequence of the details of the physical model, not of the symmetry of the model.

The analysis of the hyperfine splitting, if present, gives the most helpful clues in the construction of defect models. The number of nuclei of various species contributing substantially to the hyperfine interaction and the relative strength of these interactions give an immediate qualitative measure of the distribution of the unpaired electrons over the ions contributing importantly to the defect. The anisotropy of the hyperfine coupling constants typically indicates the contribution of *p*-like orbitals to the electronic functions and the orientation of the principal axes of the A tensors reflects the orientation of these orbitals with respect to the crystallographic axes. These data, when fully available, usually lead rather directly to a model for the defect. For the V_K-center, the large hyperfine coupling with two fluorines and the absence of resolvable lithium splittings suggest immediately a molecular model involving a bound pair of fluorine ions, without substantial participation by the lithium. The large anisotropy of the A tensors suggests a large admixture of p character in the wave functions and the presence of the [110] symmetry axis for this A tensor is consistent with either of the two models illustrated in Fig. 4. In Fig. 4(a), the unpaired electron lies in a molecular σ orbital, in Fig. 4(b), in a π orbital. In either case, both the g and A tensors would show [110] symmetry and the maximum principal

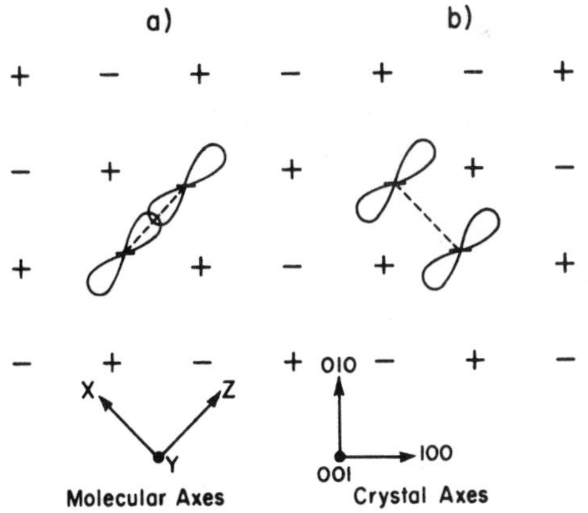

Fig. 4. Two possible models for the unpaired electron orbital of the V_K-center. (a) The accepted model with molecular axes as defined by Känzig and used in text; (b) a tentative alternative, consistent with the gross but not the fine details of the EPR spectra.

value of the A tensor would be in the [110] direction for the defect orientation as illustrated. The very closely axial symmetry of the observed spectra would suggest Fig. 4a as the more likely model, but this is more a quantitative than a qualitative conclusion.

Appeal is often made to the sign of the g-shift, or difference between the observed g-value and the free-electron value, to determine whether the defect is a trapped electron or trapped hole. Although this may sometimes give useful insight, it is a procedure to be treated cautiously since, in some sense, most defects correspond to the situation of the half-filled shell where there are often comparable contributions to the g-shift of both signs. Further, there exist systems for which the sign of the g-shift is in fact opposite to that expected by the simplest arguments. However, as soon as one has a specific model of the electronic structure of a defect, the g-shift information does give a rough qualitative test. For the V_K-center in LiF, the $\delta g_{\perp}$ is substantial (0.02), compared with $\delta g_{\parallel}$ (0.0011), and positive. As discussed in greater detail in the following section, g-shifts of molecular species tend to be larger for the magnetic field in directions perpendicular to the direction containing the lobes of the unpaired electron p function, and particularly large for π orbitals with the field parallel to the molecular axes. The model of Fig. 4(b) would imply $|\delta g_{1\bar{1}0}| > |\delta g_{001}| > |\delta g_{110}|$, which is inconsistent with the data, while the experimental axial g with $|\delta g_{001}| \approx |\delta g_{1\bar{1}0}| > |\delta g_{110}|$ fits neatly with the model of Fig. 4(a).

Further qualitative evidence to distinguish between the two models of Fig. 4 is provided by ENDOR results.[10] Note that for case (a) of Fig. 4, there is a fluorine ion on the [110] axis defining the major axis of the A tensor, while for case (b), this [110] axis contains only lithium ions. Analysis of the ENDOR data shows the presence of a fluorine nucleus with this [110] symmetry axis but no lithium, giving added evidence in support of model (a).

For these particular spectra, one is led rather quickly to an apparently satisfactory model, namely an F_2^- molecule ion with the molecular axis parallel to [110] directions, the charge state of minus one chosen because this results in an electronic configuration with a single unpaired electron in the highest antibonding σ-type molecular orbital, consistent with the hyperfine symmetry implied in model (a) and with the qualitative g variation.

2.4. Summary of Results

Such qualitative arguments combined with a wealth of data using other techniques, the most important of which is optical absorption, have led to quite specific models for a number of centers. These are summarized very briefly below. The reader is referred to other reviews[3,4] for a more extensive catalogue of the various centers which have been identified from EPR results.

2.4.1. *F*-Center

The *F*-center[11,12] is firmly established to be an electron trapped in the Coulomb potential of the positive effective charge of a negative-ion vacancy. The electron wave function is diffuse, extending out of the vacancy with substantial overlap onto the neighboring ions. In the alkali halides, the hyperfine interaction with the nuclei of the neighboring ions gives a line breadth varying from 45 to 700 G. In a few systems, LiF, NaF, RbCl, and CsCl, some resolved structure is present, but the more usual result is a single, broad, structureless line. In CaO[13] and MgO[14], on the other hand, the absence of nuclei with magnetic moments gives relatively narrow lines, with weak satellites in the case of MgO produced by the 10% abundant ^{25}Mg.

2.4.2. *R*- and *M*-Centers

The optical bleaching of crystals containing *F*-centers leads to the formation of centers which are found to be pairs and triples of *F*-centers. The *M*-center, a pair of *F*-centers with [110] orientation, has a diamagnetic ground state but a magnetic metastable excited state which has been studied by EPR.[15] The *R*-center, an equilateral triangle of *F*-centers lying in a (111) plane, has been observed both in a spin doublet ground state[16] and a spin quartet metastable excited state.[15]

2.4.3. *V*-Centers

The electron centers, *F*, *M*, and *R*, are qualitatively understood in terms of simple continuum models and as analogs of atomic and molecular hydrogenic systems. In contrast, the hole centers are much more strongly localized and are pictured in terms of the formation of molecules or molecular ions.

At liquid nitrogen temperature, for instance, a hole in the valence band does not wander freely through the crystal, but induces a strong localized distortion of the ion structure, which in turn stabilizes the localization of the hole. The final consequence of this self-trapping process in KCl, for example, is the formation of a Cl_2^- molecule ion or V_K-center by the pulling together by the hole of two adjacent Cl^- ions. The hole is thought of as being strongly localized in a σ_u antibonding orbital on this molecule ion and as interacting only weakly with the surrounding neighbors. As discussed earlier in this section, the dominant feature in the EPR spectrum of these

V_K-centers[6,17] is the well-resolved hyperfine splittings of these two nuclei. The H-center[18] is similar in structure, but arises from the inclusion of an interstitial chlorine atom into the crystal. Again there is the formation of a Cl_2^- molecule ion with a [110] axis, now substitutional for a chlorine ion. There are a variety of closely related species involving the association of these halide molecular ions with other defects, either vacancies or impurities.

2.4.4. Molecular Impurities

In crystals with other halides as impurities, one frequently sees spectra associated with mixed halogen molecule ions,[19–22] $ClBr^-$ or FCl^- in KCl, etc. Most such mixed centers show essentially the same features as the normal V_K-centers. The mixed centers containing fluorine, however, such as FCl^-, FBr^-, and FI^- in KCl and KBr show [111] symmetry and are identified as more closely analogous to the H centers, i.e., a center with an extra halogen ion, but now with a [111] orientation. A variety of other molecular impurities have also been observed in KCl, including N_2^-,[23] O_2^-,[24] HCN^-,[25–27] and FCN^-.[26] These are included as examples in the subsequent discussion because one's viewpoint in analyzing the spectrum is so similar to that taken in discussing the V-centers.

2.4.5. Atomic Impurities

As noted in the introduction, there are a number of paramagnetic systems which are naturally included in a discussion of color-center problems for historical reasons. The U-center, the hydride ion substituted for a halide, has long been studied in laboratories concerned with color-center problems and the photochemistry of systems containing this center has been studied using EPR to monitor the presence of atomic hydrogen[28] produced by optical irradiation of this defect. Neutral as well as the more conventional divalent silver[29] has been studied in alkali halides using EPR. Finally, one should mention the investigations of the divalent manganese[30] EPR in the alkali halides, the emphasis being in the study of vacancy interaction effects rather than intrinsic interest in the properties of the manganese ions.

3. QUANTITATIVE INTERPRETATION

3.1. The g-Tensor

A final step in the analysis of the color-center EPR and ENDOR spectrum is the quantitative comparison of the constants of the spin-Hamiltonian with predictions of theory based upon the qualitative model for the center. With few exceptions, the color centers do not show orbital degeneracy in the ground state. Excepting these special cases, the g-shifts, or difference between the principal values of the g-tensor and the free-electron value of 2.0023, result from the admixture by spin–orbit interaction of some orbital moment into the ground state. The distribution of the wave function of the unpaired electron(s) over a number of ions rather than a single ion gives the most important difference between the g-shift calculation for color centers and for the usual paramagnetic ions. The extreme examples of such delocalization are, of course, the paramagnetic donor and acceptor impurities in semiconductors.

For defects of orthorhombic or higher symmetry the principal axes of the g-tensor are unambiguously defined by the symmetry operations of the defect. For defects of triclinic or monoclinic symmetry, either none or only one of the axes is determined by symmetry. Kneubühl[31] has discussed the implications of crystal symmetry in detail and in particular the ambiguities in the definition of the g-tensor which result from the arbitrariness in definition of the spin basis states for low-symmetry centers.

Slichter[32] and Stone[33] have discussed with some care the quantitative treatment of the g-shift problem for the case of the multicenter defect. For example, the zz principal value of the g-shift tensor is given in second-order perturbation theory by

$$\delta g_{zz} = \sum_k \{(m/\hbar^2)\langle 0 \mid (x_k^2 + y_k^2)\xi_k(r_k) \mid 0\rangle + 2 \sum_{n \neq 0} [\langle 0 \mid l_{kz} \mid n\rangle\langle n \mid \xi_k(r_k) l_{kz} \mid 0\rangle/(E_0 - E_n)]\} \tag{7}$$

where k denotes the various nuclei over which there is appreciable value of the electron wave function and x_k, y_k, r_k, etc. the electron position measured with respect to the kth nucleus. $\xi_k(r_k)$ is given by

$$\xi_k(r_k) = (e\hbar^2/2m^2c^2)E_k(r_k)/r_k$$

where E_k is the electric field, assumed radial, experienced by the electron when on the kth ion. Here, $\mid 0\rangle$ is the orbital of the unpaired electron, and $|n\rangle$ the various orbitals to which this electron may be promoted or from which a normally paired electron may be promoted to the orbital $\mid 0\rangle$. E_0 and E_n are the one-electron energies of these various orbitals. The first term represents the spin interaction with the induced diamagnetic circulation of the electrons about the nuclei, and is usually neglected; the second is the interaction with the paramagnetic currents.

As in the conventional single-ion problem, the contribution of various excited states to the g-shift involves the product of two terms, one a matrix element of the spin–orbit interaction, the other a matrix element of the orbital Zeeman interaction. Since the spin–orbit interaction is appreciable only in regions of large electric field, and hence only when the electron is near a nucleus, the matrix element of the spin–orbit term may always be broken into a sum of contributions from various centers, with very little error. This is not, in general, a valid approximation in the case of the orbital Zeeman matrix elements, because there will typically be substantial contributions to that matrix element from those regions where the electron is not clearly to be associated with a particular one of the ions, and, as noted later, these overlap contributions may even result in a reversal of sign of the g-shift.

The result of Eq. (7) gives clear qualitative implications with regard to the g-shift for the V_K-center, as already mentioned in Section 2. Figure 5 indicates a qualitative picture of the lowest set of one-electron states for a diatomic molecule in an orthorhombic field. The states are essentially the same as for the free molecule except that the orbital degeneracy of the π states is removed by the crystalline field. The V_K-center corresponds to the configuration indicated in which all of the states are occupied by pairs of electrons except for a single hole in the highest σ_u antibonding orbital.

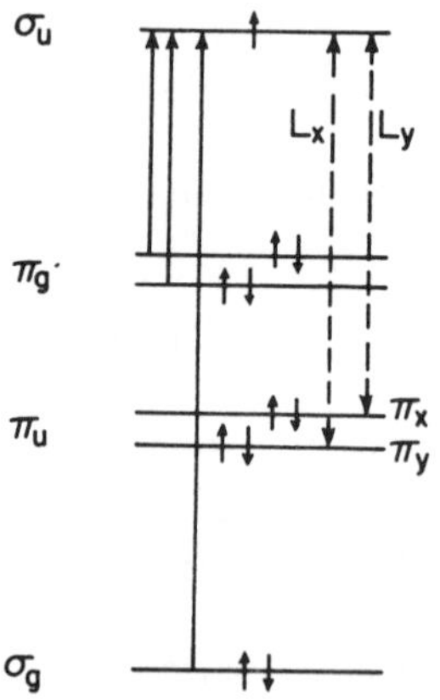

Fig. 5. Schematic of energy level diagram for a diatomic molecule. Small arrows indicate the occupation of these levels by spin-up and spin-down electrons for the V_K-center. Solid lines indicate allowed optical transitions. Dashed lines indicate states coupled by the orbital moment operator which contribute to the g-shift.

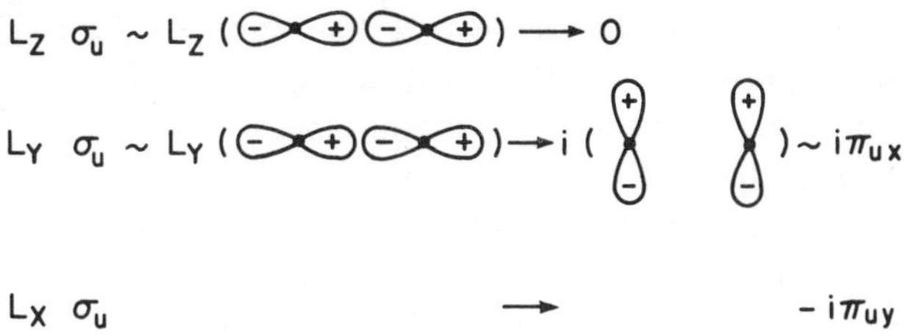

Fig. 6. A schematic representation of the coupling among the molecular orbital states of the V_K-center by the orbital angular momentum.

Because, to a good approximation, the σ_u wave function of the unpaired electron (or hole) is axially symmetric, there will be no matrix elements of the L_z operator and hence the δg_{zz} is expected to be zero. The small value of the observed $\delta g_{\parallel} \sim +0.001$ is consistent with this, the nonzero value being contributed by terms of higher order in $\lambda/(E_0 - E_n)$ than included in (7). There will also be a contribution to the $\delta g_{\parallel}$ from the admixture into the ground state of Δ symmetry molecular states by the crystal field. Because of this admixture, the operator L_z can then couple weakly to other states giving a small contribution to $\delta g_{\parallel}$. The operator L_y, on the other hand, when applied to the p-like functions centered on the individual centers, couples to a wave function of π_u symmetry as indicated in Fig. 6, as will the operator L_x. Thus one expects substantial contributions to δg_{xx} and δg_{yy}. To the extent that the two π_u states remain degenerate despite the nonaxial crystalline field, these g-shifts should be the same. A determination of the sign of the departure from axial symmetry, in turn, determines which of the two π_u states lies higher. In KCl, for instance,[6] $\delta g_{xx} = 0.0405$, $\delta g_{yy'} = 0.0424$, implying that the energy denominator for the δg_{yy} is smaller or that the x-symmetry π_u state lies higher than the y-symmetry π_u state by $\sim$15% of the π_u–σ_u splitting. The sign of this splitting is apparently inverted in NaCl for reasons that are not understood. Approximating the electronic states by linear combinations of atomic orbitals with coefficients estimated from the hyperfine coupling constants, neglecting overlap, and using free-atom spin–orbit coupling constants allows a quantitative estimate of the energy of the π_u state relative to the σ_u. The results of these calculations are presented by Castner and Känzig[6] and are qualitatively consistent with optical data of Delbecq *et al.*[34]

For the F-center, the problem is formally similar, but intuitively less transparent. The problem has been treated in detail by several authors[35–37] and results are reviewed by Schmid.[38] The observed g-shifts are small, varying from 5×10^{-4} in the light alkali halides to $\sim 500 \times 10^{-4}$ in the heavier ones. The spin–orbit mixing of excited states occurs on the neighboring ions, both the positive and negative ions giving important contributions. The magnitude is estimated from the atomic spin–orbit constants and the amount of *p* character of the wave functions on these ions as determined from the anisotropic part of the hyperfine coupling to these nuclei. The theory overestimates the δg for the Li salts by a factor of two; for the heavier salts, the theory is somewhat more successful, giving excellent agreement, 10% or better, for three systems, but understimating the magnitude of the shift by up to 40% for the remaining seven examples. In view of the use of the hyperfine coupling constants to eliminate the largest uncertainty, one would have hoped for somewhat better agreement, particularly for the Li salts.

The sign of the *g*-shift frequently gives a critical test of a model, or suggests a crucial modification of a model. However, the glib remark that a positive δg implies a hole, a negative δg implies an electron, though a useful rule of thumb, must be treated with some caution. The *R*-center is clearly an electronlike center, but shows a positive δg;[16,39] the N_2^- molecule ion has its molecular orbitals more than half full and hence might be considered to be a closed shell plus five holes rather than an empty structure plus seven electrons, yet it shows a negative δg;[40] and a sensible argument has been constructed[37] to explain an incorrect measurement of a positive *g*-shift for the *F*-center in LiF, the classic example of an electron center.

The basis of the prediction of the sign of the *g*-shift is in the structure of Eq. (7). Consider the usual case of a single unpaired electron in an orbital singlet. There are contributions to δg from a number of excited states, corresponding either to the promotion of the unpaired electron to a higher, unoccupied one-electron state—an electronlike excitation—or to the promotion of a paired electron in a lower state up to the same orbital as the unpaired electron—a holelike excitation. Because the energy denominator in Eq. (7) has the opposite sign for these two types of contributions, the contributions of the two kinds of excitation to the *g*-shift will be opposite in sign. In general, there are contributions of both signs to the δg, the sign of the δg depending upon which is the dominant term. For the *F*-center, there will be positive contributions to δg from excitations of ion core electrons on the neighbors to pair with the *F*-center electron as well as the excitations, which in fact dominate the δg, of the *F*-center electron to higher excited *F*-states. In N_2^-, the dominant contribution to the δg is from the excitation of an unpaired electron from one to the other of the π_g molecular states; although the full set of bonding and antibonding orbitals is more than half full, the crucial point is that the particular subshell of two π_g

orbitals is less than half full, giving an electronlike negative δg. The complement to this example is the O_2^- with one π_g hole and a positive δg. Finally, one should note that arguments concerning the sign of the g-shift also depend tacitly upon the assumption that the one-electron matrix elements of the orbital Zeeman and spin–orbit terms involved in evaluating (7) have a unique relative sign. In the single-force-center problem, this is always the case, but not for the multicenter problem. In the multicenter problem, there will be contributions to the orbital Zeeman term both from circulation of electron current about the individual ions and from a gross circulation about the center as a whole, whose magnitude will be substantial to the extent that there is significant overlap of the atomic orbitals on different centers. The relative sign of the first contribution to the orbital moment is the same as for a single-center problem, but the second term may have either sign and presents an additional ambiguity in interpretation if it dominates the orbital Zeeman term. Examples of this anomalous sign of δg appear in the R-center[16,39] and in a related problem, the optical Faraday rotation and magnetic dichroism of the F-center.[41] Another interesting example is the resonance of Cr^{3+} in SbI_3. The Cr^{3+} ion normally has a negative g-shift, but strong covalent interactions with the neighboring I^- ions are sufficient that the positive spin–orbit coupling of the atomic iodine configuration dominates to give a net positive g-shift.[41a]

3.2. Hyperfine Coupling

Whereas quantitative interpretation of the g-tensor gives information about the coupling to and positions of excited states of the defect, the magnitudes of the hyperfine coupling constants give quite a specific picture of the spatial distribution of the unpaired electron in the ground state of the defect and hence provide the most critical test of defect models. In absence of significant contribution of orbital moment to the g-value of the resonance, the hyperfine tensor for the interaction of a single electron with a nucleus i is given by

$$\mathsf{A}^i = g\beta g_N{}^i\beta_N \int \left\{ \frac{8\pi}{3}\,\delta(\mathbf{r}-\mathbf{r}_i) + \left[\frac{3(\mathbf{r}-\mathbf{r}_i)(\mathbf{r}-\mathbf{r}_i)}{|\mathbf{r}-\mathbf{r}_i|^5} - \frac{\mathbf{1}}{|\mathbf{r}-\mathbf{r}_i|^3} \right] \right\} |\psi(\mathbf{r})|^2\, d^3\mathbf{r} \tag{8}$$

where $\mathbf{r}_i$ is the vector position of the ith nucleus with nuclear g-value $g_N{}^i$, and $\psi(\mathbf{r})$ is the orbital of the unpaired electron. The first or Fermi contact part of the hyperfine interaction gives an isotropic contribution to the hyperfine tensor, while the second or dipolar term contributes a traceless component to the A tensor.

If the hyperfine coupling A^i is measured for a number of nuclei in the vicinity of the defect, either through ESR or ENDOR, then the contact or isotropic term, Tr A^i, together with the known values of the nuclear moments gives a value of $|\psi(\mathbf{r}_i)|^2$ at each of these nuclei and hence a rough sketch of the wave function. Further, in many instances, the integral giving the anisotropic part is dominated, through the $\langle 1/r^3 \rangle$ term, by the wave function near the nucleus and can be interpreted in terms of the amount of p character in the wave function.

If there is substantial contribution of orbital moment to the Zeeman interaction as evidenced by the departure of the g from 2, there will also be a contribution of the orbital moment to the hyperfine coupling.[42] The relative contributions of orbital and spin moment to the anisotropy of the hyperfine coupling are roughly equal to the δg, but the details will depend upon the electronic structure in question. This effect is beautifully illustrated in the work of Zeller and Känzig[43] on the O_2^- impurity in a number of alkali halides in which the variation of the oxygen-17 hyperfine anisotropy in different host crystals is entirely accounted for by the differences in orbital moment contributions as calculated from the g-shifts. It is important to remember also that the orbital contribution to the hyperfine tensor will not have zero trace and hence, in the presence of substantial g-shifts, the trace of the hyperfine tensor does not give directly the magnitude of the Fermi contact term.

In the discussion of molecular-type defects, such as the V_K-center, it is instructive to express the molecular orbitals in terms of LCAO's. If this procedure is followed, and if the atomic hyperfine coupling constants are available, one is lead immediately from the hyperfine coupling constants to the coefficients of the LCAO expansion of the electronic wave function. This approach is discussed by Castner and Känzig[16] for the V_K-center. Das *et al.*[44] have attempted a calculation from first principles of these coefficients for the V_K-center in LiF, taking into account both the electronic structure of the F_2^- molecular ion and the influence of the surrounding ions upon the F_2^-. The results are not satisfactory in that they give too low an energy, by a factor of two, in the $\sigma_g \rightarrow \sigma_u$ excitation energy and too small a value for the isotropic part of the hyperfine interaction. They speculate that their deduced value for the fluorine–fluorine separation is probably too large since a smaller nuclear separation would be expected both to increase the σ_g–σ_u splitting and to increase the degree of s hybridization in these orbitals.

The most extensive effort in predicting the hyperfine coupling constants for *F*-centers is the work of Gourary and Adrian.[36] Their approach is to construct a smooth variational wave function which is then orthogonalized to the ion cores of the surrounding ions and the constants chosen to minimize the ground-state energy using a potential appropriate to a point-ion model. The orthogonalization of the smooth function to the core functions is crucial in obtaining the correct order of magnitude of the hyperfine coupling. Holton and Blum[45] and Seidel[46] have made detailed comparisons of the experimental hyperfine coupling constants with those calculated by Gourary and Adrian. For the isotropic part of the coupling, which is proportional to the smooth wave function with an "amplification factor" determined by the orthogonalization procedure, the agreement is only fair. For the near shells, the theoretical coupling is roughly twice that observed experimentally, while for the farther shells, the trend is reversed, the theoretical amplitudes falling off faster with distance than the experimental constants. Wood and Korringa[47] suggest on the basis of a detailed calculation for LiCl that the near neighbors of the *F*-center are displaced outward from their positions in the unperturbed lattice, thus reducing the theoretical estimate of Gourary and Adrian. On the other hand, Feuchtwang,[48] in a detailed analysis of the anisotropic hyperfine coupling and of the nuclear quadrupole coupling, has concluded from the signs of the electric field gradients that the nearest-neighbor ions displace toward the *F*-center in KCl and LiF. Thus, although the theory of Gourary and Adrian gives a satisfactory general picture of the ground-state *F*-center wave function, there still remain ambiguities and uncertainties in the detailed interpretation of the ENDOR results for the *F*-center.

4. EFFECTS OF EXTERNAL FIELDS

Although EPR is basically concerned with the influence of a static magnetic field upon the low-lying levels of a degenerate system, the technique is naturally extended by the study of the influence of other externally applied fields upon these levels. Electric fields and uniaxial stress fields may both modify the constants of the spin-Hamiltonian and, to the extent that they lower the symmetry of the center, may introduce additional terms. The effect of hydrostatic stress, since it does not change the symmetry of the defect, is only to modify the constants of the spin-Hamiltonian. A further dramatic effect of uniaxial stress, not involving the constants of the usual spin-Hamiltonian, is in the removal of orientational degeneracies which, for an orientationally mobile species, such as O_2^- in the alkali halides, results in a preferential distribution of the defects in certain of the originall equivalent orientations.

The experimental parameters determined from these studies are relevan for several reasons. In some instances, one has a sufficiently detailed theo retical picture of the defect to have hopes of predicting the magnitudes o the effects of external fields; the experiment then becomes a test of thes predictions. In other cases, these coupling coefficients may enter as unde termined constants in a theoretical description of relaxation processes The experimental determination of the coupling constants is then require to test the success of the semiphenomenological theory of the relaxatio process. In the measurement of the effects of hydrostatic stress, not onl may both of the points above be relevant, but these results are also require to interpret the temperature dependence of the constants of the spin Hamiltonian as discussed in the following section.

Reichert and Pershan[49] have measured by ENDOR the electric-field induced shift in one component of the hyperfine coupling to the nearest neighbor positive ions of the *F*-center in KCl. The application of an elec tric field polarizes the *F*-center by admixture of a small amount of th excited *p* state of the *F* center. This results in a slight increase in electro density on the alkali nucleus on one side of the center, and a correspondin decrease on the other, whose magnitude may be calculated using the *F* center wave functions of Gourary and Adrian.[36] Although the Stark splitting of the ENDOR lines remain within the linewidth, the coefficient describin the change in density on the neighboring nuclei is experimentally determine from the changes in line shape to within 5%. The observed fractional shif in hyperfine coupling of 1/4% at a field of 56 kV/cm is in gratifying agree ment with a simple theoretical prediction. It will be interesting to see whethe this agreement is maintained if the results are extended to other crystals

Blum has measured the influence of hydrostatic pressure on the hyper fine coupling for *F*-centers in LiF[50] and for hydrogen atom interstitial im purities in CaF_2.[51] In both cases, one gains a reasonable picture of th magnitude of the effect by assuming that the bulk of the contact hyperfin interaction with the defect neighbors results from the orthogonalization o the defect function to the core function of the neighbors. The change in thi overlap with pressure then gives the observed change in hyperfine coupling The weakest point in the argument is the determination of the local io displacements assuming the local compressibility at the defect to be the sam as in the bulk. In CaF_2:H for instance, this assumption leads to discrepanc between experiment and the simple theoretical model of about 2.5. Onl a model which is able to estimate reliably the local ion displacements for given macroscopic stress would be expected to improve this agreement. It would be useful in this regard to analyze for the changes in anisotropic hyperfine and quadrupolar coupling constants as well.

A dramatic example of the influence of stress upon an EPR spectra is the *R*-center in KCl.[16,39] This center consists of three *F*-centers arranged in an equilateral triangle in a (111) crystal plane. The ground state has both a twofold spin degeneracy and a twofold orbital degeneracy. The strain field couples strongly to the orbitally degenerate states, the apparent g of the EPR depending sensitively upon the magnitude of the strain at the center unless the strain is large, in which case the first-order orbital contribution to the g is effectively quenched. In typical crystals, the distribution of random internal strains broadens the resonance beyond observability. The ground-state *R*-center resonance is observed only when an external uniaxial strain, large compared with internal strains, is applied to stabilize the g to a value characteristic of the large strain limit.

For defects, such as the V_K, whose symmetry is lower than that of the host crystal, the equivalence of the various orientations of the defect is removed in the presence of applied electric or uniaxial stress fields. For achievable stress fields, the differences in energies for different orientations is frequently of the order of a few to a few tens of degrees kelvin, while for electric fields, one may expect a lifting of the orientational degeneracy by the order of a few degrees kelvin for centers with large permanent electric dipole moments. For those centers which can rotate from one orientation to another in times less than or the order of minutes, the effect of the applied field is to change the relative populations of the different orientations. If the field splitting of the orientational degeneracy is large compared with the ambient temperature, one can obtain substantially complete alignment of the centers.

This field-induced alignment can be of help in the formulation of defect models through the dependence of the details of the alignment statistics upon the defect symmetry. This would be of particular relevance for a defect whose EPR spectrum shows little or no resolvable structure. In other cases, a very complex spectrum, where the multiplicity is due in part to the orientational degeneracy, may be greatly simplified by the alignment of substantially all of the defects into a single orientation. Conversely, the EPR technique is a very convenient means of monitoring the relative populations of the various orientations if the EPR spectrum of the individual center has substantial anisotropy. This technique has been applied to a detailed study of the reorientation kinetics of the O_2^- center in several alkali halides.[24]

5. DYNAMIC EFFECTS

Frequently, an EPR spectrum shows significant and occasionall marked changes as a function of temperature. A detailed understanding o these changes yields information about interactions with the phonon fiel or about reorientation or motion of the defect.

5.1. Spin–Lattice Relaxation

The most important of these effects is the spin–lattice relaxation proces which establishes thermal contact between the spins and the vibrationa degrees of freedom of the crystal. The problem of the spin–lattice relaxatio rate of the *F*-center has been treated extensively both theoretically and ex perimentally. The dominant coupling mechanism is the modulation of th hyperfine coupling by the phonons, and the theory of Kravchenko an Vinetskii,[52] using experimental hyperfine coupling constants and estimatin the coupling to the phonons using Gourary and Adrians wave functions gives quite reasonable agreement with the very long times, ~2 hr, measure by Feldman *et al.*[53] at helium temperature.

These results have recently been extended by Panepucci and Mollen auer[53a] using optical pumping techniques rather than EPR, to magneti fields ranging from 0 to 50 kG. The field dependence at high field implie relaxation via modulation of the crystal field and the small orbital admixtur to the magnetic moment rather than the modulation of the hyperfin coupling which dominates the relaxation at lower field (see pp. 393–394).

Since the general theoretical problem is essentially the same as for th usual paramagnetic ions, the T_1 problem will not be discussed here in detail.

5.2. Thermal Shifts

In the construction of the spin–Hamiltonian, one normally thinks o the defect as incorporated in a static environment, although in fact the ion are in constant motion, with the amplitude of this motion increasing wit increasing temperature. This vibrational motion may influence the constant which enter the Hamiltonian as well as providing a relaxation mechanism.

In principle, one can imagine determining the constants of the spin Hamiltonian as a function of the time-dependent positions $x_i(t)$ of al ions which interact with the defect,

$$\mathscr{H} = \mathscr{H}(\mathbf{x}_i(t)) \approx \mathscr{H}(\mathbf{x}_{i0}) + \sum_i (\partial\mathscr{H}/\partial\mathbf{x}_i) \cdot \delta\mathbf{x}_i(t) + \tfrac{1}{2}\sum_{ij} (\partial^2\mathscr{H}/\partial\mathbf{x}_i\,\partial\mathbf{x}_j) \cdot \delta\mathbf{x}_i(t)\,\delta\mathbf{x}_j(t) + \cdots \qquad (9$$

where $\mathbf{x}_{i0}$ are the low-temperature equilibrium positions and $\delta\mathbf{x}_i(t)$ the time-dependent displacements from that equilibrium. While the time-dependent parts of this Hamiltonian give rise to the spin–lattice relaxation of the defect, the resonance spectrum is determined by the time average of this Hamiltonian, since most of the time variation is rapid compared with the spin precession frequency. For an anharmonic crystal, both the linear and quadratic terms will contribute to the thermal shifts. The $\delta\mathbf{x}_i$ will have a nonzero average value related simply to the thermal expansion of the crystal, while the quadratic terms, of course, will have average values determined by the zero-point motion at absolute zero and proportional to temperature at high temperatures.

Some results are available from ENDOR data[46] on the temperature dependence of the hyperfine coupling constants for the *F*-center which are in reasonable accord with a fundamental theory of Kravchenko and Vinetskii.[54] Unfortunately, the effects are small and one has little more than an order-of-magnitude check on the theory.

Dreybrodt[55,56] has investigated very carefully the temperature variation of the hyperfine coupling for a number of *V*-centers, both intrinsic, F_2^- and Cl_2^-, and of the mixed type, FCl^- and FBr^-. Here, the contribution of the molecular vibrational coordinate is treated separately from the effects of the lattice modes and the relative contributions from each source separated experimentally by fitting to predicted temperature dependences. Because the hyperfine interactions are large, small fractional changes in hyperfine coupling give substantial shifts, ~10 G, in line position as the temperature is varied between helium and room temperature. From these results, Dreybrodt finds values for the molecular vibrational frequencies which are substantially lower (by two to eight times) than the free-molecular frequencies for the mixed centers, but are not greatly changed (~20%) for the intrinsic V_K-centers. Extensive discussion is given both of the reason for this frequency shift and of the mechanisms determining the dependence of the hyperfine coupling constants upon local distortions.

5.3. Rotation

Much more striking than the very small change in constants effected by the lattice vibrations are the dramatic changes in EPR spectra induced by rapid defect rotation. If the defect is able to reorient among several or all of the equivalent orientations described by the Hamiltonian $\mathscr{H}^i$, there are two simple limiting cases. If the reorientation frequency ω_R is much less than the frequencies characteristic of the differences between the eigenvalues of the different Hamiltonians $\mathscr{H}^i$, the spectrum will show separately the spectra associated with each of the $\mathscr{H}^i$ as discussed earlier. This is referred to as the static spectrum and is expected in the low-temperature limit. If ω_R is much larger than the orientational splittings, one sees a motionally averaged spectrum, or a spectrum described by the time-averaged Hamiltonian

$$\overline{\mathscr{H}} = (1/N) \sum_{i=1}^{N} \mathscr{H}^i \tag{10}$$

where N is the degree of orientational degeneracy. In the transition region between the static and motionally narrowed limits, the lines are typically broad, the details depending upon the particular nature of the Hamiltonians $\mathscr{H}^i$.

In NH_4Cl, the NH_4^+ ion is normally in the body-center position in a simple cubic cell with chlorine ions at the corners. The hydrogens are tetrahedrally arranged about the nitrogen and point toward four of the eight neighboring chlorine ions. At low temperature, an ordered structure is stable in which the NH_4 tetrahedra are regularly arranged so that each chlorine has just four hydrogen near neighbors. X-irradiation produces in this material a center whose very complex low-temperature spectrum (see Fig. 7a) is interpreted in terms of an NH_3Cl center.[57] During the irradiation, a hydrogen atom is removed; the resulting NH_3^+ ion then displaces toward that chlorine which was previously the neighbor of the missing hydrogen. Upon warming, the complex low-temperature spectrum loses its structure (see Figs. 7b and 7c) but at higher temperature, ~200°K, a well-defined structure (Fig. 7d) reemerges which shows full cubic symmetry. This spectrum may be neatly interpreted as a motionally averaged spectrum in which there is a tumbling motion of the NH_3 group so that it bonds first to one and then to another of the *four* chlorine ions toward which the hydrogen atoms were originally directed. It is clear from the analysis of the spectrum that the ordering of the surrounding is felt by the defect in as far as only four rather than all eight of the chlorine nuclei participate in the motional averaging.

A second, particularly dramatic, example of this motional averaging is the HCN^- molecule ion incorporated as an impurity in KCl.[27] A static spectrum is observed at liquid helium temperature showing low symmetry. It is interpreted in terms of the schematic diagram of Fig. 8 in which the bent HCN^- molecule lies in a $(1\bar{1}0)$ plane with the CN axis roughly along the [110] direction and the CN–CH bond angle is about 130°, placing the hydrogen interior to the cubic cell of the figure. Within any one of the six (110) planes, there are four possible orientations of the molecule, giving an overall 24-fold orientational degeneracy of the system.

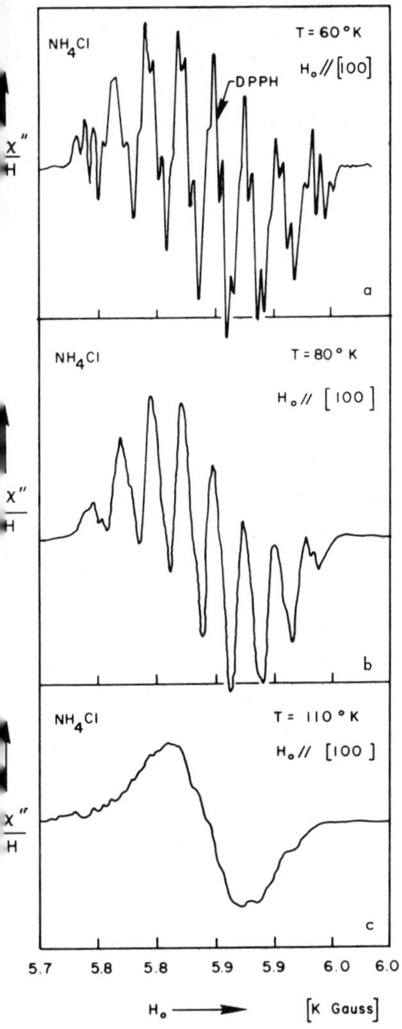

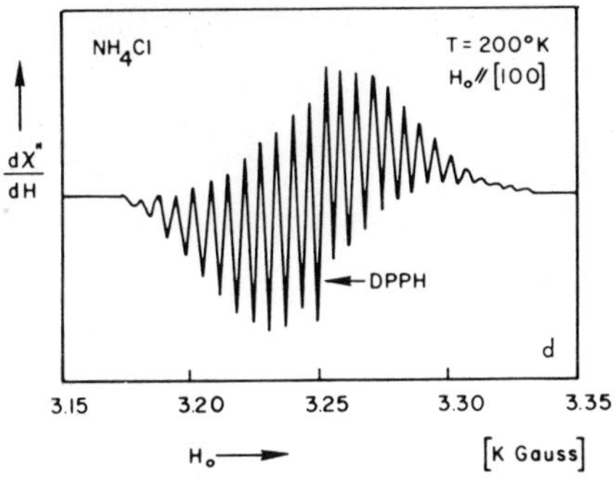

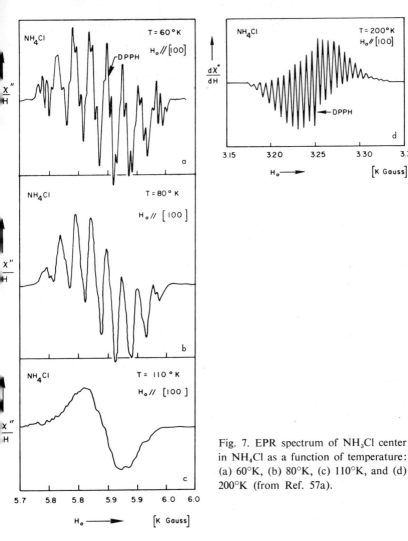

Fig. 7. EPR spectrum of NH_3Cl center in NH_4Cl as a function of temperature: (a) 60°K, (b) 80°K, (c) 110°K, and (d) 200°K (from Ref. 57a).

As the temperature is raised, a second well-defined spectrum is observed iear 40°K in which the *g* and hyperfine tensors have [111] symmetry. This pectrum is interpreted as a motional averaging of the low-temperature spectrum in which the molecule jumps rapidly among 3 of the 24 available orientations. For the defect of Fig. 8, the other two available orientations at 40°K would be those obtained by rotation of the molecule in that figure about the [111] axis either plus or minus 120°; one of these orientations is shown in Fig. 8. It is amusing that this seems to be a case of the tail wagging he dog, since in this restricted rotation, the CN axis changes direction by

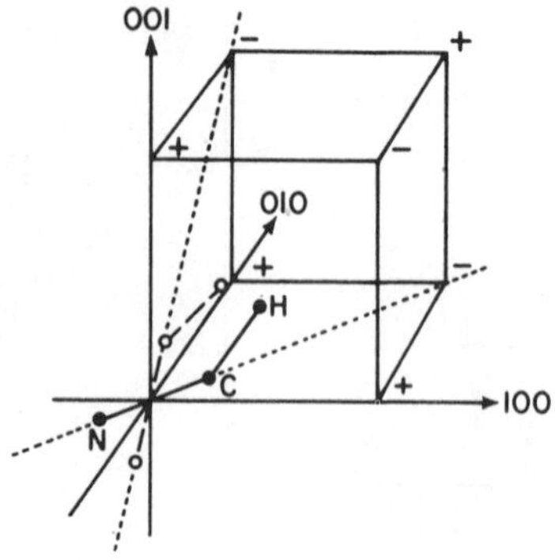

Fig. 8. Proposed model for the HCN^- center in KCl. The solid circles represent one of 24 possible configurations of the molecule. The open circles represent one of the other two configurations to which the original configuration may reorient at 40°K.

60°, involving considerable motion of the carbon and/or nitrogen, while the hydrogen stays substantially fixed in the interior of the same cell.

A further increase in temperature gives finally, at liquid nitrogen temperature, a fully isotropic spectrum, indicating that at this temperature, the molecule is free to jump rapidly among the full set of 24 possible orientations. Rough estimates of the two barriers for rotation may be made from observations of the line broadening as a function of temperature as the spectrum changes from one case to the next.

These examples involve classical, thermally activated motion of a defect among several possible orientations. A different problem arises if there is significant tunneling of the defect among the different orientations. Typically, the lowest eigenstate of the rotational motion is a symmetric linear combination of the localized states, with an associated EPR spectrum showing high symmetry. One is tempted to speak of an averaging by the tunneling. If higher pseudorotational states are thermally excited, these may have lower-symmetry wave functions and hence anisotropic EPR spectra. The narrowing of the NO_2 EPR in KCl[57a] is probably due to the depopulation of these states as the temperature is lowered from 4 to 2°K.

5.4. Translational Motion

In principle the onset of translational diffusion of defects should have consequences similar to those of rotation. Rapid diffusion of the *F*-center, for instance, should in principle result in a narrowing of the hyperfine contribution to the line width. Wolf[58] has studied the saturation behavior of the *F*-center resonance in KCl at temperatures up to 550°C. In this work, he observed the contribution of a thermally activated process to the apparent relaxation time of the *F*-center, but no significant changes in linewidth or line shape. The basic step in the diffusion of the *F*-center is the jump of a nearest-neighbor chlorine ion into an *F*-center, leaving an unpaired electron at the old chlorine site. In this process, the electron is transferred to a new nuclear environment and will contribute to a different portion of the inhomogeneously broadened line. This *F*-center diffusion process thus provides a particularly simple mechanism of spectral diffusion.[59] The rapid variation in relaxation time observed by Wolf is the consequence of this strongly temperature-dependent spectral diffusion process. An extrapolation based on these results indicates that narrowing of the line by this diffusion process should be observed only at temperatures higher than those of Wolf's experiments.

This measurement of the *F*-center diffusion rate should be clearly distinguished from those involving the observation of macroscopic drift or diffusion of *F*-center clouds. These latter experiments are dominated by processes in which an electron is thermally ionized from an *F*-center, drifts in the conduction band over distances large compared with a lattice constant, and is trapped at another negative-ion vacancy. These events happen rarely, but dominate the macroscopic diffusion because of the long step lengths involved. The EPR measurement, on the other hand, is sensitive to that step process which occurs most often, irrespective of step length, and that happens to be the short-step process involving the *F*-center chlorine ion interchange, at least in pure crystals. Thus the EPR measurement is able to characterize the rate of a diffusion process which is inaccessible to conventional techniques because it is masked by a separate mechanism.

6. DEFECT INTERACTIONS

The bulk of EPR data concerns the properties of individual defects or ions. A natural extension is to the study of defect interactions including both interactions among paramagnetic species and the influence of diamagnetic defects upon the spectra of the paramagnetic centers.

The greatest amount of work in this second area is in the study of the influence of charge-compensating defects upon the EPR spectrum of paramagnetic ions of valence different from that of the host crystal. The classic example of this type of investigation is that of Watkins,[30] who studied the Mn^{2+} spectrum in several alkali halides over a range of temperatures. The most extensive study was in NaCl, where four distinct spectra were observed corresponding to isolated manganese ions in a variety of surroundings. One expects charge compensation of the divalent ion in the monovalent alkali halide, for example, by pairing with a divalent anion impurity or with a cation vacancy.

Watkins interpreted two of the observed spectra as corresponding to manganese–cation vacancy pairs. One, with [110] symmetry, suggests a compensating vacancy in the nearest-neighbor position of the cation lattice; a second, with [100], suggested a vacancy in the next-nearest site on the cation lattice. The relative intensity of these two spectra gives the relative population of these two alternative sites for the vacancy. Since the cation vacancy still has some mobility at room temperature, one would expect thermal equilibrium to be established, in which case these relative populations give the difference in binding energies of the divalent ion–vacancy complex for the two configurations. This difference in binding energy is found to be small, only ~0.03 eV for three chlorides, and whereas in LiCl and NaCl, the nearest-neighbor [110] site is the more stable, in KCl it is the next-nearest neighbor site which is the more stable. These results are in qualitative agreement with detailed theoretical predictions, although the KCl result is in startling disagreement with simple arguments based on Coulomb energy alone. In addition to establishing plausible models for the static spectra, Watkins also observed the broadening of the lines with increasing temperature resulting from the motion of the vacancy, and was able successfully to correlate these dynamical effects on the EPR with measurements of the dielectric loss of the same material.

In addition to charge compensation by cation vacancies, another spectrum of low symmetry was noted which may have been associated with charge compensation by divalent anions. In this case, the nature of the defect giving the low-symmetry field at the manganese was not clearly established.

Finally, a comparison at a number of temperatures of the intensity of the spectra of the manganese–vacancy complexes with the intensity of an additional resonance showing full cubic symmetry, assumed that of an isolated Mn^{++} ion, gives a measure of the binding energy of these complexes. The measured value of 0.4 eV is in reasonable accord with theoretical estimates.

The EPR of the usual color centers may also be perturbed by the deviations of their surroundings from the normal structure. The F_A-center in the alkali halides is such an example in which one of the nearest-neighbor cations is replaced by a different alkali. The introduction of a sodium or lithium ion as nearest neighbor of an F-center in KCl, for instance, gives rise to a new resolved optical absorption band on the long-wavelength side of the normal F-band.[60,61] Because of the absence of resolved hyperfine structure in the F-center EPR in KCl, the effect on the EPR spectrum is not evident. Mieher,[62] however, has studied the ENDOR of the lithium F_A-center in KCl and obtained the hyperfine coupling of the electron to the lithium nucleus and to the two types of potassium nuclei, the one directly opposite the lithium, and the four in the plane parpendicular to the tetragonal symmetry axis of the F_A-center. The theoretical details of the distortions resulting from the presence of the lithium are not worked out; the ENDOR results imply that it will not be sufficient to consider simply distortions of the electron wave function, but that effects of ion displacements will also have to be taken into account.

Bass and Mieher[63] have also studied, using both EPR and ENDOR, the V_K-center in NaF as perturbed by a lithium ion substituted for one of the nearest-neighbor sodium ions of the center. Hyperfine data are presented for eight distinct nonequivalent nuclei which give a highly specific description of the distortion of the V_K-center produced by the substituted ion.

An example of strong interaction between pairs of paramagnetic defects is the F_t-center in CaO[64] and MgO.[65] This center consists of a pair of F-centers in next-nearest-neighbor oxygen sites probably with a cation vacancy in the intervening site. This center can also be thought of as a linear trivacancy oriented along the [100] direction with two trapped electrons. The spectrum is that expected for a pair of exchange-coupled F-centers with an exchange energy of $\sim$50°K in CaO and $\sim$80°K in MgO. The triplet state lies above the singlet,* so that at 4°K, no resonance is observed, while at higher temperatures, the $S = 1$ state gives a resonance with an axial field splitting of the triplet of somewhat over 300 G in both host crystals. This splitting is naturally explained as a consequence of the anisotropy of the dipole–dipole interaction between the two F-centers.

These examples represent cases where the interaction between defects is strong enough to give marked qualitative changes in the EPR or ENDOR spectra. There are numerous examples where the effect of interactions is observed only in changes in the linewidth of the resonance produced by magnetic, strain, or exchange coupling among defects. Stoneham[66] has treated in some generality the problem of broadening of resonance lines by defect interactions, using for illustrative purposes a number of defects in MgO crystals.

* See Fig. 11 in Chapter 5.

The particular problem of magnetic interactions in a system of randomly distributed dipoles has been treated by Kittel and Abrahams.[67] They predict a Lorentzian line shape, if this is the dominant broadening mechanism, with a linewidth proportional to the local defect concentration. Wyard[68] has applied these results to the study of the inhomogeneity of damage in electron-irradiated solid solutions of H_2O_2 and D_2O_2 in frozen H_2O and D_2O. The linewidth of the EPR in unannealed samples gives a measure of the local density of paramagnetic species as produced by the irradiation, while the linewidth after a mild anneal, under the assumption of a randomization without loss of the paramagnetic species, is proportional to the mean density. The ratio of these widths is then a measure of the extent to which the initial damage is localized to regions near the path of the ionizing particle. Although the data are not extensive, they seem consistent with the ideas of Kittel and Abrahams, and they demonstrate the feasibility of measuring local rather than average defect densities by careful studies of linewidth.

A qualitatively different behavior is noted in study of *F*-centers in KCl.[69–72] In freshly prepared samples, the EPR of the *F*-center is 46 G wide and shows a saturation behavior characteristic of an inhomogeneously broadened line, though not identical to the idealized case described by Portis.[73] If the crystal is exposed to light at room temperature, some of the *F*-centers are converted to *M*-, *R*-, and *N*-centers, the EPR of the remaining *F*-centers narrows to 35 G, and the saturation bahavior approaches more nearly that expected for a homogeneously broadened line. Although no entirely consistent picture has been proposed, these changes are probably due to the clustering of *F*-centers, the necessary mobility being provided by the optical excitation. The remarkable feature of these results is a well-defined stable final configuration in which further optical irradiation neither alters the relative concentrations of the *F*-centers and the various *F* aggregate centers, *R*, *M*, and *N*, nor influences the linewidth and saturation behavior of the *F*-center EPR. Schwoerer and Wolf[72] have proposed that a stable distribution develops in which the mean distance between *F*-centers is 4–5 interionic distances. The exchange interaction at this separation would then be sufficient to give some, but not extensive, narrowing of the hyperfine broadening and would similarly lead to a more homogeneous behavior of the line under saturation, as observed. Ehret and Wolf[74] have conducted a similar study of the *F*-center in RbCl, for which the hyperfine structure is partially resolved. In this case, the clustering leads to a loss of resolution of the structure and is suggestive of a smaller value of the exchange than in KCl, or perhaps simply broadening of the individual hyperfine components due to dipolar interactions. Although this final clustered state of the *F*-

centers, produced by optical bleaching, is not yet completely characterized, it may represent a useful system in which to study the dynamics of spectral diffusion.

7. CONCENTRATION DETERMINATIONS

The most useful results of the EPR technique derive from the qualitative and quantitative features of the spectrum without regard to the absolute intensity of the spectrum. Occasionally, however, measurements of the absolute intensity of the EPR or changes with external parameters of relative intensities are the crucial point of an experiment.

The most common experiment of this sort, involving measurements only of relative intensities, are those intended to establish the paramagnetic defect as the source of some other property of the crystal, typically an optical absorption band. The experiment is trivial in principle. The concentration of the paramagnetic defect is simply varied over as wide a range and by as many different mechanisms as possible, and an attempt made to correlate the changes in EPR intensity with the changes in the suspected optical absorption band.

Although trivial in concept, in practice, the results are often misleading and a number of incorrect conclusions have been reported in the literature. Broad EPR lines are not always distinguishable, nor are broad optical absorption bands. If two different species are contributing intensity to a single EPR line or optical band because of accidental superposition, it is easy to draw an incorrect negative conclusion in attempting to correlate intensity changes. On the other hand, it is fairly common in the annealing of centers that two centers will anneal simultaneously. This would occur, for instance, if the anneal consisted of a thermally activated transfer of an electron from one defect to another. If this simultaneous anneal occurs, it is all too easy to conclude incorrectly that the EPR of one of the two types of centers is to be associated with the optical absorption of the other. One's confidence in the use of EPR correlations in the assignment of models for the centers associated with optical bands obviously must depend upon the number of independent ways in which an apparent correlation is established.

The group at Argonne has clearly demonstrated that this technique can be reliably applied in their studies of a wide variety of systems including a number of *V*-centers[34,75,76] and the silver center in KCl,[29] which can be driven to either the Ag° or Ag^{++} paramagnetic ionization state. These results in turn extend one's picture of the center involved to the extent that the positions of the optical bands clarify the assignment of the energies of the various states of the center which contribute to the g-shift. In particular, for the V_K-center, two optical bands are observed corresponding to the transitions indicated in Fig. 5, the energies and relative intensities fitting in naturally with the level scheme proposed for the interpretation of the EPR spectrum.

This idea is often extended beyond a mere correlation of optical and EPR results to the use of relative EPR intensities to monitor the kinetics of photochemical reactions or other process involving paramagnetic species. Delbecq *et al.*[28] have studied in detail the photochemistry of a number of hydrogen centers, the U (a substitutional hydride ion), the U_1 (an interstitial hydride ion), and the U_2 (an interstitial hydrogen atom), and of the $U \rightarrow F$ conversion process in KCl, using combined EPR and optical results.

In work on the O_2^- ion in KCl, KBr, and KI, Känzig[24] has used the relative intensities of EPR lines to monitor the relative populations of the different orientations of the O_2^- molecule. When the thermal distribution is altered by changing an applied stress field, the detailed kinetics of the reorientation process may be conveniently studied by following the time variation of the different line intensities. The dependence of this reorientation rate upon both stress and temperature from 1.7 to 4.2°K is in moderate[77] but not complete agreement with theoretical calculations.[78,79] The reorientation mechanism in the low-temperature limit seems to be established as the process suggested by Sussman, which involves a tunneling of the molecule from one orientation to the other, with the associated absorption or emission of a phonon.

Occasionally, the absolute measurement, of an EPR signal intensity is used to determine an absolute rather than relative concentration of the defect. Because the theoretical treatment of the EPR absorption strength is straightforward, compared, say, with optical absorption, for which there are ambiguities both in local field corrections and in the calculation of the optical oscillator strengths, in principle, the EPR method gives a direct measurement of center concentration. In practice, uncertainties in field geometry within the microwave cavity and sometimes uncertainties in the contribution of wing intensities to a resonance, a particularly severe problem for Lorentzian-shaped lines, may limit the precision of the method to 10 or 20%. It is usual to measure the unknown intensity relative to that of a material with a known paramagnetic ion concentration, such as a paramagnetic salt, although one can in fact conveniently calibrate a spectrometer sensitivity with fair precision by measuring the response to a known frequency modulation of the microwave source, which response is simply related to the dispersion sensitivity. Such a calibration does not, for instance, require either a knowledge of the cavity Q nor of the absolute detector sensitivity. The optical oscillator strength of the F-band in KCl and NaCl has been determined by a comparison of the EPR and optical line strengths,[80] in reasonable accord ($\sim 5\%$) with most of the results which measured the center concentration using chemical methods or static magnetic susceptibility.

8. COLOR CENTERS AS MODEL RESONANCE SYSTEMS

There have been a number of EPR studies which do not fall neatly into the classifications above but which are worthy of mention because they illustrate an important point, that often color centers form useful model systems with which to test or illustrate ideas of resonance physics.

8.1. Inhomogeneous Broadening and Spectral Diffusion

The F-center, for instance, has long been the classic example of the inhomogeneously broadened line and there have been extensive studies of its saturation properties in an attempt to test the simple ideas of Portis[73] concerning the saturation of such a system. The Portis model is a set of spins characterized by longitudinal and transverse relaxation times T_1 and T_2 and a distribution of local fields as determined via hyperfine interactions by the distribution of possible orientations of the neighboring nuclear moments. The single, broad resonance is thought of as the unresolved superposition of a multitude of lines, each with a width characterized by the parameter T_2. The predicted saturation behavior is that shown in Fig. 9, the value of H_1 at which the knee occurs giving a value of the product T_1T_2.

The observed behavior deviates from the theory as shown by Fig. 9. A number of investigations have attempted to measure T_1 and T_2 independently and to determine the cause of this discrepancy. These include direct measurements of T_1 at low temperatures by Feldman *et al.*,[53] indirect measurements of T_1 by Blumberg[81] at room temperature, as well as low-temperature transient[82,83] and steady-state experiments[84] at room temperature designed to measure the width and shape of the individual unresolved resonance lines. These latter experiments involve measurements of the shape

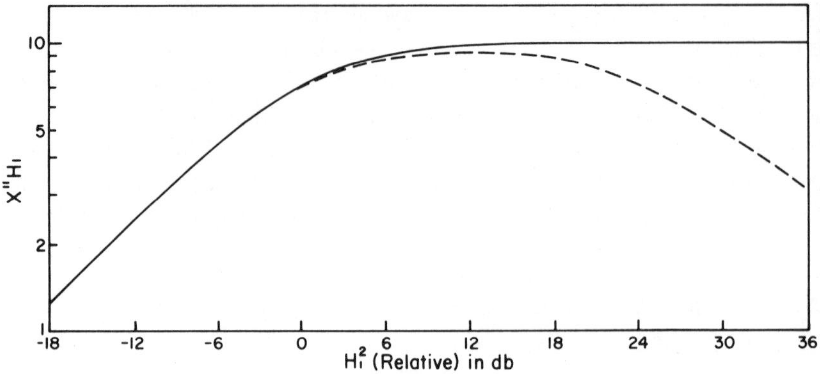

Fig. 9. Saturation plots for an inhomogeneously broadened line: solid curve—theory of Portis; dashed curve—experimental results for the *F*-center in KCl.

of the hole burned in the broad *F*-center line by a saturating microwave field. The steady-state experiments of Moran revealed a complex shape for this saturated region of the line and demonstrated clearly for the *F*-center in KCl the importance of a spectral diffusion mechanism involving the rapid relaxation of certain of the nearby nuclei and of including contributions to the line intensity of partially forbidden transitions involving simultaneous nuclear and electronic spin flips. The approximate agreement of the two curves of Fig. 9 is in fact due in large measure to the contribution of these forbidden transitions. Using an effective T_2 deduced from these experiments, Moran was able to extract a value of T_1 in good accord with the extrapolation of Feldman *et al.'s* low-temperature T_1 results. The results were also in agreement with conclusions drawn from transient observations of the behavior of the "saturated hole" by Seidel[82] and Noble.[83]

Two additional mechanisms of spectral diffusion were identified by Wolf[58], in the experiments already discussed, at high temperatures. The spatial diffusion of the *F*-center by interchange with a neighboring chlorine ion provides the unpaired electron with a new environment of nuclear spins, the hyperfine field in the new position being almost independent of the original. This jump then "scatters" the electron to a new position in the inhomogeneous line. The same effect may be achieved by exchange collisions with thermally excited conduction electrons if these conduction electrons do not experience spin–lattice relaxation during the time between successive exchange collisions with *F*-center electrons. These spectral diffusion mechanisms are particularly easy to include in a generalization of the Portis theory of saturation and the resulting predictions of saturation behavior are in excellent agreement with the experimental results.

8.2. Rapid Passage Effects

EPR line shapes in the limit of low microwave power are determined by the interaction of the paramagnetic species with its environment, including the lattice vibrations, nearby nuclear moments, and elastic strain fields. At power levels sufficient to saturate the resonance, these line shapes become distorted in a variety of ways depending on the relative magnitude of a number of experimental parameters involving the homogeneous and inhomogeneous widths of the resonance, the T_1 of the magnetic species, the frequency and magnitude of the modulation field normally used to display the resonance, and the magnitude of the microwave field.

These distortions in line shape in EPR were discussed briefly by Portis[73] in the first papers on the *F*-center resonance and then later in more detail using again measurements on the *F*-center as illustrative examples.[85] A more complete summary of these effects has been given by Weger.[86] Extensions of these ideas provide means for measuring the mean relaxation rate T_1T_2.[87,88]

8.3. Jahn–Teller Effect

In defect or impurity systems with orbital degeneracy of the ground state, the energy of the system can be lowered by a distortion of the defect or its environment which lifts the orbital degeneracy. This tendency of degenerate systems to distort to lower symmetry was first noted for molecular systems by Jahn and Teller, but most of the discussions for molecules are relevant as well to defects or impurities in solids.[89]

If the lowering of the energy by this distortion is comparable with the vibrational energies of the system, the case of the dynamical Jahn–Teller effect, the coupled electronic and vibrational motion will be rather complex. A particularly simple model for such a system was treated theoretically by Longuet-Higgins *et al.*,[90] but without reference to a specific physical system.

The *R*-center in the alkali halides, a cluster of three *F*-centers in a triangular array, has orbital degeneracy in the ground state and should be susceptible to the Jahn–Teller effect. The expected behavior is indeed observed and the strength of the Jahn–Teller coupling is appropriate to the interesting dynamical range. The work of Krupka[39] on the EPR of the *R*-center is nicely consistent with the appropriate extension of the Jahn–Teller theory and in particular shows dramatically the suppression of the orbital moment of the electron by the dynamic interaction of the electron with the displacements of the neighboring ions.[89] (See the chapter in this volume by F. Ham for other examples similar to the *R*-center.)

8.4. Optical–EPR Experiments

The combined use of optical and EPR techniques is standard in color-center problems insofar as optical absorption is often used as a convenient monitor of relative concentrations of various defects, or optical irradiation is used to alter the relative concentrations of different centers or of different ionization states of the same center through various photochemical processes. There are some experiments, however, in which the optical and EPR techniques are more intimately connected.

One such experiment is in the study of excited metastable states of color centers. In centers containing more than a single electron, such as the M (two electrons) or the R (three electrons), there may be a series of states of spin multiplicity higher than that of the ground state. Indeed, for the M-center, there is a series of triplets and for the R, of quartets. Optical irradiation in the UV allows one to establish rather large populations in the lowest of these series which have quite long life times, on the order of a minute, against decay to the ground state. Seidel *et al.*[15] have observed both the EPR and the ENDOR of these metastable states, the triplet of the M and the quartet of the R, with results confirming the picture of these centers as a pair and a triple of F-centers, respectively.

An amusing effect is observed in similar experiments on M-like centers in CaO[64] in which the shorter lifetime of the triplet state, $\sim$10 msec, requires continuous illumination of the sample to maintain a significant population of the triplet state at liquid helium temperature. The surprising result is that some of the EPR lines are observed in emission rather than absorption the required preferential populations being provided by the selection rules for the transition from the singlet to triplet manifolds. For the applied field parallel to the tetragonal axis of the center, for instance, only the $M_s = 0$ state is populated in the decay from the optically excited state, so that the $M_s = 0 \rightarrow +1$ transition is seen in absorption, the $M_s = 0 \rightarrow -1$ is seen in emission. The fluorescence lifetime is assumed short compared to the triplet state T_1 so that the triplet-state populations are determined entirely by the dynamics of the pumping process. The selection rules required to give the observed populations are consistent with predictions based upon the defect symmetry. Similar effects have been observed by Schwoerer and Wolf by optical pumping into the triplet state of certain organic molecules[89a]. See Section 3.2.1 and Fig. 11 in the chapter by S. Geschwind in this volume for a discussion of these selection rules.

Mollenauer, *et al.*[91] have performed a very sophisticated optical–microwave double resonance experiment in which the spin resonance of the relaxed excited state (fluorescence lifetime $\sim 10^{-6}$ sec) of the *F*-center is observed in KCl, KBr, and KI. The technique takes advantage of the spin memory in the optical excitation and fluorescence cycle of the *F*-center. By exciting with circularly polarized light, the ground state may be optically pumped, taking advantage of an unresolved spin–orbit splitting of the unrelaxed excited state to couple the angular momentum into the spin system. The degree of spin polarization in the ground state is determined by monitoring the magnitude of the magnetic circular dichroism as measured with an optical monitor. Because of the spin memory through the excitation–luminescence cycle, the spin polarization of the ground state may be altered by inducing transitions among the spin states with a microwave field at any point of the cycle. In particular, driving the spin resonance in the excited state will alter the ground-state spin polarization. Further details may be found in Section 4.3 of the chapter by S. Geschwind in this volume.

Mollenauer *et al.* have used this technique to discover the EPR of excited state of the *F*-center. The linewidth of the excited-state resonance is from one to three times greater than for the ground state as one goes from KCl to KBr to KI. The *g*-shift is from four to ten times that of the ground state and is of the same sign.

The detailed nature of this relaxed excited state remains unclear. The *g*-shifts, in KCl and KBr at least, are small enough to suggest that this excited state may be of predominantly *S* symmetry, in agreement with conclusions of Fitchen *et al.*[92,93] based on measurements of electric and magnetic field induced dichroism of the *F* emission spectrum. However, a naive attempt to reconcile the *g*-shift measured by Mollenauer with the parameters proposed by Fitchen for the excited state is not successful.

REFERENCES

1. J. H. Schulman and W. D. Compton, *Color Centers in Solids*, Macmillan, New York (1962).
2. W. B. Fowler, *Physics of Color Centers*, Academic Press, New York (1968).
3. G. Bacquet, J. Dugas, and P. Gautier, *J. Phys.* **27**, 97 (1966).
4. H. Seidel and H. C. Wolf, *Phys. Status Solidi* **11**, 3 (1965).
5. T. O. Woodruff and W. Känzig, *J. Phys. Chem. Solids* **5**, 268 (1958).
6. T. G. Castner and W. Känzig, *J. Phys. Chem. Solids* **3**, 178 (1957).
7. F. K. Kneubühl, *Phys. Kondens. Materie* **1**, 410 (1963).
8. G. F. Koster and H. Statz, *Phys. Rev.* **113**, 445 (1959).
9. A. Bieri and F. K. Kneubühl, *Phys. Kondens. Materie* **4**, 230 (1965).
10. R. Mieher and R. Gazinelli, *Phys. Rev. Letters* **12**, 644 (1964).

11. C. A. Hutchison, *Phys. Rev.* **75**, 1769 (1949).
12. A. F. Kip, C. Kittel, R. A. Levy, and A. M. Portis, *Phys. Rev.* **91**, 1066 (1953).
13. J. E. Wertz, J. W. Orton, and P. Auzins, *Disc. Faraday Soc.* **1961**, 140.
14. J. E. Wertz, P. Auzins, R. A. Weeks, and R. H. Silsbee, *Phys. Rev.* **107**, 1535 (1957).
15. H. Seidel, M. Schwoerer, and D. Schmid, *Z. Phys.* **182**, 398 (1965).
16. D. C. Krupka and R. H. Silsbee, *Phys. Rev. Letters* **12**, 193 (1964).
17. W. Känzig, *Phys. Rev.* **99**, 1890 (1955).
18. W. Känzig and T. O. Woodruff, *J. Phys. Chem. Solids* **9**, 70 (1958).
19. J. W. Wilkins and J. R. Gabriel, *Phys. Rev.* **132**, 1950 (1963).
20. M. L. Meistrich and L. S. Goldberg, *Solid State Commun.* **4**, 469 (1966).
21. W. Dreybrodt and D. Silber, *Phys. Status Solidi* **16**, 215 (1966).
22. D. Schoemaker, *Phys. Rev.* **149**, 693 (1966).
23. W. Sander, *Z. Phys.* **169**, 353 (1962).
24. W. Känzig, *J. Phys. Chem. Solids* **23**, 479 (1962).
25. (a) A. Hausmann, *Z. Physik* **192**, 313 (1966); (b) K. D. J. Root, M. C. R. Symons, and B. C. Weatherly, *Mol. Phys.* **11**, 161 (1966).
26. E. L. Cochran, B. C. Weatherly, V. A. Bowers, and F. J. Adrian, *Bull. Am. Phys. Soc.* **II13**, 357 (1968), and to be published.
27. S. Othmer and R. H. Silsbee, *Bull. Am. Phys. Soc.* **II13**, 661 (1968), and to be published.
28. C. J. Delbecq, B. Smaller, and P. H. Yuster, *Phys. Rev.* **104**, 599 (1956).
29. C. J. Delbecq, W. Hayes, M. C. M. O'Brien, and P. H. Yuster, *Proc. Roy. Soc.* **A271**, 243 (1963).
30. G. W. Watkins, *Phys. Rev.* **113**, 79, 91 (1959).
31. F. K. Kneubühl, *Helv. Phys. Acta* **38**, 358 (1965).
32. C. P. Slichter, *Principles of Magnetic Resonance*, Harper and Row, New York (1963), pp. 179–189.
33. A. J. Stone, *Proc. Roy. Soc. London* **A271**, 424 (1963).
34. C. J. Delbecq, W. Hayes, and P. H. Yuster, *Phys. Rev.* **121**, 1043 (1961).
35. B. Kojima, *J. Phys. Soc. Japan* **12**, 918 (1957).
36. B. S. Gourary and F. J. Adrian, *Phys. Rev.* **105**, 1180 (1957); *Solid State Phys.* **10**, 127 (1960).
37. F. J. Adrian, *Phys. Rev.* **107**, 488 (1957).
38. D. Schmid, *Phys. Stat. Sol.* **18**, 653 (1966).
39. D. C. Krupka and R. H. Silsbee, *Phys. Rev.* **152**, 816 (1966).
40. E. Gelerinter and R. H. Silsbee, *J. Chem. Phys.* **45**, 1703 (1966).
41. J. Mort, F. Lüty, and F. C. Brown, *Phys. Rev.* **137**, A566 (1965).
41a. R. W. Bené, *Phys. Rev.* **178**, 497 (1969).
42. A. Abragam and M. H. L. Pryce, *Proc. Phys. Soc.* **A205**, 135 (1951).
43. H. R. Zeller and W. Känzig, *Helv. Phys. Acta* **40**, 845 (1967).
44. T. P. Das, A. N. Jette, and R. S. Knox, *Phys. Rev.* **134**, A1079 (1964).
45. W. C. Holton and H. Blum, *Phys. Rev.* **125**, 89 (1962).
46. H. Seidel, *Z. Phys.* **165**, 218, 239 (1961).
47. R. F. Wood and J. Korringa, *Phys. Rev.* **123**, 1138 (1961).
48. T. E. Feuchtwang, *Phys. Rev.* **126**, 1616, 1628 (1962).
49. J. F. Reichert and P. S. Pershan, *Phys. Rev. Letters* **15**, 780 (1965).
50. H. Blum, *Phys. Rev.* **140**, A1998 (1965).

51. H. Blum, *Proc. 1st Internat. Conf. on Physics of Solids at High Pressures, Tucson, 1965*, Academic Press, New York, p. 409.
52. V. Ya. Kravchenko and V. L. Vinetskii, *Fiz. Tverd. Tela* **6**, 2075 (1964); *Soviet Phys.—Solid State* **6**, 1638 (1965).
53. D. F. Feldman, R. W. Warren, and J. Castle, *Phys. Rev.* **135**, A470 (1964).
53a. H. Panepucci and L. F. Mollenauer, *Phys. Rev.* **178**, 589 (1969).
54. V. Ya. Kravchenko and V. L. Vinetskii, *Opt. i Spektroskopiya* **18**, 73 (1965); *Opt. Spectry. (USSR)* **18**, 37 (1965).
55. W. Dreybrodt, *Phys. Letters* **19**, 274 (1965).
56. W. Dreybrodt, *Phys. Status Solidi* **21**, 99 (1967).
57. (a) L. Vannotti, H. R. Zeller, K. Bachmann, and W. Känzig, *Phys. Kondens. Materie* **6**, 51 (1967); (b) F. W. Patten, *Phys. Letters* **21**, 277 (1966).
57a. J. R. Brailsford and J. R. Morton, *J. Mag. Res.* **1**, 517 (1969).
58. E. L. Wolf, *Phys. Rev.* **142**, 555 (1966).
59. A. M. Portis, *Phys. Rev.* **104**, 584 (1956).
60. F. Lüty, *Z. Phys.* **165**, 17 (1961).
61. K. Kojima, N. Nishimaki, and T. Kojima, *J. Phys. Soc. Japan* **16**, 2033 (1961).
62. R. L. Mieher, *Phys. Rev. Letters* **8**, 362 (1962).
63. I. L. Bass and R. Mieher, *Phys. Rev. Letters* **15**, 25 (1965).
64. D. H. Tanimoto, W. M. Ziniker, and J. C. Kemp, *Phys. Rev. Letters* **14**, 645 (1965).
65. B. Henderson, *Brit. J. Appl. Phys.* **17**, 853 (1966).
66. A. M. Stoneham, *Proc. Phys. Soc.* **89**, 909 (1966).
67. C. Kittel and E. Abrahams, *Phys. Rev.* **90**, 238 (1953).
68. S. J. Wyard, *Proc. Phys. Soc.* **86**, 587 (1965).
69. H. Gross, *Z. Phys.* **164**, 341 (1961).
70. P. R. Moran, S. H. Christensen, and R. H. Silsbee, *Phys. Rev.* **124**, 442 (1961).
71. W. E. Bron, *Phys. Rev.* **125**, 509 (1962).
72. M. Schwoerer and H. C. Wolf, *Z. Phys.* **175**, 457 (1963).
73. A. M. Portis, *Phys. Rev.* **91**, 1071 (1953).
74. P. Ehret and H. C. Wolf, *Phys. Status Solidi* **15**, 239 (1966).
75. C. J. Delbecq, B. Smaller, and P. H. Yuster, *Phys. Rev.* **111**, 1235 (1958).
76. C. J. Delbecq, J. L. Kolopus, E. L. Yasaitis, and P. H. Yuster, *Phys. Rev.* **154**, 866 (1967).
77. R. H. Silsbee, *J. Phys. Chem. Solids* **28**, 2525 (1967).
78. J. A. Sussman, *Phys. Kondens. Materie* **2**, 146 (1964).
79. R. Pirc, B. Zeks, and P. Gosar, *J. Phys. Chem. Solids* **27**, 1219 (1966).
80. R. H. Silsbee, *Phys. Rev.* **103**, 1675 (1956).
81. W. E. Blumberg, *Phys. Rev.* **119**, 1842 (1960).
82. H. Seidel, *Int. Symp. über Farbzentren in Alkalihalogeniden, Stuttgart 1962* (unpublished).
83. G. A. Noble, *Phys. Rev.* **118**, 1028 (1960).
84. P. R. Moran, *Phys. Rev.* **135**, A247 (1964).
85. A. M. Portis, *Phys. Rev.* **100**, 1219 (1955).
86. M. Weger, *Bell Syst. Tech. J.* **39**, 1013 (1960).
87. J. S. Hyde, *Phys. Rev.* **119**, 1483 (1960).
88. A. A. Bugai, *Soviet Phys.—Solid State* **5**, 700 (1963).
89. M. D. Sturge, *Solid State Phys.* **20**, 92 (1967).

89a. M. Schwoerer and H. C. Wolf, *Proc. XIVth Colloque Ampère, 1966*, ed. by R. Blinc, North-Holland Publ. Co., Amsterdam (1967).
90. H. C. Longuet-Higgins, U. Öpik, M. H. L. Pryce, and R. A. Sack, *Proc. Roy. Soc.* (*London*) **244**, 1 (1958).
91. L. F. Mollenauer, S. Pan, and S. Yngvesson, *Phys. Rev. Letters* **23**, 683 (1969).
92. L. D. Bogan, L. F. Stiles, Jr., and D. B. Fitchen, *Phys. Rev. Letters* **23**, 1495 (1969).
93. M. P. Fontana and D. B. Fitchen, *Phys. Rev. Letters* **23**, 1497 (1969).

Chapter 8

Covalent Effects in EPR Spectra—Hyperfine Interactions*

E. Šimánek[†] and Z. Šroubek[‡]

Department of Physics
University of California
Los Angeles, California

1. INTRODUCTION

Considerable progress has been made in recent years on the understanding of covalent bonding effects in transition metal salts and complexes. Electron paramagnetic resonance (EPR) measurements have now been made on a great variety of magnetic ions in diamagnetic insulating crystals revealing the effects of covalent bonding between the magnetic ion and ligands. We shall not attempt to give a systematic description of all possible covalent effects in EPR because this material has been covered in the recent review by Owen and Thornley[1]; instead we will be primarily concerned with the effects of covalency on the hyperfine interaction. It is the latter which is most directly linked to the unpaired electron spin density and thus gives us information on the actual wave function of the magnetic ion in a crystal.

The term covalency generally refers to the effects of electron transfer between the magnetic ion and its diamagnetic surroundings. The electron

* Supported in part by the National Science Foundation and the Office of Naval Research, Grant No. NONR 233(88).

† Present address: Department of Physics, University of California Riverside, California 92502.

‡ On leave from the Institute of Radio Engineering and Electronics, Czechoslovak Academy of Sciences, Prague, Czechoslovakia.

transfer from the ligand to the half-filled shell of the magnetic cation is the most important transfer mechanism for producing an unpaired spin density at the originally diamagnetic ligand. The interaction of the unpaired electrons with the nuclear moment of the ligand gives rise to an anomalous hyperfine structure in the EPR spectrum, known usually as superhyperfine structure (shfs). The observation of this shfs in transition metal complexes and salts[2,3] has been regarded as one of the most convincing pieces of evidence for the presence of covalent charge transfer.

However, it must be emphasized that the nonorthogonality between the magnetic and ligand electrons also contributes to shfs. This mechanism is called overlap or Pauli distortion effect and it has been first considered by Gourary and Adrian[4] in their theory of *F*-centers. It turns out that the theoretical expression for shfs contains both overlap and charge transfer effects mixed up in an indistinguishable way. In view of this fact, shfs cannot be directly linked to the covalency in the pure sense of electron charge transfer. Also, covalent effects on other types of hyperfine interaction, discussed below, show a similar entanglement of overlap and charge transfer. Throughout most of this chapter, we shall adopt the configuration interaction[5] method in the framework of the Heitler–London model to calculate the unpaired spin densities involved in the various hyperfine effects. The task before us is not to attempt an *a priori* calculation of the degree of covalency. Rather, we wish merely to establish the links between the experimentally observed hyperfine parameters and the electronic state of the system by regarding the degree of covalency as an empirical parameter.

We begin Section 2 with a short description of the idea of the overlap effect in a purely ionic system. We then illustrate this effect by taking a simple example of magnetic system; the hydrogen atom in CaF_2 crystal. The previous treatment of this case[6] is improved by taking into account the ligand–ligand overlaps.

In Section 3, we discuss the effects of electron charge transfer on the distribution of unpaired spin density in a two-atom (metal–ligand) model. Three basic types of transfer are considered:

1. Ligand → metal transfer, mentioned above. The configuration interaction approach is used and the results are related to the molecular orbital description. The well-known example[7] of shfs in Ni^{2+}:$KMgF_3$ is discussed.

2. The second type of transfer is the metal–ligand transfer, involving empty ligand orbital (back-bonding).

3. Finally, we consider the ligand–metal transfer involving empty metal orbitals. The latter are split by the exchange interaction with the magnetic

electron and we are therefore dealing with an "exchange polarized transfer." This process is illustrated by an example[7] of isotropic shfs in $Cr^{3+}:KMgF_3$.

In Section 4, we consider the effects of covalency on the central hyperfine interaction parameter A of Mn^{2+} in various hosts. The experimentally observed decrease of A with the degree of covalency is interpreted in terms of a model which improves the previous treatment[8] by taking into account the effect of exchange on the radial $4s$ function involved in the charge transfer excited state. In addition to this, the $4s$ contribution to A is significantly amplified by including the overlaps between ligands and inner s shells of Mn^{2+}. The dependence of A on metal–ligand distance is also discussed.

Section 5 is devoted to the problem of transmission of unpaired spin from the magnetic cation through ligand to the diamagnetic cation. This process gives rise to a hyperfine interaction with the nucleus of the diamagnetic cation which is known as cation–cation or "supertransferred" hyperfine interaction. As an example, we discuss recent ENDOR studies[9] of ^{27}Al hyperfine interaction in $Fe^{3+}:LaAlO_3$ and $Cr^{3+}:LaAlO_3$.

Finally, we shall mention some problems arising in the first-principle calculations of the effects discussed above.

2. OVERLAP EFFECT IN AN IONIC MODEL

2.1. Unpaired Spin Density in a Three-Electron System

The basic idea of the overlap effect can be described using a simple two-center system with three electrons $m\uparrow$, $l\uparrow$, and $l\downarrow$ as shown in Fig. 1. Here, $m\uparrow$ is the magnetic spin-orbital with spin up, and $l\uparrow$ and $l\downarrow$ are the ligand spin-orbitals.

The overlap effect produces a redistribution of the charge density of the electrons with parallel spin ($m\uparrow$, $l\uparrow$) resulting in an unpaired spin density

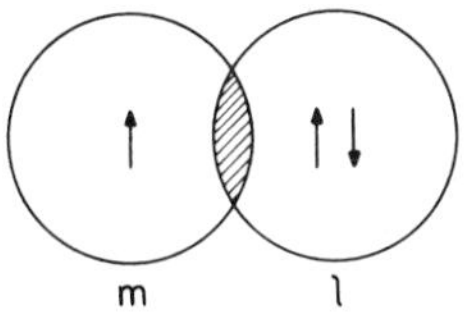

Fig. 1. The orbitals of a magnetic atom (ion) and a ligand are shown. The electrons $m\uparrow$ and $l\uparrow$ keep out of the overlap region (shaded area) by the action of Pauli exclusion principle.

at the nominally closed shell ($l\uparrow$, $l\downarrow$). It will clarify the matters if we first consider the subsystem $m\uparrow$, $l\uparrow$ described by a normalized Slater determinant

$$\Psi(1,2) = \frac{1}{[2(1-\langle m \mid l\rangle^2)]^{1/2}} \{m\uparrow, l\uparrow\} \tag{1}$$

where $\langle m \mid l\rangle$ is the overlap integral and the curly brackets indicate a determinantal wave function. The electron charge density of this subsystem is

$$\begin{aligned} \varrho(1) &= 2\int \Psi^2(1,2)\, d\tau_2 \\ &= \frac{1}{1-\langle m \mid l\rangle^2} [m^2(1) + l^2(1) - 2\langle m \mid l\rangle m(1)l(1)] \end{aligned} \tag{2}$$

This result differs from the sum of the charge densities of free constituents, $m^2(1) + l^2(1)$, in two respects:

1. The overlap charge with density $2\langle m \mid l\rangle m(1)l(1)$ is removed from the overlap region, where the Pauli exclusion principle is violated.
2. This charge is redistributed so that the charge density $m^2(1) + l^2(1)$ is enhanced elsewhere by a factor $1/(1-\langle m \mid l\rangle^2)$.

It is important to note that (2) may be derived alternately[4] from $\Psi(1, 2)$ constructed with the use of orthogonalized one-electron orbitals. The usual procedure is to orthogonalize the magnetic orbital by the Schmidt process. The determinant (1) thus becomes

$$\Psi(1,2) = \frac{1}{\sqrt{2!}} \left\{ \left(\frac{m - \langle m \mid l\rangle l}{(1-\langle m \mid l\rangle^2)^{1/2}} \right)\uparrow, l\uparrow \right\} \tag{3}$$

With determinant (3), the total electron charge density is a sum of the one-electron contributions, giving us the result in Eq. (2). The charge density due to $l\downarrow$ is

$$\varrho_\downarrow(1) = l^2(1) \tag{4}$$

Subtracting (4) from (2), we find for the difference between charge densities of electrons having upward and downward spins, respectively,

$$\varrho_\uparrow(1) - \varrho_\downarrow(1) = \frac{m^2(1)}{1-\langle m \mid l\rangle^2} + \langle m \mid l\rangle^2 l^2(1) - 2\langle m \mid l\rangle m(1)l(1) \tag{5}$$

Expression (5) determines the unpaired spin density, which can be measured around the nucleus of the magnetic ion or ligand by means of hyperfine interaction.

We shall now show that the problem can be reduced to that of three electrons ($m\uparrow$, $l\uparrow$, $l\downarrow$) if the ligand $2p_\sigma$ orbitals are combined to form molecular orbitals belonging to the representations of the cubic group. Since the magnetic orbital is a $1s$ hydrogen orbital, we have to consider the molecular orbital belonging to the A_{1g} representation. All other orbitals give zero overlap with the magnetic orbital and can be omitted. The properly normalized A_{1g} ligand orbital is

$$l = (p_1 + p_2 + \cdots p_8)/[8(1 + \sum_{j=2}^{8} \langle p_1 \mid p_j \rangle)]^{1/2} \tag{6}$$

where $p_1, \ldots, p_8$ are the $2p_\sigma$ orbitals shown in Fig. 2, and $\langle p_1 \mid p_j \rangle$ is one of the ligand–ligand overlaps.

In view of the multicenter form of (6), $\langle m \mid l \rangle$ becomes what is called the group overlap integral. Using (6), we can express $\langle m \mid l \rangle$ in terms of the atomic overlap $\langle m \mid p_1 \rangle$ as follows

$$\langle m \mid l \rangle = [8/(1 + R)]^{1/2} \langle m \mid p_1 \rangle \tag{7}$$

where

$$R = \sum_{j=2}^{8} \langle p_1 \mid p_j \rangle \tag{8}$$

We see from (5) that the overlap $\langle m \mid l \rangle$ has two basic consequences for the distribution of unpaired spin in a purely ionic system described by a Heitler–London model:

1. The spin density of the magnetic ion is enhanced compared to the free-ion value by a factor $1/(1 - \langle m \mid l \rangle^2)$.

2. The originally closed shell ($l\uparrow$, $l\downarrow$) has an unpaired spin density $\langle m \mid l \rangle^2 l^2(1)$. It is to be noted that this is not true in the overlap region, where the positive spin density is counteracted by the negative overlap term.

2.2. *Hyperfine Interactions in* $H{:}CaF_2$ *System*

In a real crystal, the magnetic ion is usually surrounded by a number of ligands and the calculation of the unpaired spin densities is complicated by the ligand–ligand overlaps. To clarify this problem, we shall consider a very simple system consisting of a hydrogen atom situated interstitially in the CaF_2 lattice. The basic experimental data on spin densities in this system have been obtained by Hall and Schumacher[6] with the use of EPR.

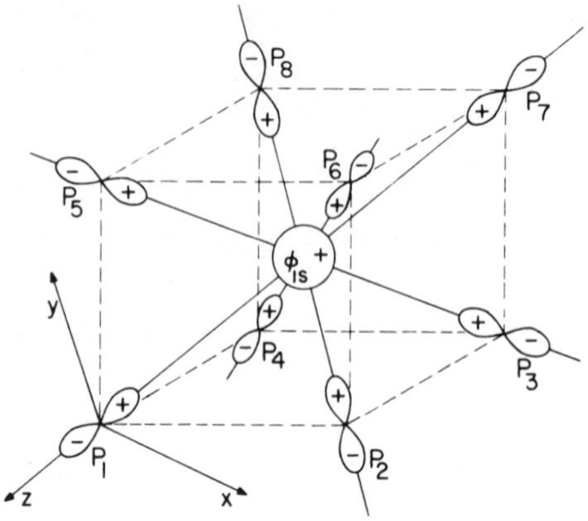

Fig. 2. The orbital of the hydrogen atom ϕ_{1s} surrounded by eight $2p_\sigma$ orbitals of the F^- ions. The fluorine hyperfine interaction Hamiltonian has axial symmetry about the axis z, coinciding with the F^-–H bond axis.

In this case, the magnetic atom is surrounded by eight F^- ions in a cubic arrangement as shown in Fig. 2. The fluorine orbitals leading to a dominant overlap with the unpaired hydrogen electron are the $2p_\sigma$ orbitals directed toward the proton. Each of those orbitals is occupied by two electrons and we are consequently dealing with a system of nine electrons. The problem of calculating the spin densities may be reduced to that of a superposition of eight H–F^- bonds if the $2p_\sigma$ orbitals are mutually orthogonalized. Gourary and Adrian[4] used Löwdin's[10] symmetric orthogonalization procedure to take into account cation–cation overlaps in the *F*-center problem.

Now, we can use (5) to calculate the unpaired spin density at proton and fluorine positions. Combining (7) with (5) and writing

$$\langle m \mid p \rangle = \langle \phi_{1s} \mid p \rangle,$$

the unpaired spin density around proton becomes

$$\varrho_P\uparrow(1) - \varrho_P\downarrow(1) = \frac{\phi_{1s}^2(1)}{1 - [8\langle \phi_{1s} \mid p \rangle^2/(1 + R)]} \tag{9}$$

Comparing (9) with (5), we see that the enhancement of the hydrogen atom spin density is $8/(1 + R)$ times larger than it would be for single H–F^- bond. If we neglect the ligand–ligand overlaps ($R = 0$), the result is a superposition of the effects of eight independent bonds. This is the essence of the *independent bond* model[5], in which one considers each metal–ligand pair separately and finally adds up the effects due to all pairs.

The spin unpairing in the $2p_\sigma$ orbitals is, according to (5) and (7),

$$\varrho_F\uparrow(1) - \varrho_F\downarrow(1) = [\langle\phi_{1s} \mid p\rangle^2/(1+R)^2]p^2(1) \tag{10}$$

2.2.1. Proton Hyperfine Interaction

The isotropic spin density (9) interacts with the proton magnetic moment via Fermi contact hyperfine interaction which is described by an isotropic hyperfine spin-Hamiltonian

$$\mathscr{H}_p = A\mathbf{I} \cdot \mathbf{S} \tag{11}$$

where $\mathbf{I}$ is the proton spin, $\mathbf{S}$ the electron spin, and A the hyperfine constant, which is related to the hyperfine field H_p at the proton as follows:

$$H_p = -AS/g_n\beta_n \tag{12}$$

where g_n is the nuclear gyromagnetic ratio and β_n the nuclear magneton. For the contact interaction, H_p is related to the unpaired spin density at the proton by the well-known formula

$$\begin{aligned} H_p &= \frac{8\pi}{3}\beta_e[\varrho_p\uparrow(0) - \varrho_p\downarrow(0)] \\ &= \frac{8\pi}{3}\beta_e\frac{\phi_{1s}^2(0)}{1 - [8\langle\phi_{1s} \mid 0\rangle^2/(1+R)]} \end{aligned} \tag{13}$$

where β_e is the Bohr magneton. If we assume that ϕ_{1s} is the $1s$ orbital of the free hydrogen atom, expression (13) indicates the enhancement of H_p of the latter produced by the overlap effect.

2.2.2. Fluorine Hyperfine Interaction

The anisotropic unpaired density in the $2p_\sigma$ orbital interacts with the fluorine nuclear moment through the dipolar interaction. It is to be noted that *two* electrons, $2p_\sigma\uparrow$ and $2p_\sigma\downarrow$, are involved in (10); but it is convenient to express the interaction in terms of the hyperfine interaction of one unpaired p_z electron which is written (see Ref. 1)

$$\mathscr{H}_{\text{dip}} = A_{2p}(2I_z s_z - I_x s_x - I_y s_y) \tag{14}$$

where $s = \frac{1}{2}$ is the unpaired electron spin, x, y, and z refer to the local bond coordinates shown in Fig. 2, and

$$A_{2p} = \tfrac{4}{5}g_N\beta_N\beta_e\langle r^{-3}\rangle \tag{15}$$

where $\langle r^{-3} \rangle$ is the radial average for the $2p_\sigma$ orbit of the F^- ion. According to (10), $\varrho_F\uparrow - \varrho_F\downarrow$ is equivalent to a fraction $\langle \phi_{1s} \mid p \rangle^2/(1+R)^2$ of one unpaired p_z electron. Therefore the fluorine hyperfine interaction in $H{:}CaF_2$ can be written

$$\mathscr{H}_F = A_F(2I_z s_z - I_x s_x - I_y s_y) \tag{16}$$

where

$$A_F = [\langle \phi_{1s} \mid p \rangle^2/(1+R)^2] A_{2p} \tag{17}$$

2.2.3. Comparison with Experiment

By putting $R = 0$, (13) and (17) go over into the "independent bond" expressions used by Hall and Schumacher.[6] They calculated $\langle r^{-3} \rangle$ and $\langle \phi_{1s} \mid p \rangle$ using a free F^--ion Hartree–Fock $2p$ wave function and a free hydrogen $1s$ wave function and obtained a value of A_F which was higher by about 30% than the corresponding experimental value. It is interesting to note that including the ligand–ligand overlaps produces a decrease of (17) by about 30% ($R = 0.15$ according to our estimates), giving a good agreement with experiment. It is to emphasize that ligand–ligand overlaps are less important for d electrons, for which the phasing of ligand orbitals makes R substantially smaller than for A_{1g} symmetry.

As pointed out by Hall and Schumacher,[6] a serious discrepancy exists between (13) and the observed proton hfs. The experimental value of A for $H{:}CaF_2$ is about 40 MHz larger than that for the free hydrogen atom $\sim$1420 MHz). On the other hand, (13) predicts about 180 MHz (after R is included). This discrepancy illustrates the incompleteness of the Heitler–London description used above. The most peculiar feature of these experimental results is that the overlap effect on the ligand is fully present, whereas the overlap enhancement of the density on the magnetic site is strongly reduced. The explanation offered in Ref. 6 is that the cubic crystal field of surrounding F^- and Ca^{2+} ions admixes $l = 4$ states into the $1s$ hydrogen ground state, decreasing $\phi_{1s}^2(0)$ at the proton.

Another documentation for the tendency of the free-ion Heitler–London model to overestimate the overlap effects will be presented in the discussion of covalency effects on the central hyperfine parameter A in Mn^{2+} Section 4.

Before leaving this example, we should note that Blum[11] interpreted the g-shift in $H{:}CaF_2$ by taking into account a charge transfer from F^- ions to hydrogen.

3. ELECTRON CHARGE TRANSFER

In this section, we describe the basic electron charge transfer processes which play an important role in determining the distribution of unpaired spin density in transition metal salts and complexes. The energy levels involved in these transfers are those belonging to the metal ion M and the diamagnetic surroundings L, as shown in Fig. 3. In order to keep the description as simple as possible, we consider one magnetic electron in a half-filled orbital m_1. For $3d$ ions in octahedral surroundings, the latter is one of the e_g or t_{2g} states and m_2 is an empty $3d$ or $4s$ orbital. The inner s shells of the magnetic ion are disregarded here, but they will be taken into account in calculating the hyperfine constant A in Section 4. As in the case of H:CaF_2, the ligand orbital l_1 is a linear combination of ionic orbitals transforming according to the same representation as m_1 or m_2 For ligands such as F^- or O^{2-}, the pertinent ionic orbitals are $2p$ or $2s$. Then, l_2 is constructed from empty $3p$ or $3s$ orbitals. As pointed out by Owen and Thornley,[1] the latter can participate in covalent bonding (metal–ligand transfer). However, there is no positive spectroscopic evidence of the existence of these states in transition metal salts. The empty ligand orbitals are definitely important in strong covalent cyanides, where the CN^- ligands have empty π orbitals participating in metal–ligand transfer (back-bonding).

In what follows, we discuss the three transfer processes depicted by the arrows in Fig. 3.

3.1. Ligand–Metal (Half-Filled-Shell) Transfer

3.1.1. Configuration Interaction Method

As mentioned above, this process is undoubtedly the basic covalency mechanism in transition metal salts of predominantly ionic character (fluorides, oxides). In order to calculate the effects of this transfer on the

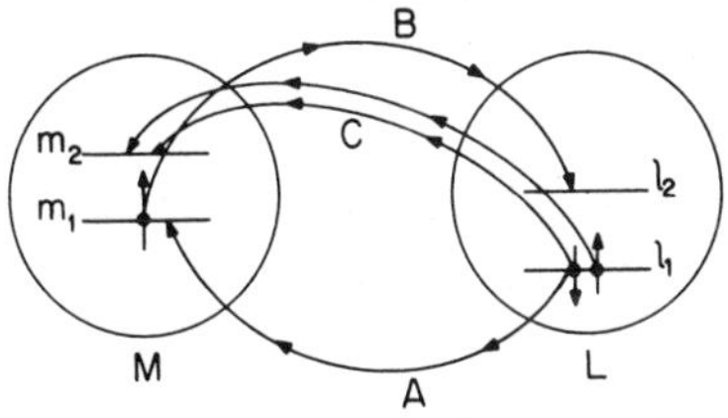

Fig. 3. Diagram showing three transfer processes, A, B, and C, between magnetic atom (ion) M and the diamagnetic surroundings L.

distribution of unpaired spin density, we use the Heitler-London model with configuration interaction. The unperturbed ground state Ψ_G is the purely ionic state described by a Slater determinant

$$\Psi_G = \frac{1}{\sqrt{3!}} \left\{ \left(\frac{m_1 - \langle m_1 \mid l_1 \rangle l_1}{(1 - \langle m_1 \mid l_1 \rangle^2)^{1/2}} \right) \uparrow, l_1\uparrow, l_1\downarrow \right\} \tag{18}$$

In writing (18), we have used the orthogonalized magnetic orbital [see (3)]. The excited (charge transfer) configuration is

$$\Psi_E = \frac{1}{\sqrt{3!}} \left\{ \left(\frac{m_1 - \langle m_1 \mid l_1 \rangle l_1}{(1 - \langle m_1 \mid l_1 \rangle^2)^{1/2}} \right) \uparrow, l_1\uparrow, m_1'\downarrow \right\} \tag{19}$$

The superscript prime indicates that the electron is transferred into a state with different radial wave function than that of the occupied d state. This reflects the fact that m_1 acts to screen out the Coulomb potential seen by m_1' (see Ref. 12). The interaction between M and L admixes Ψ_E into Ψ_G so that the total energy of the system is lowered. The physical essence of this mixing is that electron $l_1\downarrow$ is *virtually* transferred into the $m_1'\downarrow$ state.

The perturbed ground state can be written as

$$\Psi_G' = N(\Psi_G + \gamma \Psi_E) \tag{20}$$

where N is the normalization constant.

The parameter γ is a measure of the degree of covalency and can be calculated using the Ritz variational principle with (20) as a trial function. To first order in the small quantity γ, one finds

$$\gamma = \frac{\langle \Psi_G \mid \mathcal{H} \mid \Psi_E \rangle - \langle \Psi_G \mid \Psi_E \rangle \langle \Psi_G \mid \mathcal{H} \mid \Psi_G \rangle}{\langle \Psi_G \mid \mathcal{H} \mid \Psi_G \rangle - \langle \Psi_E \mid \mathcal{H} \mid \Psi_E \rangle} \tag{21}$$

where $\mathcal{H}$ is the total Hamiltonian of the many-electron system.

3.1.2. Discussion of the Covalency Parameter

The numerical evaluation of the matrix elements in (21) has been recently carried out for $KNiF_3$ and $KMnF_3$ (see Hubbard *et al.*,[13] for example). We will not be concerned with these *a priori* calculations of γ, but rather with the relation between γ and the "chemical" properties of the studied system. Unfortunately, no reliable "chemical rule" can be established for the numerator of (21), which depends sensitively on various computational details.[13] One possibility (accepted mainly by chemists) is to assume that the numerator is proportional to the overlap integral $\langle m_1' \mid l_1 \rangle$.

On the other hand, the denominator of (21) is more amenable to interpretation in terms of semiempirical parameters. It is equal to the change in energy of the system as it passes from Ψ_G to Ψ_E. In evaluating this change, one must recognize that in a crystal, the magnetic ion M and ligand L are subjected to the Madelung potential[5] due to the rest of the lattice. Then the denominator is essentially made up of (a) the energy involved in removing an electron from the ligand (electron affinity), (b) the energy involved in adding an electron to the magnetic ion (ionization energy), (c) the change of the Madelung potential, and (d) the contribution from electric polarization of surrounding ions in the excited state.[12,13]

We wish to emphasize the importance of Madelung energy in the denominator of (21). It happens sometimes that the ionization energy is larger than the electron affinity.[13] In the absence of Madelung potential, this would lead to an instability of the ground state. The dependence of the Madelung potential on the interionic distance is important for the interpretation of the volume and pressure dependences of covalency (see Section 4).

3.1.3. Distribution of Unpaired Spin Density

For calculating the effect of electron transfer on the distribution of unpaired spin density, it is convenient to rewrite (20) as follows:

$$\Psi_G' = \frac{1}{\sqrt{3!}} \left\{ \left(\frac{m_1 - \langle m_1 \mid l_1 \rangle l_1}{(1 - \langle m_1 \mid l_1 \rangle^2)^{1/2}} \right) \uparrow, \quad l_1 \uparrow, \left(\frac{l_1 + \gamma m_1'}{(1 + \gamma^2 + 2\gamma \langle m_1' \mid l_1 \rangle)^{1/2}} \right) \downarrow \right\} \tag{22}$$

Since the orbitals in (22) are mutually orthogonal, the total spin density is simply a sum of the one-electron contributions

$$\varrho_\uparrow - \varrho_\downarrow = \frac{1}{1 - \langle m_1 \mid l_1 \rangle^2} (m_1^2 - 2\langle m_1 \mid l_1 \rangle m_1 l_1 + \langle m_1 \mid l_1 \rangle^2 l_1^2) + l_1^2 - \frac{l_1^2 + 2\gamma l_1 m_1' + \gamma^2 (m_1')^2}{1 + \gamma^2 + 2\gamma \langle m_1' \mid l_1 \rangle} \tag{23}$$

Although physically more correct, the assumption $m_1' \neq m_1$ causes formal complications and below we will assume $m_1' = m_1$. Up to second order in small parameters $\langle m_1 | l_1 \rangle$ and γ, we then obtain from (23)

$$\varrho_\uparrow - \varrho_\downarrow = (1 + \langle m_1 | l_1 \rangle^2 - \gamma^2) m_1{}^2 + (\langle m_1 | l_1 \rangle^2 + 2\gamma \langle m_1 | l_1 \rangle + \gamma^2) l_1{}^2 - 2\langle m_1 | l_1 \rangle m_1 l_1 - 2\gamma m_1 l_1 \quad (24)$$

According to (24), the transfer process A has the following consequences for the unpaired spin density:

1. The spin density of the magnetic ion is decreased (proportionally to γ^2), showing that transfer is counteracting the effect of overlap.
2. The spin density near the ligand is enhanced. It is important to note that the net ligand spin density is proportional to $(\langle m_1 | l_1 \rangle + \gamma)^2$, making transfer and overlap effect experimentally undistinguishable. For that reason, one usually introduces a new covalency parameter $\lambda = \langle m_1 | l_1 \rangle + \gamma$ to describe the unpairing of the spin on the ligand.
3. The additional overlap density $-2\gamma m_1 l_1$ is due to the delocalization of $l\downarrow$. It is important for chemical bonding, but not directly in EPR.

3.1.4. Molecular Orbital Picture—Importance of Bonding Orbitals

The three-electron wave function (22) can be rewritten to a form where two electrons of the ligand character with opposite spin have identical orbital wave functions so that the unpaired spin density of the system is due to *one* electron only. Using the invariance of the determinant with respect to linear combination of orbitals with the same spin and putting $m_1' = m_1$, we obtain from (22)

$$\Psi_G' = \frac{1}{\sqrt{3!}} \left\{ \left(\frac{m_1 - \lambda l_1}{(1 - 2\lambda \langle m_1 | l_1 \rangle + \lambda^2)^{1/2}} \right) \uparrow, \left(\frac{l_1 + \gamma m_1}{(1 + 2\gamma \langle m_1 | l_1 \rangle + \gamma^2)^{1/2}} \right) \uparrow, \left(\frac{l_1 + \gamma m_1}{(1 + 2\gamma \langle m_1 | l_1 \rangle + \gamma^2)^{1/2}} \right) \downarrow \right\} \quad (25)$$

where

$$\lambda = \gamma + \langle m_1 | l_1 \rangle \quad (26)$$

The paired orbitals in (25) are called the bonding molecular orbitals, and the unpaired one is the antibonding molecular orbital. Determinant (25) forms the essence of what is called the molecular orbital description of covalency.[14] It has the advantage that all magnetic properties can be described in one-electron way, using the *antibonding* orbital

$$\varphi_A = (m_1 - \lambda l_1)/(1 - 2\lambda\langle m_1 \mid l_1\rangle + \lambda^2)^{1/2} \tag{27}$$

It should be emphasized, however, that omitting the bonding orbitals can sometimes lead to wrong conclusions. First, there is an established misconception according to which one considers the d electron covalency as a delocalization of the magnetic electron toward ligands. This naturally arises from the one-electron description (27), when the charge contribution of bonding orbitals is neglected.

The bonding orbitals contribute in an important way to various effects on the magnetic site. The reader is referred to Section 4 for an example where the inner s shells and the $4s$ orbital of Mn^{2+} are admixed into the ligand orbitals, thus contributing to the hyperfine field at the nucleus. We would like to mention briefly other examples of the importance of bonding orbitals. Let us consider bonding orbitals containing admixtures of the $2p$, $3p$, and $4p$ central magnetic ion orbitals. In a distorted surrounding, these admixtures can produce an electric field gradient at the nucleus of the magnetic ion, leading to a quadrupole interaction. The admixtures of the orbitals of the magnetic ion also contribute to the reduction of spin–orbit parameter on account of electrostatic screening. The latter effects cannot be properly accounted for by considering the antibonding orbitals only.

3.1.5. *Fluorine Hyperfine Interaction in* Ni^{2+}:$KMgF_3$

We will now illustrate the transfer process A of Fig. 3 by taking the example of Ni^{2+} in $KMgF_3$ studied previously by EPR.[7] In this case, Ni^{2+} is surrounded by a regular octahedron of F^- ions. The same type of bonding occurs in $KNiF_3$, where fluorine hyperfine interaction has been studied using nuclear magnetic resonance.[15] Moreover, several first-principle calculations have been carried out for $KNiF_3$.[13,16–18] The importance of this example lies in the fact that it involves Ni–F bonds of a pure σ-bonding type.

There are two unpaired electrons on Ni^{2+} in orbitals $d_z{}^2 = m_u$ and $d_{x^2-y^2} = m_v$ belonging to the e_g representation of the cubic group. The intraatomic exchange orients the spins of these electrons parallel to each other. The degenerate orbitals m_u and m_v become admixed with ligand molecular orbitals l_{1u} and l_{1v} of corresponding symmetry to form degenerate antibonding orbitals

$$\varphi_{Au} = N(m_u - \lambda_u l_{1u}) \tag{28}$$

$$\varphi_{Av} = N(m_v - \lambda_v l_{1v}) \tag{29}$$

where, in view of the degeneracy,

$$\lambda_u = \gamma_u + \langle m_u \mid l_{1u} \rangle = \lambda_v = \lambda \tag{30}$$

and N is the normalization constant [see (27)].

Both $2s$ and $2p$ florine orbitals participate in the bonding with the nickel ion. However, the Hartree–Fock calculations[16] show that the energy of the $2s$ state is much lower than that of the $2p$ state. Recalling that this energy (electron affinity) is involved in the denominator of (21), we can conjecture that the degree of transfer from $2s$ orbital is small compared to that from the $2p$ orbital. Sugano and Shulman[16] show that in $KNiF_3$ the parameter for $2s \rightarrow 3d$ transfer is $\gamma_s = 0.031$, whereas the parameter for $2p \rightarrow 3d$ transfer is $\gamma_p = 0.285$.

For simplicity, we omit the $2s$ orbitals and consider only the $2p$ orbitals illustrated in Fig. 4. If the ligand–ligand overlap is neglected, the ligand molecular orbitals are

$$l_{1u} = \tfrac{1}{2}(p_1 + p_4 - p_2 - p_5) \tag{31}$$

$$l_{1v} = (1/\sqrt{12})(2p_3 + 2p_6 - p_1 - p_4 - p_2 - p_5) \tag{32}$$

where the subscripts i of the p_i ligand atomic orbitals refer to the ligand positions whose numbering is shown in Fig. 4.

The total unpaired spin density is a sum of the contributions due to φ_{Au} and φ_{Av}. Using (28)–(32), the unpaired electron density in the p orbital of one fluorine is

$$\varrho_F\uparrow - \varrho_F\downarrow = \tfrac{1}{3}N^2\lambda^2 p^2(1) \tag{33}$$

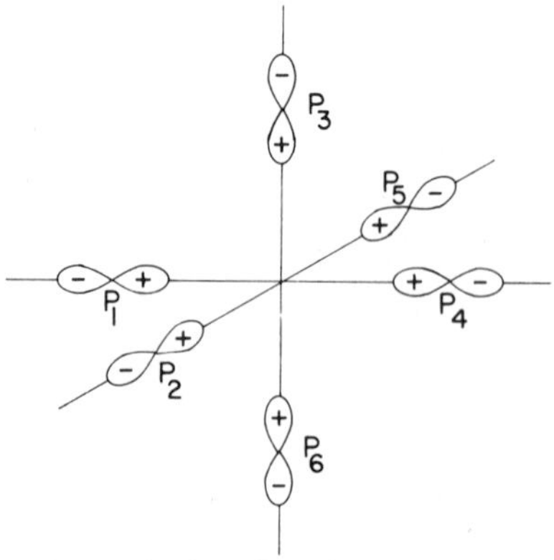

Fig. 4. Numbering of the fluorine $2p_\sigma$ orbitals in a regular octahedron.

This is equivalent to a fraction $f_\sigma = \frac{1}{3}N^2\lambda^2$ of *one* unpaired electron in a p orbital. Then, according to (14) and (15), the hyperfine interaction due to (33) is

$$\mathscr{H}_F = f_\sigma A_{2p}(2I_z s_z - I_x s_x - I_y s_y) \tag{34}$$

where $s = \frac{1}{2}$. Since the true spin of Ni^{2+} is $S = 1$, we must rewrite (34) in terms of S. For general magnitude of S, we have

$$\mathscr{H}_F = A_F(2I_z S_z - I_x S_x - I_y S_y) \tag{35}$$

where

$$A_F = (f_\sigma/2S)A_{2p} \tag{36}$$

This equation relates the experimental parameter A_F to the "atomic" parameter A_{2p} and the fraction of unpaired electrons f_σ It is to be noted that A_F also contains the direct dipole term from the spin density of the Ni^{2+} $3d$ orbitals, which can be eliminated by a simple calculation.[14]

A_F has been determined experimentally by Hall *et al.*[7] A_{2p} can be calculated from (15) using Hartree–Fock $2p$ wave function of F^-. In this way, $f_\sigma = 4.5\%$ has been obtained,[7] corresponding to about 14% of the nickel spin delocalized over the fluorine octahedron. As stressed in the introduction, this delocalization is not a direct measure of the degree of transfer, because of the entanglement with overlap $\langle m \mid l_1 \rangle$ (see $\lambda = \gamma + \langle m_u \mid l_{1u} \rangle$). The latter can be calculated if the form of the wave functions of Ni^{2+} and F^- are known in the crystal. Usually, one assumes that the free-ion Hartree–Fock wave functions represent a good approximation in a predominantly ionic crystal such as $KNiF_3$. Sugano and Shulman[16] used Watson's[19] solution for $3d$ orbital in Ni^{2+} and Froese's[20] $2p$ wave function of F^- and calculated $\langle m \mid l_1 \rangle = 0.11$. With this value, one obtains $\gamma \cong 0.2$, indicating that charge transfer gives a dominant contribution to the fluorine shfs.

However, it is to be realized that there is always certain amount of arbitrariness in such a determination of γ. The one-electron wave functions may undergo various deformations in a crystal modifying the contribution of overlap and changing the predictions about γ (see, for example, Ref. 14). Related to this problem is a question:[21] To what extent does the overlap effect based on the Heitler–London model correctly account for the Pauli exclusion principle? Several other examples (see Sections 1 and 4) involving overlap effect demonstrate the generality of this question.

Before leaving the charge transfer process A, let us mention its importance in trivalent (and higher-valent) transition metal salts. For these, the ionization energy of the magnetic ion involved in the denominator of (21) is substantially larger than in the divalent case, making the degree of covalency larger. This has been confirmed, for example, by EPR studies of fluorine shfs in Fe^{3+}:$KMgF_3$ and also by neutron diffraction studies[22] of Fe^{3+} salts.

3.2. Back-Bonding*

It is well known that this charge transfer process plays a dominant role in iron group cyanides. The CN^- ligand has empty π orbitals which can accept electrons from the metal ion. A molecular orbital analysis of iron-group cyanides has been presented by Shulman and Sugano.[23]

One interesting feature pointed out by authors is the observation that divalent iron group cyanides show a larger degree of back-donation than trivalent ones, contrary to the transfer process (A) discussed in the previous section. This can be understood again in terms of the effects of the metal ionization energies on γ. Since the transfer process (B) is toward the ligands, it is easier to transfer an electron from a divalent ion (with smaller ionization energy) than from a trivalent one.

Bleaney and O'Brien,[24] analyzing the experimental results of Baker *et al.*,[25] obtained the orbital reduction parameters $k = 0.87$ and 0.74 for $Fe^{3+}(CN^-)_6$ and $Mn^{2+}(CN^-)_6$, respectively. It is well known that k is a measure of covalency (see Ref. 1), and hence this result qualitatively confirms the above statement. Shulman and Sugano[23] have presented additional evidence for this by comparing the Mössbauer isomer shifts of ferrocyanides and ferricyanides. The difference in back-bonding makes the number of $3d$ electrons left on the iron site approximately equal for the latter compounds. This explains the experimental fact that the isomer shifts in ferrocyanides and ferricyanides are almost the same.

We shall not discuss the formulation of the spin-density distribution due to back-bonding. It is obvious from the diagram in Fig. 3 that the unpaired spin density on the ligand should be positive for transfer process B. However, Kuska and Rogers[26] studied the ^{13}C shfs of the $Cr^{3+}(CN^-)_6$ complex in $K_3Co(CN)_6$ single crystal and found negative isotropic shfs.

As shown below (see process C), the negative sign can be accounted for by a transfer from occupied carbon orbital (such as $2s$) to unoccupied chromium orbital (such as $3d_{z^2}$, $3d_{x^2-y^2}$). The CN^- ligand contains both occupied both occupied π orbitals and empty π^* orbitals. While the latter participate in back-donation, the former are involved in the transfer pro-

* Back-bonding has been recently considered to explain the anomalous temperature dependence of hfs of Mn^0 atoms in KCl (E. Šimánek and Th. von Waldkirch, *Proceedings of the XVIth Colloque Ampère, Bucharest, 1970,* ed. I. Ursu, Publishing House of the Academy of the Socialist Republic of Romania (1971), p. 243).

cess A. The observed anisotropic shfs is a result of both types of transfer and hence not amenable to a straightforward analysis in terms of single back-bonding transfer parameter.

3.3. Ligand–Metal (Empty-Shell) Transfer

Orgel[27] seems to have been the first to have suggested that electron transfer from ligand to an empty orbital of metal produces a negative spin density on the ligand.

Recognition of the importance of this process in transition metal fluorides is due to Shulman and Knox,[28] who found, using NMR, a negative isotropic fluorine hyperfine interaction in K_2NaCrF_6. In other words, the observed unpaired fluorine s density, presumably due to $2s$ orbitals, is antiparallel to the spin of the Cr^{3+} ion. Obviously, the $2s$ orbitals have no interaction with the t_{2g} orbitals (see Fig. 5) since the overlap integral $\langle 2s \mid t_{2g} \rangle = 0$. However, charge transfer is possible from the $2s$ ligand orbital to the empty metal orbitals, such as $3d(e_g)$, $4s$, $4p$. The latter can accept ligand electrons of both spin orientations, but the intraatomic exchange with the $(t_{2g})^3$ subshell causes the transfer probability to be greater for electrons with spin parallel to the chromium spin. Thus a negative spin density is produced in the $2s$ orbital of the F^- ion.[28,29] It is natural to expect a similar effect for the $2p_\sigma$ electrons. In fact, as discussed in Section 5, substantial negative spin density in oxygen $2p_\sigma$ orbitals has been recently suggested by Owen and Taylor[9] to explain the aluminum hyperfine interaction in Cr^{3+}:$LaAlO_3$. Another example of transfer process C occurs in discussing the central hyperfine parameter A (see Section 4).

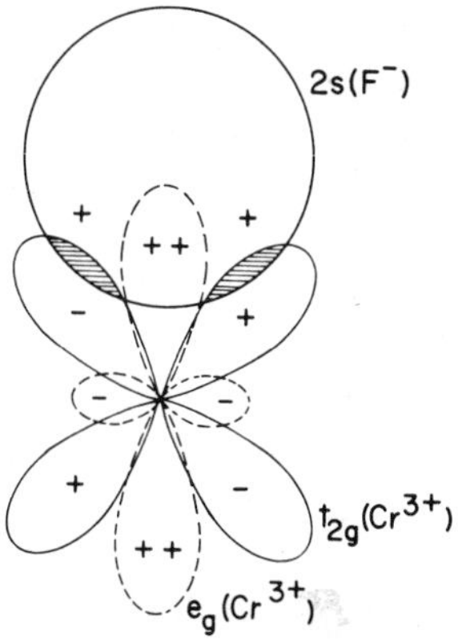

Fig. 5. The nature of the overlap between the fluorine $2s$ orbital and the t_{2g} and e_g (empty) orbitals of Cr^{3+}.

We now turn to the derivation of the ligand spin density due to this process, using the configuration interaction method. For simplicity, we take again a simple three-electron system ground state

$$\Psi_G = (1/\sqrt{3!})\,\{m_1\uparrow, l_1\uparrow, l_1\downarrow\} \tag{37}$$

where m_1 is the magnetic orbital, such as t_{2g} in the Cr^{3+} case, and l_1 is the molecular ligand orbital belonging to the same representation as the *empty* orbital of the metal m_2! We assume that m_1 and m_2 belong to different representations so that $\langle m_1 \mid l_1\rangle = 0$. In (37), we have omitted the occupied ligand orbitals belonging to the same representation as m_1. There are two excited states differing by the *spin* of the transferred electron

$$\Psi_{Ea} = (1/\sqrt{3!})\{m_1\uparrow, m_2\uparrow, l_1\downarrow\} \tag{38}$$

$$\Psi_{Eb} = (1/\sqrt{3!})\{m_1\uparrow, l_1\uparrow, m_2\downarrow\} \tag{39}$$

It should be emphasized that the excited orbital m_2 in (38) actually differs from that in (39) because of the exchange interaction with $m_1\uparrow$. This difference does not matter in calculations of the ligand unpairing. However (see Section 4), when the excited state is a 4*s* orbital and we are interested in hyperfine effects on metal, this effect has important consequences.

The perturbed ground-state wave function can be written

$$\Psi_G' = N(\Psi_G + \gamma_a \Psi_{Ea} + \gamma_b \Psi_{Eb}) \tag{40}$$

where N is the normalization constant. The transfer parameters may be calculated using expression (21). The exchange integral $\mathcal{J}$ between m_1 and m_2 makes the energy of Ψ_{Ea} smaller than that of Ψ_{Eb}. Referring to expression (21), we can write

$$\gamma_a = b/(\Delta E - \mathcal{J}) \tag{41}$$

$$\gamma_b = b/\Delta E \tag{42}$$

where b refers to the numerator and ΔE to the denominator in (21).

The exchange integral $\mathcal{J}$ is positive and hence $\gamma_a > \gamma_b$. This produces a negative unpaired spin density on the ligand which can be calculated by rewriting (40) as follows:

$$\Psi_G' = \frac{1}{\sqrt{3!}}\left\{m_1\uparrow, \left(\frac{l_1 + \gamma_a m_2}{(1 + 2\gamma_a\langle m_2 \mid l_1\rangle + \gamma_a^2)^{1/2}}\right)\uparrow,\right.$$
$$\left.\left(\frac{l_1 + \gamma_b m_2}{(1 + 2\gamma_b\langle m_2 \mid l_1\rangle + \gamma_b^2)^{1/2}}\right)\downarrow\right\} \tag{43}$$

Determinant (43) differs from (40) by the presence of small cross-products proportional to $\gamma_a \gamma_b$ which can be neglected. The one-electron orbitals in (43) of the ligand character are bonding molecular orbitals with slightly different degree of delocalization, due to exchange. Consequently, the charge transfer process C is sometimes called the "exchange polarization" of bonding orbitals. A straightforward summation over the spin densities due to one-electron orbitals in (43) gives for the total unpaired spin density in orbital l_1

$$\varrho_L \uparrow - \varrho_L \uparrow = (\gamma_b^2 + 2\gamma_b \langle m_2 \mid l_1 \rangle - \gamma_a^2 - 2\gamma_a \langle m_2 \mid l_1 \rangle) l_1^2 \qquad (44)$$

Inserting (41) and (42) into (44) and assuming $\Delta E \gg \mathscr{I}$ (typically, $\Delta E = 10$ eV, $\mathscr{I} = 1$ eV), we obtain

$$\varrho_L \uparrow - \varrho_L \downarrow \cong -(2\mathscr{I}/\Delta E)(\gamma_b^2 + \gamma_b \langle m_2 \mid l_1 \rangle) l_1^2 \qquad (45)$$

where l_1 is the combination of $2s$ orbitals with e_g symmetry. It is interesting to compare (45) with the corresponding expression for transfer process A. According to (24), the latter produces an unpaired spin density $(\langle m_1 - l_1 \rangle^2 + 2\langle m_1 \mid l_1 \rangle \gamma + \gamma^2) l_1^2$. The form (45) leads to a conjecture that the unpairing due to process C is much weaker than that due to process A. There is no direct overlap term in (45) and furthermore we have a reduction due to $\mathscr{I}/\Delta E$.

This situation can be illustrated by comparing the isotropic fluorine shfs of Ni^{2+}:$KMgF_3$ and Cr^{3+}:$KMgF_3$. The Ni^{2+} has a half-filled e_g shell so that the unpaired fluorine $2s$ density comes from process A and overlap, whereas Cr^{3+} has empty e_g shells available for process C. The EPR experiments of Hall and Schumacher[6] yield a fraction of unpaired $2s$ electrons f_s defined as

$$\varrho_F \uparrow - \varrho_F \downarrow = f_s \cdot \phi_{2s}^2 \qquad (46)$$

The experimental values[7] are $f_s = 0.49\%$ and $f_s = -0.031\%$ for Ni^{2+}:$KMgF_3$ and Cr^{3+}:$KMgF_3$, respectively. As mentioned above, Sugano and Shulman[16] calculated for the $2s \to 3d$ transfer parameter γ_s in $KNiF_3$ a value of 0.031. For a rough estimate, we assume the same value for γ_b and put $\langle m_2 \mid l_1 \rangle = 0.1$ and $\mathscr{I}/\Delta E = 1/10$ in (45). We then obtain $(2\mathscr{I}/\Delta E) \times (\gamma_b^2 + \gamma_b \langle m_2 \mid l_1 \rangle) \cong 8 \times 10^{-4}$. Expressing the molecular orbital in terms of atomic ϕ_{2s} orbital, Eq. (46) yields for f_s a value $f_s = -(1/6) \times 10^{-4} = -0.013\%$, which is of the same order as the experimental value. It is interesting that in this expression for f_s, the term $\gamma_b \langle m_2 \mid l_1 \rangle$ is three times larger than the direct admixture γ_b^2, showing the importance of transfer-overlap cross-terms.

4. COVALENCY AND HYPERFINE CONSTANT A

In this section we will be concerned with the effects of covalency on the hyperfine parameter A in $3d^5$ ions, such as Mn^{2+} or Fe^{3+} situated in an octahedral environment.

Let us first consider the origin of A when these ions are in a free state. As discussed by Heine[30] and Wood and Pratt,[31] the hyperfine interaction for S-state ions is due to the polarization of $1s$, $2s$, and $3s$ shells by exchange with $3d$ electrons (core polarization). The exchange field of $3d$ electrons affects the inner electrons of up and down spin differently, so that the orbital wave function $\phi_{ns}(r)$ for the ns shell of up spin is different from $\phi'_{ns}(r)$ belonging to down spin. Consequently, one obtains a finite difference of the electron densities at the nucleus which leads to a hyperfine field of contact character

$$H_{ns} = (8\pi/3)\beta_e[\phi_{ns}^2(0) - \phi_{ns}'^2(0)] \tag{47}$$

This is a contribution to the core polarization field coming from unpairing in the ns shell. Then, the total hyperfine field of core polarization is

$$H^{\text{c.p.}} = \sum_n H_{ns} = (8\pi/3)\beta_e \sum_n [\phi_{ns}^2(0) - \phi_{ns}'^2(0)] \tag{48}$$

Using relation (12) between hyperfine field and parameter, one can write for the S-state ions

$$A_0 = -\left(\frac{8\pi g_n \beta_n \beta_e}{3S}\right) \sum_n [\phi_{ns}^2(0) - \phi_{ns}'^2(0)] \tag{49}$$

where the 0 indicates that we are dealing with a free ion. The individual contributions H_{ns} have been computed for transition metal neutral atoms and ions by Watson and Freeman[32] using the unrestricted Hartree–Fock method. In what follows, we will be discussing mainly Mn^{2+}, for which they calculate $H_{1s} = -30$ kG, $H_{2s} = -1400$ kG, and $H_{3s} = +740$kG. The change of sign between H_{2s} and H_{3s} is due to the different radial extensions of $2s$ and $3s$ shells.

From the EPR study of Mn^{2+} in different salts, Van Wieringen[33] has shown that there is a regular variation of A with the degree of covalency. The hyperfine field $H^{\text{c.p.}}$ depends on the octahedrally arranged ligands as follows[32]:

Ligands:	H_2O	F^-	O^{--}	S^{--}	Se^{--}	Te^{--}
$H^{\text{c.p.}}$ (kG):	695	695	580	490	460	420

We will attempt to find an explanation for the general tendency of A to decrease with covalenty. The quantitative estimates will be confined to the more ionic compounds, such as fluorides and oxides. The latter are more amenable to quantitative treatment than covalent compounds because the degree of covalency in fluorides and oxides is known from neutron diffraction measurements.[22]

The mechanisms responsible for the modification of A due to covalency can be classified as follows: First, we have an indirect effect of the covalent modified $3d$ electrons on the core polarization field $H^{\text{c.p.}}$. Second, the $1s$, $2s$, $3s$, and $4s$ orbitals can contribute directly when they become admixed by overlap ($1s$, $2s$, $3s$) and transfer ($4s$) from the ligand orbitals.

4.1. Indirect Effect

In weakly covalent compounds, the change in the $3d$ spin and charge distribution is due to ligand–metal overlap and transfer (process A). The most trivial effect is the change of the $3d$ spin density described by the first term of (24). This is, in fact, a normalization correction producing a decrease of the $3d$ spin density if the transfer parameter γ is greater than the metal–ligand overlap $\langle m_1 \mid l_1 \rangle$. It is to be stressed, however, that this "uniform" decrease refers only to that part of the $3d$ shell which is not significantly overlapped by the ligand electrons.

Since the core polarization hyperfine field is proportional to the total spin of the $3d$ shell, we expect from this normalization correction a decrease of A. To estimate this effect in Mn^{2+}, we assume that only e_g electrons are effected by covalency and overlap (t_{2g} electrons are taken as in a free ion). Using (24), one finds for the modified value of A

$$A = A_0[1 + \tfrac{2}{5}(\langle m_u \mid l_u \rangle^2 - \gamma_u^2)] \tag{50}$$

where, according to (28), m_u is the $3d_z{}^2$ orbital and l_u is the corresponding ligand combination (31) of the $2p_\sigma$ orbitals. In what follows, we neglect the ligand $2s$ orbitals. As a specific example, let us consider Mn^{2+}:MgO and $Mn^{2+}:KMgF_3$, for which the hyperfine parameter A is 81.5 cm^{-1} and 91 cm^{-1},[7] respectively (at 4°K).

It is interesting to attempt an explanation of the decrease in A on going from fluoride to oxide host. The overlap integral $\langle m_u \mid l_u \rangle$ can be calculated using Hartree–Fock wave functions, and we then obtain a value $\langle m_u \mid l_u \rangle = 0.13$ for both Mn^{2+}:MgO and $Mn^{2+}:KMgF_3$. The transfer parameter γ_u can be estimated from the neutron scattering results[22] for MnO and MnF_2. If we neglect π bonding and use $\langle m_u \mid l_u \rangle = 0.13$, these results[22] are accounted for by $\gamma_u = 0.18$ for both MnO and MnF_2. Using these numbers, we obtain from (50) $A = 0.995 A_0$. Thus, the normalization correction gives quite negligible decrease (~0.5%) of A. Obviously, this mechanism cannot account for the large (~10%) decrease of A observed on going from fluoride to oxide host.

As mentioned above, the normalization effect does not describe properly the modification of the spin density in the overlap region. The latter is important for the analysis of A because it affects the outer part of the 3s orbital. According to (24), the overlap spin density is negative and thus it may cause (by exchange potential) the down-spin orbital ϕ'_{3s} to expand, resulting, according to (47), in an increase of H_{3s}. This effect may account for a decrease of A with overlap and γ, but it is difficult to estimate quantitatively.

4.2. Direct Effect

The s electrons of the magnetic ion contribute directly to A by the hyperfine field of contact type. Their contribution can be affected by the ligand electrons as follows: First, the ligand electrons can be virtually transferred into the 4s state. Exchange polarization of the latter then produces a 4s hyperfine of the same direction as the spin of the 3d shell. In view of the negative sign of the core polarization hyperfine field, this transfer leads to a decrease of the magnitude of A. Second, the $2p_\sigma$ ligand orbitals overlap significantly with the filled inner shells 2s and 3s. Thus the latter become admixed into the ligand orbitals. Since $\phi_{ns}(0) \neq \phi'_{ns}(0)$, this admixture gives rise to a local unpairing at the nucleus, enhancing the ns shell hyperfine field H_{ns}. The enhancement is proportional to $\langle \phi_{ns} \mid 2p_\sigma \rangle^2$ and hence we expect that it will be largest for the 3s shell. Since $H_{3s} > 0$, we again exect a decrease in the magnitude of A on account of this overlap effect. Third, there are important cross-products between the overlap-admixed 2s and 3s and transfer-admixed 4s orbital.

As mentioned above, the ligand–4s transfer is an example of the transfer process C and we can use the configuration interaction method described in Section 3. Actual calculations show that the overlap admixture of the 1s state gives small effects and we will neglect it for simplicity. Then, the ground-state wave function for the Mn^{2+} ion in an octahedral surroundings is the determinant

$$\Psi_G = N\{\phi_{2s}\uparrow, \phi'_{2s}\downarrow, \phi_{3s}\uparrow, \phi'_{3s}\downarrow, \phi_{3d_1}\uparrow, \ldots, \phi_{3d_5}\uparrow, l_1\uparrow, l_1\downarrow\} \qquad (51)$$

where $\phi_{3d_1}, \ldots, \phi_{3d_5}$ describes the 3d configuration of the free Mn^{2+} ion.

We have neglected the $3d$–ligand overlap and covalency, which give a small indirect effect described in the previous section. There l_1 is a symmetric (A_{1g}) combination of the ligand ionic orbitals. For simplicity, we again neglect the $2s$ orbitals (smaller transfer to $4s$ and smaller overlap) and confine ourselves to the $2p_\sigma$ orbitals. Using these orbitals, the properly normalized orbital l_1 in an octahedron is

$$l_1 = [6(1 + R)]^{-1/2}(p_1 + \cdots + p_6) \tag{52}$$

where $p_1, \ldots, p_6$ are the six $2p_\sigma$ orbitals, shown in Fig. 4, and R is the ligand–ligand overlap correction

$$R = \sum_{j=2}^{6} \langle p_1 \mid p_j \rangle \tag{53}$$

The two excited configurations involved in the transfer process C are

$$\Psi_{Ea} = N_a\{\phi_{2s}\uparrow, \phi'_{2s}\downarrow, \phi_{3s}\uparrow, \phi'_{3s}\downarrow, \phi_{3d_1}\uparrow, \ldots, \phi_{3d_5}\uparrow, \phi_{4s}\uparrow, l_1\downarrow\} \tag{54}$$

$$\Psi_{Eb} = N_b\{\phi_{2s}\uparrow, \phi'_{2s}\downarrow, \phi_{3s}\downarrow, \phi'_{3s}\uparrow, \phi_{3d_1}\uparrow, \ldots, \phi_{3d_5}\uparrow, l_1\uparrow, \phi'_{4s}\downarrow\} \tag{55}$$

It should be stressed that the $4s$ excited state in (54) and (55) is taken so that $\phi_{4s} \neq \phi'_{4s}$ are unrestricted.

The perturbed ground state

$$\Psi_G' = N'\{\Psi_G + \gamma_a\Psi_{Ea} + \gamma_b\Psi_{Eb}\} \tag{56}$$

can be rewritten to a "molecular bonding orbital" form

$$\Psi_G' = N_m'\{\phi_{2s}\uparrow, \phi'_{2s}\downarrow, \phi'_{3s}\uparrow, \phi'_{3s}\downarrow, \phi_{3d_1}\uparrow, \ldots, \phi_{3d_5}\uparrow,$$
$$(l_1 - \sum_{n=2}^{3} \langle l_1 \mid \phi_{ns}\rangle\phi_{ns} + \gamma\,\phi_{4s})\uparrow, (l_1 - \sum_{n=2}^{3} \langle l_1 \mid \phi'_{ns}\rangle\phi'_{ns} + \gamma\,\phi'_{4s})\downarrow\} \tag{57}$$

The one-electron orbitals in (57) are orthogonal and thus the total density can be calculated in a straightforward way giving for the unpaired s density at the nucleus of Mn^{2+}

$$\varrho_\uparrow(0) - \varrho_\downarrow(0)$$
$$= \left\{\sum_{n=2}^{3} [\phi_{ns}^2(0) - \phi'^2_{ns}(0)]\right\} + \left\{\sum_{n=2}^{3} \langle l_1 \mid \phi_{ns}\rangle^2[\phi_{ns}^2(0) - \phi'^2_{ns}(0)]\right.$$
$$\left. + 2\sum_{n,m=2}^{3} \langle l_1 \mid \phi_{ns}\rangle\langle l_1 \mid \phi_{ms}\rangle[\phi_{ns}(0)\phi_{ms}(0) - \phi'_{ns}(0)\phi'_{ms}(0)]\right\}$$
$$+ \left\{\gamma_a^2\phi_{4s}^2(0) - \gamma_b^2\phi'^2_{4s}(0)\right.$$
$$\left. - 2\sum_{n=2}^{3} \langle l_1 \mid \phi_{ns}\rangle[\gamma_a\phi_{ns}(0)\phi_{4s}(0) - \gamma_b\phi'_{ns}(0)\phi'_{4s}(0)]\right\} \tag{58}$$

By the curly brackets, we divide (58) into three basic terms:

1. The first one is the unpaired s density in the free ion, which, according to (49), is responsible for the free-ion hyperfine parameter.

2. The second term is the correction of the invididual H_{ns} due to the overlap effect. It can be estimated using the unrestricted values $\phi_{ns}(0)$ and $\phi'_{ns}(0)$ calculated by Watson and Freeman.[32] For Mn^{2+}:MgO, the overlap integrals $\langle\phi_{ns} \mid 2p_\sigma\rangle$ have been calculated using Watson's ϕ_{ns} wave function[19] for Mn^{2+} and $2p$ wave functions[36] for O^{2-}. Then, the group overlaps $\langle l_1 \mid \phi_{ns}\rangle$ can be evaluated with the result $\langle l_1 \mid \phi_{2s}\rangle = 0.33$ and $\langle l_1 \mid \phi_{3s}\rangle = 0.17$. Using these values, we find that the second term of (58) reduces the magnitude of A_0 by about 2%. The latter value is composed of the diagonal $2s$ and $3s$ contributions ($\sim +3\%$) and the ($2s$, $3s$) crossproduct ($\sim -1\%$). The negative sign of this cross-product has its origin in the fact that the numbers of nodes of the $2s$ and $3s$ functions differ by one. This effect also plays a role in the mechanism of cation–cation hyperfine interaction discussed in Section 5. Though substantially larger than the indirect effect, the pure overlap effect is not sufficiently strong to account for the 10% difference of A for Mn^{2+}:MgO and Mn^{2+}:$KMgF_3$. It is also not directly linked to the electron affinity of the ligands which is shown to have so strong influence on A.

3. Let us now consider the last terms involving the $4s$ admixture.* Inspection of the unrestricted Hartree–Fock calculation[32] for the neutral Mn atom ($3d^5\,4s^2$) shows that $\phi_{4s}(0)$ is significantly larger than $\phi'_{4s}(0)$, the corresponding contact hyperfine fields being $H_{4s}\uparrow = 1750$ kG and $H_{4s}\downarrow = 1150$ kG. This is caused by the $3d$–$4s$ exchange interaction, which pulls the $4s$ orbital (with spin parallel to the $3d$ spin) inward, increasing $\phi_{4s}(0)$. It is natural to expect that the excited $4s$ orbitals in (54) and (55) exhibit a similar property. The excited configurations Ψ_{Ea} and Ψ_{Eb} contain $3d$-$4s\uparrow$ and $3d$-$4s\downarrow$ configurations of Mn^+ and hence we expect somewhat larger values for both $\phi_{4s}(0)$ and $\phi'_{4s}(0)$ than in a neutral atom. Another source of unpairing is the exchange splitting of the excited configurations, making $\gamma_a > \gamma_b$ (see transfer process C in Section 3). It is useful to consider these two unpairing effects separately to find which one is dominating.

1. First, let us assume $\gamma_a = \gamma_b = \gamma$ and calculate the $4s$ contribution [third term of (58)] to the unpaired s density:

$$\varrho_{4s}\uparrow(0) - \varrho_{4s}\downarrow(0) = \gamma^2[\phi_{4s}^2(0) - \phi_{4s}'^2(0)] - 2\gamma \sum_{n=2}^{3} \langle l_1 \mid \phi_{ns}\rangle[\phi_{ns}(0)\phi_{4s}(0) - \phi'_{ns}(0)\phi'_{4s}(0)] \qquad (59)$$

* For a detailed analysis of this effect, see E. Šimánek and K. A. Müller, *J. Phys, Chem. Solids* **31**, 1027 (1970).

As an example, we will take again Mn^{2+}:MgO with the overlaps given above and assume values of $\phi_{4s}(0)$ and $\phi'_{4s}(0)$ corresponding to $H_{4s}\uparrow = 1750$ kG and $H_{4s}\downarrow = 1150$ kG.[32] The latter values are the calculated 4*s* contributions for the neutral Mn atom. Recent analysis of the exchange polarization of bonding orbitals in a molecular Mn complex[36a] shows a reduction of atomic radial polarization by the effect of ligand → 4*s* bonding. Hence these values correspond to an upper limit of the 4*s* unpairing. Using them, the contact hyperfine field arising from (59) is calculated to be

$$H_{4s} = 600\gamma^2 + 220\gamma \quad [\text{kG}] \tag{60}$$

It should be emphasized that γ appearing in (60) is the octahedral group transfer parameter rather than the Mn–O bond parameter. With this definition, the total occupancy of the 4*s* state turns out to be $2\gamma^2$. The second term in (60) is due to the (3*s*, 4*s*) and (2*s*, 4*s*) cross-products. For small values of γ, this term provides an important enhancement of H_{4s}. As mentioned above, this is one of the few examples mentioned in this chapter where the effects of transfer are substantially modified by overlaps. A similar phenomenon occurs in the calculation of the total ("paired") *s*-electron density at the nucleus of iron and is shown[37] to be important in the interpretation of the Mössbauer isomer shift of ^{57}Fe. In order to estimate the effect of H_{4s} in (60) on the hyperfine parameter *A*, we take typically $\gamma = 0.2$ (corresponding to a configuration $3d^5 4s^{0.08}$) and obtain from (60) $H_{4s} = 78$ kG. Thus the hyperfine constant *A* decreases from the free-ion value A_0 by about 10%, the contribution of the term linear in γ being about 6%.

2. Let us now investigate the effect of the exchange splitting of the excited state by putting $\phi_{4s}(0) = \phi'_{4s}(0)$ and allowing only γ_a to be different from γ_b in the third term of (58). Neglecting the cross-terms, one obtains for the unpaired *s* density

$$\varrho_{4s}\uparrow(0) - \varrho_{4s}\downarrow(0) = (\gamma_a^2 - \gamma_b^2)\phi_{4s}^2(0) \tag{61}$$

where $\phi_{4s}(0)$ is the "restricted" Hartree–Fock solution. As in Section 3, we can relate $\gamma_a^2 - \gamma_b^2$ to the exchange splitting $\mathscr{I}$ of the excited state and to the promotion energy ΔE for the transfer $2p \to 4s$. According to Eqs. (41) and (42),

$$\gamma_a^2 - \gamma_b^2 \cong (2\mathscr{I}/\Delta E)\gamma_b^2 \tag{62}$$

Using (62), one can write for the contact hyperfine field corresponding to approximation (61)

$$H_{4s} = (2\mathscr{I}/\Delta E)H'_{4s}(0)\gamma_b^2 < 150\gamma^2 \quad [\text{kG}] \tag{63}$$

In estimating the right-hand side of (63), we have assumed $\mathcal{T} < 1$ eV, $\Delta E > 20$ eV. The "restricted" hyperfine field of the unpaired 4s-electron $H_{4s}^{r}(0)$ is taken to be 1500 kG. Recently, Henning[8] suggested a model to explain the covalent reduction of the hyperfine parameter A in Mn^{2+} and Fe^{3+} ions. His assumptions are essentially equivalent to those leading to formula (63). However, comparison of (63) with (60) shows that the latter gives rise to a dominant contribution to A. For $\gamma = 0.2$, (63) gives $H_{4s} = 6$ kG, which is less than 1/10 of the contribution due to (60).

In view of this fact, it seems to us that expression (60) could also provide an explanation of the decrease of A on going from Mn^{2+}:$KMgF_3$ to Mn^{2+}:MgO. Though the neutron scattering results[22] show no difference between the $3d$ covalency of MnF_2 and MnO, there is a general "chemical" belief that oxides possess a greater amount of charge transfer than fluorides. Hence, we also expect a larger $2p \rightarrow 4s$ transfer [and greater γ in (60)] in Mn^{2+}:MgO than in Mn^{2+}:$KMgF_3$. Referring to (60), only a small increase in 4s bonding (about 10%) is needed to explain the observed difference between hyperfine parameters. In connection with this question, let us recall the measurements[38] of the form factor of Mn^{2+} salts, which show that the manganese charge distribution is expanded relative to the theoretical Hartree–Fock charge distribution. Hubbard and Marshall[39] suggested that the covalency admixed (expanded) 4s orbital may be responsible for this observation.

4.3. Dependence of A on Interionic Distance

If we accept that the dominant covalent effects on A are those leading to formula (60), we can also expect a dependence of A on the cation–anion distance. The latter enters in the overlap integrals $\langle l_1 \mid \phi_{ns} \rangle$ making the linear term in γ increase as this distance decreases. In addition to this, we also expect that γ increases with decreasing interionic distance. By referring to formula (21), this may be explained by the increase of $\langle l_1 \mid \phi_{4s} \rangle$ overlap leading to an increase of the numerator of (21). This effect is counteracted by the increase in the Madelung energy in the denominator. Furthermore, since the 4s orbital is very spread out, we expect a slow variation of $\langle l_1 \mid \phi_{4s} \rangle$ and γ with the Mn–O distance.

In order to find the lowest limit of the variations of A with distance, we assume that γ is independent of distance and calculate the changes of (60) due to variations of $\langle l_1 \mid \phi_{ns} \rangle$. Calculation shows that increasing the Mn–O distance by about 10% leads to a decrease of H_{4s} by about 15 kG, which corresponds to a 2% increase of A.

It is interesting to compare this result with the experimental data on Mn^{2+} in various oxides. As Geschwind[40] pointed out, there is a remarkable independence of A on the cation–anion distance of the host. For MgO, CaO, and SrO, the values of A (at $T = 4°K$) are -81.5, -81.6, and -80.9 ($\times 10^{-4}$ cm^{-1}). respectively. Unfortunately, the exact Mn–O distance in these crystals is not known. Analyzing the EPR data on the cubic crystal field parameter a, Geschwind makes a conjecture that the Mn–O distance closely follows the cation–anion distance in the host (2.1, 2.4, and 2.5Å for MgO, CaO, and SrO). We then see that though the Mn–O distance presumably increases by about 20% on going from MgO to SrO, *A decreases* by less than 1%!

On the other hand, our theoretical model predicts at least 2% *increase* of A. One possible explanation of this discrepancy is that the overlap effect is smaller than calculated from the simple Heitler–London model (see also Section 2). Another possible effect which can account for this anomaly is due to the contribution[41] of zero-point lattice vibrations to the hyperfine parameter A. This effect is connected with the explicit temperature dependence of A observed by Walsh *et al.*[34] It has been suggested[42] that the dynamical noncubic crystal fields give rise to a $4s$ admixture into the $3d$ states. This admixture is responsible for a contact hyperfine field which counteracts the negative field of core polarization. The theory[13] predicts a decrease of A which is proportional to the average *square strain*. Then, the zero-point lattice vibration leads to a decrease of A compared to the rigid lattice value. A method has been proposed[41] for deducing this decrease from the strength of the temperature dependence of A. For Mn^{2+} in SrO, a much faster temperature dependence has been observed[43] than for Mn^{2+} in MgO. This result suggests also that the zero-point effect might be stronger in SrO, which could explain the above-mentioned anomaly. Actual calculation has shown that the rigid lattice values of A in MgO and SrO are exactly equal. From this result, one would conjecture that there is no distance dependence of the overlap integrals $\langle l_1 \mid \phi_{ns} \rangle$. However, it is to be emphasized that the oxygen wave function may change on going from MgO to SrO so that $\langle l_1 \mid \phi_{ns} \rangle$ remains effectively unchanged.

5. CATION–CATION HYPERFINE INTERACTION

So far, we have considered the hyperfine interaction with the nuclei of the magnetic ion and its nearest-neighbor ligands. It is natural to expect that the delocalization of the unpaired spin extends beyond the ligands, leading to a hyperfine interaction with more distant nuclei. Usually, those nuclei belong to the cations and thus we are dealing with transferred hyperfine interaction of cation–cation type (supertransferred hyperfine interaction). Its presence has been revealed in several EPR experiments[44,45] on $3d$ transition metal impurities in diamagnetic hosts. Recently, Owen and Taylor[9] reported aluminum hyperfine interaction measurements on Fe^{3+} and Cr^{3+} in $LaAlO_3$. The importance of these measurements is due to the simple structure (perovskite) of the host, where the Fe(Cr)–O–Al bonds involved are 180° and thus amenable to a theoretical treatment. Here, we will confine ourselves to a short description of the basic processes leading to the aluminum hfs. For experimental and theoretical details of this study, we refer the reader to the original paper.[9]

5.1. Fe–O–Al *Bond*

The position of the trivalent cation La^{3+} in the perovskite structure is not favorable for transmitting the spin density between Fe^{3+} and Al^{3+} cations and it is therefore assumed that the main contribution to the aluminum hfs is due to the Fe–O–Al 180° bond. The electron orbitals, belonging to a simplified σ-bond configuration, are illustrated in Fig. 6. In what follows, we will be interested in the unpaired s-electron density at the aluminum nucleus. From symmetry arguments, only $3d_{z^2}$ orbital can contribute to the latter. The Fe–O covalent σ bond is simplified by omitting the $2s$ oxygen orbitals.

There are two basic mechanisms contributing to the aluminum unpaired electron density. First, the unpaired electron density in the p orbital of O^{2-} ion can be transmitted to the $2s$ occupied orbital of Al^{3+} by $\langle p \mid 2s \rangle$ overlap. Second, the d electron of Fe^{3+} can be virtually transferred to the empty $3s$ orbital of Al^{3+}. We will now discuss these two mechanisms separately.

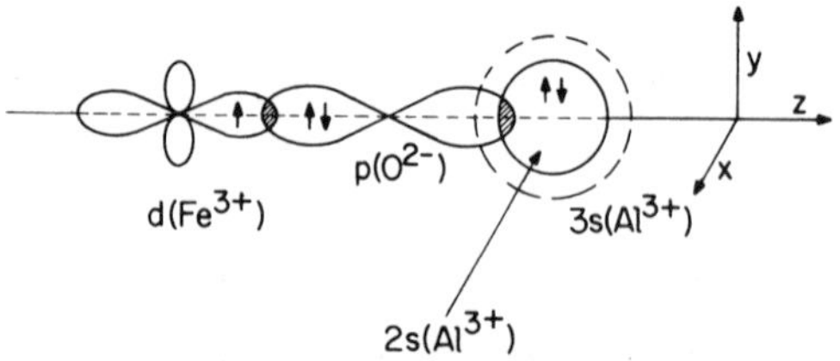

Fig. 6. The orbitals participating in the mechanisms of isotropic hyperfine interaction of ^{27}Al in an Fe–O–Al bond. d is the $3d_{z^2}$ orbital of Fe^{3+}, p is the $2p_\sigma$ orbital of O^{2-}, and $2s$ and $3s$ belong to the Al^{3+} ion.

5.1.1. Overlap Mechanism

It is interesting to see that already in the purely ionic ground state there is an unpaired s-electron density in the $2s$-aluminum orbital. Referring to Fig. 6, the ground-state determinant is

$$\Psi_G = N\{d\uparrow, p\uparrow, p\downarrow, 2s\uparrow, 2s\uparrow\} \tag{64}$$

where N is the normalization constant. To calculate the unpaired $2s$ density arising from the pure overlap effect, we first orthogonalize the one-electron orbitals in (64) using Schmidt process and obtain for Ψ_G

$$\begin{aligned}\Psi_G = N'\{&[d - \langle d \mid p\rangle(p - \langle p \mid 2s\rangle 2s)]\uparrow,\\ &(p - \langle p \mid 2s\rangle 2s)\uparrow, (p - \langle p \mid 2s\rangle 2s)\downarrow, 2s\uparrow, 2s\downarrow\}\end{aligned} \tag{65}$$

In deriving (65), we first orthogonalize p to a $2s$ and then d to the (orthogonalized) p orbital, neglecting the small overlap $\langle d \mid 2s\rangle$. According to (65), only the first (antibonding) orbital contributes to the total unpaired spin density. By neglecting the higher-order terms, one obtains from (65) the unpaired $2s$ density at the nucleus of aluminum

$$\varrho_{\mathrm{Al}}\uparrow(0) - \varrho_{\mathrm{Al}}\downarrow(0) = \langle d \mid p\rangle^2 \langle p \mid 2s\rangle^2 \phi_{2s}^2(0) \tag{66}$$

where $\phi_{2s}(0)$ has been introduced to denote the amplitude of the $2s$ wave function at the nucleus. The result (66) can be interpreted as meaning a transmission of *oxygen unpaired density* $\langle d \mid p\rangle^2$ by $\langle p \mid 2s\rangle$ overlap to the $2s$ orbital of aluminum. Then, one can make a conjecture that in the presence of oxygen–iron transfer, the unpaired $2s$ density is

$$\varrho_{\mathrm{Al}}\uparrow(0) - \varrho_{\mathrm{Al}}\downarrow(0) = (\langle d \mid p\rangle + \gamma)^2 \langle p \mid 2s\rangle^2 \phi_{2s}^2(0) \tag{67}$$

where γ is the $2p \to 3d$ transfer parameter (process A). Formula (67) can be derived[46] by including this transfer in the antibonding molecular orbital for the Fe–O σ bond.

5.1.2. Transfer Mechanism

Owen and Taylor[9] considered a transfer of the oxygen p electrons to the $3s$ state of aluminum. Since both $p\uparrow$ and $p\downarrow$ electrons are transferred with the same probability, there is no diagonal unpaired $3s$ density. However, in combination with the overlap term discussed above, there is an important ($2s$, $3s$) cross-product. A different type of transfer was considered by Huang *et al.*[46] in their study of cation–cation hyperfine interaction in $KMnF_3$. The $3d$ electron of one Mn^{2+} ion is *directly* transferred to the empty $4s$ orbital of the other Mn^{2+} ion. In view of the large cation–cation distance, the transfer probability for this process is small, but this is compensated for by the fact that unpaired electron of only one spin orientation is transferred. As suggested previously,[47] one can apply this process to the Fe–O–Al bond and consider the $d \rightarrow 3s$ virtual transfer which is equivalent to admixing into (64) an excited configuration

$$\Psi_E = N\{3s\uparrow, p\uparrow, p\downarrow, 2s\uparrow, 2s\downarrow\} \tag{68}$$

Then the perturbed ground state is

$$\begin{aligned} \Psi_G' &= N'(\Psi_G + \gamma_3 \Psi_E) \\ &= N''\{(d + \gamma_3 3s)\uparrow, p\uparrow, p\downarrow, 2s\uparrow, 2s\downarrow\} \\ &= N'''\{(d - \langle d \mid p \rangle p + \gamma_3 3s + \langle d \mid p \rangle \langle p \mid 2s \rangle 2s)\uparrow, \\ &\quad (p - \langle p \mid 2s \rangle 2s)\uparrow, (p - \langle p \mid 2s \rangle 2s)\downarrow, 2s\uparrow, 2s\downarrow\} \end{aligned} \tag{69}$$

As in (65), an orthogonalization of one-electron orbitals is used in (69). The unpaired s-electron density at the nucleus of aluminum is given by (69) as follows

$$\begin{aligned} \varrho_{\mathrm{Al}}\uparrow(0) - \varrho_{\mathrm{Al}}\downarrow(0) &= \gamma_3{}^2 \phi_{3s}^2(0) + \langle d \mid p \rangle^2 \langle p \mid 2s \rangle^2 \phi_{2s}^2(0) \\ &\quad + 2\gamma_3 \langle d \mid p \rangle \langle p \mid 2s \rangle \phi_{2s}(0) \phi_{3s}(0) \end{aligned} \tag{70}$$

This result can be generalized by incorporating again a $2p \rightarrow 3d$ transfer. This can be formally done,[46] as in (67), by replacing $\langle d \mid p \rangle$ [in (70)] by $\langle d \mid p \rangle + \gamma$. The ($2s$, $3s$) cross-term is quite important. It has the same sign as the diagonal $3s$ and $2s$ terms and thus amplifies their contribution. Generally, the sign of such cross-terms depends on the number of nodes of the wave function involved.[46] For instance, the ($1s$, $3s$) cross-term is negative and thus diminishes the resulting unpaired s density.

5.1.3. Aluminum Hyperfine Interaction

The unpaired s density (70) gives rise to a contact hyperfine interaction described by the spin-Hamiltonian

$$\mathscr{H} = A_s \mathbf{I} \cdot \mathbf{S} \tag{71}$$

where $\mathbf{I}$ is the ^{27}Al nuclear spin, $\mathbf{S}$ is the Fe^{3+} electron spin ($S = 5/2$), and A_s is the isotropic hyperfine parameter. The experimental isotropic hfs involves, besides this contact interaction, a direct dipolar interaction with the spin density of the Fe^{3+} $3d$ orbitals. The latter can be calculated and thus the contribution of the unpaired s electrons can be determined from the experiment.

Let us consider the overlap mechanism which involves the unpairing in the $2s$ orbital of aluminum. The fraction f_s of the unpaired $2s$ electrons is defined [see Eq. (33)] as

$$\varrho_{\mathrm{Al}\uparrow}(0) - \varrho_{\mathrm{Al}\downarrow}(0) = f_s \cdot \phi_{2s}^2(0) \tag{72}$$

Comparing (72) with (67), we obtain a "theoretical" expression for f_s

$$f_s = (\langle d \mid p \rangle + \gamma)^2 \langle p \mid 2s \rangle^2 = \lambda^2 \langle p \mid 2s \rangle^2 \tag{73}$$

λ^2 is the fraction of unpaired $2p_\sigma$ electrons on oxygen. From the work of Nathans *et al.*,[22] one estimates $\lambda \approx 5\%$. Owen and Taylor[9] calculate $\langle p \mid 2s \rangle = 0.094$. Using these numbers in (73), one obtains a theoretical estimate $f_s \approx 0.04\%$.

It is interesting to compare this result with the experimentally determined value of f_s. According to (12) and (13), the unpaired density (72) gives rise to an isotropic hyperfine interaction (71) with

$$A_s = (f_s/2S)A_{2s} \tag{74}$$

where

$$A_{2s} = (16\pi/3)g_N\beta_N\beta_e\phi_{2s}^2(0) \tag{75}$$

The calculation of Owen and Taylor[9] gives $A_{2s} = 51.4 \times 10^3$ MHz. Using this value and the measured[9] value $A_s = 3.26$ MHz, one deduces from (74) $f_s = 0.03\%$, which is in good agreement with the above estimate based on overlap contribution (73). Thus it seems that no $3s$ transfer is necessary to account for the experimental hfs of aluminum. However, it must be realized that in (67) the $1s$ orbital has been omitted. The latter is responsible for a negative ($1s$, $2s$) cross-term[9] which *diminishes* the resulting s density. Another ambiguity arises from the uncertainty of the overlap integrals involving the oxygen wave function. The latter may depend on the type of salts,[47] so that errors in $\langle p \mid 2s \rangle$ may arise by using a "universal" type of O^{2-} wave function.[36]

5.2. Cr–O–Al *Bond*

According to (67), the unpaired $2s$ density at the aluminum nucleus is proportional to the unpaired electron density in the $2p_\sigma$ orbital of the intervening O^{2-} ion. In Cr^{3+}:$LaAlO_3$, the magnetic t_{2g} orbitals of Cr^{3+}

have no overlap with the $2p_\sigma$ orbital of oxygen and the polarization of the latter occurs only on account of an exchange polarized transfer of $2p_\sigma$ electrons to unoccupied $3d_{z^2}$ and $4s$ or $4p$ orbitals. As indicated by (45), this type of transfer gives rise to a negative spin density in the $2p_\sigma$ orbital. Referring to (67), this implies a *negative* unpaired aluminum $2s$ density induced by $\langle p \mid 2s \rangle$ overlap. From symmetry, no direct $d \rightarrow 3s$ transfer is possible and thus $\gamma_3 = 0$. Thus the $3s$ orbital becomes admixed only through $p \rightarrow 3s$ transfer, as introduced by Owen and Taylor.[9]

These authors have observed negative A_s in Cr^{3+}:$LaAlO_3$. From the measured value of A_s and calculated[9] value of A_{2s}, the fraction of unpaired electrons in the aluminum $2s$ orbital is $f_s \simeq -0.006\%$. Putting this and the above-mentioned value of $\langle p \mid 2s \rangle$ into (73), one obtains for the oxygen $2p_\sigma$ unpaired density a value -0.75%. Referring to Eq. (45), the negative unpaired $2p$ density is roughly $-2\mathscr{T}/E = -1/10$ times the unpaired density due to σ bonding ($\simeq 5\%$). Hence, the results of Owen and Taylor[9] are also in agreement with the prediction of (45) and support the idea of a negative p_σ spin density on oxygen. The importance of these results is amplified if we realize that the negative p_σ density on the ligand of $(t_{2g})^3$ complexes leads to the same anisotropic ligand hyperfine interaction as the positive p_π density. Thus the transferred hyperfine interaction (for instance, with F^- nuclei) reflects the presence of the negative p_σ density only as an increase of the anisotropic shfs parameter.[7] Thus only *cation–cation* hyperfine interaction selectively picks up the p_σ density.

An alternate interpretation of negative f_s would be a direct transfer from aluminum $2s$ orbitals to the empty $3d_{z^2}$, $4s$, or $4p$ states of chromium (see transfer process C in Section 3). However, in view of the great cation–cation distances, the transfer integral for such a process is expected to be very small.

6. COMMENTS ON FIRST-PRINCIPLE CALCULATIONS

The material presented in this chapter shows that various hyperfine effects in transition metal salts can be satisfactorily explained using a simple configuration interaction description based on the Heitler–London model. However, as mentioned above, the quantitative treatment of the parameters of this theory, such as overlaps and charge transfer parameter γ, is still difficult.

A difficulty one already meets in describing the purely ionic state is the lack of knowledge of the "atomic" orbitals in a solid-state system. As a result of overlapping, the constituent ions start losing their individuality. Then, the overlap integrals (based on individual atomic orbitals) have no direct physical significance and should be regarded as mathematical parameters belonging to a particular approximate description of the system. We have indicated that the free-ion Heitler–London model is often an oversimplification of the real situation in ionic crystals.

The same type of problem extends into the *a priori* calculations of γ. The configuration interaction method also requires a knowledge of the excited charge transfer states. These states are usually more expanded[12,13] than the ground-state orbitals and thus more affected by the solid-state surroundings. Obviously, the calculation of the matrix element

$$\langle \Psi_G | \mathscr{H} | \Psi_E \rangle$$

in (21) is very sensitive to the exact form of the excited-state orbitals. So far, *a priori* calculations[13,15–18] of the covalency effects have been done with free-ionic wave functions for both Ψ_G and Ψ_E.

Another problem arising in the first-principle calculations of γ is connected with the perturbation treatment of the configuration interaction method. Physically, the first-order formula (21) corresponds to a *one-electron* charge transfer process, all other electrons remaining in the ground state. However, it has been shown[12,13] that other, higher-order processes are important for calculating the degree of covalency. For instance, the ligand–metal transfer is accompanied by excitations of other electrons which correspond to electric polarization of the surroundings and redistribution of $3d$ electrons due to Coulomb interaction.[13] These excitations have been shown to increase the unpaired spin density at the ligand compared to the one-electron process (21). Hence, it appears that the many-electron aspects of virtual charge transfer have to be taken into account in an accurate treatment of covalency effects.

7. COMMENTS ON EXPERIMENTS

In conclusion, we should briefly mention how in practice one deduces the hyperfine parameters of interest from the measured EPR spectra. In accordance with the general concepts exposed in the sections we are interested in the central hyperfine interaction described by the spin–Hamiltonian (11) and discussed in length in Section 4 and in the ligand (fluorine) hyperfine interaction described by the spin-Hamiltonian (16) or (35).

We return to the resonance of H in CaF_2 which was studied in Section 2 to illustrate the procedures leading from the recorded spectra to the microscopic parameters for central and ligand hyperfine interaction. The spectrum of CaF_2:H is particularly suitable for the illustration because the individual resonance curves are very well resolved, all magnetic moments involved (electron and nuclear) have the smallest possible value $\frac{1}{2}$, and the overall symmetry is cubic. Because of the cubic symmetry, for some particular orientations of the magnetic field with respect to the crystal axes, the resonance pattern and consequently also the interpretation of the spectra simplify enormously. Furthermore, since there is only one magnetic electron, there are no complications due to crystal field effects.

The EPR spectra of CaF_2:H have been obtained by Hall and Schumacher[6] at room temperature with a high-sensitivity 9-kHz (X band) microwave spectrometer. The resonance (absorption derivative) for the magnetic field parallel to the [111] axis is shown in Fig. 7. The spectrum apparently consists of two main groups of 11 lines each. The large spacing between the centers of the two patterns is due to the electron–proton magnetic interaction, the smaller spacing between the 11 lines of each group is caused by the hyperfine interaction with the 8 fluorine spins.

The electron–proton interaction is formally described by the isotropic spin-Hamiltonian $A_p\mathbf{S} \cdot \mathbf{I}$ [Eq. (11)] which should be added to the electron Zeeman energy $g\beta\mathbf{S} \cdot \mathbf{H}$ and to the nuclear Zeeman energy $g_p\beta_N\mathbf{I} \cdot \mathbf{H}$. In the case when the hyperfine interaction and the nuclear Zeeman energy are much smaller than the electron Zeeman energy, $A\mathbf{S} \cdot \mathbf{I}$ can be treated as a perturbation and $g_p\beta_N\mathbf{H} \cdot \mathbf{I}$ can be neglected. Then, the observed doublet

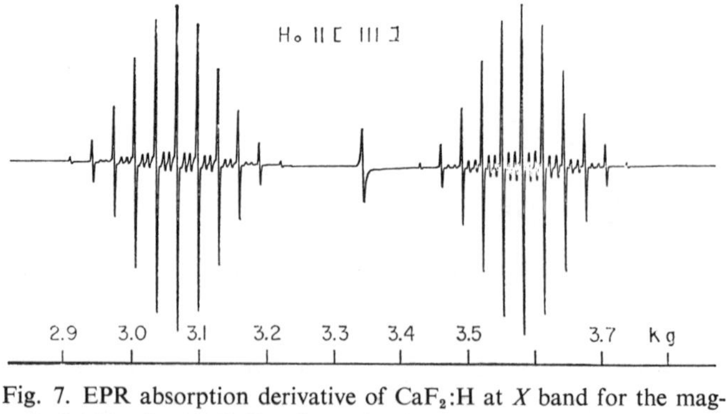

Fig. 7. EPR absorption derivative of CaF_2:H at X band for the magnetic field in the [111] direction. The doublets between the main lines are forbidden transitions and the line near 3.35 kG is a DPPH g-marker. (After Hall and Schumacher.[6])

splitting δH of the resonance pattern (expressed in terms of the magnetic field) is related to the hfs constant A by a simple formula

$$\delta H = A/g\beta \tag{76}$$

The g-factor can be found from the relation $h \quad = g\beta H$, when ν is the frequency of the microwave spectrometer and H is the magnetic field which corresponds to the center of the overall resonance pattern.

Whenever the constant A is not small in comparison to $g\beta H$, the exact Breit–Rabi formula[48] has to be applied. This formula involves g and A parameters implicitly and should be solved numerically for each particular case. The way to proceed is to measure the resonance magnetic field H_m (the field which corresponds to the central resonance of the upper group in Fig. 7) and H_n (the field which corresponds to the central resonance of the lower group in Fig. 7) of the two doublet components and substitute in the equations

$$\begin{aligned} h\nu &= \tfrac{1}{2}A(1 + X_n) + \tfrac{1}{2}A(1 + X_n{}^2)^{1/2} \\ h\nu &= \tfrac{1}{2}A(1 + X_m{}^2)^{1/2} - \tfrac{1}{2}A(1 - X_n) \end{aligned} \tag{77}$$

where $X_n = g\beta H_n/A$ and $X_m = g\beta H_m/A$. The two equations (77) can be solved for two unknown parameters g and A. It turns out in our case of CaF_2:H that the parameter A determined from the relation (76) is about 1% off in comparison with the exact A calculated from (7).

The hyperfine interaction with fluorine nuclei must be generally described by a tensor interaction of the form

$$\mathscr{H}_F = \sum_{\alpha=1}^{8} \mathbf{S} \cdot \mathsf{T}_\alpha \cdot \mathbf{I}_\alpha \tag{78}$$

where the sum over α goes over all eight nearest-neighbor fluorine ions. The interaction energy must not be changed by any of the symmetry operations which are allowed by the model. Namely, for one particular fluorine ion (for example, the ion p_1 in Fig. 2), the Hamiltonian must be axially symmetric about the line connecting the fluorine ion with the center of the cube since this line coincides with the body diagonal and has threefold rotation symmetry. The body diagonal must be also one of the principal axes of T_1. According to the Fig. 2, the body diagonal is denoted by z and we can use the following notation:

$$T_1^{zz} = T_\parallel, \qquad T_1^{xx} = T_1^{yy} = T_\perp \tag{79}$$

The spin-Hamiltonian for the hf interaction with the fluorine P_1 can then be written as

$$T_{\parallel} I_z S_z + T_{\perp}(I_x S_x + I_y S_y)$$

It is convenient to decompose $T_{\parallel}$ and $T_{\perp}$ into isotropic (a) and anisotropic (b) components as follows:

$$T_{\parallel} = a + 2b, \qquad T_{\perp} = a - b \tag{81}$$

Only the anisotropic part of the interaction is discussed in Section 2.2. Then the constant A_F in the Eq. (11) is equivalent to the constant b defined by the relations (81). This is because only $2p$ orbitals on F^- ions are considered, the $2s$ ones being neglected.

For the magnetic field along the [111] direction, there are three pairs of fluorine nuclei for which the angle θ between the tensor axis and the magnetic field is $\theta = 71°$ and one pair for which $\theta = 0°$. To proceed further, we should therefore find the energy levels due to the Hamiltonian (80) for a general direction of the magnetic field. The problem is not straightforward because generally the orientations of the external magnetic field of the electron spin and of the nuclear spin may not coincide. But if the electron Zeeman energy is substantially larger than the hyperfine interactions, the electron spin is practically aligned in the direction of the external magnetic field and the electron spin projection numbers are good quantum numbers. A large operator $A\mathbf{S} \cdot \mathbf{I}$ may admix states of different projection numbers, but in the case of CaF_2:H, the neglect of such an admixture introduces negligible error.

On the other hand, the direction of the external magnetic field and therefore the direction of the electron spin may not necessarily be parallel with the effective magnetic field acting on the fluorine nucleus. The effective magnetic field is usually much larger than the external one and it is more convenient to quantize the fluorine spin along the local effective field. Such procedure was suggested by Castner and Känzig[49] and gives exact energies for $\theta = 0$ and 90° and energies to within a few per cent for arbitrary angle θ. The quantity which is observed in EPR experiments is the energy difference between two different orientations of the electron spin. According to the paper of Castner and Känzig,[49] the hyperfine interaction with one fluorine α contributes to this energy by

$$\Delta E(M_F{}^{\alpha}) = M_F{}^{\alpha}[T_{\perp}{}^2 + (T_{\parallel}{}^2 - T_{\perp}{}^2)\cos^2\theta^{\alpha}]^{1/2} \tag{82}$$

where $M_F{}^{\alpha}$ is the projection quantum number of the fluorine spin in the direction of the effective magnetic field.

Consequently, due to the interaction with the fluorine spin α, each EPR line should be split into two components separated by

$$\delta H^{\alpha} = (1/g\beta)[T_{\perp}^{2} + (T_{\parallel}^{2} - T_{\perp}^{2})\cos^{2}\theta^{\alpha}]^{1/2} \tag{83}$$

In our case, we have two fluorines (total maximum spin $I = 1$) with $\theta = 0$, which should split each EPR line into three components separated by $T_{\parallel}/g\beta$. Each of these lines is further split into seven components due to the interaction with the remaining six fluorine nuclei with $\theta = 71°$ (having total maximum spin $I = 3$) and we should expect generally $3 \times 7 = 21$ different absorption lines. The experimental spectrum in Fig. 7 is seen to contain only 11 principal lines which can be synthesized only by assumption that the splitting due to the 0° group is twice that due to the 71° group. The assumption yields the equation

$$T_{\parallel}/2g\beta = (1/g\beta)[T_{\perp}^{2} + (T_{\parallel}^{2} - T_{\perp}^{2})\cos^{2}71°]^{1/2} = \delta H \tag{84}$$

where δH is the separation between two nearest principal lines. From Eq. (84), the unknown $T_{\parallel}$ and $T_{\perp}$ can be calculated.

It must be emphasized Eq. (84) or any other similar equations for other directions of the external magnetic field cannot give the relative sign of $T_{\parallel}$ and $T_{\perp}$. This may cause serious difficulties mainly if we try to decompose $T_{\parallel}$ and $T_{\perp}$ into the components a and b [see Eq. (81)]. Examination of Fig. 7 reveals there is a prominent doublet between each pair of principal lines. These transitions are forbidden transitions which correspond to the selection rule $M_S = \pm 1$, $M_F = \pm 1$. They occur because the total magnetic field acting on the nucleus is reoriented neither by exactly 0° nor by exactly 180° when an electronic transition occurs. Their physical significance lays mainly in the fact[50,51] that the field interval between an allowed line and an adjacent forbidden line depends on the relative sign of $T_{\parallel}$ and $T_{\perp}$. It turns, out, for example, that the observed forbidden field interval can be calculated to within 10% by the assumption the $T_{\parallel}/T_{\perp} \lessgtr 0$ and within 0.01% by the assumption that $T_{\parallel}$ and $T_{\perp}$ have the same sign. Assuming $T_{\parallel}$ and $T_{\perp}$ of the same sign, the parameters a and b can be calculated unequivocally from the equations (81). This is also how the parameter $A_F = b = +11.3 \times 10^{-4}$ cm^{-1} has been deduced from experimental data. The choice of the positive sign is based on the physical arguments analyzed in Section 2.

As another example, we mention the paramagnetic resonance of Cr^{3+} in K_2NaGaF_6 first reported by Helmholtz *et al.*[152] The resonance spectrum is not so well resolved as the resonance CaF_2:H, but if forms a typical example of transition ion metal resonances.

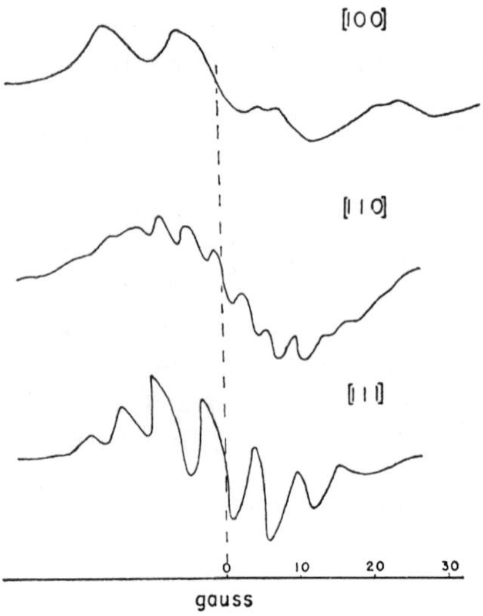

Fig. 8. EPR absorption derivative of CrF_6^{3-} dissolved in K_2NaGaF_6 with the magnetic field in the indicated directions (after Helmholz *et al.*[52]).

The Cr^{3+} replaces Ga and is surrounded by six fluorine ions which form a perfect octahedron. Because the most abundant chromium isotope has no nuclear moment and the Cr^{3+} electron spin 3/2 is not affected by the cubic crystal field, only one electron spin resonance line is usually observed. The fine structure of the resonance in Fig. 8 is, therefore, solely due to the hyperfine interaction with fluorine ions. The interaction is again described by the Hamiltonian (79) with the only difference being that, instead of $8F^-$ ions in CaF_2, there are six fluorine ions around the chromium ion (see Fig. 4) and the symmetry axes of the hfs tensors are along [100] axes.

The method of fitting of the experimental data follows the same line as in the previous case. The best fit is obtained for

$$|T_\parallel| = 10.1 \times 10^{-4}\ \text{cm}^{-1}, \qquad |T_\perp| = 3.4 \times 10^{-4}\ \text{cm}^{-1}$$

Substituting into (81), we get four possible pairs of parameters a and b:

$$a = \pm 5.7 \times 10^{-4}\ \text{cm}^{-1}$$
$$b = \pm 2.2 \times 10^{-4}\ \text{cm}^{-1} \tag{85}$$

or

$$a = \pm 1.1 \times 10^{-4}\ \text{cm}^{-1}$$
$$b = \pm 4.5 \times 10^{-4}\ \text{cm}^{-1} \tag{86}$$

The "forbidden transitions" are not resolved, so that there is no experimental way to distinguish between the four possibilities. We should recall our discussion in Section 3 of the physical origin of the hyperfine structure in similar systems. The isotropic fluorine hfs is due to the ligand–metal (empty-shell) transfer and must be weak and negative. The anisotropic fluoring hfs (b) is due to the p–$d_{t_{2g}}$ transfer, is relatively strong, and also unambiguously negative. The only combination which fulfill these requirements is $a = -1.1 \times 10^{-4}\ \mathrm{cm}^{-1}$ and $b = -4.5 \times 10^{-4}\ \mathrm{cm}^{-1}$.

Before closing this chapter, we would like to point out the fact that transfer hyperfine interaction in insulators has furnished exact information about delocalization of unpaired density and covalency in a vast number of cases. On the other hand, one has not observed a single example of shfs of transition metal impurities in diamagnetic metals. This is due to the large linewidth and special features of s–d interactions which wash out, for example in the case of CuMn, even the central Mn hfs.

ACKNOWLEDGMENTS

We would like to express our sincere thanks to Nai Li Huang, R. Orbach, S. Geschwind, and M. Tachiki for their illuminating discussions. We are also indebted to Karel Zdánsky for his contribution to our understanding of this subject.

REFERENCES

1. J. Owen and J. H. M. Thornley, *Rep. Progr. Phys.* **29**, 675 (1966).
2. J. H. E. Griffiths, J. Owen, and I. M. Ward, *Proc. Roy. Soc. A*, **219**, 526 (1953).
3. M. Tinkham, *Proc. Roy. Soc. A*, **236**, 535 (1956).
4. B. S. Gourary and F. J. Adrian, *Phys. Rev.* **105**, 1180 (1957).
5. F. Keffer, T. Oguchi, W. O'Sullivan, and J. Yamashita, *Phys. Rev.* **115**, 1553 (1959).
6. J. L. Hall and R. T. Schumacher, *Phys. Rev.* **127**, 1892 (1962).
7. T. P. P. Hall, W. Hayes, R. W. H. Stevenson, and J. Wilkens, *J. Chem. Phys.* **38**, 1977 (1963).
8. J. C. M. Henning, *Phys. Letters* **24A**, 40 (1967).
9. J. Owen and D. R. Taylor (to be published).
10. Per-Olov Löwdin, *J. Chem. Phys.* **18**, 365 (1950).
11. H. Blum, in *Proc. Int. Conf. on Physics of Solids at High Pressures, Tucson, Arizona 1965*, Academic Press, New York, p. 409.
12. E. Šimánek, Z. Šroubek, and M. Tachiki, *J. Phys. Soc. Japan* **22**, 547 (1967).
13. J. Hubbard, D. E. Rimmer, and F. R. A. Hopgood, *Proc. Phys. Soc.* **88**, 13 (1966).
14. W. Marshall and R. Stuart, *Phys. Rev.* **123**, 2048 (1961).
15. R. G. Shulman and S. Sugano, *Phys. Rev.* **130**, 506 (1963).

16. S. Sugano and R. G. Shulman, *Phys. Rev.* **130**, 517 (1963).
17. R. E. Watson and A. J. Freeman, *Phys. Rev.* **134**, A1526 (1964).
18. E. Šimánek and Z. Šroubek, *Phys. Status Solidi* **4**, 251 (1964).
19. R. E. Watson, MIT SSMIG Technical Report, No. 12 (1959).
20. C. Froese, *Proc. Cambridge Phil. Soc.* **53**, 206 (1957).
21. F. J. Adrian, *J. Chem. Phys.* **32**, 972 (1960).
22. R. Nathans, G. Will, and D. E. Cox, in *Proc. Int. Conf. on Magnetism, Nottingham 1964*, Institute of Physics and Physical Society, London, p. 327.
23. R. G. Shulman and S. Sugano, *J. Chem. Phys.* **42**, 39 (1965).
24. B. Bleaney and M. C. M. O'Brien, *Proc. Phys. Soc.* (*London*) **B69**, 1216 (1956).
25. J. M. Baker, B. Bleaney, and K. D. Bowers, *Proc. Phys. Soc.* **B69**, 1205 (1956).
26. H. A. Kuska and M. T. Rogers, *J. Chem. Phys.* **41**, 3802 (1964).
27. L. E. Orgel, *J. Chem. Phys.* **30**, 1617 (1959).
28. R. G. Shulman and K. Knox, *Phys. Rev. Letters* **4**, 603 (1960).
29. R. G. Shulman, in *Seminar on the Magnetism in Metals and Alloys*, American Society for Metals, Cleveland, Ohio (1959), p. 56.
30. V. Heine, *Phys. Rev.* **107**, 1002 (1957).
31. J. H. Wood and G. Pratt, Jr., *Phys. Rev.* **107**, 995 (1957).
32. R. E. Watson and A. J. Freeman, *Phys. Rev.* **123**, 2027 (1961).
33. J. S. Van Wieringen, *Disc. Faraday Soc.* **19**, 118 (1955).
34. W. M. Walsh, Jr., J. Jeener, and N. Bloembergen, *Phys. Rev.* **139**, A1338 (1965).
35. E. Šimánek and Nai Li Huang, *Phys. Rev. Letters* **17**, 699 (1966).
36. R. E. Watson, *Phys. Rev.* **III**, 1108 (1958).
36a. E. Šimánek and K. A. Muller (to be published).
37. E. Šimánek and Z. Šroubek (to be published).
38. J. M. Hastings, N. Elliot, and L. M. Corliss, *Phys. Rev.* **115**, 13 (1959).
39. J. Hubbard and W. Marshall, *Proc. Phys. Soc.* **86**, 561 (1965).
40. S. Geschwind, in *Hyperfine Interactions* (Proc. NATO Advanced Summer School, Aix-en-Provence, 1966), eds. A. J. Freeman and R. B. Frankel, Academic Press, New York (1967), p. 225.
41. R. Orbach and E. Šimánek, *Phys. Rev.* **158**, 310 (1967).
42. E. Šimánek and R. Orbach, *Phys. Rev.* **145**, 191 (1966).
43. J. Rosenthal, L. Yarmus, and R. H. Bartram, *Phys. Rev.* **153**, 407 (1967).
44. N. Laurance, E. C. McIrvine, and J. Lambe, *J. Phys. Chem. Solids* **23**, 515 (1962).
45. I. Chen, C. Kikuchi, and H. Watanabe, *J. Chem. Phys.* **42**, 186 (1965).
46. Nai Li Huang, R. Orbach, and E. Šimánek, *Phys. Rev. Letters* **17**, 134 (1966).
47. Nai Li Huang, R. Orbach, E. Šimánek, J. Owen, and D. R. Taylor, *Phys. Rev.* **156**, 383 (1967).
48. G. Breit and I. I. Rabi, *Phys. Rev.* **38**, 2082 (1931).
49. T. E. Castner and W. Känzig, *J. Phys. Chem. Solids* **3**, 178 (1957).
50. J. M. Baker, W. Hayes, and M. C. M. O'Brien, *Proc. Phys. Soc.* (*London*) **A254**, 272 (1960).
51. A. M. Clogston, J. P. Gordon, V. Jaccarino, M. Peter, and L. R. Walker *Phys. Rev.* **117**, 1222 (1960).
52. L. Helmholtz, A. V. Guzzo, and R. N. Sanders, *J. Chem. Phys.* **35**, 1349 (1961).

Index

F

G

H

Q

R

T

U

V

Y

Z